CENTERBOOK

The Center for Advanced Visual Studies and the Evolution of Art-Science-Technology at MIT

CENTERBOOK

The Center for Advanced Visual Studies and the Evolution of Art-Science-Technology at MIT

Elizabeth Goldring and Ellen Sebring

SA+P Press
School of Architecture + Planning
Massachusetts Institute of Technology
Cambridge, Massachusetts

ZKM | Center for Art and Media Karlsruhe
Karlsruhe, Germany

The MIT Press
Cambridge, Massachusetts, and London, England

A publication by
SA+P Press, MIT School of Architecture and Planning, and
ZKM | Center for Art and Media Karlsruhe
In collaboration with
The MIT Press, Cambridge, Massachusetts, and London, England.

This book was set in Garamond and Avenir.
Printed and bound by Zarbock GmbH & Co. KG, Frankfurt am Main, Germany.
Distributed by The MIT Press, Cambridge, Massachusetts, and London, England.

ISBN 978-0-9981170-5-8
Library of Congress Control Number: 2018938606

Visual Narrative, Graphic Design, and Layout: Ellen Sebring
Cover, Front Matter and Final Print Version Design: Paul Montie
Design Consultant and Print Liaison: Bernd Kracke, President, Hochschule für Gestaltung / University of Art and Design Offenbach
Archivist: Jeremy Grubman, CAVS Special Collection, ACT/MIT
Editing: Sarah Olson, Patsy Baudoin
Index: Tobiah Waldron
Reproduction Rights: the Authors and the CAVS Special Collection, ACT/MIT

Production support provided by
MIT School of Architecture and Planning, Hashim Sarkis, Dean
MIT Museum, John Durant, Director
Philip S. Khoury, Associate Provost at MIT
ZKM | Karlsruhe, Peter Weibel, CEO and Chairman

Contents

Foreword

In 2011 the Massachusetts Institute of Technology (MIT) celebrated its sesquicentennial anniversary. When the MIT Museum asked the extended MIT community to contribute to the celebrations by helping us to select 150 artifacts representing the spirit of the Institute, it was striking that artworks featured prominently in the lists. As a result, the museum's *MIT 150 Exhibition* devoted an entire section to "Artistic MIT," featuring works by Alexander Calder, Aaron Copland, György Kepes, Richard Leacock, Otto Piene, and many others.

Art has always been part and parcel of the life of the Institute. Indeed, MIT's founding document, William Barton Rogers' "Objects and Plan of an Institute of Technology" (1860), refers specifically to "A Society of Arts, a Museum of Arts, and a School of Industrial Science." While the arts were pursued from the very outset, it was not until a full century later that the Institute may be said to have found its distinctive artistic voice. For it was during the turbulent years of the 1960s, amidst protests about civil rights and the Vietnam War, and the 1970s, with the emergence of the environmental and women's movements, that MIT first began actively cultivating artistic endeavors that depended on some of the distinctive attributes of a first-rate research university.

Decisive here was the initiative of the Hungarian-born artist György Kepes, who persuaded the Institute to establish the Center for Advanced Visual Studies (CAVS). After joining the MIT faculty in 1947, Kepes had become steadily more interested in the dialogue between art and science; and upon its founding in 1967, CAVS quickly established a reputation as a place for interdisciplinary work, a place where artists who were interested in science and scientists who were interested in art were assured a warm welcome. Early among those who were attracted to CAVS as fellows was the co-founder of the influential Group ZERO, German artist Otto Piene; and under the successive directorships of Kepes (1967-1974) and Piene (1974-1994), CAVS established a distinctive and enviable reputation for art in dialogue with science and technology.

In this volume, Elizabeth Goldring and Ellen Sebring provide a rich visual history of CAVS on the occasion of the Center's 50th anniversary. As their work reveals, CAVS stimulated an enormous body of artistic work, much of it (as with the extraordinary *Centerbeam*, 1977, superbly well documented here) profoundly collaborative. The quality of this work established CAVS's and MIT's reputations in the visual arts, and—together with its near contemporary, the New York-based Experiments in Art and Technology (E.A.T.) initiative—helped to put art, science, and technology firmly on the artistic and cultural map internationally. The success of CAVS demonstrated the multiplicity of the mutual interconnections between art, science, and technology, as well as the very great value of providing faculty and students with opportunities to engage at the multiple interfaces among them.

This lesson has been well learned at MIT. CAVS itself merged with the Visual Arts Program in 2009 to create the MIT Program in Art, Culture and Technology (ACT); today, ACT works alongside many other departments, labs, and centers—Comparative Media Studies/Writing, the Glass Lab, the Media Lab, Music and Theatre Arts, and the Center for Art, Science and Technology (CAST), to name a few—to provide an extraordinary variety of creative opportunities in the arts.

The former faculty, fellows, and students of CAVS can all be justly proud of what they have accomplished.

John Durant, Mark R. Epstein Director, MIT Museum

Preface

The question, Why do we need a book about the Center?, is partly answered by the response to the question, Why did MIT need CAVS? Founding Director György Kepes saw in CAVS a chance for humankind to retrieve an inner compass through artists' interventions with science and technology.

> Today we need a compensating inner guide more than ever before. For there is no doubt that we face, individually and collectively, a profound crisis brought about by the increasingly dynamic complexity of our social fabric. The ideas, images, and feelings of a smaller world now largely past cannot keep pace with what is developing before our eyes. It is as if our human capacities were growing by linear increments and our problems by exponential increments. We need to take an exponential leap to the new level. (Kepes, *The Center for Advanced Visual Studies,* inaugural catalog, 1968, 7)

CAVS was the first such program at an academic institution—an institution renowned not for art, but for science and engineering. CAVS was MIT's response to the socio-political environment of the 1960s and to new forms of communications technology. It was a heady time, full of promise and potential, idealism and intensity, where artists took the chance to make a difference. CAVS Director Otto Piene characterized the tendency toward unfettered experimentation, emboldened curiosity, and the quest for newness:

> If I want to see what it looks like when a hydrogen-inflated plastic hose burns and explodes, I must go to a place where this experiment will not hurt anybody; but the experiment as such is necessary to turn a thought into an experience. Almost all art forms are still more sensual than cerebral. (Piene, *More Sky*, 1970, 55)

In 2018, CAVS celebrated its fiftieth anniversary with an exhibition at the MIT Museum. The CAVS archive—begun in the 1980s as a series of exhaustive folders and files at the Center's Building W11 home—has at last been launched as the "CAVS Special Collection" by the MIT Program in Art, Culture and Technology (ACT). Now memorialized, while being reduced to an archive, CAVS is only beginning to be understood. Young artists and academic researchers are attempting to retrieve the world that was the Center for Advanced Visual Studies.

It is fitting that the first Center book to be published is written by insiders, and produced at MIT. Those of us who were there felt the burn. We knew the passion, dedication, and hard work that sustained us, and kept the program alive. This is not the story of individual artworks by individual artists. It is the story of the group work and collaboration that formed the core of CAVS activities. The large-scale public environmental artworks, exhibitions, symposia, and conferences that we participated in as a community inspired an enduring sense of friendship and family, while producing an impressive body of work. The community still exists. Together, we have made this first record—by no means encyclopedic, but with luck, a springboard to further research and publications. The title can be traced to efforts within CAVS—as early as the 1970s —to create a book describing the Center's mission and work, and mirrors the collective spirit of group projects like *Centerbeam* and *Centervideo*.

Centerbook concentrates on the years 1967-1994, ending with Otto Piene's retirement to become CAVS Director Emeritus. In addition to chapters on the founding and first directors, and exhibitions and symposia, there is a chapter on education and the creation of the first graduate program of its kind, which served as a dynamic model internationally. The chapter on Sky Art and the *Sky Art Conferences*, while a tribute to Piene, is significant for CAVS group work and continuing outreach activities.

Centerbook's key themes of collaboration, environment, new media, education, and group and personal projects reflect the timeless impulse of artists to create and contribute enduring value to society. In the spirit of Kepes' Bauhaus roots, CAVS artists broke boundaries through transdisciplinary thinking and experiments with genre not previously associated with artwork. The creative palette included anything going on at MIT. This positivism continued with Piene, exemplified in a text written during his Group Zero days:

> Yes, I dream of a better world.
> Should I dream of a worse?
> Yes, I desire a wider world.
> Should I desire a narrower?
> (Otto Piene, "Paths to Paradise,"
> *ZERO 3*, Düsseldorf, July 1961)

Piene's frequently cited words are followed by a paragraph which, together with Kepes' founding vision, helps to explain CAVS as a collaborative community willing to risk—succeed or fail:

> My dreams are different from songs and sagas. I am not pining away from longing and resignation because no patron will give me smoke and light. I already have my 12 searchlights, they belong to me. But they are just the beginning, for I would like 12 times 12, and then more, and they must be strong enough to light up the moon. (Piene, "Paths to Paradise")

Every CAVS participant, including many not specifically covered within the limitations of this first book, contributed to an art-science-technology DNA.

As publications by Kepes and Piene demonstrate, visual narrative can communicate history and ideas as powerfully as written narrative. *Centerbook* balances both approaches. Digitization of the paper files, slides, and notebooks in the CAVS Special Collection and Otto Piene Collection (fueled in part by this book), resulted in an archive of nearly 30,000 entries. The collection of sources colligated across genre and time revealed patterns and complexities, practice and process, recurring themes, and unpublished behind-the-scene moments. Sebring, who has worked large visual databases, did the design and layout of *Centerbook* to permit immersive looking. The image sequences and juxtapositions, captions and quotes, complement the written text. The "visual narrative" authoring mode made possible by the dynamics of the digital archive and production methods opens a panoramic yet intimate view on twenty-seven years of working life at CAVS.

Centerbook is the story of art and artists living and breathing the often-alien worlds of MIT. CAVS's visual legacy drives the narrative and design, which echoes the size and layout of Kepes' books and reflects the opening words of his groundbreaking work, *The New Landscape in Art and Science* (1956): "This book is meant to be looked at more than read." *Centerbook* may be both looked at and read as the visions and voices of CAVS past tell a prescient tale.

Elizabeth Goldring and Ellen Sebring

Acknowledgments

Centerbook, a visual narrative by Elizabeth Goldring and Ellen Sebring, relies on images provided by CAVS fellows, the CAVS Special Collection, and Otto Piene's slide archive housed at the MIT Museum. Texts are drawn from CAVS catalogs, MIT publications, archival materials and Joan Brigham's interviews. The outline of the book emerged from meetings with Joan Brigham, Keiko Prince, Otto Piene and Elizabeth Goldring. They were later joined by the head of the CAVS Special Collection, Jeremy Grubman, and Ellen Sebring, with initial consultation from Lowry Burgess, former Professor and Dean at the College of Fine Arts at Carnegie Mellon University. With the support of the late Dean William Mitchell, earlier ideations of the book were developed by Debra Edelstein, and later by Elizabeth Finch, Lunder Curator of American Art at the Colby College Museum of Art. Otto Piene championed the project until his death in 2014.

A number of former CAVS fellows, in particular Vin Grabill, Seth Riskin, Beth Galston, Sally Weber, Bill Seaman, Christopher Janney, and Todd Siler provided commentary as readers. Hashim Sarkis, Dean of the School of Architecture and Planning (SA+P); John Durant, director of the MIT Museum; Philip Khoury, Associate Provost at MIT; Gediminas Urbonas, professor at ACT; Peter Weibel, Chairman and CEO of ZKM | Center for Art and Media Karlsruhe; and Bernd Kracke, President of the Hochschule für Gestaltung (HfG) Offenbach am Main, backed the book for publication. The authors are grateful to contributors Durant, Weibel, Kracke, Prince, and Urbonas for adding context to the narrative.

Generous support for the publication came from SA+P at MIT; the Office of the Provost at MIT; the MIT Museum; ZKM | Center for Art and Media Karlsruhe; and printing support from The Hochschule für Gestaltung (HfG) Offenbach am Main. We also want to acknowledge the efforts of Melissa Vaughn, Director of Communications at SA+P, and Justin Kehoe at MIT Press. We thank Sarah Olson for her thoughtful editing of the manuscript, and Patsy Baudoin and Stacie Slotnick for editorial fine tuning. Heather McCune, Carl Canner, Alexandra Folts, and Deborah Douglas, Director of Collections & Curator, Science & Technology at the MIT Museum, offered both encouragement and insightful feedback as non-CAVS readers. Deborah Douglas and Marion Cunningham, Administrative Officer of ACT, were instrumental in the intake of the Otto Piene Collection which involved digitization of 7,550 slides, many being seen for the first time in this publication. We are especially grateful to Jeremy Grubman for his research, advice and commitment to the project.

There is no one story of CAVS. There are as many stories as there are fellows and graduates. We are very appreciative to each and every CAVS alum who contributed to this seminal movement in art, science, and technology at MIT.

Finally, the authors remain thankful to each other for their flexibility, perseverance and enthusiasm rendering this publication in the true collaborative spirit that epitomized CAVS.

Image Credits

Unless otherwise noted, the images in this book come from two collections at MIT: the CAVS Special Collection housed at ACT; and the Otto Piene Collection housed at the MIT Museum.

The Interviews

Our book was a process. The formal interviews of fellows came about as Elizabeth, Keiko, and I were doing research for the chapter on the social climate of the Center. Together, we decided that it would be best to hear as many voices as possible. I volunteered to conduct interviews with former CAVS fellows and alumni. For over a dozen years I traveled far and wide to carry out these interviews not only throughout New England, New York, and California, but also in Canada, Spain, Greece, Germany, England, and France in pursuit of the Center's international community. The recorded conversations, which often lasted hours, even days, have been excerpted here to preserve the diverse timbre of CAVS voices. Their answers were often lengthy, and there was a great deal of conversation back and forth. I transcribed the tape-recorded interviews, and then sent the transcripts to each interviewee to revise and correct.

I asked only three questions. First, how did you come to be at CAVS? Second, what did you do when you were there, including everything about your life there? And third, how do you understand that experience?

Joan Brigham, former CAVS Fellow

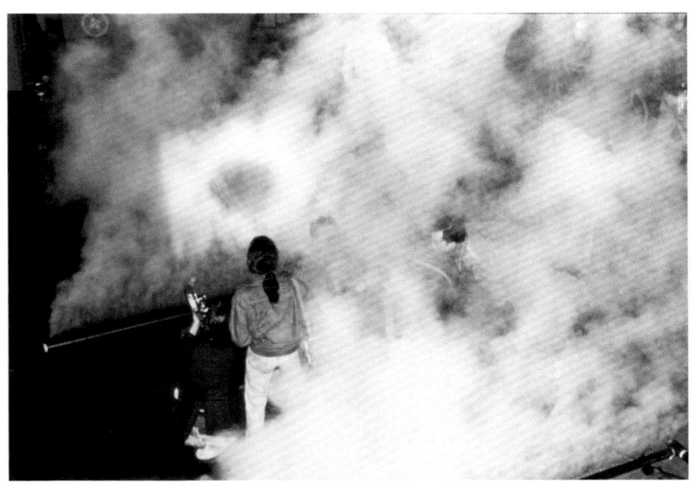

Joan Brigham and Stan VanDerBeek, "Steam Screens," Brigham pictured looking at VanDerBeek's computer-generated film imagery projected onto her steam installation, Whitney Museum of American Art, 1979
Photo: Francene Keery

Chapter 1

Founding and the Directors

> The scientist may be an extra brain to the artist, and the engineer may be an extra arm to the artist, whereas the artist can be an extra eye to the scientist and engineer. (Otto Piene, *Centerbeam* film directed by Richard Leacock and Jon Rubin, 1978)

In the late 1960s, a time of widespread anti-war protests, MIT did the unexpected and established the first academic center for research and collaboration in art, science, and technology. The Center for Advanced Visual Studies would bring artists to the MIT campus to explore innovative, peaceful expressions of a rapidly evolving technological era. As a breeder of intermedial art forms, CAVS became a model for media labs worldwide.

Centerbook focuses on CAVS during the 27 years spanning 1967 to 1994 under its first two directors. Chapter 1 introduces the tenure of Founding Director, György Kepes, and long-term Director, Otto Piene, and touches on the Center's post-Building W11 nomadic years helmed by Co-Directors Paul Earls and Elizabeth Goldring, and Directors Steve Benton and Krzysztof Wodiczko.

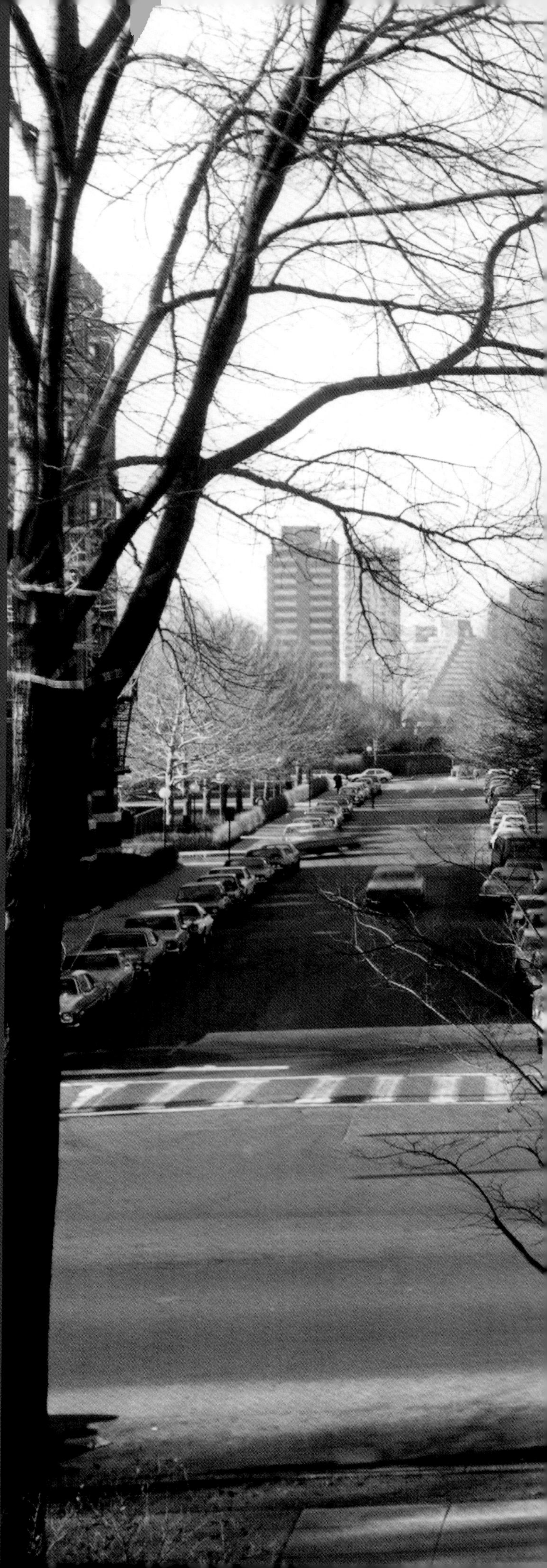

Right: Peaked roofs draw light into the artist studios in MIT Building W11, the home of CAVS at 40 Massachusetts Avenue and Amherst Street, Cambridge, Massachusetts
Photo: Nishan Bichajian

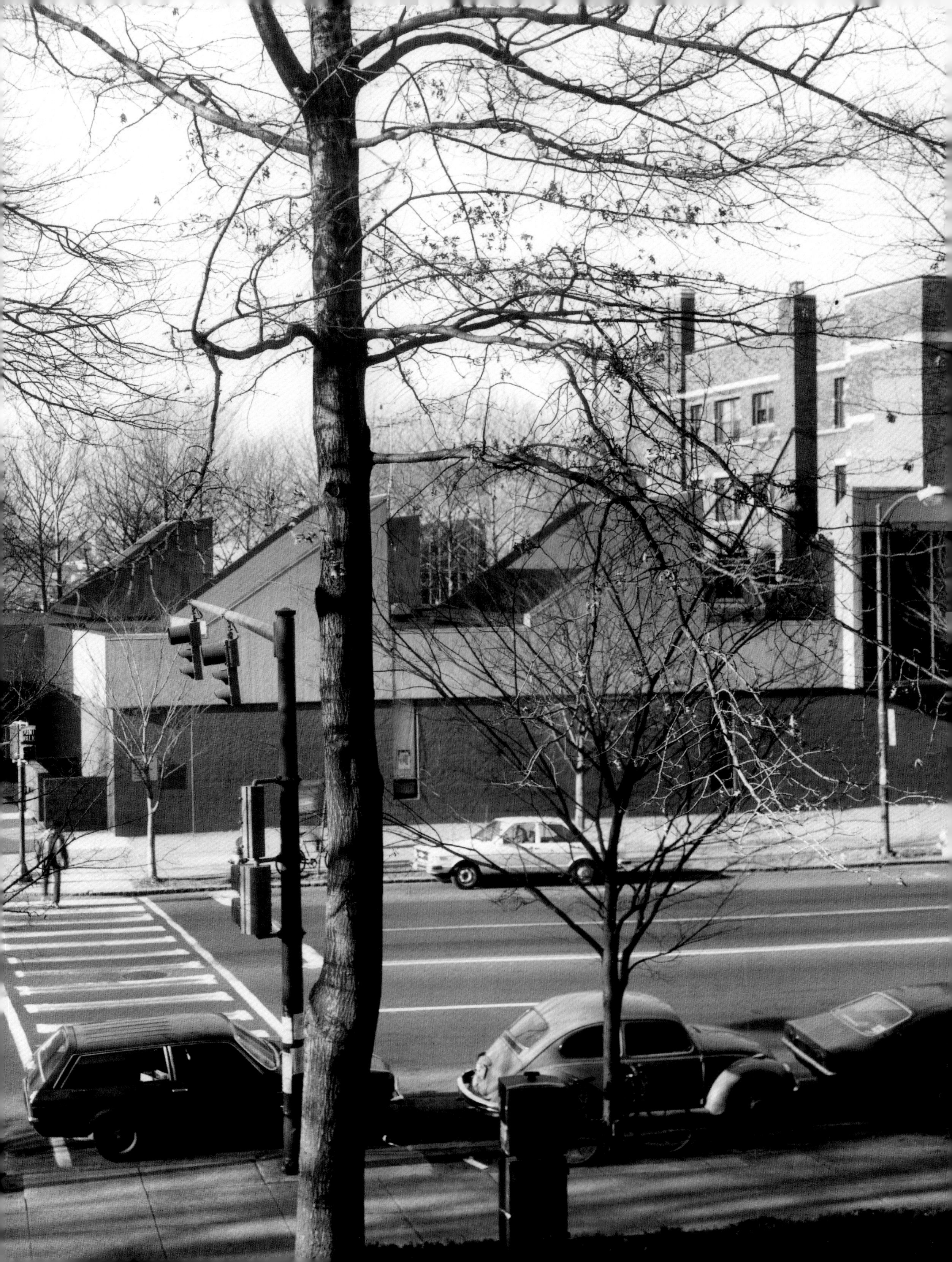

György Kepes, CAVS Founding Director, 1967-1974

On March 21, 1968, the year following its founding, the Center for Advanced Visual Studies was dedicated at the Massachusetts Institute of Technology along with the Center for Theoretical Physics.

In his address, MIT President Howard Johnson concluded that the occasion allowed MIT "a rare opportunity to celebrate the wholeness of man—his learning, his experimental approach to life—his appreciation of what he is and what he may become."

CAVS (pronounced C.A.V.S.) was the brainchild of Hungarian-born MIT professor György Kepes (1906-2001). Kepes had long espoused the unity of science and art, and the urgent need to repair their fission:

> Science and the arts have steadily grown apart. The division has grown into something like open warfare in the last century, when many artists, repelled by the physical and social blight that

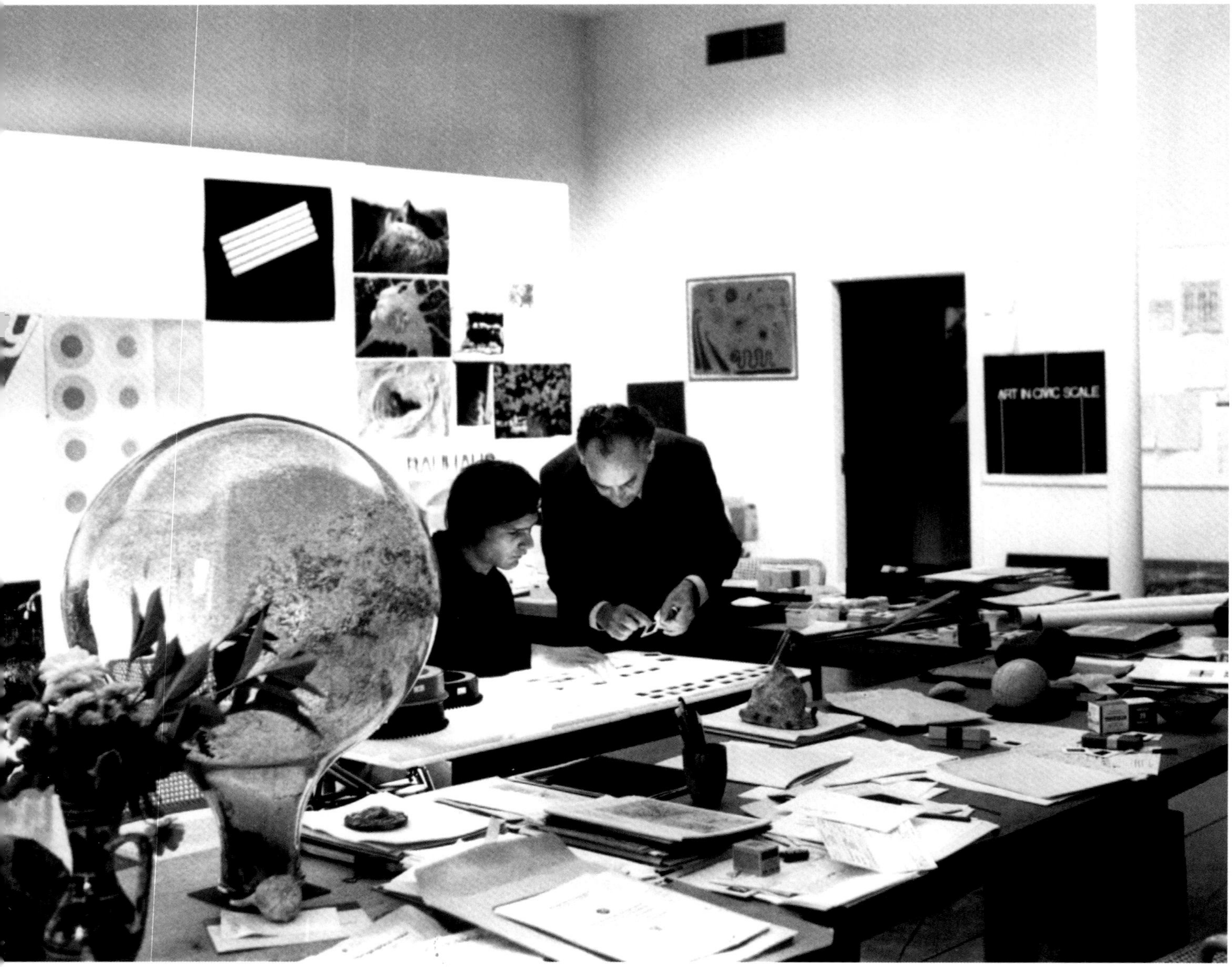

György Kepes working in his office/studio at CAVS, MIT, 1971
Photo: Nishan Bichajian

accompanied the industrial revolution, turned their backs on the technical and scientific world. Thus, majestic achievements of the human mind were seen exclusively in their disruptive effects on man and society. Our need to harmonize our inner and outer world became both patent and urgent. (György Kepes, *The Center for Advanced Visual Studies,* inaugural catalog, MIT, 1968, 2)

He advocated an increased contact between disciplines that would make MIT a meaningful base for artists:

> Art and science complement each other; together they constitute the full horizon of creative mental life. Hence their mutual importance. Just as scientists and engineers fill out their horizons through communication with artists, so do artists through communication with scientists and

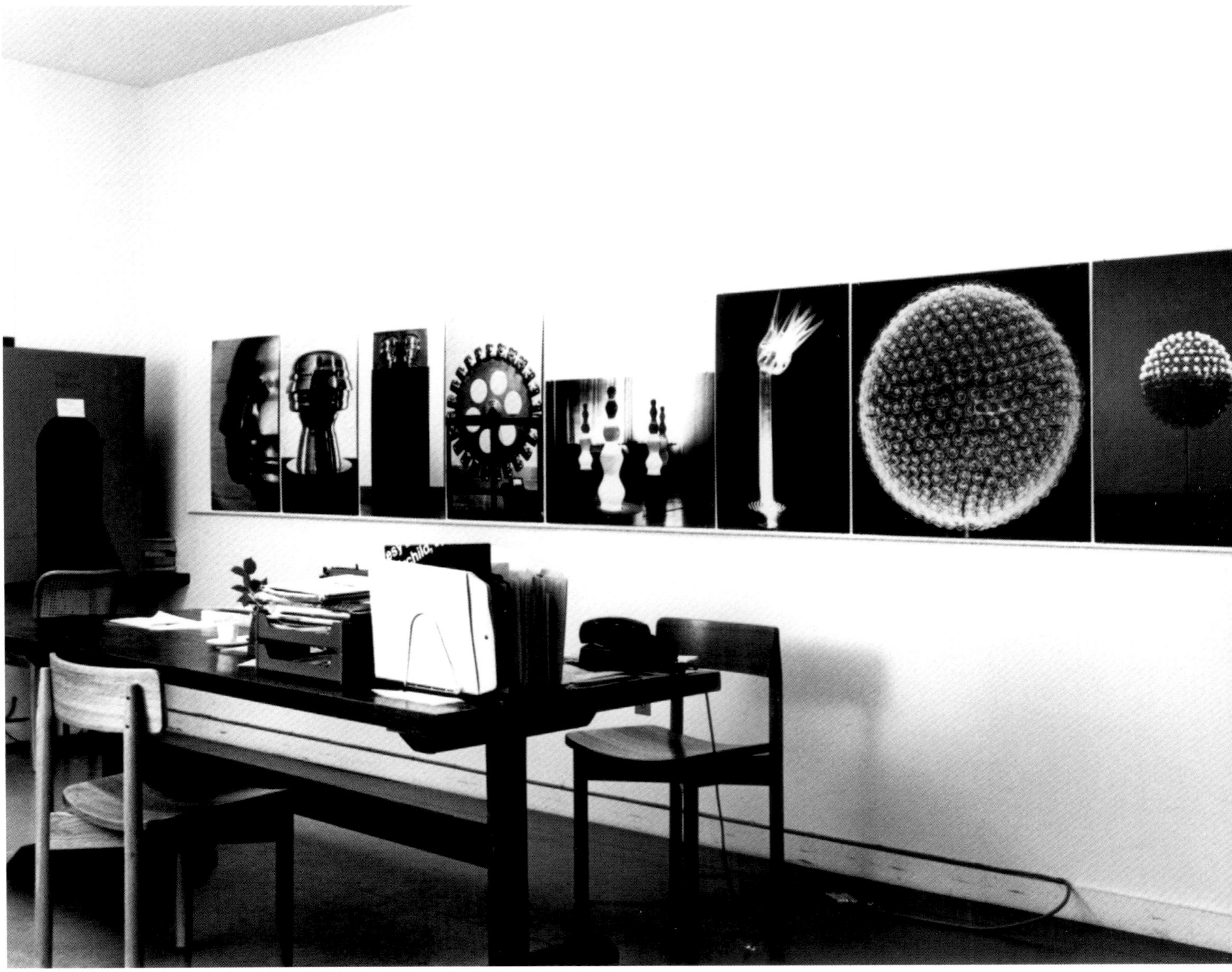

Otto Piene studio, during the *Art in Civic Scale* symposium, CAVS, MIT, May 19, 1971
Photo: Nishan Bichajian

engineers. When artists find science so forbidding or so withdrawn a realm that they will not or cannot enter, they restrict themselves to half the available spectrum of thought and feeling; half the world slips through their fingers. (György Kepes, *The Center for Advanced Visual Studies*, inaugural catalog, MIT, 1968, 3)

As a young filmmaker, Kepes worked with Bauhaus teacher László Moholy-Nagy and later joined Moholy-Nagy's New Bauhaus in Chicago as head of the light department.

> [Kepes] moved to Berlin in 1930 at the invitation of a fellow Hungarian, Moholy-Nagy, a charismatic former Bauhaus teacher, who was at the forefront of experimentation with light and moving imagery. He was as much an extrovert as Kepes was an introvert, but they were both so driven and intellectually inquisitive that they forged a dynamic, if occasionally explosive partnership. When Moholy-Nagy fled Nazi Germany in 1935, Kepes followed him, first to Amsterdam, then London and, finally, Chicago, where they founded a new design school, the short-lived New Bauhaus. (Alice Rawsthorn, "A Master of Image and Information," *The New York Times*, Oct. 10, 2010)

In 1946, Kepes became Associate Professor of Visual Design at MIT's School of Architecture and Planning. It took two decades for his vision of an MIT workshop where artists would collaborate with scientists and engineers to become a reality, coming to fruition during what Fellow Lowry Burgess described as an explosive socio-political climate:

> György Kepes created CAVS at MIT in the midst of socio-political-cultural explosions. Cambridge, Massachusetts was a hotbed of strife and change. Everywhere, both in the immediate context of MIT, Cambridge and Boston, as well as in Kepes' native Eastern Europe and Asia, Latin America and Africa, massive revolutions of the human spirit were manifest in upheavals and the turmoil of deep passions [and] the utter commitment to transformation. All those movements surrounded and impacted Kepes and his creation of the Center for Advanced Visual Studies. (Lowry Burgess interview)

Part of the rationale for creating CAVS as a bastion of humanism was to appease Vietnam War protesters and counterbalance MIT's involvement in military-related research in the wake of World War II:

> Certainly, György Kepes was well aware that MIT was in the middle of this surrounding transformative and creative fire. The deep internal and external controversies concerning MIT's role in expanded war research and sponsorship by the US government and the 'military-industrial complex'—an awareness that ultimately culminated in the formation of the Union of Concerned Scientists in 1969—were intensely debated on the MIT campus and elsewhere. Historian William Irwin Thompson (*At the Edge of History: Speculations on the Transformation of Culture*, 1971) was teaching at MIT where he articulated the internal necessities of MIT to deflect attention from itself by creating humanistically oriented research and centers to camouflage its defense research. The creation of CAVS was one of the MIT actions Thompson named. Nonetheless, at the time, Kepes was a well-known anti-war and pro-civil rights advocate originating from the many historic situations he actively engaged with. (Lowry Burgess interview)

Facing page: György Kepes, director, with fellows Otto Piene and Harold Tovish at CAVS, MIT, 1969
Photo: Nishan Bichajian

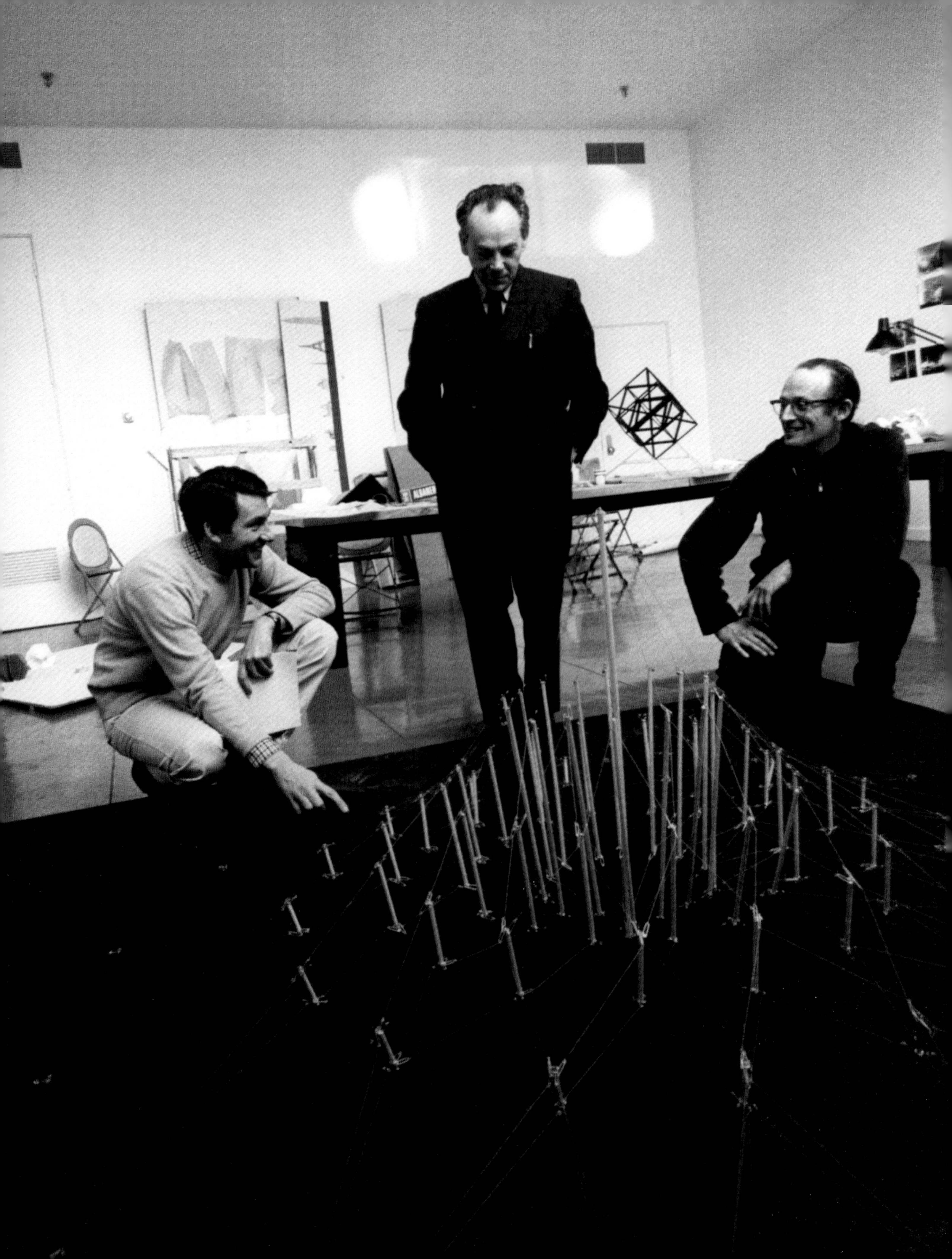

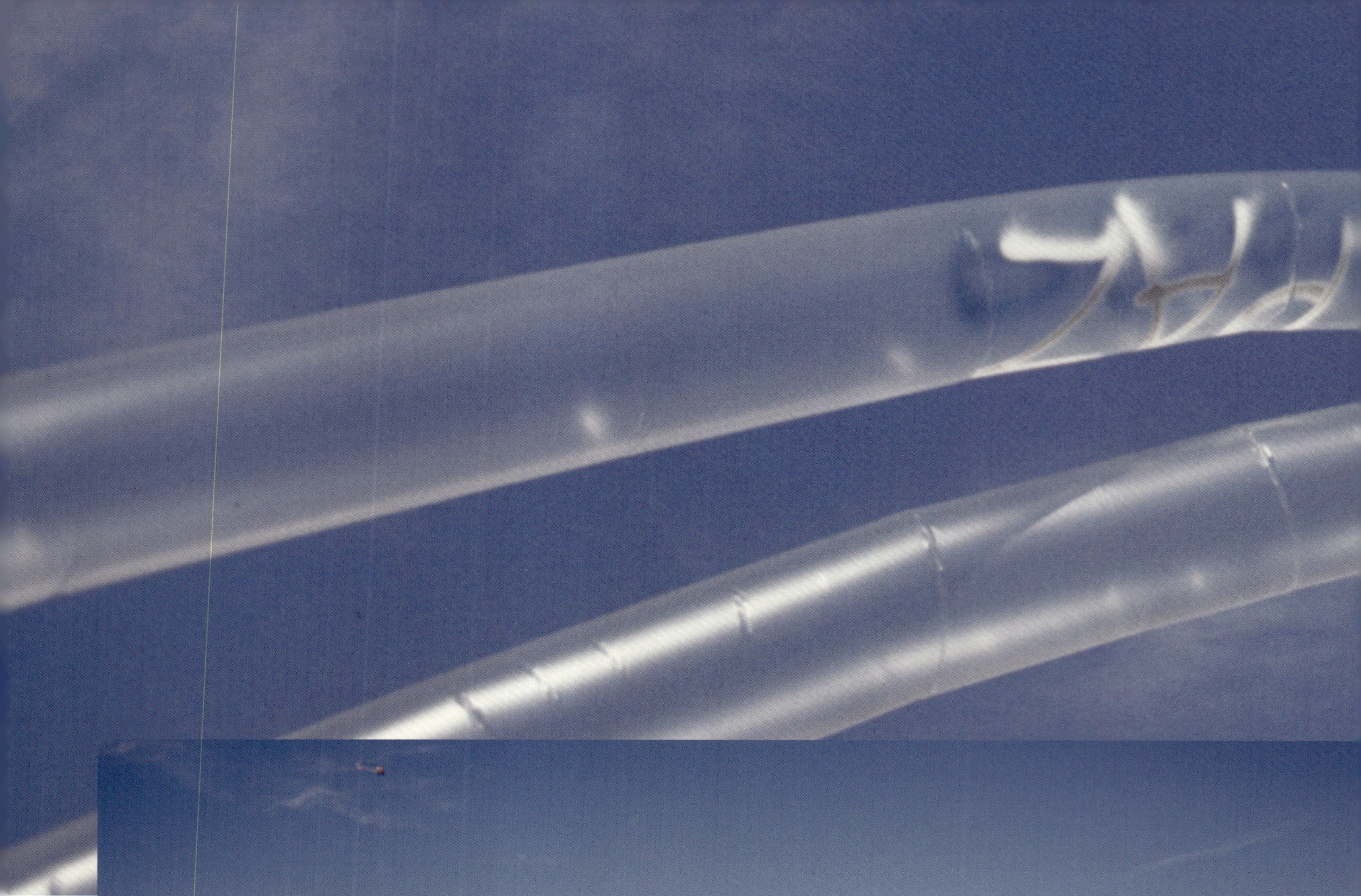

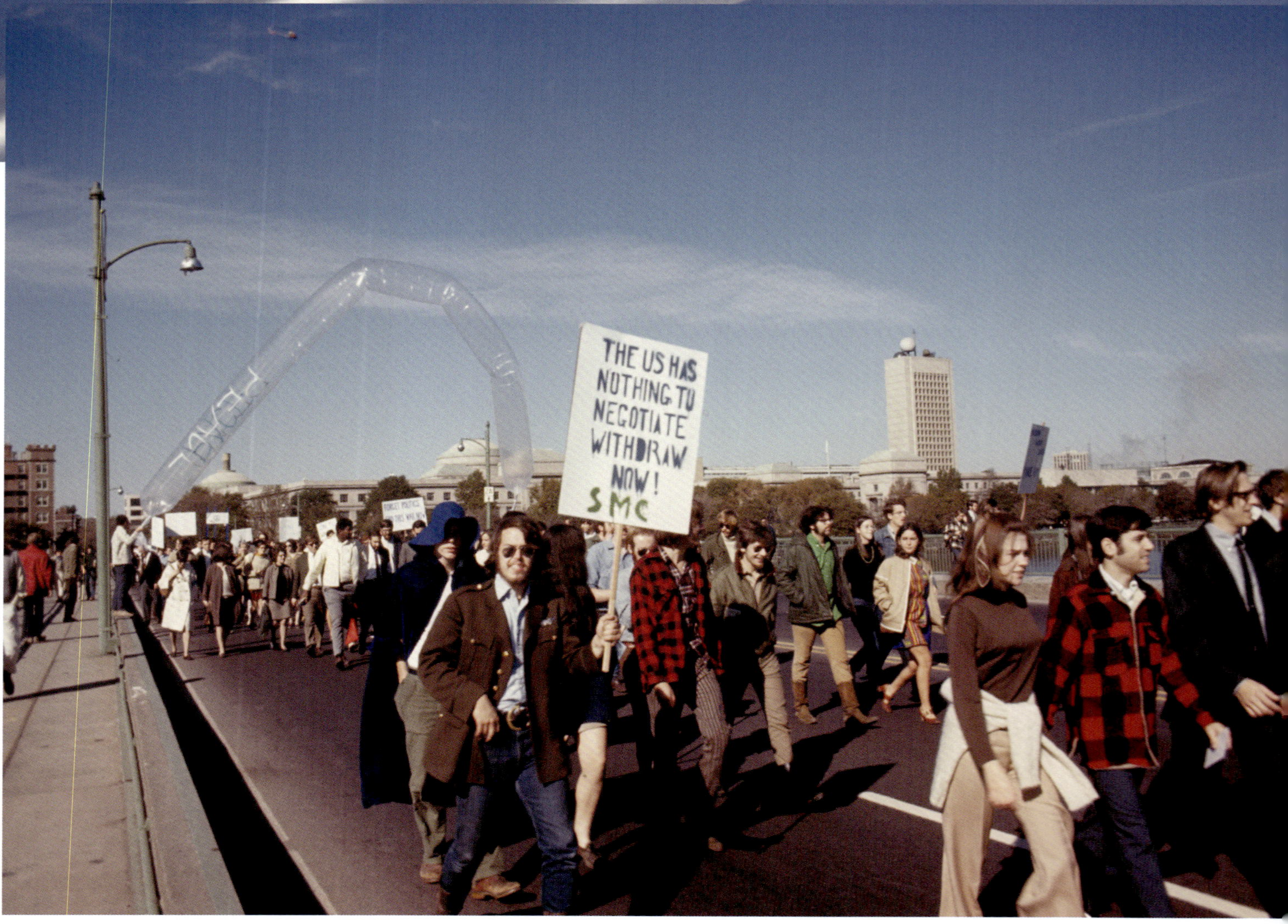
THE US HAS
NOTHING TU
NEGOTIATE
WITHDRAW
NOW!
SMC

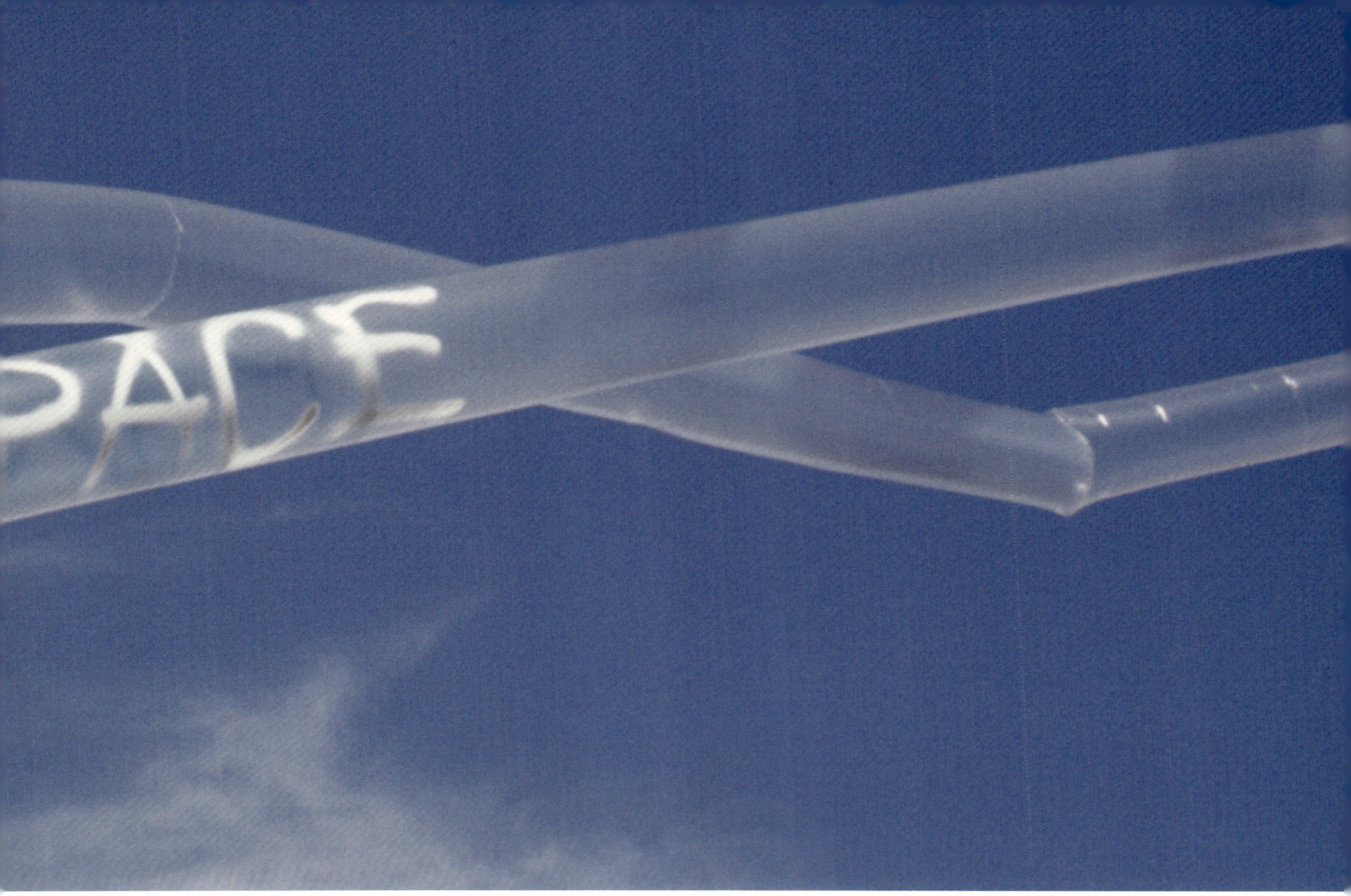

In October 1969, when 100,000 antiwar protesters formed the largest crowd ever to occupy Boston Common, MIT's contingent stood out because of [MIT President] Jerome Wiesner's solidarity and Piene's "insane balloons." (Matthew Wisnioski, "Arts / Science / Technology" in *A Second Modernism: MIT, Architecture, and the 'Techno-Social' Moment*, 2013, 188)

The Center is definitely . . . speaking out and trying to act in the name of the right values. That is, love and peace, and interaction among well-meaning creative people. Not only in the arts but also in the sciences, and among the engineers. . . . And these peace-loving people were not only among the artists, they were also among the scientists, particularly among those at MIT who had been major players in World War II, such as those who were involved in the concept and development of nuclear bombs. (Otto Piene interview in *A Second Modernism*, 772-773)

When the demonstrators crossed the Harvard Bridge into Boston, they raised their arms in peace signs at police and press helicopters swooping low over them. (*The Harvard Crimson*, 1969)

Above and facing page: students fly Otto Piene's "peace" inflatables while crossing the Harvard Bridge (MIT in the background) en route to the Boston Common during the national protest, the Moratorium to End the War in Vietnam, October 15, 1969
Photo: Otto Piene Collection

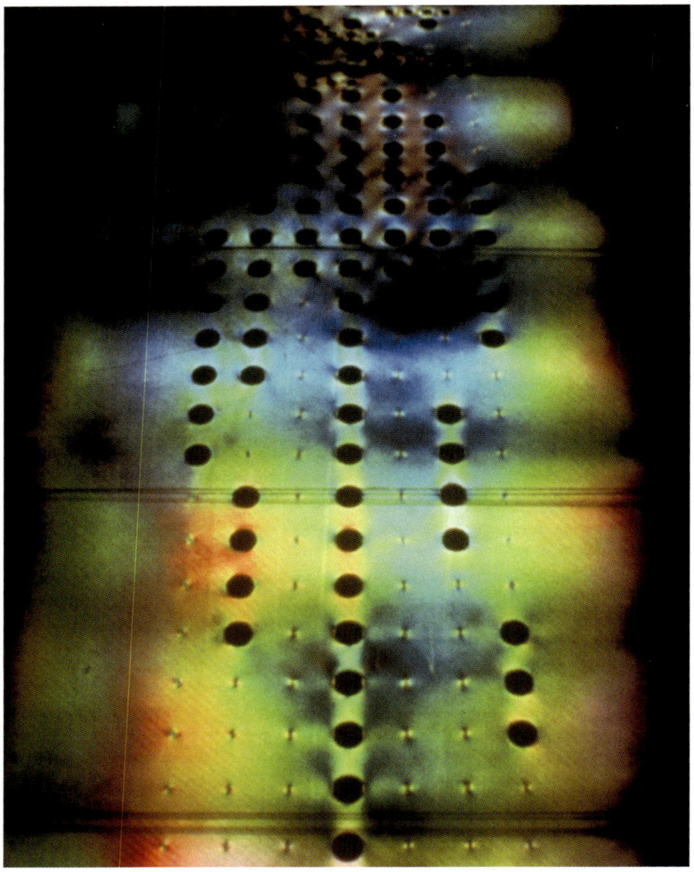

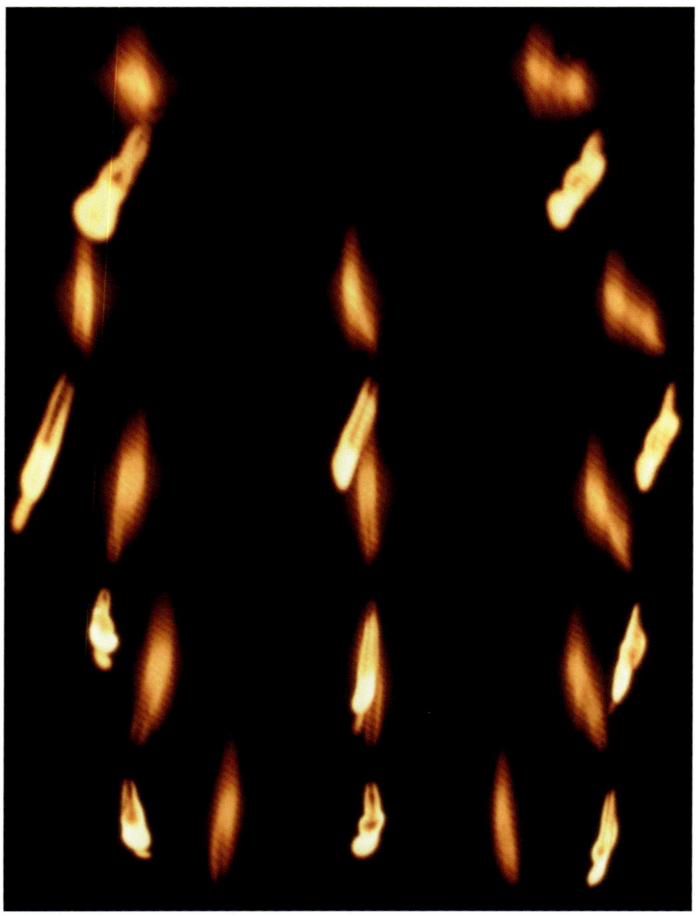

Immediate precursors in the broader art-and-technology movement included Bell Labs and New York City-based Experiments in Art and Technology (E.A.T.), founded in 1967 by engineers Billy Klüver and Fred Waldhauer and artists Robert Rauschenberg and Robert Whitman. E.A.T. artists joined forces and employed engineers to create interdisciplinary performances and installations. E.A.T. differed from CAVS in being industry rather than university based, and in generally emphasizing engineering over science. Kepes envisioned CAVS as an artists' equivalent to Princeton University's Institute for Advanced Study, which hosted such scientific luminaries as Albert Einstein, Kurt Gödel, and J. Robert Oppenheimer, and which, in part, inspired the Center's name. Jack Burnham, appointed as a CAVS fellow in 1968, commented on the naming of the Center:

> To many, the Center's formal name seemed academically pretentious—Center for Advanced Visual Studies. Yet, essentially we were artists playing with our computers, light environments and holographic realities. No tweedy academics in horn-rimmed glasses and white lab coats. (Jack Burnham interview)

Historically related to CAVS, the Bauhaus in Germany (1919-1933) and Black Mountain College in North Carolina (1933-1957) were artist collaboratives, both with the unspoken mandate to not be art schools in the traditional sense:

> When John Rice established Black Mountain College in 1933, he sought to create a school that dissolved distinctions between curricular and extracurricular activities, [and] conceived of education and life as deeply intertwined. (Ruth Erickson, *Leap Before You Look*, 2015, 77)

Black Mountain artists included Robert Rauschenberg, Joseph and Anni Albers, John Cage, Charles Olson, and others wishing to work interdisciplinarily, combining poetry and dance, painting and music, and so forth. Kepes was familiar with this model, and invited a number of Black Mountain teachers,

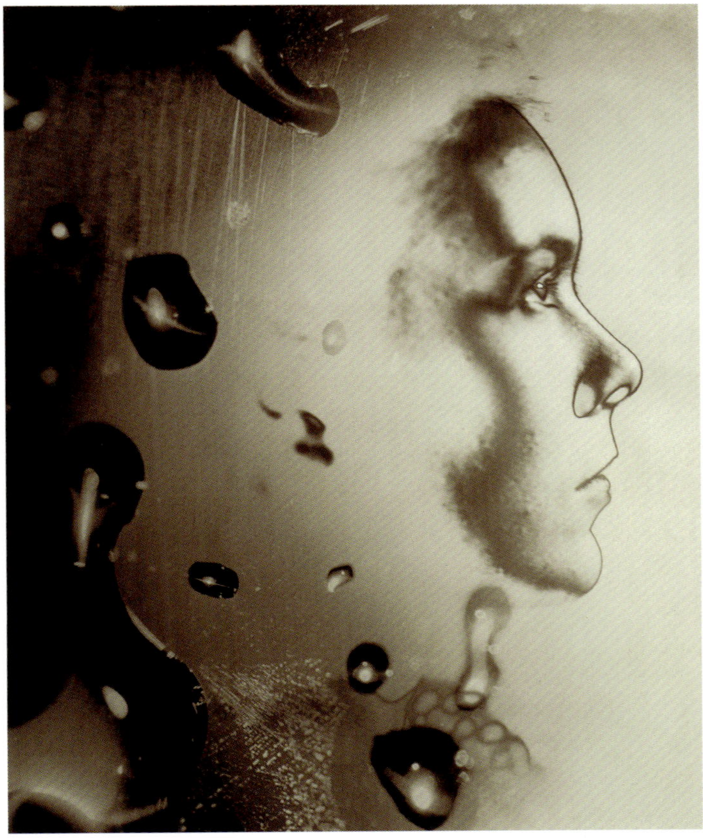

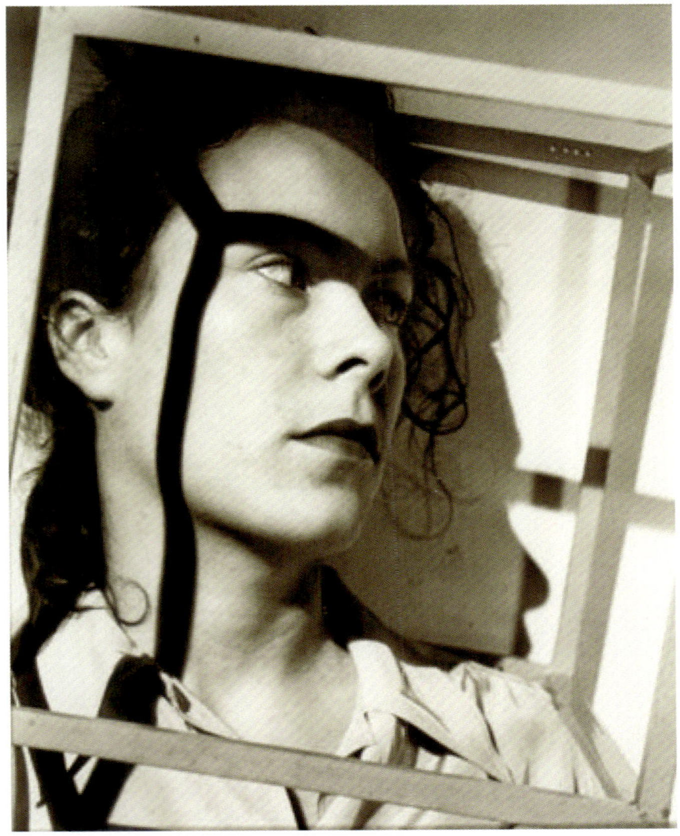

including John Cage and Buckminster Fuller, to contribute to his publications and symposia.

Kepes was part of the academic intelligentsia in Cambridge, Massachusetts. He and his wife, Juliet, an artist and children's book author, hosted culturally rich evenings at their summer home, designed by close friend Marcel Breuer, on Long Pond in Wellfleet, as well as at their home at 90 Larchwood Drive in Cambridge's Brattle Street area, a favorite residential quarter of Harvard academics.

The Archives of American Art at the Smithsonian Institution holds photographs and letters exchanged by Kepes and his famous contemporaries, including Ansel Adams, Josef Albers, Rudolf Arnheim, Harry Bertoia, Marcel Breuer, Alexander Calder, Henry Dreyfuss, Charles Eames, Max Ernst, Buckminster Fuller, Naum Gabo, Walter Gropius, Jean Helion, Aldous Huxley, Le Corbusier, Margaret Mead, Piet Mondrian, Robert Motherwell, Pier Luigi Nervi, Robert Oppenheimer, Herbert Read, Hans Richter, George Rickey, Jonas Salk, David Smith, and Saul Steinberg.

Above: György Kepes, photographs of his wife, Juliet, left, "Solarization," c. 1938; right, "Juliet's Shadow Caged," 1939/41
Courtesy of the Estate of György Kepes

Facing page top: György Kepes, with William Wainwright, "Photoelastic Walk," interactive light installation, 1969
Photo: Nishan Bichajian

Facing page bottom: György Kepes, with Paul Earls and Mauricio Bueno, "Flame Orchard," sound-responsive flame field, 1972
Photo: Nishan Bichajian

Creating a Center at MIT

In correspondence and conversations with leading MIT scientists and engineers over many years, Kepes shared his vision of a groundbreaking interdisciplinary center. He was a master consensus builder, as can be seen in a letter to MIT President Julius Stratton:

> I talked with Dr. Townes, and Professors Wiesner, Zacharias, C.S. Smith, Teuber, Rossi, Bitter, Morrison, Edgerton, Cerrillo, Lynch, Rosenblith, Fleisher, George Clark, and others. . . . [My meetings] were rewarding and most encouraging . . . I received valuable suggestions and promises of collaboration—qualified, naturally, by the availability of time. (Kepes letter to President Stratton, June 17, 1965)

Once the faculty and administration embraced Kepes' idea, attention turned to finding an appropriate location for the Center. Two buildings headed the list: the MIT Coop (Cooperative Society) bookstore and an industrial building next to the Daggett complex. Kepes evaluated their suitability as artist studios:

> The Coop has a convenient, central location, close to the School of Architecture and to Kresge [Auditorium], whose small theater could be used for a seminar and projection room. . . . The Coop's shortcomings are its characterless space arrangement and the need, imposed by its complicated structure, to doctor it considerably in order to transform it into a congenial working space. The industrial plant near the Daggett building has a simple, honest, rustic strength. But the unfinished nature of its space presents several liabilities; vibration, from the nearby printing plant, that can be harmful for photographic and film work; and an almost uncontrollable dirt and dust problem, which is equally menacing to film and photographic work. This building, like the Coop, is just a little meager in space, and will not give us sufficient storage area, which is almost as important in studios as work space itself. (Kepes letter to President Stratton, June 17, 1965)

The Coop was selected despite its limitations. MIT announced the renovation of a new center for "the interplay among artists and people from other fields" (MIT press release, July 13, 1967). It was renovated for some $250,000 and divided into five studios with vaulted skylights tailored to a community of artists. MIT Building W11 at 40 Massachusetts Avenue on the corner of Amherst Street would become the emblematic home of CAVS for the next 29 years.

> My first visit to Cambridge was in the mid-seventies, a new city strongly defined by the river as an element of its context, and creating the border between Boston and Cambridge. I saw Cambridge as a specialized place, a college town, a large intellectual community, an historical spin-off from 1968. It was searching for alternative educational situations but it was also a kind of celebration city. MIT was there with its big buildings with Greek columns, industrial construction, real estate, and recycled barracks. Later I remember seeing Noam Chomsky in his office, which looked like a barracks. The MIT campus looked like a temporary facility, more like an ex-colony or a military compound, an accumulation of things that created a mosaic from many different sources, making different kinds of contradictory, interesting patterns. (Antonio Muntadas interview)

For the initial budget and space requirements of the new Center, Kepes proposed $157,000 for equipment, and $180,000 as an annual operating budget that included salaries for a select number of artist fellows. Later, these figures were increased to $220,000 for equipment and $225,000 for the operating budget.

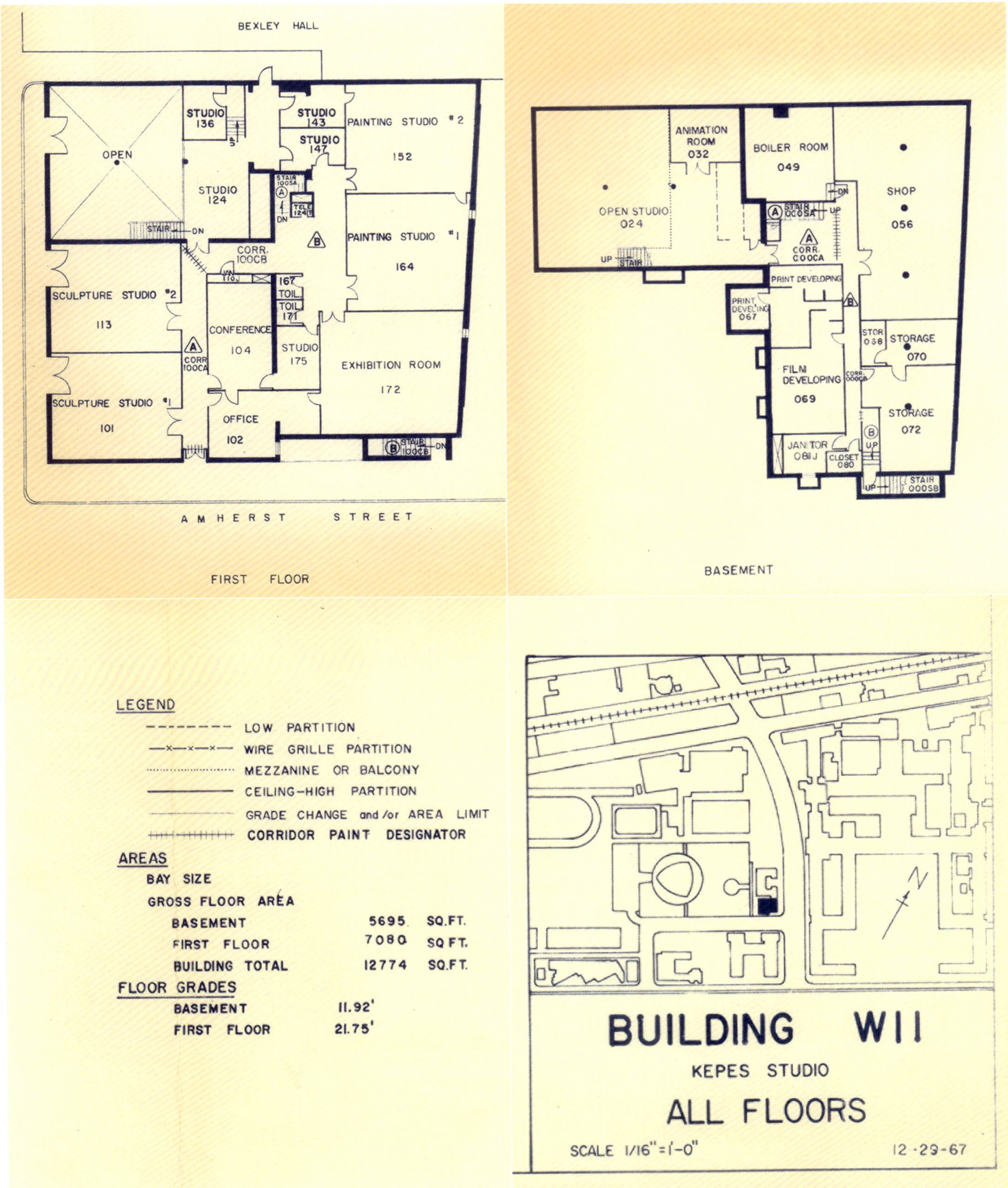

Blueprints for the new CAVS at MIT Building W11 on the corner of 40 Massachusetts Avenue and Amherst Street, Cambridge, Massachusetts; concept and design, Marvin Goody and György Kepes, 1967
Source: CAVS Special Collection

First Artist Fellows

Harold Tovish, a Boston artist who was also an ardent pacifist, became CAVS's first appointed fellow:

> Kepes and I were in the same gallery, The Alpha Gallery. I had an exhibition there and he came and saw two pieces which interested him. One was an illusion piece. The other involved a three-dimensional movie. A whole series of heads were mounted on a wheel—there were holes under each head—which triggered camera-like, high-speed photographs, so they disappeared and returned. The other piece used parabolic mirrors, to make an illusion and people were invited to touch the sculpture. When they did, it wasn't there. Kepes was intrigued, called and asked if I would be interested in joining CAVS and I said, "of course." (Harold Tovish interview)

Kepes met German-born artist Otto Piene, together with Charles Eames and Harry Bertoia, at Piene's 1965 Light Ballet exhibition at the Howard Wise Gallery in New York City. He told Piene about his plans to form CAVS. In ensuing correspondence, Piene responded to Kepes' invitation:

> I thought about your plan to suggest my being invited as a fellow to work on the light tower and found that I like your project very much. . . . Your suggestions about studios, about distribution of time for mutual as well as individual purposes and about the possibilities of using and exploring the facilities and the expert advice at the Institute seem very sensible to me. The work on the great light project could be a challenge to everyone involved. My conclusion is that I would like to accept if all your plans work out well. (Piene, letter to Kepes, Jan 28, 1966)

Kepes asked Piene what he would need in his studio, to which Piene answered pragmatically in the same letter, "space, gas, water, electricity, and fireproof walls." He recollected that he arrived to find lots of outlets, but not much power and no fireproof walls. Piene had co-founded Group Zero in 1959, as an international group of like-minded artists, galvanized by the all-out destruction of World War II, who often exhibited and worked together.

> From the beginning we looked upon the term [ZERO] not as an expression of nihilism—or as a dada-like gag, but as a word indicating a zone of silence and of pure possibilities for a new beginning as at the countdown when rockets take off—zero is the incommensurable zone in which the old state turns into the new. (Piene, "The Development of the Group Zero," *Lightworks*, Volume 13, 1980)

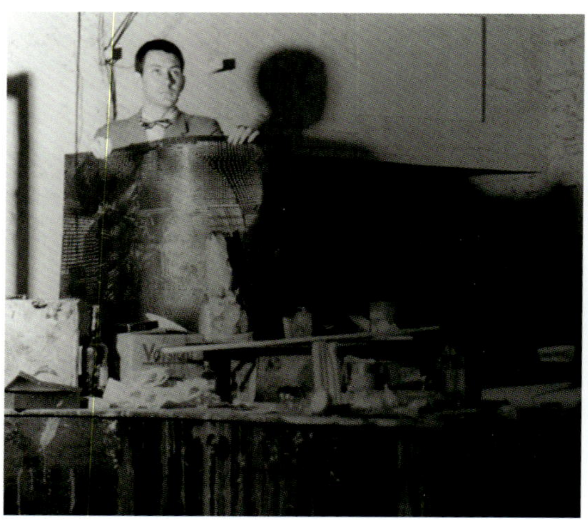

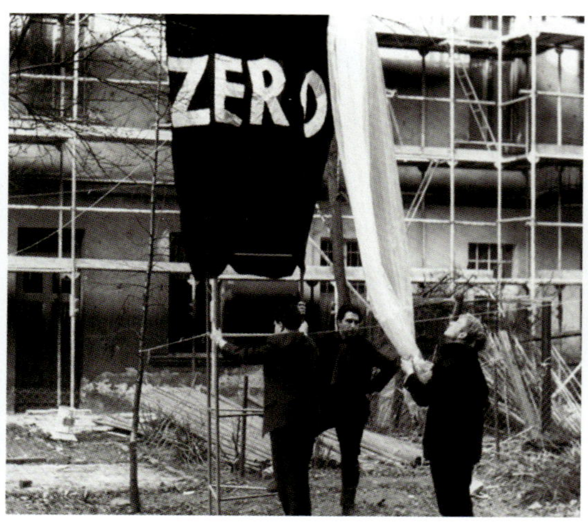

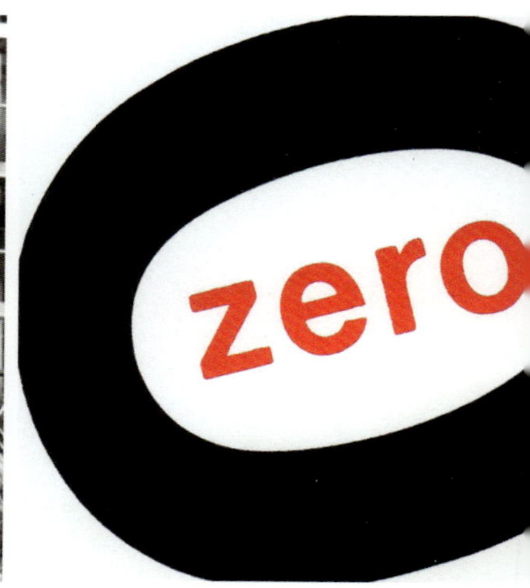

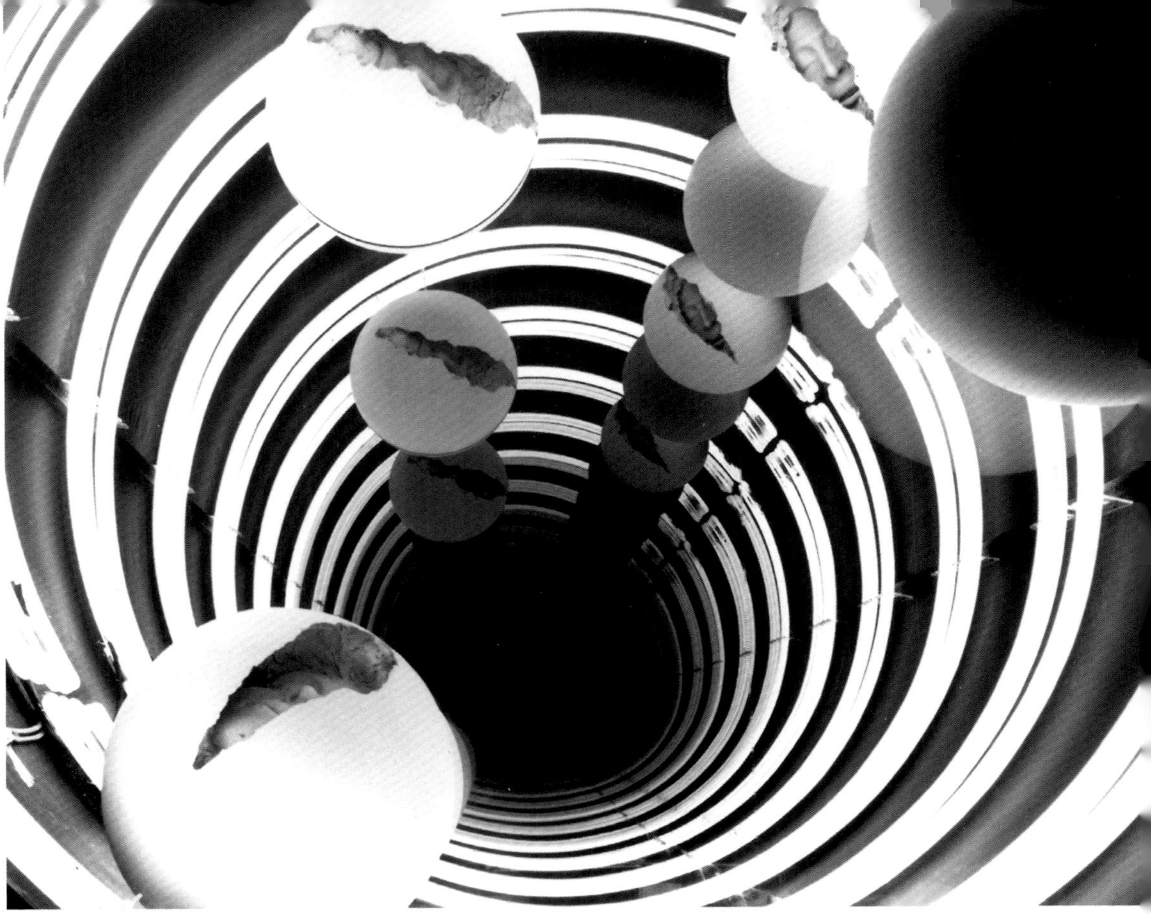

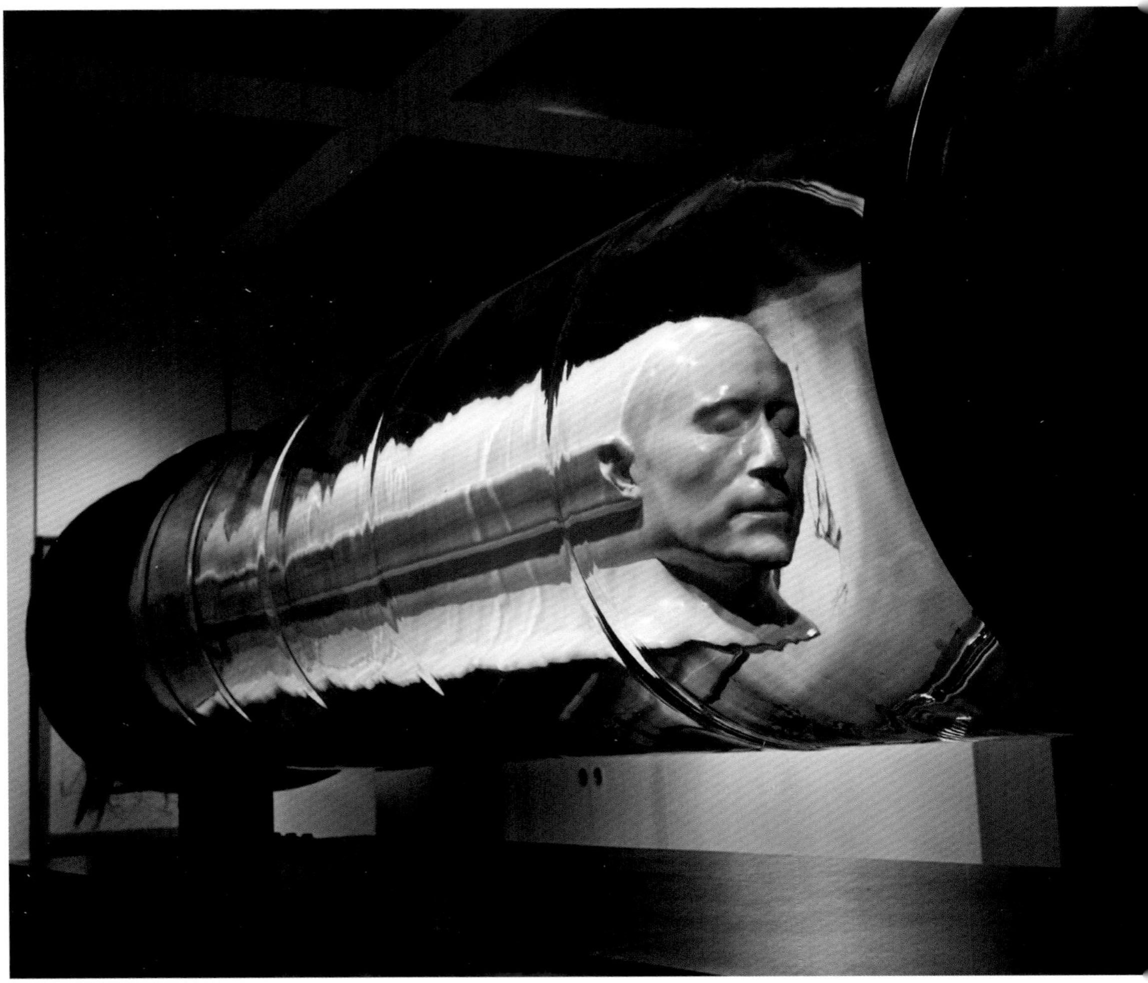

Above right: Harold Tovish, "Downfall," mixed media, 1984
Photo: CAVS Special Collection

Bottom right: Harold Tovish, "Accelerator," sculpture in glass tube, inspired by Cambridge Elextron Accelerator (Harvard and MIT), 1968
Photo: CAVS Special Collection

Facing page far left: Otto Piene with stencil in his studio at Gladbacherstrasse 69, Düsseldorf, Germany, 1959 (detail)
Photo: Paul Brandenburg

Facing page middle: Hermann Goepfert, Gotthard Graubner, and Heinz Mack hoisting ZERO Flag, *ZERO—der neue Idealismus*, Galerie Diogenes, Berlin, Germany, 1963 (detail)
Photo: Hilde Zenker

Facing page right: Group Zero logo (detail)
Source: CAVS Special Collection

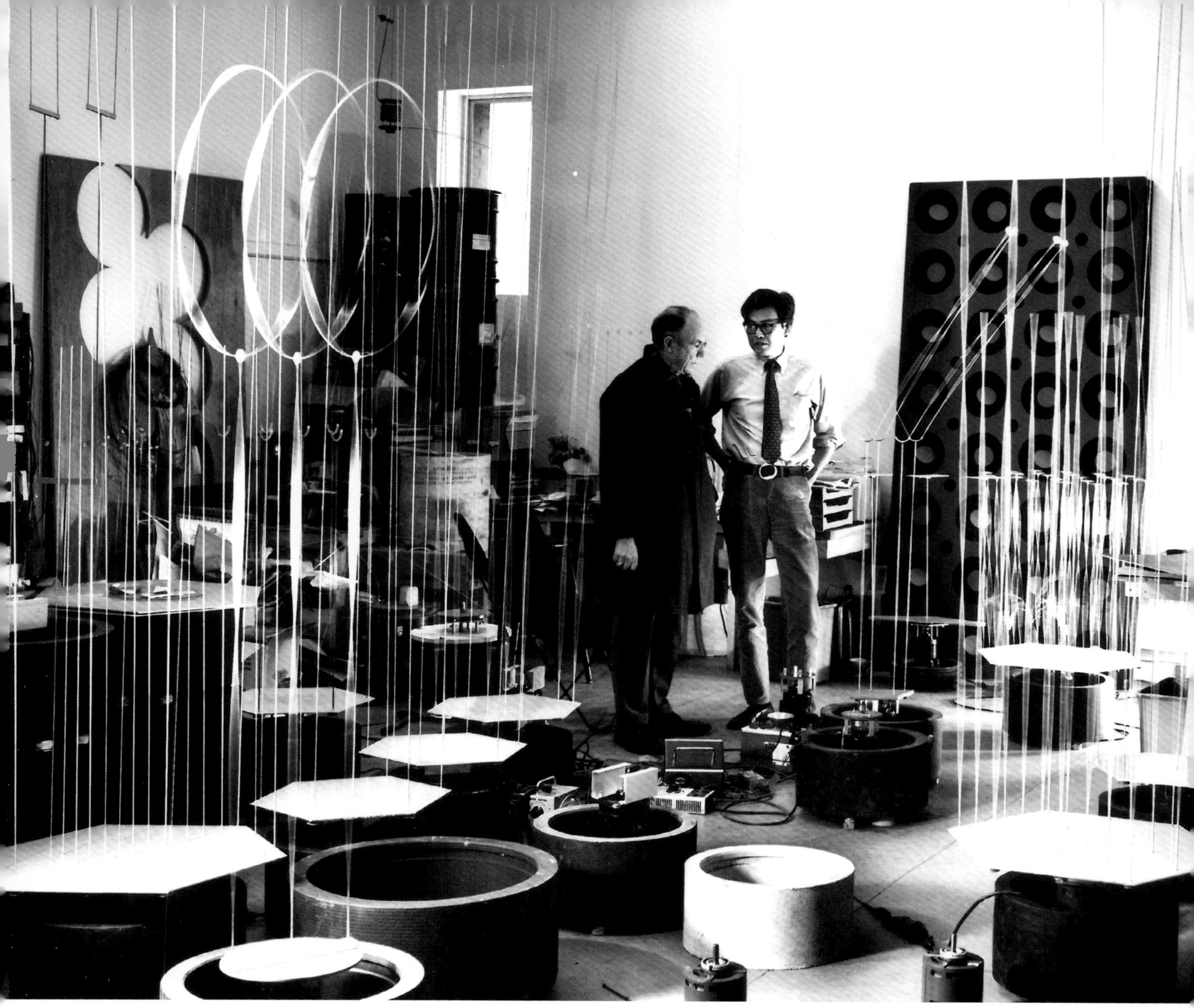

Between 1968 and 1969, the intense, highly theoretical international workshop and think tank of the Kepes years hosted the initial fellows—Burnham, Ted Kraynik, Piene (the first CAVS fellow from outside Boston), Vassilakis Takis, Tovish, Wen-Ying Tsai, and Stan VanDerBeek—who were established kinetic and light artists drawn to Kepes, MIT, and the idea of a world community of artists. Reflected in the Center's mandate were Kepes' widely published views on art, science, technology, the environment, and art on a civic scale. Kepes envisioned artists and engineers working together to bring water and sky back to urban environments, and toward the aesthetic design of technical apparatus.

CAVS's mission should be seen as a continuum of the interchange between art and science. What makes the Center so fascinating is that it sits historically at a critical moment in that timeline when science and technology reached a turning point. Science no longer searches for the law of nature. It began to manipulate the laws of nature. CAVS came into existence at this moment in time. This could not have happened without MIT. (Friedrich St. Florian interview)

Takis, a fellow who used magnetism to create sculptures both illustrative and tactile, invented a machine to produce electricity out of the undulations of the sea:

> If you collaborate with a scientist, he will try to collaborate with you to make a machine to send SOS signals. This machine was made by me, an artist, it was a lovely thing. (Vassilakis Takis interview)

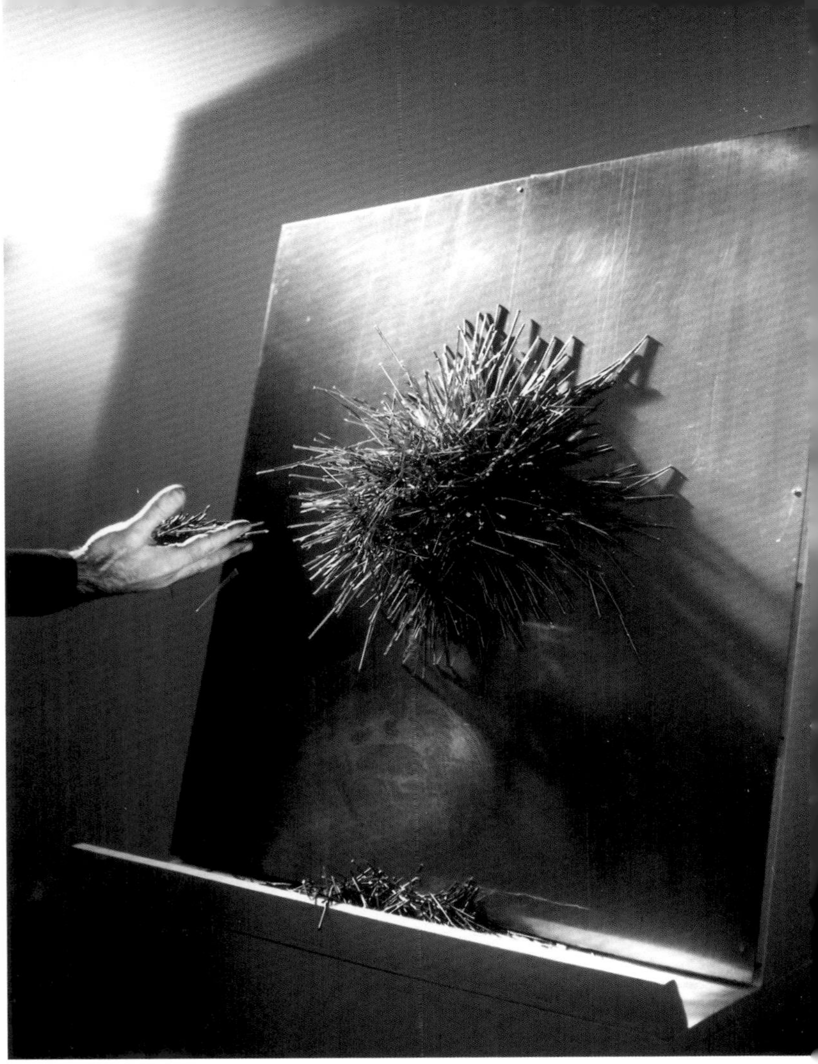

Right: Vassilakis Takis, "Anti-Gravity," Takis exhibition, Hayden Gallery, MIT, November 15-December 8, 1968

Bottom: Takis' studio (detail), in foreground, *Dials* (unfinished), and bases of *Signals*, CAVS, MIT, 1968

Facing page: Wen-Ying Tsai in his studio at CAVS with Harold "Doc" Edgerton, an MIT professor and electrical engineer who frequented CAVS and crossed the boundaries of art and science with his famous stroboscopic photographs, CAVS, MIT, December 1969
Photos: Nishan Bichajian

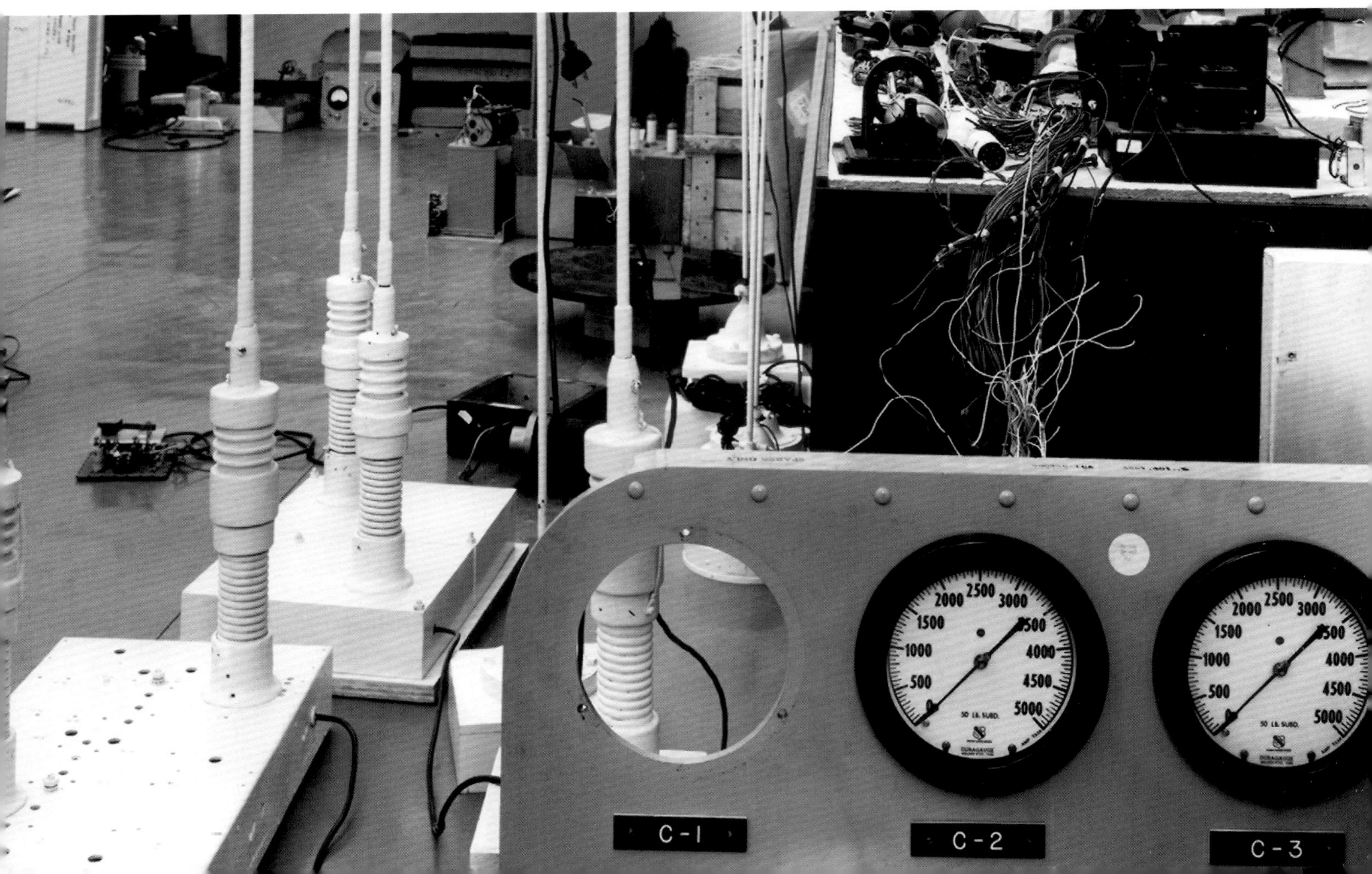

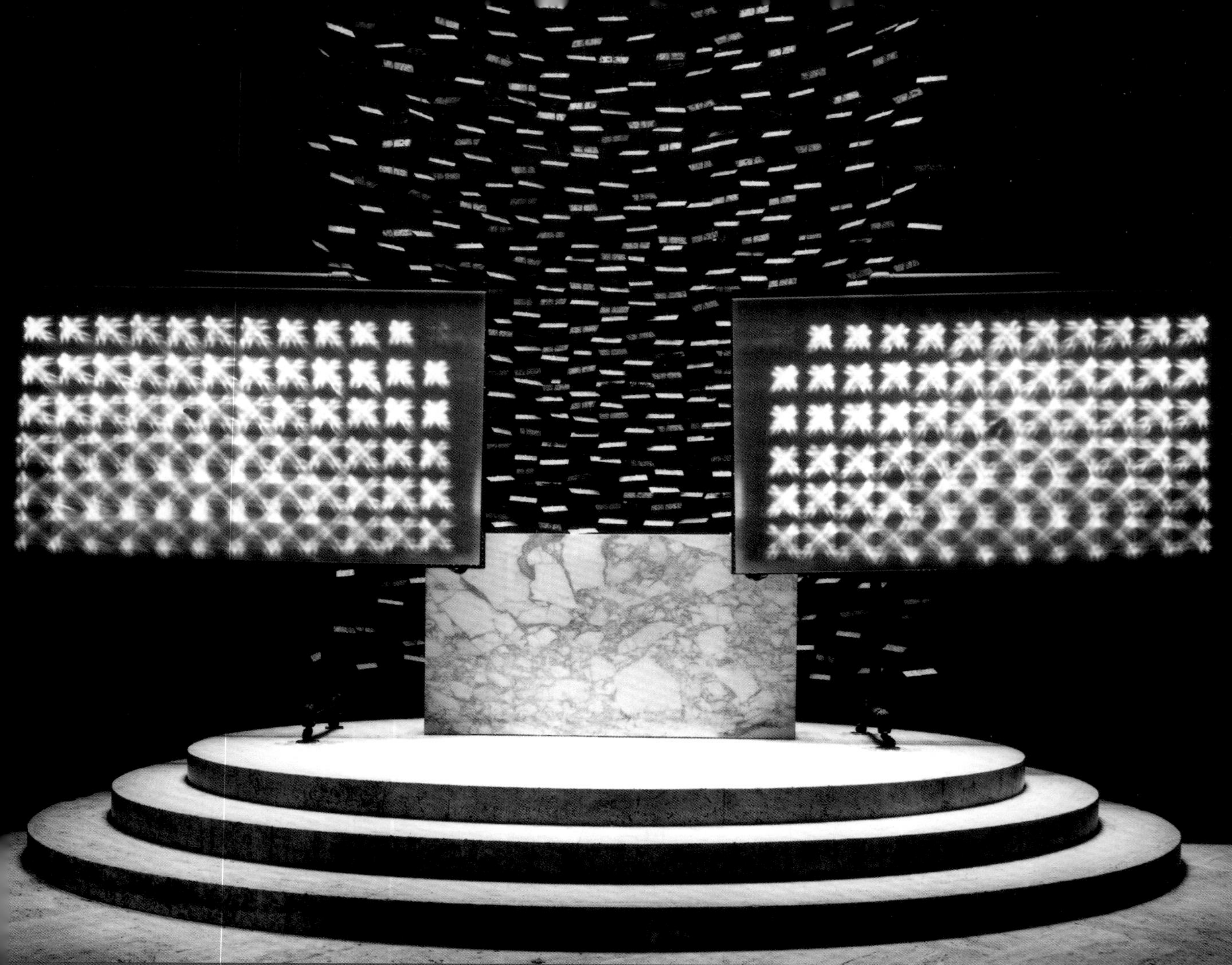

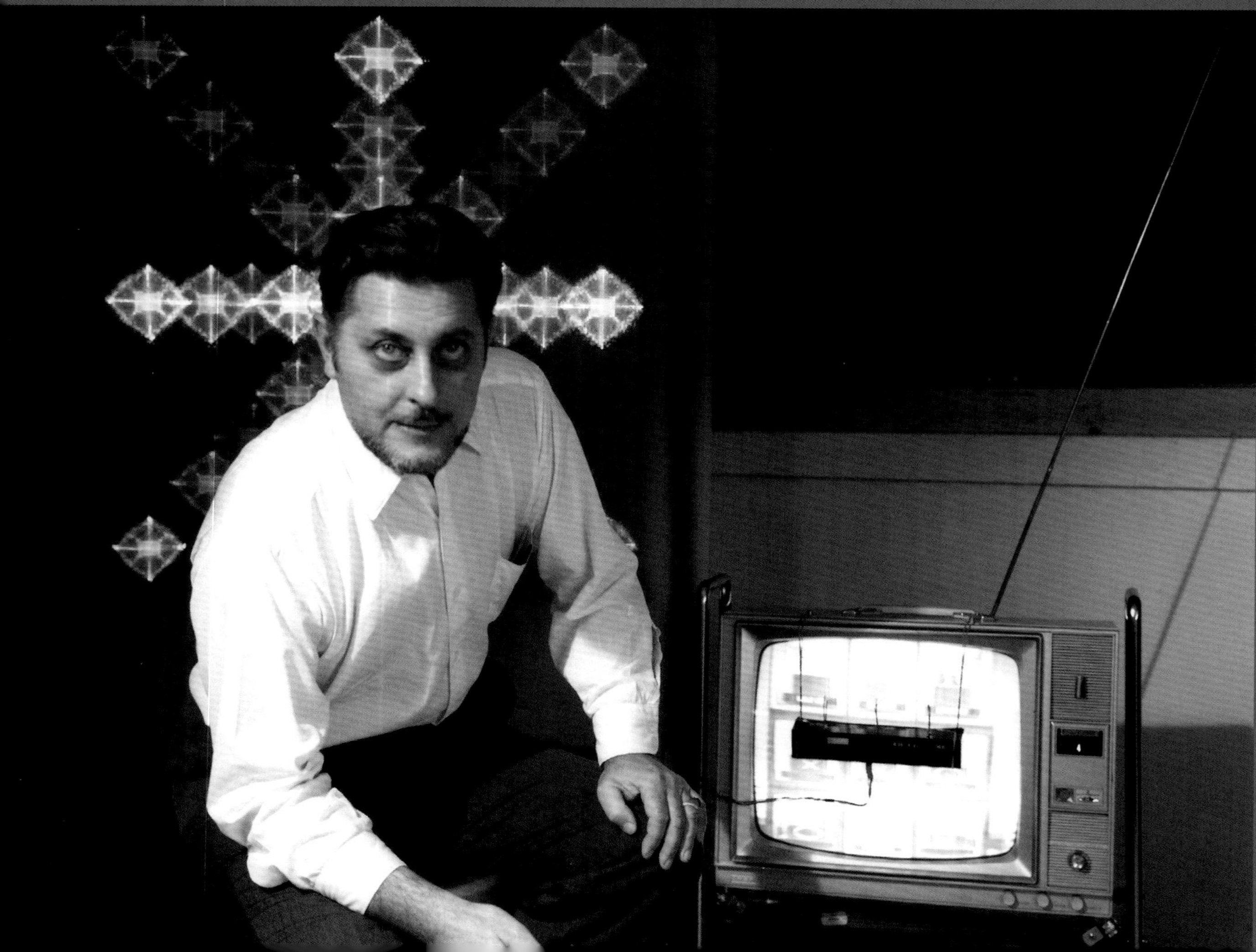

The year 1968-69 was the second year of the Center's existence. Activities initiated in the previous year began to develop in three interconnected areas. The first is that of individual creative efforts, the artist Fellows concentrating primarily on finding new technical means to express their personal aspirations. The second is the area of cooperative ventures, tasks of major dimensions calling for collaboration of the artists of the Center with scientists and engineers from the faculty and student body of the Institute. The third area involves communicating ideas generated at the Center to the general public, through publications, forums and symposia. . . . The Center was invited in February by the Institute of Contemporary Art to present a three day light spectacle for the dedication of the new Boston City Hall. Ted Kraynik, Otto Piene, and Gary Rieveschl created a sequence of light events using a wide range of simple and complex lights. . . . The Fellows were involved in a number of events which presented the Center's ideas to the community. On January 15, Jack Burnham organized a forum in the Little Theatre of Kresge Auditorium on the theme "The Artist and the Scientific Community." The Director and the Fellows also participated in a major symposium presenting the ideas of the Center at the annual conference of the College Art Association.
(Kepes, *MIT Report of the President*, 1969, 435-437)

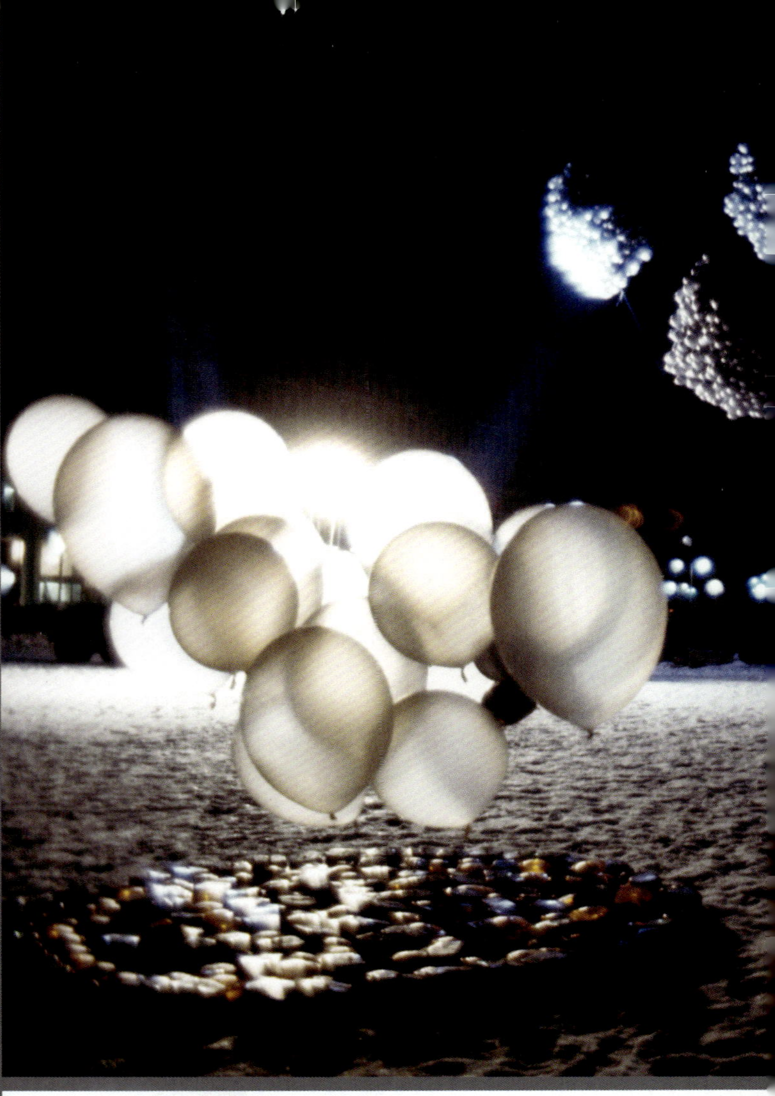

Above right: Otto Piene, CAVS, "Silver Balloon Event," inauguration of new Boston City Hall, February 16, 1969

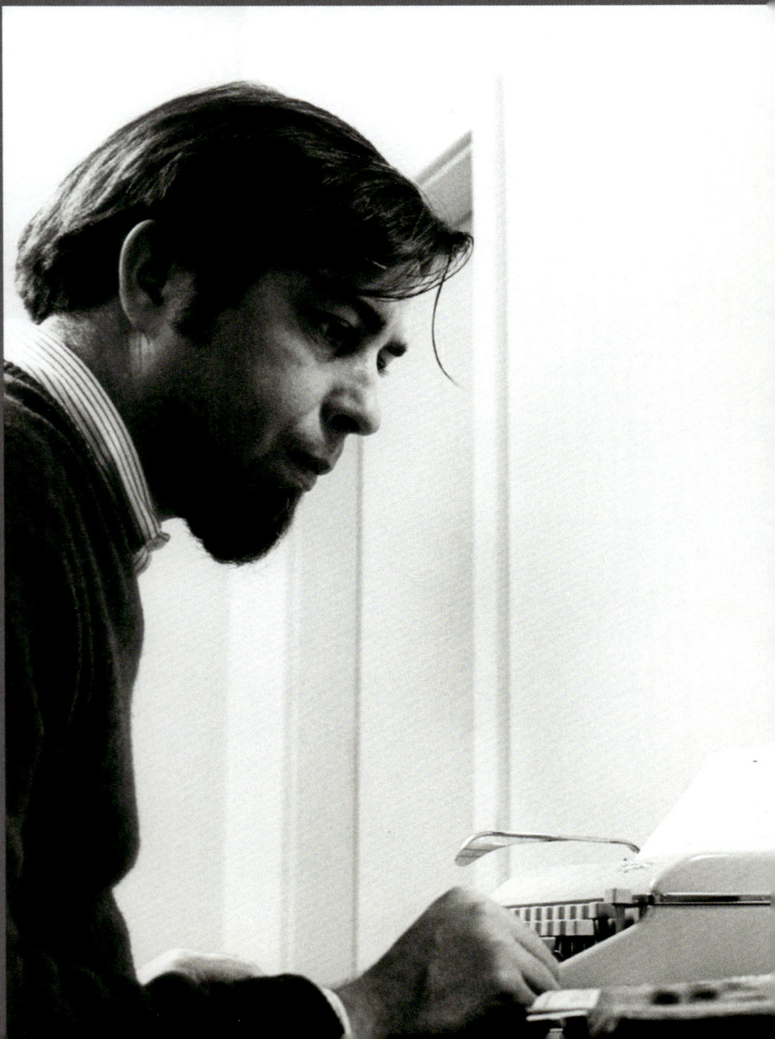

Bottom right: Jack Burnham, author of *Beyond Modern Sculpture: The Effects of Science and Technology on the Sculpture of This Century*, published in 1968

Facing page top: Ted Kraynik, light installation, MIT Chapel, 1972
Facing page bottom: Ted Kraynik, "recently-appointed Fellow . . . with his Video-Luminar No. 3," portrait and press release from MIT Office of Public Relations, 1968

Photos: Nishan Bichajian

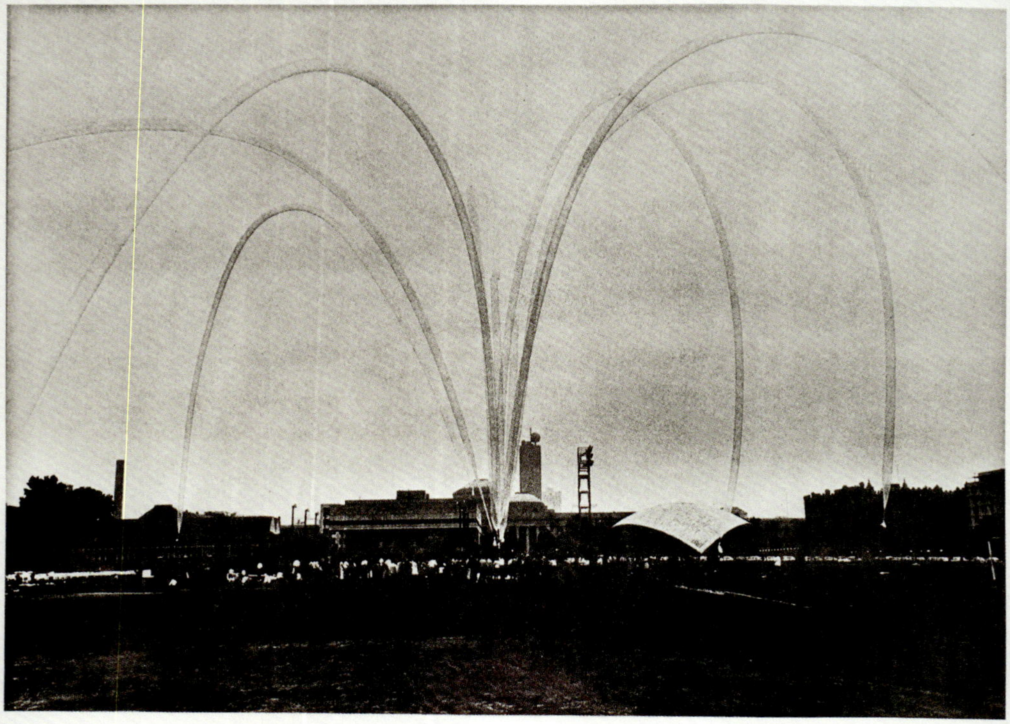

A SEQUENCE OF EQUILIBRIUM
(AND A PROFILE OF M.I.T.)

"LIFT AND EQUILIBRIUM" WAS A DAY OF SKY-FLYING SCULPTURAL EXPERIMENTS THAT WENT ALOFT BETWEEN 4^30 A.M. AND 11 P.M., JUNE 25, 1969, FROM BRIGGS ATHLETIC FIELD, MASSACHUSETTS INSTITUTE OF TECHNOLOGY, CAMBRIDGE/MASS. . AT SUNRISE THREE 120-FOOT-HIGH LETTERS, Y-O-U, FORMED OF 29-INCH-IN-DIAMETER SECTIONS OF TRANSPARENT POLYETHYLENE TUBING, WERE FILLED WITH HELIUM, TIED TOGETHER AND SENT UP TO AN ALTITUDE OF 600 FEET, WHERE THEY FLOATED UNTIL NOON. "Y-O-U" WAS CONTROLLED FROM THE GROUND BY LIGHTWEIGHT LINES. THE MAJOR EVENT OF THE DAY, "EQUILIBRIUM", A MANNED HELIUM SCULPTURE, BEGAN SOON AFTER 2 P.M. . IT CONSISTED OF EIGHT 240-FOOT-LONG, HELIUM-FILLED, TRANSPARENT POLYETHYLENE SECTIONS, EACH 41 INCHES IN DIAMETER, JOINED AT ONE END TO A PARACHUTE HARNESS AND RIDDEN TO AN ALTITUTE OF 100 FEET, FIRST BY SUSAN PETERS, 16-YEARS-OLD, 95 LBS., OF IPSWICH/MASS., THEN BY VERA SIMONS, ARTIST AND BALLOON PILOT, FROM HOUSTON/TEXAS. THEIR FLIGHTS WERE CONTROLLED BY A GROUND CREW OF COLLABORATORS AND COLLABORATING SPECTATORS HOLDING ROPES ATTACHED TO THE PARACHUTE HARNESS. WALTER LEWIN, ASTROPHYSICIST AT THE CENTER FOR SPACE RESEARCH, M.I.T., COLLABORATED WITH ME ON PLANS AND CALCULATIONS FOR "EQUILIBRIUM". OTHER COLLABORATORS WERE: EDWARD FRANKLIN, SCULPTOR, PARTIAL CREATOR OF "LIFT", A GROUND-ANCHORED HELIUM MOBILE CONSISTING OF SEVERAL HUNDRED TIED-TOGETHER BALLOONS OF DIFFERENT SIZES, SHAPES AND COLORS; FRANK CARLTON, ARTIST AND FILMMAKER, WHO IN THE EVENING PLANTED THIRTY ULTRA-VIOLET-LIT STALKS OF FLUORESCENT RED AND BLUE POLYETHYLENE TUBING IN THE GROUND, AND GARY RIEVESCHL, ARTIST, WHO IN ADDITION TO CO-ORGANIZING THE DAY, PRESENTED ICE-AND-FIRE SCULPTURES AFTER SUNDOWN.

PHOTO: BOB LYON

With the cooperation of over thirty participants, most of whom were MIT students, Otto Piene organized two major outdoor sculptural air experiments on the MIT Athletic Field. Using in one event hot air-filled balloons and in the other helium-filled balloons, hundreds of feet in length, the artist and his collaborators produced a vital choreography of dancing sky sculptures. (Kepes, *MIT Report of the President*, 1969, 436)

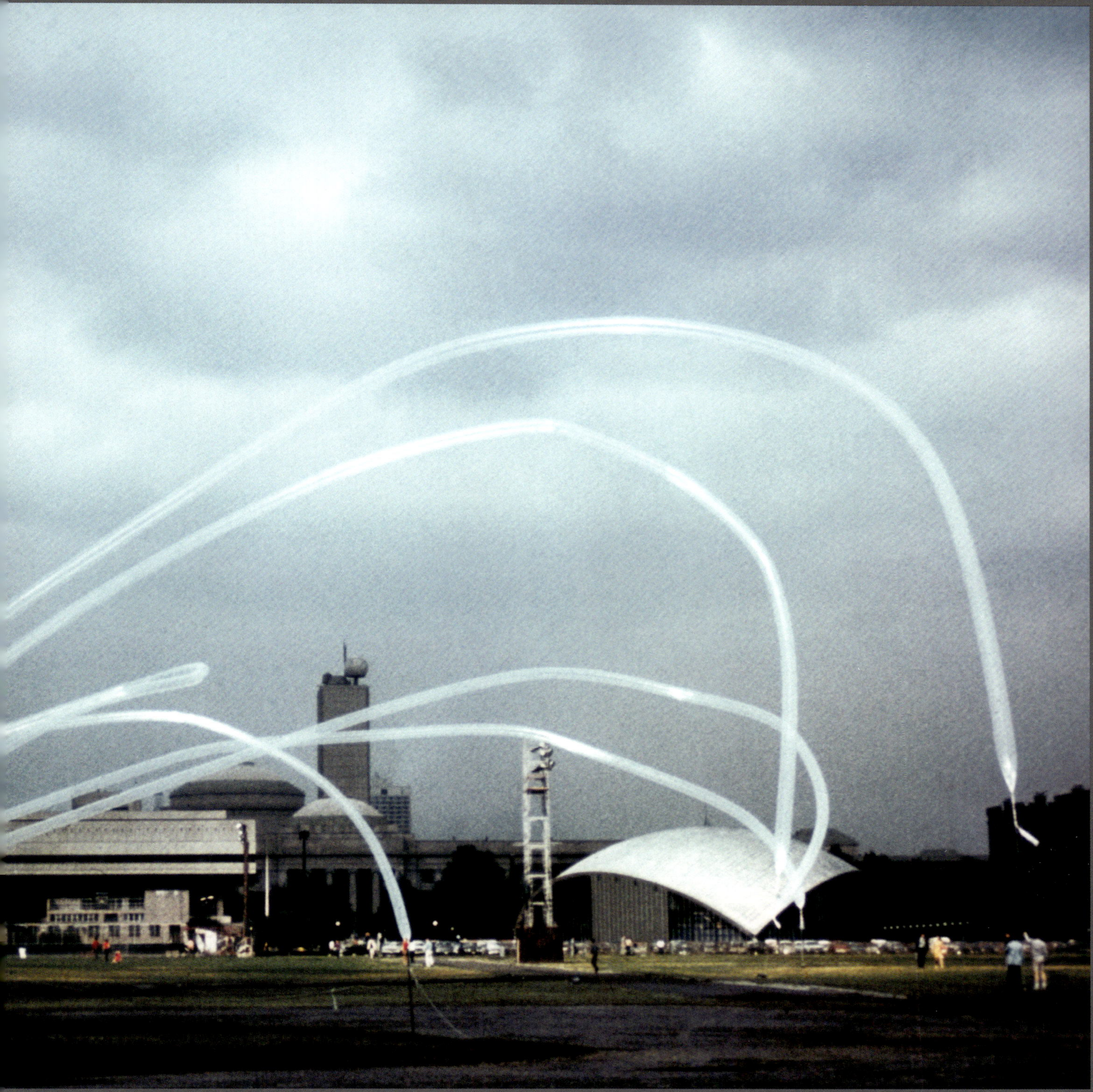

Above: Otto Piene, "Lift," from CAVS sponsored all day event, *Lift and Equilibrium*, Briggs Athletic Field, MIT, June 25, 1969
Photo: Bob Lyon, MIT

Facing page: Otto Piene, "A Sequence of Equilibrium (and a profile of MIT)," lithograph, from the *Sky Art* portfolio of 25 prints, 1969
Photo: Ante Glibota

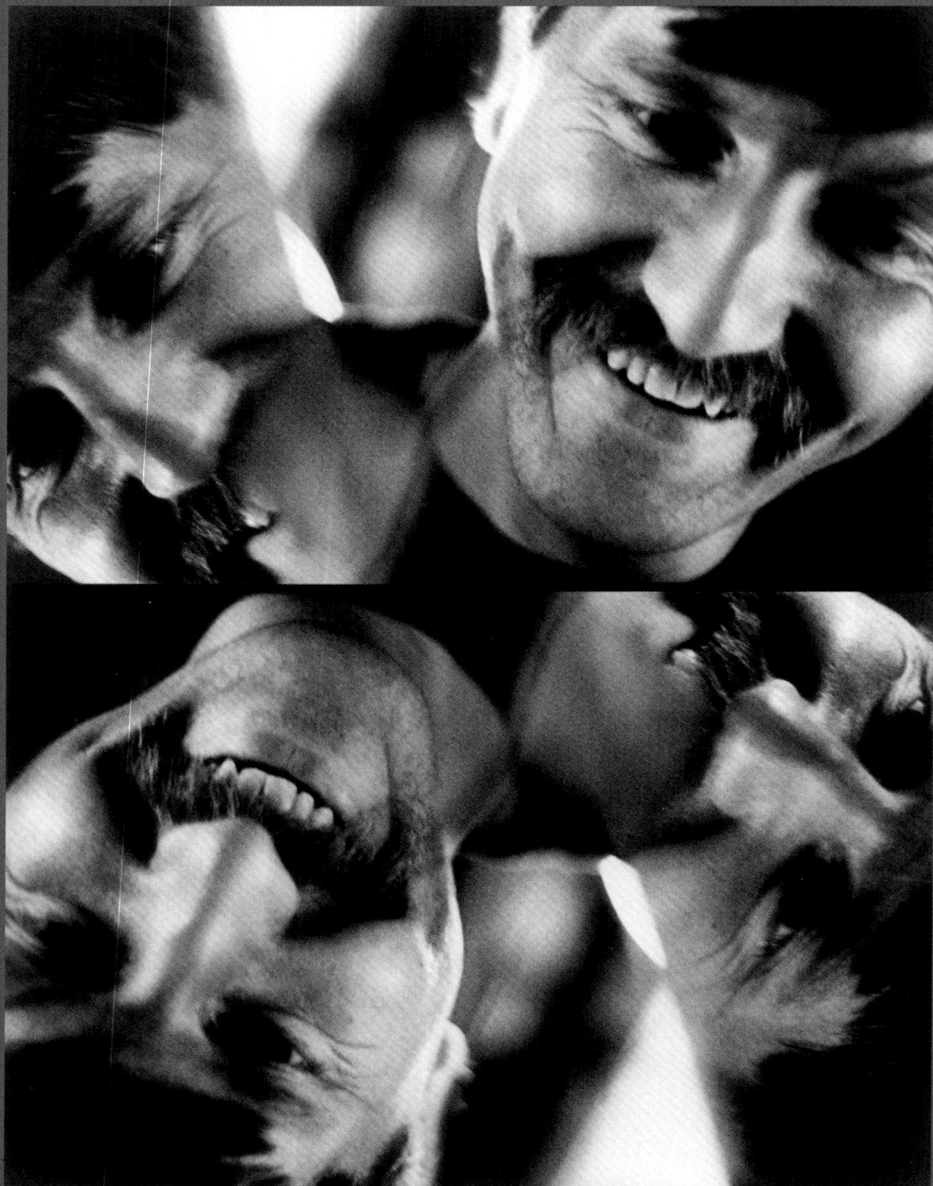

I like to think that life is a dissolve. . .

. . . and that seeing is the real illusion, that a sense of reality
is a sense of the senses. . .
that a sense of reality is a sense of non-sense. . .
that movies should delight the eye and rearrange the senses. . .
that movies are changing the art of seeing. . .
that movies are an art of seeing. . .
that movies are an illusion. . .
that seeing is believing.
(Stan VanDerBeek, excerpt from "Re: Vision," *The American Scholar,* Vol. 35, No. 2, Spring, 1966, 335)

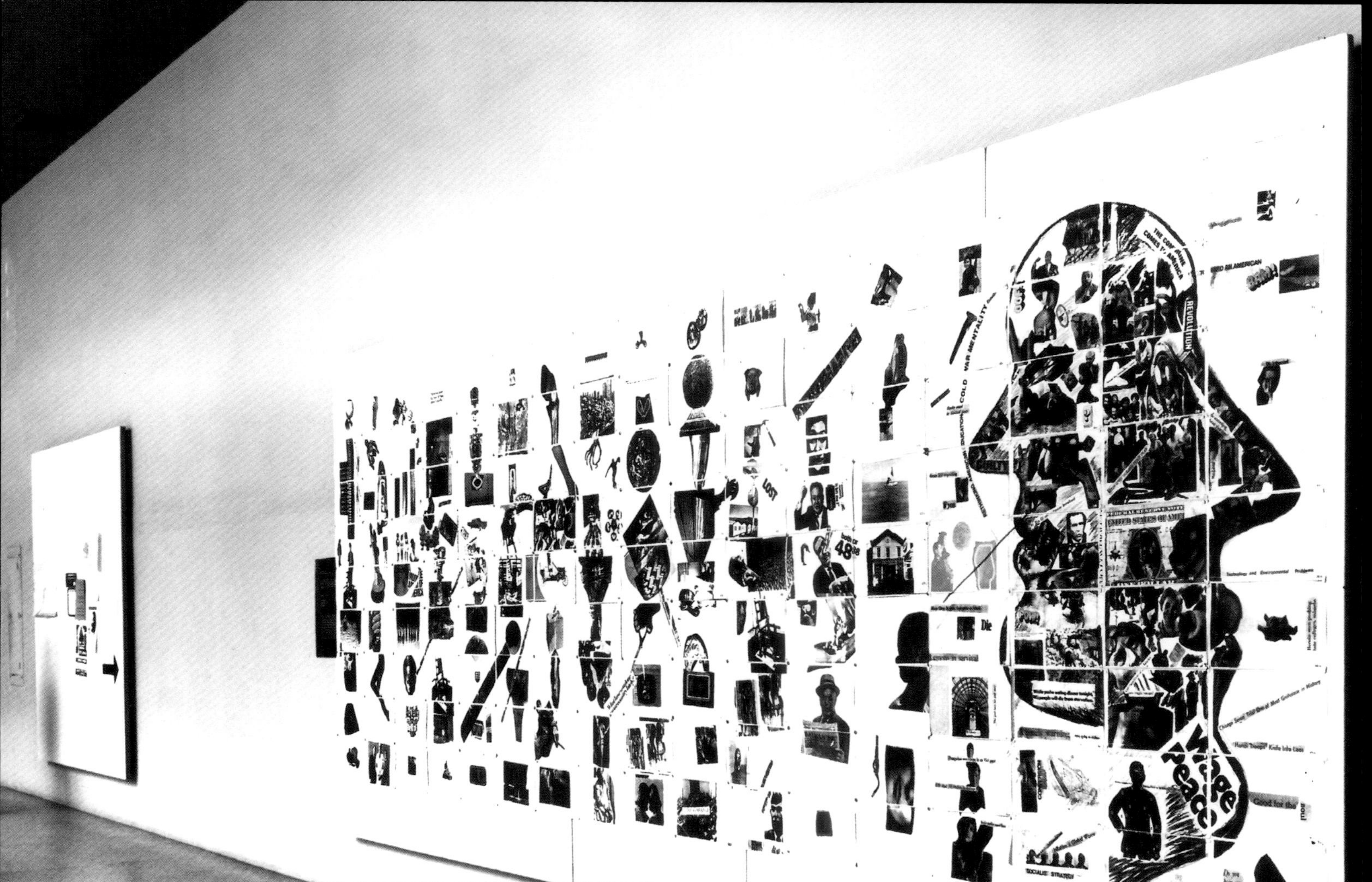

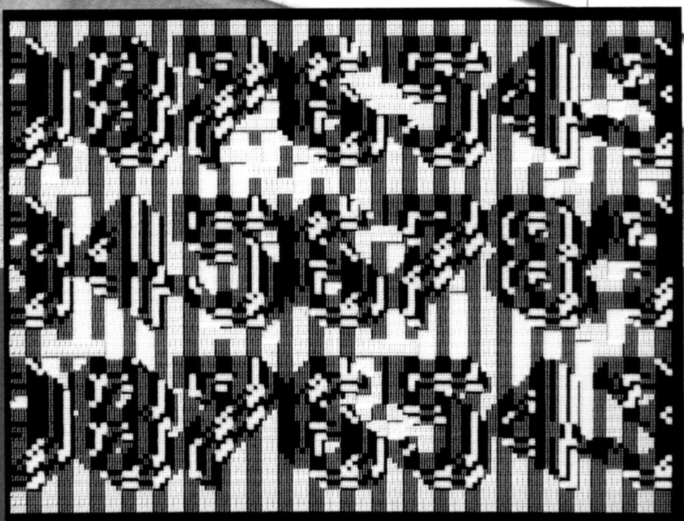

Above: Stan VanDerBeek, "Walls of the World," a mural assembled from images transmitted by telephone, *Exploration*, CAVS exhibition, Hayden Gallery, MIT, 1970
Photo: Nishan Bichajian

Inset: Stan VanDerBeek, "Tapestry: No. 3," *Poemfield Series*, computer-generated typographic poetry created with the 8-bit programming language BEFLIX written by collaborator Ken Knowlton, 1964-69
Source: CAVS Special Collection

Facing page: Stan VanDerBeek, portrait c. 1968
Source: CAVS Special Collection

Stan VanDerBeek, who is constructing an enormous mural (23x6 feet) by telecopier from Cambridge, said, "My phone hasn't been plugged in yet and my exhibit won't work without it." . . . by making use of a telecopier (which transmits photographs by telephone) an artist can send the same mural simultaneously to several different museums. ("Art Steeped in Technology," *The Washington Post*, April 4, 1970)

Boston Harbor Project

The first group project of CAVS fellows was the predominantly conceptual *Boston Harbor Project.* Kepes described it as "a kinetic light form on an imposing scale, hopefully to be realized in Boston Harbor."

> Its purpose is to provide the urban environment with a focal hearth, a monumental gateway matched to the age of flight. But this projected form is intended to be more than an exciting formal acrobatic utilizing sophisticated technical means; it aspires to have a vital symbolic power. If our ambitions can be realized, it could play a significant role in the Boston celebration of the 200th anniversary of 1776. The development of such a complex form challenges not only the artist, but also the structural, electronic, computer and systems engineer, the city planner, the psychologist and sociologist. It will involve the solution of a sequence of demanding problems, for this intense animated light form has to be visually intelligible and effective from land, sea, and air, from varying distances, and in changing weather; and it has to satisfy existing and potential needs of our urban life. (Kepes, *MIT Report of the President*, 1968)

According to Piene, the 1976 US Bicentennial celebrations were a lure for fellows, but the project would ultimately remain conceptual, as the celebrations were moved from Boston to Washington, DC.

> They were all confident that something could be done, although it was not clear what it would be, because Kepes's carrot for people to come here was that there was going to be the US Bicentennial to be celebrated in Boston. . . . They came mostly intrigued by this grand plan of Kepes to be the major motor and the major drive in celebrating the US Bicentennial in 1976. . . . However, the site of the celebrations was moved to Washington, and it wasn't happening in Boston. . . . It was like the air went out of the great balloon of expectations. (Otto Piene, Matthew Wisnioski, "Arts / Science / Technology" in *A Second Modernism: MIT, Architecture, and the 'Techno-Social' Moment*, 2013, 772)

Charles River Project

The *Boston Harbor Project* was followed by the somewhat messianic *Charles River Project.*

> The river in the urban environment: its life and meaning. Since 1971 Fellows at the Center for Advanced Visual Studies have been thinking and working on the problems of and possibilities for the Charles River. Different approaches and results have included educational and informational systems, multimedia, reforestation of the banks of the river, park-playgrounds, sound gardens, light theatres, civic sculptures that could mask noise and exhaust, float on the surface of the river, expand into a huge walkway park, or become a fountain. Participating in the project this year were Lowry Burgess, Juan Downey, Friedrich St. Florian, Michio Ihara, Athena Tacha, Stan Resnicoff, Joe Moss, Maryanne Amacher, Keiko Prince and Bill Parker. (NEA Portfolio, 1974, CAVS Special Collection)

> The American environment—that is what Kepes talked about—the revitalization of the Charles River. We understood it completely. We wanted to do it! Kepes said it was a task—our task. Go ahead, you just have to do it, raising our consciousness. I don't know where the word "public art" came from. Kepes consciously used the term "Civic Art." Well, "civic" with humanity not just the public. As a citizen you have work to do. Our imaginations could go anywhere. (Keiko Prince interview)

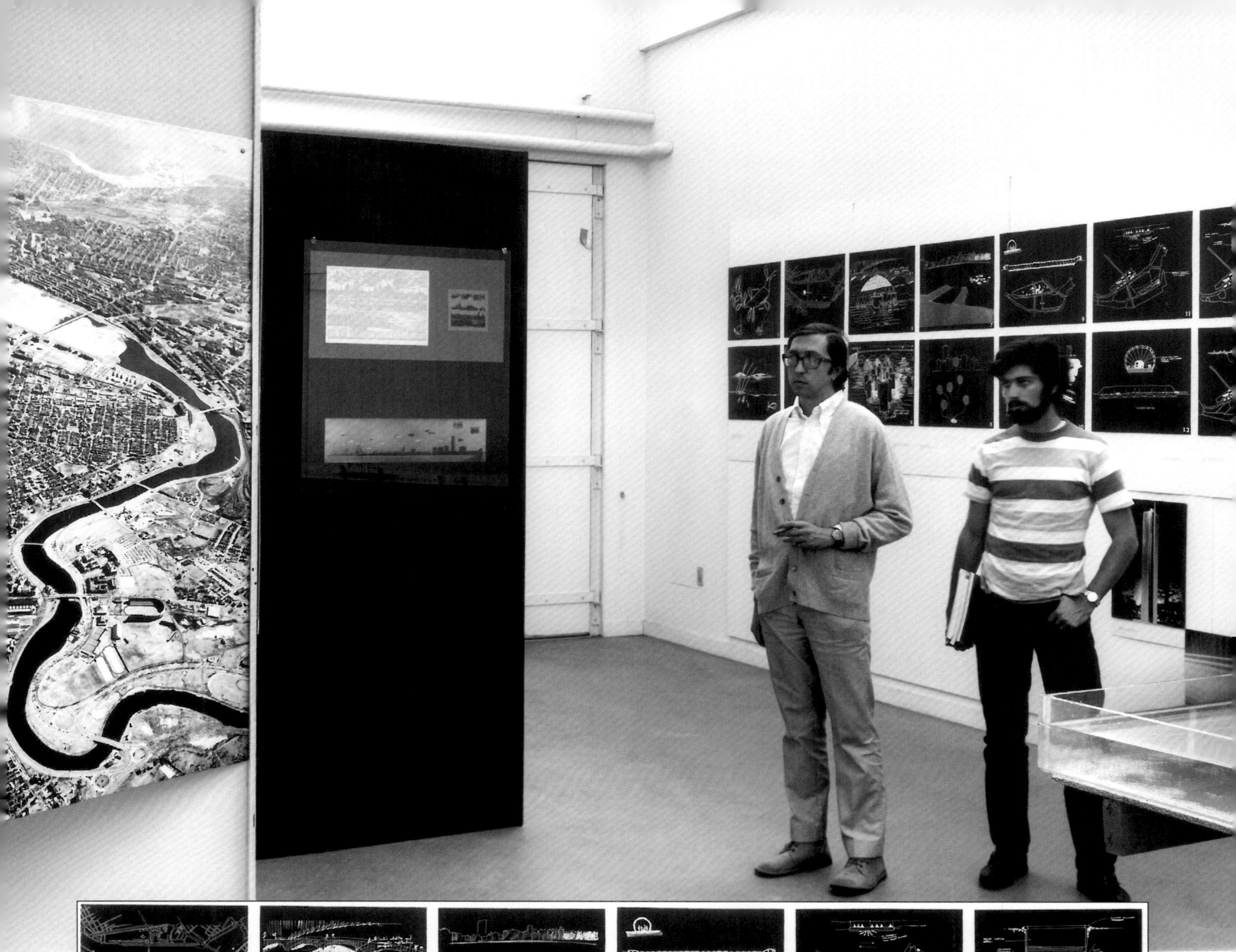

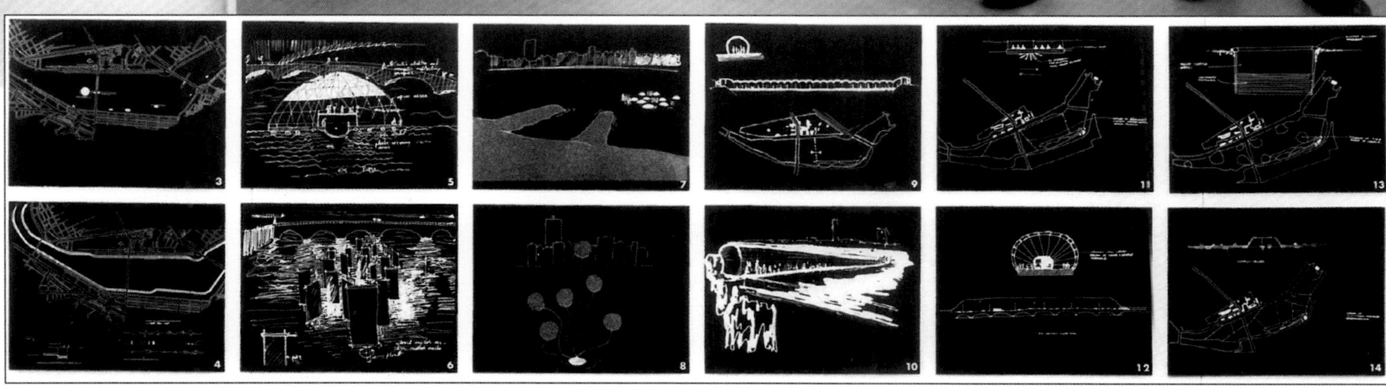

The purpose of the project is to refocus civic awareness on the potentially immense role of a river in urban life in general, and the Charles River in the life of Boston and Cambridge in particular. The project includes a great number of not yet existing uses of the River, as well as the utilization of some of the age-old ways in which a river can enrich the urban life. (Kepes, *MIT Report of the President*, 1971, 33)

Above: *Charles River Project*, 1971-73, Juan Navarro Baldeweg (left) and colleague view proposals displayed at CAVS, MIT, October 3, 1971

Inset: detail of *Charles River Project* proposals, 1971-73

Photos: Nishan Bichajian

Kepes' vision and the ideas and work of CAVS artists engaged the public and broader intellectual community through group exhibitions and symposia. New fellows began to arrive. Lowry Burgess described the transition to the next generation of fellows, a pattern of changing faces within a continuum that would persist over the lifetime of the Center, building a connected community and sustained legacy.

> When I showed up it was a transitional time, in hindsight, between the time when the first so-called "big fellows" came—1968-69, Takis, Otto, Tsai, and all those folks—and a group of much younger people, much less formed artists, and I think the great instigator for that was John Goodyear. He felt that a younger force was missing in his time at the Center. He actually cornered Kepes and asked him to think rather differently about how he appointed the fellows. . . . What John had commented on about the early Center was that [there was], to put it brutally, not much positive interaction among the fellows. Each was working in his own studio. (Lowry Burgess interview)

The growing vitality of CAVS was unmistakable. Kepes' vision had taken root, helped by an aberrant critical moment of societal upheaval, and the unexpected became reality in a Center devoted to discourse between the arts and sciences at MIT.

> Spun out of our heads, science and art remain anemic and without root, and need strengthening contact with nature once again. The natural world remains the common basis for all of us, even though it is changed beyond recognition from the world of nature known to our fathers. It still starts for us where we come in contact with it—through our senses. Science has opened up resources for new sights and sounds, new tastes and textures. If we are to understand the new landscape, we need to touch it with our senses and build the images that will make it ours. For this we must remake our vision. (Kepes, *The New Landscape in Art and Science*, 1956, 20)

In 1974, in accordance with MIT retirement policies, Kepes stepped down as CAVS director. A symposium, *Arts and the University: The Visual Arts at MIT*, was held in his honor. Following his retirement, Kepes participated as an artist in collaborative Center projects. He was a revered guiding light, whose October 4 birthday was celebrated each year by the fellows, until his death in 2001.

Facing page: György Kepes, simulated light effects, floating mirroring buoys, proposed for the Bicentennial celebration, Boston Harbor, 1967
Photos: Nishan Bichajian, 1969

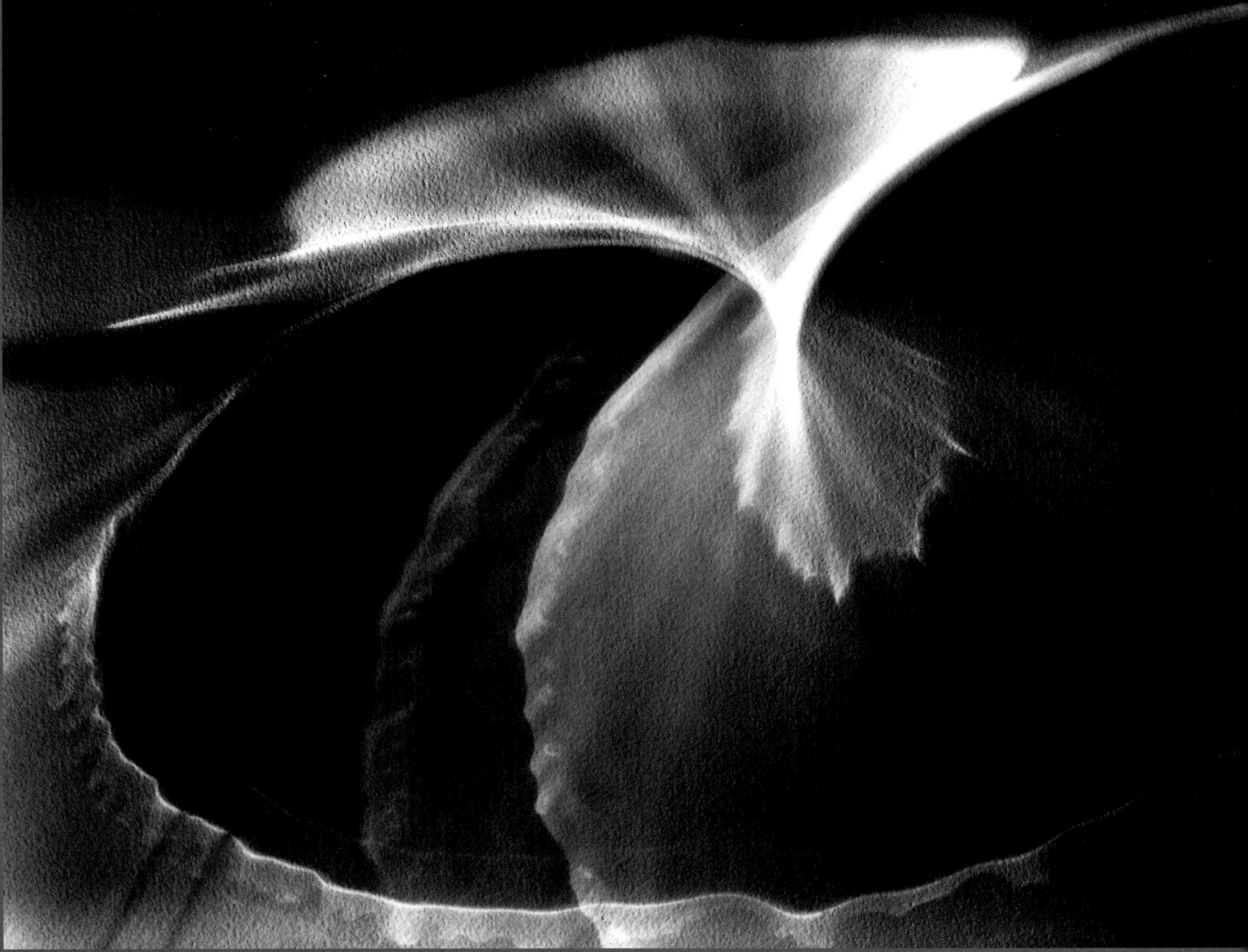

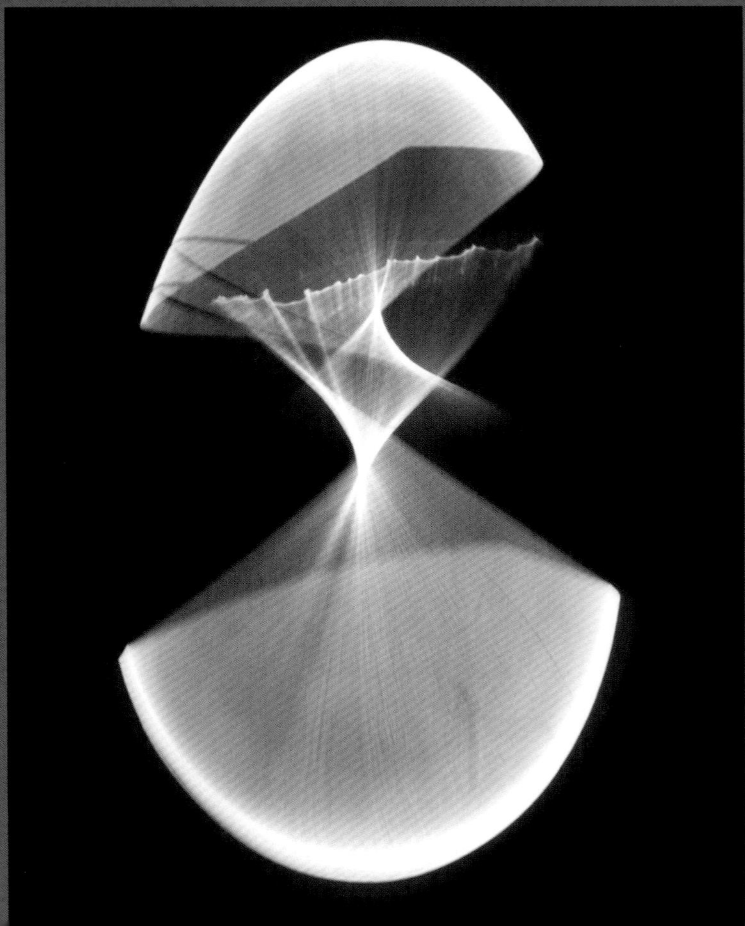

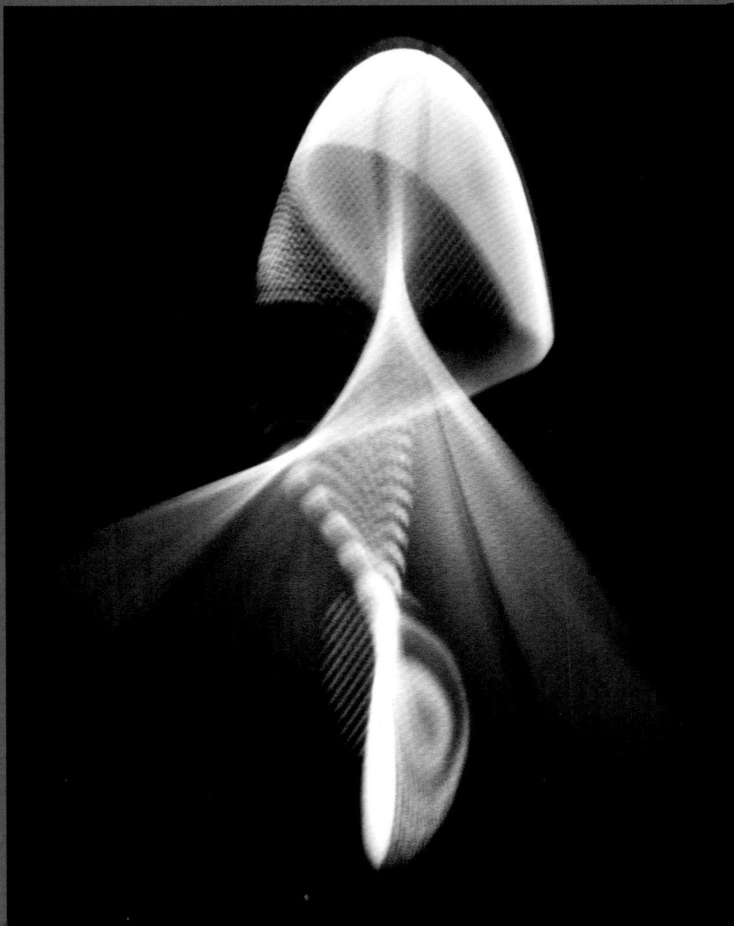

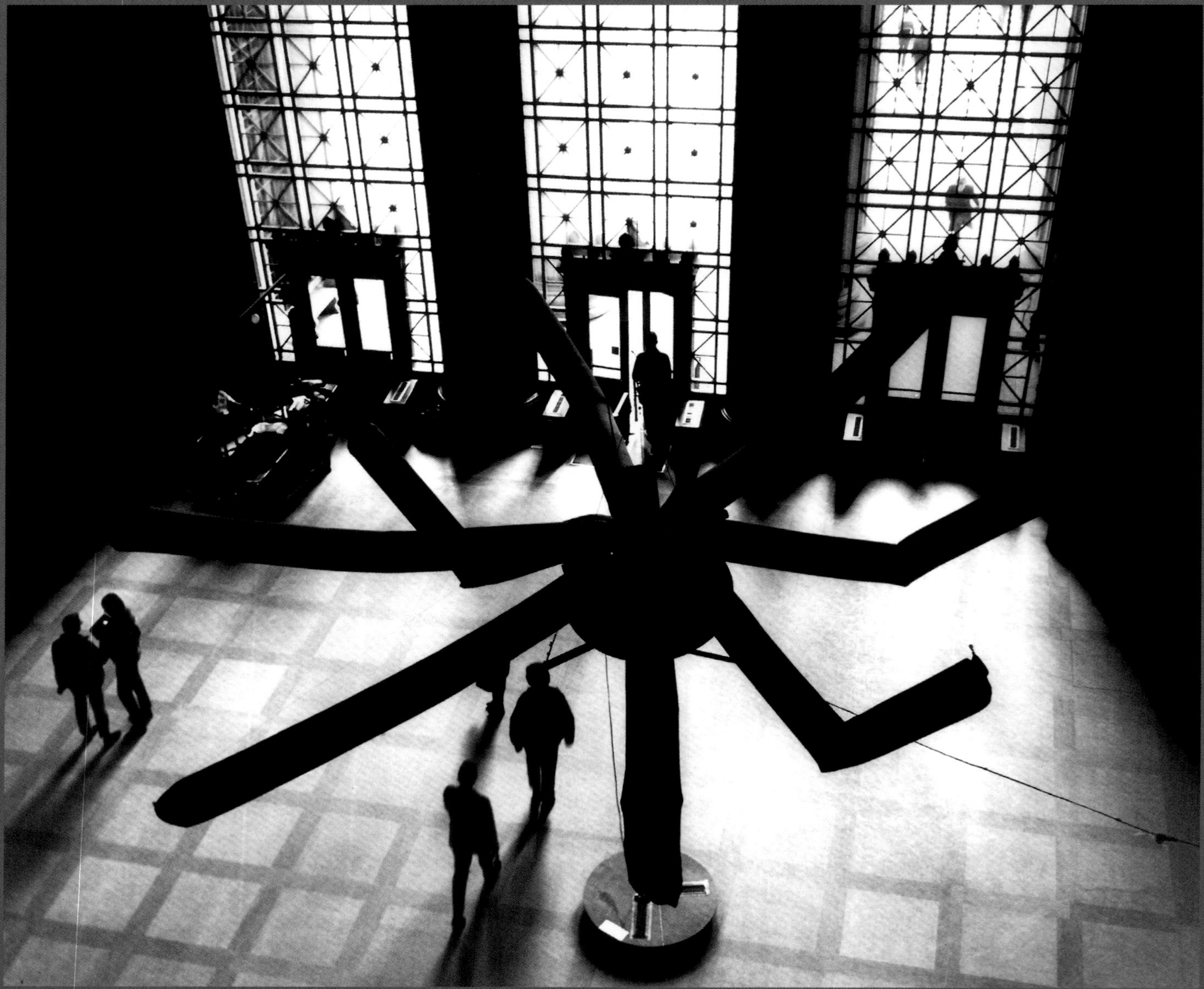

Otto Piene, "Octopus," with Cliff Henricksen, timed inflatable sculpture, Lobby 7, MIT, 1969
Photo: Nishan Bichajian

Otto Piene, CAVS Director, 1974-1994

On October 1, 1974, Otto Piene (1928-2014) officially became the director of CAVS, with, he remarked, "several mouths to feed and $700 in the till." Out of necessity, and from a conviction that artistic reality is the highest form of artistic achievement, Piene directed Center activities toward tangible manifestations.

> Whereas Kepes viewed the creative force of artists and scientists as complementarily similar, Piene saw essential dissimilarities in both the processes and intentions. He enjoyed melding this diversity into collaborations of unlikely members. It was the tension of new associations and configurations that generated the dynamism at the core of the CAVS Group. Piene's group was a collaborative octopus which stretched in all directions to find new attachments. (Elizabeth Goldring, "Otto Piene and the Center for Advanced Visual Studies," *Otto Piene*, Ante Glibota, editor, Delight Edition, 2011, 309)

Piene focused on building a celebratory ideal as a vital part of the Center's teaching and practice. In addition to being director of CAVS, Piene held a faculty position in the Department of Architecture as Professor of Design and Environment. One of Piene's major contributions to CAVS was the founding of the Master of Science in Visual Studies (SMVisS) program together with MIT Professors Richard Leacock and Nicolas Negroponte, joined later by Professor and CAVS Fellow Muriel Cooper. The program offered a pedagogical vision for young artists, and its operating budget helped to stabilize the Center's financial base at MIT.

> With Otto there was a new philosophic, almost pedagogical, concern in terms of teaching—the presence of the teaching process, which had not been primary at the Center at all. Otto brought that in a major way, and influenced a lot of the new young artists, because there was a pedagogical function now at the Center. A master's program came in, and that brought a lot of creativity with it . . . activity, thinking, interesting projects and events. Certainly, it attracted and informed, and brought in a broad layer of extremely competent students who have since gone out and done remarkable things everywhere. (Lowry Burgess interview)

Large-scale collaborations and group works flourished with full participation from younger, unformed artists. Students in the SMVisS program often became fellows, ensuring a multigenerational, multinational, gender-balanced program, rare at that time for MIT. Piene generally embraced Kepes' philosophy regarding art, science, technology, and environment, but there were differences. Piene was not in favor of the *Denkmodell* approach, believing instead in actions and evidence. He also espoused a fully democratic *modus operandi*, at the same time believing in the supremacy of art.

CAVS Administrative Officer, Astrid Hiemer, observed that "[Kepes'] sizable director's office and meeting room became the CAVS exhibition room and meeting space. Piene and staff moved into small inside offices and over time Fellows doubled and tripled to share studios with graduate students." (Hiemer, "New MIT Program ACT Inauguration," *Berkshire Fine Arts*, April 4, 2010) Collaboration at CAVS took many forms, from seminars, conferences, and group exhibitions to the development of new artistic media in such technical areas as holography, laser, video, and the execution of large-scale multimedia performances and events like *Centerbeam* and *Desert Sun/Desert Moon*. Piene himself worked in these fields both as an individual artist and as a collaborator, and served as coordinator for these large-scale events. With students and fellows he engaged in several European architectural competitions, such as Constance University, Tête de Défense in Paris, and the Bonn Government Center.

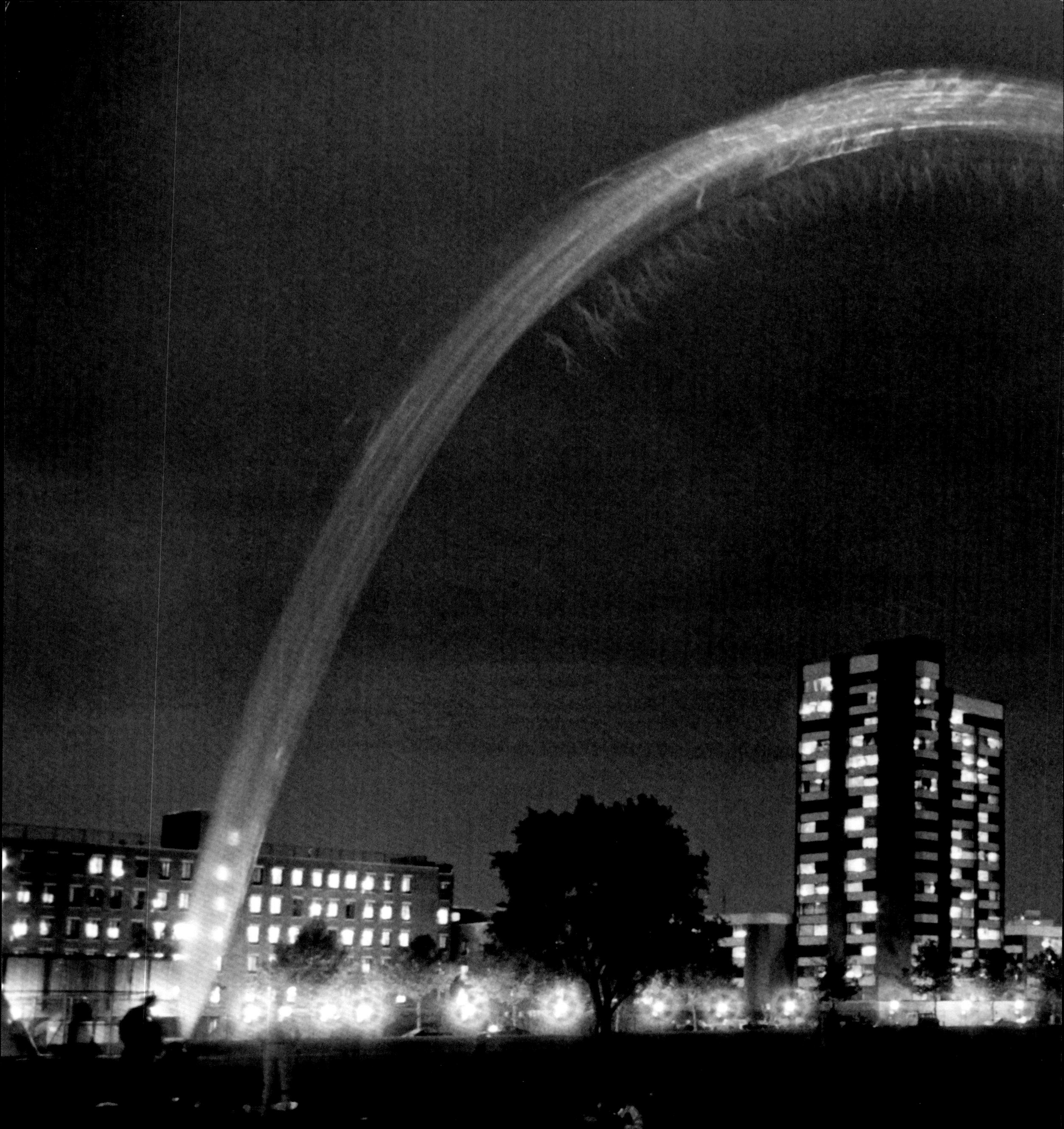

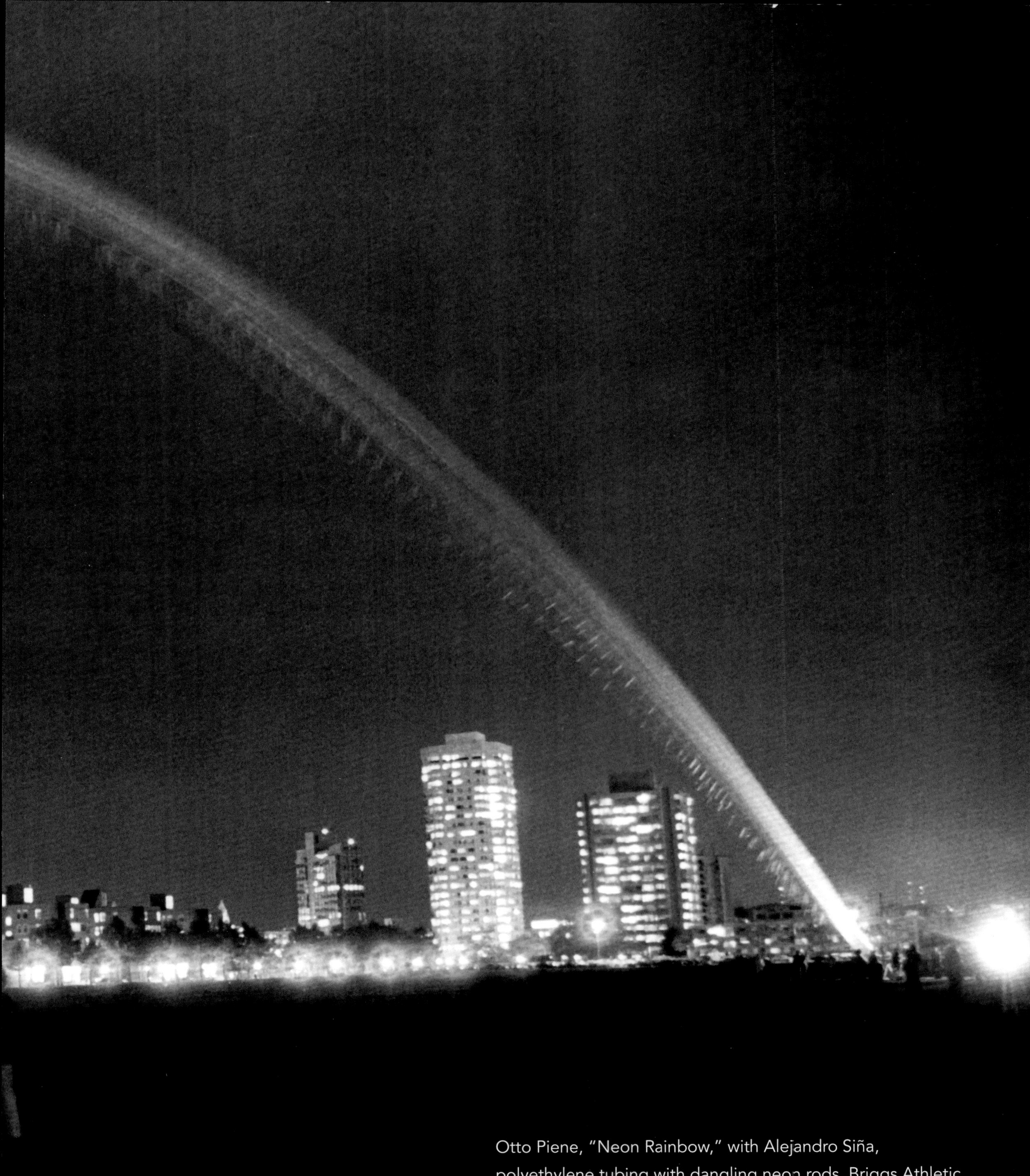

Otto Piene, "Neon Rainbow," with Alejandro Siña, polyethylene tubing with dangling neon rods, Briggs Athletic Field, *Arttransition* conference, MIT, October 17, 1975
Photo: Roger N. Goldstein

Elizabeth Goldring and Paul Earls, CAVS Acting Co-Directors, 1994-1995

Following Piene's retirement, Dean William J. Mitchell appointed Paul Earls and Elizabeth Goldring, both CAVS Senior Fellows, as acting co-directors. While the search began for a permanent director, Goldring and Earls were charged with maintaining the status quo and "not rocking the boat."

Earls and Goldring continued Piene's model of bimonthly fellows' meetings as well as work-in-progress exhibitions and presentations by fellows and invited guests. Goldring remembered Kepes' words, "The Center has had a great past. I'm certain it will have a great future."

Krzysztof Wodiczko, CAVS Director, 1995-1996; 2004-2008

During the directorship of Polish-born Krzysztof Wodiczko, CAVS moved up to renovated third-floor spaces in MIT Building N52, with small clean studios and glass doors distributed around a central core. Facilities for digital arts, an emerging focus at CAVS, were emphasized.

> The Center has consolidated its new identity and is re-establishing itself as a potential leader in the field of critical art and design thanks to its new emphasis on integrating ethical, social and cultural study with the innovative experiment inspired by and challenging contemporary scientific and technological developments. In doing so, the Center for Advanced Visual Studies has begun to redesign its educational and research methods and programs. (Wodiczko, *MIT Reports to the President*, 1995, 219)

Wodiczko revived the antimilitaristic focus present at the founding of CAVS. The first initiative of CAVS developed jointly by Mitchell and Professor Wodiczko, was an open invitation to the art and scientific communities entitled "The Demilitarization of Technology."

> Among many responses to the Demilitarization of Technology proposal which are currently under review is the "Metaphorming Military Technology" project by Todd Siler, a graduate of CAVS and the only artist who received a PhD from MIT. Siler says that "I share the belief of Professors Wodiczko and Mitchell who maintain that designers and artists can become catalysts, opening a dialogue between the human and technical sciences in directing the transfer of defense technologies into critical/cultural projects." (Wodiczko, *MIT Report of the President*, 1995)

At that time, Wodiczko also considered changing the CAVS name from the Center for Advanced Visual Studies to the Center for Art, Culture and Technology, a theme that would later be picked up by Professor Ute Meta Bauer, head of the Visual Arts Program (VAP) at MIT.
Wodiczko described the work of his Interrogative Design Group at CAVS, which also outlined his philosophy:

> Modern design is often supposed to resolve a problem but in reality it becomes a facade behind which the actual problem is hidden so that there is only a superficial answer to it. There seems to be a similar expectation with technology: each new gadget and new system is supposed to resolve some problem. Those who work on scientific or technological research have some playful ideas, they are working on applications for those ideas in order to sell them or connect them with the expectations of the industry or consumers. The basic idea is to further improve the existing world. In fact, both technology and design—technology as a technical opportunity and design as the relation between this opportunity and the world of needs—can reveal and clarify needs that should not exist. They both have

the potential to expand on our access to the complexity of the present world rather than just projecting into the future.

Technology and new media and design could articulate and expose the problematic world in which we live and at the same time provide emergency service and help to people who really need it. With media and technology we can achieve two things at the same time. We can do cultural work and provide access to the circulation of power for those who are least likely to have it. That way we provide emergency help and transform the understanding and awareness of the world. Those who are marginalized, displaced, and misfortunate could in fact become agents of this new and prophetic way of understanding the world. They can provide an image of the world from the point of view of those who are in trouble and a vision of a better world. There are a lot of things we could achieve if the cultural world would connect technological design and artistic research in the same place at the same time. Most of the time we are dealing with disconnected organizations and institutions. (Wodiczko from an interview by Christiane Paul, "The Prophet's Prothesis: An Interview with Krzysztof Wodiczko," *Sculpture Magazine*, May 1999)

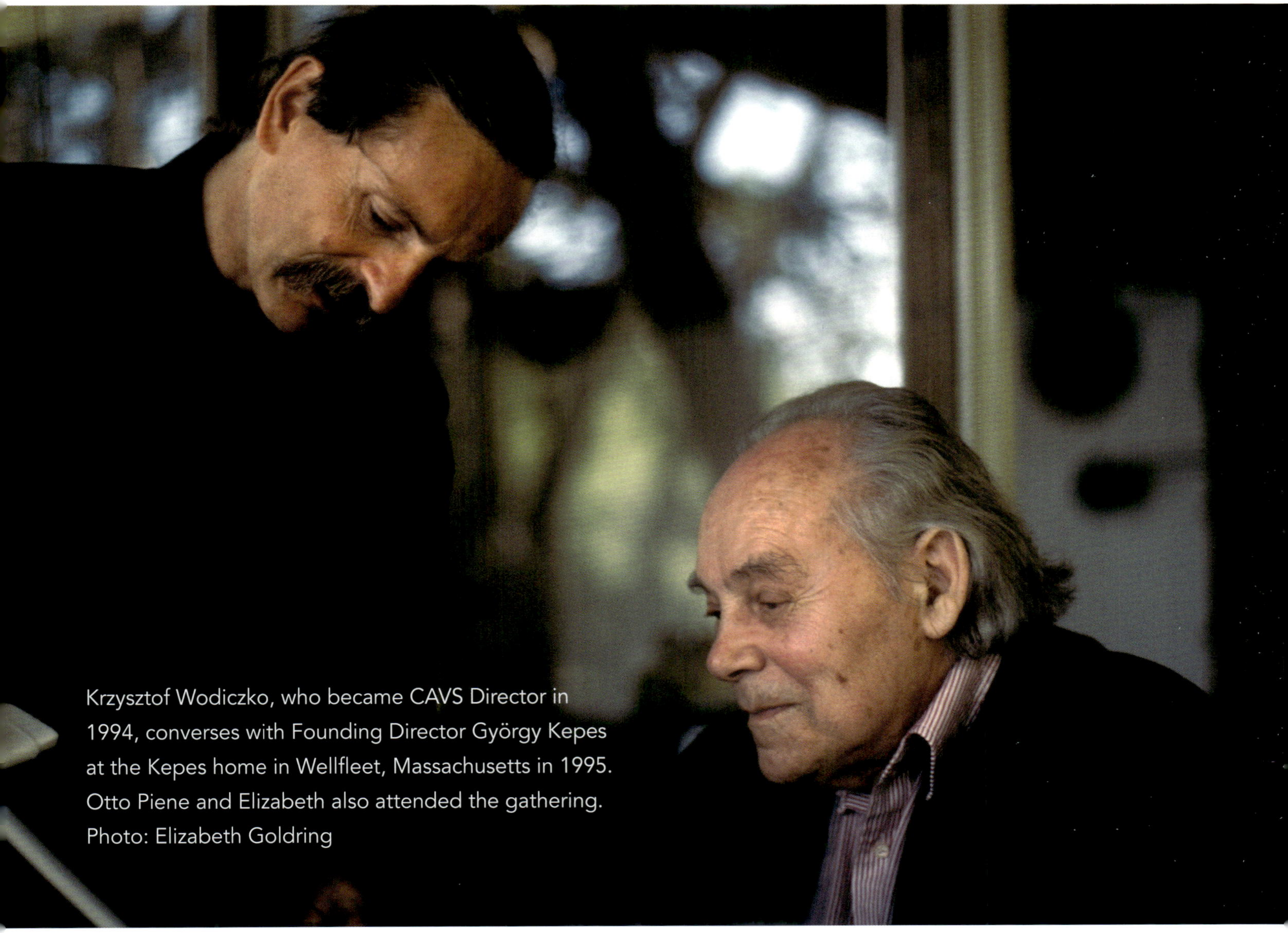

Krzysztof Wodiczko, who became CAVS Director in 1994, converses with Founding Director György Kepes at the Kepes home in Wellfleet, Massachusetts in 1995. Otto Piene and Elizabeth also attended the gathering. Photo: Elizabeth Goldring

Steve Benton, CAVS Director, 1996-2003

Professor Steve Benton, who led MIT's Spatial Imaging Group, and who had collaborated with CAVS fellows on holography projects since 1975, served as director from 1996 until his untimely death in 2003. He believed that, "an emerging mission of the CAVS is the exploration of the digital arts as a common ground for collaborative projects. Our goal is the creation of important art that could not or would not be possible except at MIT." (Benton, *MIT Reports to the President*, 1998, 245)

> I like Kepes' statement about three tasks which he intended for artists, but which relate to everyone, especially everyone at MIT. We must build a bridge between man and nature, which is the genesis of concern with environmental art; we must build bridges between man and man, which I see as the genesis of the collaborative style between art, science and technology to bring these different cultures together; and we have to build bridges inside ourselves, which involves the science of this new century. (Benton, *Sky Art Conference 2002* catalog, 18)

During the latter years, CAVS modeled itself on the early Center, with fellows working in individual studios on individual projects. There were specific groupings within CAVS headed by fellows, such as Wodiczko's Interrogative Design Group, and Goldring's Vision Group. Several research fellows were appointed during this time, and people came from outside to give presentations and lectures. Often, well-funded fellows lived in New York, and commuted to MIT during the week for teaching and other responsibilities. Without the W11 Building, there was no cohesive, full-time working community as there had been with Kepes' invited fellows and during Piene's tenure.

MIT Program in Art, Culture and Technology (ACT), 2009-present

Each of the CAVS directors was an accomplished artist, with the exception of Steve Benton, who was a noted scientist. Each directorship was guided by vision rather than administration. The demise of CAVS was clearly an act of reorganization and administration, rather than a statement of vision. Ultimately, CAVS became an archive—the CAVS Special Collection—housed at the MIT Program in Art, Culture and Technology (ACT). ACT was an amalgam of the Visual Arts Program—the undergraduate teaching program established in 1989—and the CAVS research program, helmed by founding director Ute Meta Bauer from 2009 to 2012, Professors Renée Green, from 2012 to 2014, Gediminas Urbonas, from 2014 to 2018, and Judith Barry, from 2018.

> "We see the visual arts as an essential component of offerings," says Dean Adèle Naudé Santos. "The creation of this new program, combining the efforts of two historic endeavors, reflects our commitment to the growing importance of the visual arts at MIT. In addition to giving the arts a higher profile on campus and in the city, the program's signal position in our new building will put its faculty, students and researchers in daily contact with their counterparts at the Media Lab, and at the many other centers and programs in the complex, enhancing everyone's efforts." ("A New Program in Art, Culture and Technology," *Plan 75*, MIT School of Architecture + Planning, 2010)

The fact that the Center ultimately found a home in proximity with the Media Lab didn't ensure collaboration. Cross-talk took place during the early years of the Media Lab when artists bridged CAVS with Leacock and Glorianna Davenport's Film/Video section, Benton's Holography Lab, Cooper's Visible Language Workshop (VLW), and Vercoe and Machover's experimental music studios. György Kepes' far-reaching advocacy for a "new landscape" manifested with the founding of the Center for Advanced Visual Studies. As CAVS vanished, its legacy became more present as something precious—a singular, meteoric force.

Inside the Studios

The following section looks inside the studios at the working life during the early years (1969-1974) of CAVS. The artists, their process, and materials are often captured in the black and white photographs of Nishan Bichajian, who was handpicked by Kepes to document the tools and methodologies of art-science-technology practice. Texts come from artist statements, proposals, summaries written by Kepes and Piene for the annual *MIT Report of the President*, fundraising portfolios, news articles written at the time, and exhibition wall texts and catalogs.

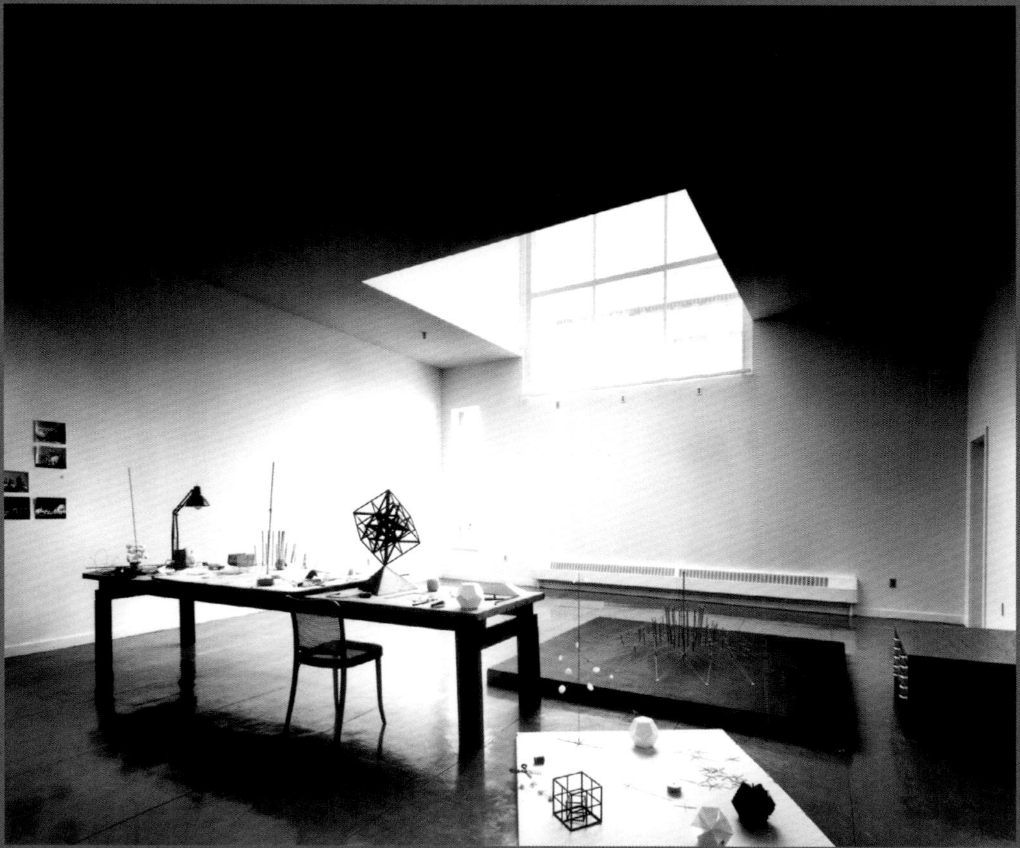

Studio at CAVS, facade of the main MIT building across Massachusetts Avenue visible in the window, 1969
Photo: Nishan Bichajian

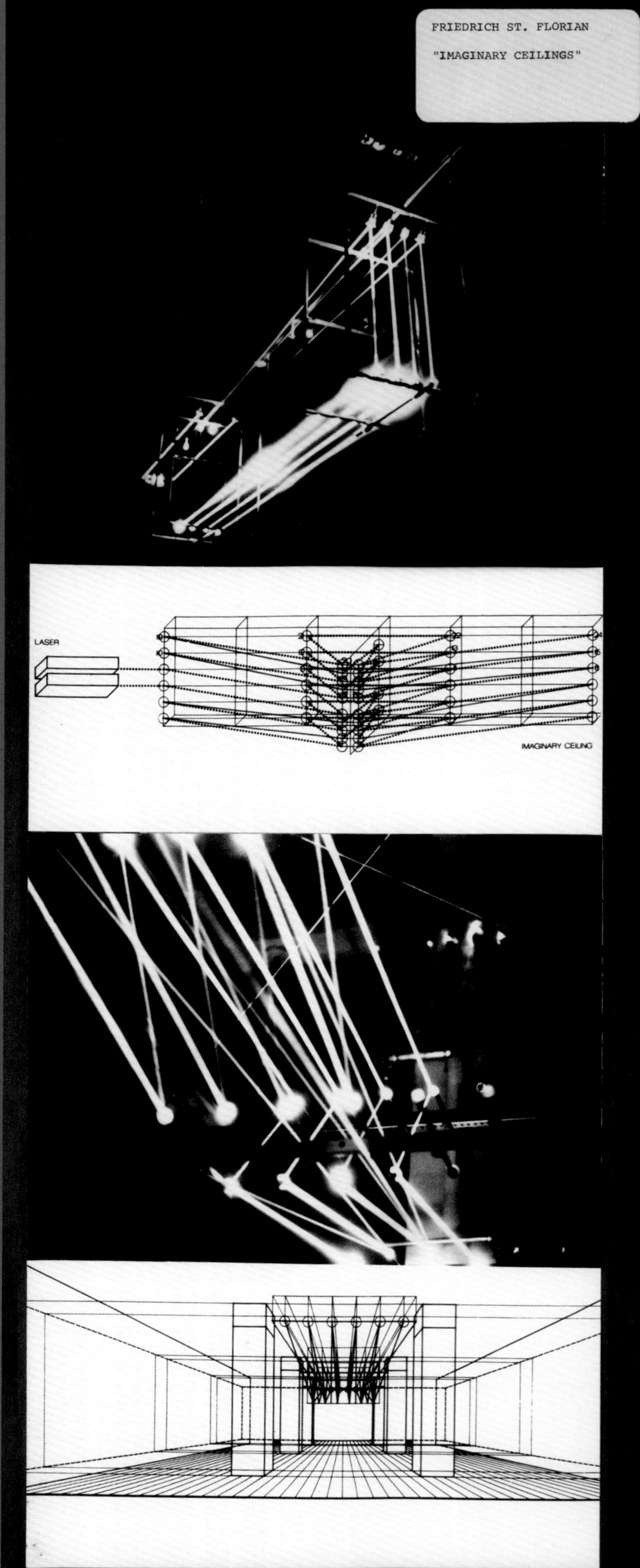

I wanted to construct an imaginary room so that when people entered the room they would go through a set of electric eyes, the laser would be activated and its beam would be guided by mirrors to create an imaginary space. We needed smoke in the room in order to make the laser beam visible. . . . When I began working at the Center, I linked up with the Department of Architecture and their students helped me construct a revised version of an imaginary space that was shown at the Hayden Gallery in 1973. (Friedrich St. Florian interview)

Facing page: Friedrich St. Florian in his studio at CAVS, MIT, 1973

Right: Friedrich St. Florian, "Imaginary Ceilings," laser constructions, from "Individual Work by Fellows at the Center for Advanced Visual Studies, 1973-74," CAVS portfolio

Photos: Nishan Bichajian

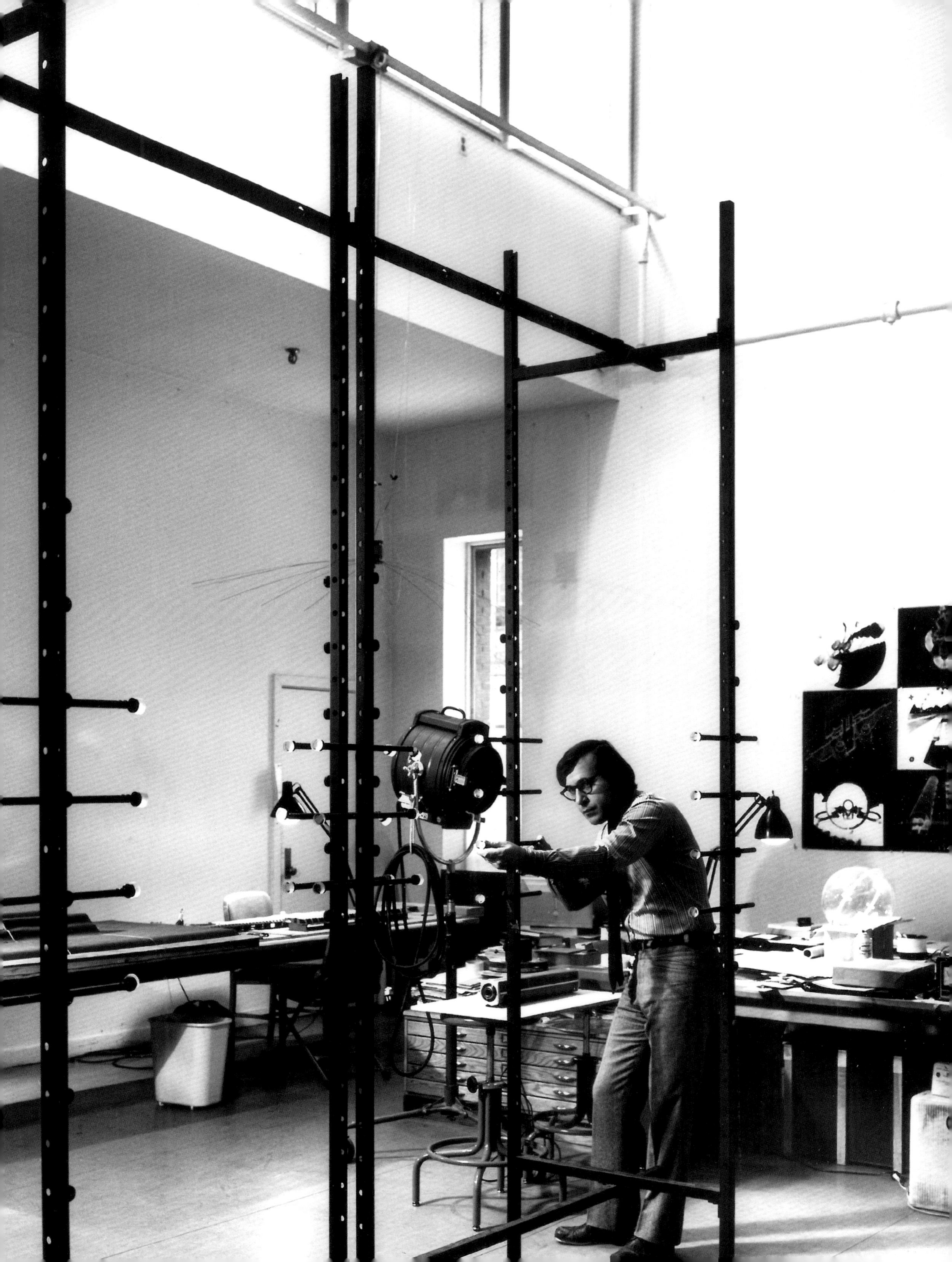

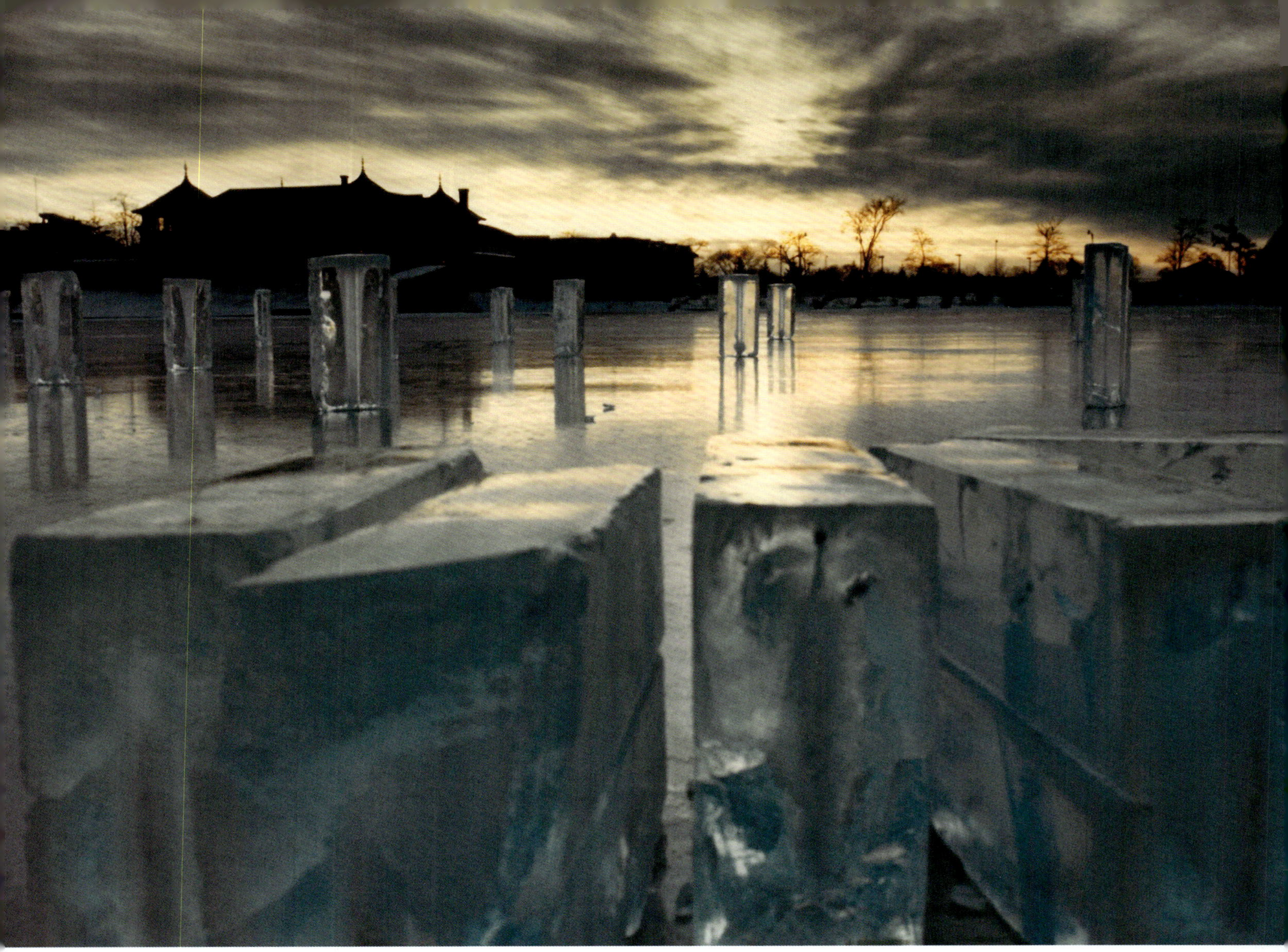

Above: Lowry Burgess, "Listening for Light-Hinge," ice sculpture on the Charles River, part of *Earth, Air, Fire, Water: Elements of Art* show, Museum of Fine Arts, Boston, 1971
Photo: George Cope

Facing page: Burgess in his studio, CAVS, MIT, 1972
Photo: Nishan Bichajian

Facing page inset: Burgess, "The Heptagonal Crystal," drawing for an earthwork in Mexico
Photo: Nishan Bichajian

Burgess arranged large blocks of ice, amounting to some 24 tons altogether, on the surface of the river and set crocus bulbs in them which will eventually drift to shore and take root. Time, temperature, and states of matter were connected in this piece in a somewhat poetic way, but the beauty of it came with the use of the light, particularly evening light, which the river seems to attract, and the use of that peculiarly inviolate space just above the river's surface . . . (Kenneth Baker, "Review, Earth, Air, Fire, Water: Elements of Art," *Artforum*, v. 9, March, 1971, 72-74)

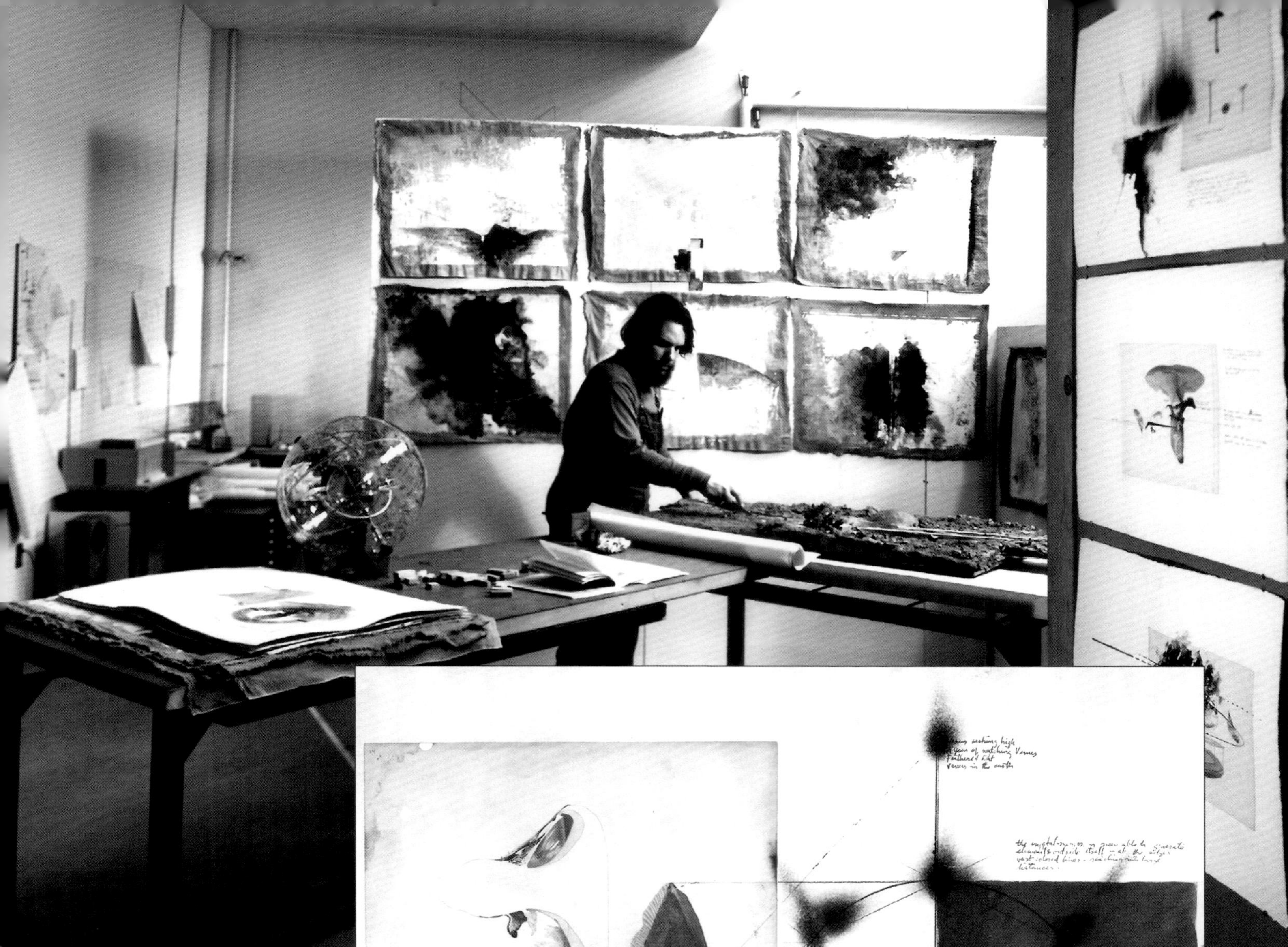

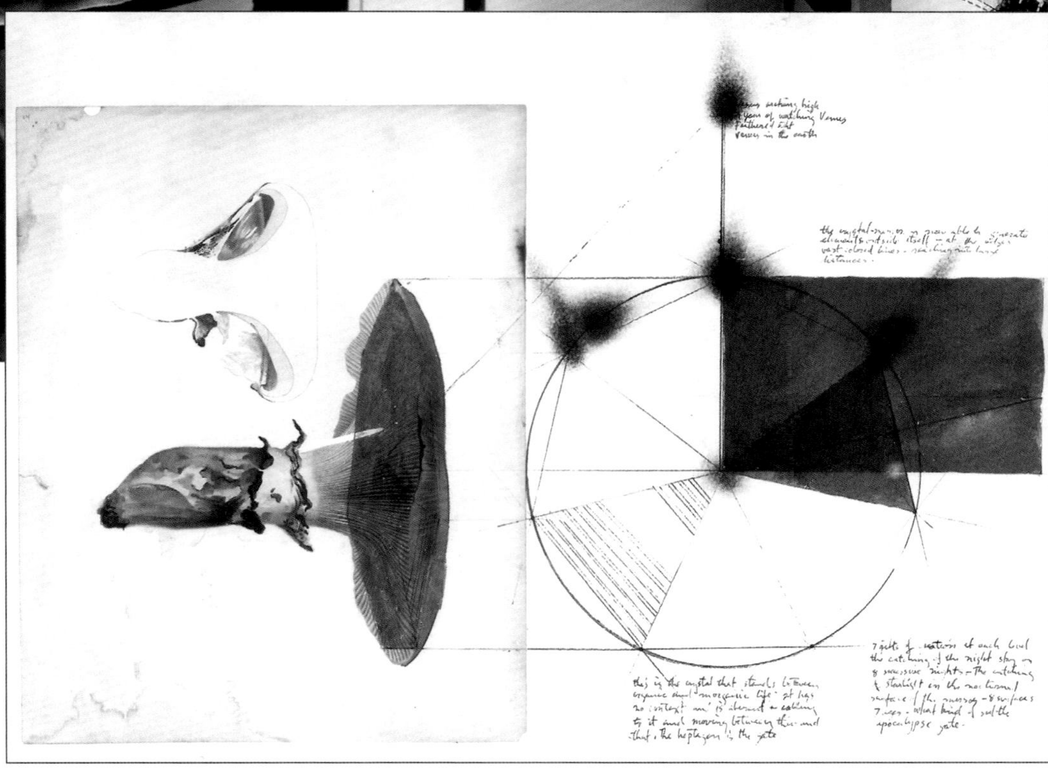

STUDIO

I have continued experimentation in the studio developing ideas for new large works as well as investigating holograms as drawings and various new printing techniques. . . . I am particularly interested in daylight holograms that can be seen with simple sunlight. I have focused on the curious area between the diffraction patterns and the image. I am interested in the way an image appears and disappears with change in light or angle. Otherwise my studio investigations have dealt with the registering of cosmic space in an interior chamber and particularly with areas of contradiction, irony and substitution. (Burgess, from "Individual Work by Fellows at the Center for Advanced Visual Studies, 1973-74," CAVS portfolio)

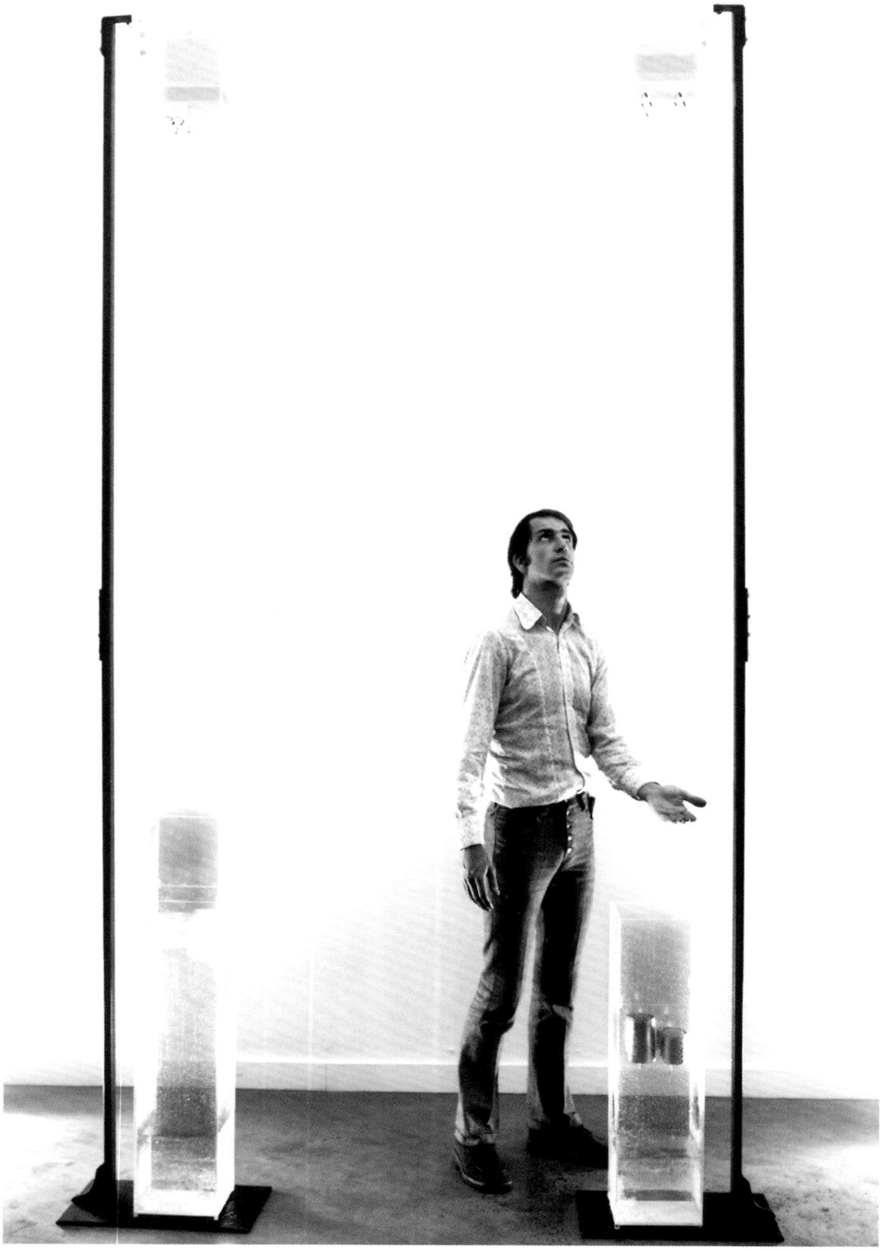

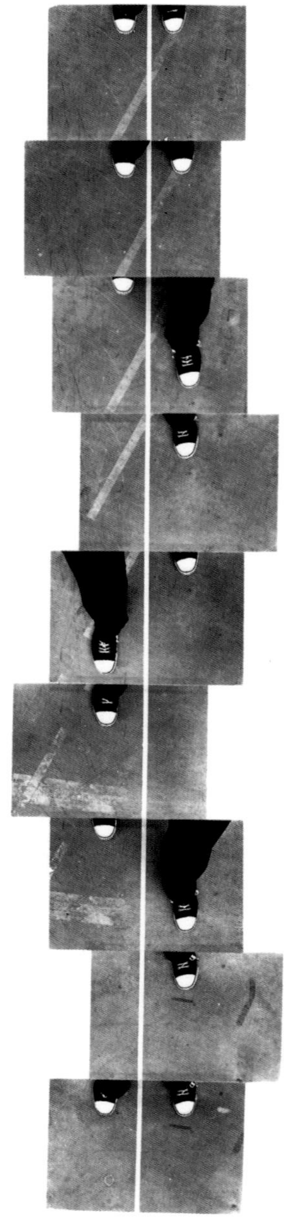

Above left: Luis Frangella, "Music Box," CAVS, MIT, 1973
Photo: Nishan Bichajian

Above right: Luis Frangella, "Gravity Center in a Trajectory," photo work, 1974
Source: CAVS Special Collection

MUSIC BOX This music box was presented at the Wadsworth Atheneum Tactile Gallery in Hartford, Connecticut as part of a Center exhibition. Later developments in the form of columns of sound were done following the same sound-producing effect. Water drips from the upper boxes, on top of the column, into the glass pieces of the lower part.

GRAVITY CENTER IN A TRAJECTORY A series of photographs shows the shifting of the gravitational center of the body of a person walking on a straight line. The camera has recorded this movement by focusing on the feet while located at the gravity center. (Frangella, from "Individual Work by Fellows at the Center for Advanced Visual Studies, 1973-74")

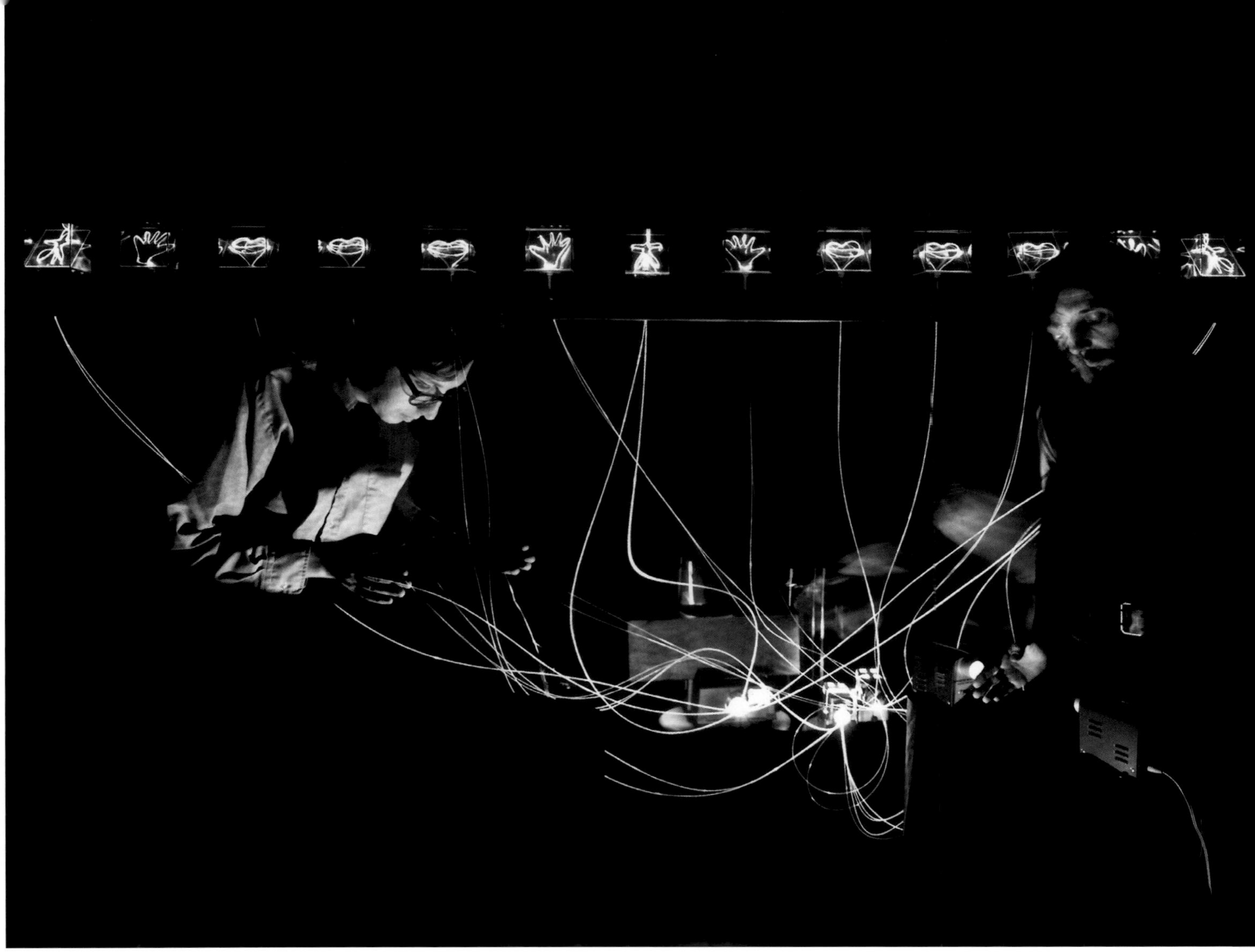

- shaking hands, greeting - more closed interaction, greater commitment - median rhythm
  -(H (---) H) - (H (---) H)-, four images

- pressing, kissing - big commitment - closed rhythm
  - (- (LLL) -) - (- (LLL) -) -

This piece has the power of visualizing this intangible ("hidden architecture," as it is called by E. T. Hall), that is being generated at any moment and place by the communication process. Moreover, it would permit experimentation with the possible equivalences between some conventions associated with the interpersonal communication and a visual convention created over the physical space. . . . Because each action is being transferred by other electric means of communication, the result is identifiable with the omnipresence of an integrated social temperament. (Baldeweg, from "Individual Work by Fellows at the Center for Advanced Visual Studies, 1973-74")

Above: Juan Navarro Baldeweg, "Gates from the Body," the artist (left) and Mauricio Bueno, CAVS, MIT, 1972
Photo: Nishan Bichajian

Bottom: Juan Navarro Baldeweg, typed excerpt from, "Individual Work by Fellows at the Center for Advanced Visual Studies, 1973-74"
Source: CAVS Special Collection

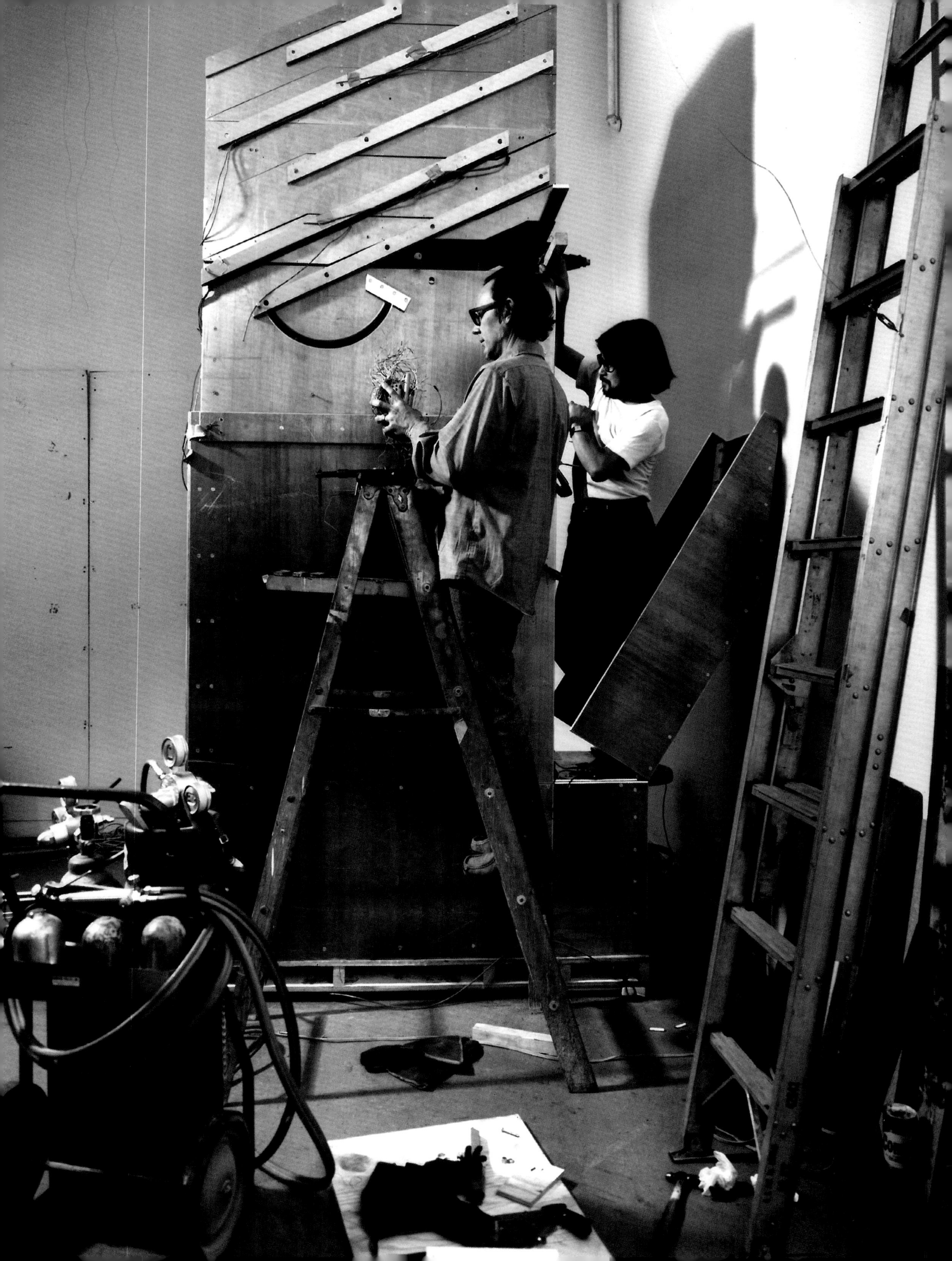

While at CAVS, Paul Earls and Michio Ihara collaborated on a performing sculpture titled "Pan":

As Mr. Ihara developed, he said, he found that screwdrivers and metal were more useful tools for executing his personal visions than brushes and canvas. Mr. Earls was fascinated by the possibilities of electronic music. The result is "Pan"—which, both artists agree, has come to have a life of its own. At rest, sculpturally, it looks simply like a large, rectangular aluminum and steel packing case with a lopsided house on top. But when plugged in it takes on various configurations in all sorts of movement patterns as it unfolds to its full 16-foot height and stretches out its "neck," much like a giraffe going after treetop leaves, to its full 27-foot extension. . . . " It has been programmed," Mr. Earls explained, "to react, in inverse proportions, to the sound and movement around, as well as to its own internal sound system. It was made with a complete logic. In other words, I know exactly what its components are. It is just that there are so many possible combinations it can react to that it is impossible to predict what it will do and when." Indeed, he readily admitted, he has created something with, more or less, an independent existence. "It Is a god that I can't completely control," he said, "and, of course, it's kind of scary because it's not what I thought it would be." (Phyllis Funke, "Science Is Blended With Art in Display," *The New York Times*, April 7, 1974, 97)

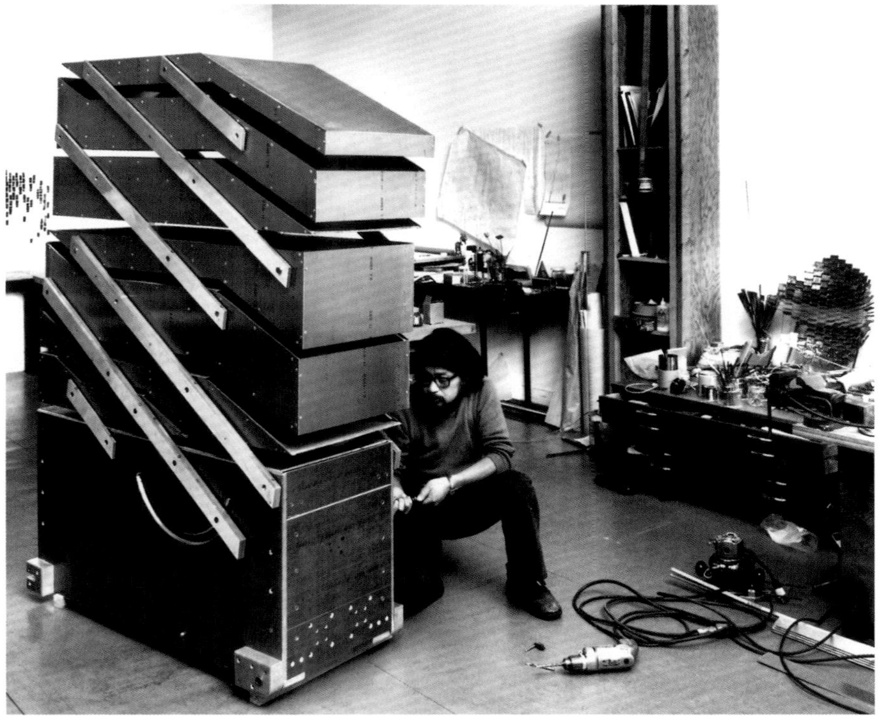

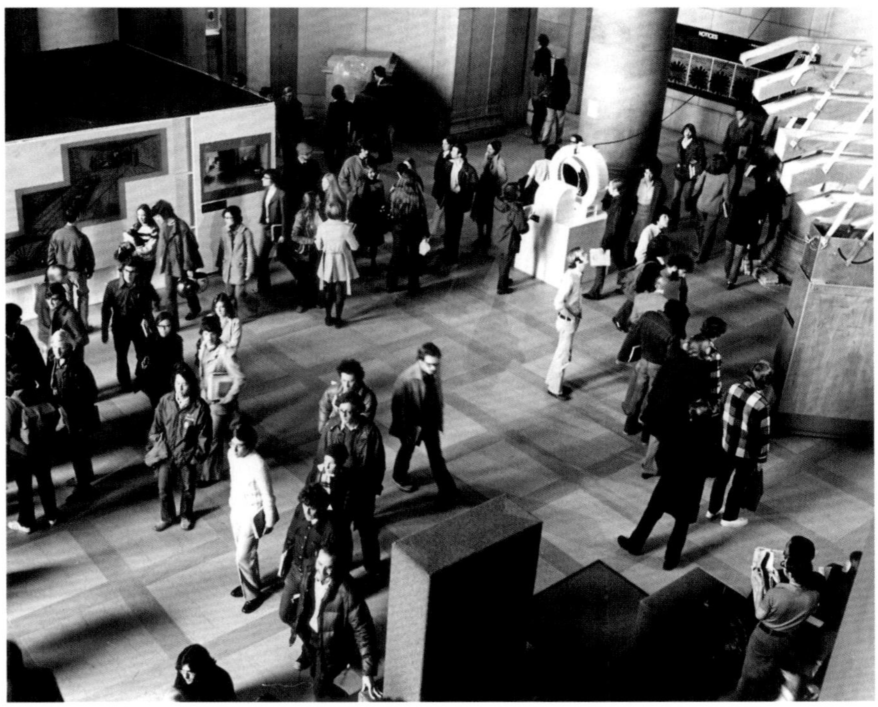

Above: Michio Ihara works on "Pan," interactive sculpture, CAVS, MIT, 1972
Photo: Nishan Bichajian

Bottom: "Pan," **Asterisk* show, *Mulitiple Interaction Team*, Lobby 7, MIT, November 7-15, 1974
Photo: Calvin Campbell, MIT

Facing page: Paul Earls and Michio Ihara work on "Pan" at CAVS, MIT, 1972
Photo: Nishan Bichajian

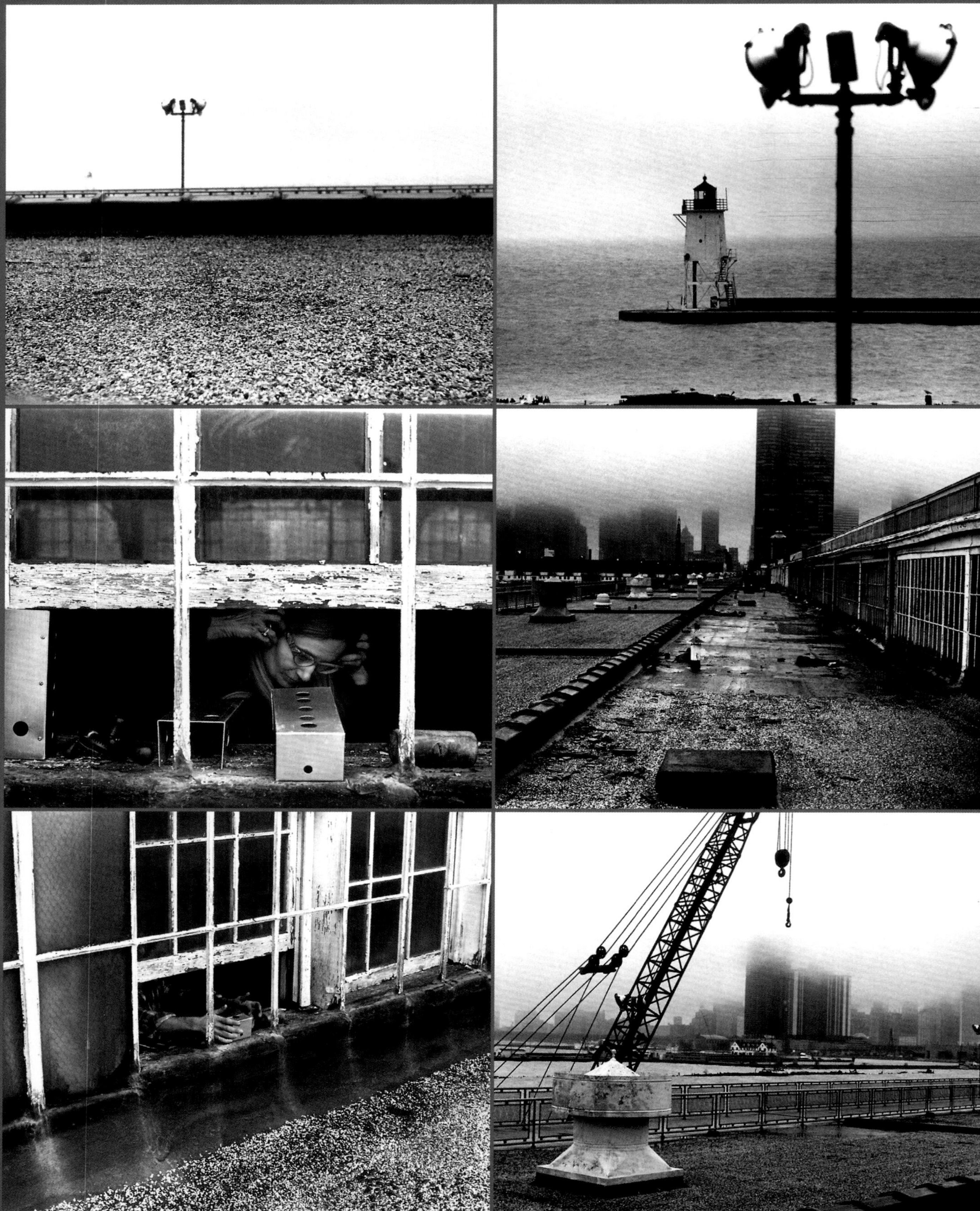

Maryanne Amacher, sounding environments recorded for performances of "Everything in Air," May 11, and "City-Links," May 12, Gallery at Museum of Contemporary Art, Chicago, 1974. Photos: CAVS Special Collection

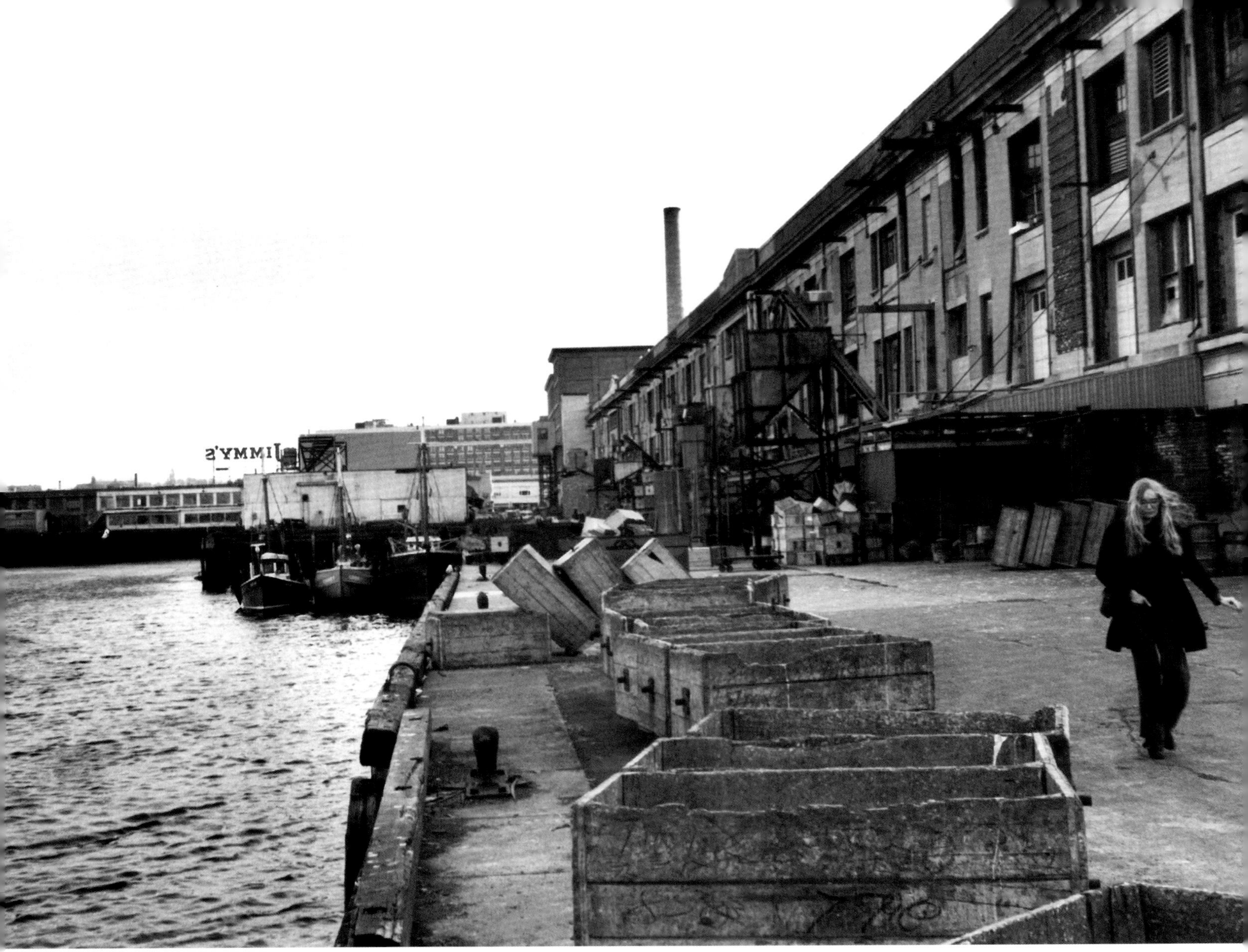

I selected tapes from my collection "Life, Time and its Music" to make a sounding environment for the particular interior space of the Gallery at the Museum of Contemporary Art, Chicago. I mixed the selected material, 6 tracks of recorded sound, in a continuous live performance. The composition explored familiar acknowledgements of boundary and illusion: auditory dimension and perspective existing between selected sounding material and the physical possibilities within and outside the immediate structure of the performing space. . . . An ambience. Apparent by sound. Exploring familiar acknowledgements of boundary and illusion. Activating certain mental possibilities. (Amacher, "Individual Work by Fellows at the Center for Advanced Visual Studies, 1973-74," CAVS portfolio)

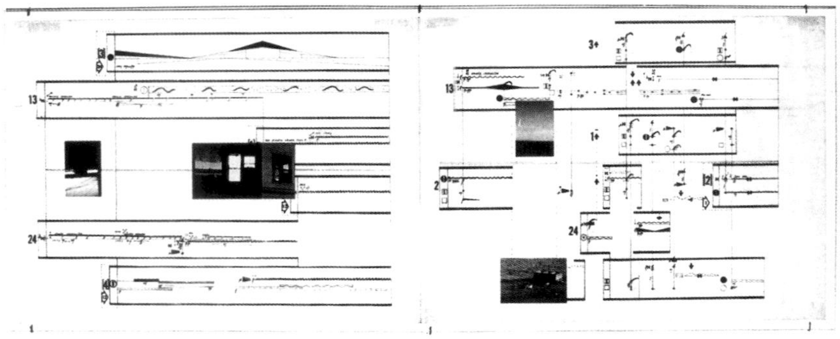

Above: Maryanne Amacher, "The composer checking microphone installation at Boston's pier No. 6" (*Chicago Guide*, May 1974). Photo: Mike Guran

Bottom: Maryanne Amacher, musical score, "Hearing the Space Day by Day, 'Live'," *Interventions in Landscape* exhibition, Hayden Gallery, MIT, 1974
Source: CAVS Special Collection

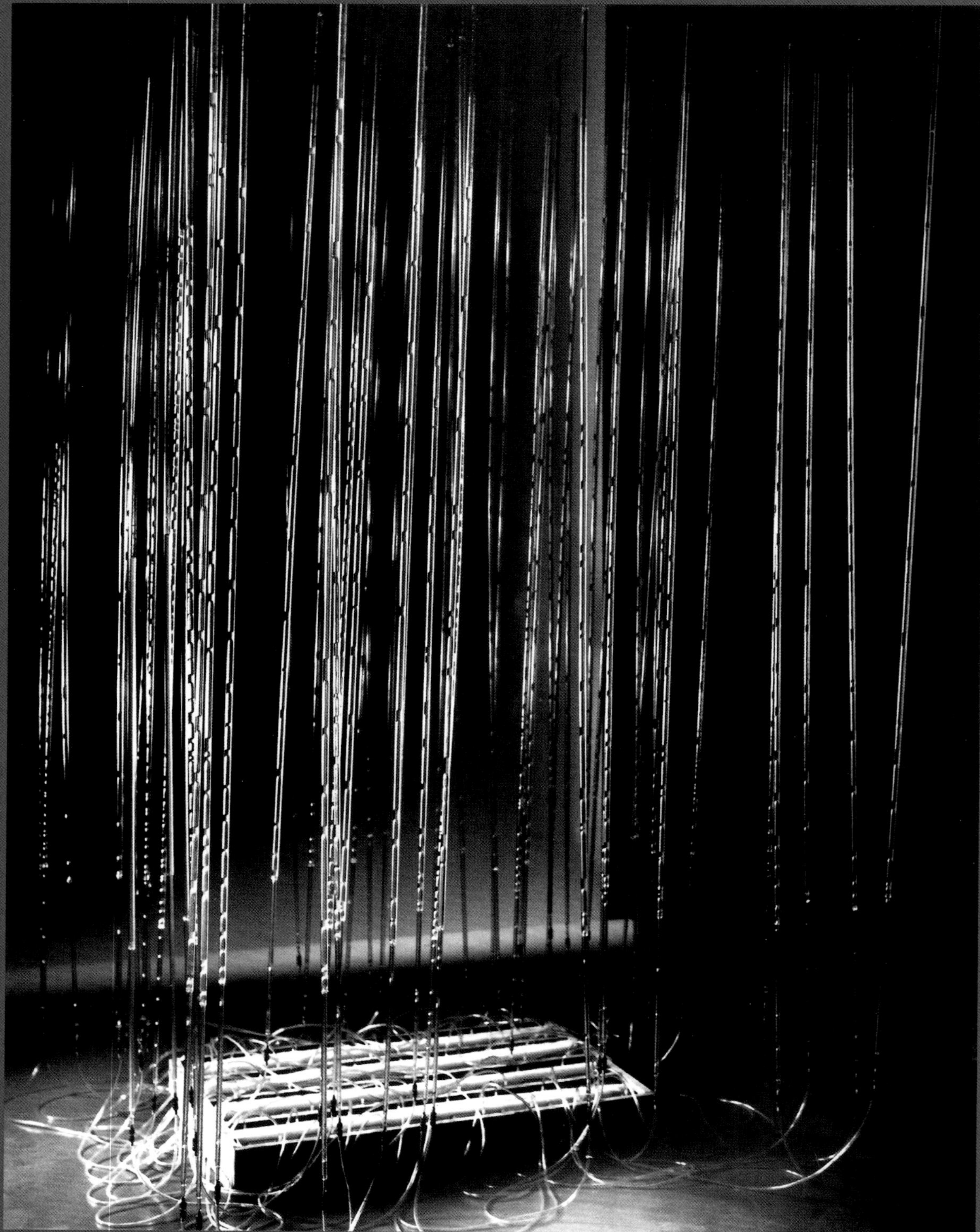

Mauricio Bueno, "Forty-nine Tubes,"
CAVS, MIT, 1972
Photo: Nishan Bichajian

Mauricio Bueno, on a Fulbright fellowship from Bogota, Colombia, is engaged in the design of environmental sculpture. . . . Mr. Bueno is now exploring the possibility of an urban scale light windmill which utilizes a rotating cylinder, mirror, and laser to create varying geometric patterns. (Kepes, *MIT Report of the President*, 1970, 37)

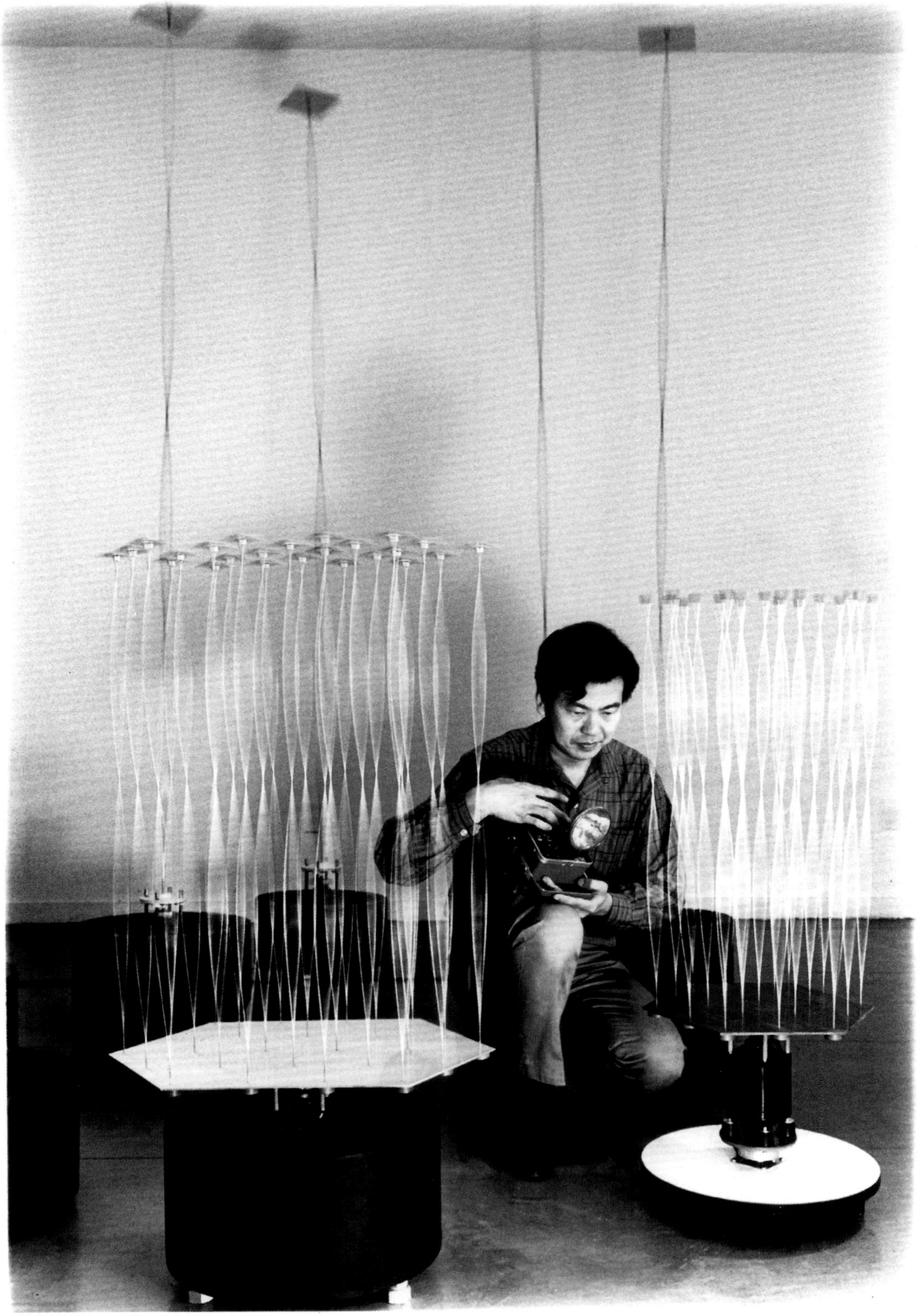

I think it was during the period of art at its peak, creatively. Everyone was searching for ideas, different groups of people all over the city, the country, the world. Who is doing something? . . . When I heard about CAVS, I wanted to find out what was happening. I have many books about kinetic art by Kepes. . . . When we met it was like old friends right away. He talked about Chinese culture and his friends at Harvard, Asian specialists. . . . (Wen-Ying Tsai interview)

Wen-Ying Tsai, "Cybernetic Scuptures," the artist in his studio at CAVS, MIT, 1969
Photo: Nishan Bichajian

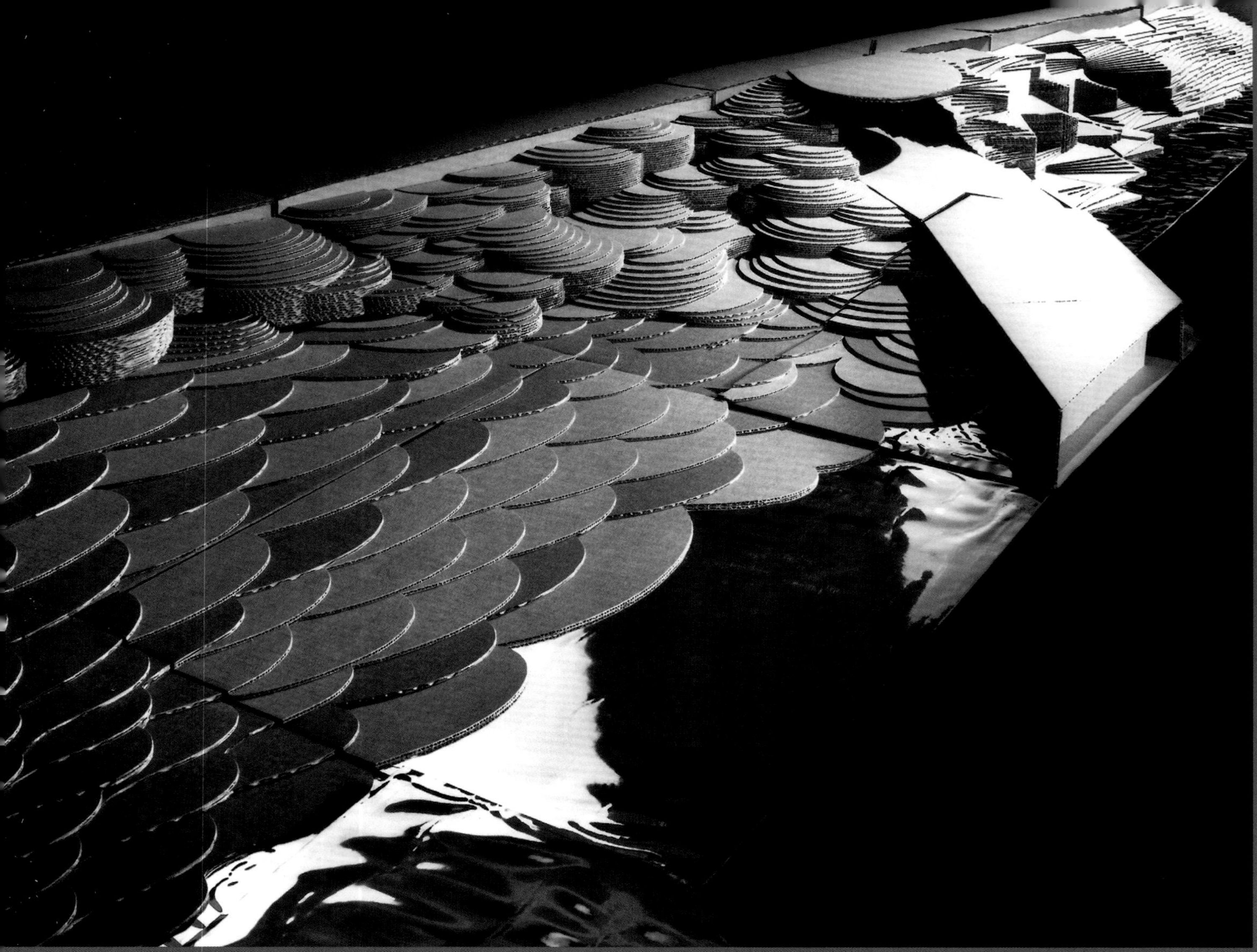

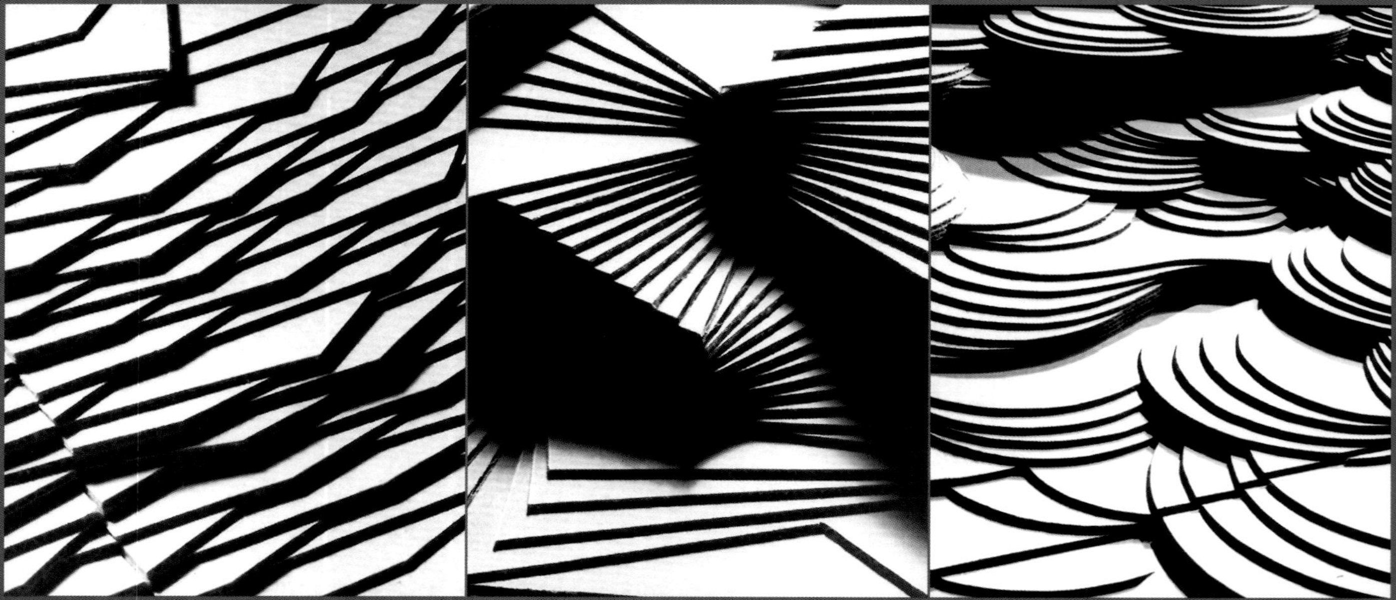

Athena Tacha, "Step Sculpture," model for step-and-terrace environmental sculpture for banks of the Charles River, CAVS, MIT, 1974

Photo, top: Nishan Bichajian
Photos, bottom: Athena Tacha (details)

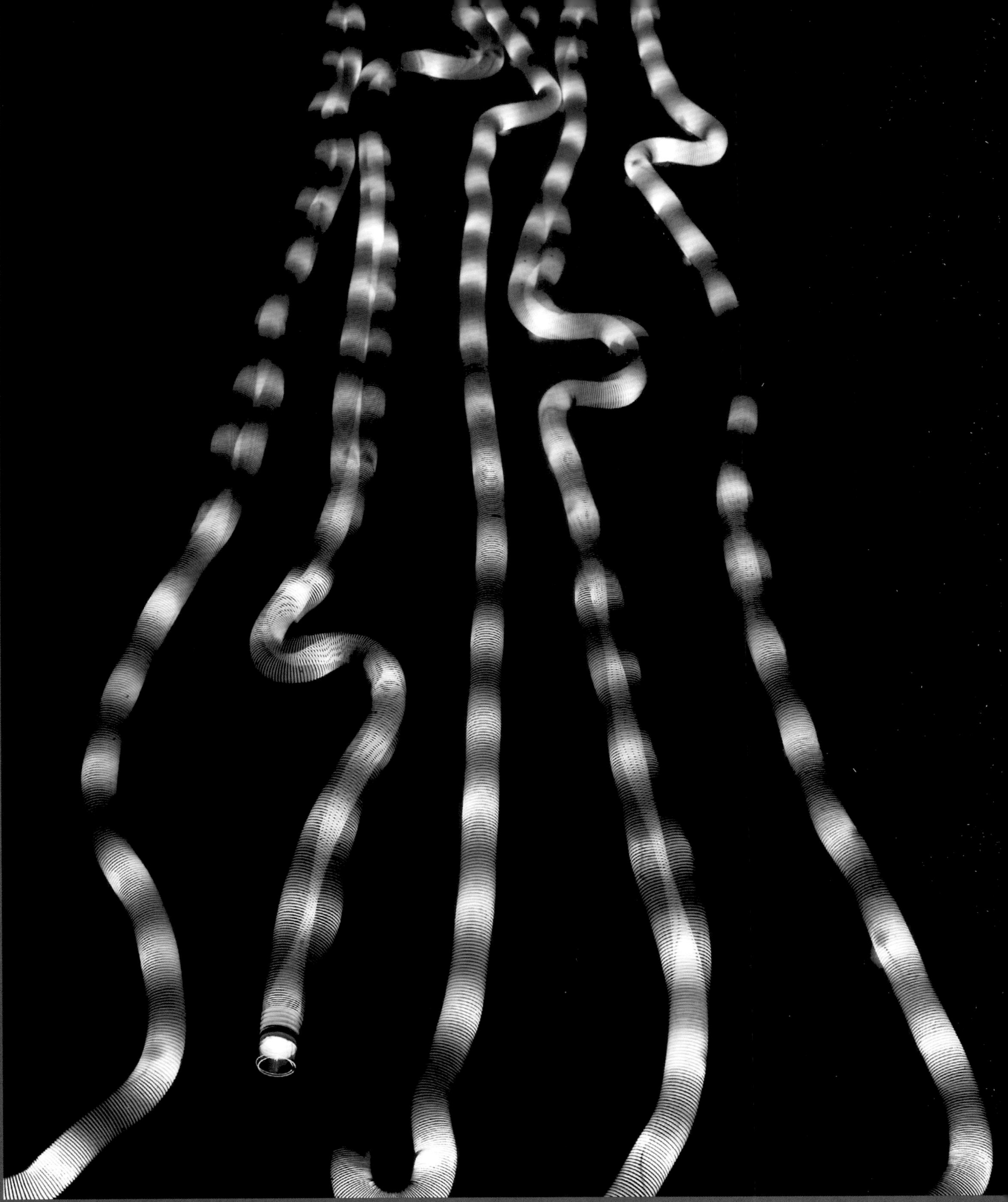

Howard York, "Charles River," sculpture, CAVS, MIT, 1972
Photo: Nishan Bichajian

Chapter 2

Art in the Environment: Education at CAVS

Education at CAVS focused on future orientation in the arts, exemplified in György Kepes' *The New Landscape in Art and Science* (1956) and the multi-volume *Vision + Value* series (1965-1966) with essays by artists, architects, educators, psychologists, and scientists on an "education of vision." Piene wrote of the need for a new approach to education in his abecedarium-style book, *More Sky*.

> The artist-planner is needed. He can make a playground out of a heap of bent cans, he can make a park out of a desert, he can make a paradise out of a wasteland, if he accepts the challenge to do so. . . . In order to enable artists of the future to engage in large-scale planning and shaping of tasks, art education has to change completely. (Piene, *More Sky*, MIT Press, 1970/1973, 4)

From the beginning, artist-fellows taught classes as part of their participation in the CAVS community. In 1976, MIT launched the Master of Science in Visual Studies (SMVisS) program, creating an early model for art-science-technology degree programs. Students and fellows worked together and shared studios, extending CAVS research and practice into global projects and a future creative vision.

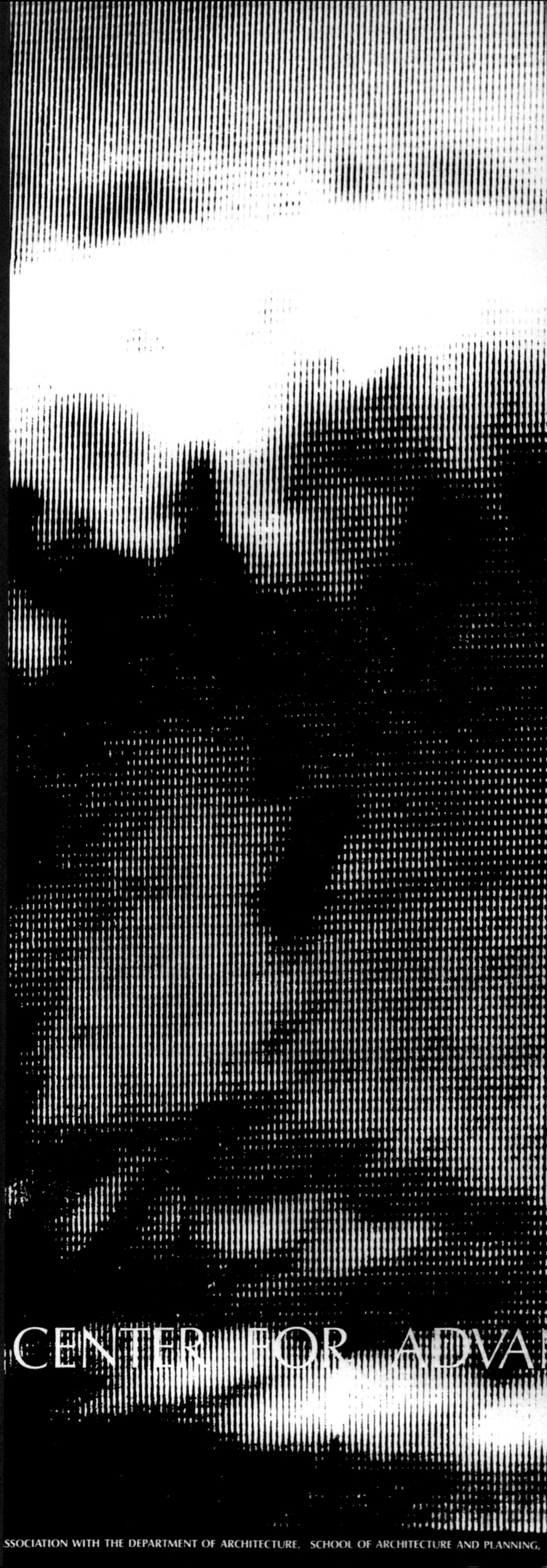

Right: poster (detail), Subjects, fall semester, Center for Advanced Visual Studies, MIT, 1975
Source: CAVS Special Collection

subjects , fall semester 197

4.801 ART AND THE ENVIRONMENT (3-0-6)
Otto Piene, Friedrich St. Florian

Continuation of lecture series on environmental art plus presentations on foreign cultures by Center Fellows: assignments for investigations by students. Initial meetings at 3-133, further meetings at the Center for Advanced Visual Studies, Building W11.

First meeting, Tuesday, September , 7-10 p.m., 3-133.

4.845 ADVANCED VISUAL DESIGN: GLASS, GAS & ELECTRICITY (0-4-8)
Otto Piene, Alejandro Sina, Pam Vandiver; Dan Dailey, Massachusetts College of Art

Theory and practice of some aspects of kinetic and light art. Electricity and glass as tools and means for indoor and outdoor sculptures and projects. Use of C.A.V.S. facilities, Department of Metallurgy glass shop and Mass. College of Art glass-blowing facility. Design and presentation of works in C.A.V.S. exhibit. General art and workshop experience required.

First meeting, Thursday, September 11, 9 a.m., 7-437.

4.841 ENVIRONMENTAL LIGHT AND COLOR (2-4-6)
Robert Preusser, Nishan Bichajian

Optical phenomena and psychological responses inherent in the perception of light and color serve as the basis for their applications within the context of architectural and urban settings. Emphasis on the dynamic and qualitative attributes of artificial illumination as the means for creating expressive environments. Photography utilized as an analytic tool and for recording experiments. Studio, E21, second floor. Mondays and Wednesdays, 2-5 p.m.

First meeting, Wednesday, September 10, 2 p.m., E21.

10

4UR or 4.899 SPECIAL PROBLEMS IN ENVIR
(OPTION I) MULTIMEDIA WORKSHOP (Units
Paul Earls, Otto Piene, Robert Preusser

A joint projects course in the product
media performances. May be taken eithe
Problems subject or a UROP project. Re
Wednesday afternoons, at the Center for
Studies, Building W11, 4 p.m.

First meeting, Wednesday, September 10,

4.895 PROJECTS IN ENVIRONMENTAL ART: "PASSAGES" (2-0-4)
Paul Earls and Fellows, Center for Advanced Visual Studies

The transformation and creation of a temporary passage/corridor, including selection of site, thematic formulation, planning, design, execution, documentation, and evaluation, together with presentation-events of the work of the Fellows of the Center for Advanced Visual Studies. Designed for artists interested in environmental work and for advanced undergraduates ready to realize collaborative projects. Regular meetings Tuesdays from 10 a.m. to noon, at the Center for Advanced Visual Studies, Building W11, with additional laboratory hours to be scheduled.

First meeting, Tuesday, September 9, 10 a.m., W11.

September 9	Introduction: Otto Piene, Paul Earls
September 16	"Conceptual Art", Whitney Chadwick
September 23	"Environmental Art", Whitney Chadwick
September 30	"Multi-gates", Lowry Burgess
October 7	"Processional Spaces", Friedrich St. Florian
October 21	"Memory Connection", Keiko Prince
October 28	"Immediate Environments", Luis Frangella
November 4	"Video Environments", Peter Campus
November 11	"Live Space", Maryanne Amacher
November 25	Project
December 2	Project
December 9	Project Evaluation

4.899 SPECIAL PROBLEMS IN ENVIRONMENTA

1. THE AVANT GARDE FILM, 1965-1975
Jon Rubin

Viewing and analysis of some of the mos
narrative, non-documentary films made d
decade. Discussion and elucidation of
artist does with his/her materials. P
on filmic structures. Filmmakers whos
studied include: Frampton, Nelson, Ge
Warhol, Conrad, Herbert and others.

The class will meet every other week o
Thursdays, 7:30-10:00 p.m.

2. STUDIOS IN COLOR OR SOUND
Lowry Burgess, Paul Earls

(In conjunction with Massachusetts Col

4.861 GRAPHICS LABORATORY I (0-3-2)
Nishan Bichajian

Free hand drawing of still life forms i
from simple lines to light and shadow m
Control of line, value, color, texture
with the use of pencil, pastel, conte c
charcoal and brush. Also drawing from
to develop accurate observation in acti
line and form.

First meeting, Tuesday, September 9, 1

VISUAL STUDIES

The Evolution of Education at CAVS

György Kepes was a professor for nearly thirty years in MIT's School of Architecture and Planning. His *Vision + Value* series and *The New Landscape in Art and Science* still serve as the unofficial textbooks for art- and environment-related classes. Kepes' fellow professors Robert Preusser and Richard Filipowski taught visual design and, in 1974, Preusser became the Center's director of education. While Kepes envisioned the Center as a haven for pure research, he articulated through his teaching, writing, and symposia his vision of a new paradigm for art and science.

> What has happened in art is itself a part of a very broad movement in which science has made the major contribution. Through its dynamics of rigorous logic twentieth-century scientific understanding has come to conclusions not unlike those of artists. (György Kepes, *Arts of the Environment*, 1972, 7)

The 1970 edition of the annual *MIT Report of the President* stated that under Kepes the Center was "not a teaching unit of MIT," but that his course, Advanced Visual Design, "served to introduce students to the Center's goals and to the nature of the projects developed here." Graduate students from MIT and Harvard University "were able to utilize the Center's work space and to consult freely with Professor Kepes while exploring the no-man's land between free artistic expression and socially urgent environmental tasks."

> There is the legacy of what Kepes did, the books he has written about his time in Chicago with Moholy-Nagy and later. What is so important about the books is that they have inspired almost every generation since. . . . He began to see what was just over the horizon, which really helped us all to envision the future. Part of the story is that those who passed through Cambridge went on in various ways, but carried on these ideas. . . . CAVS at MIT is a place where all that got focused in its 25 year period. (Greg Garvey interview)

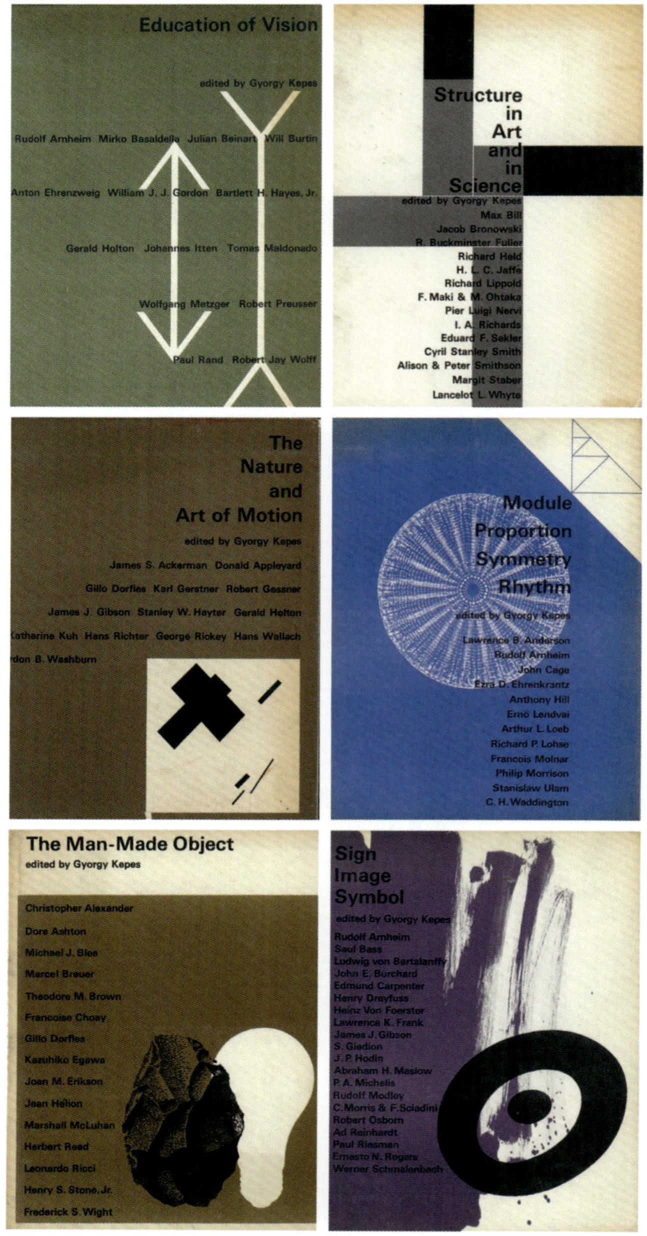

IV The New Landscape

Magnification of optical data
Expansion and compression of events in time
Expansion of the eye's sensitivity range
Modulations of signals
(Kepes, *The New Landscape in Art and Science*, 1956, 100)

Above: György Kepes, editor, *Vision + Value* series of six books, in order of publication (left to right from top left), George Braziller, New York, 1965-66
Image source: Bill Seaman

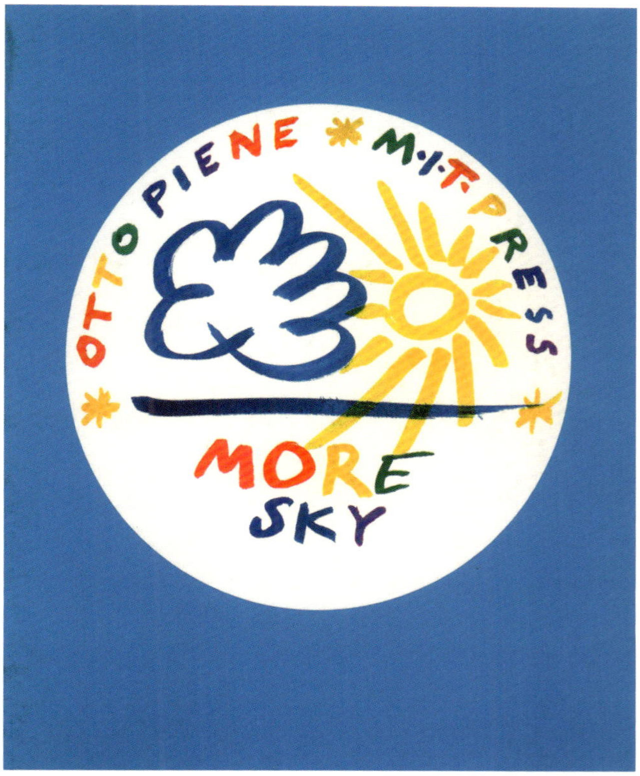

"Ideas" 108

in such number for and about art have been discussed, especially since the intellectualization of art (in modern art) that occurred following Chevreul, Apollinaire, Worringer, Marinetti, Malevich, that I feel that a deadline should be set, maybe August 1, 1973, after which the use of art language should be restricted to content, art itself, as opposed to comment or interpretation. Word ideas will no longer be relevant. When properly set out, reality is strong enough to permit the measurement of things by their presence and intensity, as opposed to their potential. After all, not that many good ideas have been added since Jules Verne, Paul Scheerbart, Laszlo Moholy-Nagy.

Given the advanced age of technology, it is deeds that count, not babble. We have to realize that visual or sensual evidence can lead us out of the dark ages of an archaic intellectualism in which catchwords like patriotism, America, German, or white were the pied-piped tunes of mankind's herds. You hang a sculpture in the wind, and the wind becomes visible. You talk about the wind, and the wind becomes a word. You talk about art and technology, and they turn into a phrase easily overcome or neutralized at least by words, a process also known as sloganeering.

Again: one way to do things is to do them.

Above: Otto Piene, *More Sky*, book cover drawn by Piene, and page 108, "Ideas," where Piene underlined one of his favorite comments: "one way to do things is to do them," MIT Press, 1970
Source: CAVS Special Collection

Kepes' volumes educated through essays by luminaries such as Norbert Wiener, Bruno Rossi, Buckminster Fuller, Louis Kahn, Margaret Mead, Katharine Kuh, and others. Piene's *More Sky* complemented Kepes' writings with a more personal pedagogical approach.

> The artist catering to the tastes and expectations of the chosen few is a pitiable creature. Artists' obligations lie elsewhere. Villages, cities, regions, states, countries, continents have been turning ugly since the beginning of the industrial revolution. Planning of large change in the provisions for physical living has been left to architects with limited sensual imaginations. The architecture of the mind that is produced by 'the media'—newspapers, radio, television—is the work of 'professionals' with equally limited creative capacities. (Piene, *More Sky*, 1970, xii)

Both Kepes and Piene were master teachers with divergent pedagogical visions. As co-founder of Group Zero in post–World War II Germany, Piene was already developing his *Luftschule* concept—a Sky Art school without boundaries or borders. Piene spent approximately thirty hours a month flying between the US and Europe. His emerging *Luftschule* was an extended human organism that flies, all senses prickled. Piene described his integrative approach to teaching, community, and art making:

> Teaching has been the catalyst for many other things that happened and turned into art and new art forms. In my view to this day teaching is a good thing to do because it helps not only people grow but also helps minds clear and organize themselves. Not only for the teachers but also for the students of whatever institutions. It wasn't so much methodical teaching, it was the practice, the working that was done together between generations and conquered new territory. (Otto Piene, interview by Bill Seaman, 2012)

The Master of Science in Visual Studies

Many former CAVS fellows and students considered the graduate program leading to the Master of Science in Visual Studies (SMVisS) degree the grandest achievement of CAVS. The program passed MIT's academic council to become official in 1974, and in 1976, its first graduates were Derith Glover and Michael Moser. As director, Piene focused on education at CAVS in a way that would change the Center's complexion.

> During the 1974-75 academic year, the Center greatly expanded its educational program at MIT, attracting students from many departments at the Institute and from art schools and universities in the Greater Boston area. . . . Unsolicited applications for a graduate program at CAVS from students across the country and from special students who have been attending Center subjects, has resulted in the formation of an Education Group at CAVS this past spring. The present purpose of this group is to investigate and to propose the incorporation of an interdisciplinary Master of Science degree in Art as an MIT course of study. This preprofessional training program would be the first of its kind claiming and requiring interdisciplinary skills and activities among art, science, engineering, and humanities. It is the Education Group's conviction that MIT is the optimal place for such a program to emerge. (Piene, *MIT Report of the President*, 1975)

Piene, with Professors Richard Leacock, head of the Film/Video Section, and Nicholas Negroponte, head of the Architecture Machine Group (ArchMach), conceived of a degree program. The program was designed by artists for artists working in areas of art related to science and technology—those for whom it made sense to be at MIT. Piene said, "it was an art program, but because it was housed at MIT the word 'art' could not be used." Piene suggested that the program was immediately successful "because MIT students were starved for musical values and practice—by musical I mean the muses." (Otto Piene, Matthew Wisnioski, "Arts / Science / Technology," *A Second Modernism: MIT, Architecture, and the 'Techno-Social' Moment*, 2013, 781)

> When I was in college in the 70s, the first VCR came to an art campus. The idea of blending the new media technologies and sciences into a very fresh curriculum in Kansas City was suggested by Dale Eldred, my mentor, very forward-looking—the first person I ever saw doing something like that. He claimed the place to go, the ONLY place to go was CAVS. When I got to CAVS they made it clear they weren't an art school—fascinating—it was a paradigmatic shift for me as an artist. Now I had to think about the definition [of art] in a much broader social way. (Jennifer Hall interview)

> I knew I was in the right place for what I wanted to do. It felt comfortable. If you were the type of person who needed to be shown how to do things, it would be disorienting, but if you realized you had to become a sponge and absorb everything you could get, you could negotiate what areas didn't apply to you and work independently. It was a total academy for learning. (Walter Dent interview)

The SMVisS program provided pre-professional education with a large number of the graduates going on to work professionally as art educators and artists— holographers, video artists, sculptors, painters, and multimedia artists/performers—with their feet solidly planted at the edges of the art, science, and technology landscape. Students rubbed shoulders with fellows as they collectively worked toward the realization of major projects, including *Centerbeam*, *Monocle*, *Desert Sun/Desert Moon*, and the *Sky Art Conferences.*

The airy light-filled studios at the front and the darker spaces at the back of Building W11 were shared by students and fellows alike, facilitating creative interactions, discussions, and experiments in new media and old, from computer-coded installations to drawing. With rapid changes in technology, each SMVisS class and generation of fellows developed a shared identity visible in their work.

Interactions with the other sections of the SMVisS program—the Architecture Machine Group, Film/Video Section, and the Visible Language Workshop (VLW)—were integral to graduate study. Students also worked at the MIT Creative Photography Laboratory, directed by Starr Ockenga, the successor to Minor White who had initiated the lab in 1965. The degree program became a cohesive umbrella for the various art-science-technology groups. When many of these sections joined the new Media Lab, CAVS remained separate as a chartered center with its own designated building. CAVS students routinely vanished into the broader Institute to pursue holography, electronic music, animation, and early computer-based media, and work in labs from glass fabrication to biology and brain sciences. MIT specialists in scientific and technical areas were generally receptive to the ideas of these artist-students.

> One advantage of CAVS being at MIT was in reducing the fear factor around technology. The assumption was that you could figure it out, and liked the challenge of doing just that, especially with something brand new.
> (Ellen Sebring interview)

MIT undergraduates who were engaged in coding, engineering, and science took classes at CAVS, providing technical know-how for collaborative projects, and in some cases, becoming artists themselves. Students also cross-registered from local colleges, including Harvard, the School of the Museum of Fine Arts Boston, and Massachusetts College of Art, further diversifying the community. In 1986, CAVS student Todd Siler became the first visual artist to earn a PhD at MIT, integrating brain science, psychology, and art.

> I seriously needed to find a safe place in the world . . . a place that encourages this natural confluence between the arts and sciences, rather than discouraging it. I was hungry to huddle with anyone who was open to these ideas, research them further, and when I came to the Center I felt like I'd made it to Mecca.
> (Todd Siler interview)

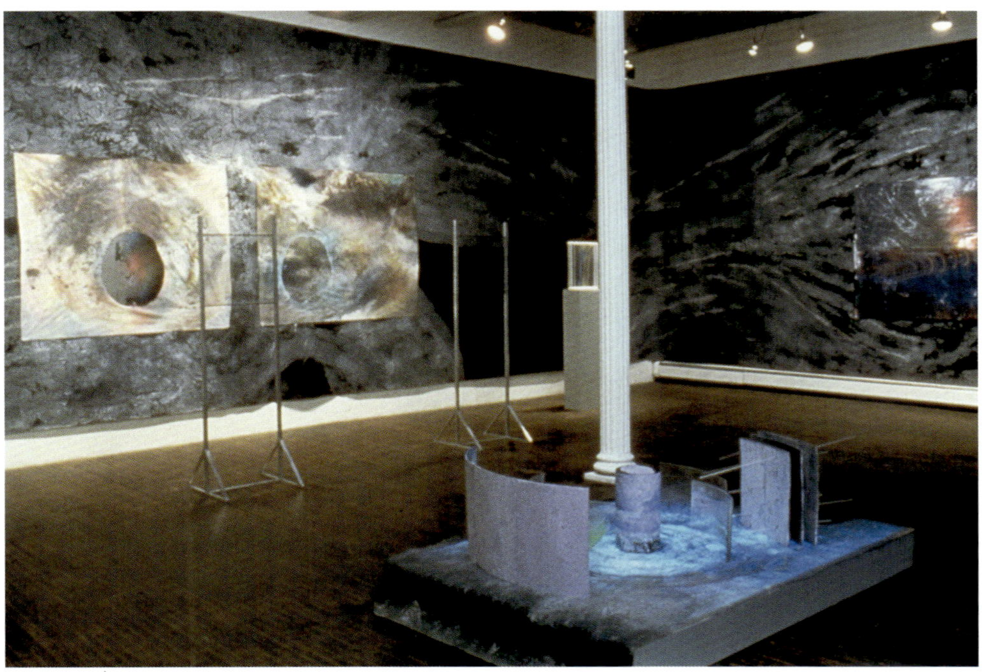

Todd Siler, "The Brain Theater of Mental Imagery," mixed media, 1983; "Cerebrarium" model, 1983-85, Ronald Feldman Fine Arts, NYC
Photo: Todd Siler

Massachusetts Institute of Technology Department of Architecture

1978-79

master of science in visual studies

The Program

The Master of Science in Visual Studies program (S.M.Vis.S.) was initiated in 1978 in response to a long history of diversity in areas of study offered by the Department of Architecture. Previously, part-time, non-degree studies in the visual arts have offered the Institute a repertoire of general studies, the Department a sequence of distributed requirements, and the occasional student a program of concentration. The S.M.Vis.S. is formalization of the latter with the three-fold purpose of: offering students professional opportunities in areas of scientific inquiry and visual study outside the architectural professional track; contributing to undergraduate and architectural programs by having a more critical mass of teaching and research; stimulating advanced research programs by attracting graduate students committed to a specific line of inquiry. The S.M.Vis.S. is an experimental program without the precedents of specific subjects sequences, or career opportunities. As such, prospective applicants are urged to make personal contact with the faculty and staff associated with the specific area of concentration.

Currently, three programs are offered within the S.M.Vis.S. They are: film/video, environmental art, and computer graphics. Their selection results from an existing momentum in teaching and/or research, formidable enough to justify concentrated study. Each group has a very specific bent on its topic. Film/video has a long-standing reputation in *cinéma verité* or *direct film*, punctuated by research and development in sync sound super-8. Environmental Art is marked by a tradition of combining art and technology launched by the early work of Gyorgy Kepes. Computer graphics is flavored by the quest for interactive and intelligent systems.

Degree requirements have been fashioned in a manner to encourage students to capitalize upon other areas in the Department and the Institute at large. In contrast to apprenticing with a mentor, candidates are expected to outline their own area of interest. Our educational goals are tinted by a self-interest in nurturing fields of study available uniquely at MIT. Consequently, prospective applicants should see their own educational objectives as research-based and professionally oriented toward acquiring skills and insights, while at the same time participating in the building of a graduate program.

The faculty are well-known innovators in their fields. However, the academic staff is not limited to the advisors in charge of each area of concentration, but stretches across the whole Department, including research associates directing very specific research projects. The S.M.Vis.S. program sees itself as a collage of colleagues, rather than a classroom of exercises.

Curricula

The Master of Science in Visual Studies degree is awarded upon satisfactory completion of an approved program of study. This includes a minimum of 100 units plus a thesis acceptable to the Department.

Thirty-two (32) credit units per term may be considered a full-time load. However, a student without a teaching or research assistantship can normally average between 45 and 50 units per term at the Institute. This means that a student with an assistantship should expect the program to take a minimum of four semesters. An incoming student without an assistantship should anticipate a duration of three semesters, devoting most of the third semester to thesis.

Curricula will depend very much upon the background and the selected area of concentration of the individual. Students are not bound by the areas described in this brochure but may also consider joint programs and combined areas of study. Examples of such study are video art, computer art and media research.

During the first semester students will be expected to examine all the resources which are available at MIT. Moreover, 40 of the required 100 units must be taken outside the field of concentration. While precision of what is and is not in the field of concentration proves impossible, candidates should recognize and expect to conform with the intention of the requirements. That is to say, students will be expected to take subjects, conduct laboratory experimentation, participate in seminars, or work in studios, outside both the area of their specific concentration and their group of co-workers.

Graduate Programs

The Department of Architecture offers four graduate programs of study: the Master of Science in Visual Studies (S.M.Vis.S.) program; the Master of Architecture (M.Arch.) program, which is the first professional degree for students interested in the professional practice of Architecture; the Master of Science in Architecture Studies (S.M.Arch.S.) program in the fields of environmental design, housing and settlement design, and building systems design; and the Doctor of Philosophy (Ph.D.) program in Architecture, Art and Environmental Studies.

The Department also offers a four-year undergraduate program that leads to the degree Bachelor of Science in Art and Design (B.S.A.D.).

Film/Video

Film and video at MIT draw from a rich and distinguished tradition of documentary production. The experimental approaches and the range of subject matter which characterize recent work at the Film/Video Section are rooted in *cinéma verité* forms which emerged during the 1960s. Methods of spontaneous, unscripted moviemaking with a hand-held camera and sync sound recording techniques remain at the focus of our teaching and research. However, theory and application have extended far beyond any vision of the genre ten years past. Through work produced by staff and students, the documentary form has gained new definition in autobiography, cable television, science communication, performance and mixed media (opera, theatre, dance).

Since its inception in 1969, the Film/Video Section has maintained a persistent involvement in the design of evolving media technologies and techniques. Through the pioneering efforts of Richard Leacock, a sync sound super-8 filmmaking system developed at MIT revolutionized that medium for independents, schools and small production companies across the country. Filmmakers working with Edward Pincus have extended the parameters of personal cinema and explored innovative approaches to one-person sync sound 16mm production. More recently, video portapak cameras have been modified for shoulder mount and low-light shooting. Research in high quality, low cost production technologies has led to the design of systems which interface super-8 and video, super-8 and 16mm, and experiments with live performance. Respective attributes of various media, as well as new forms which result from their combination, are being tested on cable television systems at MIT and elsewhere.

The S.M.Vis.S. program in film/video now numbers about seven students. Demonstrated competence in film or video production is prerequisite to admission, and the submission of a film or videotape is required for application. Prospective applicants are urged to visit and discuss their special interests with people involved in the program.

Through intensive training and practice in the techniques of unscripted sync sound moviemaking, the program emphasizes the development and refinement of moviemaking skills. Camerawork, editing, sound, film/video concepts and dissemination strategies are treated as parts of a whole rather than as narrow craft skills like the professions of editor, camera operator, director.

The Film/Video Section is well equipped, and the limited number of students assures ample access to sophisticated shooting and editing equipment in double system 16mm and super-8 film, in ½" and ¾" video. Other facilities include a cablecasting station, a color video studio, film to video transfer, and electronic image processing.

S.M.Vis.S. candidates are expected to initiate, finance, and execute several short film/video productions plus a longer film or tape which will qualify as thesis material. Even more than technical proficiency, the curriculum stresses exploration of individual perception using the most appropriate movie-making tools. In a rewarding association with fellow moviemakers and through constant viewing and discussion of works in progress, students formulate and evolve their own projects.

As a complement to regular subject offerings, workshops such as those recently conducted by Jonas Mekas and David Cort, and frequent guest presentations lend significant breadth to the film/video program. Students are encouraged to enrich their concentrations through selection from a vast array of overlapping and interdisciplinary resources at MIT. The Film/Video Section also maintains a strong tie with professional film, video and television communities. Individual work is sometimes supplemented by collaborative projects with outside sponsors.

Director: Richard Leacock
E21-010, MIT

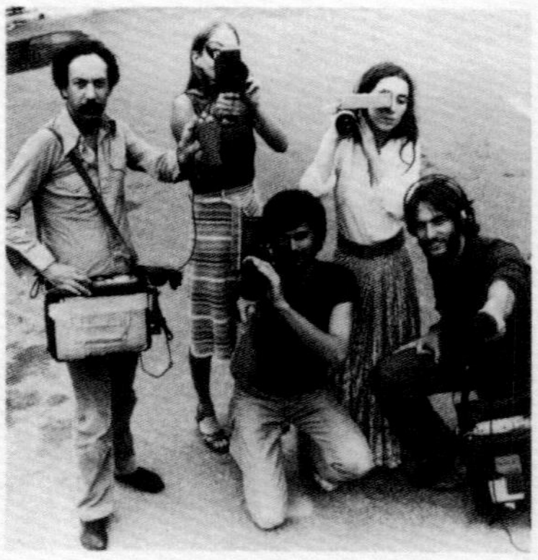

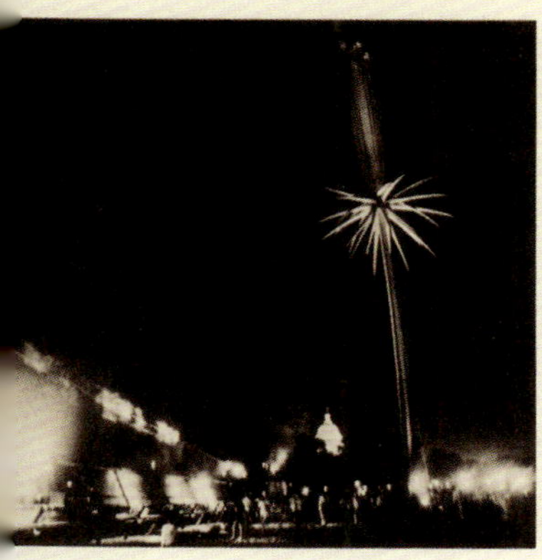

Environmental Art

The Environmental Art area of concentration of the Master of Science in Visual Studies program results from the willingness of the artists at the Center for Advanced Visual Studies to share their experiences with students working towards related objectives. Among the primary fields of specific interest to CAVS are: environmental sculpture and painting and sculptural architecture of all scales; public leisure installations; celebrations; elemental, kinetic art and aesthetic/psychological ecology.

Further disciplines of art and art research pursued by fellows of CAVS have become, consequently, and with increasing emphasis, developmental media work—especially progressive video art—and new forms of kinetic art, such as holography, laser research and multi-modal performance.

The founding philosophy of the Center's efforts is based upon the elementary human need of expression in all available media—hence, the Center's place at MIT and its continuing commitment to communicating and integrating efforts in art and technology. Twenty resident and visiting artist fellows work at the Center on personal and communal projects which enhance the collaboration of artists, scientists, scholars and engineers. At the same time the artists contribute to CAVS' teaching program under the direction of Otto Piene, head of CAVS since 1974, and Professor Robert Preusser, CAVS' director of education. The Center was founded in 1967 by Gyorgy Kepes and it has maintained its unique character as a place for the synthesis of artistic, scientific and technological contributions to contemporary social and cultural needs.

Among past fellows of CAVS were Maryanne Amacher, Karin Bacon, Douglas Davis, Juan Downey, Ron Hays, Rockne Krebs, Otto Piene, Jon Rubin, Friedrich St. Florian, Alan Sonfist, Vassilakis Takis, Harold Tovish, Wen-Ying Tsai, Stan VanDerBeek.

Present fellows include Lowry Burgess, Peter Campus, Harriet Casdin-Silver, Paul Earls, Paul Matisse, Antonio Muntadas, Carl Nesjar, Alejandro Sina, Aldo Tambellini.

Because of its specific character which demands intricate attention, the Environmental Art graduate program is currently limited to a few students. This permits the development of interdisciplinary programs for each student according to his or her individual need and vision. The layout of artistic, scientific and technological studies follows the students' projected objectives, and a balance between studio work and intellectual studies will be established. Students of the program will have studio spaces in the CAVS building, and other MIT facilities so that they will partake in the Center's activities and be in constant personal touch with the resident artists. Besides vivid cooperation with the MIT faculty at large, strong ties are at work among CAVS artists, students, and faculty of the School of Architecture and Planning's Department of Architecture with which CAVS is associated.

Director: Otto Piene
Building W11, MIT

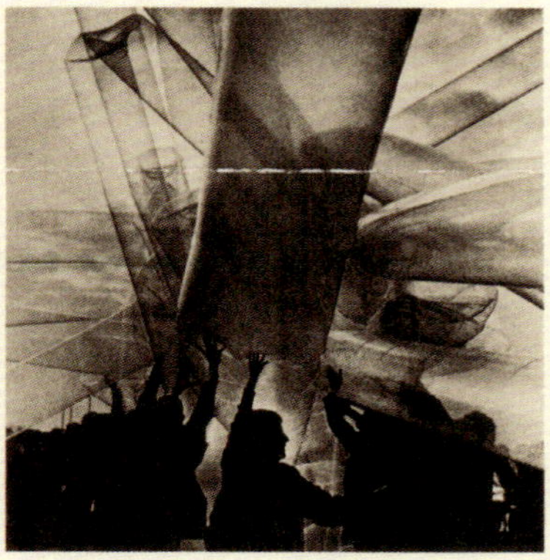

Computer Graphics

The future of computer graphics is clearly rooted in raster scan display technology, in general, and video, in particular. As such, the previously separate disciplines of computer graphics, image processing and broadcast are coming together with a major focus on digital television.

This area of concentration in the S.M.Vis.S. program is for students with previous computer programming experience, interested in research into new qualities of computer interactions. The area of study is very much on and at the interface of people and computers, with a heavy emphasis toward engaging many senses at the same time. Current research includes: sound sync computer graphics, eye-tracking, and the recognition of body movements.

Computer animation is a large part of the program. Simulation techniques, 3-D scripting languages and interest in flight simulator-like presentations are exemplary. Candidates should view these both as facilities to use and as technologies to develop. A characteristic of this area of concentration is that the line between user and system developer is thin or non-existent.

Lastly, there is digital television itself. This includes new hardware designs, advanced software support, and the incorporation of new technologies like the optical videodisc, thin film transistor displays, and picture processors. Students can engage in the design as well as the use of these machines.

The Computer Graphics area of concentration is housed within the Architecture Machine Group, which itself is loosely tied to the Laboratory for Computer Science. Yet, more informally, we find our students collaborating with the Artificial Intelligence Laboratory and the Division for Study and Research in Education. Prospective applicants should read the MIT Bulletin in order to get a more general overview of computer science activities at MIT.

The Department's own facilities include six large minicomputers, interconnected by an elaborate operating system. The graphic facilities include three large frame buffers and an array of small displays. In addition, there is a full video studio with both magnetic and optical videodiscs. Other facilities include sound and voice output, 3-D digitizers, and the so-called "media room."

Candidates should apply for this area of concentration with the understanding that it is part of a large and changing research program. Financial support is available through research grants and contracts. However, it is that very work in progress which sets the tune of computer graphics on a semester by semester basis. At this writing, research programs include: advanced mapping techniques, automated maintenance and repair, sound sync computer animation, new input technologies, and multi-media management information systems. Candidates looking for financial assistance should be aware of the fact that research funds are awarded according to the government's fiscal year and hence announcements of specific support may be delayed or last minute.

Director: Nicholas Negroponte
Room 9-518, MIT

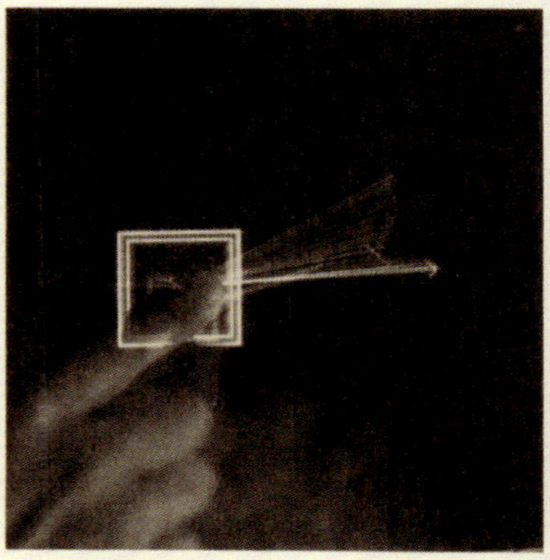

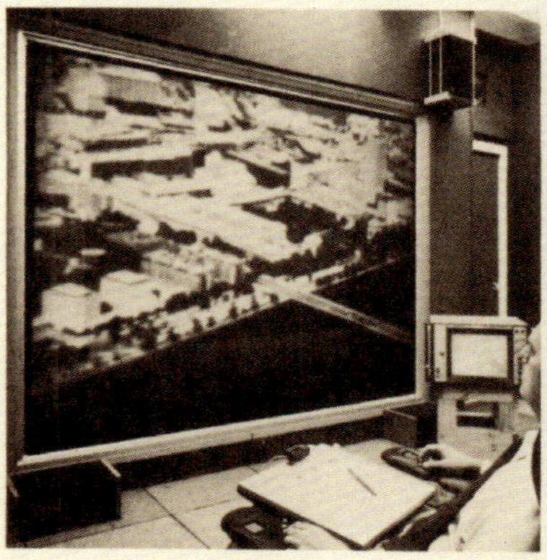

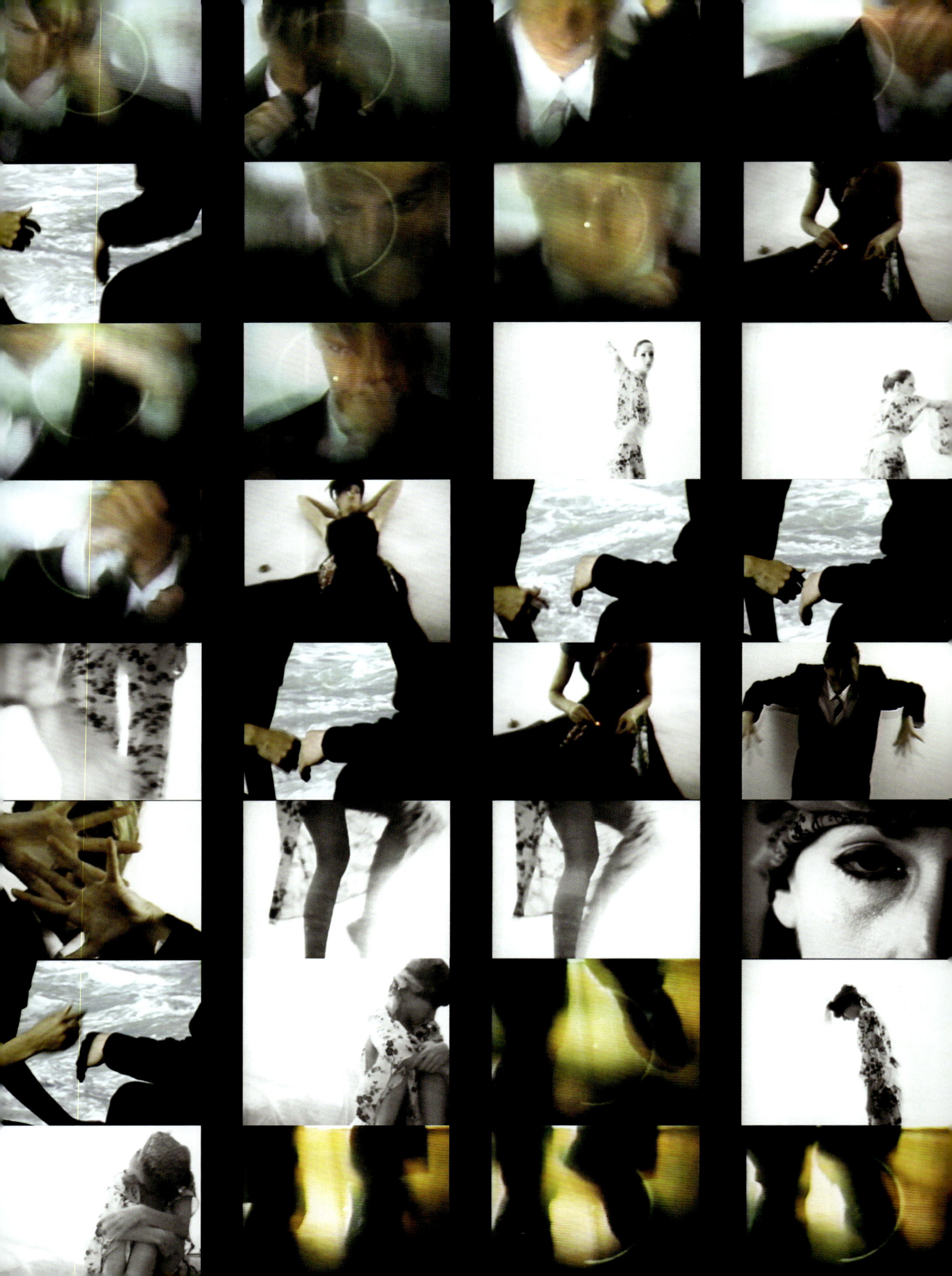

Beyond Building W11

MIT generally responded to emerging media by creating labs or resource centers to gather information on user interactions before establishing institutional policy. For example, video as a consumer medium—video that would flow two ways rather than as an impenetrable stream from professional broadcasters toward viewers—was the focus of Educational Video Resources (EVR), under the direction of physicist Edwin F. Taylor. CAVS artists distorted the video signal to generate aesthetic innovations in EVR's multi-camera Experimental 'X' Studio. EVR made broadcast-quality cameras and edit rooms, then expensive and unwieldy, freely available.

Seminal innovations in personal filmmaking both in the technology of small-format film cameras and in the *cinéma vérité* style of Leacock, Film/Video's charismatic director, were introduced to CAVS students who shot with Super 8 film, edited on flatbeds, and interacted with Film/Video's student documentary filmmakers.

The artists disrupted narrative norms: reconfiguring the established forms of time-based media; remixing image, music, and sound; and turning television cameras on themselves to personalize the medium. CAVS artists Luc Courchesne, Ellen Sebring, and Bill Seaman contributed pieces to *Elastic Movies* (1984), probably the earliest computer-controlled interactive artists' videodisc (Grau, 2017), produced by Glorianna Davenport and Benjamin Bergery at the Film/Video section as a series of experiments in interactive storytelling.

> EVR provided a major upgrade of the video format that was otherwise available at the Film/Video Section. It was great to learn video by using 1/2" reel-to-reel systems, but the precision of EVR's 3/4" systems allowed for a complex form of collage video art, and CAVS students took advantage of EVR to pursue what became an altogether new way to produce video art. (Vin Grabill interview)

To name just some of the greater MIT resources, CAVS artists worked at Harold Edgerton's MIT Strobe Lab, Csound inventor Barry Vercoe's Experimental Music Studio, Tod Machover's pioneering MIDI studio, and Stephen Benton's Spatial Imaging Research Group. Project Athena, an early campus-wide distributed computing network founded at MIT in 1983 (still active today), employed fellows Rus Gant and Ben Davis.

> Of all the sections, I felt that CAVS gave me the greatest flexibility. First it was about making art, a type of art that was not bound by any conventions or traditions. Lots of possibilities fit under a small umbrella. Having my base at CAVS I worked at the Experimental Music Studio, I programmed in PL/1 to make digitally processed images of Beth Soll, MIT choreographer for the Architecture Machine Group, and did a little work at the VLW. (Greg Garvey interview)

In 1985, many of the SMVisS sections moved into the new Media Lab housed in the white-paneled Wiesner Building designed by I. M. Pei & Partners. CAVS students were welcomed by professors like Glorianna Davenport, researching interactive documentary, and Muriel Cooper, head of VLW, to continue collaborations and projects under a new roof. For a transitional period, they produced experimental artworks using 1-inch video, MIDI interfaces, computer graphics work stations, and the Cube, a forty-foot square black box space.

Facing page: Bill Seaman and Ellen Sebring, "The Boxer's Puzzle," still frames from videotape, 1986
After graduating from the SMVisS program, the artists worked at Film/Video Section's 1-inch studio in the Media Laboratory to blend Sebring's performance imagery with Seaman's poetic text and music, evoking the "puzzle" of multiple personae within a single being.
Photo grid: Bill Seaman

Center For Advanced Visual Studies

SUBJECTS, FALL SEMESTER, 1980

CENTER FOR ADVANCED VISUAL STUDIES
MASSACHUSETTS INSTITUTE OF TECHNOLOGY
BUILDING W11, 40 MASSACHUSETTS AVENUE
CAMBRIDGE, MASSACHUSETTS 02139
TELEPHONE: 253-4415, 253-6849

IN ASSOCIATION WITH THE DEPARTMENT OF ARCHITECTURE
MASSACHUSETTS INSTITUTE OF TECHNOLOGY

PROFESSOR OTTO PIENE, DIRECTOR
PROFESSOR GYORGY KEPES, DIRECTOR EMERITUS
PROFESSOR ROBERT PREUSSER, DIRECTOR OF EDUCATION

ELIGIBLE FOR REGISTRATION IN THESE SUBJECTS ARE UNDERGRADUATE AND GRADUATE STUDENTS FROM ALL SCHOOLS AND DEPARTMENTS AT M.I.T. AND OTHER SCHOOLS WITH WHICH THE INSTITUTE HAS CROSS-REGISTRATION ARRANGEMENTS (MASSACHUSETTS COLLEGE OF ART AND MUSEUM OF FINE ARTS SCHOOL STUDENTS CONSULT THE C.A.V.S. OFFICE)

4.801 ART AND THE ENVIRONMENT (3-0-6)

Otto Piene, Elizabeth Goldring

Contributions by artist/Fellows of the Center for Advanced Visual Studies and M.I.T. scientists, engineers, and administrators.

Lecture and investigation series on the history, theory and practice of environmental art; assignments for environmental investigations in various media by students.

Fall Term, 1980
Emphasis: Sky Art

The upper crust of this planet and the lower layers of the atmosphere where earth and sky blend (and where we live); ant piles, bird maps, star scrapers; sky art; migrant apparitions ancient and contemporary.

Tuesdays, 7-10 pm, 3-133.

4.831 ENVIRONMENTAL ART (3-3-6)

ENVIRONMENTAL MUSIC
Paul Earls, Chris Janney

An analysis of musical events, celebrations, and participative collaborative sound sculpture that extends beyond the concert hall tradition. The creation of original instruments, works, and events for architectural performance.

Wednesdays, 9 am-1 pm, W11.

4.841 ENVIRONMENTAL LIGHT AND COLOR (A) (2-4-6)

4.842 ENVIRONMENTAL LIGHT AND COLOR (A) (2-9-9)

Robert Preusser, Nishan Bichajian

A study of the dynamic and qualitative attributes of light and color within the context of architectural and urban settings. Emphasis on innovative alternatives to prevailing practice in the illumination of man-made environments and the use of light as an environmental art medium. Projects include additive and subtractive color interaction, minimal lighting, reflected color, complementary color shadows, and the coordination of light and movement. Photography utilized as an analytical tool for recording experiments.

Mondays and Wednesdays, 2-5 pm, E21-202.

4.845 ADVANCED VISUAL DESIGN (A) (0-8-12)

Otto Piene

Theory and practice of some aspects of kinetic, light, and media art; environmental art and celebrations; planning of public art complexes: sculptural architecture and architectural sculpture. Work on public commissions and exhibitions of environmental scale. Mostly individual assignments and tasks.

Thursdays, 9 am-1 pm, W11.

4.861 LIFE DRAWING OR STILL LIFE (0-3-2)

Nishan Bichajian

LIFE DRAWING

Free hand drawing for beginners. Students may choose to draw from live models to develop skills in observation of action, proportion, line, texture, and form using conté crayon. Kinetic drawing as well as detail and perspective rendering with emphasis on technique and spontaneity. May be repeated once.

STILL LIFE

A beginning drawing course designed to enhance accurate observation and eye-hand coordination. Emphasis on basic drawing principles and formal elements (line, form, texture, design, composition, perspective, movement) using forms from nature, landscapes, and interiors. Experimentation with pen/ink, charcoal, pencil, and wash.

Students consult instructor. Tuesday, September 9, between 12-1 pm, W11. Subsequent meetings to be scheduled.

4.895 SPECIAL PROBLEMS IN ENVIRONMENTAL ART (A)

THE ARTISTS SPEAK — WORK OF THE FELLOWS OF THE CENTER FOR ADVANCED VISUAL STUDIES (0-9-0)
Paul Earls, Joan Brigham, Mark Mendel (coordinators)

A series of individual presentations by the artist/Fellows of the Center for Advanced Visual Studies of their recent and current work. Media and topics include sky art, inflatables, lasers, holography, multi-media, film, video, photography, telecommunications, environmental art and poetry, music, light and color, time displacement, steam, celebrations, and events. Each session will be devoted to the work of one artist; a detailed schedule will be available at the first session.

Thursdays, 8-10 pm, W11.

4.896 SPECIAL PROBLEMS IN ENVIRONMENTAL ART (A)

VIDEO-PERFORMANCE SEMINAR (0-9-0)
Wendelin Glatzel

The seminar will explore practical and theoretical problems of performance and staging in the medium of artistic video (b&w and color).

Participants will design a performance-production around a selected theme.

Wednesdays, 5-9 pm, W11.

4.897 SPECIAL PROBLEMS IN ENVIRONMENTAL ART

INTERACTIVE VISUAL TELECOMMUNICATION (0-9-0)
Aldo Tambellini

"What we are building is a nervous system of mankind which will link together the whole human race for better or for worse in a unity which in an earlier age could never have been linked." —A. Clark

The new merging telecommunication systems will restructure our social life, produce social change,and allow new creative concepts to evolve.

The workshop is designed as a general overview, history and theory, of contemporary and future interactive telecommunication media and alternative systems with the intent of understanding them and utilizing them towards a "global creative event".

Interactive visual media are used to accommodate visual "conference" communication over a distance. The workshop will be concerned with the new technologies: satellite, interactive cable, the videodisk, teletext, picturephone, slow scan and facsimile transmission as new concepts in communications with unique characteristics. We will discuss how they work, how we can realistically use them and the way in which they have a direct influence on the production, distribution, and reception of information and messages.

Past workshops have organized and collaborated with existing media groups to produce a series of interactive media events connecting M.I.T. with centers in other countries. We have used slow scan television, computer conferencing and set up models for an interactive two-way cablecast. Essential to the workshop is the documentation of events with photography, slides, videotapes, followed by post production and editing.

We will work on developing a model for a new global telecommunications event. Assignments in creative research in telecommunications and the investigation of various media. Guest speakers and field trips are also a part of the workshop.

This workshop is directed towards the larger concept of COMMUNICATIONSPHERE, a network to develop channels between artists, engineers, scientists, social anthropologists and all those concerned with the impact of telecommunications on contemporary society.

Wednesdays, 4-6 pm, W11.

4.898 SPECIAL PROBLEMS IN ENVIRONMENTAL ART (A)

THE ART OF HOLOGRAPHY (0-9-0)
Harriet Casdin-Silver

Research in holographic techniques directed toward the realization of artistic concepts.

Student enrollment will be limited. Consult instructor at first meeting.

First meeting, Thursday, September 11, 4 pm, W11.

4.899 SPECIAL PROBLEMS IN ENVIRONMENTAL ART (A)

VIDEO ANIMATION (0-9-0)
Betsy Connors

Workshop in video animation to explore the developing processes through the production of works. Techniques of the independent editing and mixing of sound and image will be emphasized. Specific preparations for broadcast television systems will be discussed. Some experience with video or animation is desirable.

Mondays, 3-6 pm, 9-355.

4.909 SPECIAL PROJECTS IN VISUAL ARTS (A)

BASIC PHOTOGRAPHY FOR ARCHITECTS (0-12-0)
Nishan Bichajian

An introduction to basic photography such as film, processing, printing technique, copy work filtration, metering, etc.

A study of the role of light — daylight and artificial — and its relationship to forms, tone, and texture in architectural photography. Exploring ways to use light creatively on various forms such as reliefs, sculptures, and architectural models.

The 35mm camera and a 4 x 5" studio camera will be used for clasroom work as well as architectural subject matter. A 35mm camera is required.

First meeting, Wednesday, September 10, 12-1 pm, W11.
Subsequent meetings to be scheduled.

4 UR UNDERGRADUATE RESEARCH IN ARCHITECTURE

MULTI-MEDIA RESEARCH (UNITS TO BE ARRANGED)
Paul Earls, Robert Preusser

Research in multi-media techniques and production. Independent or collaborative work. In particular, proposals encouraged which assist in the current work with micro-processor generated laser projections. Consult instructors for more details and to initiate individual proposals.

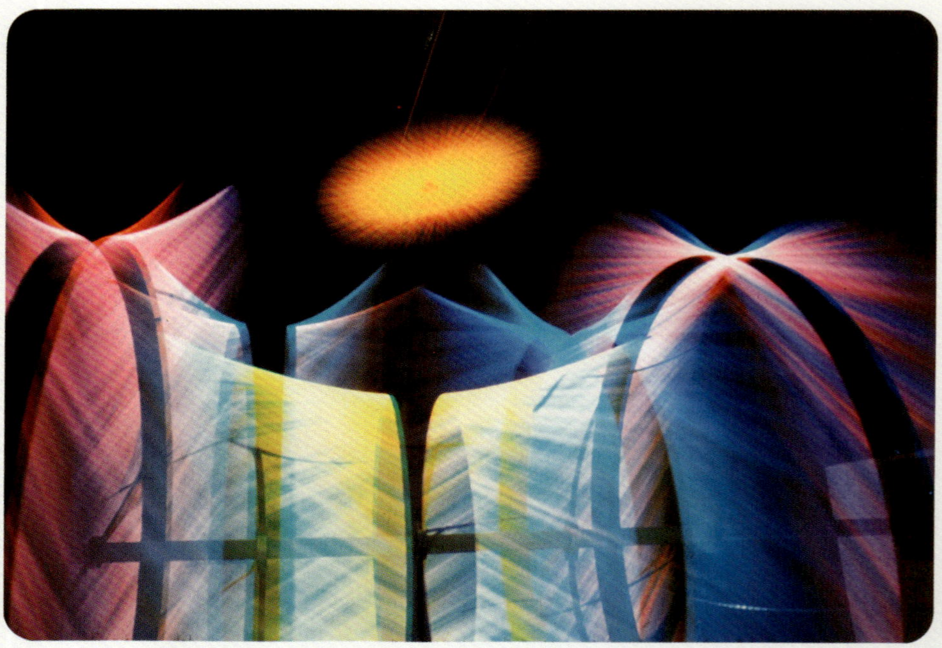

Alejandro Sina, Installation of moving neon sculptures: "Spinning Shaft" and "Night Butterfly", "5 Artists/5 Technologies", Grand Rapids Art Museum, Grand Rapids, Michigan, 1979
Photo: Ed Saunders

Classes

New students entering the Center would quickly discover the "exhibition room" filled with white director's chairs where classes and meetings were held, and in the opposite corner of the building, the multistory experimental space known as the Pit. Over the course of their studies, students met with Piene in his office to discuss their work, theses, problems, and ideas as he sat in a blue wingback chair against the backdrop of a large dark painting—the metaphorical center of the Center.

> The Center was the conceptual center. The Center got me thinking on a much deeper intellectual level about things in a holistic way. You would talk about James Joyce one minute, listen to Paul Earls' music, and then Paik would walk in. It was about how things can inform each other in very rich ways. That's how the concept of interdisciplinary and trans-disciplinary got started. (Bill Seaman interview)

> The CAVS seminar would be quite informal. We typically sat with Otto who would tell us about his recent trips, his experience of the art world and the challenges of the new approaches and spaces for a contemporary art practice rooted in technology. I found it fascinating to hear this European artist in an American culture and how he could maintain an active presence in such a large territory. He introduced us to a certain geography of the world and to a formidable scale in which to build a career as an artist. (Luc Courchesne interview)

> I had been accustomed to an educational structure that had a great deal of handholding. When I arrived I was initially appalled: what kind of education is this? You are not telling me what to do in these classes. There are just a bunch of people talking, there didn't seem to be any requirements. It took me about six weeks to realize that this is freedom; this is not a traditional school; this is a research environment. (Don Ritter interview)

From the beginning, fellows shared CAVS teaching responsibilities, trying out topics from their current research. "The Artist Speaks" series offered a forum for fellows to present work in progress for discussion.

> Maryanne [Amacher] and I ended up teaching a class at least two semesters called "Live Space." The focus was mostly on how we see depth and hear in depth—an interesting balance of our perceptions and how we are immersed in our everyday spaces. (Scott Fisher interview)

Piene's class in Advanced Visual Design was the backbone of the program, offered each semester and open only to SMVisS graduate students. He also offered an undergraduate/graduate class in Art and Environment, which he taught with CAVS fellows including Friedrich St. Florian, Elizabeth Goldring, and Paul Earls. The topics ranged from "Art and Technology" to "A Matter of Taste" to "Art/Architecture and Performance." Each semester the classes were listed on CAVS posters and distributed on campus.

> "Art and the Environment" classes sometimes lasted until midnight, whereas they were scheduled from 7-10 or something like that. A spirit of curiosity, of invention, a sense of joy of discovery, and sharing of what "had just been found," gave the classes that special flavor, or taste of voluntary excitement. Many MIT science and engineering teachers, researchers, and architects would come to speak in the Art and the Environment class. (Otto Piene, interview by Bill Seaman, 2012)

Facing page: CAVS poster of fall semester 1980 subjects, with image of Alejandro Siña's neon sculptures, CAVS, MIT, 1980
Source: CAVS Special Collection

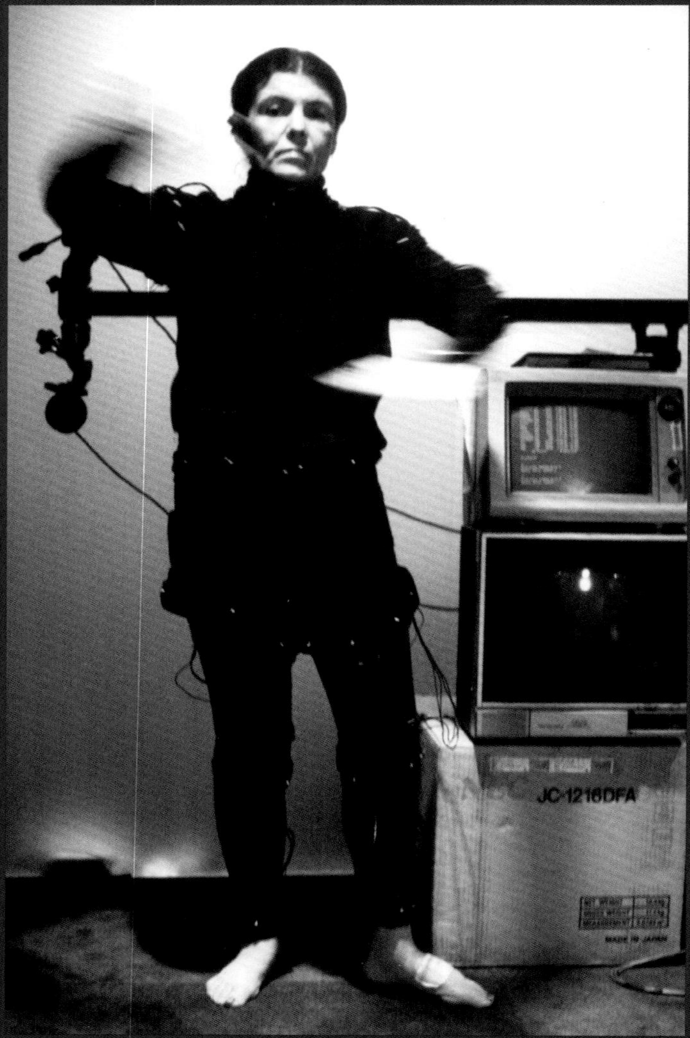

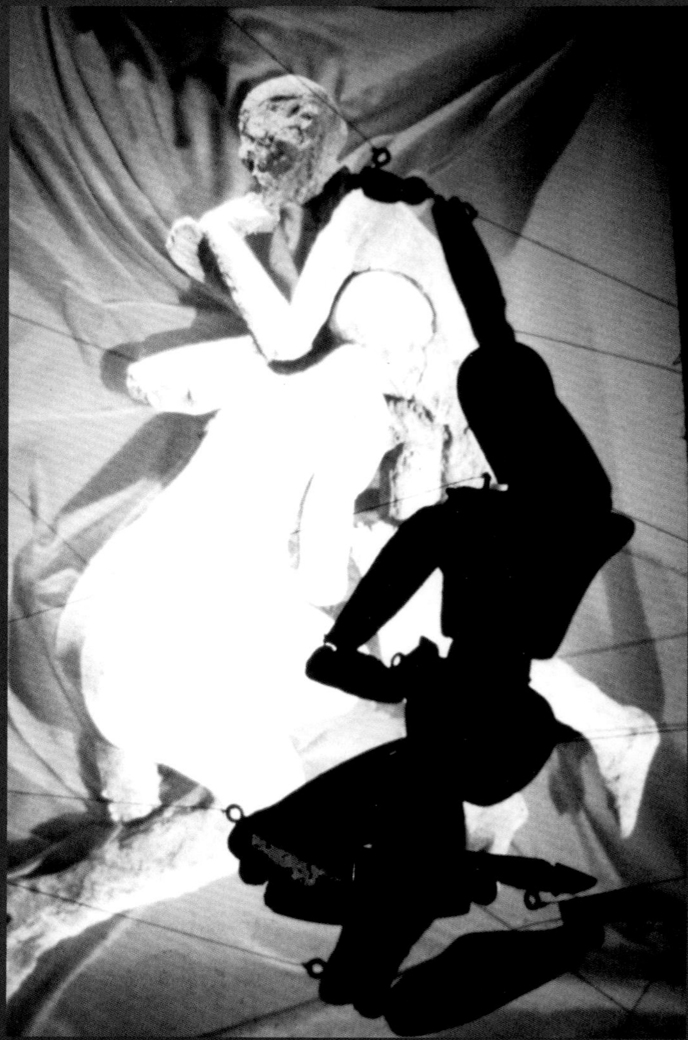

Above left: Jennifer Hall, "animation-by-enactment," artworks for the project "Body Tracking," infrared and electromagnetic tracking system to capture human motion for driving animated characters. Tracking suit by Hall, pictured worn by performer Sarah Dickinson, both CAVS graduate students. The Computer Marionette Group began at the Architecture Machine Group, and moved to the Media Lab under the Graphical Marionette Group. These pictures were taken just a few months after the Media Lab opened. Computer Marionette Group, MIT Media Lab, 1985
Photo: Marc Bayard Locascio

Above right: Jennifer Hall, "Boarding Pompei," artworks for the project "Computer Controlled Marionette," optical tracking system with edge detection, with MIDI controlled marionette. Performances of projections and shadow puppets, 1985-86
Photo: Jennifer Hall

> I used to have a once-a-week workshop at CAVS. I was teaching the whole concept of media. I had some students from the Museum School, some became part of my group I called Communicationsphere. They were punk rockers. (Aldo Tambellini interview)

> With Elizabeth I co-taught a course, "Visual Language," using the environment of war, the Gulf War of 1991. CAVS's roots are in the literature of Emerson, Thoreau's transcendentalism, utopian, visionary, connected with German Romanticism. (Panos Kouros interview)

Despite the Center's focus on innovation, as a fellow during the Kepes period, Piene instituted popular evening figure drawing classes. Later, Nishan Bichajian's figure drawing classes attracted students from across MIT.

> If you need technology, there are always going to be reasons why it's not possible for you to create what you want to create. But if you can draw, that means you can save up all your ideas for the time when they may be possible to happen. It is the most fantastic way of notating all your intellectual activity. So once I found how fantastic it was to draw, knowing all the frustrations of technology, I decided to teach drawing and I did. (Paula Dawson interview)

Twenty-one women and thirty-six men earned the SMVisS degree from CAVS under Piene's tutelage. After his retirement, the Visual Arts Program took over the SMVisS program. However, nontraditional and multidisciplinary project-oriented education lived on for CAVS fellows who taught classes during MIT's winter Independent Activities Program (IAP) and worked with students through the Undergraduate Research Opportunities Program (UROP) until 2010.

> During my four years working on Goldring's Seeing Machine Camera project, I designed my first digital retina print, helped teach my first class, delivered my first conference presentation, ran my first clinical trial, published my first journal paper, and won my first innovation competition. I was also introduced to the human factor in engineering design. I came to understand that true innovation should not only seek to help people survive but also strive to help them reach their fullest potential despite physical, social, environmental, or economic constraints. (Faye Wu, PhD candidate, Department of Mechanical Engineering, MIT, email to Elizabeth Goldring, October 4, 2016)

The education program resulted in the proliferation of art-and-technology programs worldwide. Former students and fellows went on to teach in and even found art-science-technology and new media programs in leading academic institutions.

> At one point I was at a meeting at CAVS and we listed all the institutions across the world that had media programs. The majority, 90%, were being run by CAVS compatriots. CAVS goes beyond just that building. The first generation has now come to mid-career, and we are just now seeing the second generation, including my students, going out and being educators exponentially, not just teaching but developing institutional curriculum. The philosophies they are using are CAVS philosophies. The very fact that CAVS was there set a precedent. (Jennifer Hall interview)

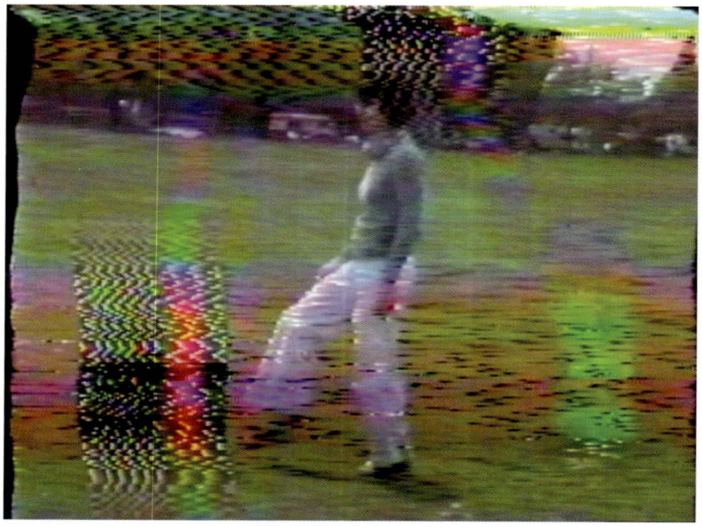

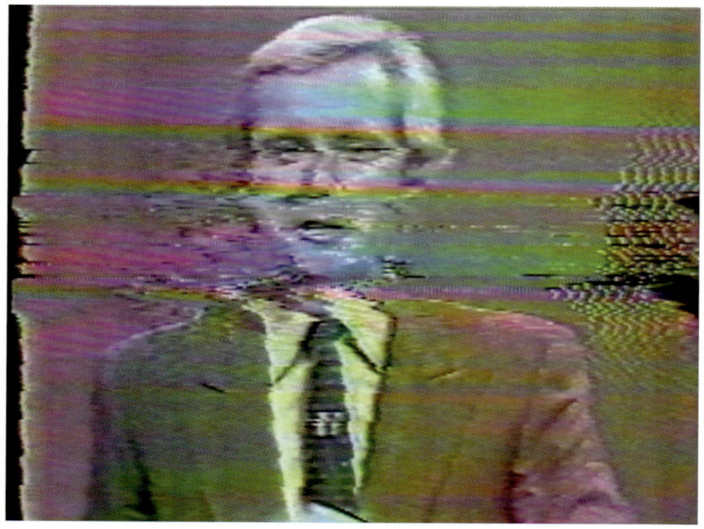

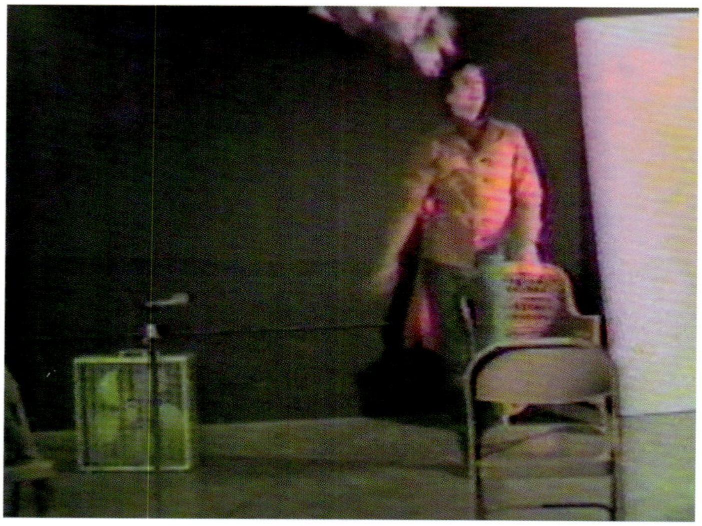

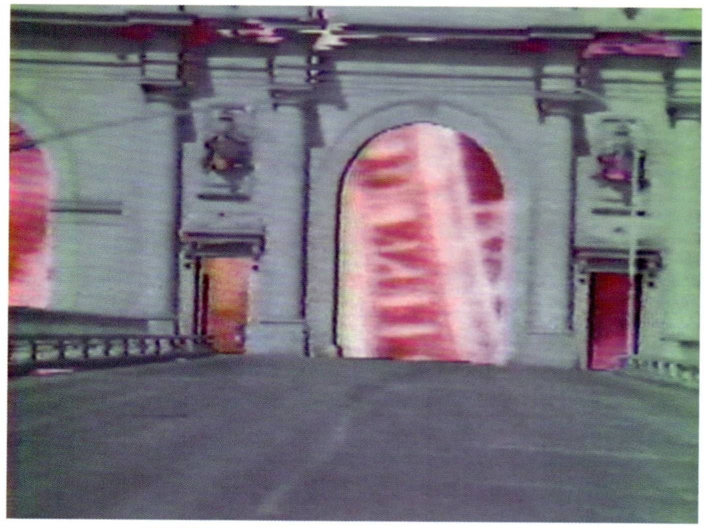

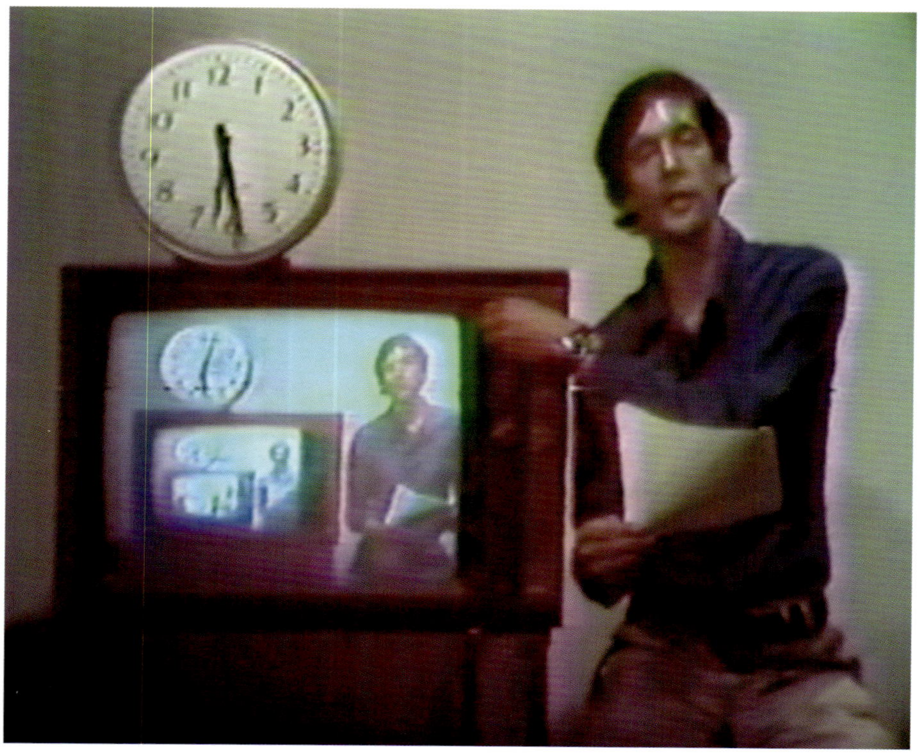

Top: Vin Grabill, "8 Short Processes," videotape exploring musical composition through percussive timelapse, signal processing, and modular actions, 1980-83

Left: Vin Grabill, "Performance in Color Video and Live Time," performance and videotape, 1979
Photos: Vin Grabill

Both pieces were part of Grabill's graduate studies at CAVS. Grabill blurred the boundaries between art and documentary approaches, collaborating with Center artists, and videotaping many CAVS projects.

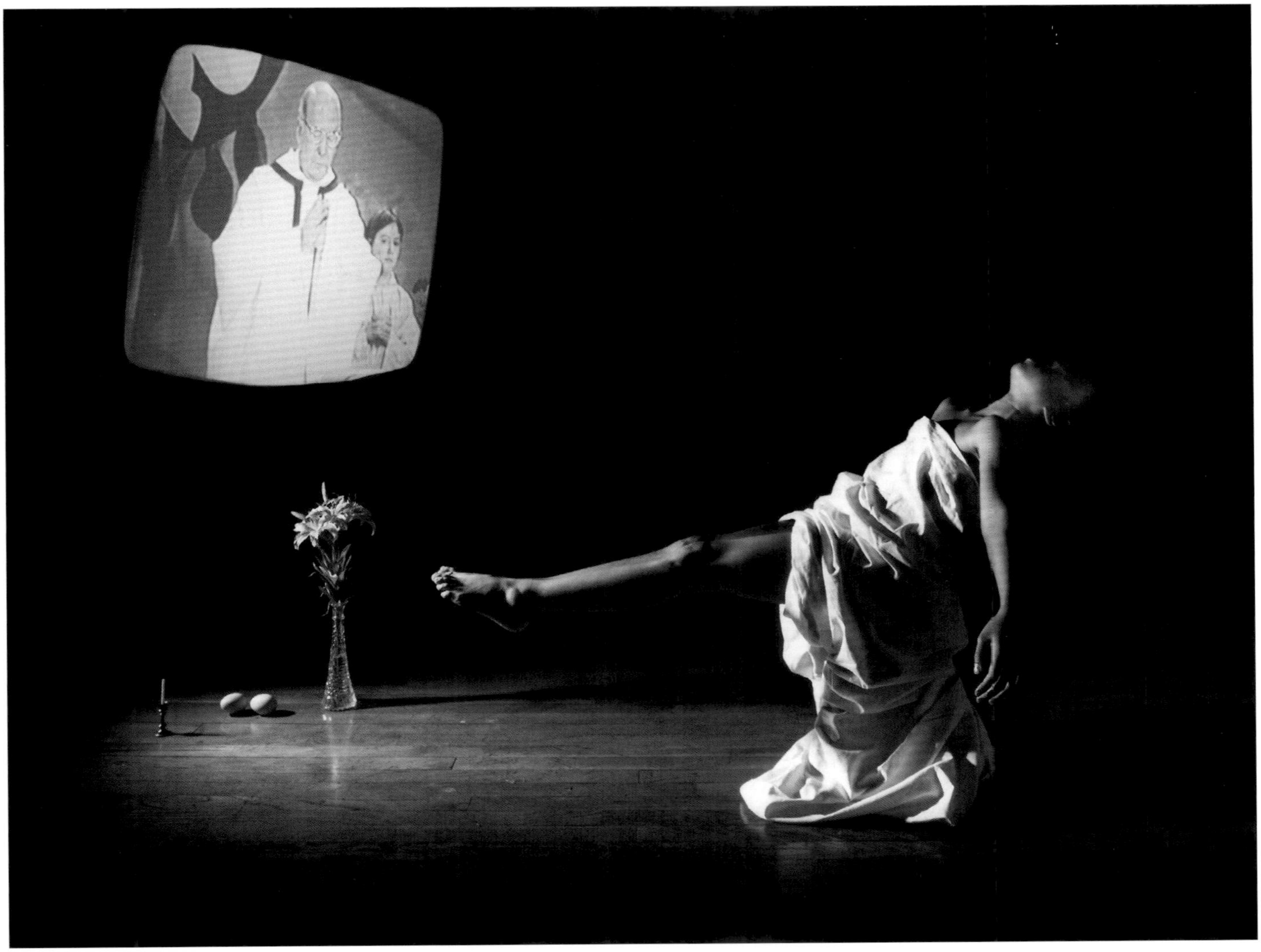

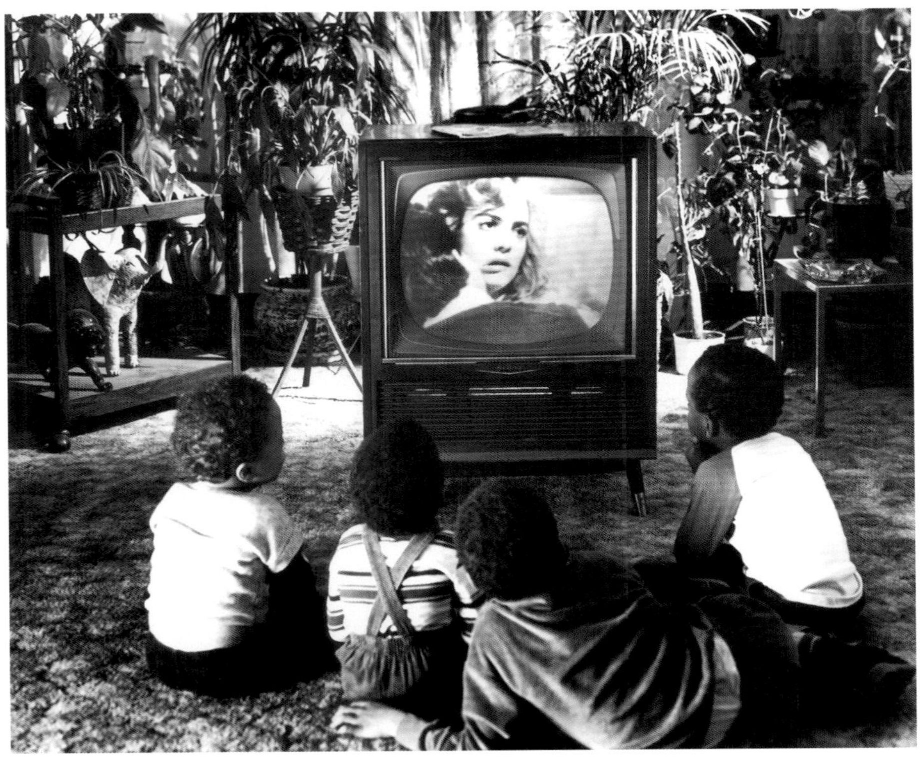

Above: Walter Dent, "Annunciation," photograph, 1983
Photo: the artist

Right: Walter Dent, "Massacre of the Innocents," photograph, 1988
Photo: the artist

Dent completed the SMVisS degree and was often hired by Piene to photograph CAVS events.

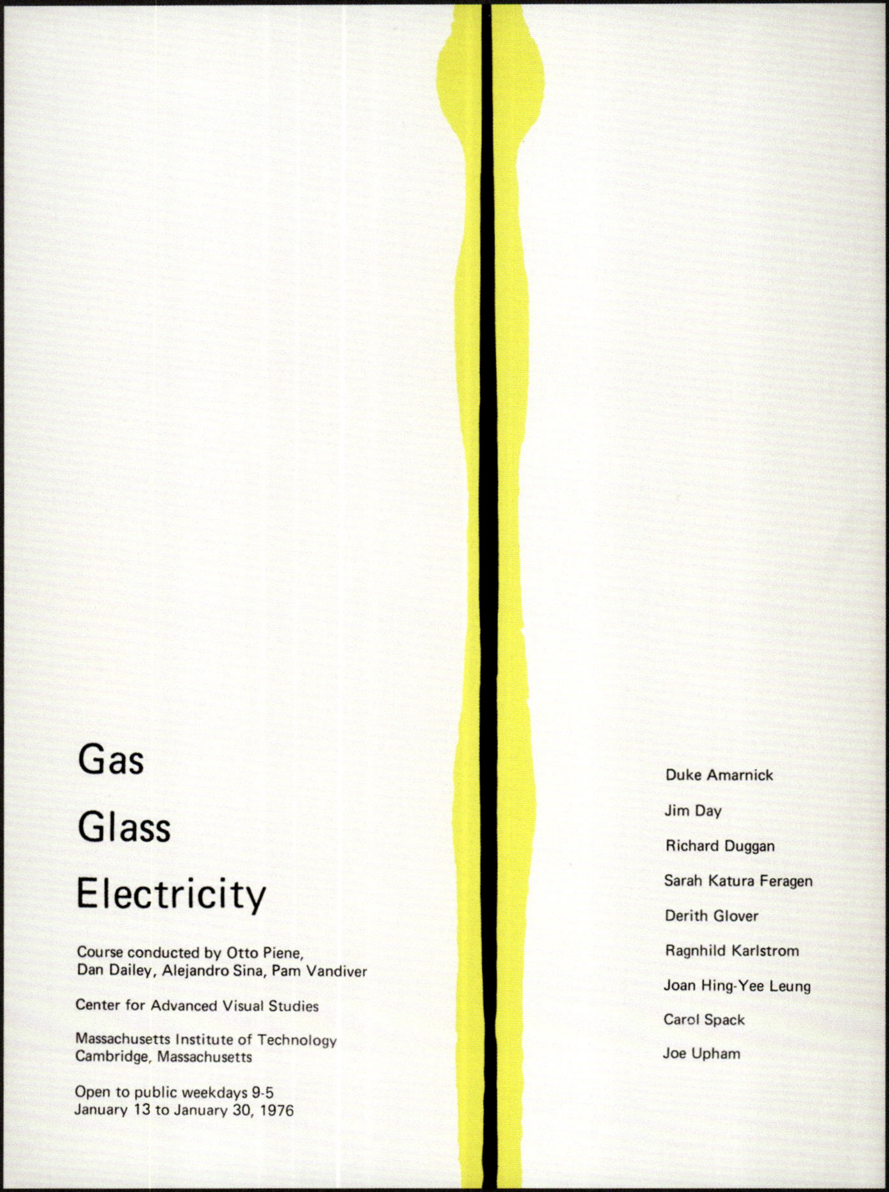

Above: "Gas Glass Electricity," poster for an Independent Activities Period (IAP) course offered by CAVS artists Otto Piene, Dan Dailey, Alejandro Siña, and Pam Vandiver, January 1976
Image: CAVS Special Collection

Facing page: Bernd Kracke, "Data Network," performance, the artist burns a rolodex of "physical" data, *Sky Art Conference '83*, BMW Headquarters, Munich, 1983
Photo: Otto Piene Collection

Collaboration. Curious word. Fifty years ago it was a condemnation, especially with a German. Twenty-five years ago, it was a guiding principle in György Kepes' formation of the Center for Advanced Visual Studies at MIT, as environmental art requires multiple levels of cooperation and collaboration among artists, designers, scientists, engineers, administrators, and patrons. Artists at two different stages of their careers are most likely to engage collaboratively: the very young (and inexperienced), and the mature (very experienced). (Paul Earls, "On Collaboration," *Otto Piene*, Glibota, editor, 2011, 396)

Art-Science-Technology Education and Inventing the Future

I met Otto Piene and CAVS for the first time at *documenta 6* in 1977. I was just about to finish art school in Hamburg, Germany. Centerbeam as a collaborative art piece blew me away not only because of the integrated effort of so many different artists but through the experimental and innovative use of very different tools and technologies. After talking to Otto about the options of joining the postgraduate program at the Center my mind was set to get there as soon as possible. It took me two years to prepare all the necessary applications and funding to finally arrive at CAVS in late 1979. The transition from cold war and still-divided Germany to MIT and the post Vietnam War-traumatized US was quite challenging and awesome. But I dug deeply into the Master's program, and into all venues of MIT on a 24/7 basis. Great artists, scientists and students to learn from and to collaborate with were all over the place. Besides my home base at CAVS, I discovered wonderful islands of individual research and artistic experiments like the film section headed by Ricky Leacock, the Visible Language Workshop headed by Muriel Cooper, and the Architecture Machine Group headed by Nicholas Negroponte. I didn't leave after finishing my Master's degree and stayed on as a fellow to participate in various large-scale, collaborative projects like the *Centervideo* exhibition, telecommunication events, and Sky Art conferences. In 1989, I became a Professor for Media Design at the newly founded and CAVS-inspired Academy of Media Arts Cologne in Germany. Ten years later, in 1999, I started my Cross Media Lab as a Professor for Electronic Media at the University of Art and Design Offenbach (Hochschule für Gestaltung). I became the President in 2006 establishing a widespread agenda of art, design, science, and technology which is still alive and active. Our *B3 Biennial of the Moving Image* has become an international platform since 2013 for the dialog of all artistic and scientific disciplines working with time-based and immersive narration. Now we are planning to build a complete new campus for the University of Art and Design of the 21st Century in 2024. In all, it will be forty-five years inspired and powered by CAVS and MIT.

— Bernd Kracke

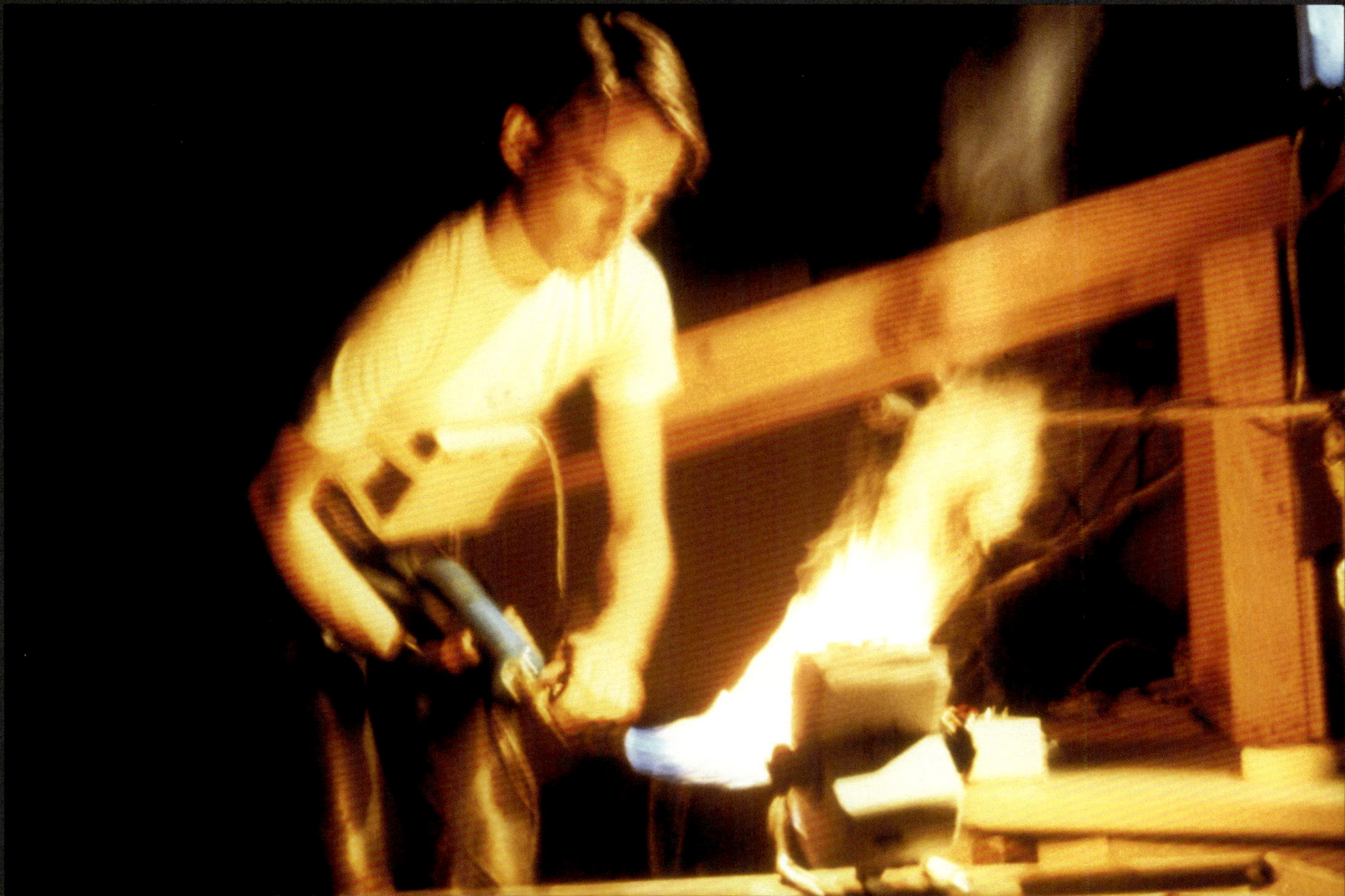

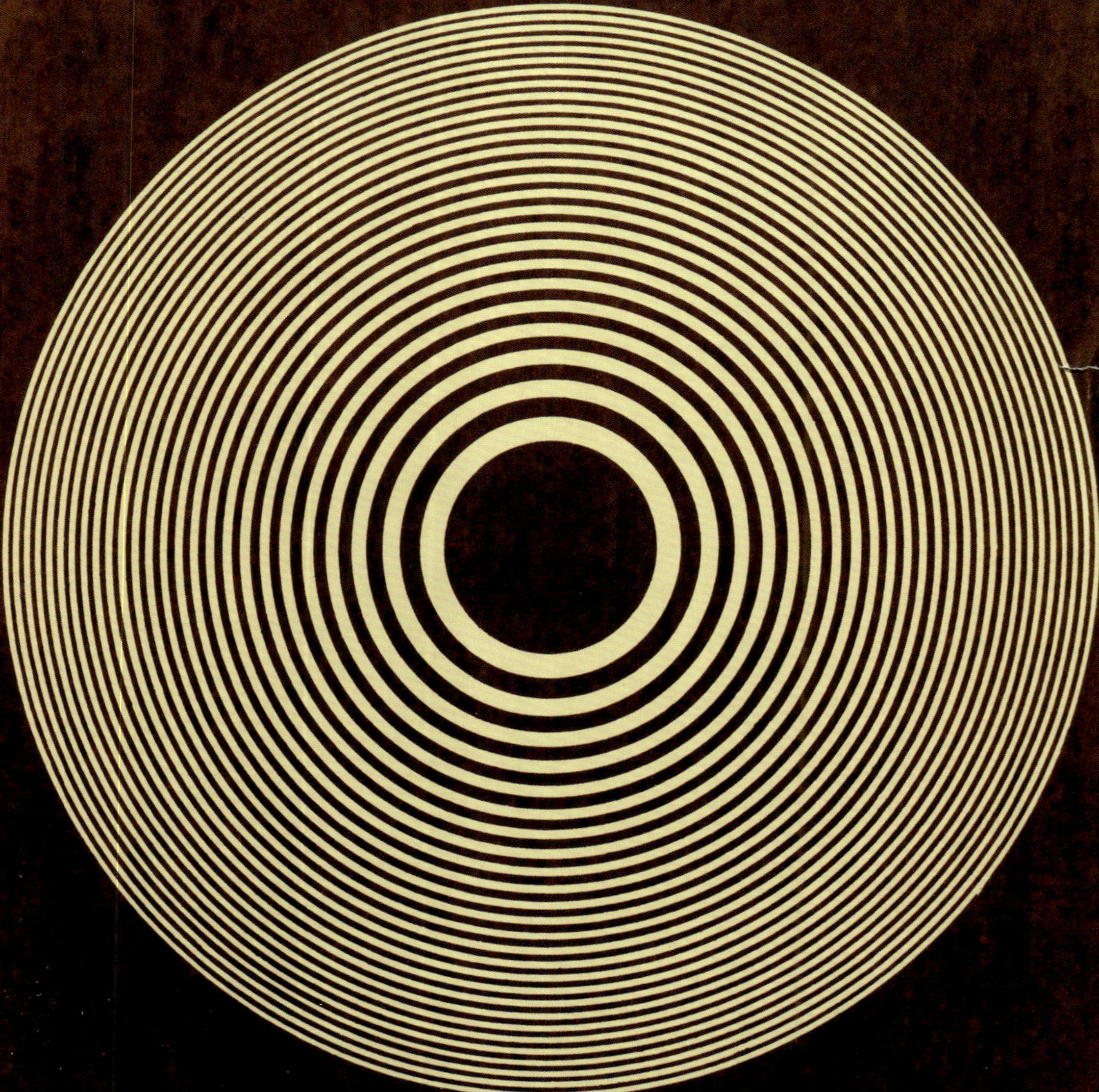

Ring the Lobby

An electronic celebration of the
resonances of the MIT Building 7
Lobby and notable historical events
of the Halcyon Advent Season
by Paul Earls, Composer,
Fellow of the Center for Advanced
Visual Studies, with Charles E. Miller,
Stroboscopist.– An evolving multi-
channeled recorded/live/electronic/
environmental/participatory
musical system operating weekdays
at noon, 2, and 4 pm, with special
performances at 5 pm, from December
4 through 15, 1972 – Special thanks
to Bose Corporation, Tech HiFi,
Pat Lee, Jim Kates, Steve Ringle,
Bart Johnson, Roy Harvey – Presented
by the Lobby 7 Committee

Lobby 7

Lobby 7, just inside the main entrance to MIT at 77 Massachusetts Avenue, was for a while the site of collaborative exhibitions and happenings that intersected with visitors and the MIT community at large. These activities, organized by Suzanne Weinberg, included holiday concerts, experimental music and dance performances, and exhibitions. CAVS Lobby 7 projects were produced by students, fellows, and invited artists. The following section features work shown in Lobby 7, including the topical interactive shows, *Weather* (1973), and *Food* (1975) with the participation of Julia Child. *Love* and *Death* were planned shows, but never produced.

Facing page: *Ring the Lobby*, poster: "An electronic celebration of the resonances of the MIT Building 7 Lobby and notable historical events of the Halcyon Advent Season by Paul Earls, composer, Fellow of the Center for Advanced Visual Studies, with Charles E. Miller, Stroboscopist—An evolving multi-channeled recorded/live/electronic/environmental/participatory musical system operating weekdays at noon, 2, and 4 pm . . . Presented by the Lobby 7 Committee,"
Lobby 7, MIT, 1972
Source: CAVS Special Collection

The lobby is 100 feet high and has a reverberation time of over 7 seconds, which Earls plans to take advantage of with the use of a feedback system as well as music. In addition, there will be an electronic piano, films, strobe lights, participatory systems and a kinesthetic sculpture. (MIT Press Release, November 30, 1972)

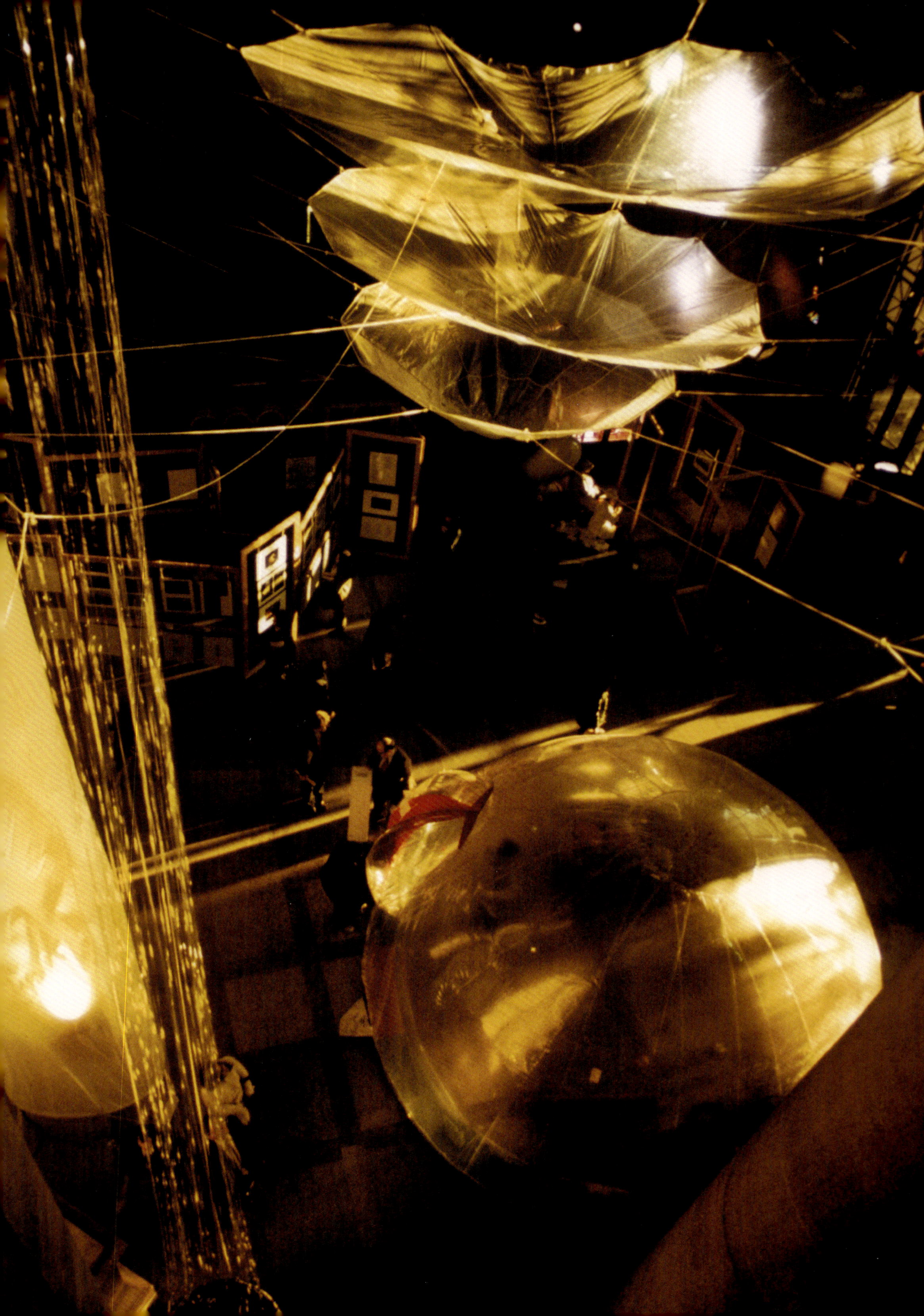

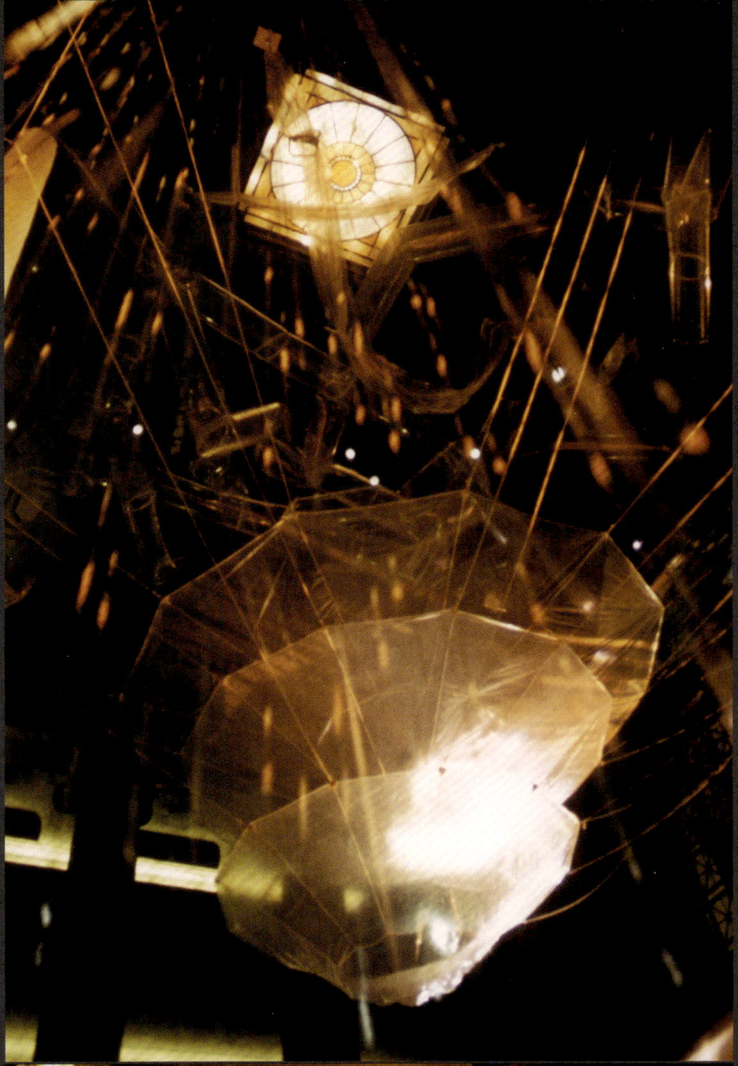

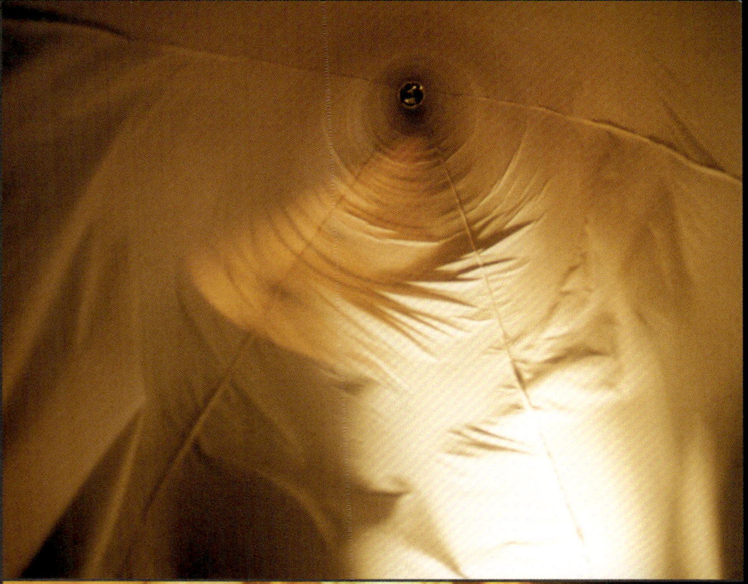

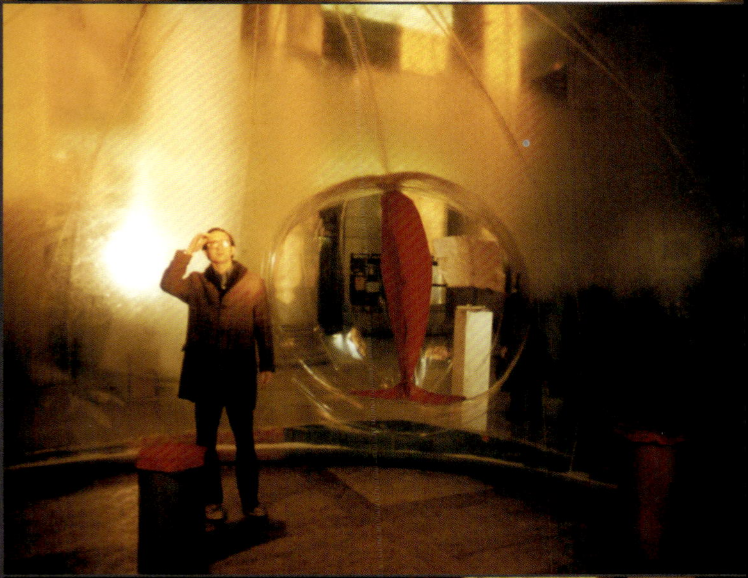

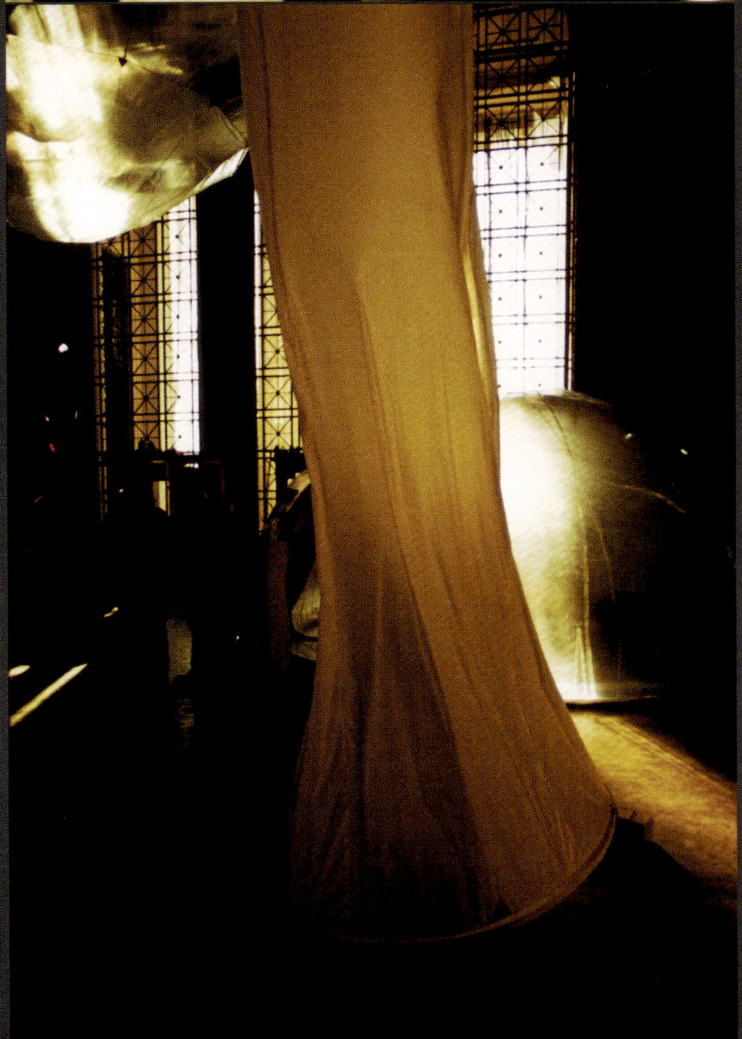

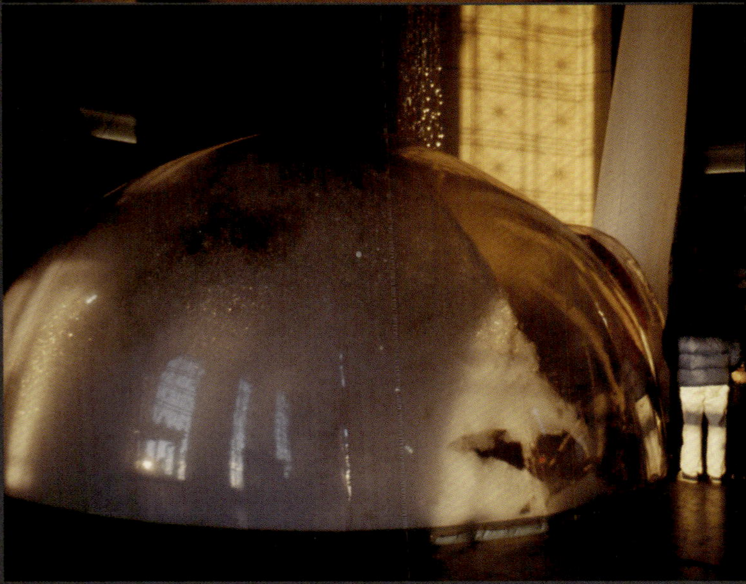

Weather show, CAVS fellows and students, tiered sky work, "RAIN" by Otto Piene and Mitch Benoff, Lobby 7, MIT, November 14-16; 19-21, 1973
Photos: Roger N. Goldstein, MIT architecture student

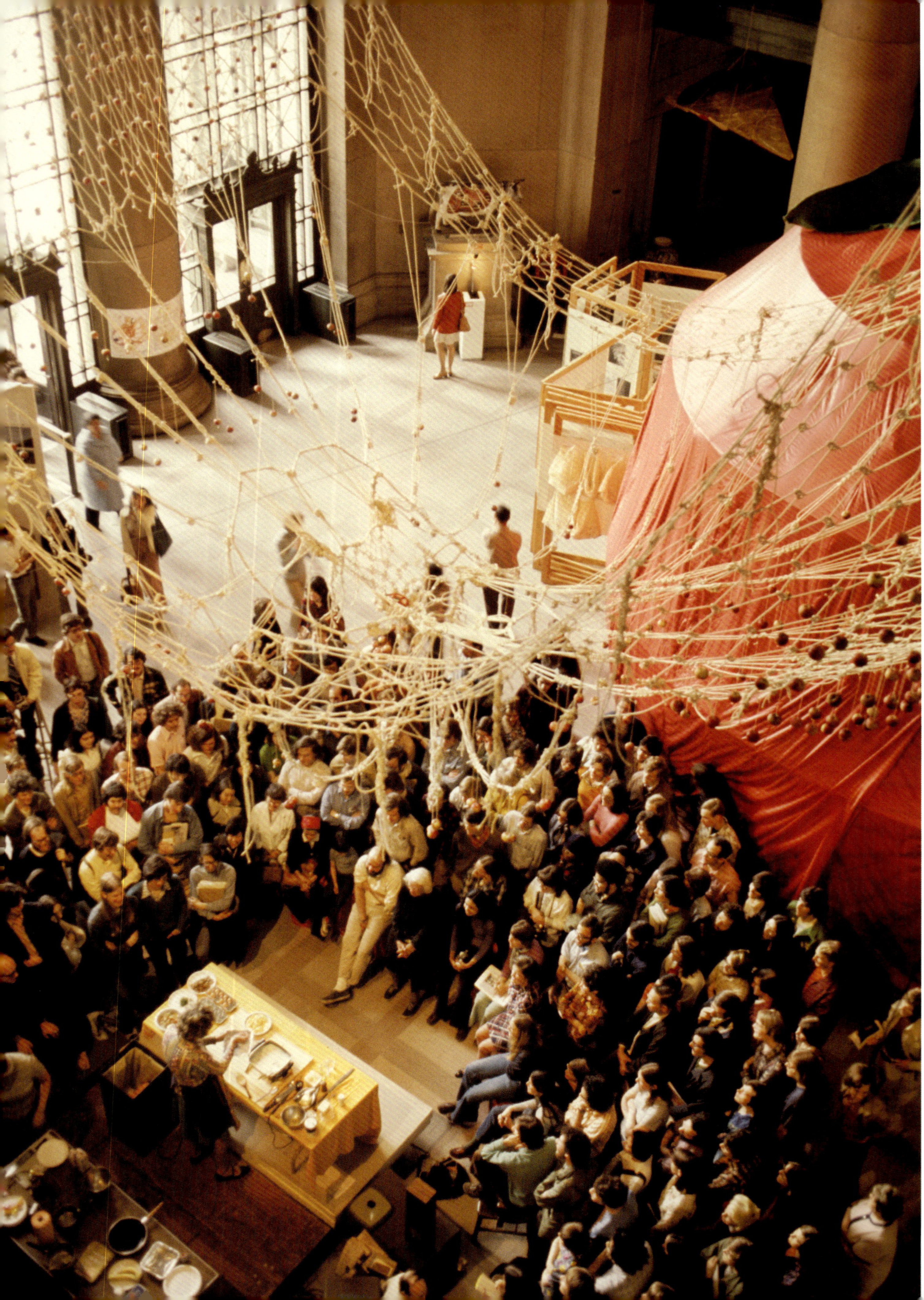

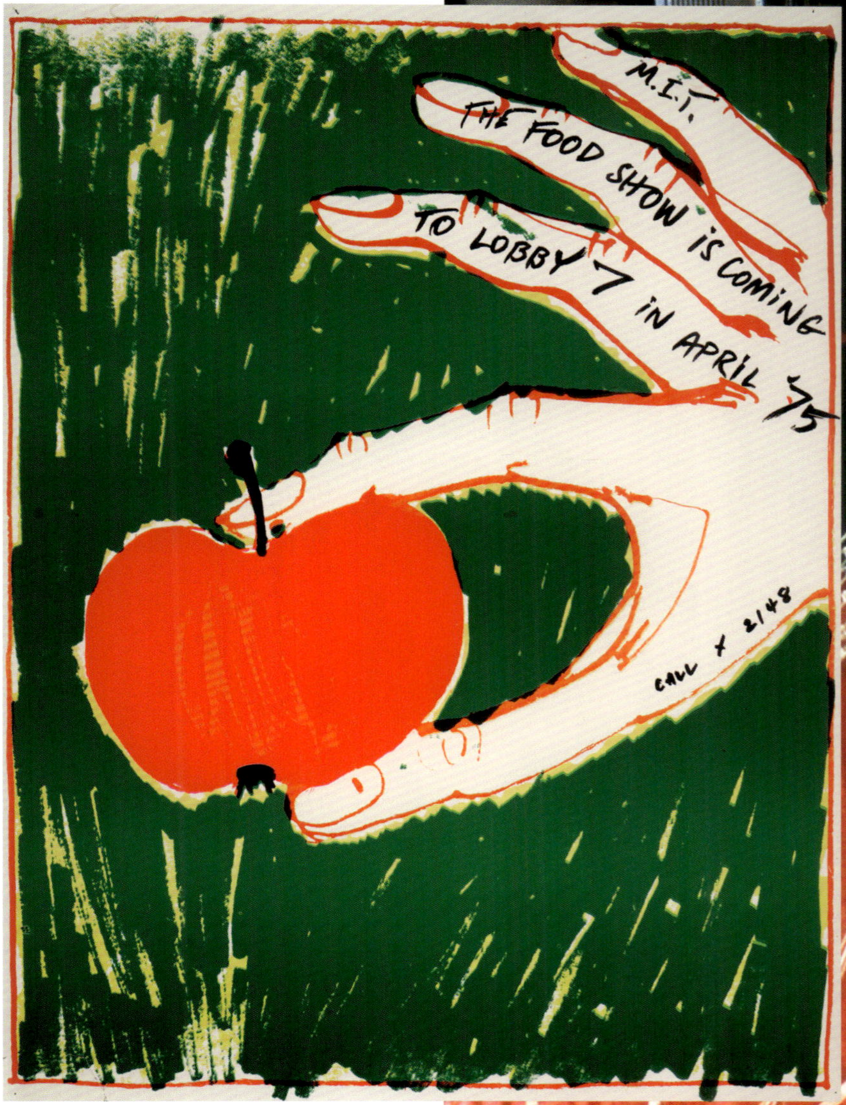

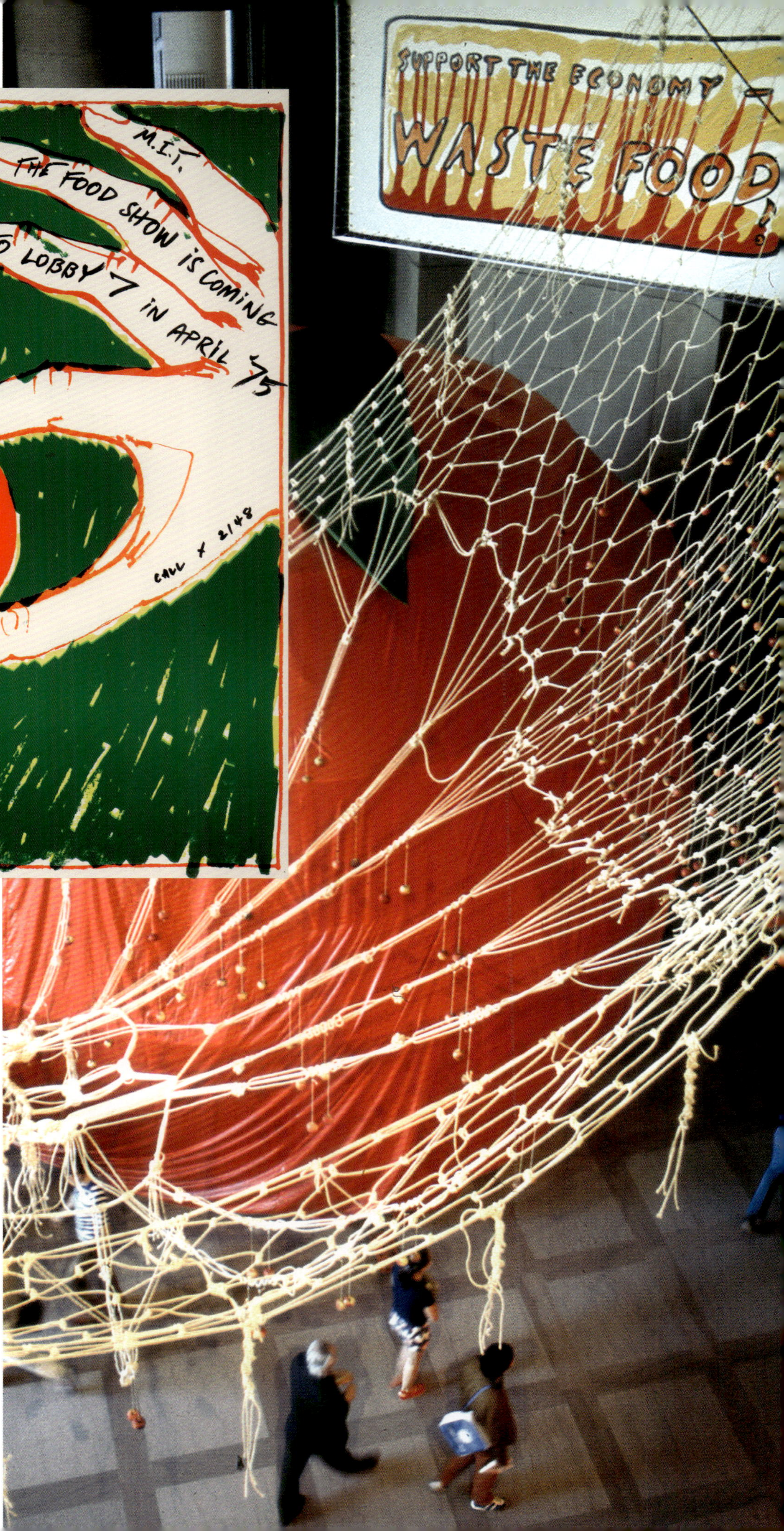

Top and right: *Food* show, CAVS fellows and students, with posters by Otto Piene (above inset and upper-right corner), Lobby 7, MIT, April 7-18, 1975

Facing page: Julia Child, French cooking master, Cambridge local, and star of a WGBH cooking television series, gives a demonstration for the *Food* show, 1975

Photos: Roger N. Goldstein

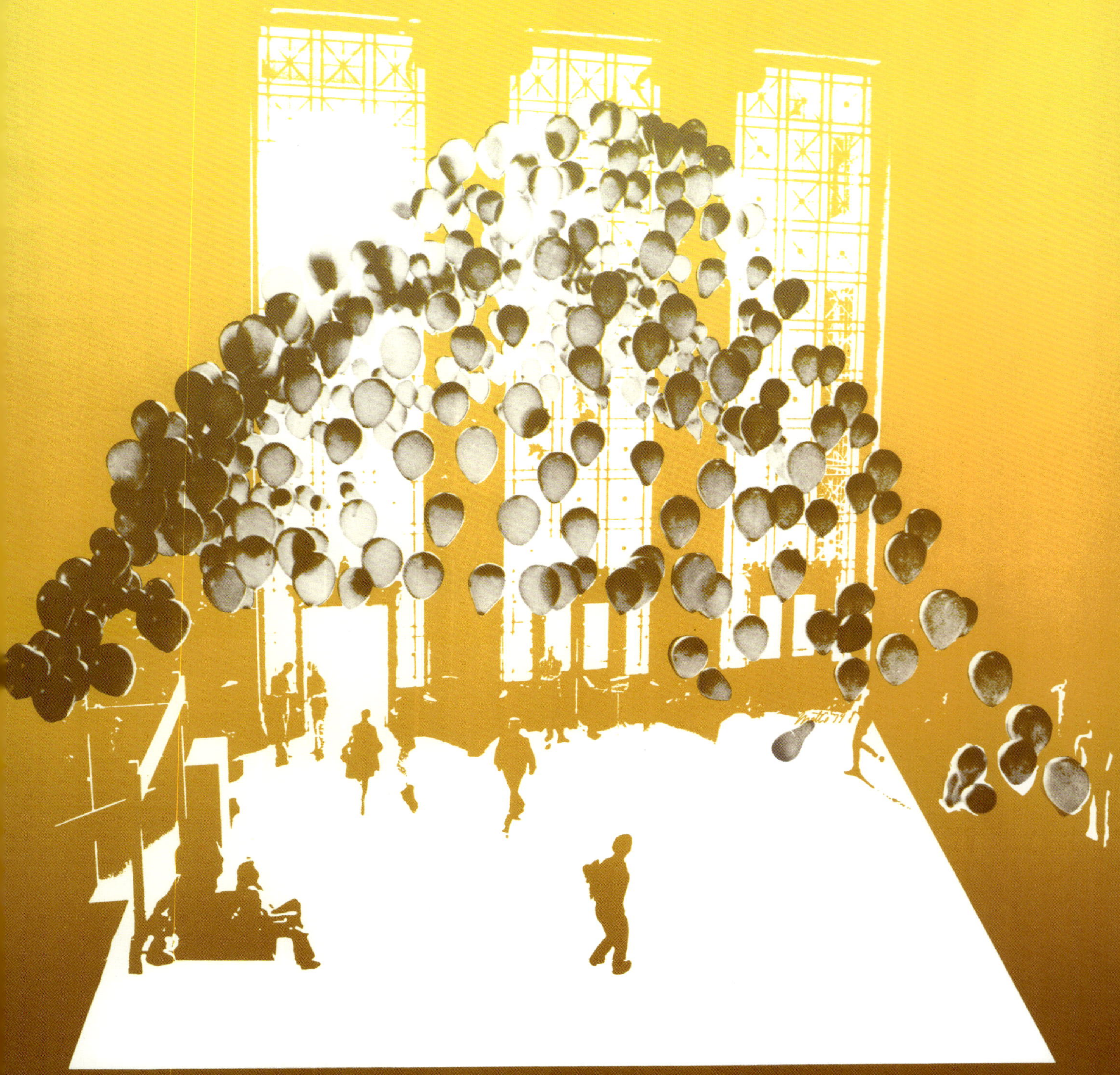

son of balloon carpet bldg 7 lobby wed may 1 8:00 pm–10:00 pm

presented by otto piene & the lobby 7 committee mit sponsored by saarländisher rundfunk | german television

In 1973, students working with Otto Piene launched the project "Silver Balloon Carpet," and the following year Piene created "Gold Balloon Carpet." Both site-specific projects took advantage of Lobby 7's height and access to the cupola through upper balconies, as did all of the Lobby 7 projects that used inflatable and other media to extend the canvas for upward viewing.

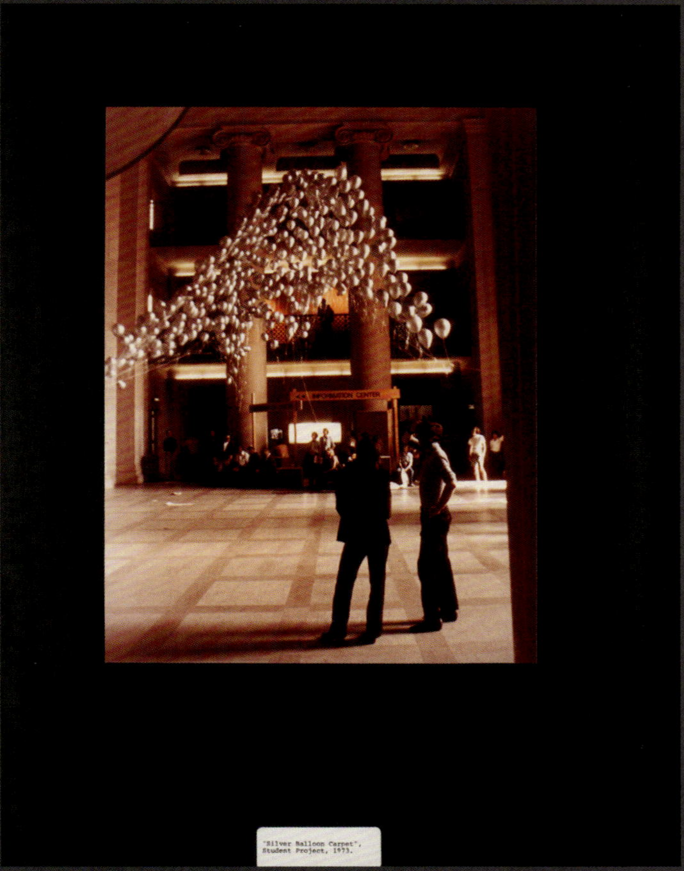

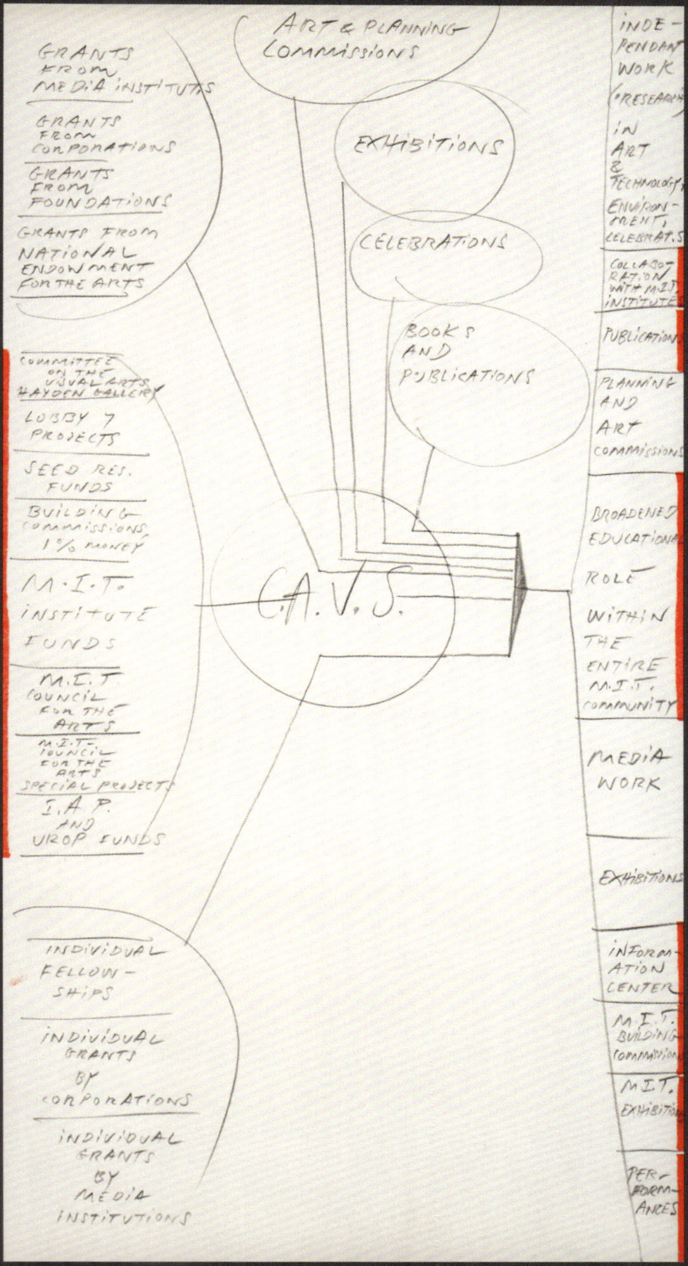

Top left: "Silver Balloon Carpet," student project, Lobby 7, MIT, 1973, page from *Plans and Perspectives*, CAVS portfolio, MIT, 1975
Source: CAVS Special Collection

Facing page: Otto Piene, "Son of Balloon Carpet," poster for MIT Lobby 7 event, May 1, 1974
Source: CAVS Special Collection

Top right: Otto Piene, planning sketch in which he put a "broadened educational role within the entire MIT community" at the center of CAVS's future, *Plans and Perspectives*, CAVS portfolio, MIT, 1975
Source: CAVS Special Collection

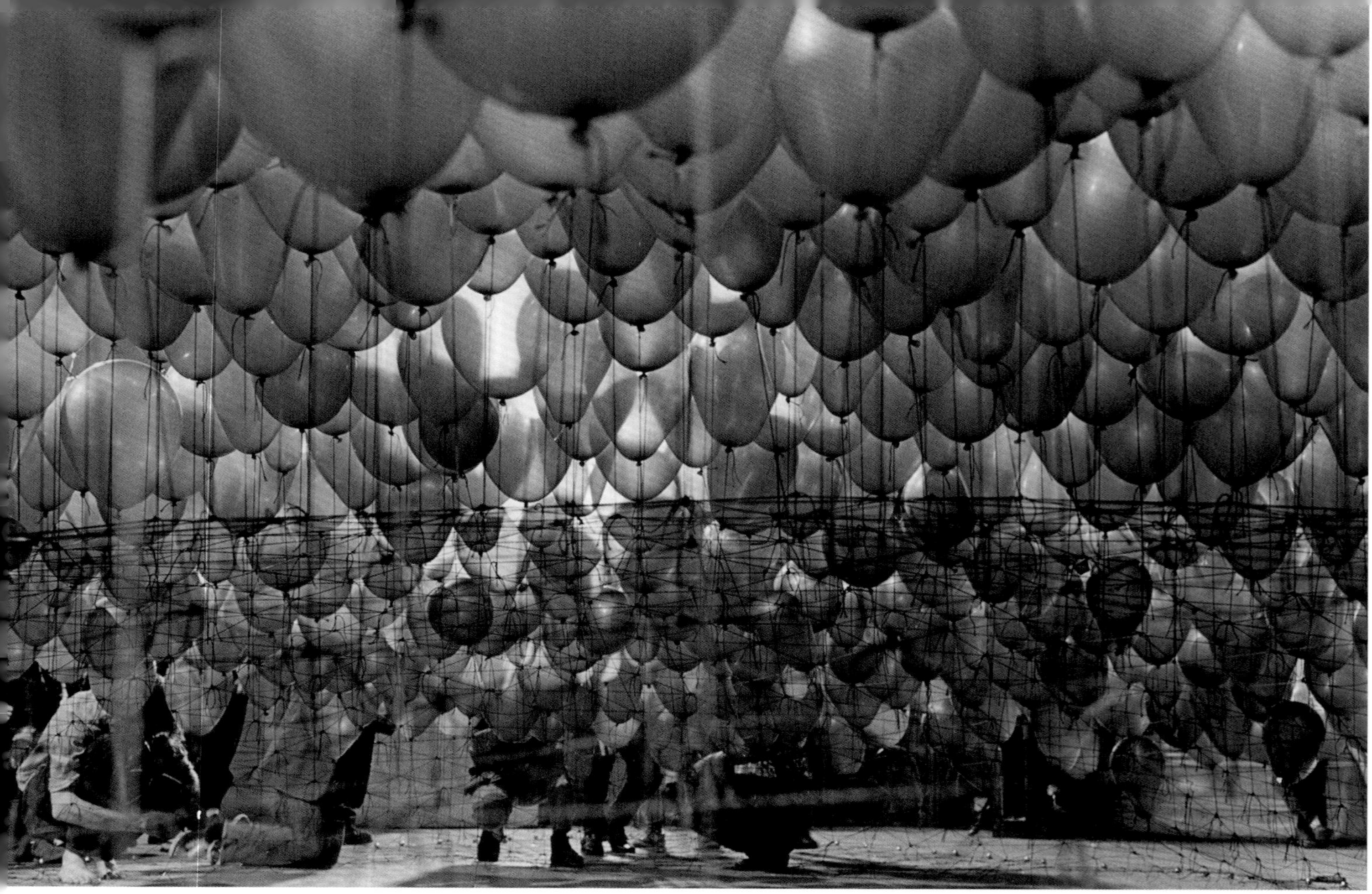

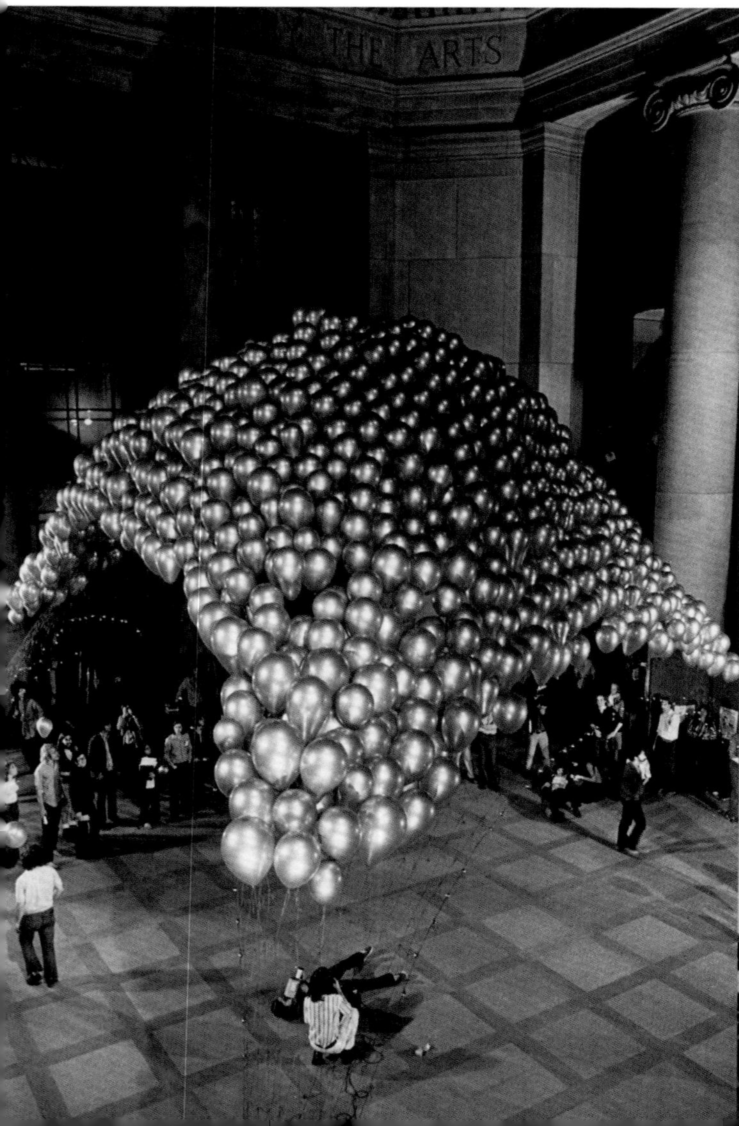
THE ARTS

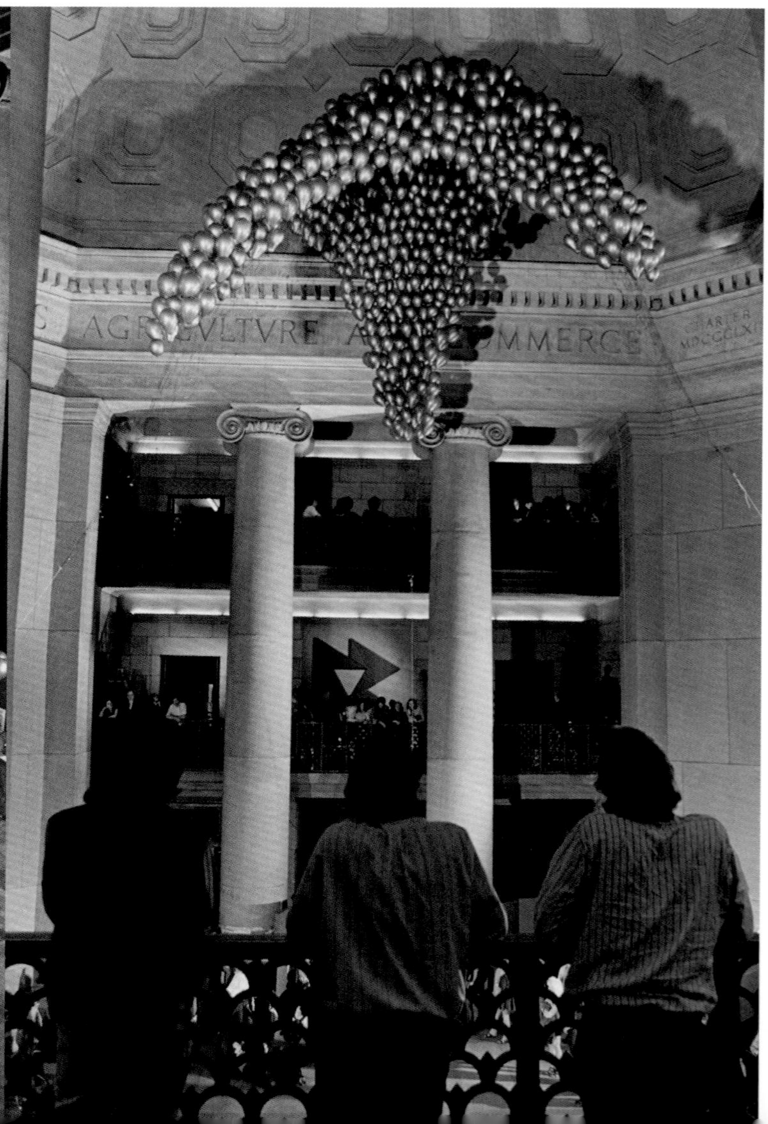
AGRICVLTVRE
MMERCE

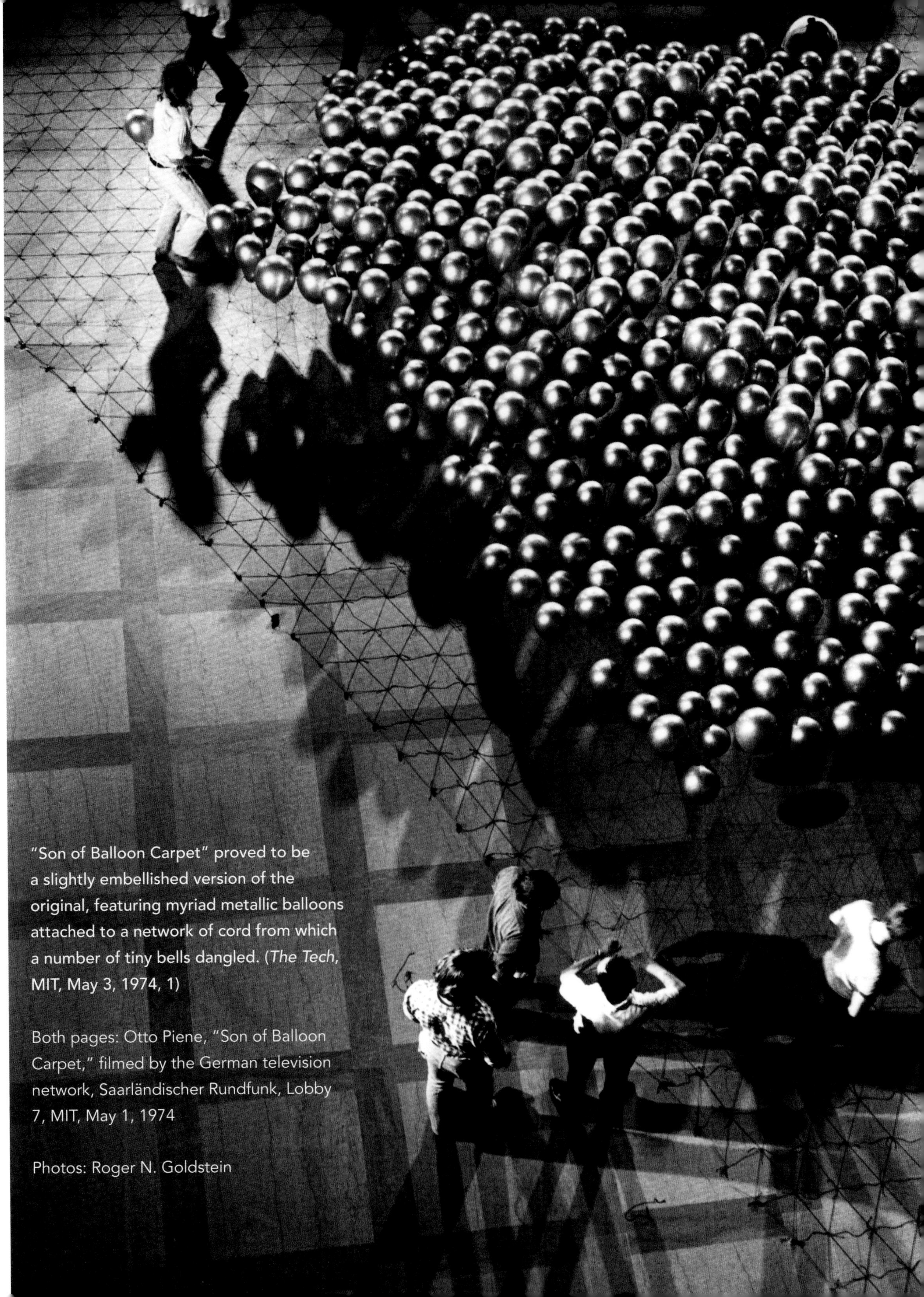

"Son of Balloon Carpet" proved to be a slightly embellished version of the original, featuring myriad metallic balloons attached to a network of cord from which a number of tiny bells dangled. (*The Tech*, MIT, May 3, 1974, 1)

Both pages: Otto Piene, "Son of Balloon Carpet," filmed by the German television network, Saarländischer Rundfunk, Lobby 7, MIT, May 1, 1974

Photos: Roger N. Goldstein

Chapter 3
In Celebration of Collaboration

Collaboration was the hallmark of the Center, its essence and its *raison d'être*. What set CAVS collaboration apart was a communal working dynamic as well as the Center's location in the heart of MIT. Collaborations forged in the hallways, exhibition room, Pit, and studios of Building W11, often galvanized by invitations for group exhibitions, formed the Center's collective artistic voice. The kinetic performing sculpture, *Centerbeam*, as a group work, defined unprecedented ways of collaboration at CAVS. Designed for *documenta 6* (1977) in Kassel, Germany, *Centerbeam* was called "An Aqueduct to the 21st Century" by Manfred Schneckenburger. (*Centerbeam* catalog, 1980, 27) The sculpture traveled to the National Mall in Washington, DC, the following year. In 2017, a 40th anniversary exhibition—*Centerbeam: A Performative Sculpture by CAVS*—was mounted by ZKM, the Center for Art and Media Karlsruhe, to celebrate the seminal art-and-technology work. This chapter explores CAVS's collaborative spirit through the works *Centerbeam* and *Monocle* as examples of the unlikely partnerships among artists working together within a research institute.

Right: *Centerbeam*, a kinetic performing sculpture created collaboratively by CAVS artists, *documenta 6*, Kassel, Germany, 1977
Photo: Otto Piene Collection

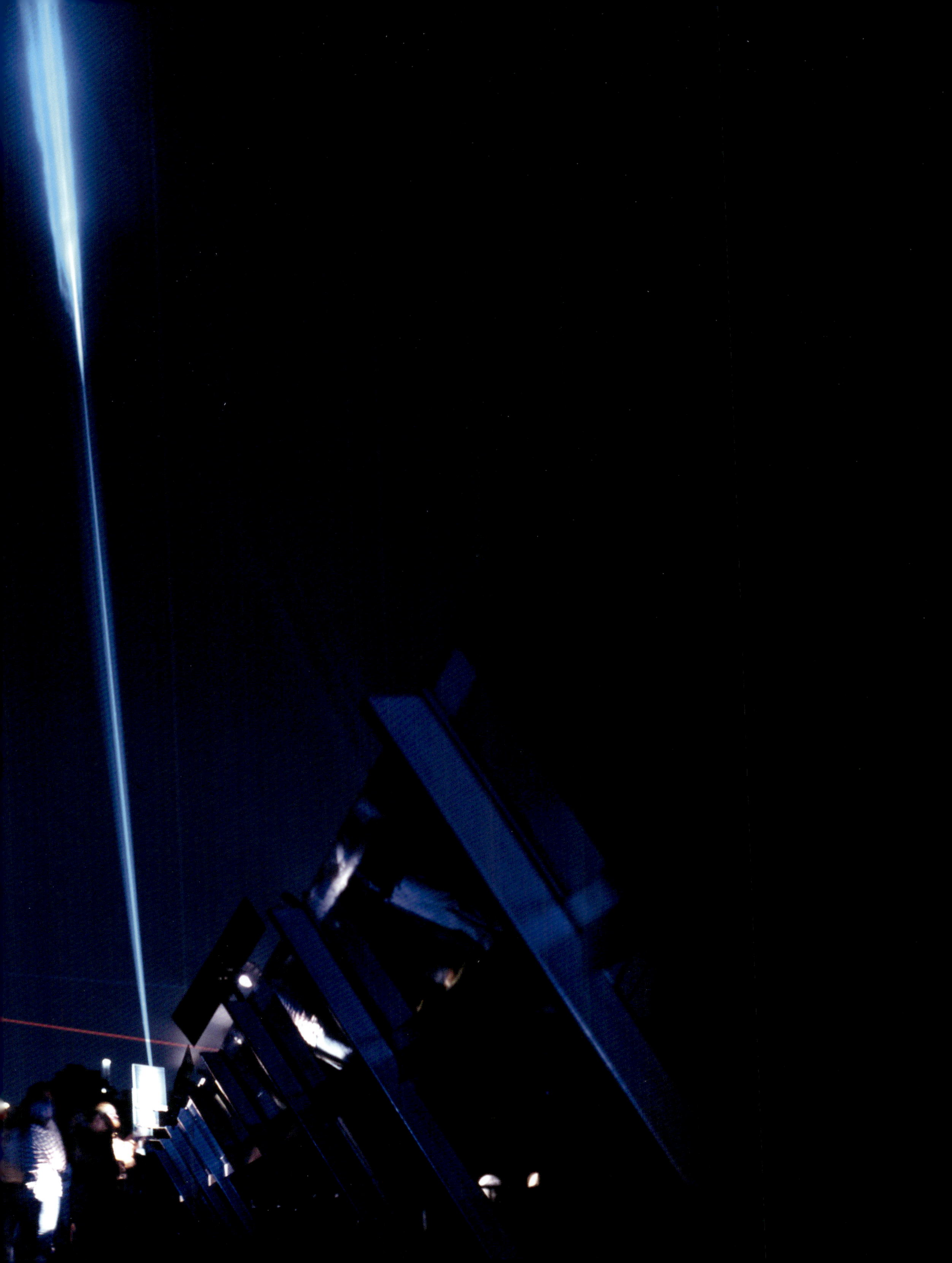

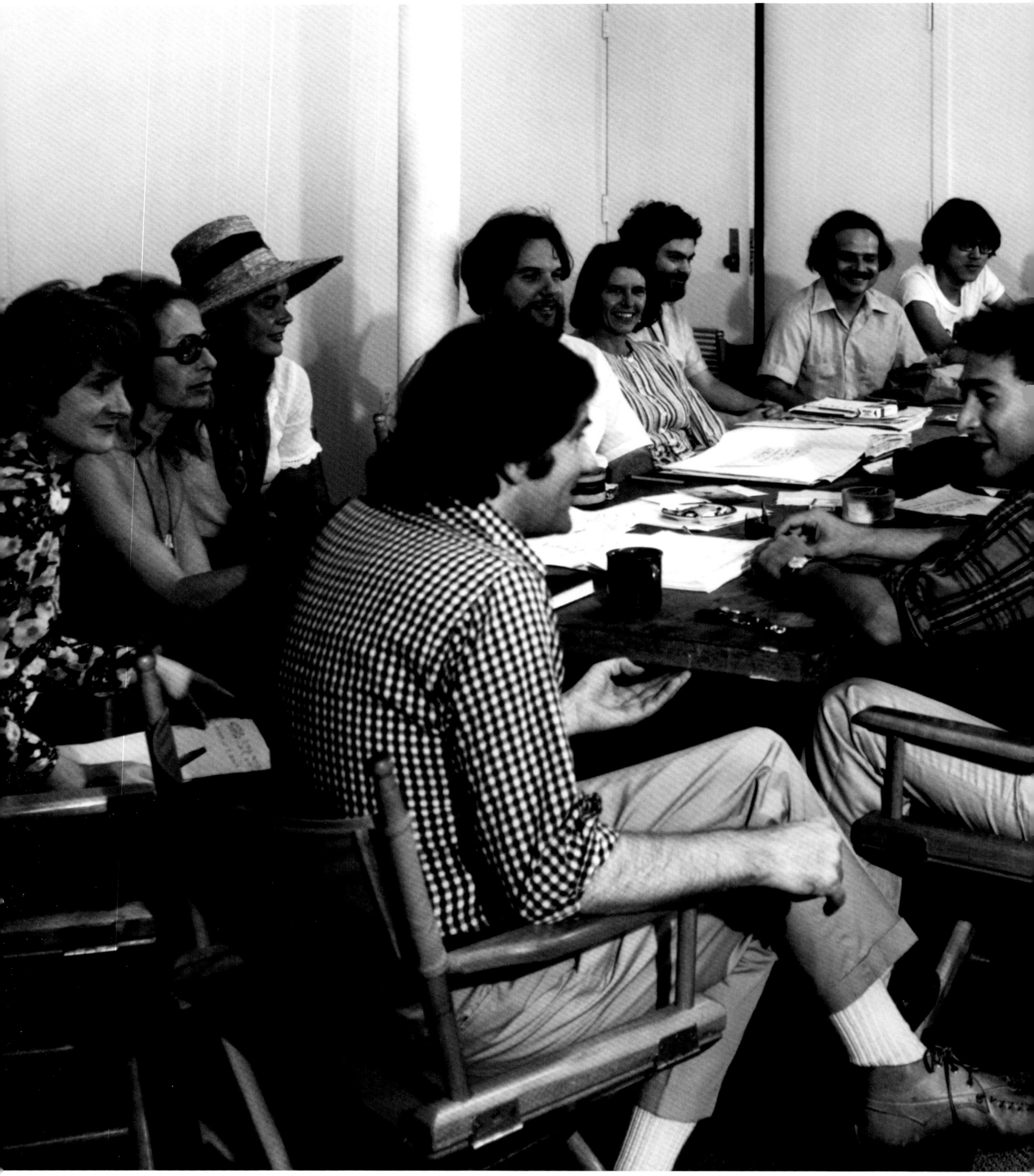

Centerbeam: a Seminal Collaborative Work

Fellows' meetings at the long table in the exhibition room, which included students, junior fellows, senior fellows, administrative staff, and invited scientists and engineers, set the stage. Manfred Schneckenburger had invited CAVS to participate as a group in the 1977 *documenta 6* in Kassel, Germany.

The CAVS community responded to Otto Piene's request for a common idea. Harel Kedem, a graduate student, proposed a habitat fabricated of abandoned computer-age hardware and software. Piene suggested a diamond structure poised in the landscape that would use various materials and media to amplify the sun's reflective energy. Lowry Burgess put forth the idea of a water prism complemented by a household of conduits, which he called *Centerbeam*. The proposals were democratically voted on. Burgess' was selected.

> I proposed a work in which a very long bundle of pipelines of elements and energies were encrusted with numerous images and informational elements (video, holograms, laser, antennae, mirrors, etc.). This woven bundle was to be absolutely straight and level running through the landscape and possibly some buildings. It appeared to me as a living staff of energies, media and phenomena like sparkling notations. (Lowry Burgess, *Centerbeam* catalog, 1980, 26)

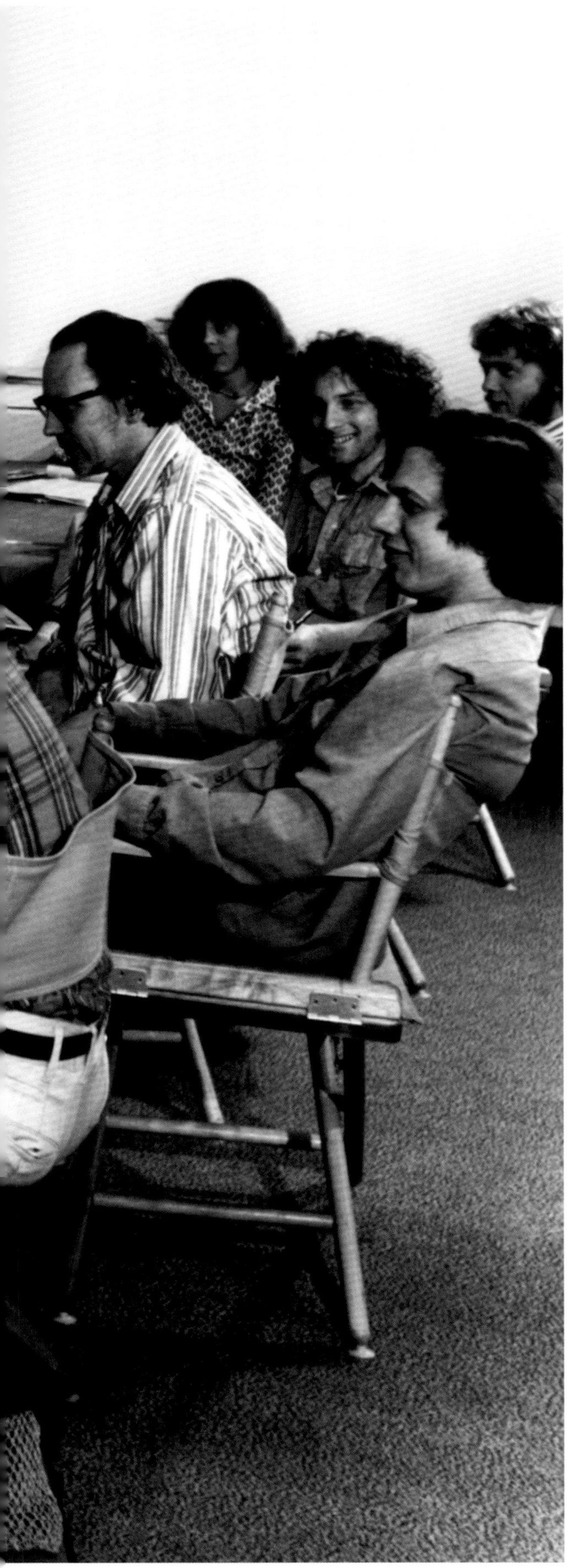

Left: Fellows' meeting, planning *Centerbeam* in the exhibition room, CAVS, MIT, 1977
Clockwise around the table, from left: Astrid Hiemer, Harriet Casdin-Silver, Elizabeth Goldring, Lowry Burgess, Joan Brigham, Ken Kantor, Alejandro Siña, Mark Chow, Karin Bacon, Nancy Doll, Werner Ahrens, Mike Moser, Paul Earls, Bill Cadogan, Mark Mendel, Otto Piene
Photo: Nishan Bichajian

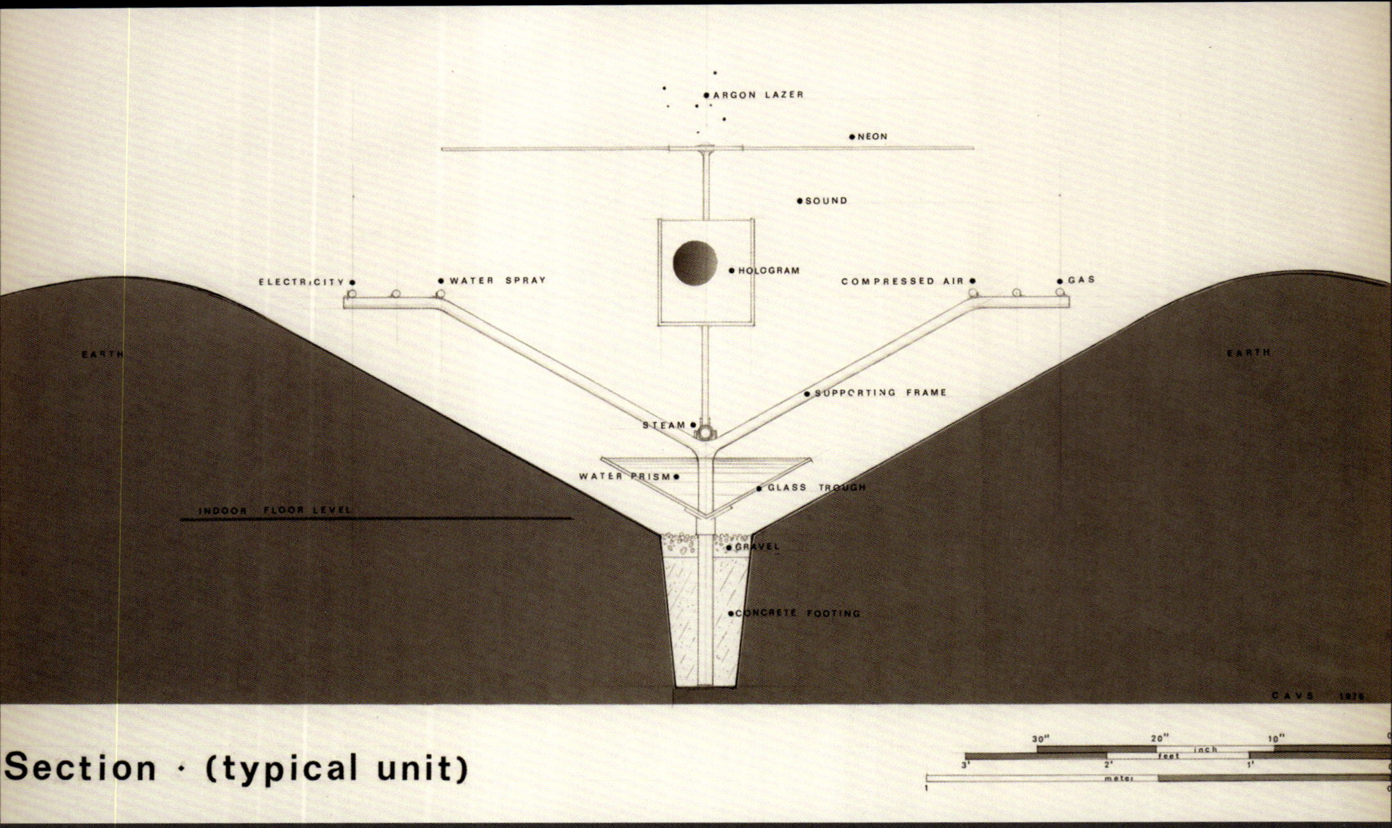

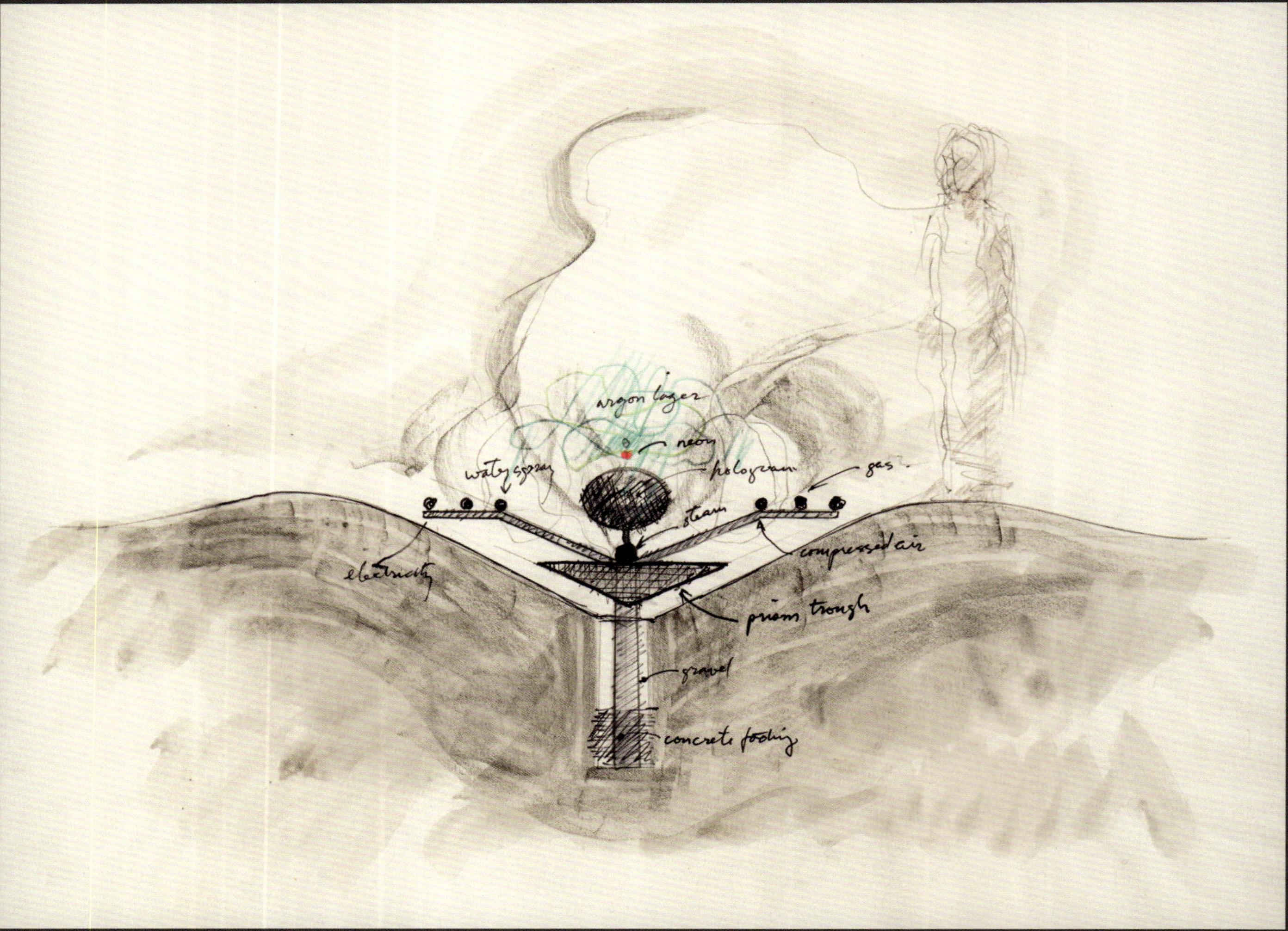

Both pages: Lowry Burgess' concept artwork for *Centerbeam*; and schematics, CAVS, MIT 1976-77
Source: CAVS Special Collection

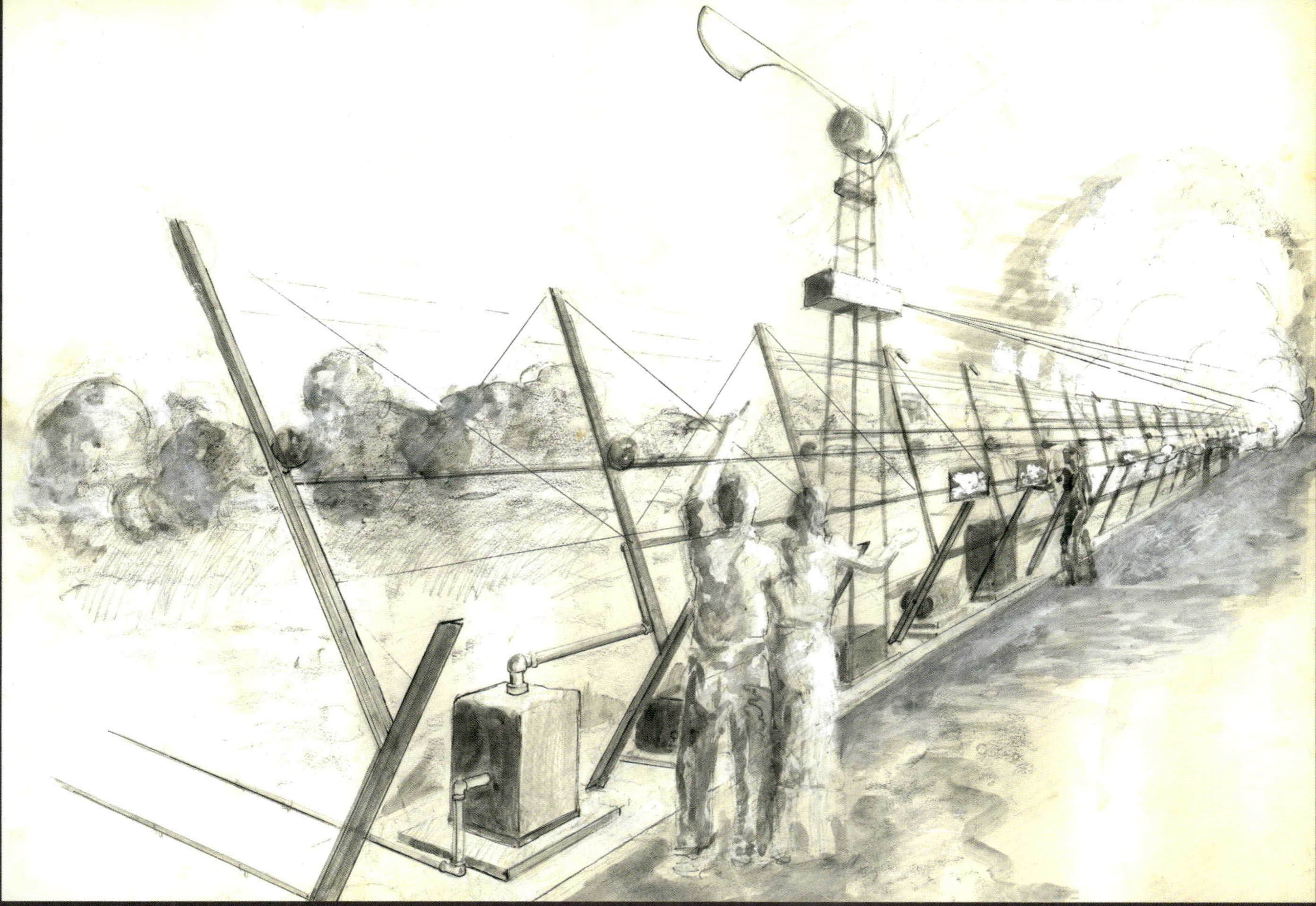

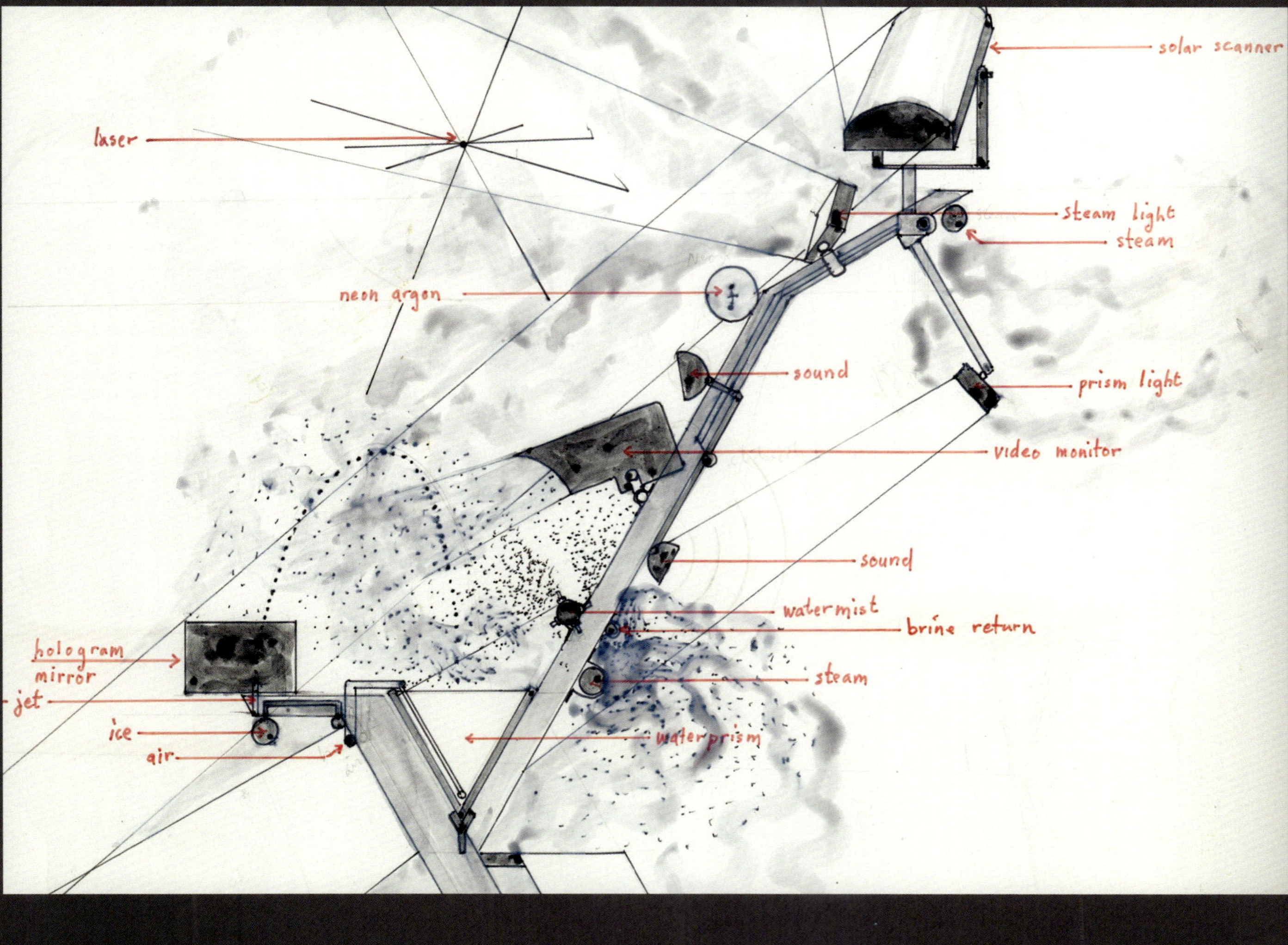

solar scanner
laser
steam light
steam
neon argon
sound
prism light
video monitor
sound
watermist
brine return
hologram
mirror
steam
jet
ice
waterprism
air

What ensued in the following six months was the design, siting in Kassel, mockups of sections, experimentation in the Pit and the alley adjacent to CAVS, and fabrication of lines and media.

> I was hanging around the Center and the next thing was getting involved with *Centerbeam*. There was a need to have Harriet's solar holograms tracked. That was during the summer of 1977. *Documenta* changed my life. Going to Europe for the first time, falling in love with a German girl . . . I had never worked on a team art project like that. We built a model of *Centerbeam* in the Pit. That was one of the most intense, exhilarating, educational, funny experiences I ever had. At a fellows' meeting, Otto looked at the pictures of the model and said we had to do it all over again, something was wrong. (Mike Naimark interview)

> 1977—I was responsible for the *Centerbeam* "ice line" and I also designed the support for the water prism. We built a scale model in the Pit. Wading around in the water we could get electrocuted, said Otto. I was sent to get a tree and brought back a big branch. It was crazy and a lot of fun. (Carl Nesjar interview)

> I hoped that we would see the integration of all our efforts into a coherent union—a single, strict, yet highly elaborated form—thereby creating an intensity and symbiosis beyond any specific technology, media or context. What appealed to me was the occasion to work with both human and environmental relationships not readily available in our individual work. In all this process it has been both remarkable and rewarding that such a diversity of minds and hearts can find a nesting structure. (Lowry Burgess, *Centerbeam* catalog, 1977, 26)

Right: *Centerbeam* test, CAVS fellows left to right, Bill Cadogan, Mark Chow, Mike Moser, Stan VanDerBeek, Aldo Tambellini, and Harriet Casdin-Silver, the Pit, CAVS, MIT, 1977
Photo: Nishan Bichajian

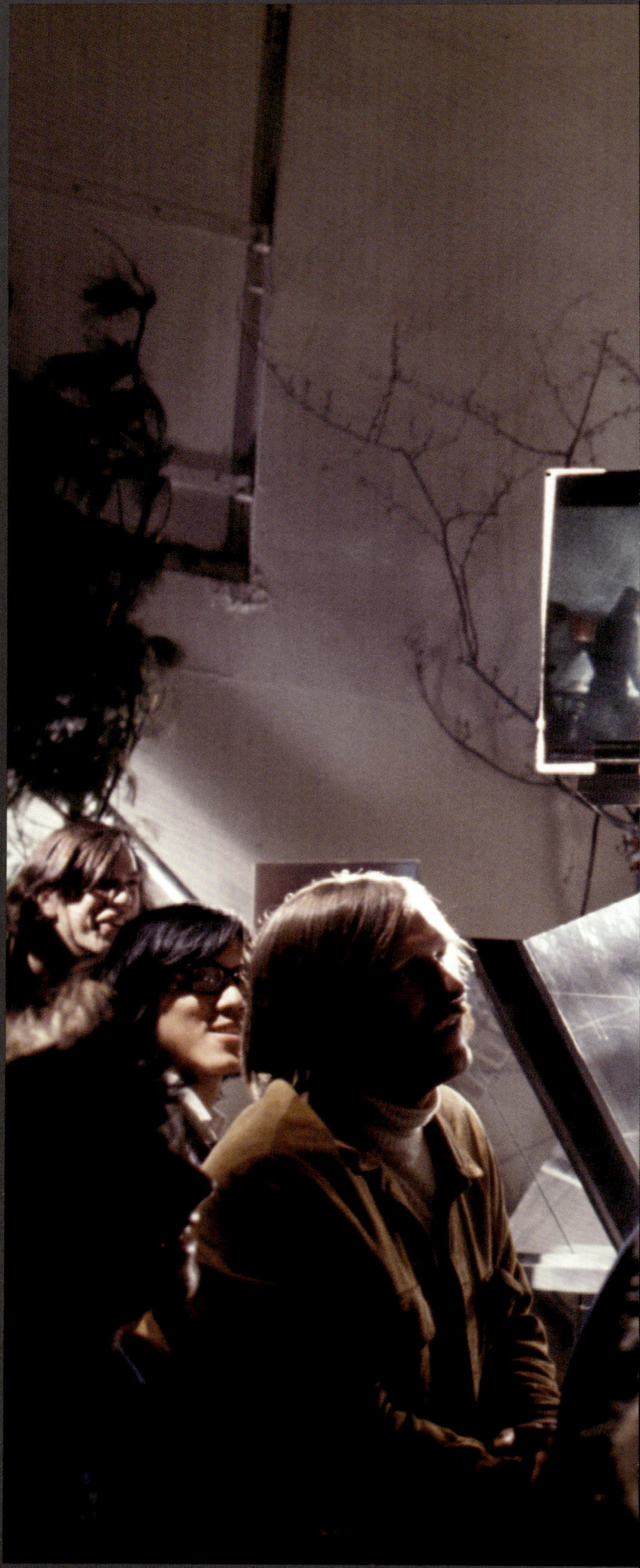

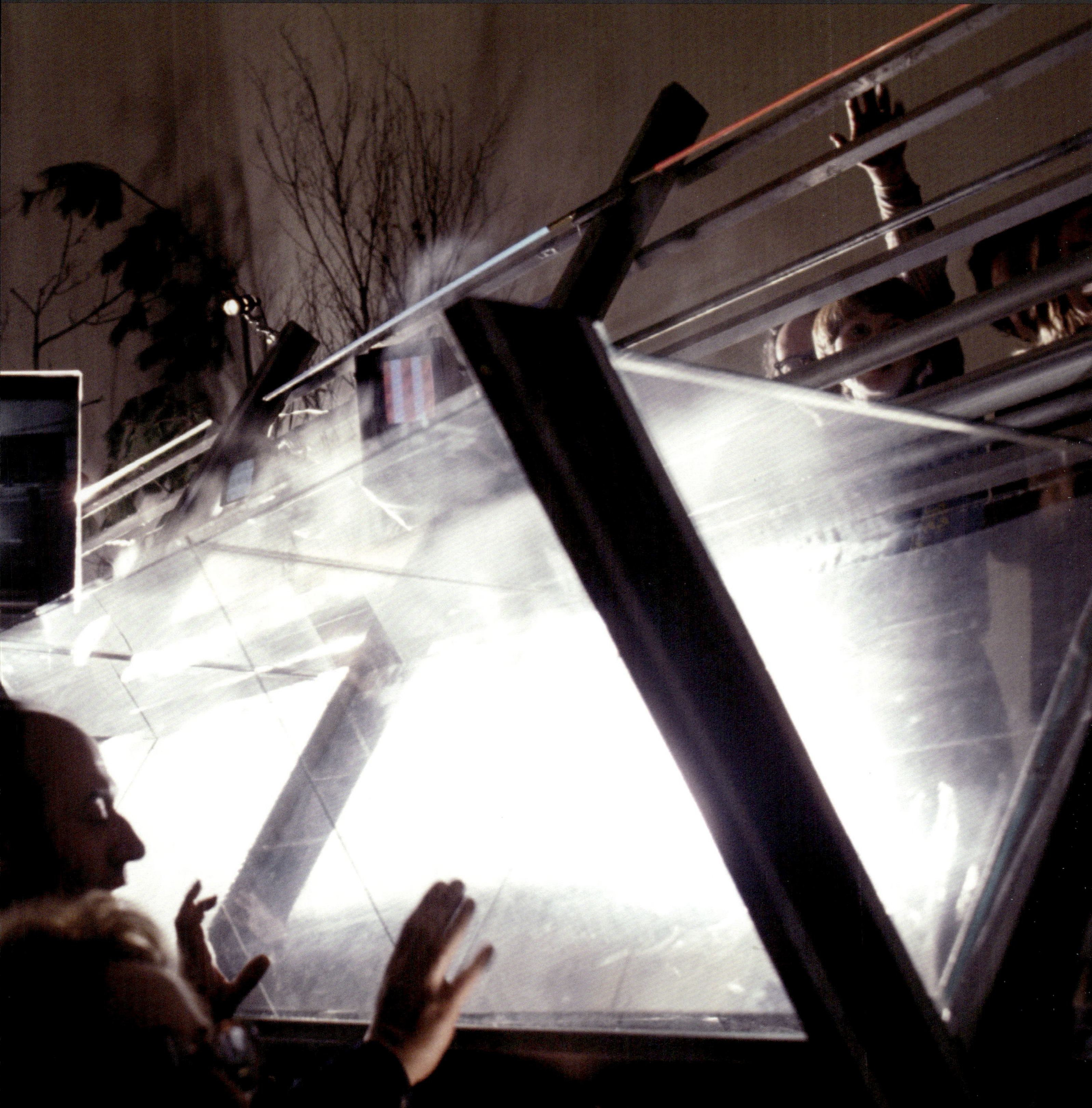

All of the artists traveled to Kassel to install and "play" the sculpture. In the *Centerbeam* catalog, Goldring described the various "lines" that converged in its central core:

Water Prism
The water prism is a trough shaped mass of water supported approximately 4 feet above the ground that absorbs, refracts and reflects the nuances of natural and artificial light surrounding *Centerbeam*. It creates haloes around anything reflected in it and straight horizontal lines appear bent or curved on it as one looks into it.

Holography Line
Eight white light transmission hologram plates project imagery frontally toward the viewer. Holography is the process of recording laser light waves, which are then diffracted by the hologram plate or film to reconstruct the 3-dimensional image in space. A hologram is in essence a light sculpture.

Steam Line
The steam system is composed of two pipes laid horizontally along *Centerbeam*, with small holes permitting the steam to exit, forming a gently spiraling cloud of steam. At maximum pressure, the combined plumes of steam reach 40 feet in height, providing a moving screen for laser projections.

Laser Line
A 4-watt mixed-gas (krypton and argon) laser that produces a columnated beam of polarized "white" green light is an essential part of *Centerbeam*. When directed through a 60° prism, at least ten different colored lines emerge, ranging from bright red to deep blue.

Electricity Line
The electricity line powers the large machinery, i.e., the steam generator, compressors, and laser as well as the sound, video, and hologram-illuminator systems.

Neon-Argon Line
Two lines—one red (neon gas), one blue (argon/mercury)—form the top section of the *Centerbeam* parallelogram of pipes. They are barely visible in daylight but brightly so at night.

Brine Line
The brine line forms a cooling cycle for various parts of *Centerbeam* and specifically creates small ice events within the work. Cool "Freon" gas (dichlorodifluoromethane) circulates through thin copper coil submerged in a brine container, gradually cooling the brine to below freezing.

Compressed Air Line
The compressed air line extends a jet stream of pressurized air throughout the structure. The compressed air plays with steam, water, and people at specific intervals along the work and, at special occasions, caters to the inflatable flower.

Water Line
Water is an element used throughout the work. The water line charges the steam generator, cools the laser, provides water to form ice on the brine line and supplies the water prism.

Sound/Communication Line
Recorded instrumental, vocal and electronic music that has been subjected to multiple time-displacement processing is played back at irregular times through a series of loudspeakers placed along *Centerbeam*. The system changes periodically and can be varied by the spectator.

Video Reflection (*Centerbeam*, Washington, DC)
Live television images of the participant/viewers of *Centerbeam* are mixed with pre-arranged, programmed images on television monitors dug into the ground. . . . The result is a collage of time shifted images with delays of varying lengths.

Poetry Line
The poetry line consists of the 102-line *Poem On*, silhouette screened with mirroring silver onto eight glass rectangles. Poem mounts are located above the water prism and when the angle of sunlight is perpendicular to the mirror face, the glass acts as a stencil, "writing" or "casting" the poem onto the water surface.

Grow Line
The grow line features fast-growing seasonal vegetation. It provides counterpoint to the technology of other lines. Vegetation grows out of open ceramic troughs.

Drawing with Your Eyes (*Centerbeam*, Washington, DC)
The Drawing with Your Eyes machine, designed as a therapeutic as well as artistic tool, is one of the participatory stations along *Centerbeam*. The apparatus uses the moving pupil to reflect a pencil beam . . . picked up by a video camera and transmitted to a video screen that displays a "continuous" drawing.

Kalliroscopic "White River" (*Centerbeam*, Washington, DC)
A "white river" of enveloped kalliroscopic fluid runs underneath the water prism reflecting sunlight and the sun spectra emanating from the water prism. It is a current of gentle turbulences, evanescent yet part of a continuum. As visitors touch the "white river" surface, they modify its flow.

Facing page: top, installing *Centerbeam*; middle, Lowry Burgess and Joe Moss plumb the water line; bottom, water prism and Carl Nesjar's frozen "Brine Line," CAVS at *documenta 6*, Kassel, 1977
Photos: Mira Cantor

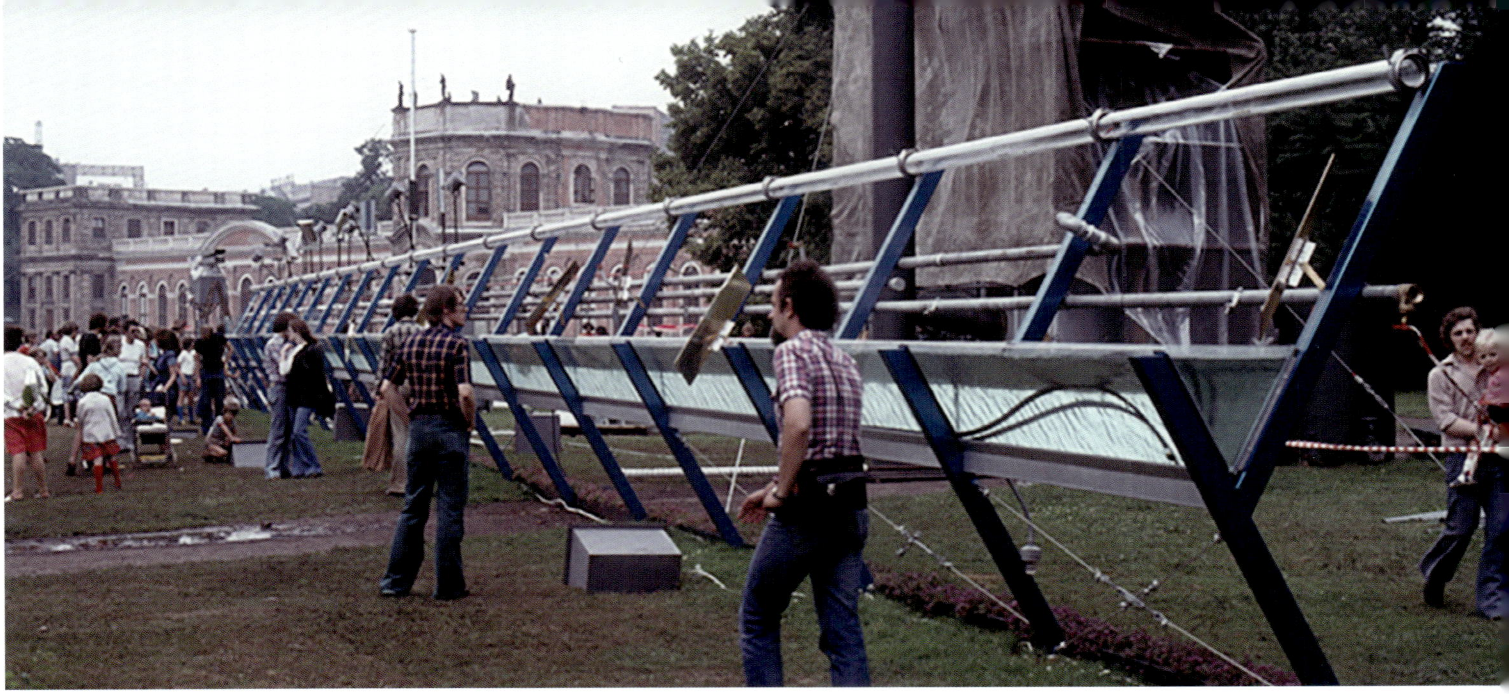

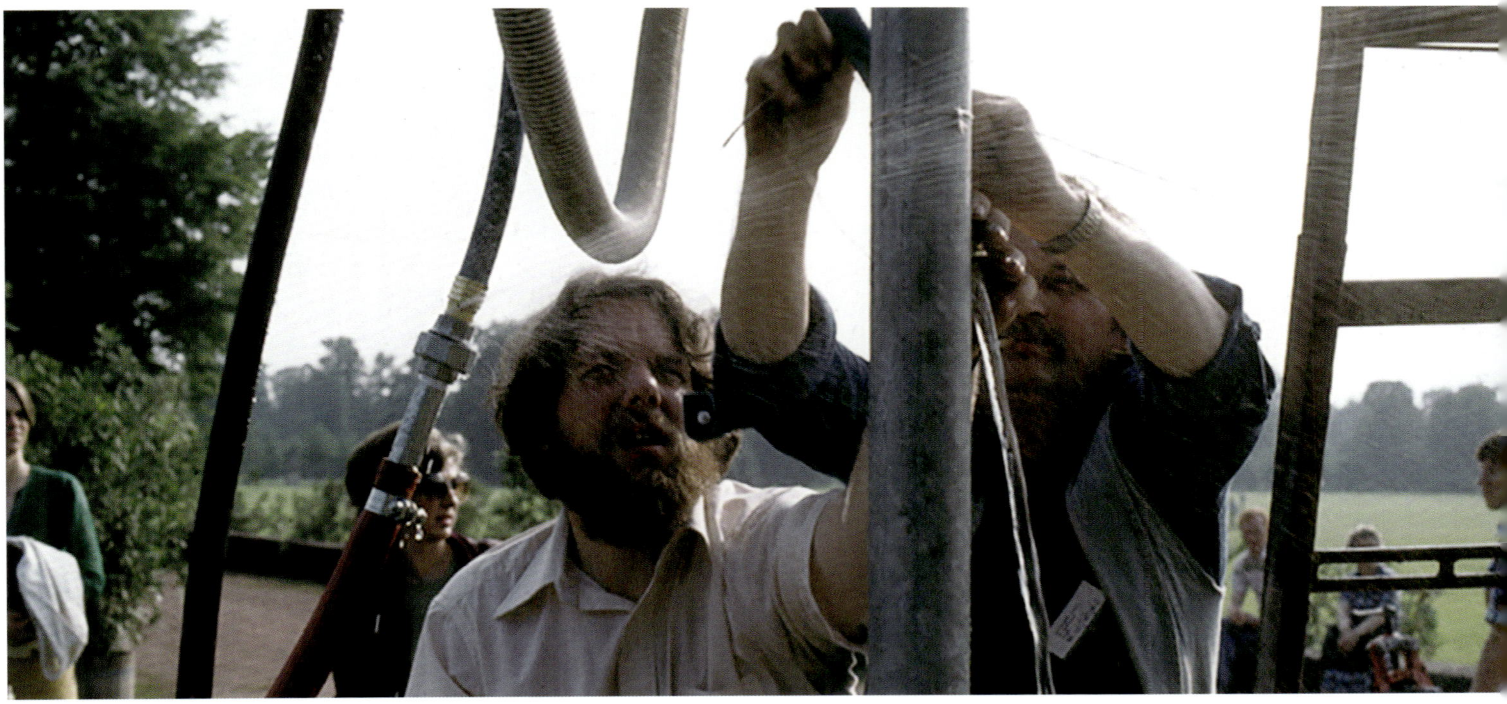

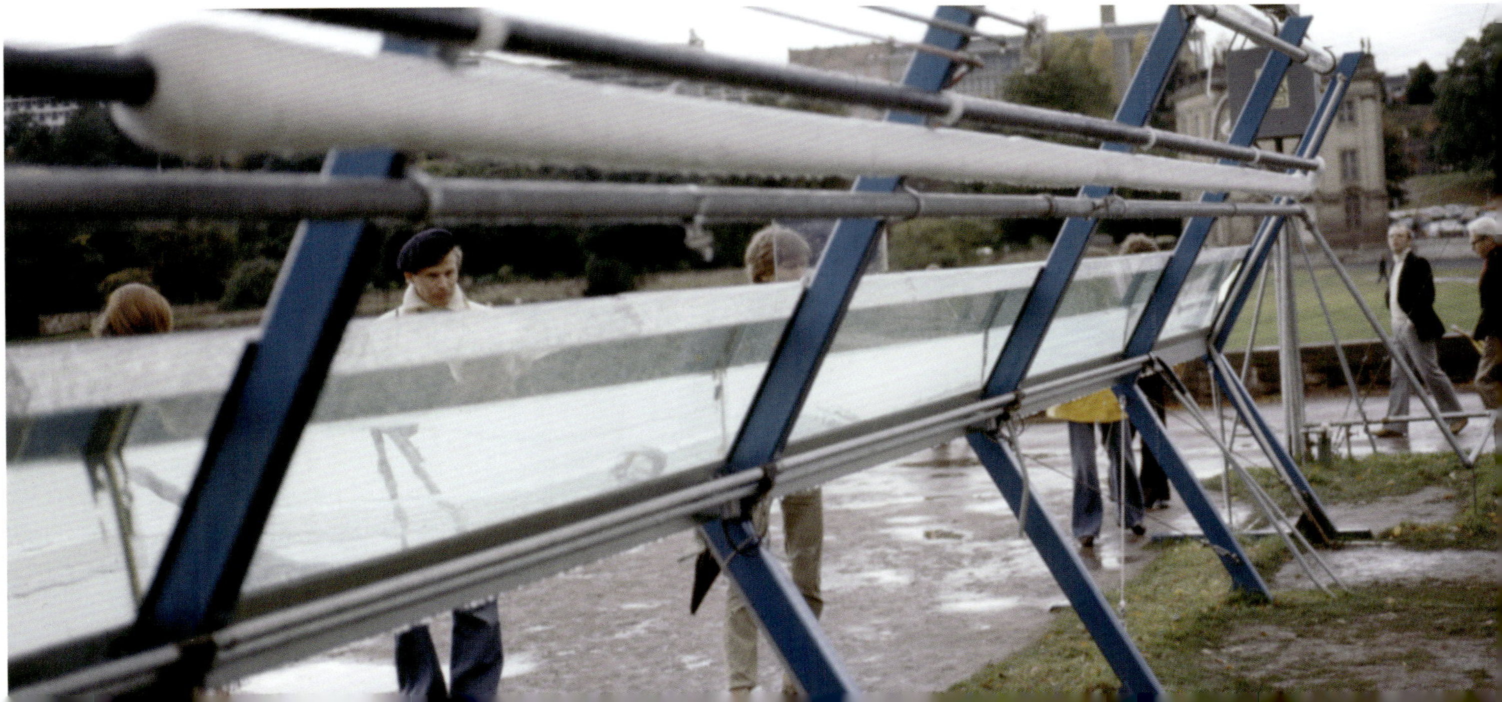

Centerbeam was 144 feet in length and was created by fourteen artists, five scientists and five engineers. It was open to the public from June 24 through October 2, 1977 in Kassel, Germany.

> Phenomena incorporated in and surrounding the work are light (sunlight, reflected light, neon, argon, xenon, laser light), sounds (live, natural, electronic, transmitted, processed), daylight holograms, ice formations, and scanned laser images. Within the larger related activities of steam, water, air and light occur numerous small-scale phenomena of sound and video. (Goldring, *Centerbeam* catalog, 1980, 54)

Centerbeam was not a "clothesline," but an integrated sculpture that went beyond individual statements. The goal was to create new work that would engage audience participation or collaboration with the public. In Piene's words, "energy transfer" between artwork and audience was the core experience, facilitated in part by the active, communicative, engagement of the artists. Piene's proposal in the 1977 *Centerbeam* portfolio was not far from its realization in Kassel:

> *Centerbeam* is an integrative work arising from the collaboration of artist/Fellows of the CAVS. Specifically, *Centerbeam* is a low straight 200 foot structure carrying elemental pipelines of compressed air, gas, steam, water, neon, sound and electricity. It is placed in a shallow furrow of earth and supported by a structure of fifty "y" shaped props. The more general phenomena incorporated in the pipe structure are light (neon, xenon, and argon), sounds (natural, live, and processed), daylight holograms, ice formations, and floating laser drawings and word images. Within this activity of steam, water, and air occur numerous small scale events of light, ice, and sound. They surround a nucleus of holographic plates supporting and paraphrasing the essential image of a horizontal, arterial beam of elemental activity. This sculptural gesture would intersect an existing architectural space and extend into the landscape. (Otto Piene, "Centerbeam '77: documenta 6" portfolio, CAVS, MIT, 1977, 3)

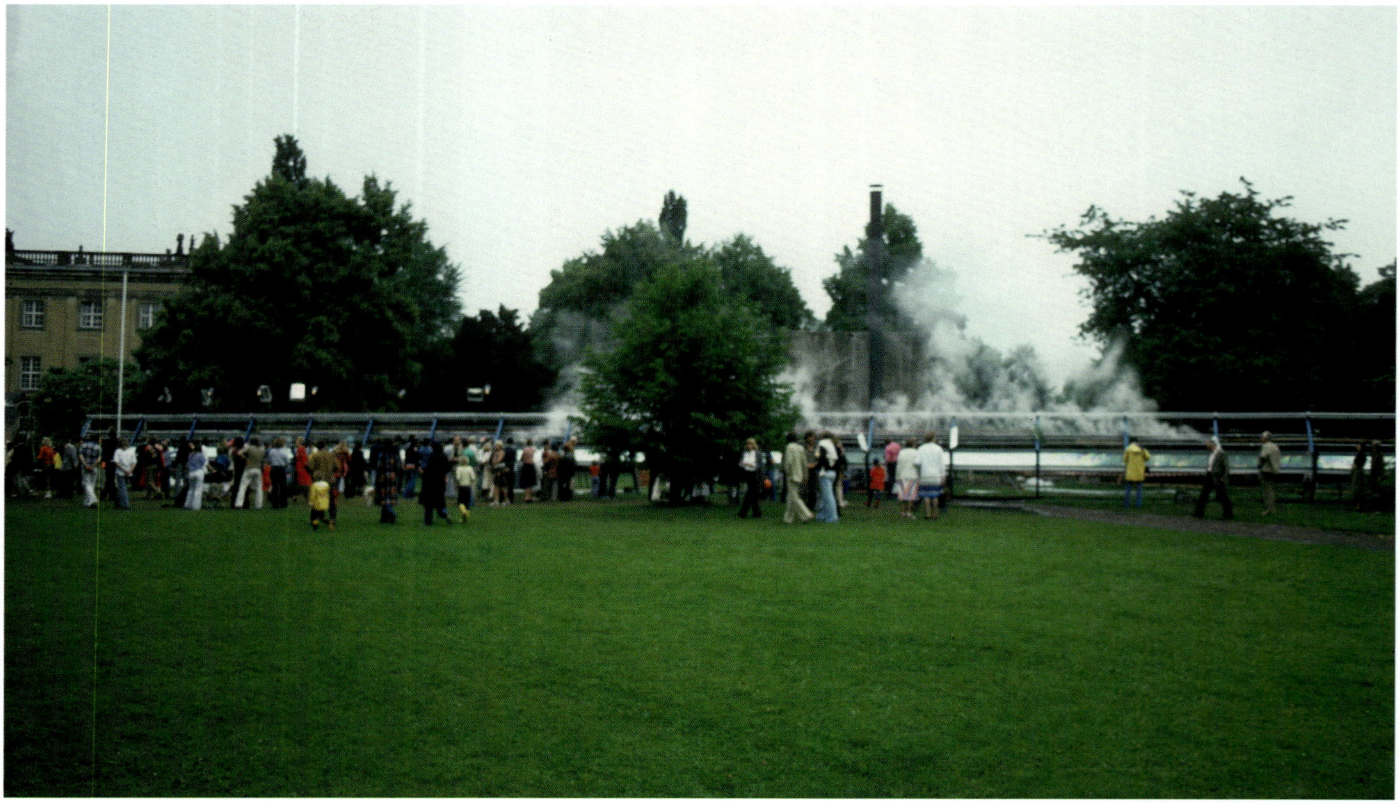

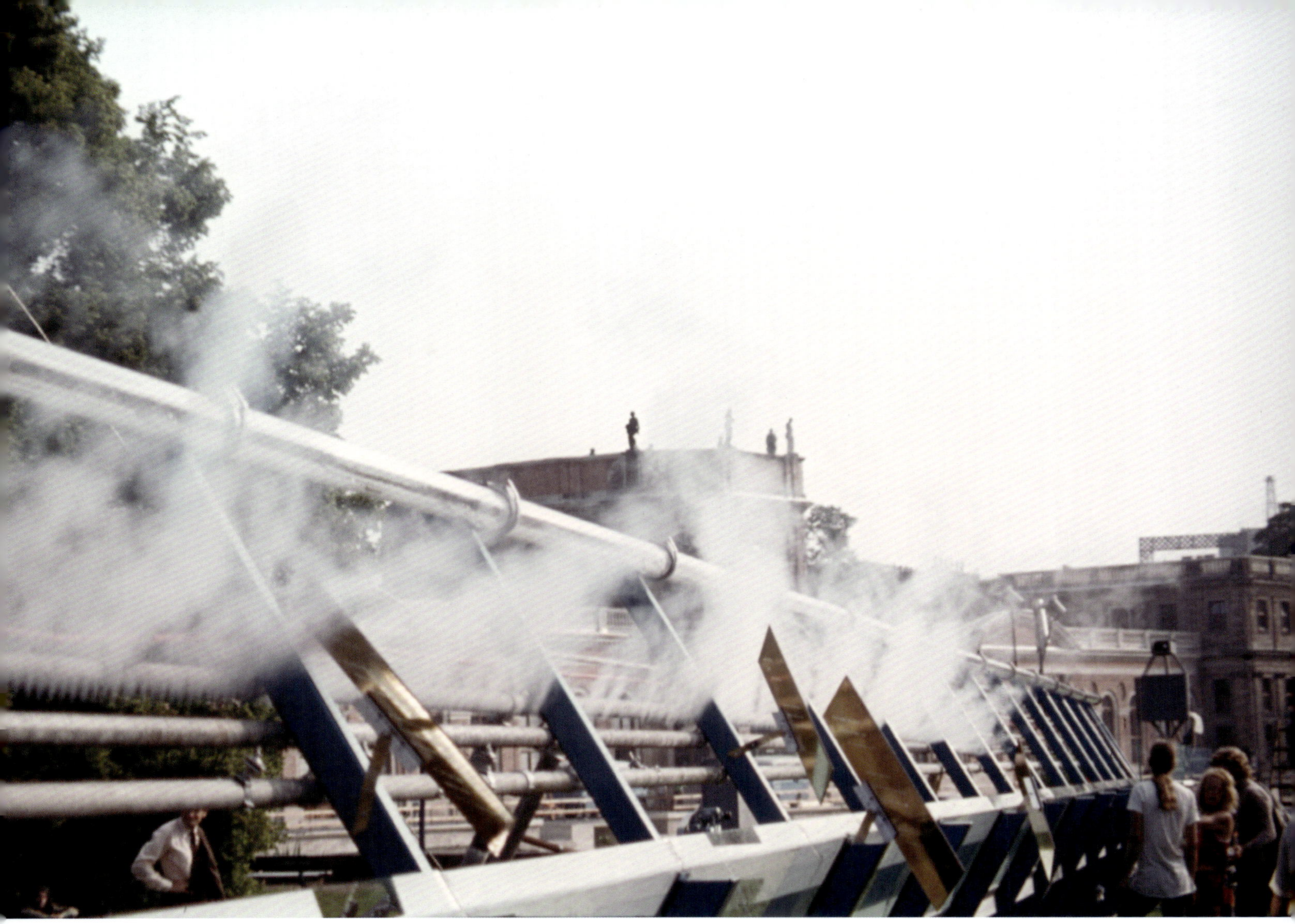

The CAVS artists who collaborated to create *Centerbeam* for *documenta 6*, Kassel, 1977:

Joan Brigham	Steam Line/Water Line
Lowry Burgess	Structural Design/Concept/Water Prism
Harriet Casdin-Silver	Holography Line
Paul Earls	Laser Line/Music/Sound Line
Elizabeth Goldring	Writer/Curator
Michio Ihara	Structural Design
Harel Kedem	Grow Line
György Kepes	"Envelope" Laser Projection
Mark Mendel	Poetry Line
Michael Moser	Video
Carl Nesjar	Brine Line
Otto Piene	Structural Design/Concept/Inflatable
Alejandro Siña	Neon-Argon Line
Aldo Tambellini	Video

I think *Centerbeam* was an example of a good collaboration. I can recognize each artist. (Jose Maria Ytturalde interview)

Above: Steam line and mirrored poetry panels during the day, *Centerbeam*, CAVS at *documenta 6*, Kassel, 1977
Photo: Dietmar Löhrl

Facing page: The 144-foot length of the structure is glimpsed in this view of *Centerbeam*, CAVS at *documenta 6*, Kassel, 1977
Photo: MIT Museum

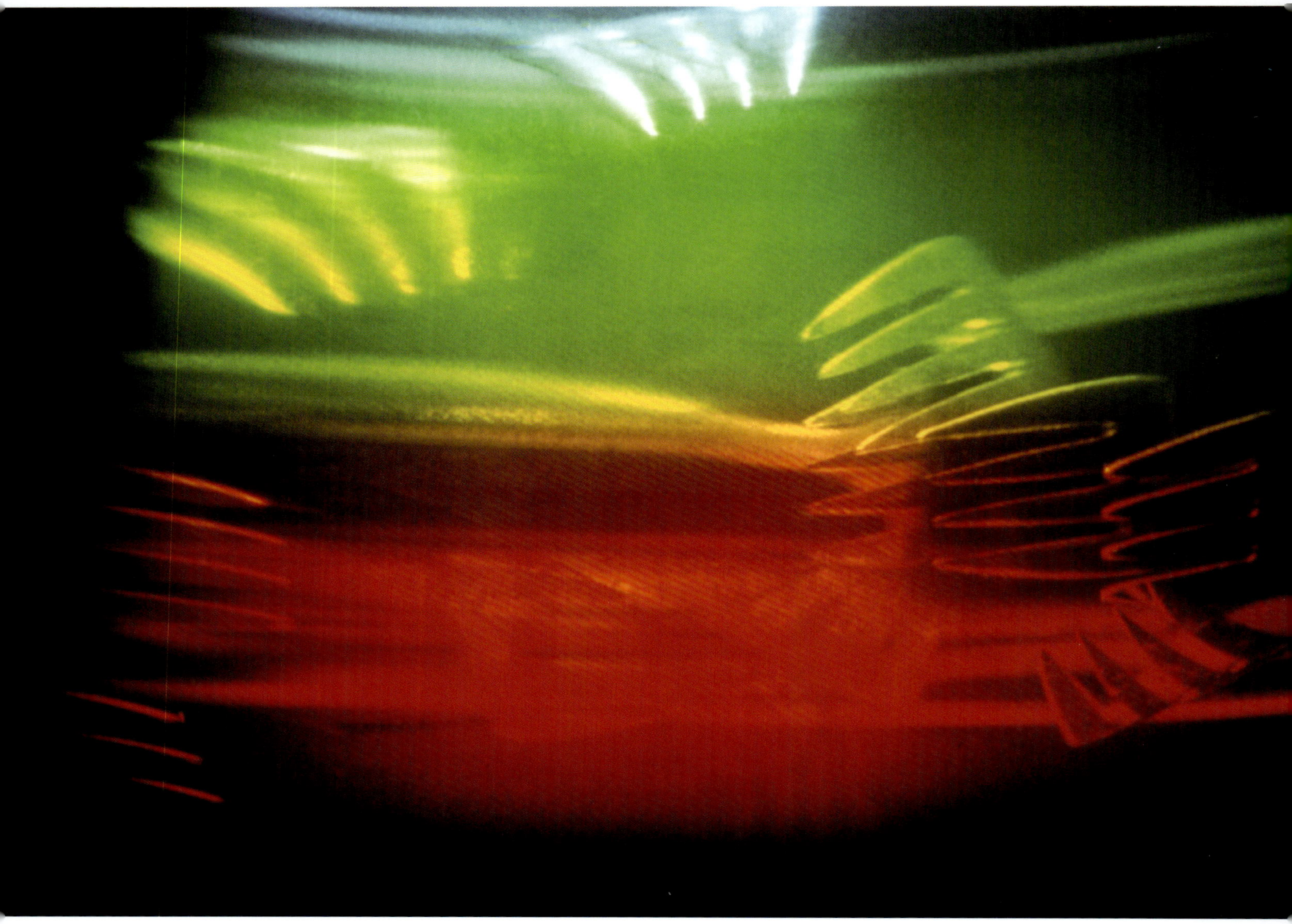

Harriet Casdin-Silver developed holograms titled, "Equivocal Forks II," for *Centerbeam*. The "Holography Line" was a series of holographic plates, 12½-inch x 16½-inch, that projected imagery frontally toward the viewer. These white light transmission holograms used the sun or illuminators as light sources. Casdin-Silver's feminist imagery drew on an everyday domestic tool—forks—piercing dangerously forward, literally, in space.

> . . . plates must be sealed to protect the emulsion from the elements. Holograms have never been exhibited outside, nor have they been illuminated by sun-tracking devices. (Casdin-Silver, *Centerbeam* catalog, 1977, 77)

Both pages: Harriet Casdin-Silver, "Equivocal Forks II," white light transmission holograms, *Centerbeam*, CAVS at *documenta 6*, Kassel, 1977
Photos: Dietmar Löhrl

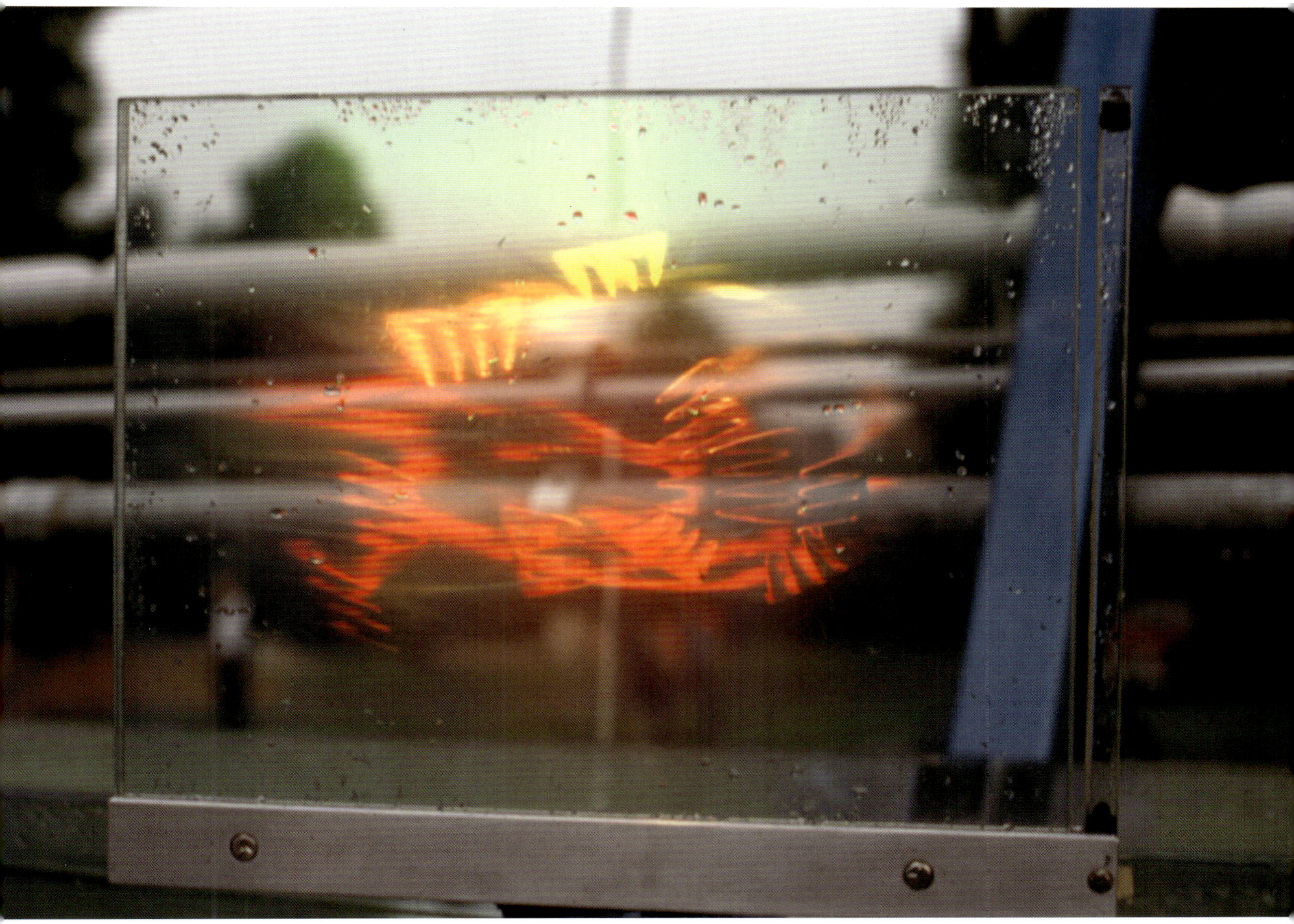

That was a huge experience for me: Ten "Equivocal Forks II" on this vast outdoor environmental sculpture. At my insistence the sun was to be the light source. Solar trackers for the solar holograms were invented by Walter Lewin, MIT astrophysicist. The artist and the scientist agreed that the aesthetics were as important as the mechanics. The holograms for *documenta 6* had been laminated but they were vulnerable to the outside with throngs of people around them every day. One hologram was stolen and another left hanging, but the thieves could not get it off *Centerbeam*. We did a lot of living! (Harriet Casdin-Silver interview)

It was important just to pose the question. *Centerbeam* was a good example of that: problems that Harriet had with the mirrors. Everybody had some problems—everywhere. But they had the dream, the vision of what they wanted. (Alejandro Siña interview)

ALL THE VERBS REWIRED MILES
WORDS TO SCAN LIKE LIVING DIALS
EVERY METER FLUCTUATES
THOUGH CONSTANT WITHIN THE DIODE'S GAZE
SOFT FLARE BYRON'S TENDER RAYS
WHEN IT RAINS OUTSIDE IT POURS
OUR LOVE IN A CLOUD OF DOORS
NEON STOCKING ROLL IN STEAM
ICE BLUE ANGEL DREAM
YOUR ART YOUR NATURE
MY SALT IN YOUR SHAKER

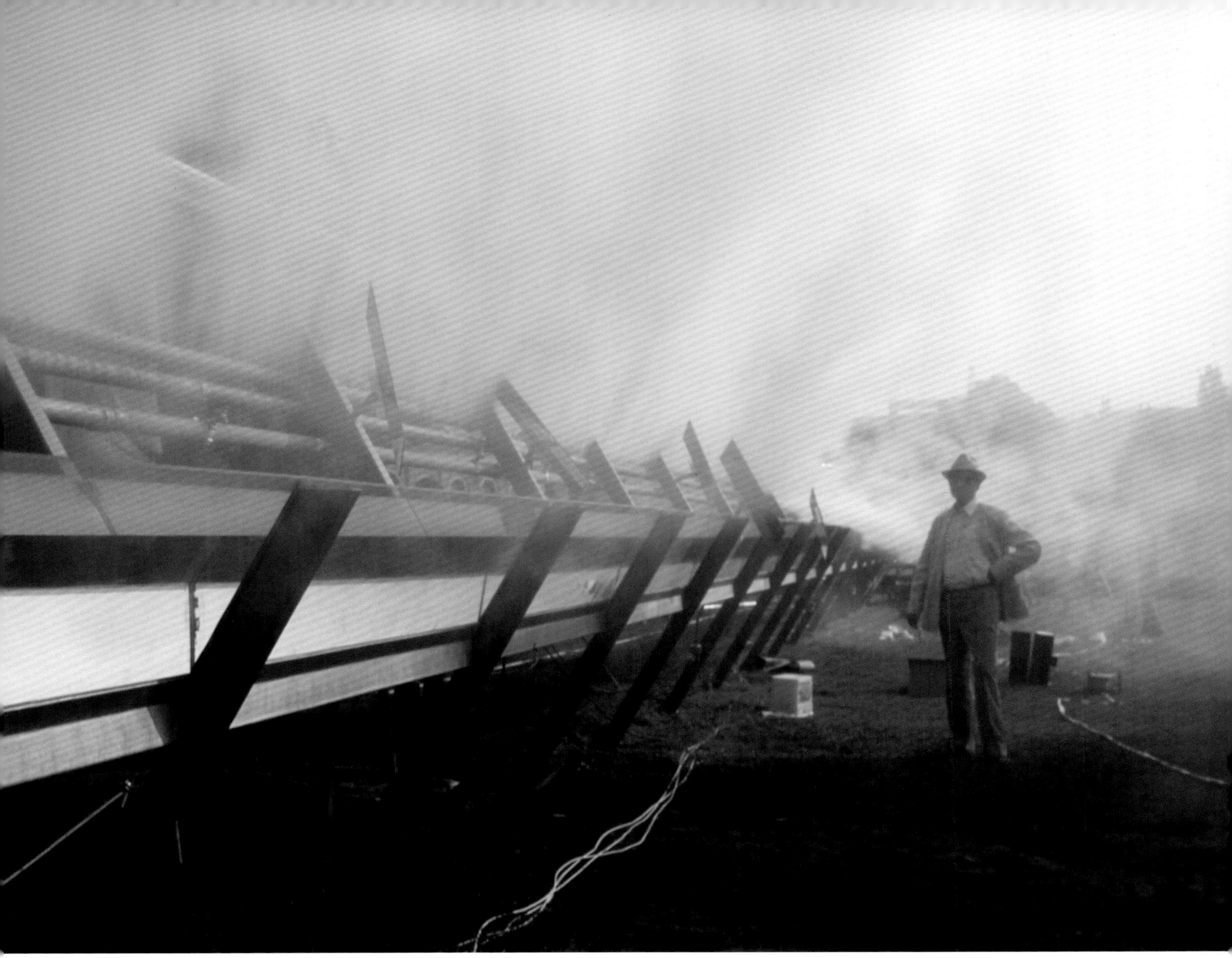

Centerbeam. I was working with the idea of trying to present a poem with no supporting structure around it. I wanted the poem to float in space, in the air. A book is a support structure for poetry; it has its own mediating influence on the way you read a poem, the ideation of a poem. I had always imagined ways of making a poem float in space, just the poem. . . . *Centerbeam* was an attempt to present in a pure way, a simple way without any structure around it that would mediate the way you saw it. You would just come upon it the way you would come upon a stream flowing or a truck driving by, part of our everyday experience. However, I think the specific poem on *Centerbeam* might be lost to people, but the idea of a poem floating on water, people could get that. (Mark Mendel interview)

Above: Joan Brigham, "Steam Line," steam animated the sculpture, and became a canvas for other media, *Centerbeam*, CAVS at *documenta 6*, Kassel, 1977
Photo: CAVS Special Collection

Facing page: Mark Mendel, "Poem On" poem etched onto eight mirrored glass rectangles, *Centerbeam*, CAVS at *documenta 6*, Kassel, 1977
Photo: CAVS Special Collection

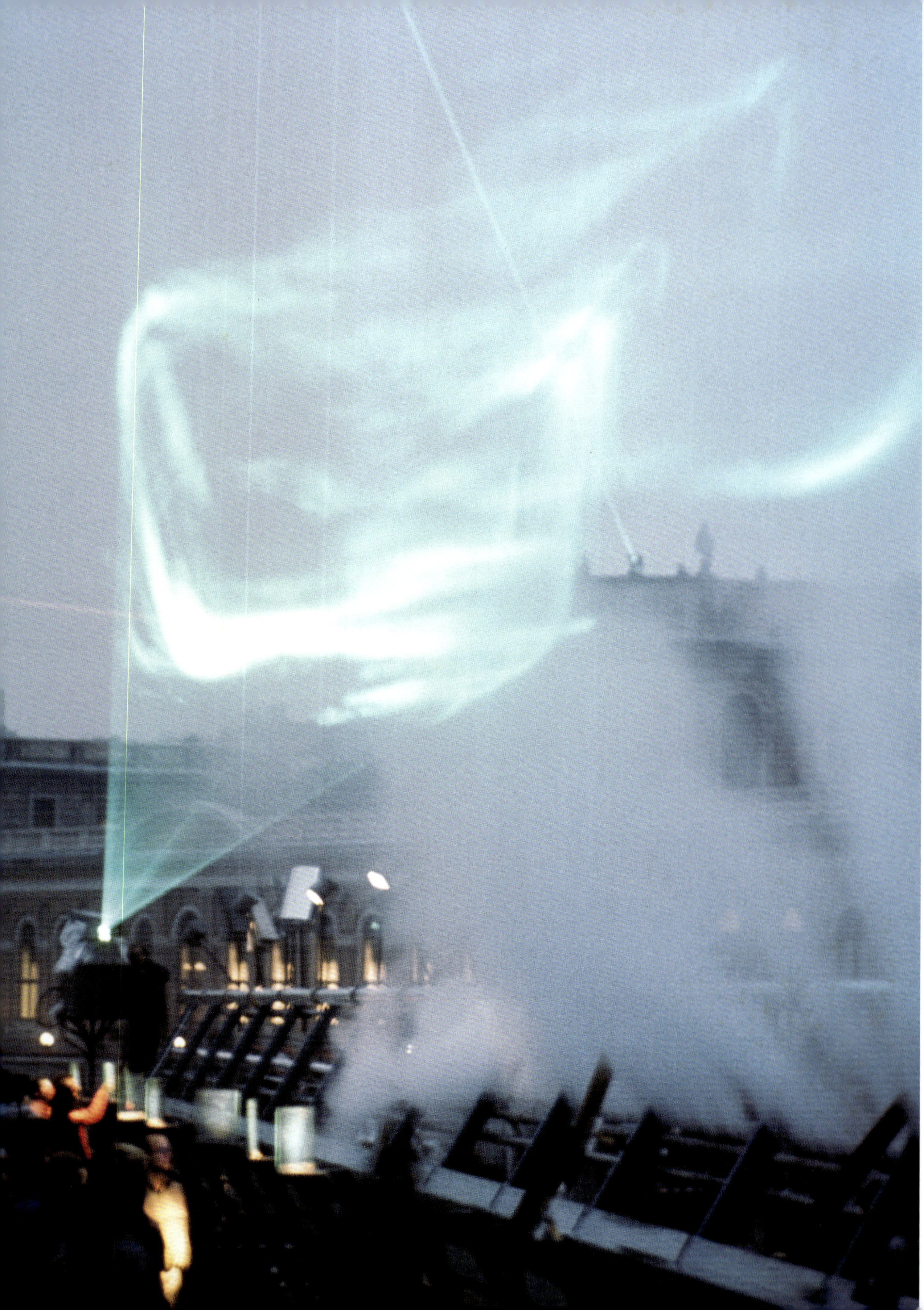

Centerbeam was a performing sculpture as well as a stage and stage set with machinery for events and performances. Events in Kassel were inter-media celebrations, such as Paul Earls' laser projections of imagery by György Kepes on steam by Joan Brigham. Spectacles included elemental events combining ice, steam, water spray, and compressed air.

> *Centerbeam* draws a kind of encyclopaedic sum from about twenty elementary powers and energies, which are orchestrated into an impressive performance of sensations and participation. . . . It is evident that without exception the emphasis is on the sensual-visual qualities of elements and energies, and it is these which have been clustered for the program. The degree of total visualization provides a touchstone of their transformation into artistic media. (Manfred Schneckenburger, *Centerbeam* catalog, 1980, 28-29)

> The form of *Centerbeam* is cyclopean, both in its vast size and in the sense that it monumentalizes a one-eyed vanishing point perspective. Viewed from the ends, it appeared to rush toward the horizon. Viewed from the sides it was an amalgam of aqueduct . . . pipeline, and wall. (Lawrence Alloway, *Centerbeam* catalog, 1980, 5)

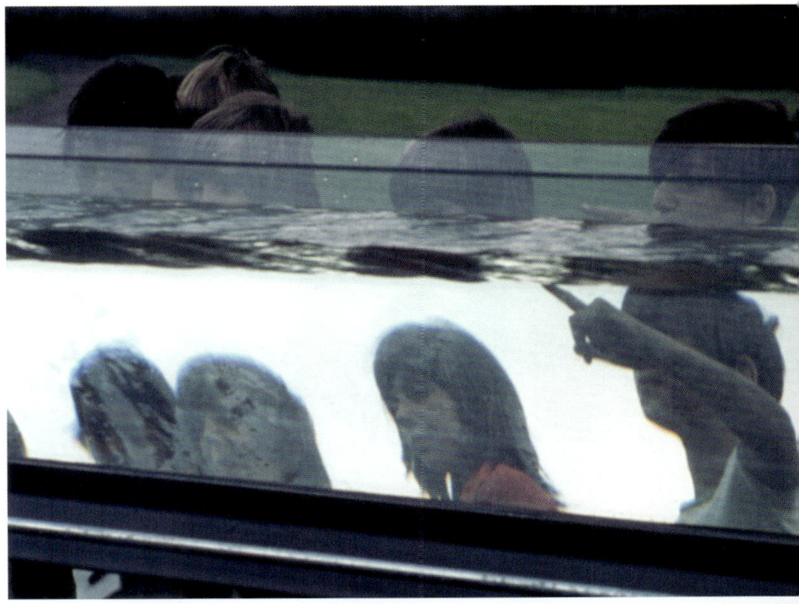

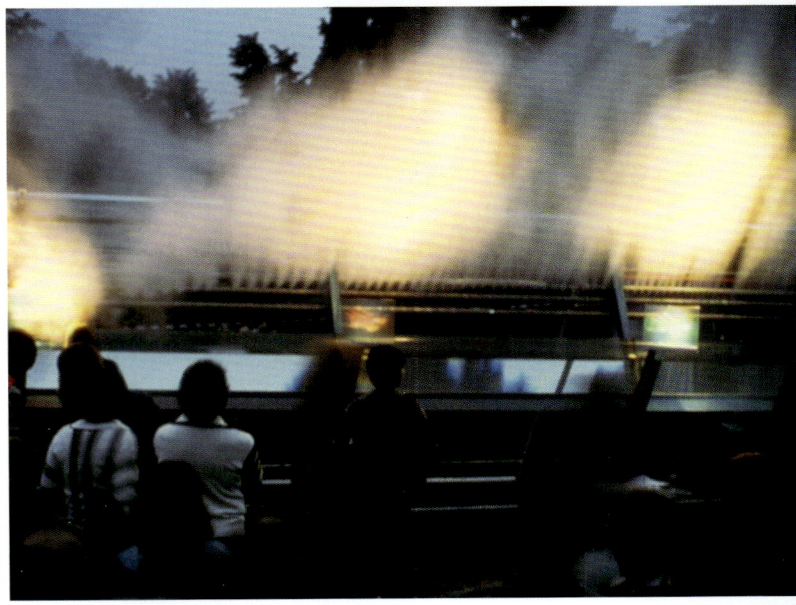

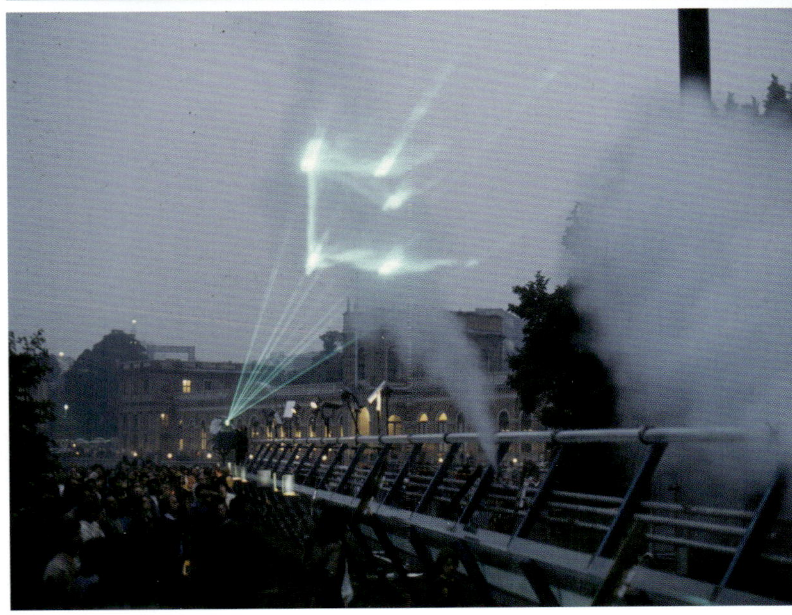

Top right: Children play with the water prism, *Centerbeam*, CAVS at *documenta 6*, Kassel, 1977

Middle right: Night brought increased illumination to the steam and holograms, *Centerbeam*, CAVS at *documenta 6*, Kassel, 1977

Facing page and bottom: György Kepes' "Envelopes" drawn by Paul Earls' lasers on Joan Brigham's steam, *Centerbeam*, CAVS at *documenta 6*, Kassel, 1977

Photos: Dietmar Löhrl

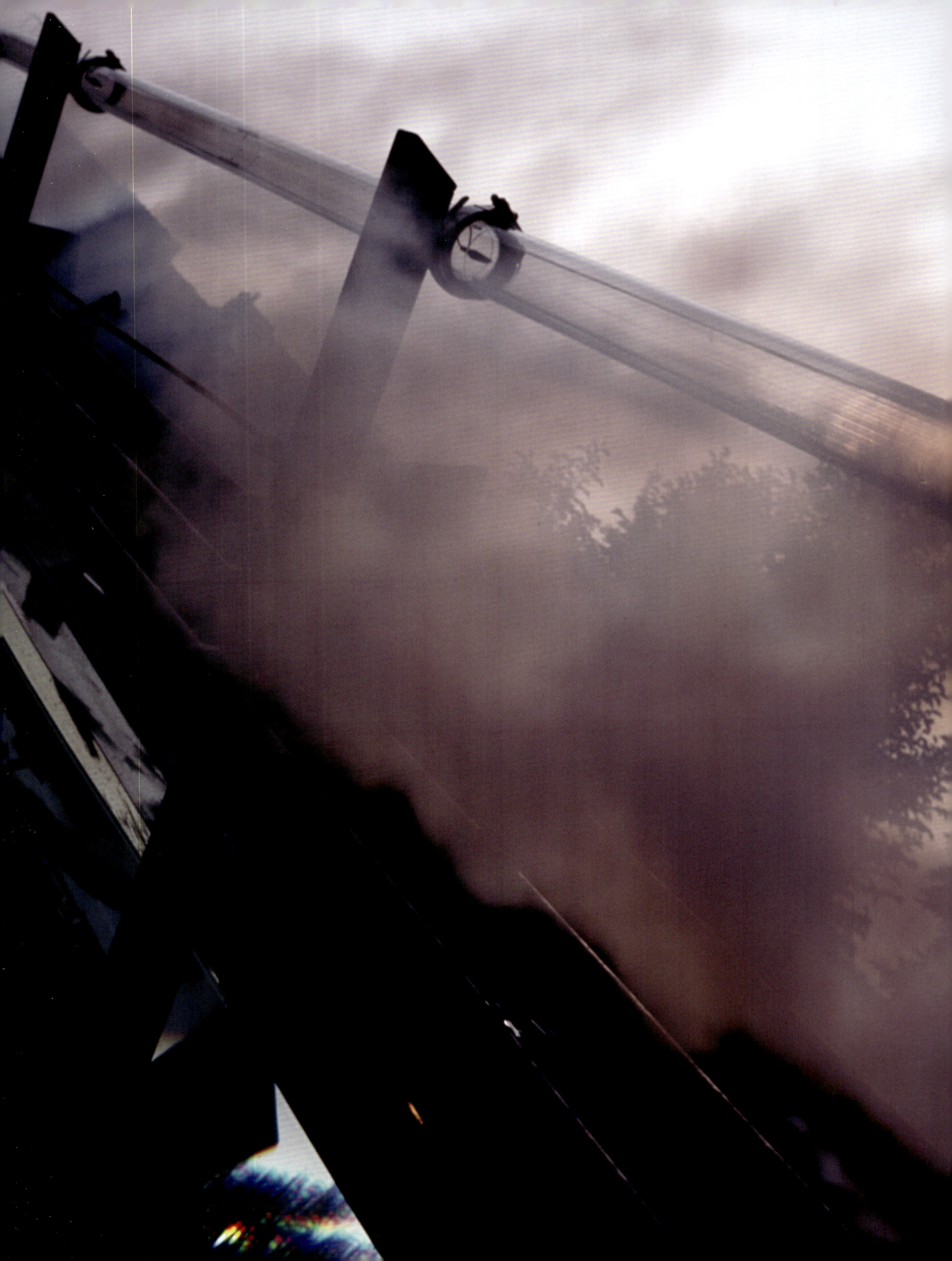

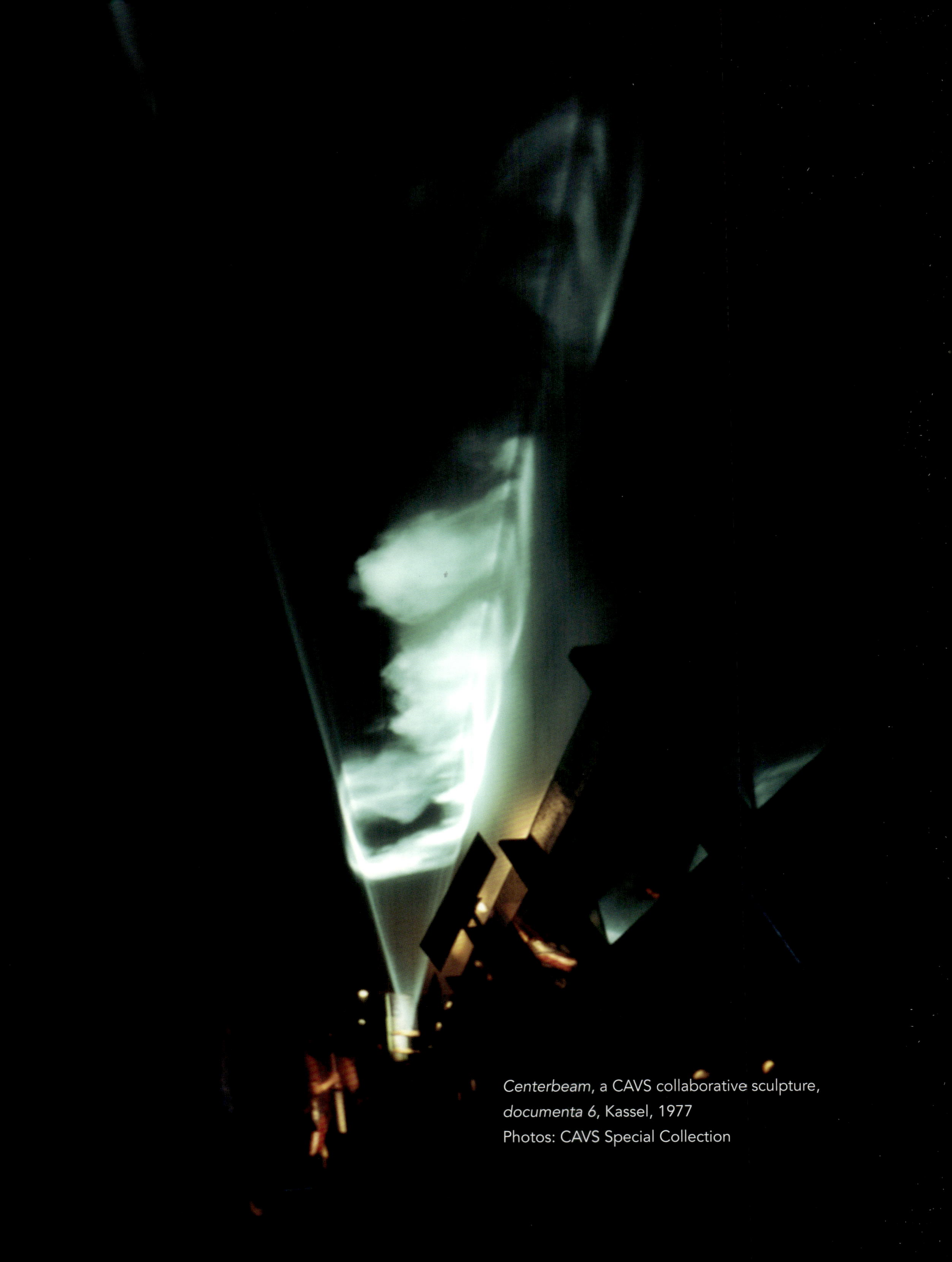

Centerbeam, a CAVS collaborative sculpture,
documenta 6, Kassel, 1977
Photos: CAVS Special Collection

Centerbeam, National Mall, Washington, DC, 1978

An expanded *Centerbeam* envisioned for the National Mall in Washington, DC, became a reality with support from the National Endowment for the Arts, the US Information Agency, the Smithsonian Institution, and the National Park Service.

For three and a half months in 1978, *Centerbeam* operated day and night. Artists working together shared the spectacle with visitors to the National Mall within sight of the US Capitol. Twenty-two artists, along with scientists and engineers, assistants, and students, participated.

Regularly scheduled performances featured laser and film projections on steam clouds and flying sculptures (sky flowers). Piene's twenty-three sky events became harbingers of the *Sky Art* conferences to come. Theatrical productions included the sky opera, *Icarus*, and the fairy tale opera, *Bremen Town Musicians*, with floating inflatable costumes, laser imagery, and singers and musicians amplified by sound and light media.

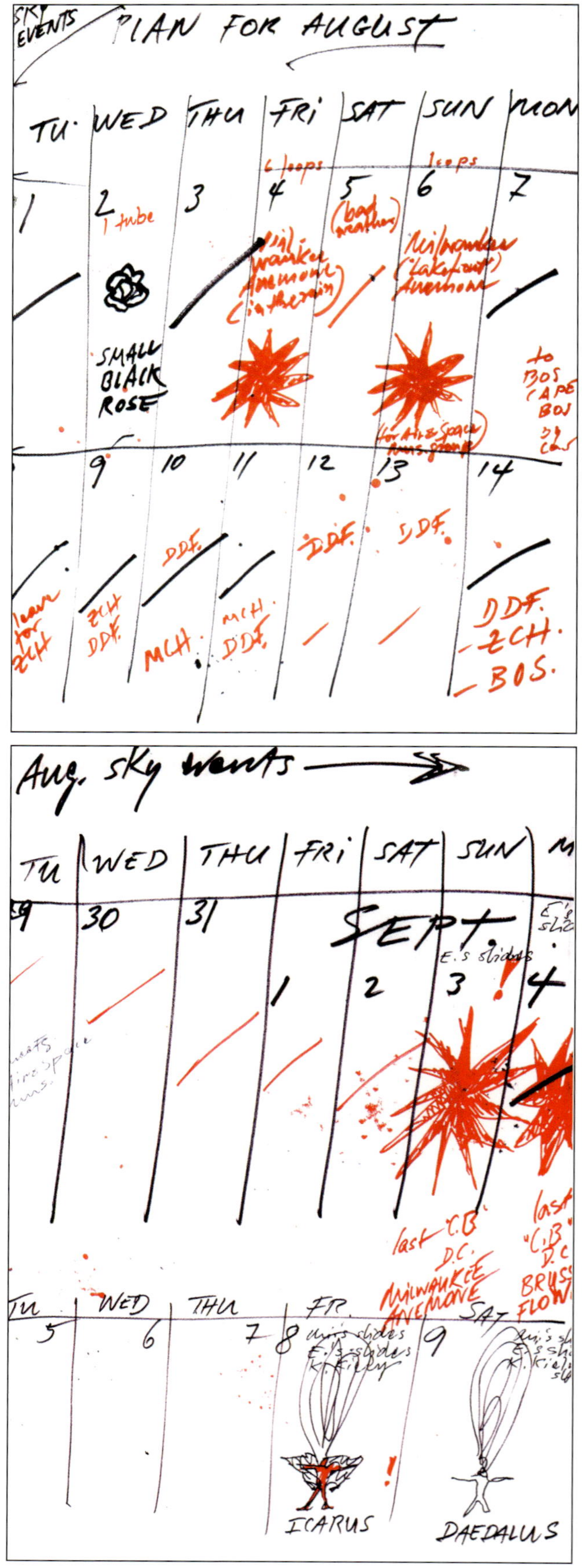

Right: Otto Piene, "Calendar for D.C. Flowers/Sky Events," two pages from Piene's sketched sky events schedule for *Centerbeam*, National Mall, Washington, DC, 1978
Photos: Ante Glibota

Facing page: Otto Piene, "Milwaukee Anemone" inflatable sculpture, Paul Earls, laser projections, and Joan Brigham, steam, performances and sky events, *Centerbeam*, National Mall, Washington, DC, 1978
Photo: CAVS Special Collection

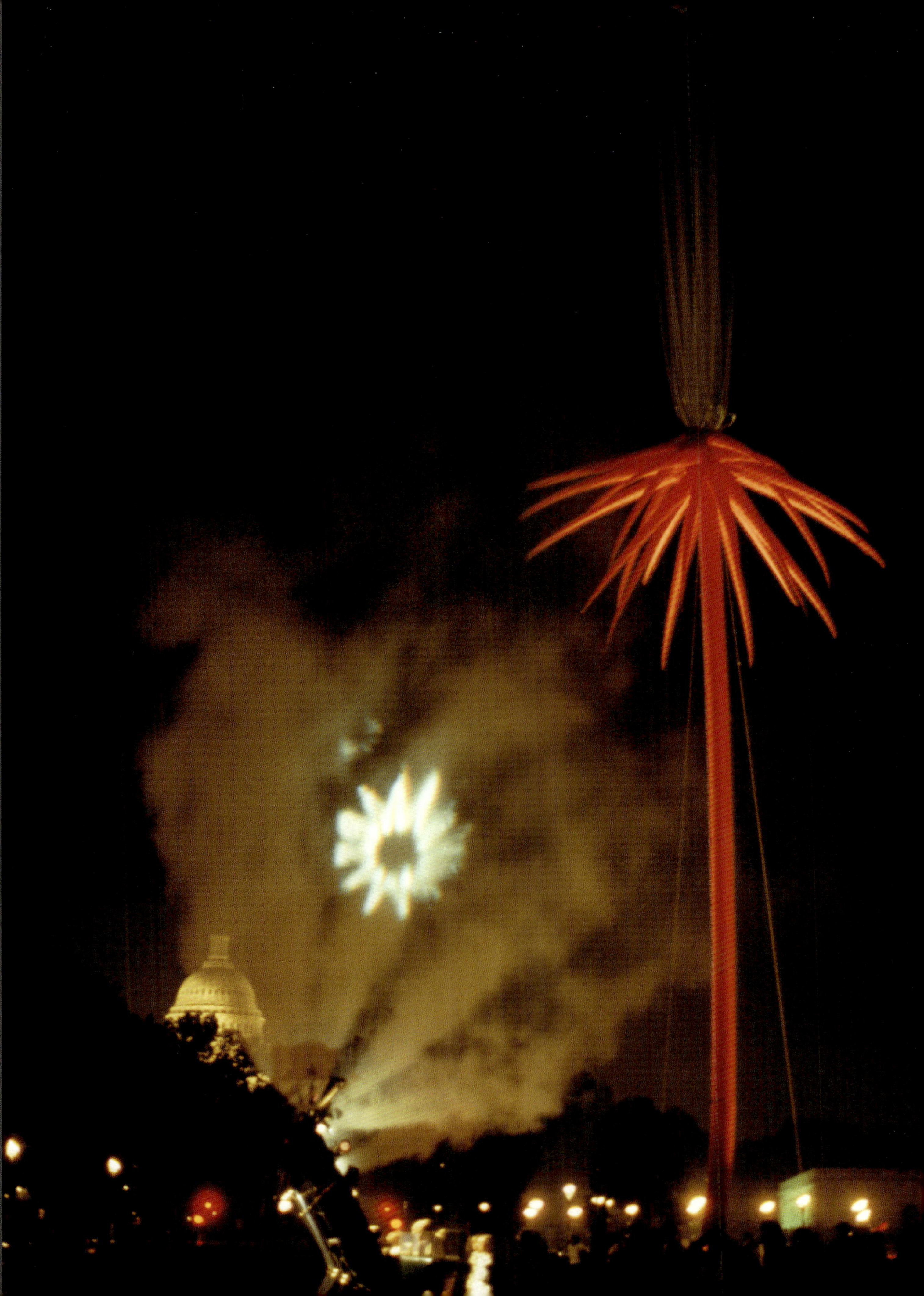

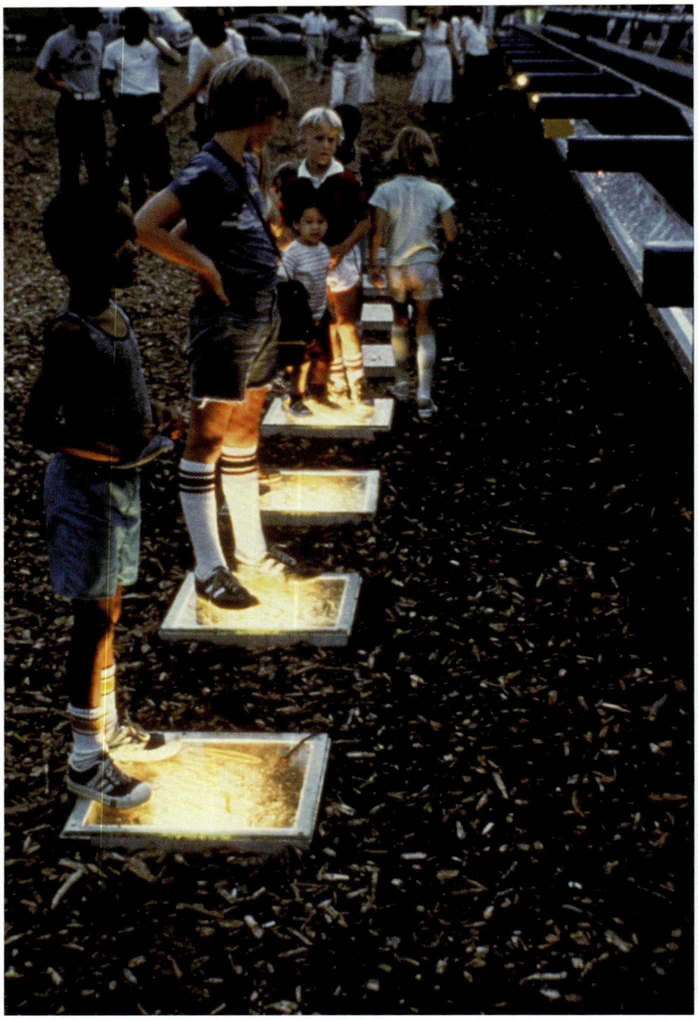

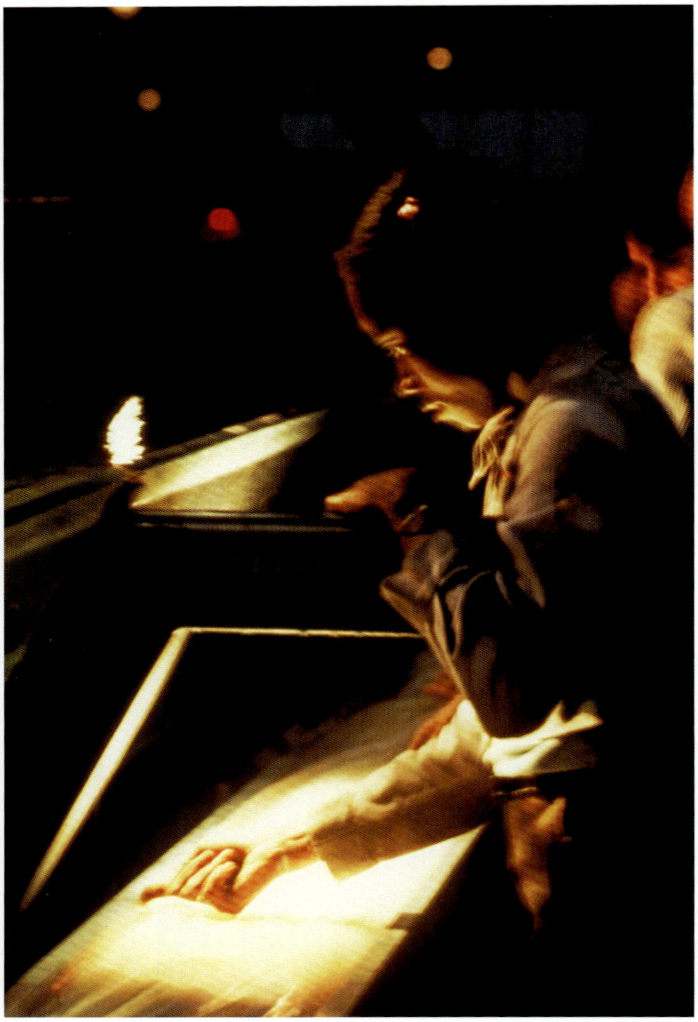

The science and technology component of *Centerbeam* remained raw and visible, process revealed, in an artwork that breathed and changed, froze and melted, and spoke with sound and words written in steam and light. It changed with the outdoor environment, responded to foot traffic and touch, and captured visitors' images on video in its generative system:

> I don't think of *Centerbeam* as being a collaborative work, I think of it as a group work. Lowry's concept for *Centerbeam* was the musical staff. He laid down five little strips of balsa wood on a table and said, "This is a music staff on which everybody gets to hang their stuff." And so his concept was ideal for the Center, because it left everybody free. Nobody had to be pigeonholed into doing something they didn't want to do, or put something where they didn't want to put it. Everybody was free to put what they wanted wherever they wanted on this structure. And the result and interactivity of all these elements was what it was. (Christopher Janney interview)

Above: Children interact with *Centerbeam*, left, Christopher Janney's "Sound Shuffle"/"Light Walk," right, Paul Matisse's kalliroscopic "White River," *Centerbeam*, National Mall, Washington, DC, 1978
Photos: Calvin Campbell, MIT

Facing page: *Centerbeam* at night, with laser projections on steam, National Mall, Washington, DC, 1978
Photos: K. M. Kiely

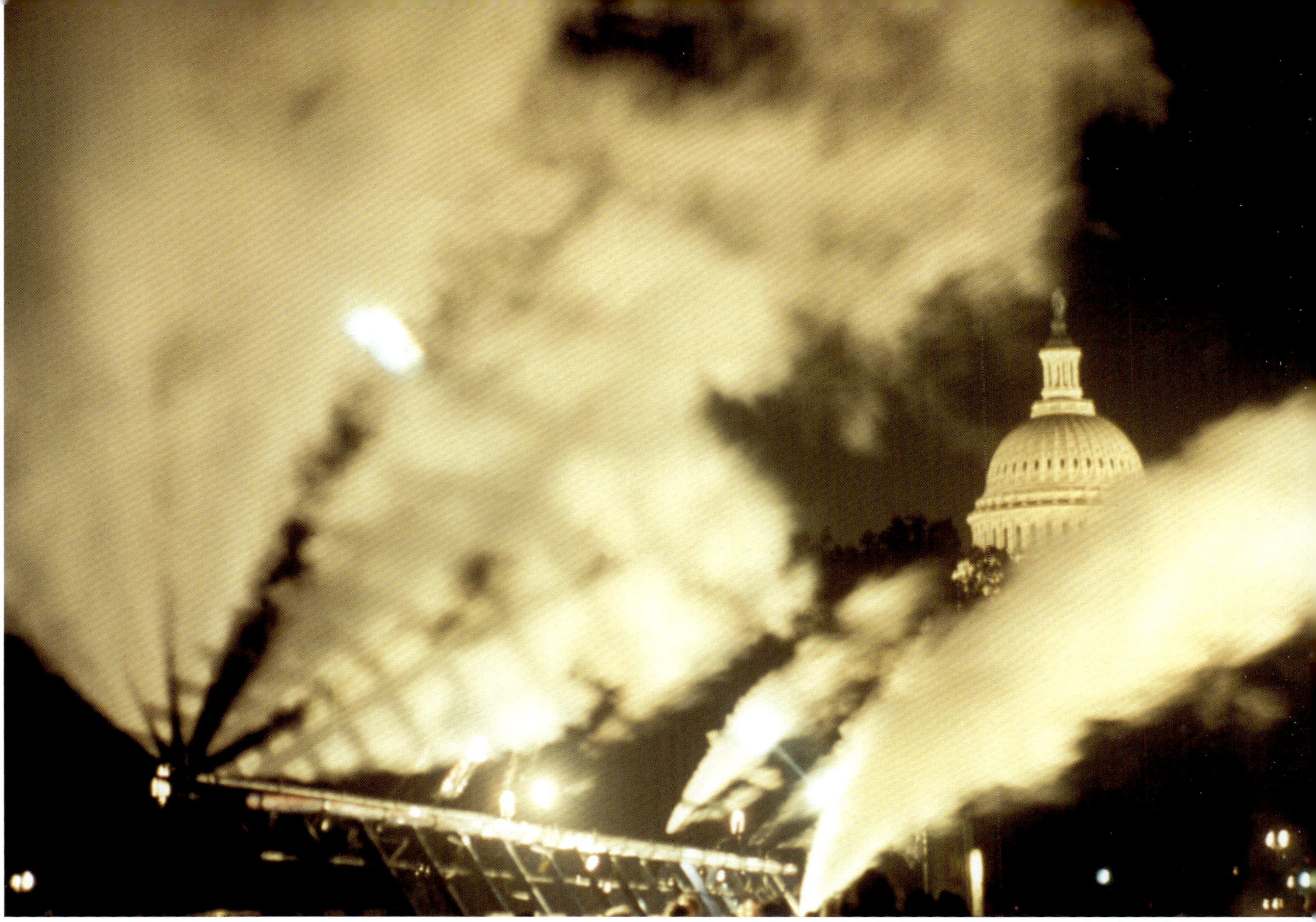

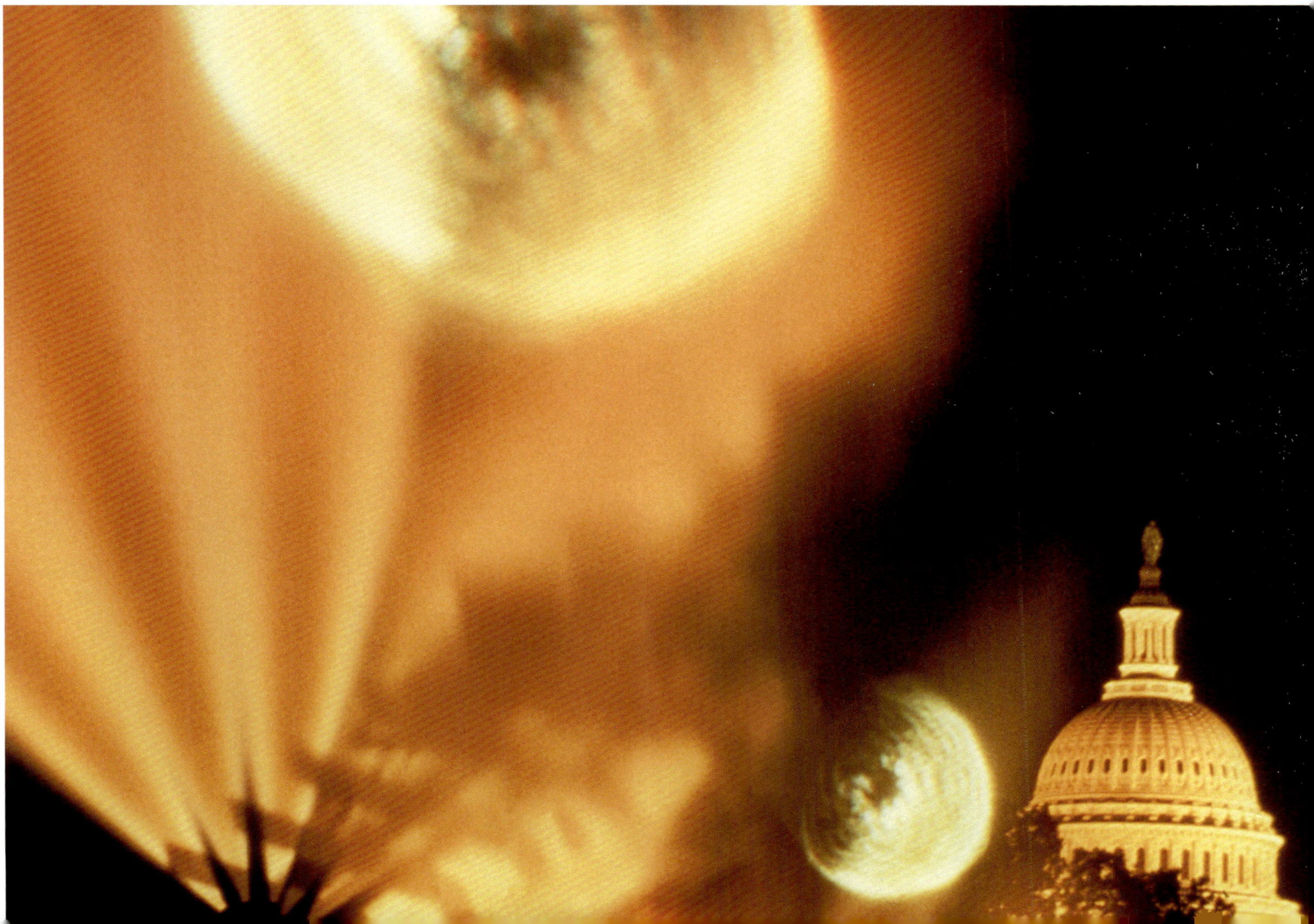

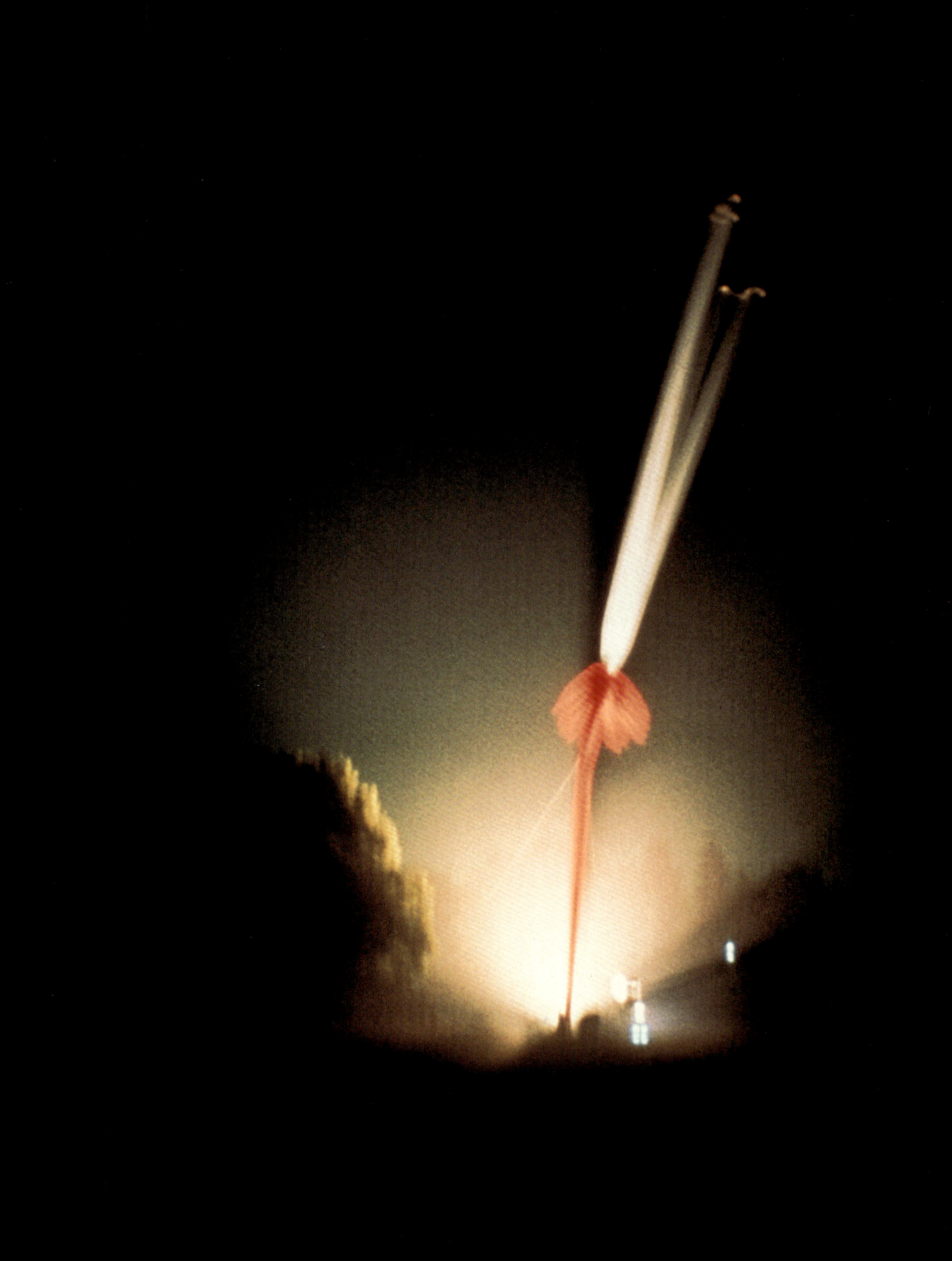

The additional artists (listed here) brought the total for *Centerbeam* Washington, DC to twenty-two:

Mark Chow	Video
Betsy Connors	Video
Alva Couch	Laser Line
Derith Glover	"Drawing with Your Eyes"
Christopher Janney	"Sound Shuffle"/"Light Walk"
Paul Matisse	Kalliroscopic "White River"
Michael Moser	Video
Muntadas	Video
Don Thornton	Holography Line

The sculpture was revisited in a 2017 exhibition and symposium, *Centerbeam. A Performative Sculpture by CAVS*, at the ZKM | Center for Art and Media Karlsruhe, which considered not only the historical weight of the work, but also the challenges for the archival preservation of such complex multimedia artworks. Under the direction of Peter Weibel, the event was curated by Lívia Nolasco-Rózsás, Morgane Stricot, and Philipp Ziegler in cooperation with the MIT Museum, MIT Program in Art, Culture and Technology, and PAMAL (Preservation & Art—Media Archaeology Lab), Avignon. The ZKM exhibition of *Centerbeam* is covered in Chapter 8 of this book.

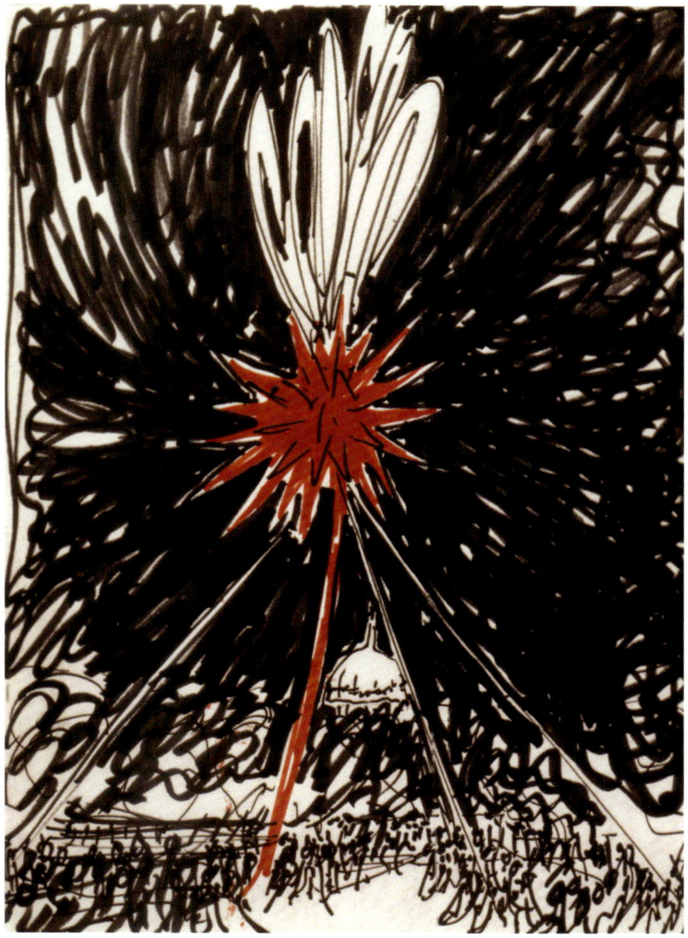

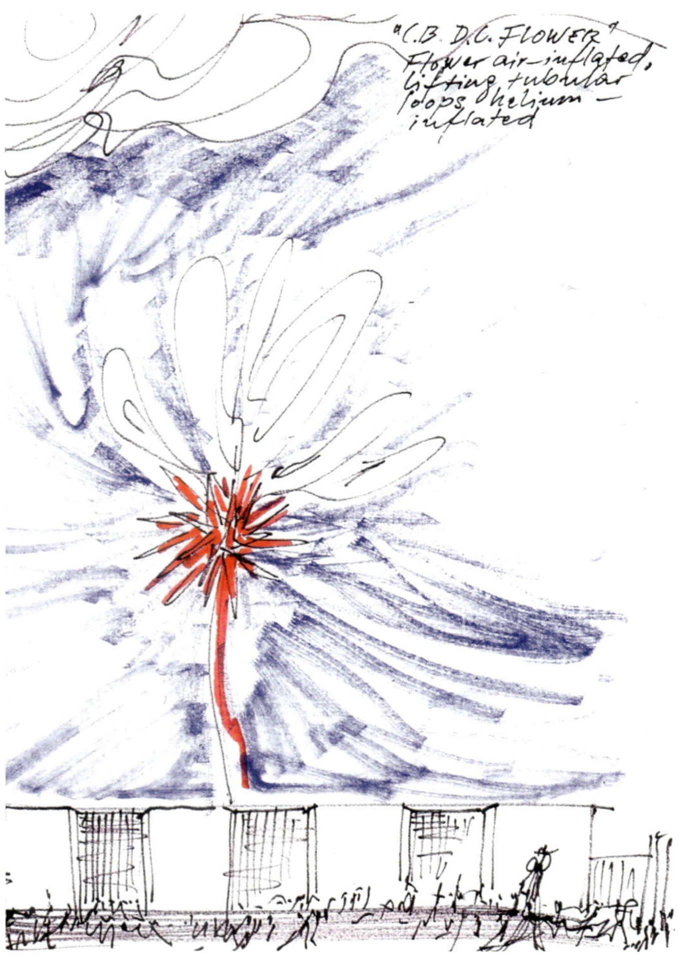

Top right: Otto Piene, "Brockton Flower," a 1979 drawing of inflatable sculpture from 1978 sky events at *Centerbeam*, Washington, DC
Photo: Ante Glibota

Bottom right: Otto Piene, "C.B. D.C. Flower," drawing (detail), dated 1978 by Piene on drawing
Photo: Ante Glibota

Facing page: Otto Piene, "Flower" inflatable sculpture, opening ceremony *Centerbeam*, CAVS at *documenta 6*, Kassel, 1977
Photo: Alejandro Siña

CAVS Artists' Interactions

Collaboration among several artists or an artist working with scientists or engineers was a founding principle of CAVS.

> The newly established Center for Advanced Visual Studies, which provides for collaborative work relationships between artist on the one hand and scientists and engineers on the other, may seem an extraordinary undertaking for an institution dedicated to science and engineering. We live in a time when many traditional conceptions have had to undergo revaluation, however, in response to emerging cultural needs. The lines of communication between art and science are too constricted, a situation as harmful for science as for art. Finding ways to increase the flow of ideas is a challenge and a necessity. (György Kepes, *The Center for Advanced Visual Studies* inaugural catalog, 1968, 1)

Group projects of the Kepes era such as the *Boston Harbor Project* and *Charles River Project* were collaborative in the sense of artists rubbing shoulders rather than creating a single work like *Centerbeam*. These projects colligated individual environmental artworks on a single theme in a public space. Throughout the history of CAVS, there were times when the conditions for collaboration were not favorable or artist fellows disinclined to collaborate.

> We had meetings and we would try to work out some group project, but never succeeded . . . I have no memory of anything more than discussion. In the first place, it was a rather hopeless idea for us to collaborate. After all, we were middle-aged people [disinclined] to change our ways. [Kepes] wanted us to collaborate, but he would have had to come up with a real project. (Harold Tovish interview)

> Collaboration was with MIT not other fellows. Kepes told me, "Well, you do your own research." I am not for collaborative work with other artists. I am extremely interested in collaborating with scientists and engineers. You learn a lot, you have answers to all kinds of things. (Piotr Kowalski interview)

Building W11's spacious studios were designed to foster creative research, and Kepes generally appointed one fellow per studio. Piene increased the opportunities for exchange by assigning two or three fellows per studio, stating, "Center artists [were] invited to be Fellows partly because of their inclination to work in groups." The Center's communal workspaces—the Pit, exhibition room, and alley—were hubs for creative interactions. The initiation of the graduate program, the influx of curious students who roamed greater MIT, and the added teaching responsibilities of fellows dramatically increased communication and exchange. Artists found enclaves to work within their fields outside of the Center, and collaborations proliferated across campus in specialized research labs, related teaching areas of the graduate program, fabrication workshops, and production studios.

> All of the projects had a single enduring quality as collaboration. In many respects the complex nature of the technologies we were using were impossible to do alone. Expertise could be drawn from any part of CAVS and MIT. Attempting to research and understand arts communication seemed to be the primary objective. What happens to works passed through different media? What impact do technologies have on image coherence? What is the documentation? Collaboration on these questions meant that anyone involved had to have a curiosity about these questions. (Ben Davis interview)

> You can hear all that stuff about collaborations, but to see it actually work . . . I remember the first time I talked to Shawn [Brixey] about collaborating. The first thing I said was, "I'm not going to be your switch! Because that's how dancers had been placed in technological environments: "If you go over here, the blue light will come on." . . . We had

to really keep going, going, going until we came to a mutual understanding . . . everybody feeling like it wasn't any one person's project, it was everybody's. It is rare. (Laura Knott interview)

Artist-inspired collaborations extended beyond MIT to non-artist organizations, as with Eric Begleiter's "Dimensional Foods," and to government institutions like NASA and the Smithsonian Institution. NASA commissioned projects by Lowry Burgess, and Gloria Brown Simmons, and supported the development of Elizabeth Goldring's seeing machine, which enabled people with little or no sight to view images of outer space. CAVS Fellow Joe Davis was the first non-scientist to address the Engineering Colloquium at the Goddard Space Flight Center:

> I was working on the shuttle project for the arts in my first year [at CAVS]. I built an electron gun for Ruby Falls at MIT's Ultra-High-Voltage laboratory and tested at the Air Force Geophysical Laboratory at Hanscom Field. (Joe Davis interview)

Light was a persistent theme and pivotal point in collaboration, as a projected and painted space, and as a transforming essence.

> My science work was becoming more and more involved with laser light. . . . My fourth year was almost exclusively working with Kepes. He gave me all the support anybody could ever get. I had my plasma work, taking glass shapes, and doing something inside them with gas and electricity [where] light interacts and moves shapes. I worked in the physics lab trying to understand this stuff from a scientific perspective. The nuts and bolts of physics—how the light is made, magnetic forces—became a physics dissertation, and then, borrowing from it for its visual phenomenology to make it into an art experience, something that could tell the story that had nothing to do with science, but which could not be made without it. (Bill Parker interview)

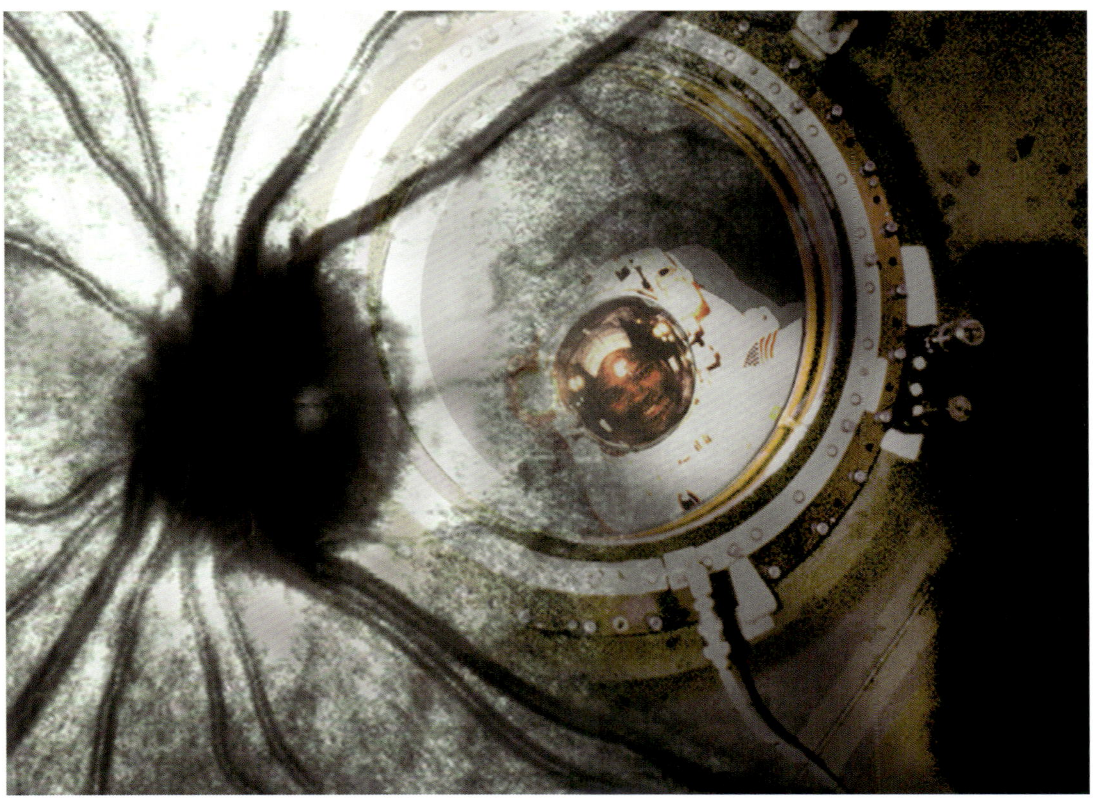

Elizabeth Goldring, "Eye to Eye (Astronaut)," retina print, commissioned by NASA, 2003
Source: the artist

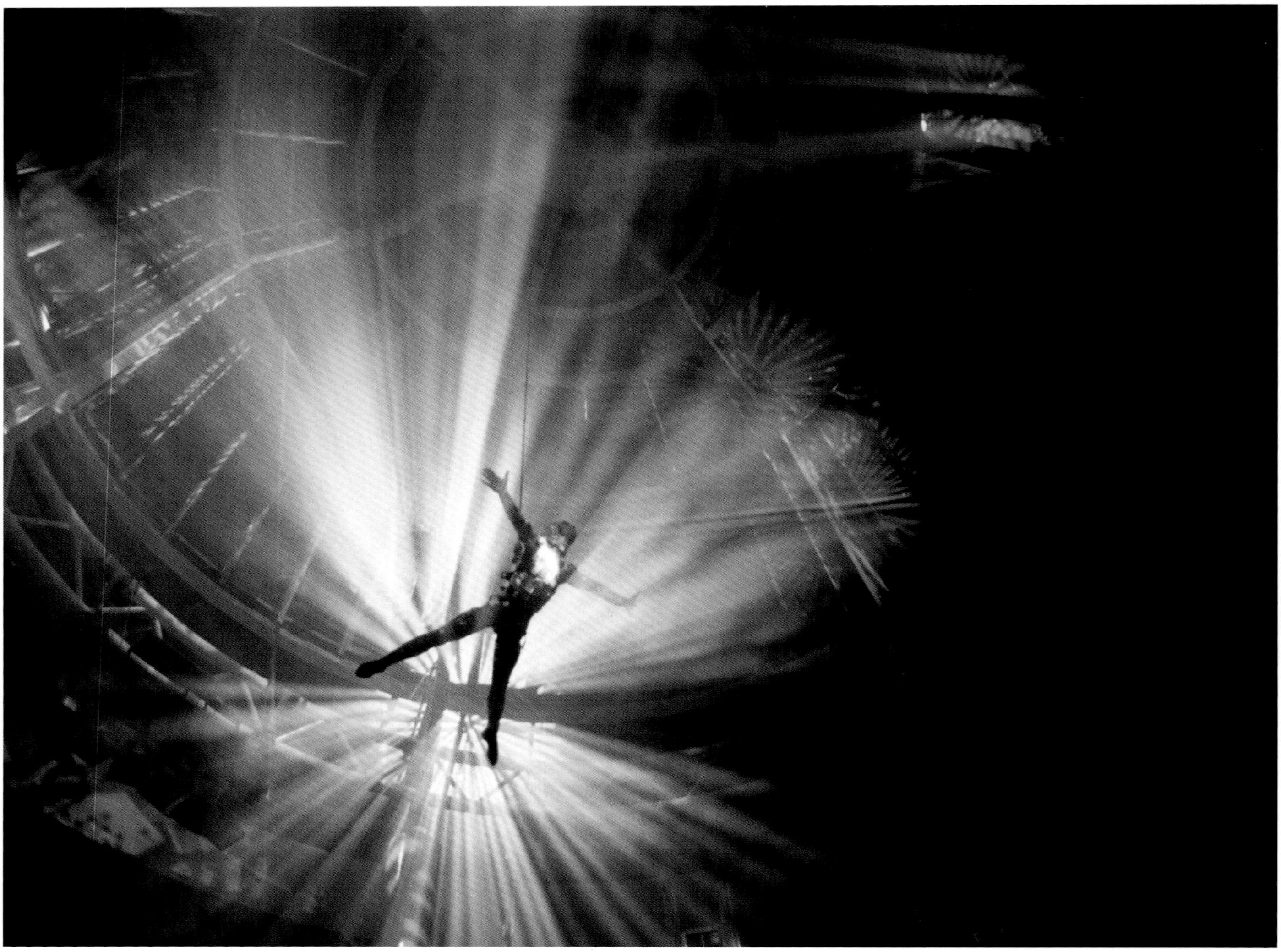

Above: Otto Piene, event with Seth Riskin performing "Light Dance," opening of Schadow Arkaden, Düsseldorf, 1995
Photo: K. Medau

What followed in my graduate career were many student projects that were elaborations of this artistic language of light, as well as more public projects that were collaborations with the residents of the Center. I worked on the parallel bars for my movements. I was using circular swinging and rotating movements, highly controlled and slowed so that the extensions of light from my body would move appropriately over a long distance. . . . I did collaborative work with Otto. We actually did flights, "Light Dances" in the night sky. . . . The flights were good, the collective experience of a light body in the sky was a kind of dream of humanity manifest. (Seth Riskin interview)

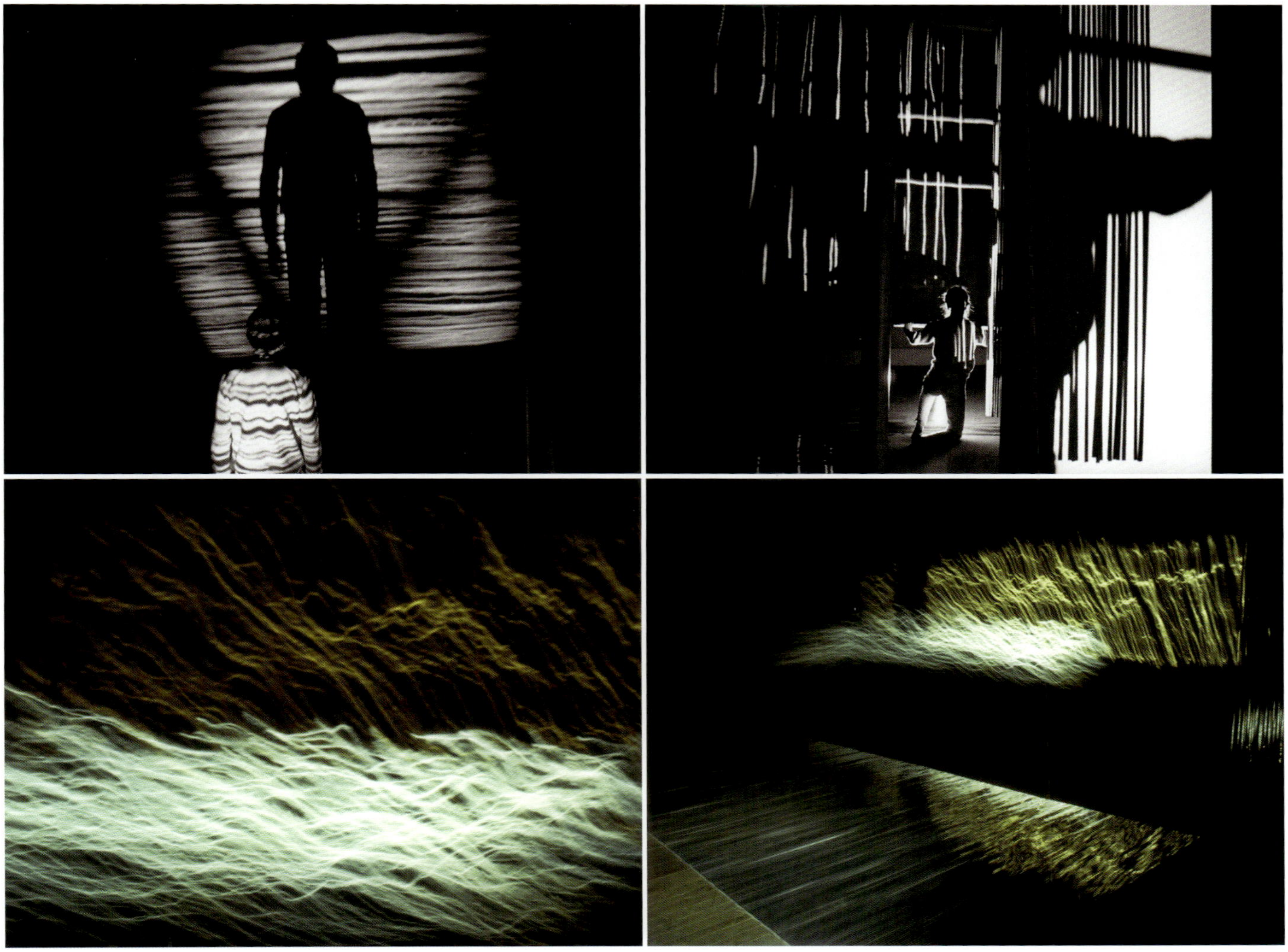

The idea of an object made by an artist that might exist within a museum or gallery space was one of the things that happened at the Center. There were connections to the art world of museums and galleries . . . the idea of a well-crafted object made for viewing pleasure was really important. The other thing was the idea of collaboration, celebration, team work and taking work out of a hermetically sealed situation and putting it out in the world and letting the viewer get involved. Once you start getting involved with the object and the space, the way light changes, the movement, the performance possibilities—BOOM!—you are into a conditional situation. Things can change, evolve and are not dependent on one person's efforts. It is only recently that art and technology caught on in a big way in the art world. (Beth Galston interview)

Top row: Beth Galston, "Mirror Light," with performers Galston and CAVS colleagues Luc Courchesne and Ellen Sebring, the Pit, CAVS, MIT, 1983

Bottom row: Beth Galston, "Lightwall" installation, Kingston Gallery, Boston, 1983
"Reflections evolve in color and intensity, evoking fire, clouds or an icy pond as wavelike light is transformed by four slide projectors that illuminate mirrored panels in slow dissolves." (Galston, description for Kingston Gallery, 1983)

Photos: Beth Galston

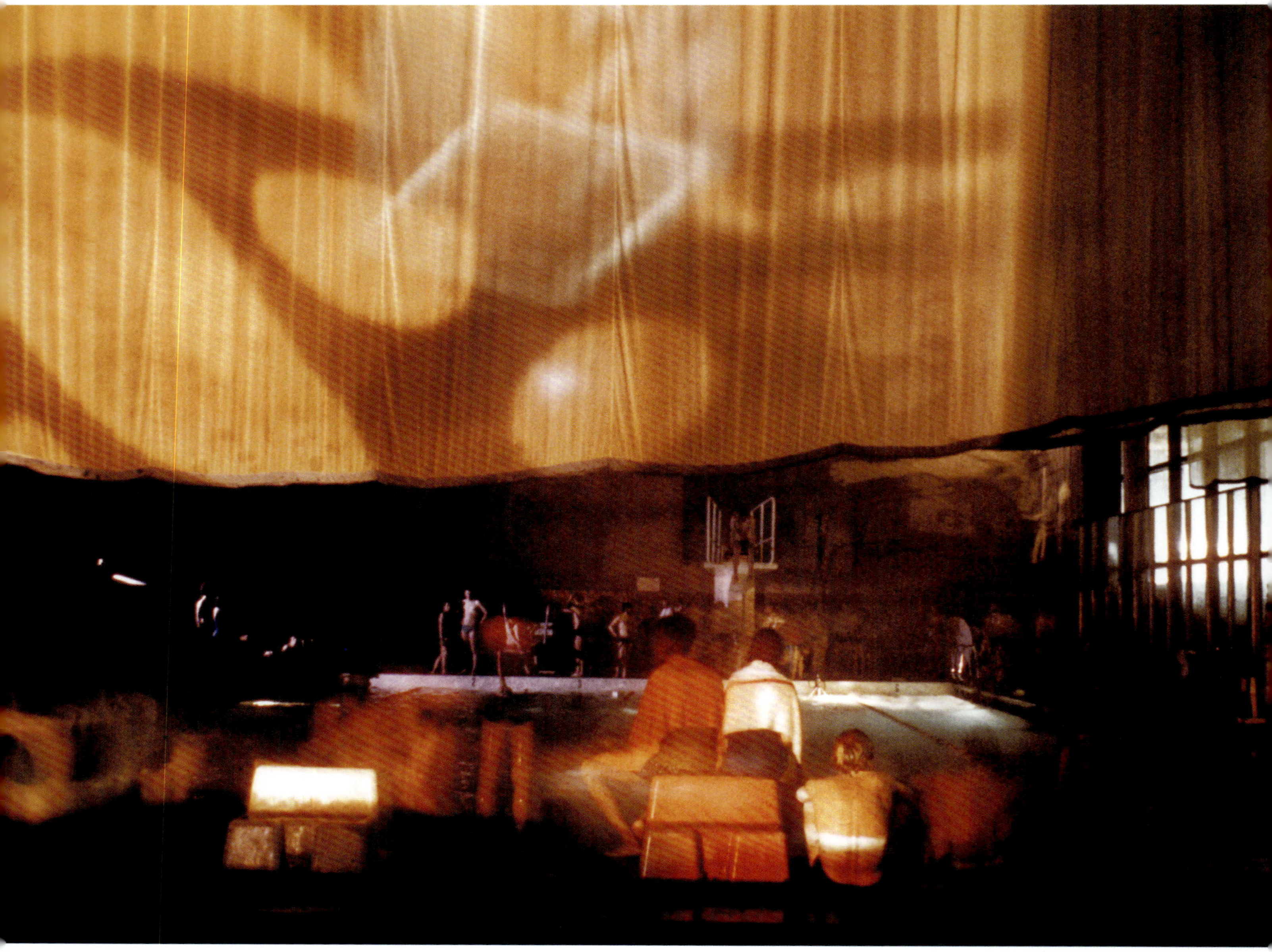

It's the behavior, it's the freedom of experimentation that went on with Black Mountain [College] . . . it was the same kind of slightly—not hostile—but, indifference towards the "professional artists" that we experienced at CAVS. It was, "Never mind your exhibits, what are you doing right now?" That kind of thing. "What are you interested in doing right now?" It was very open. It was that, and I'm sure that that's why Stan [VanDerBeek] was there. (Joan Brigham interview)

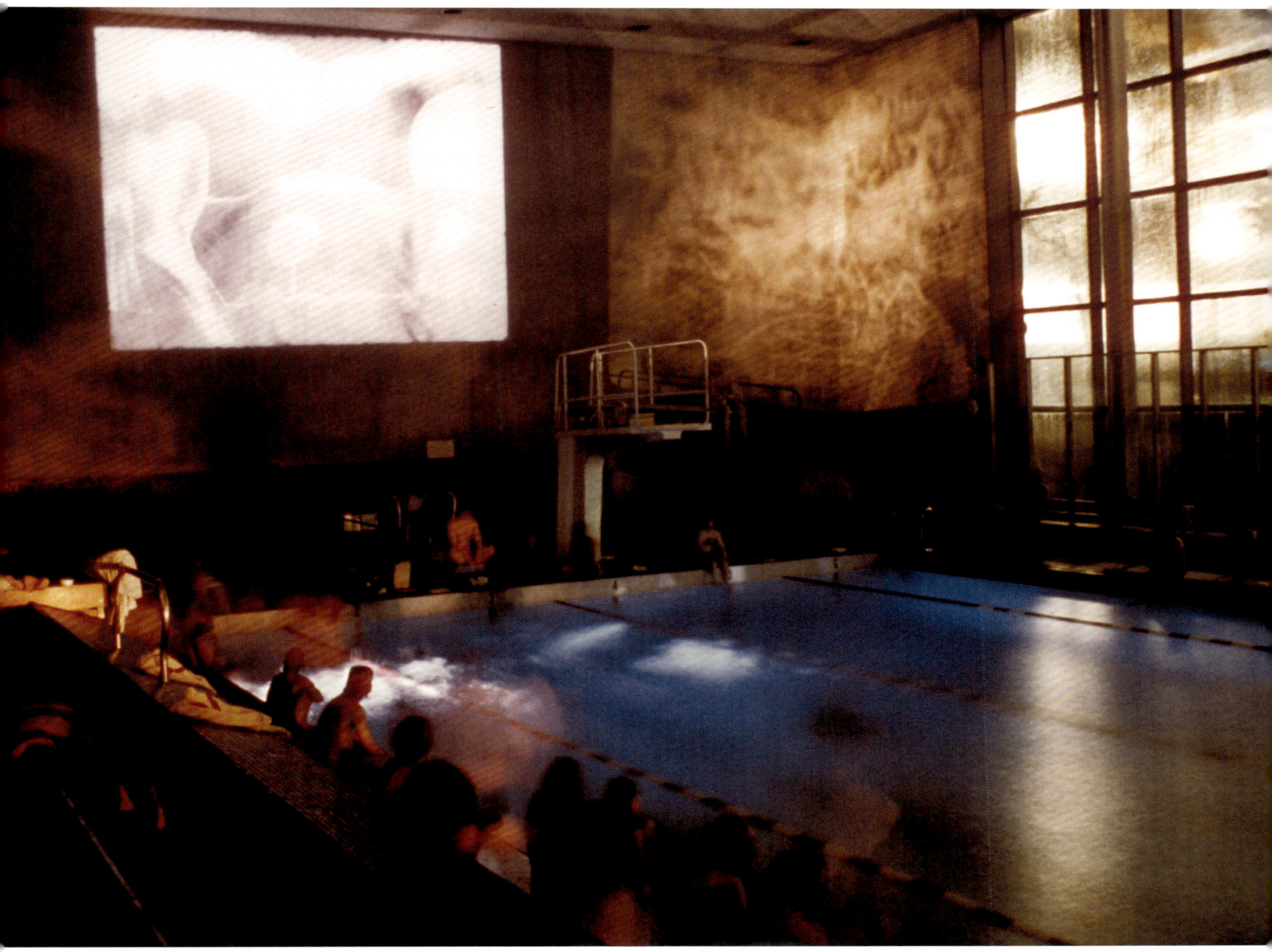

The darkened room was filled with steam on which multiple slide and film images were projected directly above the audience/swimmers while, underwater, strobe lights were installed, cyalume lights floated and were tossed among the swimmers. Underwater video and a sonar sound system completed the setting. The audience was entirely free to participate physically in any way, at any level of perception or cognition it wished, changing the conditions for its responses at will. (Joan Brigham description)

Both pages: Stan VanDerBeek and Joan Brigham, "Under Aquarius," participatory event with slides, film and video projections, strobe lights, and sound, MIT Alumni Pool, March 14, 7-10 pm, 1976
Photos: Otto Piene Collection

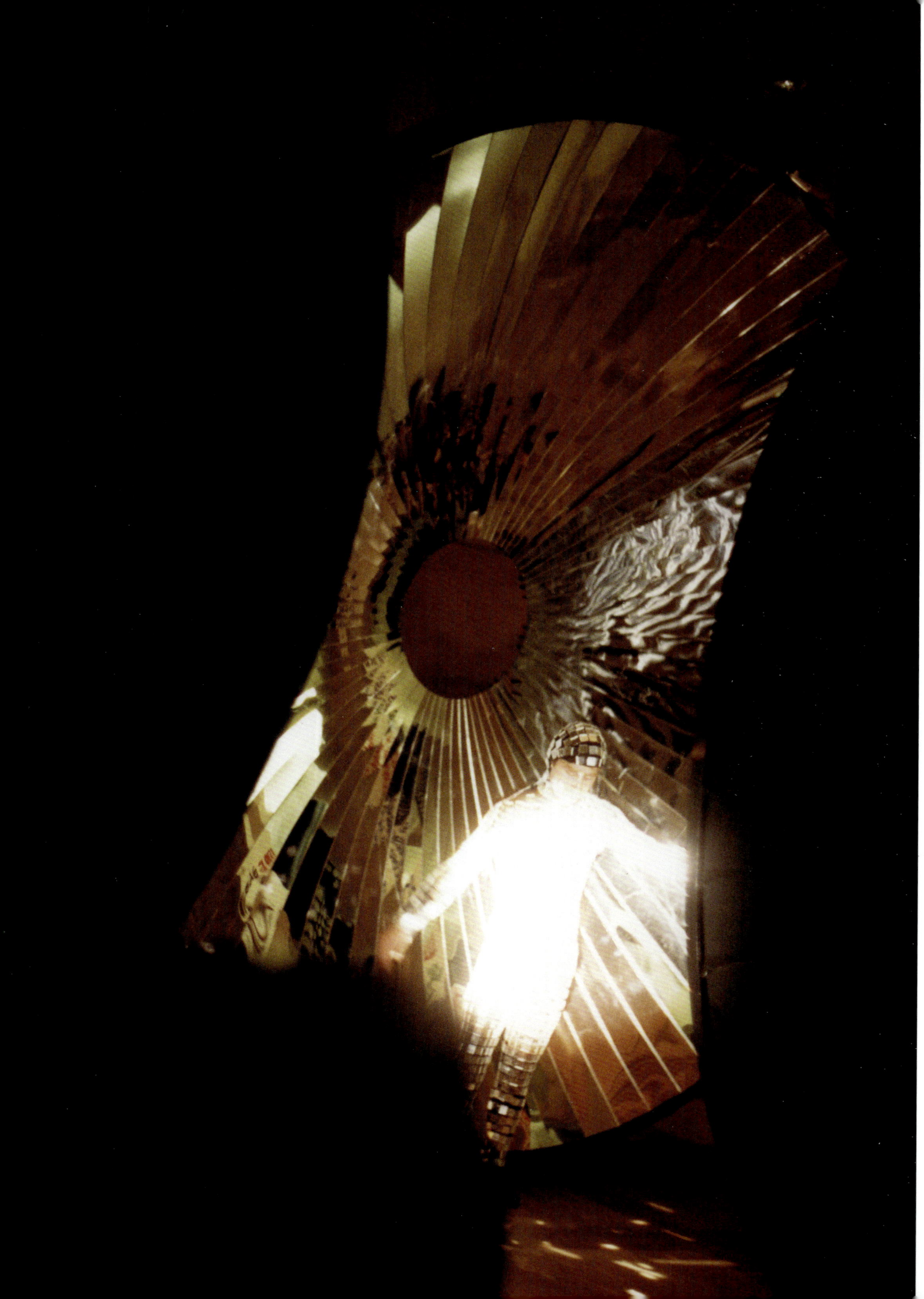

Monocle

The large-scale group projects of the 1970s and 1980s, including *Centerbeam*, *Monocle*, and *Desert Sun/Desert Moon*, were conceived as group artworks. *Monocle* (1985), a collaborative sculpture designed and executed at CAVS, formed the core of the exhibition *Mehr Licht* at the Hamburger Kunsthalle. Piene wrote in the catalog, "*Monocle* addresses primarily 'the noblest sense' (Leonardo)—the eye."

> By the time we created *Monocle* in 1985, we had done a few collaborative CAVS projects so we had a sense of how to make group projects together. The fellows had an initial meeting where Otto laid out the terms of the invitation, timetable and budget. As usual, the time was short and there wasn't a great deal of funding. But the exhibition, *Mehr Licht* ("more light," Goethe's last words) was spiritually and physically inspirational to everyone.
>
> Lowry brought in an 8-inch diameter novelty-shop item of two concave mirrors, showing how they gave the illusion of an object floating above. Everyone liked the "eye" aspect of it. Elizabeth suggested the concept/title of a "monocle." Otto then thought of former Fellow Joe Moss, and his concave parabolic sound-mirror work. Joe came up the next week and we started to muse about how we might make a set of twelve-foot parabolic mirrors. Joe said we could build them from scratch but time was an issue. We mused about using existing satellite dishes, but the budget was too tight to purchase them. Elizabeth then found a satellite dish company in North Carolina that was willing to donate two dishes. (Christopher Janney interview)

With directors Steve Benton and Krzysztof Wodiczko, CAVS collaborations took the form of affiliated groups within the Center, for example, the Interrogative Design Group headed by Wodiczko, and the Vision Group headed by Goldring. These somewhat autonomous groups included students, scientists, and engineers from both MIT and Harvard University. The experiences and experiments with collaboration held the CAVS community together even after the merger of CAVS with the Visual Arts Program, and the demise of the Center.

The legacy of CAVS often appears in speculative language written by Kepes and Piene about the need, challenge, and promise of a new type of artist and artwork positioned to open lines of communication in a technologically saturated time. *Centerbeam* and the collaborative thinking and actions of CAVS artists and their colleagues stand as tangible realizations of those aspirations.

Facing page: *Monocle*, a collaborative work by CAVS artists Otto Piene, Joe Moss, Christopher Janney, Lowry Burgess, and Paul Earls, with Elizabeth Goldring, featuring performer Seth Riskin, *Light/Space/Time: CAVS 25th Anniversary Celebration*, Lobby 7, MIT, 1994
Photo: Walter C. Dent

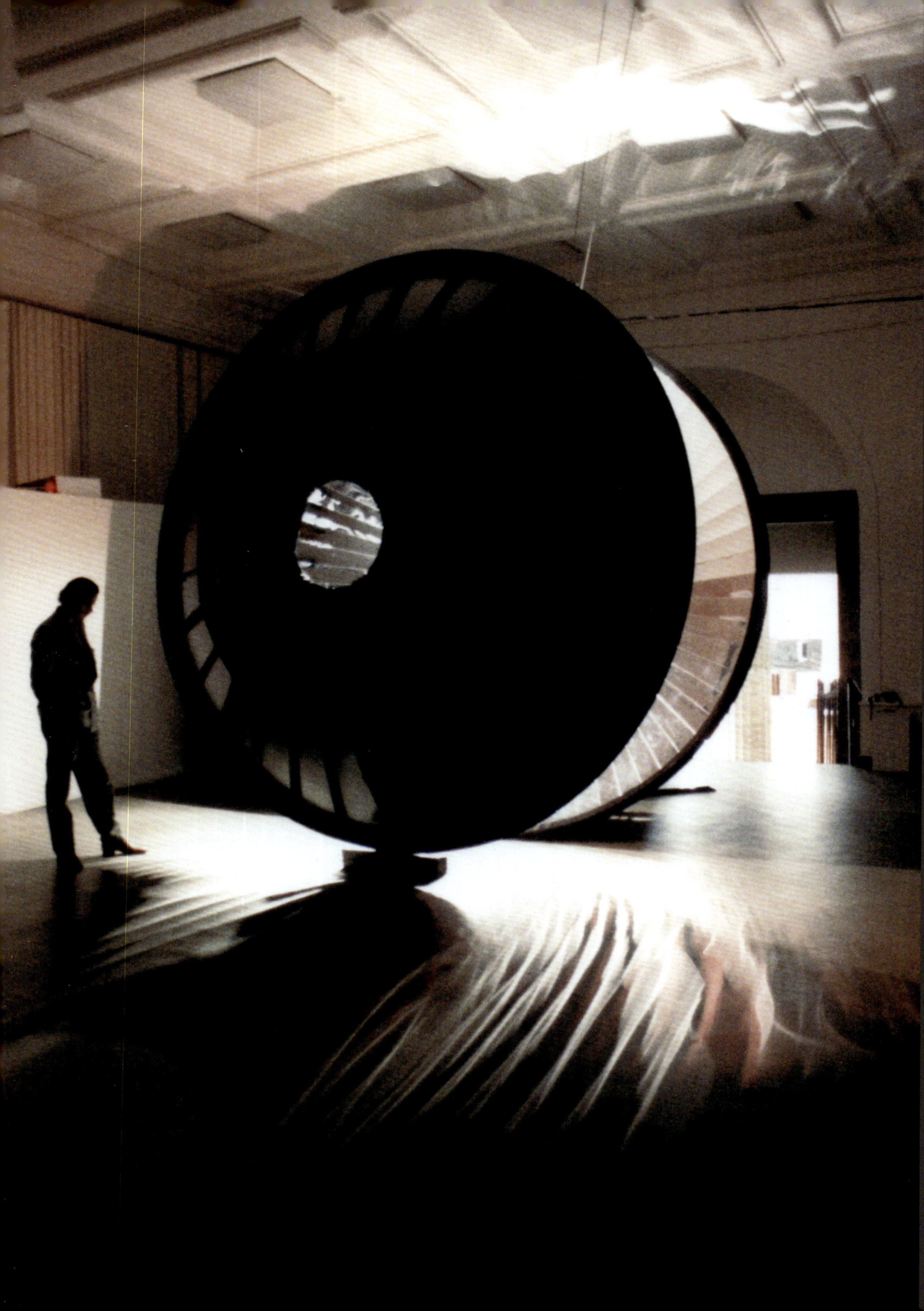

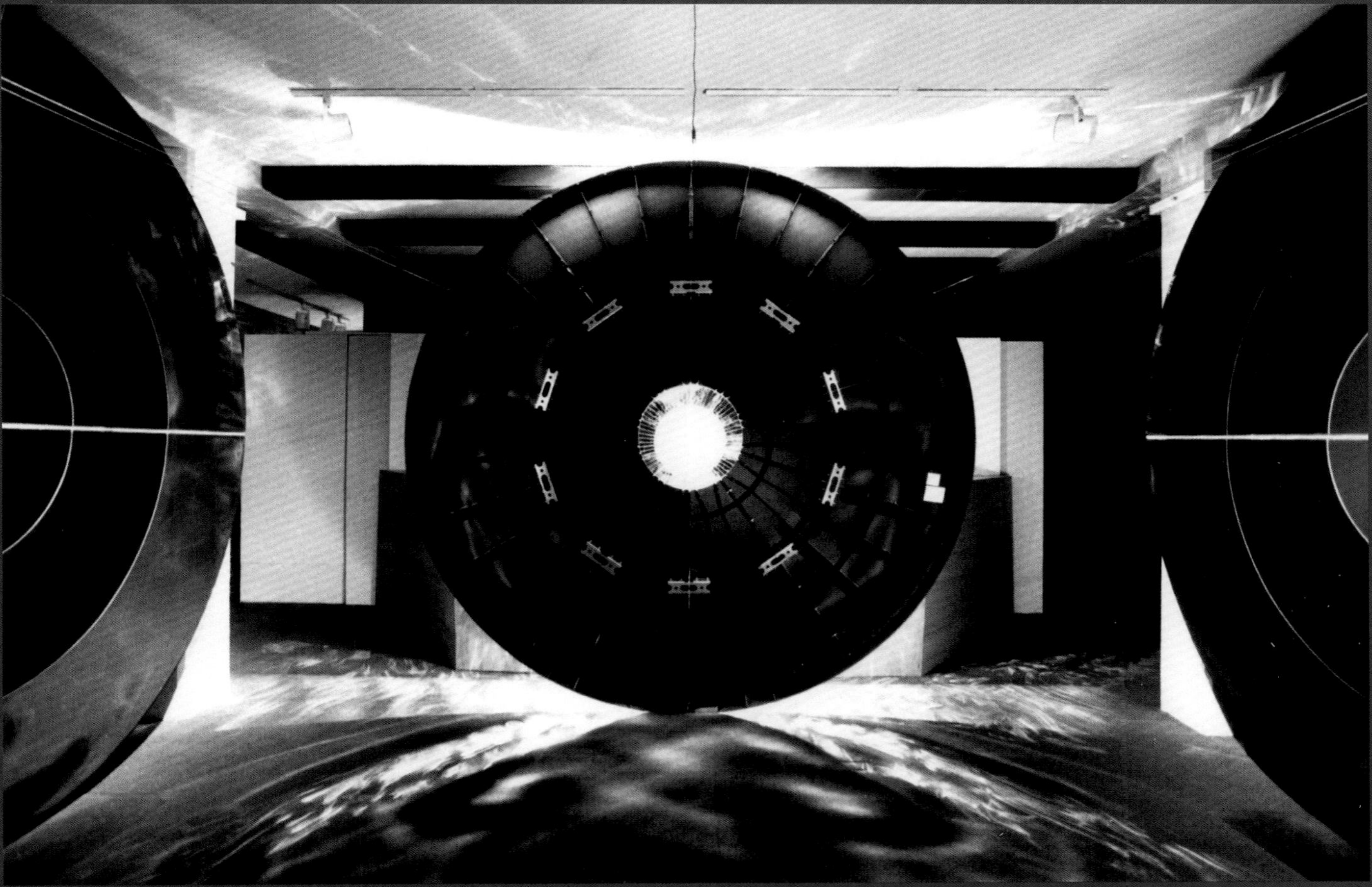

The mechanisms, including the images, are controlled by sounds made by a visitor standing on a glass plate. These statements are caught by two huge hollow acoustic mirrors and—here we ask in vain: "how?"—translated into both the laser beam movements as well as intense fluctuations of music intoned to the whole. We can touch the gigantic apparatus . . . look at ourselves in two faceted optical concave mirrors, and, if the rush is not too great, try, in all kinds of experiments, the interaction between us and the machine. (translated from article by Thomas von Randow, "Ausstellung in Hamburg: 'Mehr Licht', Bilder werden Räume," *Die Zeit*, August 9, 1985)

Above: *Monocle*, by Otto Piene, Joe Moss, Christopher Janney, Lowry Burgess, Paul Earls, with Elizabeth Goldring, *LightsOROT*, CAVS exhibition, Yeshiva University Museum, New York, New York, 1988
Photo: Walter C. Dent

Facing page: *Monocle*, by Otto Piene, Joe Moss, Christopher Janney, Lowry Burgess, Paul Earls, with Elizabeth Goldring, *Mehr Licht*, CAVS exhibition, Hamburger Kunsthalle, Hamburg, Germany, 1985
Photo: Christopher Janney

Chapter 4

The Social Climate at CAVS

The iconic Building W11 at MIT hosted a hive of artistic activity and social relationships that nurtured experiments, new artworks, and group projects. A diverse community of fellows and students kept the Center live twenty-four hours a day. "Once a fellow, always a fellow"—and most remained connected. This chapter evokes the ways community influenced individual and group work; the atmosphere of the Center as a living body; the multiple generations that formed a cohesive legacy as media and technology changed over time; and the unique "community of fellows."

Right: Charlotte Moorman performs on the cello with Paul Earls on piano; Fellows Mandi McIntyre in the background, and Michael Naimark in the foreground shooting video documentation, the Pit, CAVS, MIT, 1980
Photo: Otto Piene Collection

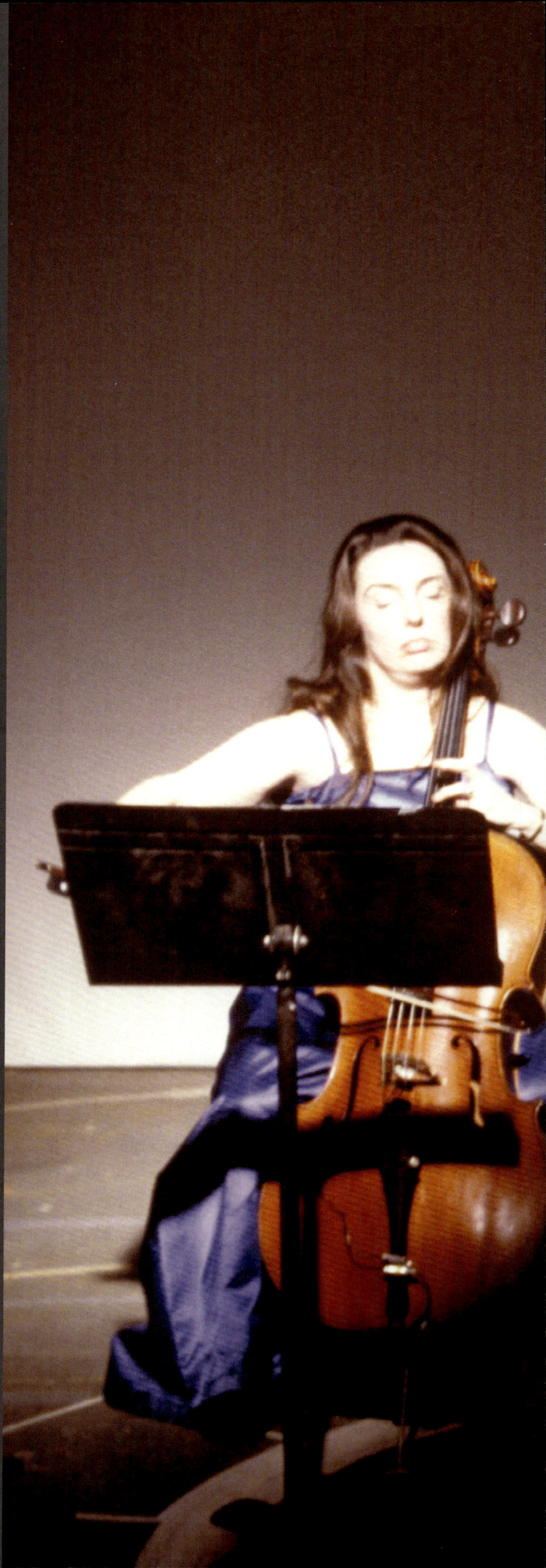

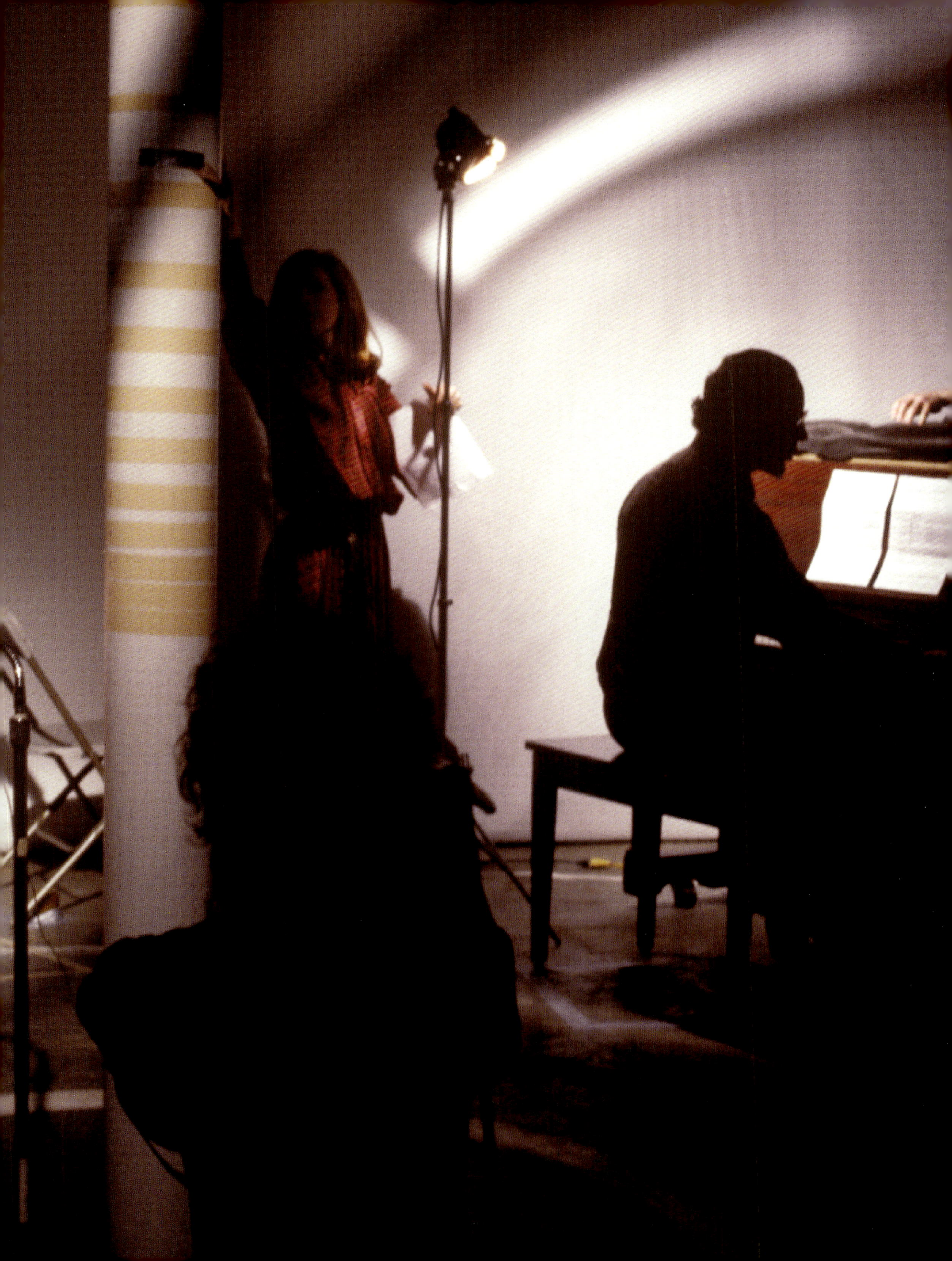

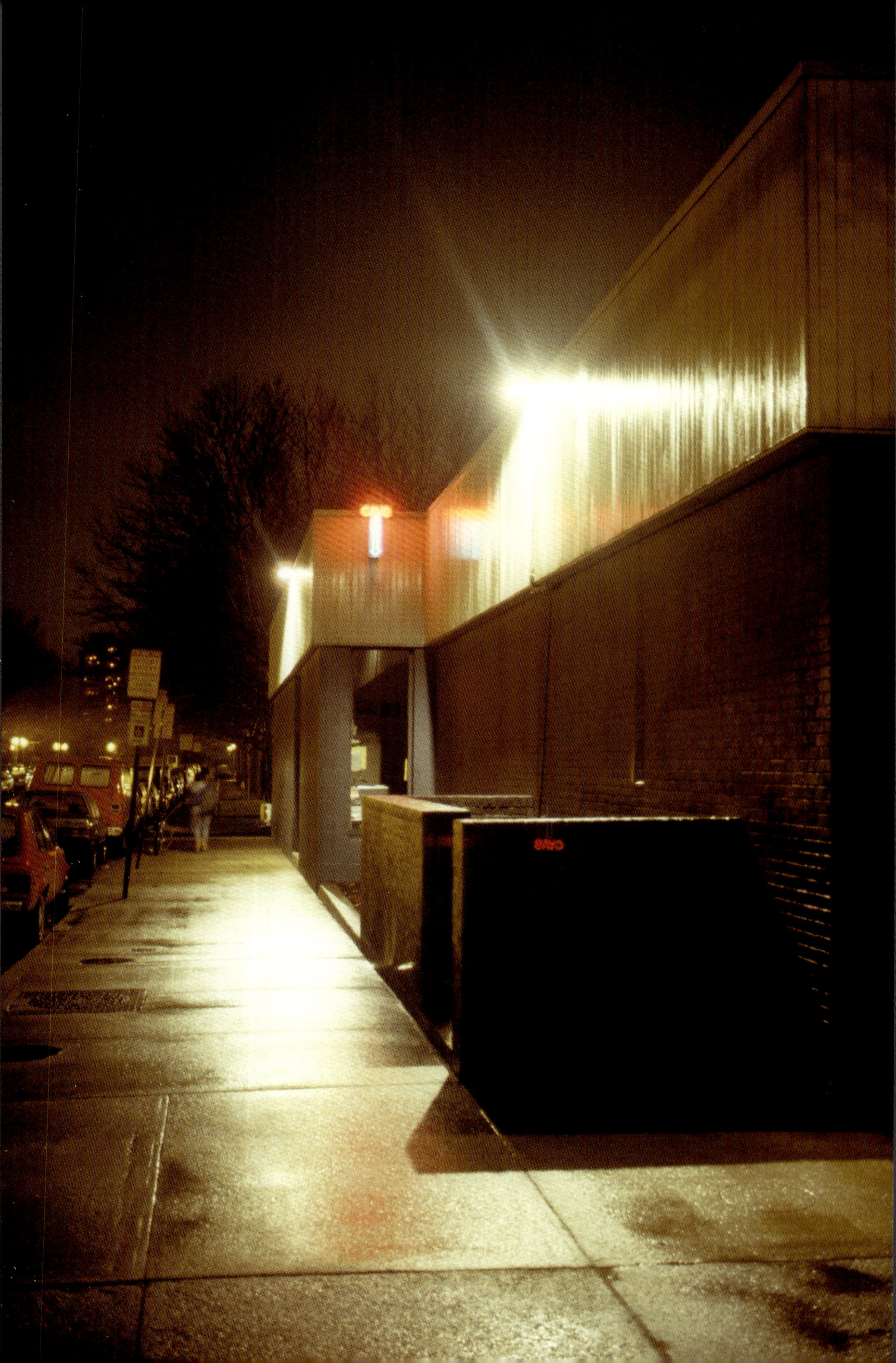

Living Together

It was the 1970s, a time of free love, intense relationships, and all-consuming communal experiences. CAVS was not immune to these tendencies. "It was a real brew of social activism, and the Center was right in the middle of it. It's impossible to disassociate the Center from this context." (Joan Brigham interview)

Cambridge dinner parties hosted by the Kepeses in their Larchwood Drive home included illustrious guests from the scientific communities of Harvard and MIT, as well as one or two fellows. These frequent, rather formal social events complemented the seminars and conferences and were part of an ongoing idea exchange at CAVS.

> Relations at CAVS were cordial but definitely hierarchic: everyone was equal but some artists were more equal than others, depending on how they fitted into György Kepes' very European system of values. (Jack Burnham interview)

> I think that the interaction of the people there—the personality of the Center when it was under Kepes—was of extremely powerful artists in their own right. Kepes would say we had at the entrance to the Center, a hat rack and an ego rack. I want you to put your ego on the right one. (Bill Parker interview)

> . . . the friendliness and openness of everyone, of sharing ideas was a very important part . . . more discussions of ideas as opposed to critiquing . . . trying to understand what Tsai or Takis were doing. . . . That is why I think of the Center as having a very healthy dialog. (Alan Sonfist interview)

Daily life and art were often indistinguishable. Celebratory occasions were artworks as well as social gatherings. These included birthdays that became events, the performances of Nam June Paik and Charlotte Moorman, Mira Cantor's *Centersheet* and *Centerbeach*, Piene's after-hours sky event parties, and Miralda's food art. Keiko Prince's feasts blended the aesthetics of food and community. Her enduring hospitality to visiting artists and sushi dinners at her Cambridge home were legendary:

> We had a fantastic social life. Keiko was always having parties. She is just like a tornado. She had a baby grand piano, which I used to play. Her house was so aesthetic. I rented her attic. (Paula Dawson interview)

> And the Center parties—I remember Nishan cooking using a soldering torch. Keiko did her food things. (Carl Nesjar interview)

Facing page: CAVS with neon signs and lighting created by John Powell at the request of Otto Piene for the 25th Anniversary celebration, MIT Building W11, 40 Massachusetts Avenue, Cambridge, 1993
Photo: John Powell

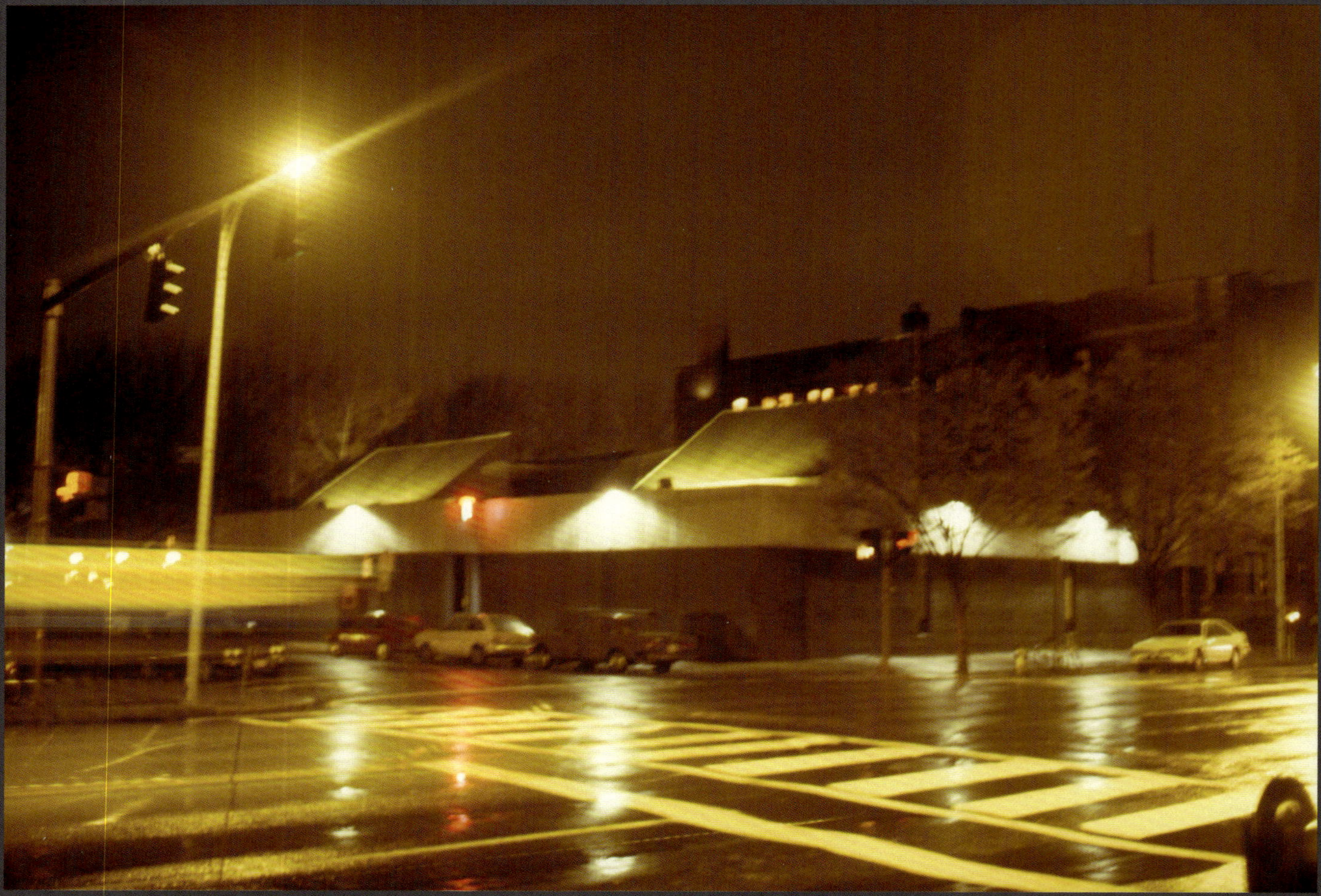

CAVS with neon signs and lighting created by John Powell at the request of Otto Piene for the 25th Anniversary celebration, MIT Building W11, 40 Massachusetts Avenue, Cambridge, 1993
Photo: John Powell

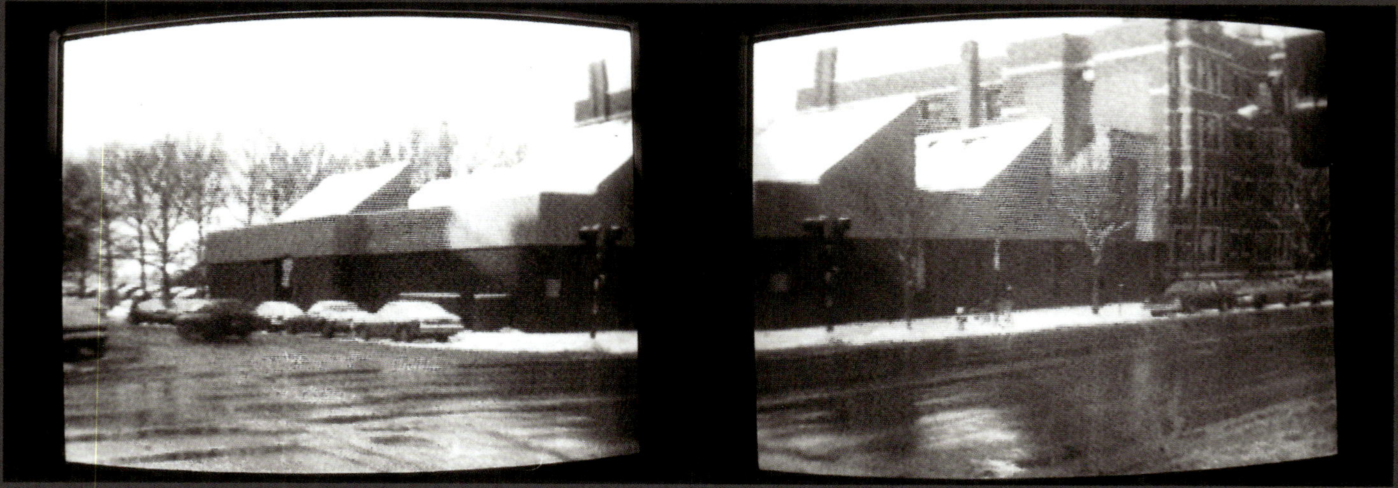

Vin Grabill, video collage, Building W11, MIT, 1980
Photos: Vin Grabill

Building W11

Building W11 was across the street from the MIT main campus, situated at the crossroads of Massachusetts Avenue and Amherst Street.

> The charter of the Center was art, not hidden under the table. It was to make art. Across the street were dozens and dozens of buildings for tech. Otto used to tell us students that the Center doesn't have a lot of money or space but this is your playground. Find collaborators and whatever you need over there. (Michael Naimark interview)

The Center building, as characterized by Dieter Jung, was "weird and fantastic; a *center* of energy." (Jung interview) The façade and roof, with protruding architectural skylights, presented a closed, nearly inaccessible view. Occasionally things would happen in the alley outside the Center or on the roof or even around the outside perimeter. However, most of what went on was an inner hub of activity not readily visible to the public.

> I remember looking at a photograph of the building, and not being able to figure it out . . . fascinating, but completely indecipherable! What was really going on in this place? Part of the reason I ended up choosing [to attend CAVS] was a sense of the unknown: it sounded interesting, and I hadn't a clue as to what it would be like. (Beth Galston interview)

> I remember the Pit, everyone was trying to do construction and I found, down in the bowels of W11, a little workspace. Paul Earls' lasers and electronic music were often splashed all over the walls and ceilings. (Todd Siler interview)

In Keiko Prince's words, the Center was "live" twenty-four hours a day. Occasionally fellows actually lived there unofficially, despite the discouragement of both Kepes and Piene. Maryann Amacher and Luis Frangella set up housekeeping behind a screen of potted plants in the first studio on the left.

> Luis Frangella had a kitchen. It was a secret. We could keep things secret because Kepes came in at nine in the morning and left at five in the evening, every day. (Keiko Prince interview)

> Life at the Center: well, nobody wanted to take the studio with Joe Davis, so I got it. It was the first year, and there were a lot of stories. Mornings at 8 am, Otto arrived in the office. He would rap on the studio door because he knew Joe was in the habit of sleeping there. Joe was living there, and I was working there twenty hours a day. We had all the facilities—coffee, smoking. I was working, still in graduate school, and I was still a gymnastics coach in the Athletic Department. I would be doing my work, and Otto would storm in looking for Joe in all the stacks of equipment—somewhere in there, Joe was tucked in. Sure enough he disturbed Joe, and a conflict would ensue, then it would all quiet down. It was a real living environment, maybe more there than in other studios. (Seth Riskin interview)

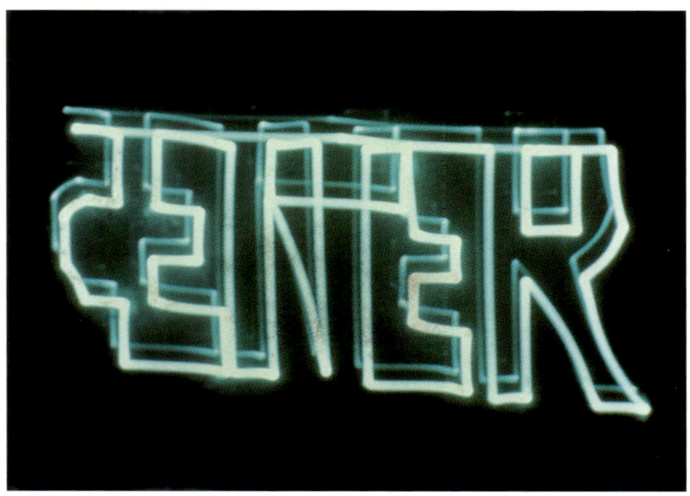

Paul Earls, "Center," laser drawing
Photo: CAVS Special Collection

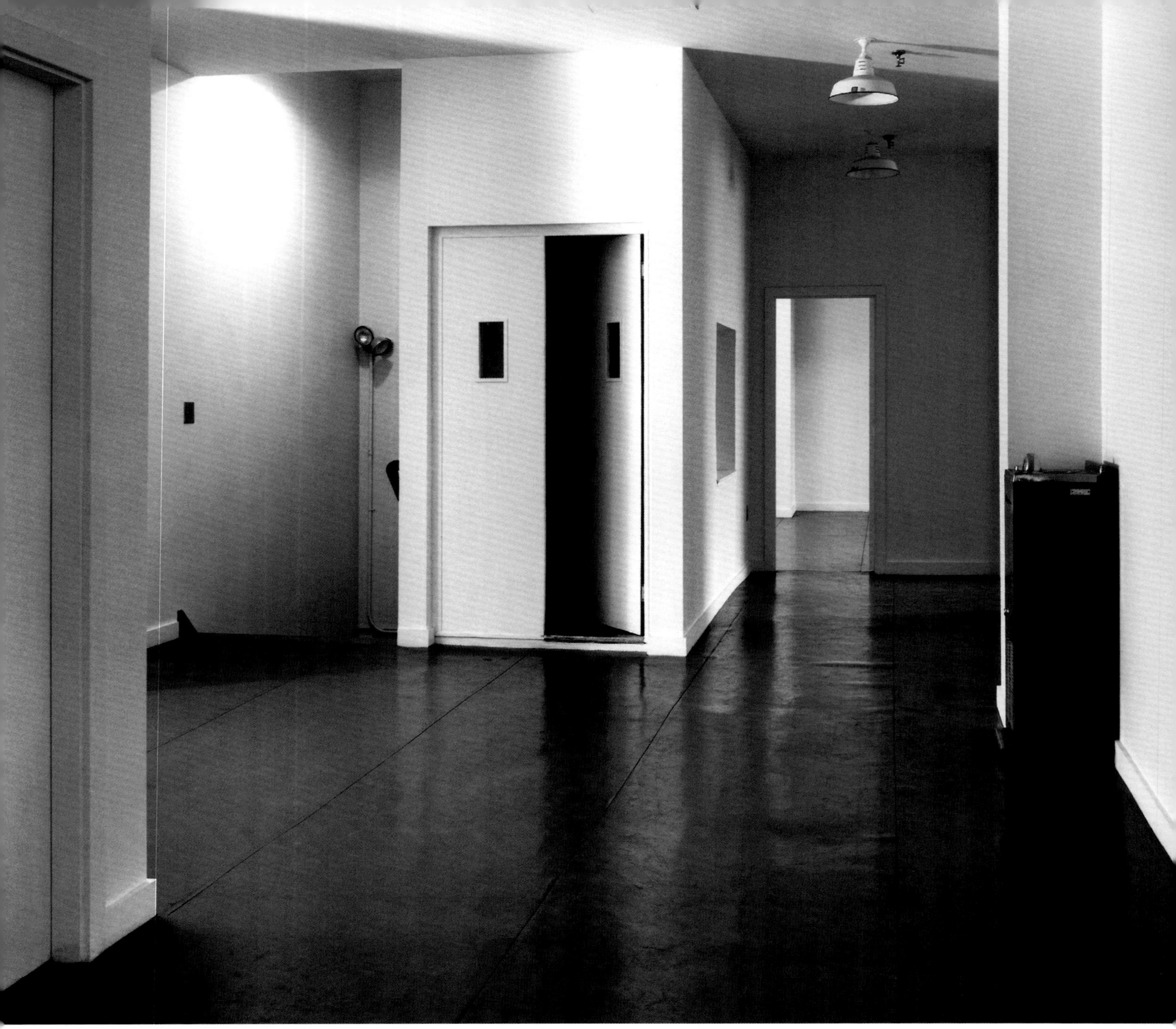

Above: Hallway on the main floor of CAVS, MIT Building W11, 1970s

Facing page: Michio Ihara (foreground left) and others at work in the basement workshop of CAVS, MIT, 1972

Photos: Nishan Bichajian

Piene described the evolution of the building over time:

> The Center was built to house five fellows and a director. The architecture reflected this structure—there were five studios and one director's office. The largest room was the director's office. The former Coop [student bookstore] was turned into a somewhat interesting building for very little money. The total budget was $250,000 for the renovation. The outside of the Center

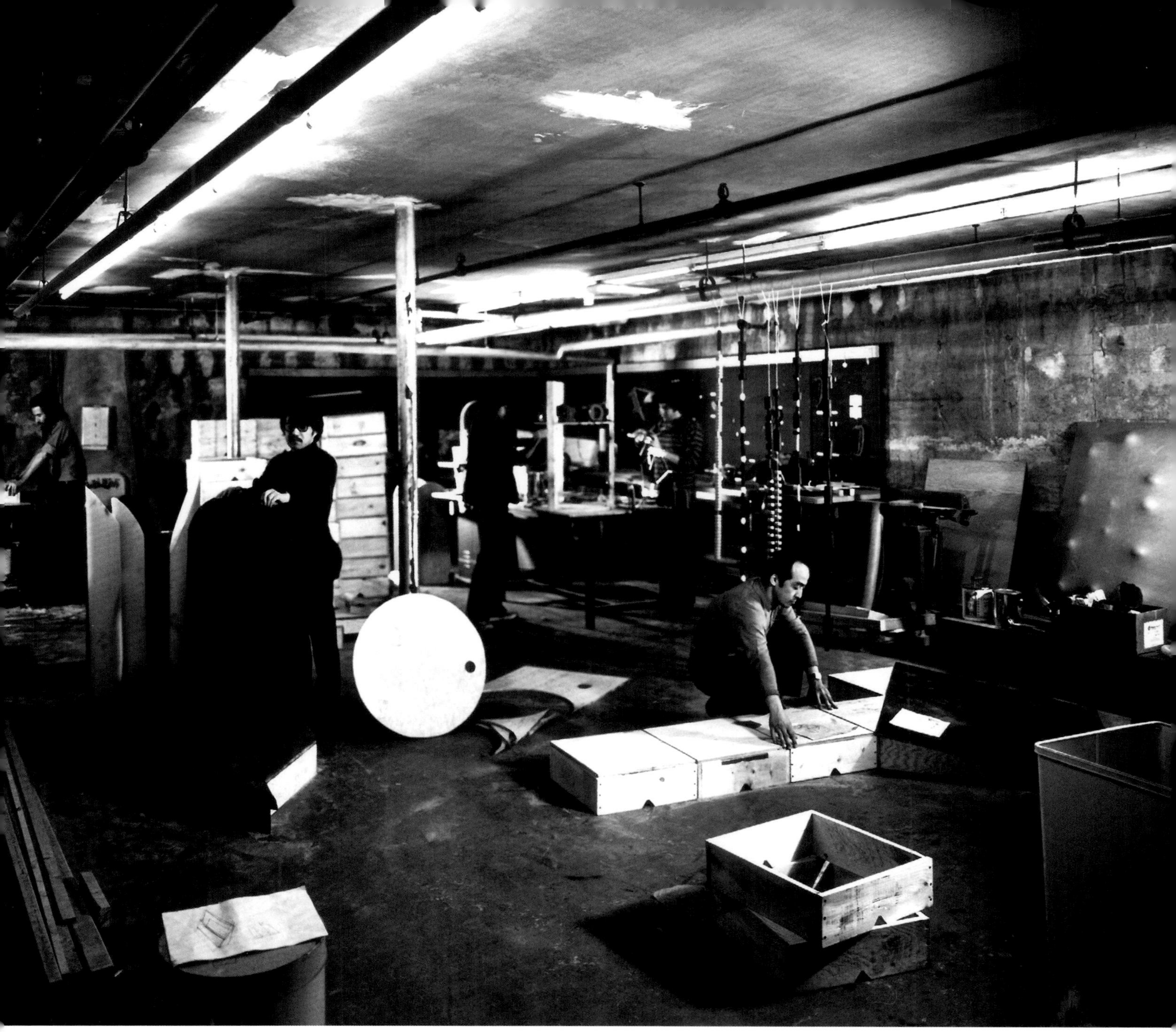

was painted in a graphite grey, but the inside was white except for the basement workshop. The studios looked much like art-school studios look today. Everything was white and kind of expressionless. There was a workshop in the basement, which was a rough space, almost like an old-fashioned sculpture workshop, except you couldn't do anything dirty. You could use a saw and hammer. Later on, much later, we got a welding license. (Otto Piene interview)

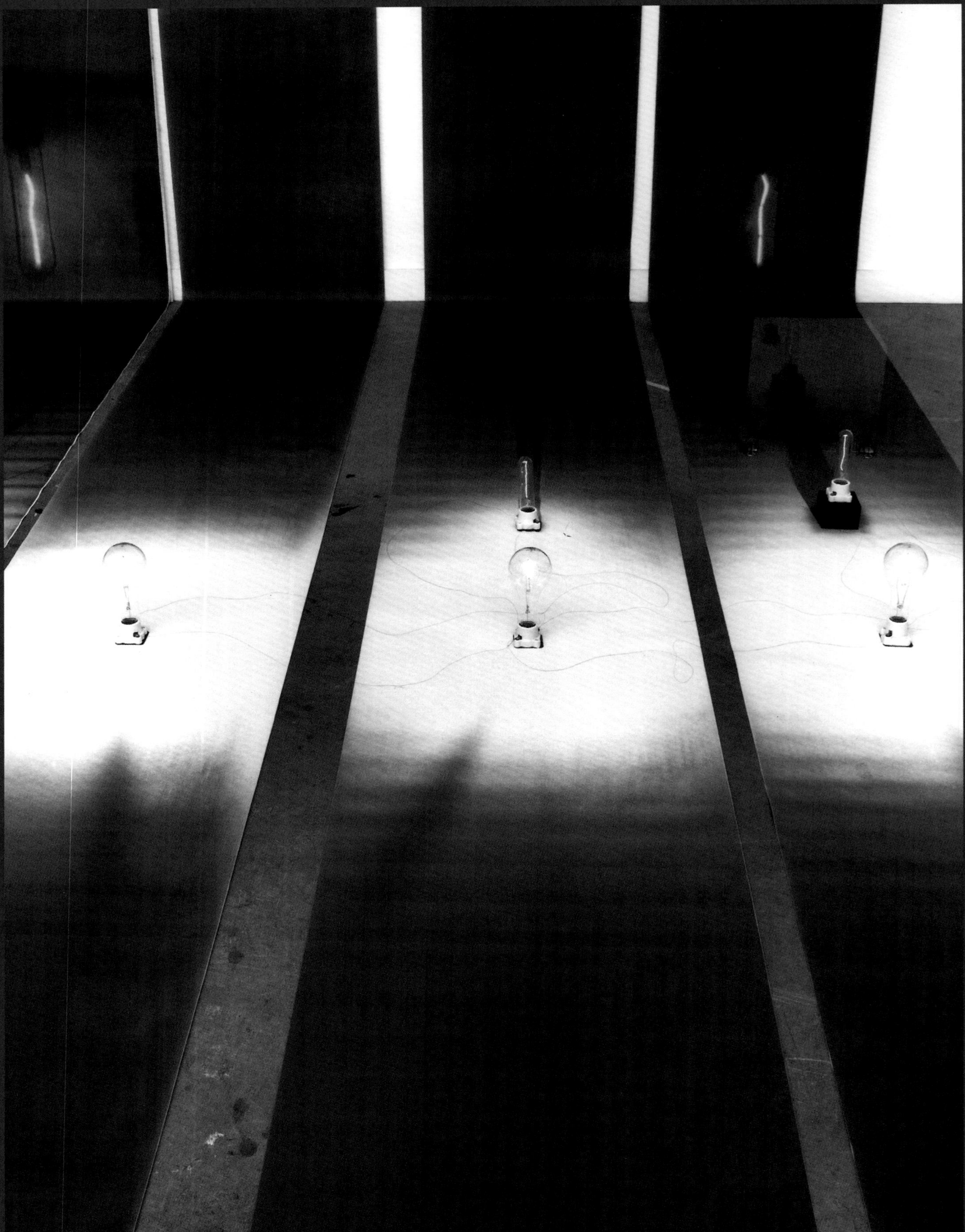

Artistic media and ideas permeated the CAVS working environment across space—through the building—and time—through multiple generations of fellows.

Right: Alejandro Siña, "Original Neon Touch Bulb," described by the artist as "red streams of lightning produced in sculptural arrangement" ("Individual Work by Fellows at the Center for Advanced Visual Studies, 1973-74" portfolio), MIT, 1973
Photo: Nishan Bichajian

Bottom left: "Filament Sculptures" (1970) by Otto Piene with Frances Masland, at *Arttransition*, CAVS, MIT, 1975
Photo: Roger N. Goldstein

Bottom right: Otto Piene with Frances Masland, "Filament Sculpture," 1970
Photo: Ed Oliver

Facing page: Juan Navarro Baldeweg's studio, CAVS, MIT, 1974
Photo: Nishan Bichajian

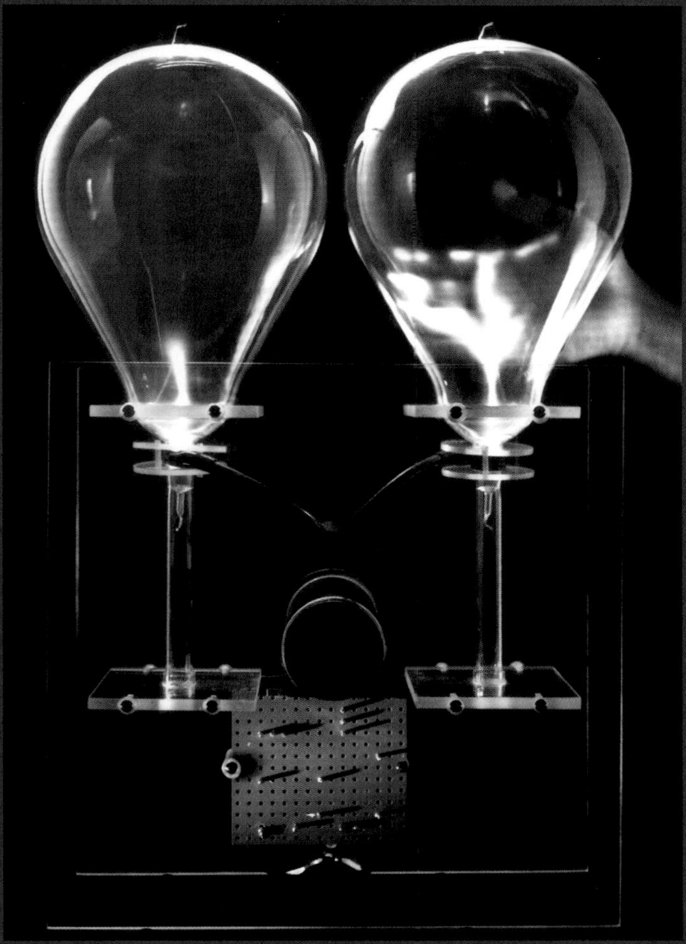

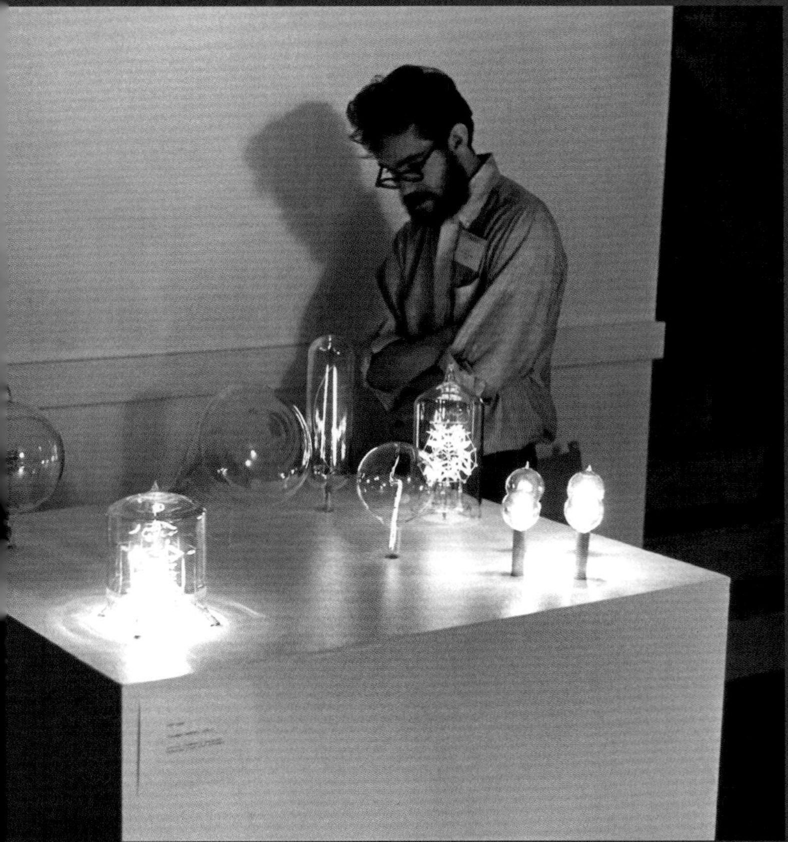

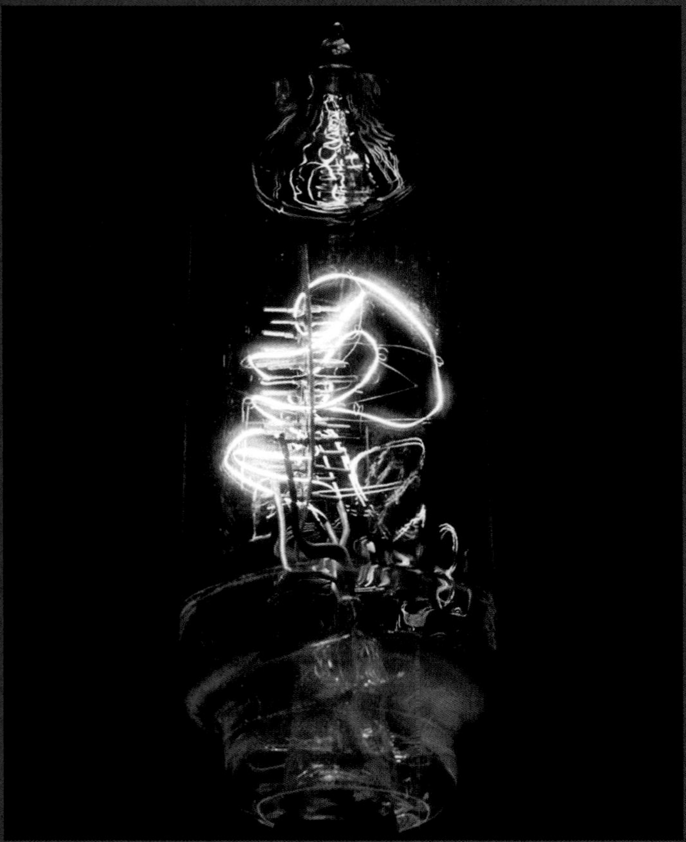

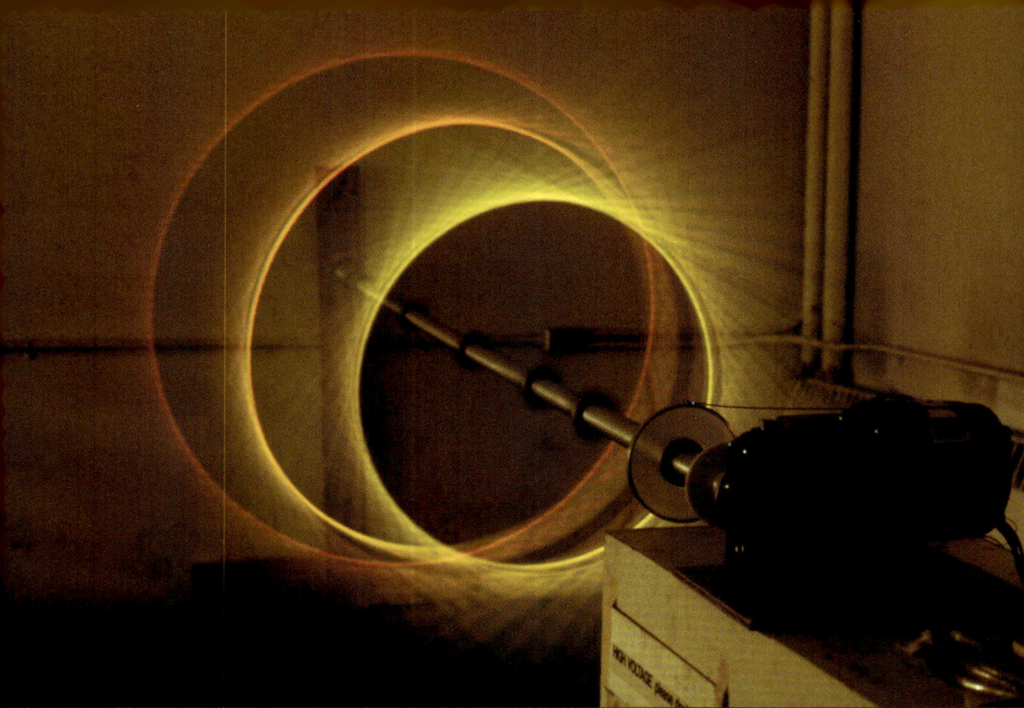

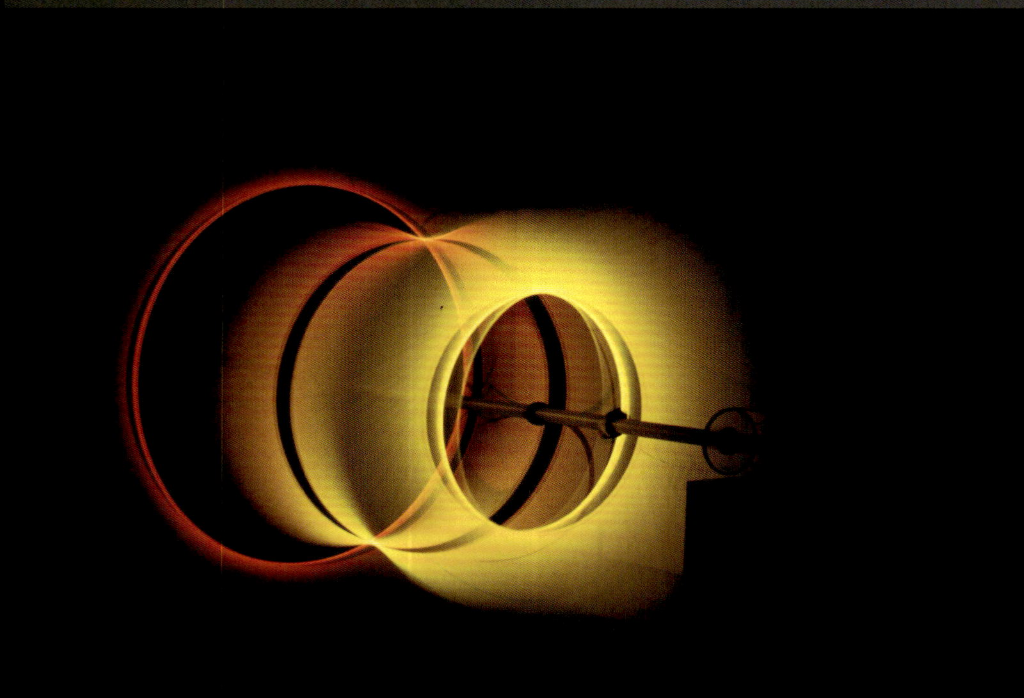

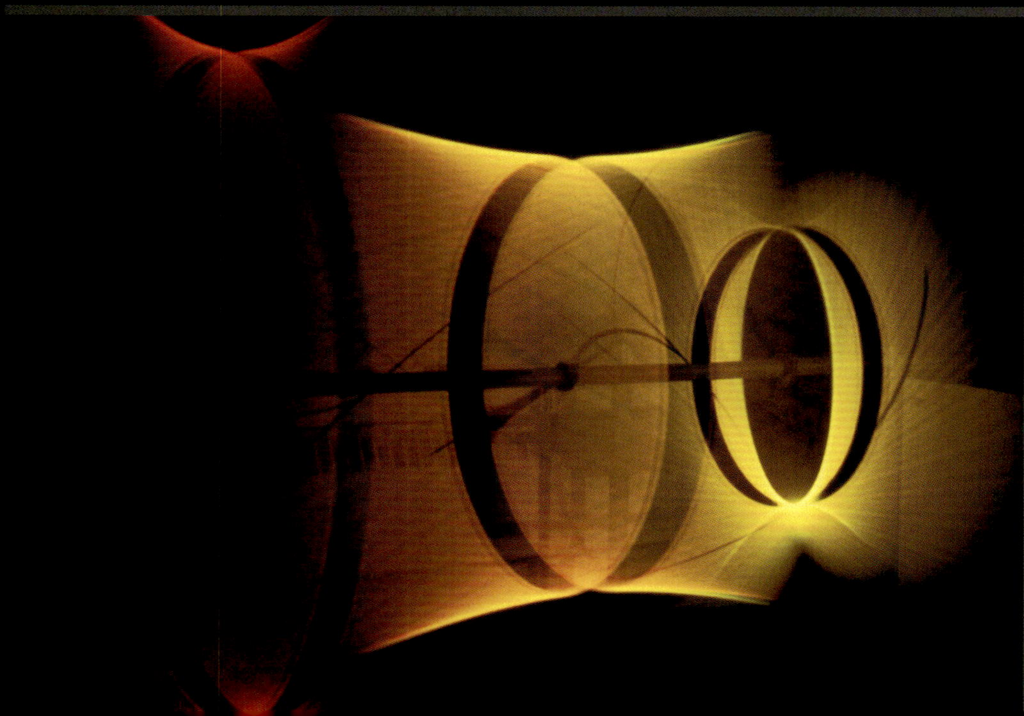

There were always light artists in residence who needed darkness to work. Alejandro Siña developed kinetic light sculptures in the Pit. His neon works became part of the *5 Artists / 5 Technologies* exhibition featuring CAVS artists at the Grand Rapids Art Museum, Grand Rapids, Michigan in 1979. Siña reminisced about CAVS's nighttime ambience:

> You couldn't find many fellows during the day. It was the nighttime when all the things were happening . . . some sense of peace at that time of night, no interruptions, and everybody liked to work at night, including me. I was doing light so I needed darkness. . . . Sometimes when you arrived at night, the place was full of smoke; other times there was a lot of water on the floor; sometimes you smelled good cooking. Always somebody was putting together and experimenting with some ideas. This stimulated the imagination and for me, these were the golden years of influence and interaction at CAVS. Now it seems like a dream. (Alejandro Siña interview)

Both pages: Alejandro Siña, "Spinning Shaft," moving neon sculptures, the Pit, CAVS, MIT, 1978
Photos: Alejandro Siña

Facing page bottom right:
Alejandro Siña, the artist with his sculpture "Corona Discharge," the Pit, CAVS, MIT, 1978
Photo: CAVS Special Collection

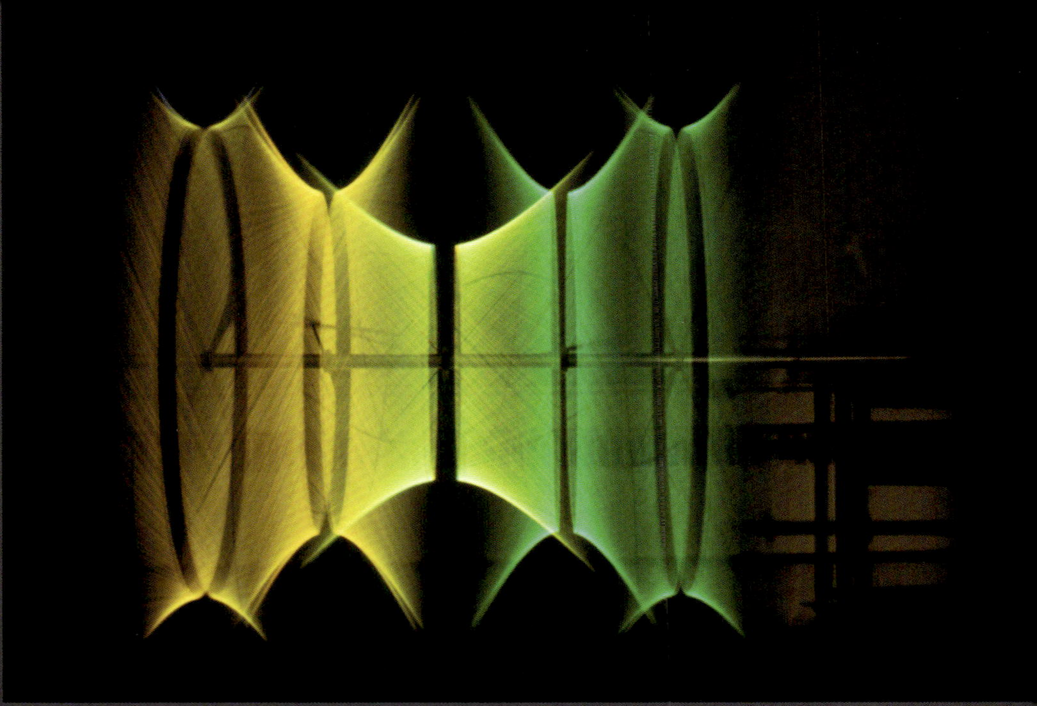

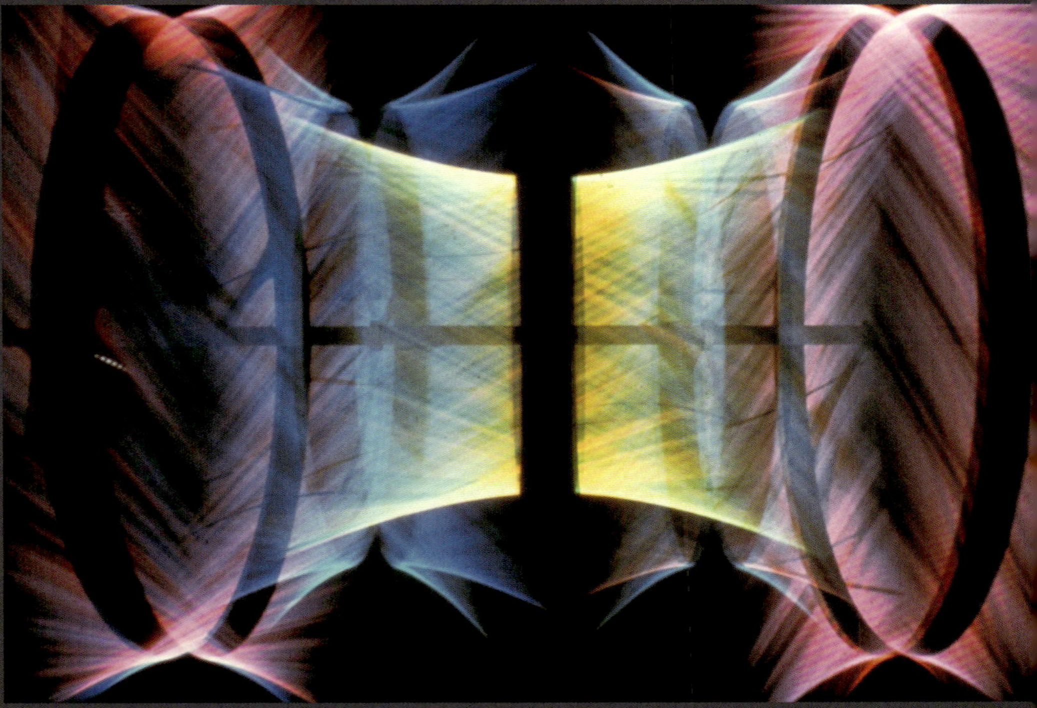

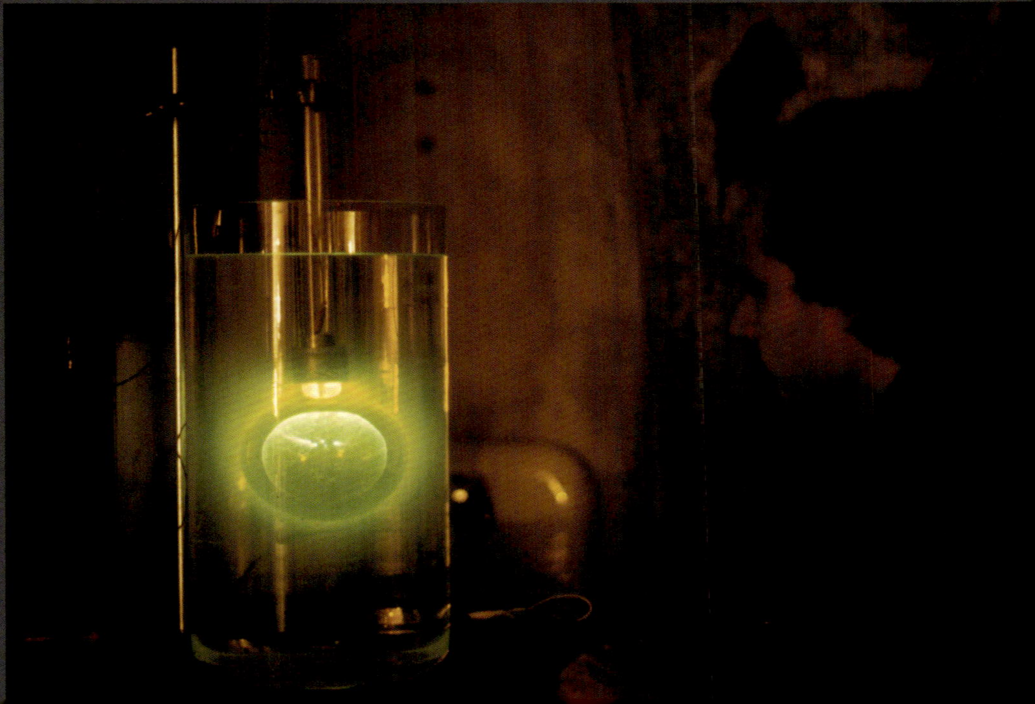

Physics, chemistry, and the human metabolism affect the visual format of electric kinetic sculptures by Alejandro Siña. . . . The gas produces the effect while other factors, such as metal, water, or mechanics, determine the programing of the patterns. Xenon gas aided by a metal bolt, for example, emits spaghetti-like spurts of electricity inter-dispersed by crackling lights. In *Corona Discharge* neon gas light flashes are controlled by water pressure within a glass tank. (article titled "Alejandro Siña," *Arts Magazine*, November, 1978, 30)

In the following impressionistic text written for *Centerbook*, long-time fellow Keiko Prince describes "the time in between"—the last years of Kepes' directorship and the initial years of Piene's tenure. For her, these years remain distinctive within the social fabric of CAVS.

Center LIVE: 1971-1975

Sound seeping through the Pit. The Velvet Underground. High volume. Midnight. Coming from Maryanne's studio? Where she's painted its floor SkyBlue? Where she's receiving sounds of Boston Harbor? Via a low corridor zip-zapping darkrooms, machine shop, storage—a tail of tadpole, the Pit (28′x28′x28′) being its head. And it was in this very Pit, in 1972, that we began to create and construct "the tactile gallery for the blind." "Dialogue for the Senses," so Kepes titled our work, and Michio Ihara conducted us working together. "The biggest hit seemed to be the sound floor. It was made up of 62 boxes, 34 of which produce elemental musical sounds when trod upon" ("Hartford Gallery Displays for the Blind," *The New York Times*, May 18, 1972). Our very first CAVS dinner happened in the Pit in 1974, or was it 1975? Flowers in hundreds of tiny beakers swayed with candles covering randomly arranged 4′ x 8′ tables. "The Pit's levitating!" a voice, maybe of Avatar Morais. But tonight, we—the Pit tribe nocturnal—start to make a slow ascent to Luis Frangella's studio. He's making a potful of soup there. ". . . wonder if Kepes knows what we are up to so late at night," murmurs Alejandro. The hall is blasting with light glaring photos of Center people. On both sides. Oh yes, who's presenting tomorrow? "At the meeting we informed each other about what we are working on." (Friedrich St. Florian) The Charles River revitalization project was a good part of each presentation. Our drawings, scaled models, and related experimental works built in the Pit, were photographed by Nishan Bichajian. Kepes bridged us "across Mass Ave" to engineers and scientists. We, in turn, received support/energy from them. Its link layered and widened . . . from mathematician Leon Shiman to mechanical engineering Professor Woody Flowers to even Sea Grant people. ". . . for me an extremely vital time . . . interacting with people from different parts of the world, background, and training." (Lowry Burgess) Guests and visitors frequented W11. Arata Isozaki, Peter Cook, Jonathan Benthall, Provost Rosenblith, Dean Porter, and many others from the School of Architecture and Planning. Particularly, Professor Kevin Lynch who wrote the book, *The Image of the City* (MIT Press, 1960). In its Preface, "The work . . . was done under the direction of Kepes and myself." We came to see the Center's two environmental projects, "Boston Harbor" (1968) and "Charles River" (1971-73), as Kepes' long-carried agenda for future urbanity, and his vision of "Civic-Art"—arts beyond gallery and museum. Kepes' programmed light mural at KLM in New York City (1959) impacted artists worldwide, Billy Klüver and Robert Rauschenberg among them, who, around that time, I was told, were organizing E.A.T. In Spring 2002, at the New York gallery showing his work, Billy Klüver told me, "I've got to come to Kepes' Memorial. I've got to. I have so much to say about Kepes!" and "I was flabbergasted!" Kepes' fervor for cityscape reached Maryanne [Amacher] in '73. "Imagine collaborating on Times Square acoustics!" She and Scott Fisher ". . . ended up teaching a class called 'Live Place'. The focus was how we see, hear, in depth . . . immersed in everyday spaces." (Scott Fisher) "Perception" was our word, and the common denominator for our work. "A big important chapter was the friendships made [at CAVS]" (Juan Navarro Baldeweg). Yes, we are hunters. Hunters for conduits connecting nature/universe, reverberating in a sensorial realm yet to be excavated in us homo sapiens, lurking in a 24/7 urban labyrinth.

— Keiko Prince

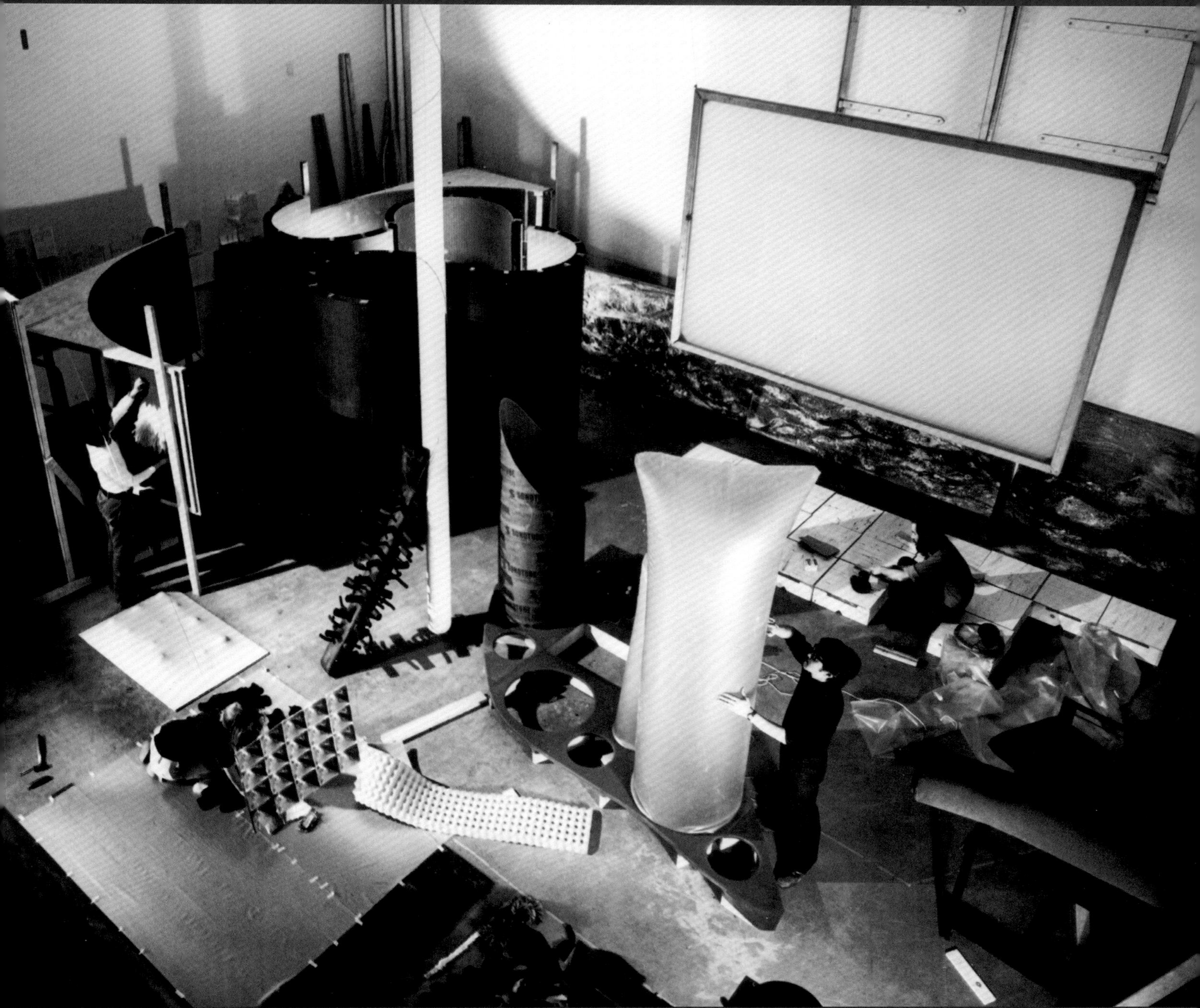

Above: CAVS fellows, including Michio Ihara (foreground), work in the Pit at CAVS to develop "Dialogue for the Senses," the premiere exhibition of the Tactile Gallery at the Wadsworth Atheneum, Hartford, Connecticut, 1972
Photo: Nishan Bichajian

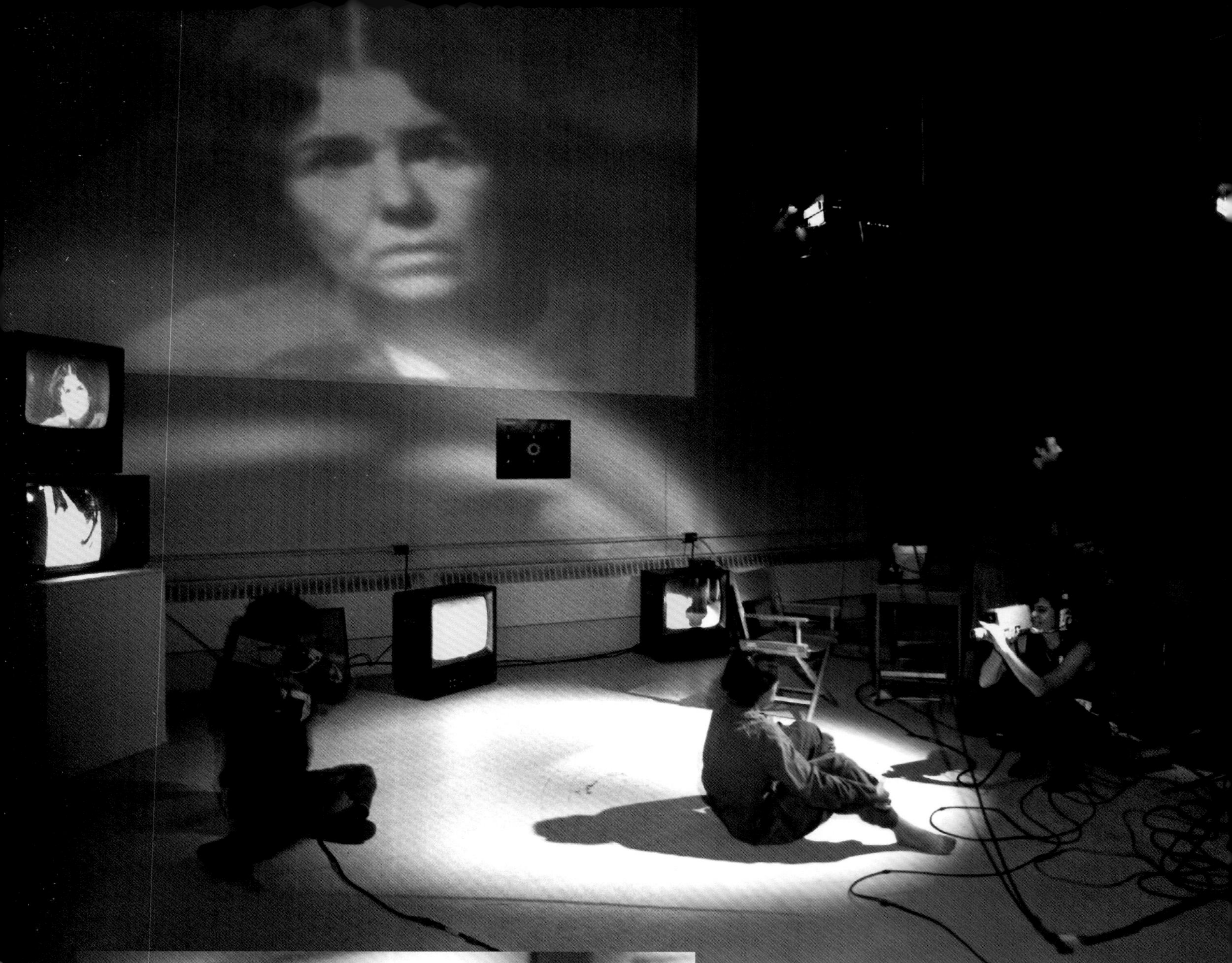

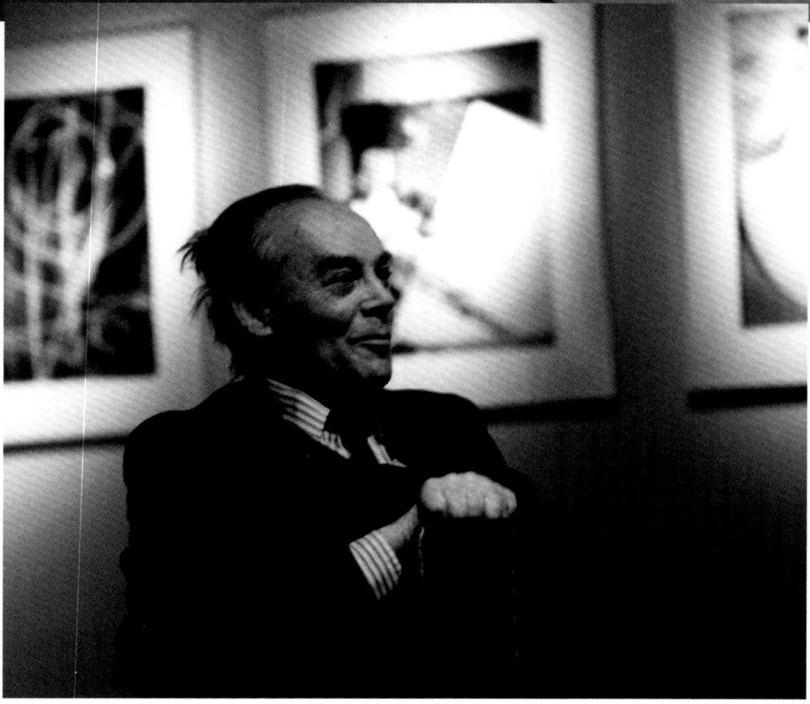

Above: Aldo Tambellini with his group Communicationsphere, "Tele-Sky" Performer Sarah Dickinson appears live and on multiple screens during a Slow Scan Television (SSTV) transmission between CAVS in Cambridge, and Eric Gidney/Metro Television in Sydney, Australia. The doors to the alley next to W11 are open (upper right) to engage additional audiences.
Sky Art Conference '81, the Pit, CAVS, MIT, September 25, 1981
Photo: Christian Lischewski

Bottom: György Kepes, 1985
Photo: Veronica Falcão

Artistic Social Collaboration

Lowry Burgess described the alliance of nature and technology combined with artistic social collaboration as Kepes' essential call to the artists at CAVS:

> To the extent that we were engaged in this positive transformative process we were a vital force inside the wider culture, believing that art can give humans and their societies a broader expressive communicative advantage. (Lowry Burgess interview)

Goldring described her close, personal relationship to Kepes, saying that he was like a grandfather and mentor:

> After my accident, when I was in bed and blind, he and Juliet called me regularly and he would always ask me, "What are you seeing?" At first I found it a frustrating question, but I truly think it took my work in a new direction. (Elizabeth Goldring interview)

The car accident had exacerbated her vision loss due to proliferative retinopathy.

Lowry suggests that, while Kepes had been disappointed by the lack of interaction and friction among the first fellows, with his arrival and others (Earls, St. Florian, Bueno, Navarro Baldeweg), a new mood emerged.

> An intense social life began and the Center was inhabited twenty-four hours a day, seven days a week. We often met in each other's studios in the afternoon for a smoke and just plain silence together. Our children came to the constant weekend parties, which often went until dawn. CAVS was on fire with experimentation and the nearly weekly exhibitions of those experiments. (Lowry Burgess interview)

The social fabric under Piene changed considerably in part to conform to the changing social climate. Piene's practice, which dominated the CAVS social life and grew in part out of what Prince described as "Center LIVE," was more organic and less formal. He believed that the work and life of artists were inseparable. Aldo Tambellini, a longtime collaborator with Piene, shared this belief in groups:

> I am a strong believer in groups, artists interconnected with each other. I don't believe in the individual artist isolated from society. Art is dependent on society and contributes works to society and the artist doesn't separate himself. He can do his individualistic work but I think he should interact with the rest of society, which I ended up doing at the Center. . . . My philosophy remains the same today. (Aldo Tambellini interview)

There were more women at the Center. Piene's directorship was more inclusive with students, junior fellows, senior fellows, scientists, and engineers gathered in a work atmosphere that gave way to late-night "Pizza and Rolling Rock." Piene's Tuesday evening "Art and Environment" class was frequently continued after hours at one of the Center's chief watering holes, the Paradise Café.

> Fellows were very supportive of each other, that comradeship, that's what people love about the Center. Almost everybody there was doing some oddball stuff; you felt that people understood what you were doing, they supported it in a big way. (Mark Mendel interview)

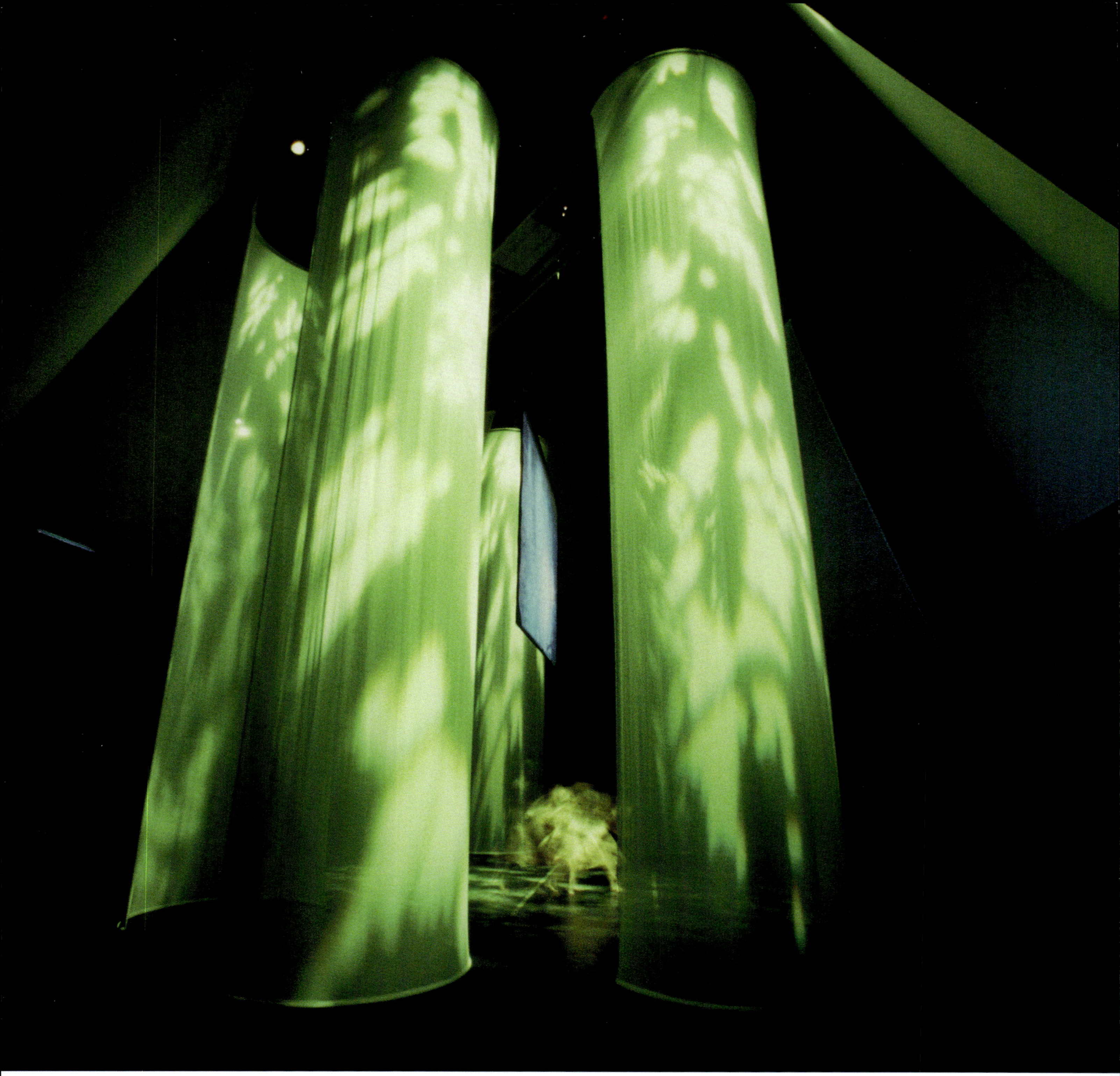

Above: Beth Galston (set) and Ellen Sebring (video and music), "Aviary," a collaborative performance with Sarah Skaggs (choreography and dance), Philippe Villers Experimental Media Facility, the Cube, MIT Media Lab, May 1988
Photo: Beth Galston

Facing page: Bernd Kracke (live video processing) and Ellen Sebring (flute performance), MIT Chapel, 1981
Photo: Bernd Kracke

At times, everyone was in love with everyone else. There were clashes, friendships, and love affairs, even marriages and an occasional divorce. There was always the feeling that one could count on anyone who was at the Center. Much of this stemmed from the work being done—new and imagined projects that were being tried out and that required more hands and brains than a single person could muster.

> CAVS was a large family with age groups and clans, the juniors and the elders, the regulars and the passers-by, but no matter what, we all belonged to a sort of movement that reached beyond what we could grasp. (Luc Courchesne interview)

Intense working relationships often led to collaborations that were not just one artist helping another to realize a project, but the forging of a mutual work. Many projects required complex setups, and fellow artists' donations of time and energy, as was the case for Ellen Sebring and Beth Galston's performance piece, *Aviary*:

> The Media Lab had just been built, and I worked there on *Aviary* . . . creating music in Tod Machover's MIDI lab at night, like many artists, when the equipment was free. David Atherton collected sounds of birds, ice, wind, and made Foley sound effects that became the keyboards. Once, I woke up over the keyboard, and listened back to music I had written in a dream state—the sampling keyboards were that intuitive—it was like listening to the unconscious mind. (Ellen Sebring interview)

The free collaboration among colleagues led to shared languages. Grabill ran with a camera held low to simulate an animal's point of view in the forest and appeared as a bear in Sebring's deconstructed fairy tales. Galston, a sculptor, became a musical object in Grabill's staccato time-lapse work. Seaman transformed a coastal train journey into a visual-musical-poetic hybrid. Kracke used feedback and chroma key to transform Sebring's flute performance in the nearby Eero Saarinen-designed MIT Chapel. Bob Rosinsky, Luc Courchesne, and Atsushi Ogata shared a quasi-comic storytelling approach. The human dimension was the intangible yet dominant feature of CAVS collaboration.

> We did a lot together, ate together, commiserated together, and collaborated, helping out on each other's projects. But we also jammed. I did some things with Ken [Kantor] and Bernd [Kracke] putting audio/video into a live jam session down in the basement. Ken was the technical whiz. Bernd had an interest in performance. Having that building was such a boon. It allowed almost any media-related experimental enterprise to take place in the Pit. (Vin Grabill interview)

> The intensity, the level of commitment and passion that existed at the Center while I was there is probably what I remember and miss the most. The Center was active around the clock, you could walk in at three in the morning and the night shift was at work. They would sleep somewhere during the day and reappear the following evening. I could walk into Michio's space and see him working with metallic structures and then go into the next studio and see Frederich working on his holographic clouds; it was another world. I think the richness came from each individual but also in the contrasts, variations. (Georges Singer interview)

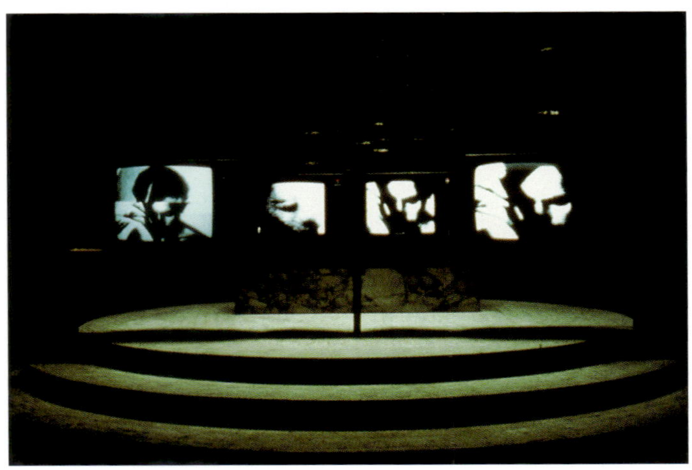

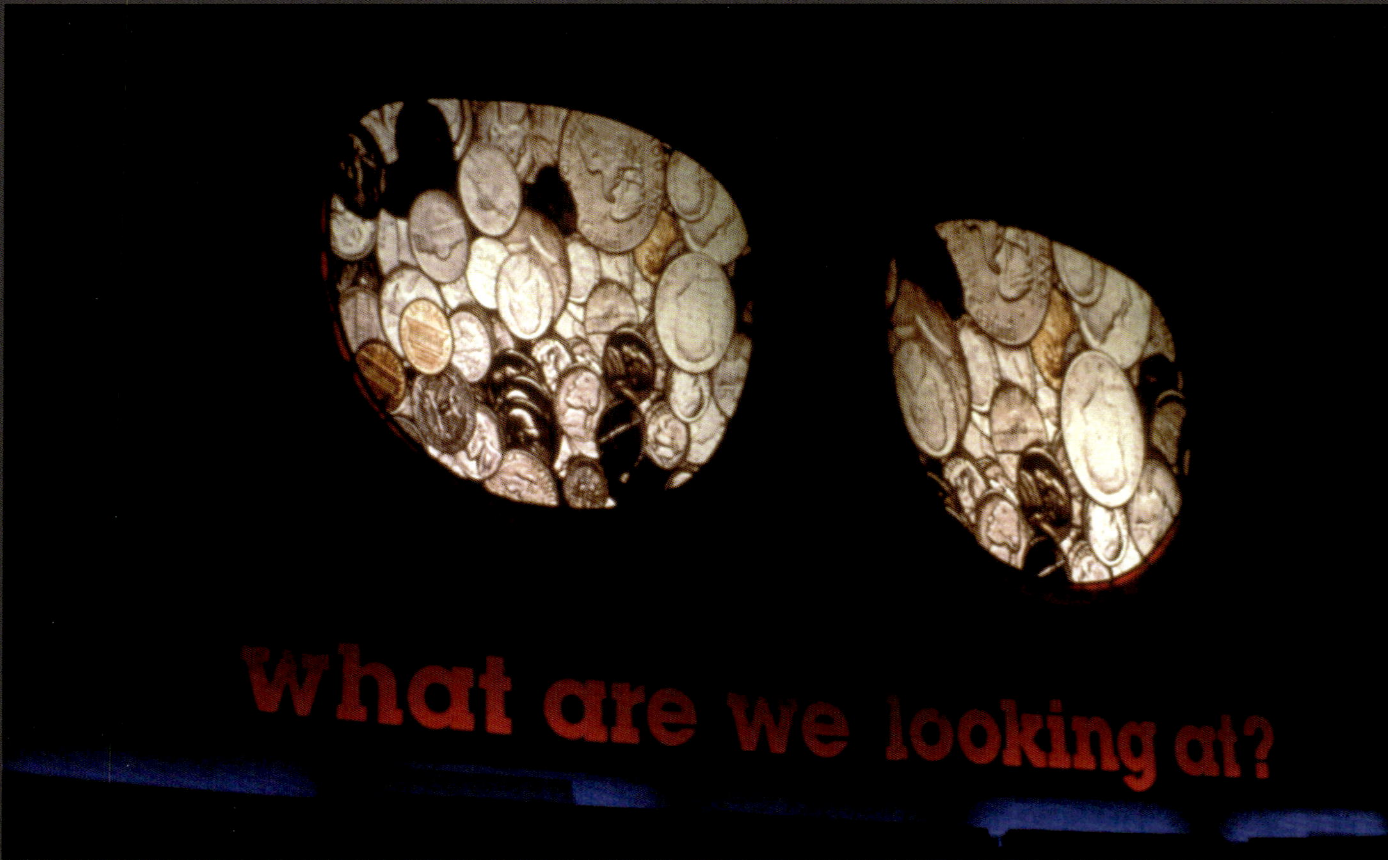
what are we looking at?

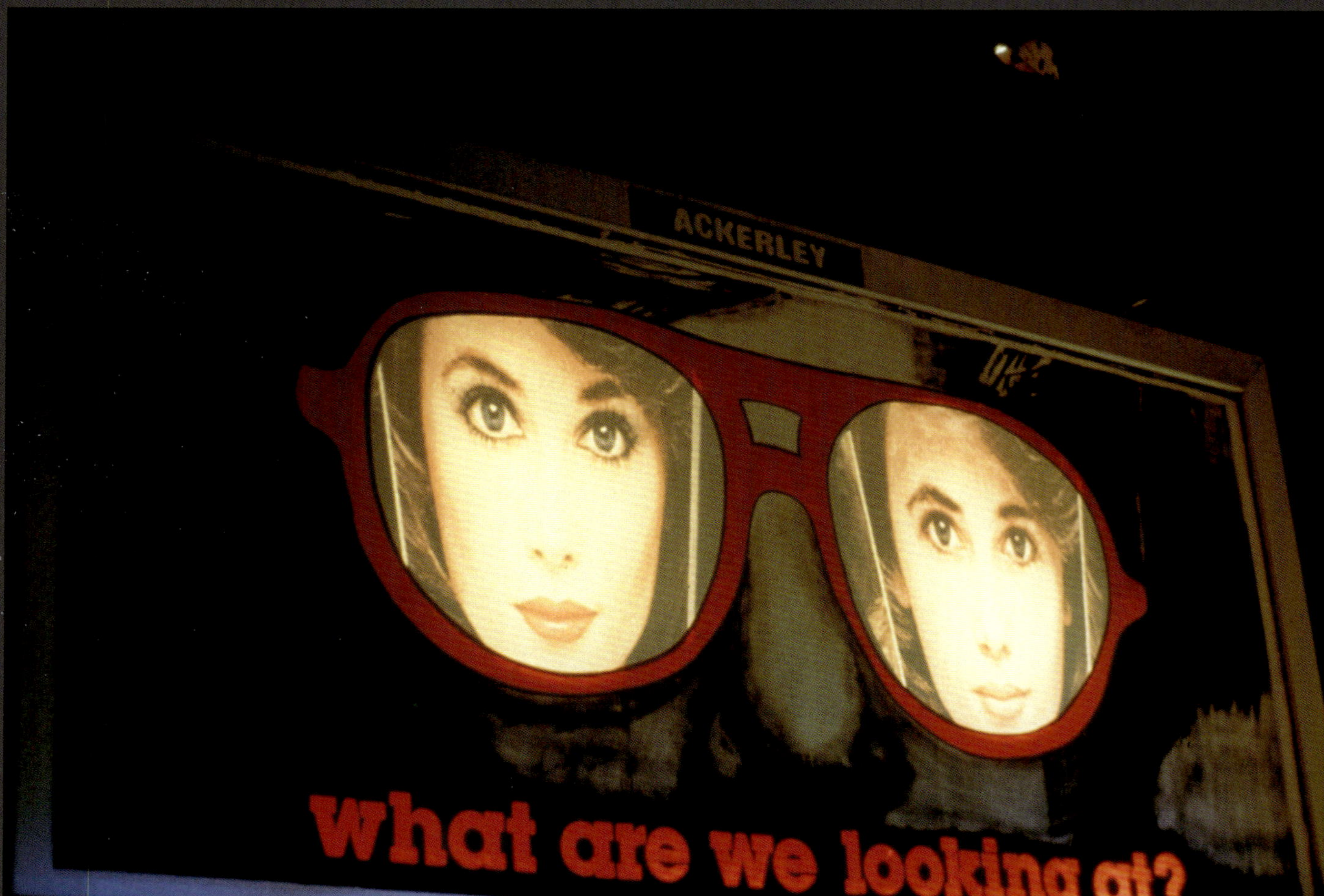
ACKERLEY
what are we looking at?

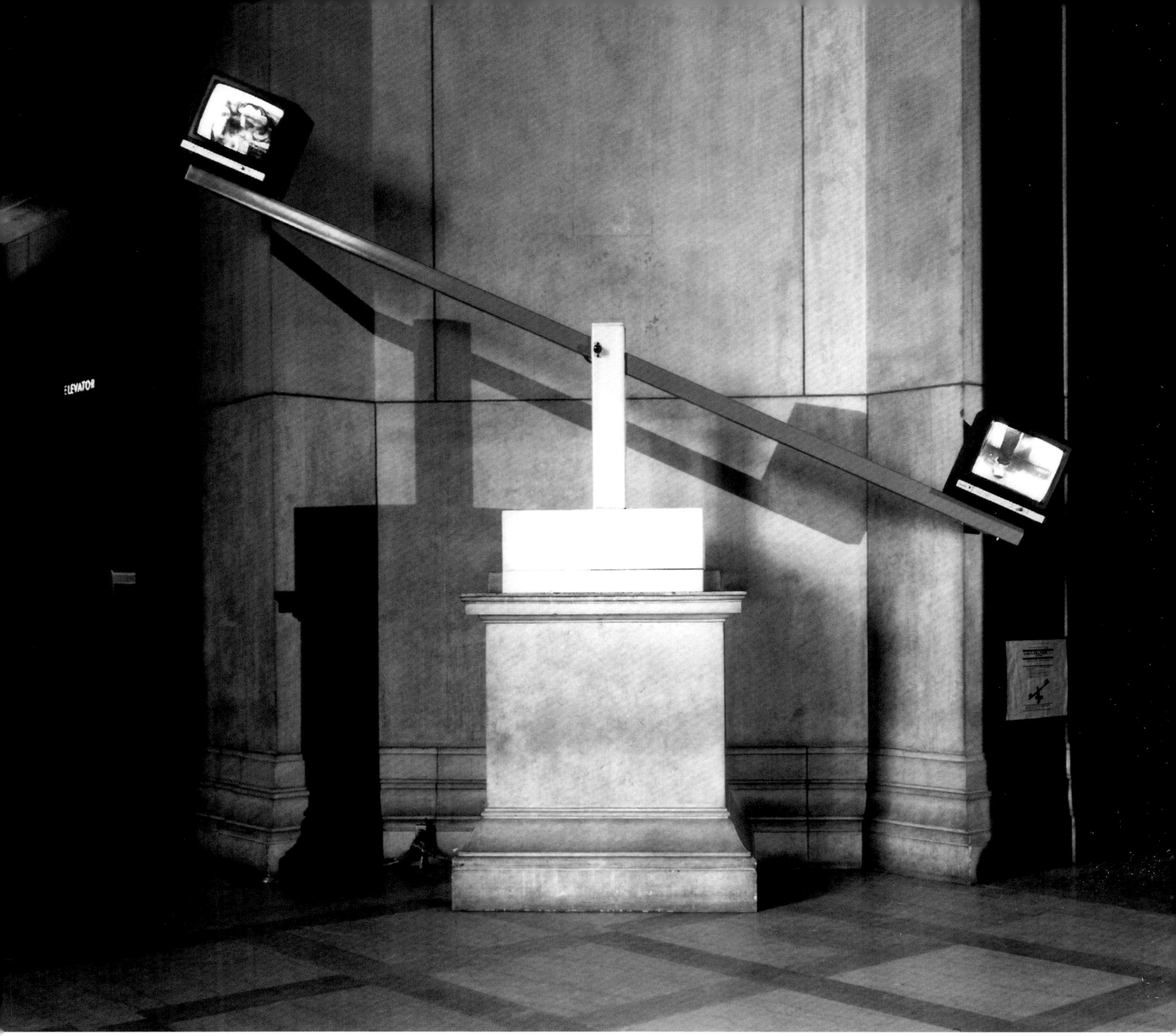

haute CULTURE is a video installation using a seesaw.
haute CULTURE is about the "shifting values, perspectives and points of view" of two public spaces —a museum and a shopping mall.
haute CULTURE is a visual commentary proposing a social/cultural rereading of these public spaces.
(Muntadas, text reproduced from special preview poster, 1984)

Above: Muntadas, "haute CULTURE (PART II)," special preview installation, Lobby 7, MIT, June 13-21, 1984
Photo: Nishan Bichajian

Facing page: Anne Bray and Muntadas, "Media Eyes," billboard installation with nightly slide projections the artists called "aerial anti-ads," above 300 Massachusetts Avenue, Central Square, Cambridge, *Sky Art Conference '81*, CAVS, MIT, 1981
Photos: Muntadas

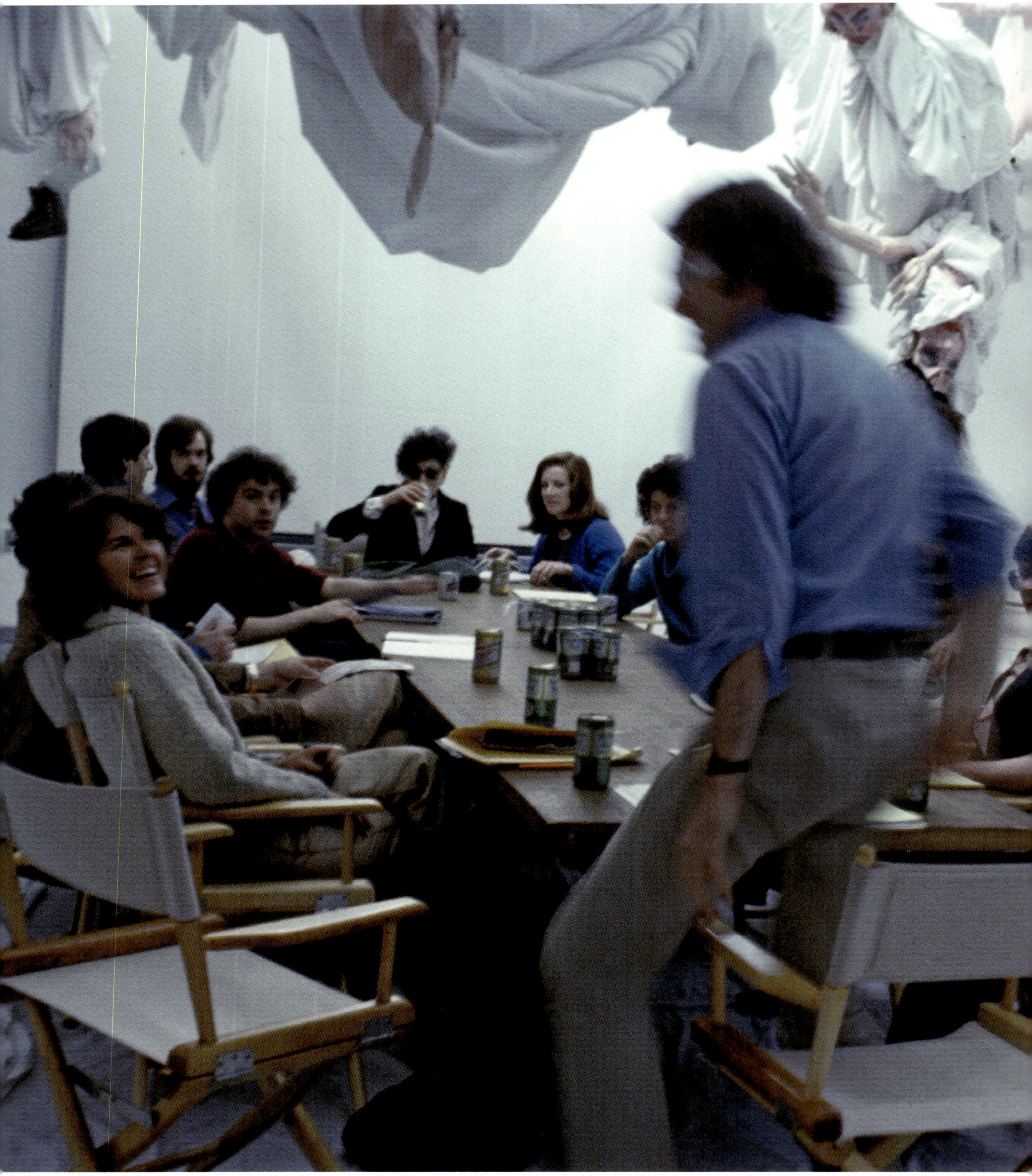

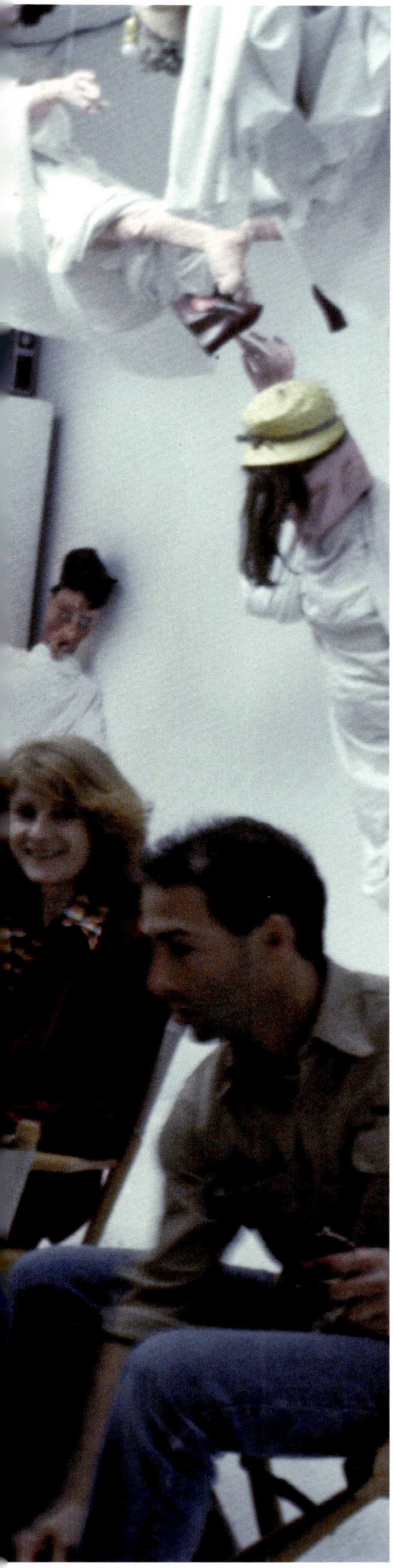

Fellows' Meetings

Large-scale group projects like *Centerbeam* and *Desert Sun/Desert Moon* (covered in the chapter on collaborations) imbued CAVS artists with a sense of living and working together. In general, the attitude of the fellows was cooperative, rather than competitive. Alejandro Siña commented on the "famous fellows' meetings" at the long table in the exhibition room.

> We would all sit around a long table with Kepes or Otto. Many subjects were discussed. Some people were reprimanded. There were all these artists with moods and strong personalities. Lots of humor and drama spiced this unique, rich interaction. (Alejandro Siña interview)

> I loved the weekly Center meetings. The first thing Otto did was to read through all the announcements, invitations, requests with the thought that one of us might be interested in participating. Almost all the meetings were about art. That was absolutely the antithesis of any educational group meeting I'd ever participated in. We were talking about art all the time, sharing ideas, raising questions. Otto had the administration in the background, never in the foreground. His meetings were not only the problems of art, but also the opportunities of art. Probably a lot of people didn't appreciate that but I thought Otto was incredibly generous, going through his mail! Information is something that gives power. If you keep people from having this information, you are in a more powerful position. (Tal Streeter interview)

> It is very difficult for artists to collaborate, to find some fertile field. Part of the fun of the fellows' meetings was watching to see how artists could "wind" together and come up with something. (John Newman interview)

Left: A fellows meeting held amidst the exhibition *Centersheet*, by Mira Cantor, in the exhibition room at CAVS, MIT, 1980. Fellows (and others), around the table left to right: Joan Brigham, Paul Earls (out of view), Aldo Tambellini (out of view), Ken Kantor, Vin Grabill, Bill Parker, Shelley Lake, Mandi McIntyre, Beth Galston, Bernd Kracke (out of view), Michael Naimark (out of view), Jane Barnes, CAVS Administrative Officer Astrid Hiemer, Mark Mendel, and Otto Piene
Photo: Elizabeth Goldring

People marvel at the prodigious amount of artwork and experiments produced at CAVS during the 1970s and 1980s. It had to do with the personal and communal bonds, and the energies, erotic at times, that motivated this creativity. CAVS was not a closed circle; rather it welcomed new blood.

> Even though many of the graduate students were working outside the Center, we all congregated at CAVS to discuss our work. What was fascinating about the Center was the life, the energy, even though people worked behind their doors, there was an openness that you could wander in and out of projects, people's lives, casually observing things, starting up conversations here and there, giving advice, but most important there was camaraderie—a sense that other artists shared similar passions and mutual interests that allowed you to see you weren't so alone in the world. (Todd Siler interview)

International travel was part of life and work at CAVS. Piene's mission included exposing graduate students and fellows to unfamiliar working conditions. While at MIT, artists who came from around the globe went out into the world to participate in shows, seminars, and events, creating a lasting international CAVS community. Physical participation rather than virtual appearances promoted friendships that survive today:

> There were different festivals all over the world where you would run into fellows. That's the sense of community about the Center. The history of the Center is embodied in all those individuals. Whatever happens is a result of the vision of Kepes and Otto to start something like that and continue it. (Peter D'Agostino interview)

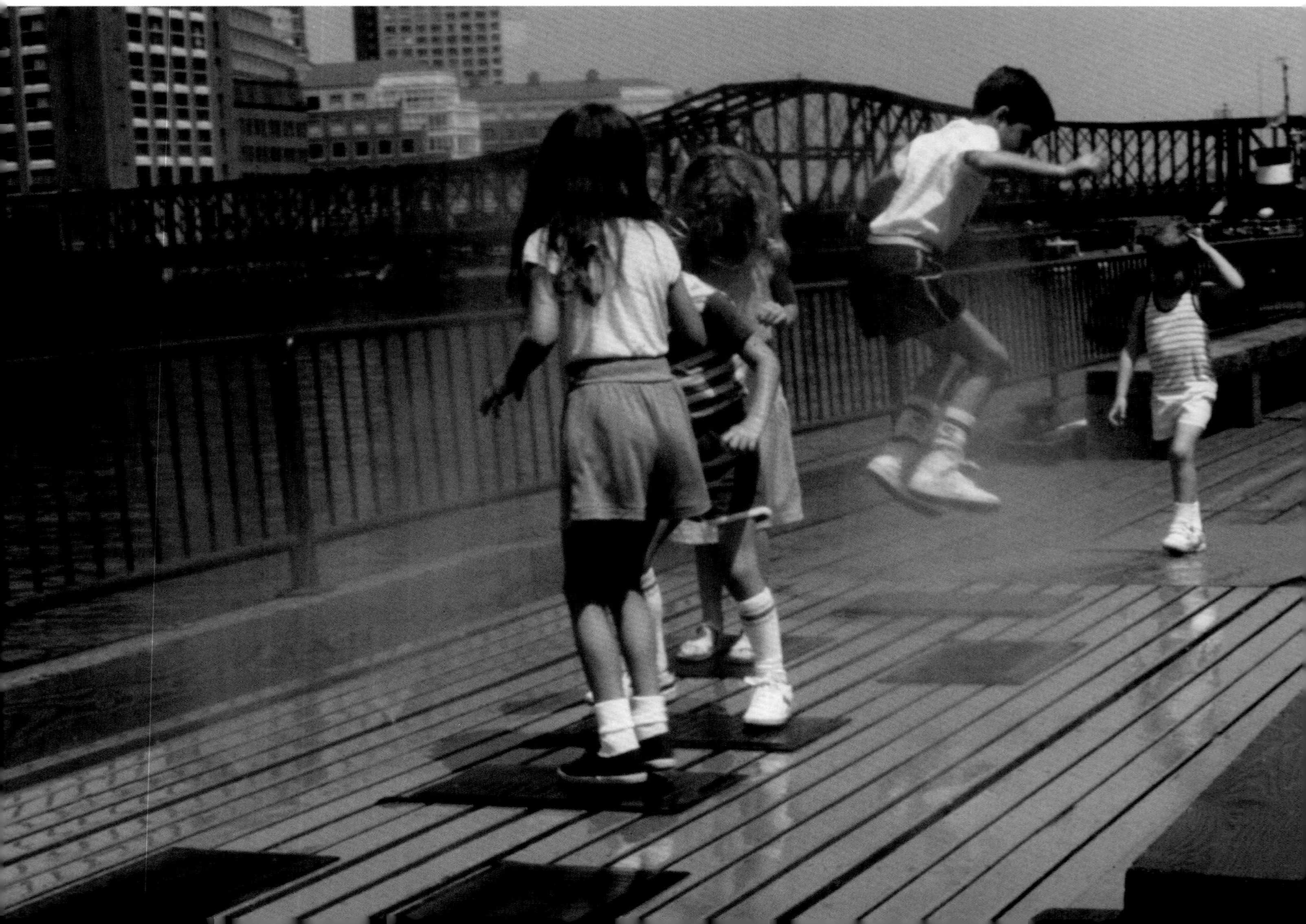

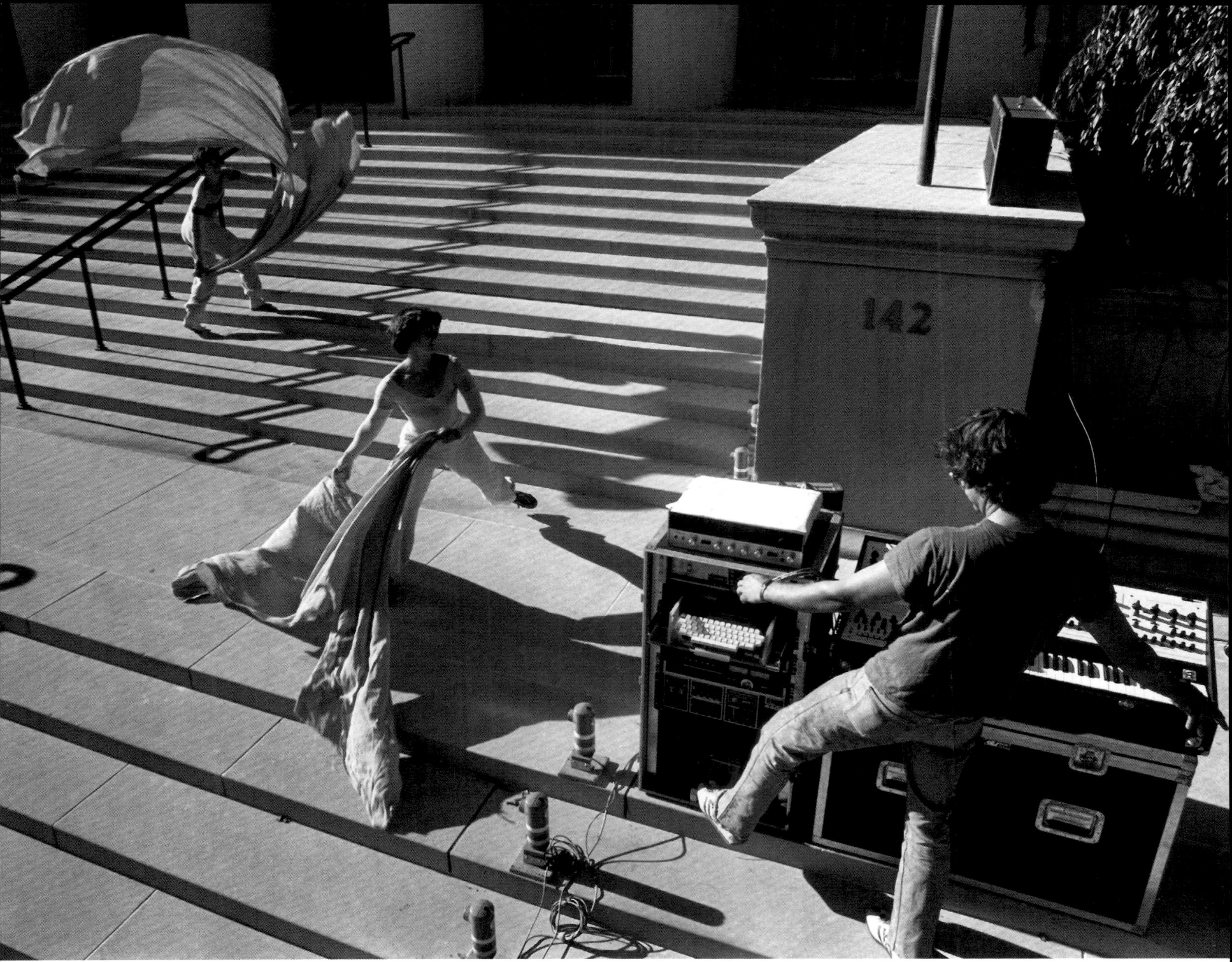

"Soundstair" transforms any staircase into a musical instrument. Sensing devices placed on the steps trigger sounds from a synthesizer. Two or more people on the stairs "play" the stairway together. . . .

"Hopscotch" is a water fountain composed of eight pressure-sensitive pads wired to a computer, synthesizer, and solenoid-activated water-mist jets. As people jump they activate the jets and sound. These phenomena follow the pattern "played," or preprogrammed patterns. In addition, there is a riddle that, when solved by stepping on the pads in a particular sequence, "Hopscotch" responds with a "dance" of its own. (Janney project descriptions, 1989)

Above: Christopher Janney, "Soundstair," the artist and dancers, Walker Memorial, MIT, 142 Memorial Drive, Cambridge, 1979
Photo: D. Dubusque

Facing page: Joan Brigham and Christopher Janney, "Hopscotch," interactive installation, Boston Children's Museum, Fort Point Channel, Boston, 1989
Photo: Christopher Janney

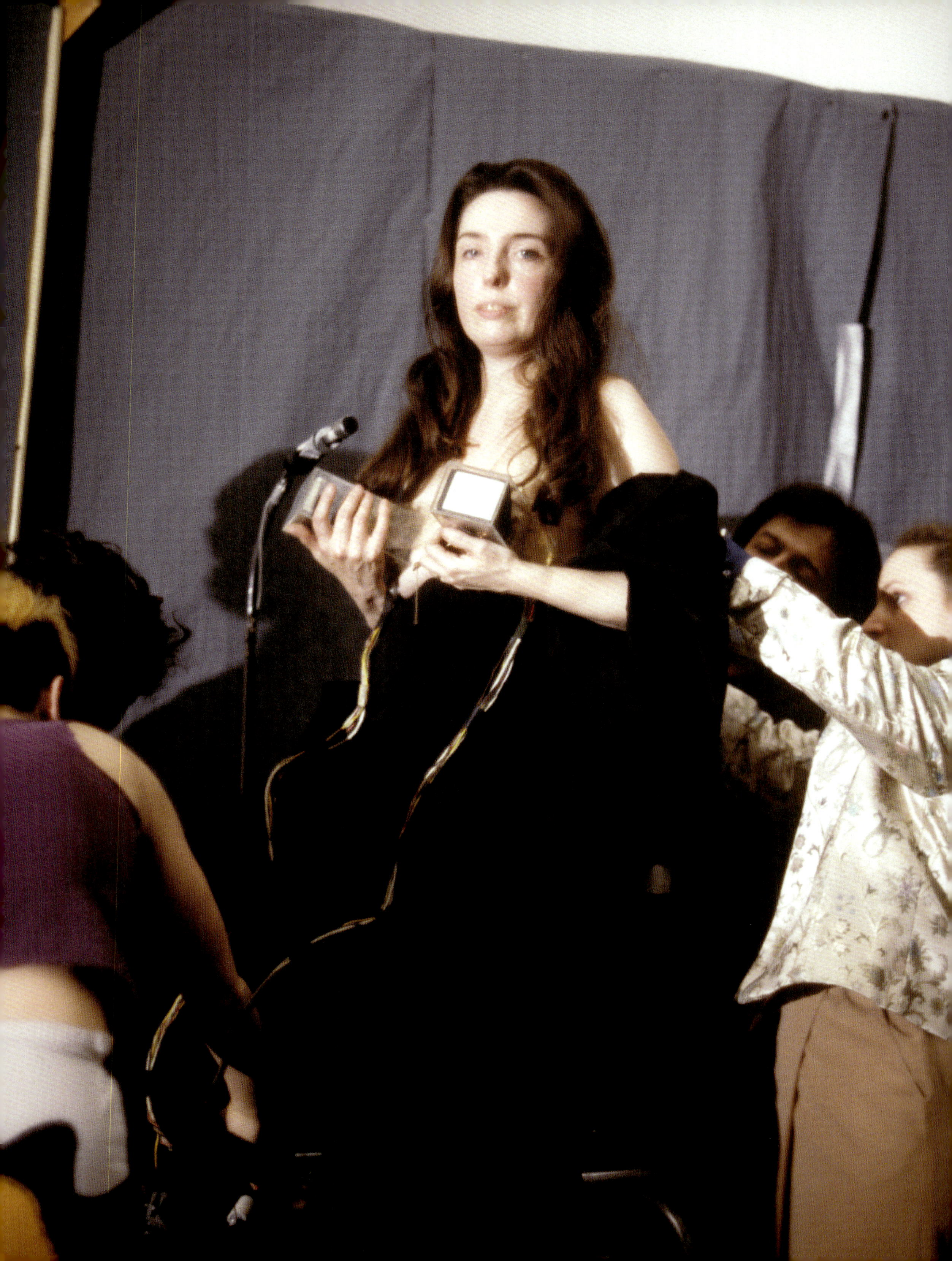

Charlotte Moorman and CAVS

This section looks at the intimate relationship between CAVS Fellow, avant-garde performance artist and cellist, Charlotte Moorman (1933-1991), and the community at the Center for Advanced Visual Studies.

Moorman was a tireless provocateur, most famous for the "TV Bra for Living Sculpture" and "TV Cello," which she performed with her long-time collaborator, CAVS Fellow Nam June Paik. Other collaborators included John Cage, Carolee Schneemann, Joseph Beuys, Takehisa Kosugi, Paul Earls, and Otto Piene. Her presence electrified the social aura of CAVS, and fellows stepped up to work with her. Keiko Prince described the iron will behind Moorman's southern beauty queen charm: "Charlotte knew what she wanted."

Moorman performed at CAVS throughout the 1980s, and, together with Otto Piene, "Sky Kiss" at each of the Sky Art Conferences. The following pages cover events on May 16-17, 1980.

> The May 16 performance will be dedicated to Howard, Barbara and Juliet Wise. Mr. Wise is president and founder of Electronic Arts Intermix, Inc., a non-profit corporation based in New York City that fosters the use of video as a means of individual expression and communication. (Paula Ruth Korn, MIT News Office, May 14, 1980)

> I would like to dedicate these performances to Howard and Barbie Wise . . . I like to let people know while I'm still alive how much they have meant to me throughout my career. Paik is especially happy about your invitation to me and we have already hired an engineer and work has started toward getting his t-v cello ready to appear in this "golden oldies" performance art series. (Charlotte Moorman, letter to Otto Piene, April 25, 1980)

Facing page: Charlotte Moorman prepares to perform "TV Bra for Living Sculpture" (1969) by Nam June Paik, CAVS, MIT, 1980
Photo: Otto Piene Collection

V 5 80

night letter

Charlotte Moorman

your letter & material have arrived stop letter in the mail stop please prepare for May 16 and 17 concerts stop please call stop love

Otto

CHARLOTTE MOORMAN 47 W. 46th n.y. 10036

new address: 62 Pearl Street
New York, New York 10004
(212) 344-6434
May 21, 1980

Provost Francis Low
Department of Physics
Massachusetts Institute of Technology
Cambridge, Massachusetts 02139

Dear Dr. Low:

Thank you so much for honoring me with your presence at my concert this past Friday night. I was especially happy that you participated in Yoko Ono's CUT PIECE, and I am enclosing a piece of cloth from the dress as a small momento.

So many people at the Center for Advance d Visual Studies were more than considerate in going so far out of their way in putting aside the pressing work of the year's end to help me set up, document, publicize, and produce this concert series, as well as actually perform some of the pieces with me. I would like to extend my hearty thanks to Otto Piene, Nishan Bichajian, Joan Brigham, Lowry Burgess, Mira Cantor, Glorianna Davenport, Paul Earls, Dan Gibson, Elizabeth Goldring, Vin Grabill, Pat Hearn, Astrid Hiemer, Ken Kantor, Paula Korn, Shelley Lake, Mandi McIntyre, Michael Naimark, Bill Parker, Keiko Prince, Brian Raila & Aldo Tambellini.

In all my years of performing around the world, I have never seen such a unique institution as C.A.V.S., where artists, academics, and community can share in the freedom and stimulation of the creative arts. M.I.T. is to be congratulated in its continuing support of the Center and its vital work.

I enjoyed meeting you and performing at M.I.T.

Sincerely,

Charlotte Moorman

Massachusetts Institute of Technology Center for Advanced Visual Studies 40 Massachusetts Avenue Cambridge, Massachusetts 02139

- CONCERT -

C H A R L O T T E   M O O R M A N

Friday, May 16, 1980

9:00 P.M.

Program

Concerto for TV Cello (1971) by Nam June Paik
or
TV Bra for Living Sculpture (1969) by Nam June Paik

One for Violin Solo (1962) by Nam June Paik

Instrumental Music (1965) by Takehisa Kosugi

Cut Piece (1961) by Yoko Ono

Intermission

Variations on a Theme by Saint-Saëns (1965) by Nam June Paik

Infiltration - Homogen für Cello (1966) by Joseph Beuys

Guest Artists: Paul Earls, Kieko Prince

Telephone (617) 253-4415

Massachusetts Institute of Technology Center for Advanced Visual Studies 40 Massachusetts Avenue Cambridge, Massachusetts 02139

- CONCERT -

C H A R L O T T E   M O O R M A N

Saturday, May 17, 1980

9:00 P. M.

Program

TV Bra for Living Sculpture (1969) by Nam June Paik

Chamber Music (1963) by Takehisa Kosugi

26 1.1499' for a String Player (1953-'55) by John Cage

Shoes of Your Choice (1964) by Alison Knowles

Intermission

That Great Cello in the Sky (1980) by Paul Earls
(World Premiere)

Sonata for Adults Only (1964) by Nam June Paik

Guest Artists: Paul Earls, Ken Kantor

Telephone (617) 253-4415

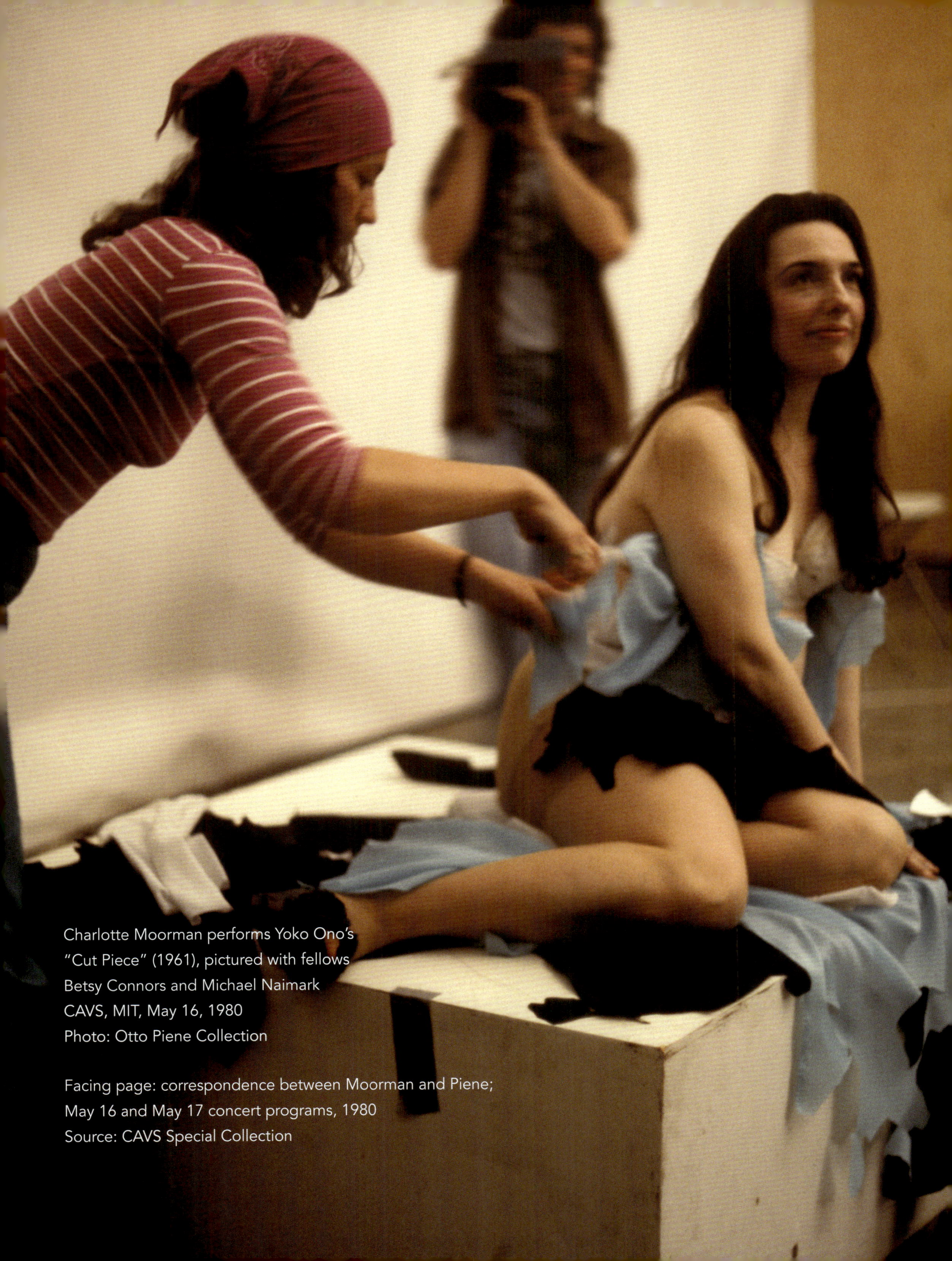

Charlotte Moorman performs Yoko Ono's "Cut Piece" (1961), pictured with fellows Betsy Connors and Michael Naimark
CAVS, MIT, May 16, 1980
Photo: Otto Piene Collection

Facing page: correspondence between Moorman and Piene; May 16 and May 17 concert programs, 1980
Source: CAVS Special Collection

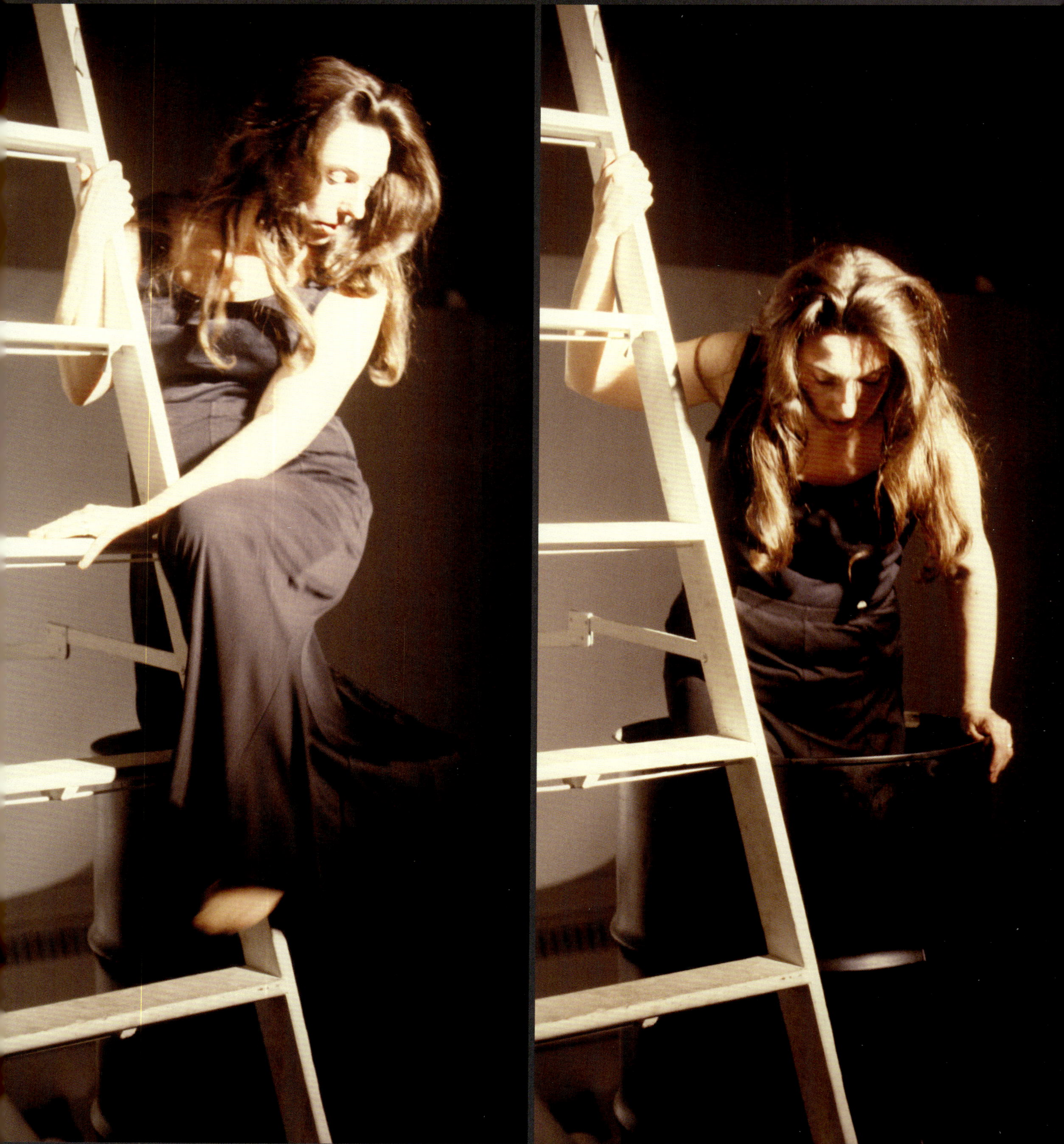

Charlotte Moorman performs "Variations on a Theme by Saint-Saëns" (1965) by Nam June Paik, the Pit, CAVS, MIT, May 16, 1980. In the work, Moorman's cello performance of Saint-Saëns' 1886 composition, "The Swan," with Paik

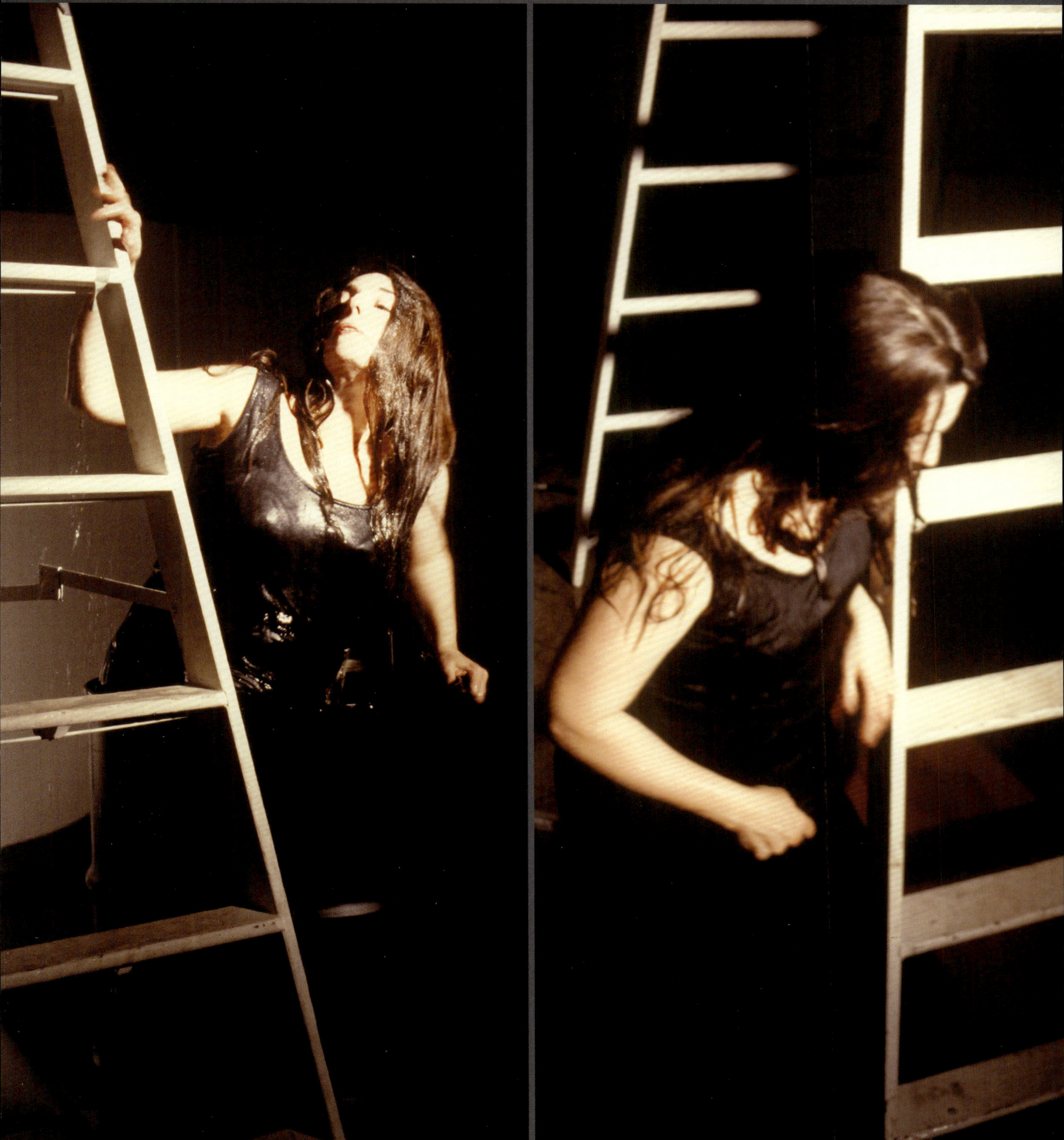

on piano, was punctuated by a section in which she stopped playing and immersed herself in a drum of water. Moorman then returned to her cello, and she and Paik finished playing the piece. Photos: Otto Piene Collection

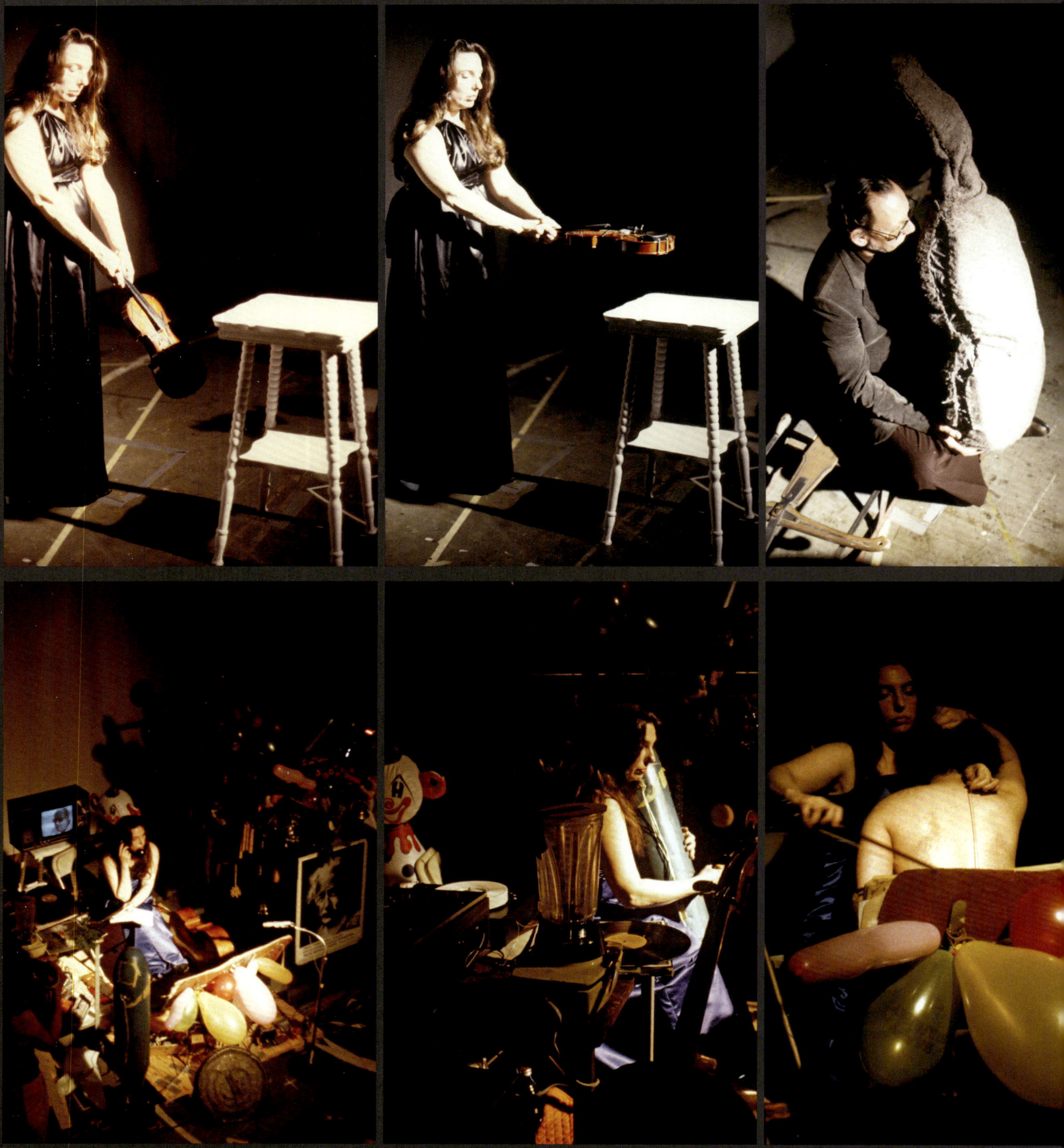

Top row, left to right across both pages: Charlotte Moorman, performance of "One for Violin Solo" (1962) by Nam June Paik; "Infiltration-homogen für Cello" (1966) by Joseph Beuys, with Paul Earls; "TV Bra for Living Sculpture" (1969) by Nam June Paik, the Pit, CAVS, MIT, May 16, 1980. All photos: Otto Piene Collection

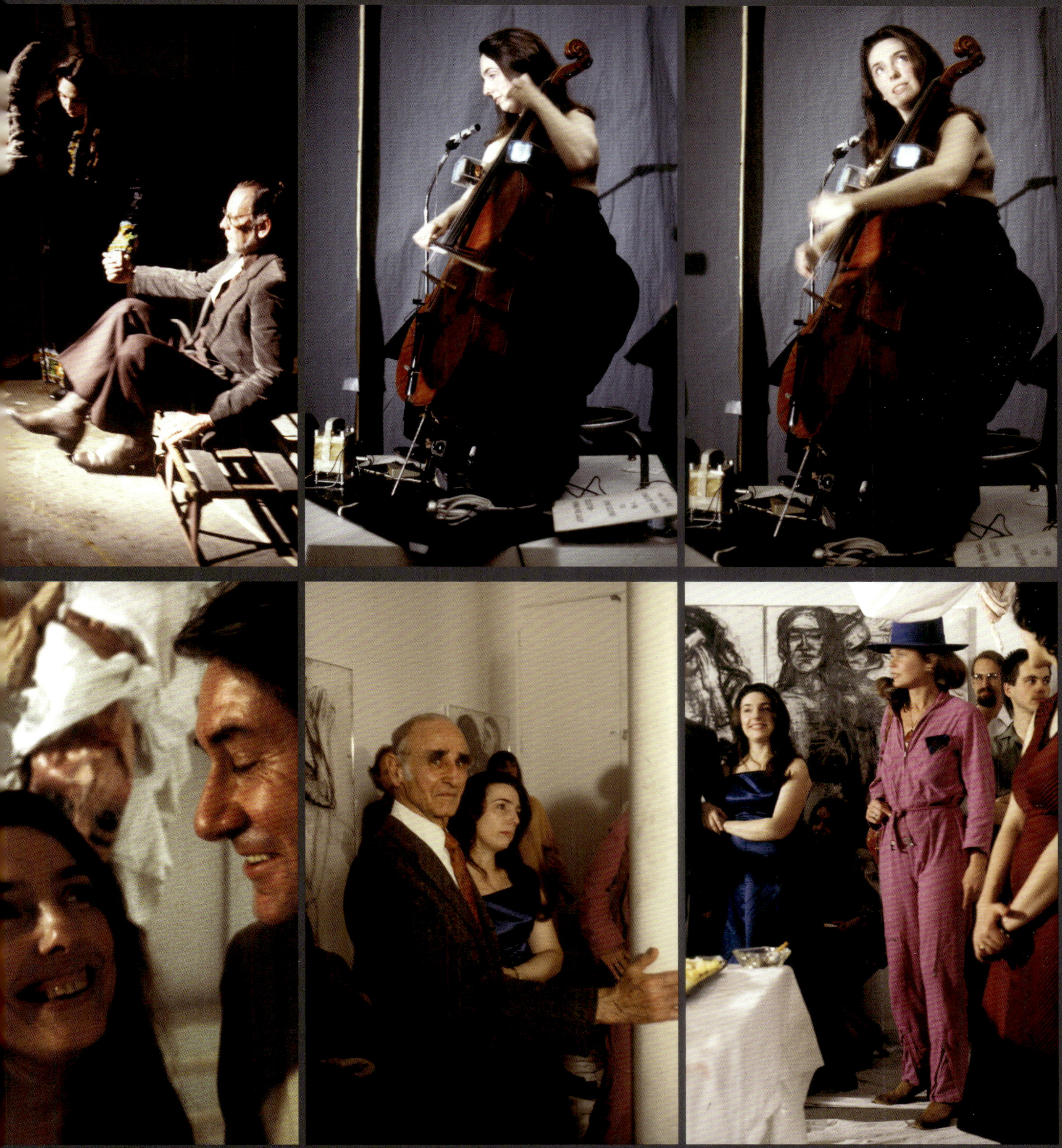

Bottom row, left to right across both pages: Charlotte Moorman, performance of "26'1.1499 for a String Player" (1953-55) by John Cage, with Ken Kantor; Moorman at the reception during the closing of Mira Cantor's *Centersheet* installation, with Otto Piene (left), Howard Wise (middle), Elizabeth Goldring (right), CAVS, MIT, May 17, 1980

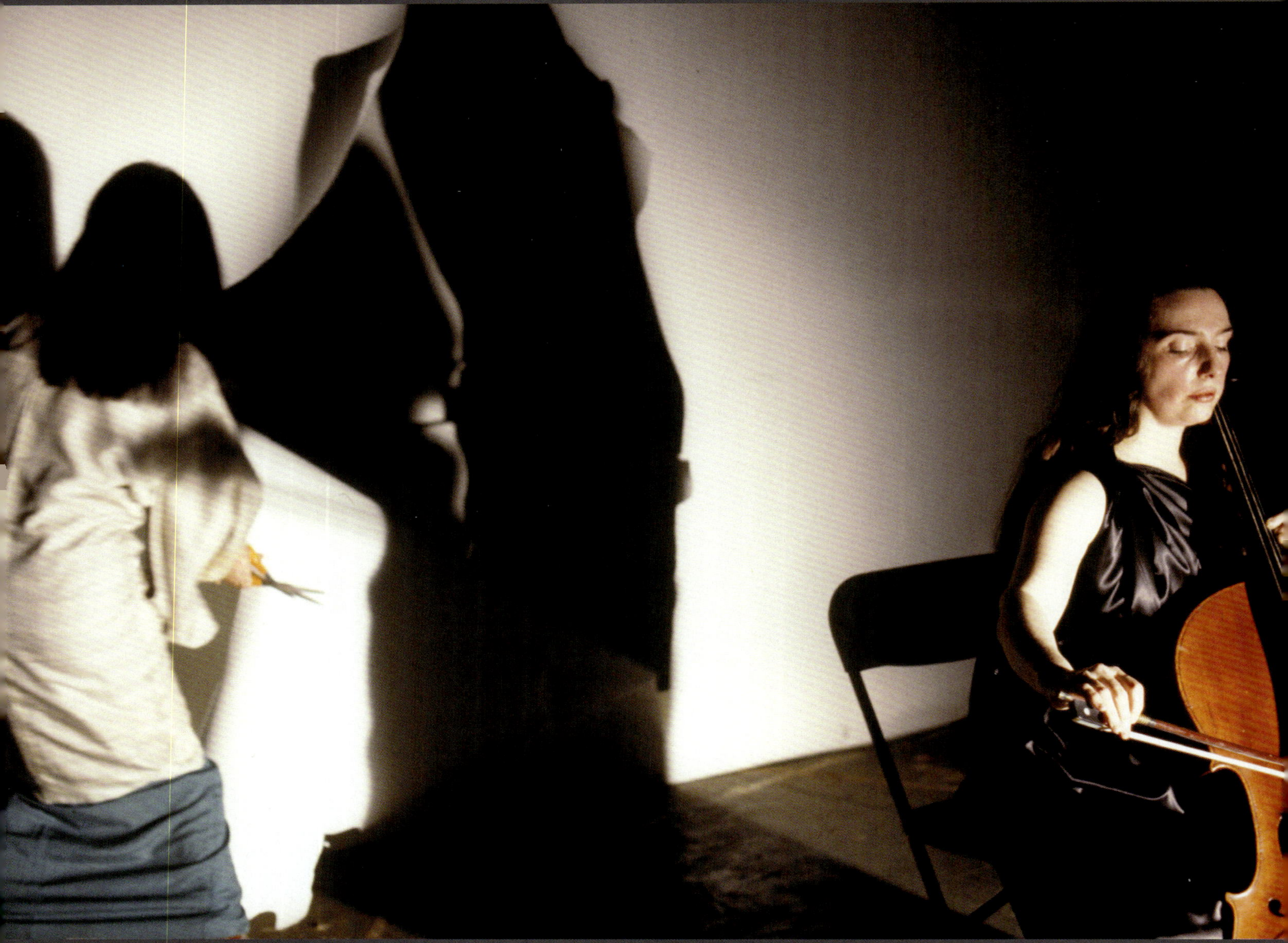

Charlotte asked me to wear kimono, which I did. She was really very demanding about the performances. We didn't worry about regulations, like fire. Otto was concerned. (Keiko Prince, 2018)

Top: Charlotte Moorman, performance of "Instrumental Music" (1965) by Takehisa Kosugi, with Keiko Prince, the Pit, CAVS, MIT, May 16, 1980
Photos: Otto Piene Collection

Facing page: Charlotte Moorman, performance of "Chamber Music" (1963) by Takehisa Kosugi, Moorman dedicated the performance to Elizabeth Goldring who, at the time, was suffering from blindness, the Pit, CAVS, MIT, May 17, 1980
Photos: Otto Piene Collection

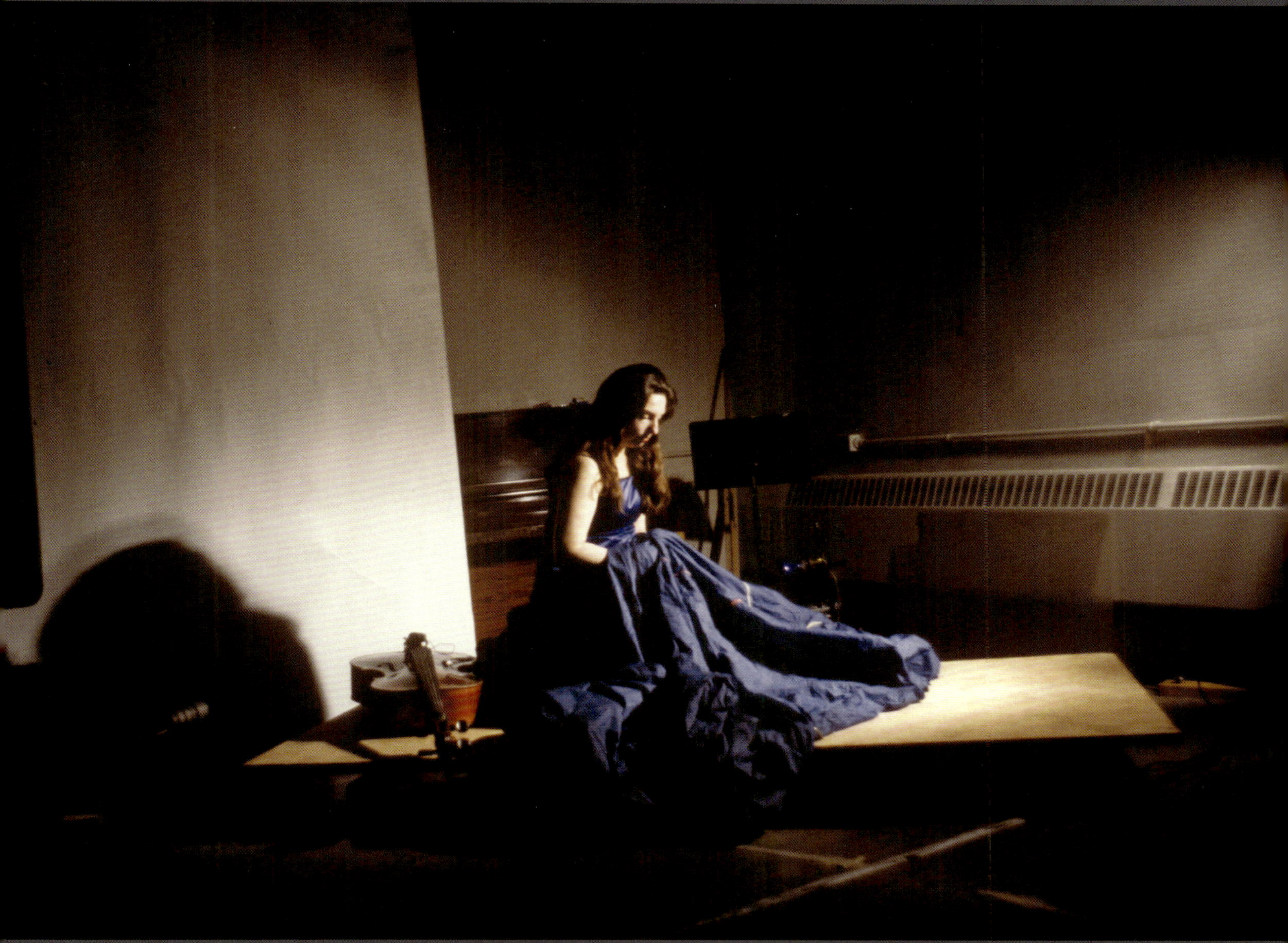

FOR CHARLOTTE (censored hero of *The Opera Sextronique**). DYING HERO IN YOUR FIGHT WITH CANCER

You played your cello wired, topless,
under water,
coming out of Kosugi's blue bag.
You made ice cellos, chocolate cellos, human cellos,
bombs, T.V.s.
You played cello fish, cellos at Guadalcanal,
red crosses pinned to Josef Beuys' felts.

When Otto Piene said he'd fly you from his helium flower,
you didn't believe your luck.
You, fastened to a grand bouquet of helium-filled,
polyethylene loops, 200 foot-long tubes bent in half.
You, a cello strumming, flying sculpture, a sky angel,
floating out across the Danube, "somewhere in Austria,"
going up to 300 feet,
maybe higher.

For the premiere of "Sky Kiss"
you wore your custom-padded parachute harness,
your green gown and white satin cape.
Otto checked the ropes himself.
He didn't want to lose you,
wouldn't let you disappear
like a Queen.

—poem by Elizabeth Goldring, from "Without Warning,"
Helicon Nine Editions, BkMk Press, University of Missouri-Kansas City, Kansas City, Missouri, 1995

**The Opera Sextronique* (1967) by Nam June Paik

November 22, 1991

Mr. Frank Pileggi
62 Pearl St. 5th floor
New York, NY 10004

Dear Frank:

For years now I have been dreading this moment telling myself it would never happen, or I wouldn't live to see it. Even when Paik—at *ARTTRANSITION '90*—said, quite softly, that Charlotte's performances would be her last with Paik, I refused to accept it; not vis-a-vis him, not to myself. Like me, many others will have to come to terms with Charlotte's death—or not. Today it is my duty and my honor to tell you, her husband of unwavering and heroic loyalty, that this place with everybody in it—fellows, staff, graduate students—will not be the same without her, will not be the same with the thought that she will not come back. I hope you will come and help us honor her—and you—with events and commemorations in the future.

All of us at CAVS—and many of those who have lived and worked here in the past—send you their compassionate embrace and their most intense greetings and wishes. We vow to nurture the greatest respect and love for Charlotte the person and Charlotte the artist forever; we will contribute our best to making the world understand and remember what a radiant woman Charlotte is in the world and the art of our confused century.

Our love to you who has helped her live and work beyond human limits.

Otto Piene
Fellows
staff
graduate students, CAVS

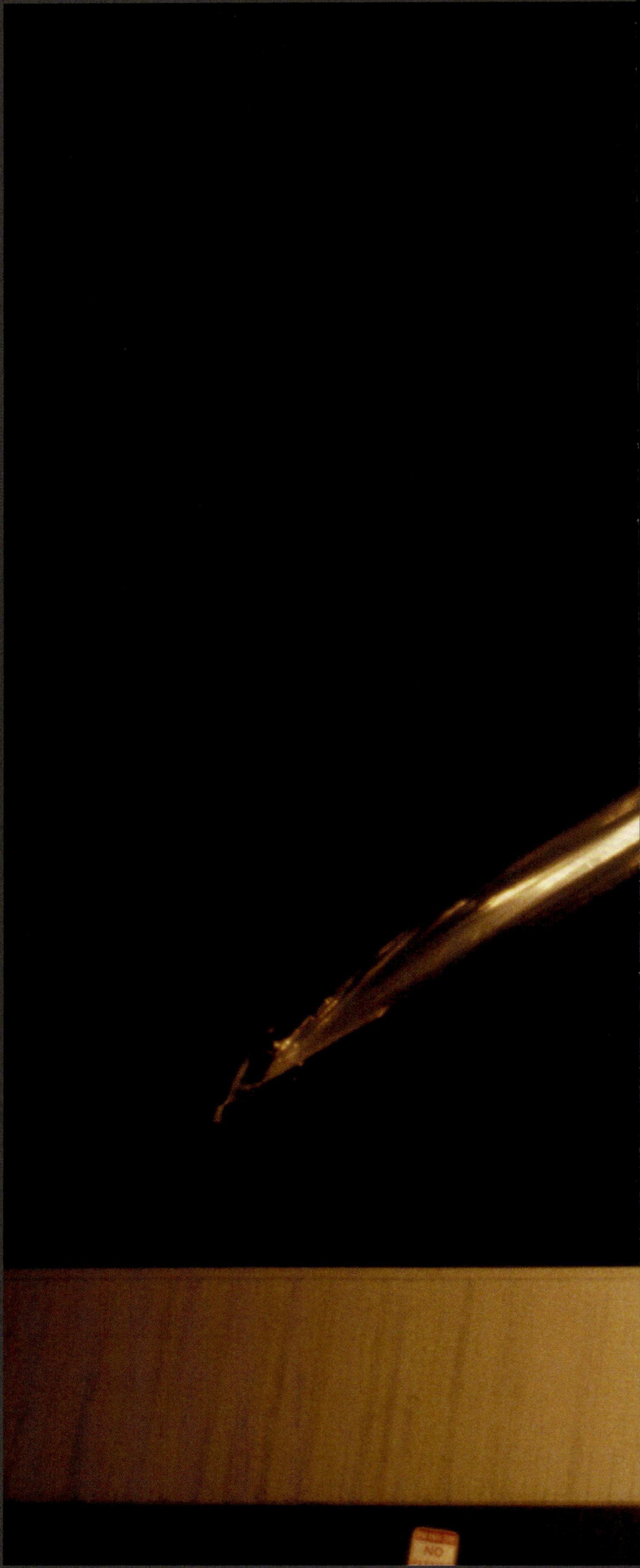

Right: Charlotte Moorman prepares to perform "Cello Sonata" by Mieko Shiomi, pictured with Otto Piene and his "Iowa Star" inflatable, 20th year reunion, CAVS rooftop, MIT, 1988
Photo: Walter C. Dent

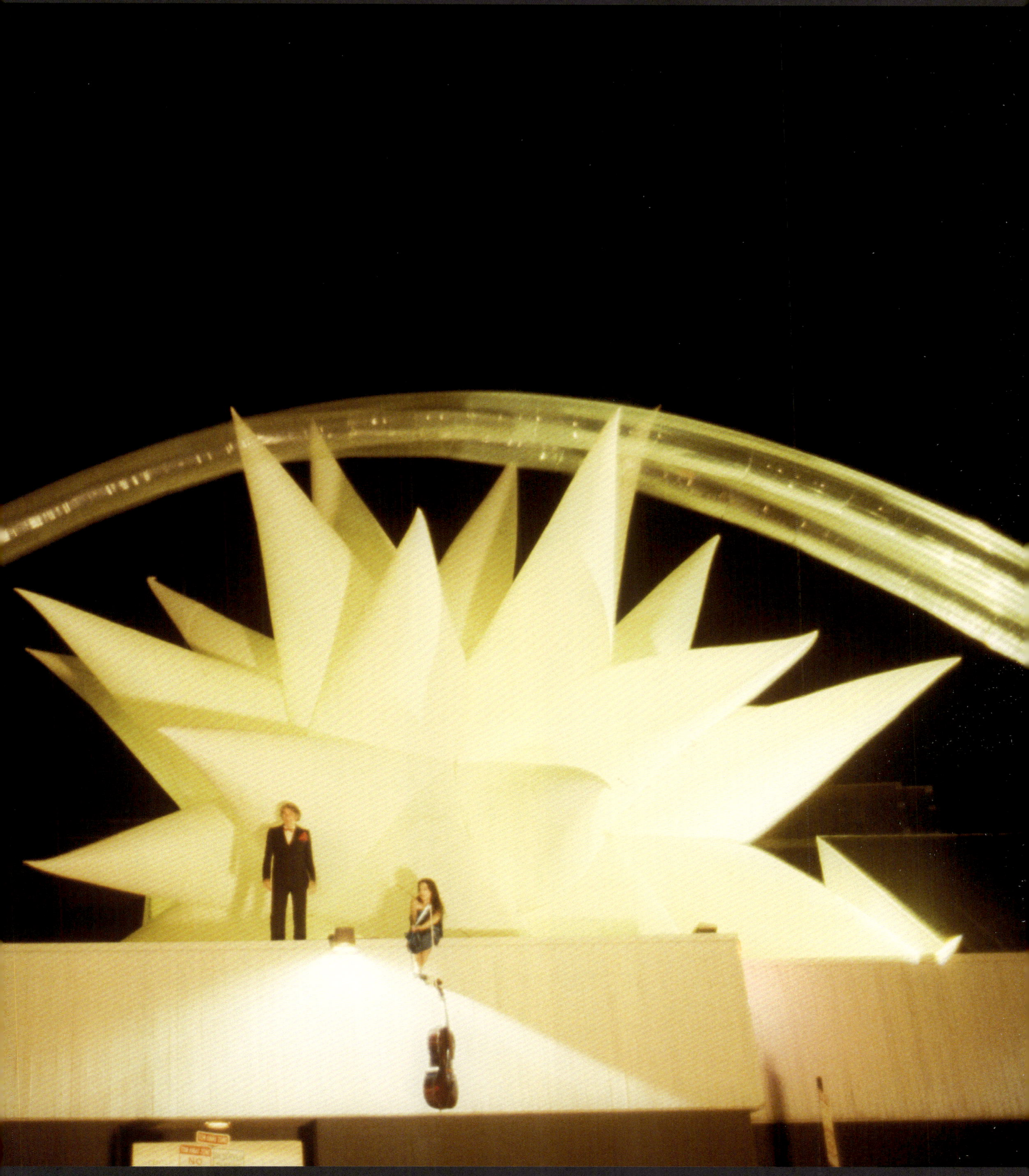

Chapter 5

Exhibitions: The Artist as Curator

The role of artist as curator, an innovative idea in the late 1950s, was adopted by Kepes at the New Bauhaus in Chicago, and by Piene with Group Zero and the Evening Exhibitions in Düsseldorf. Under their directorships, CAVS group exhibitions were curated and developed by artist fellows. The exhibitions—kinetic, interactive, and frequently informed by the theme of light—featured current projects and experiments in art, science, and technology. Just as the CAVS Fellows constituted an international group, so too were CAVS exhibitions international in scope, appearing in European, Asian, Mexican, Canadian, and South American as well as US venues.

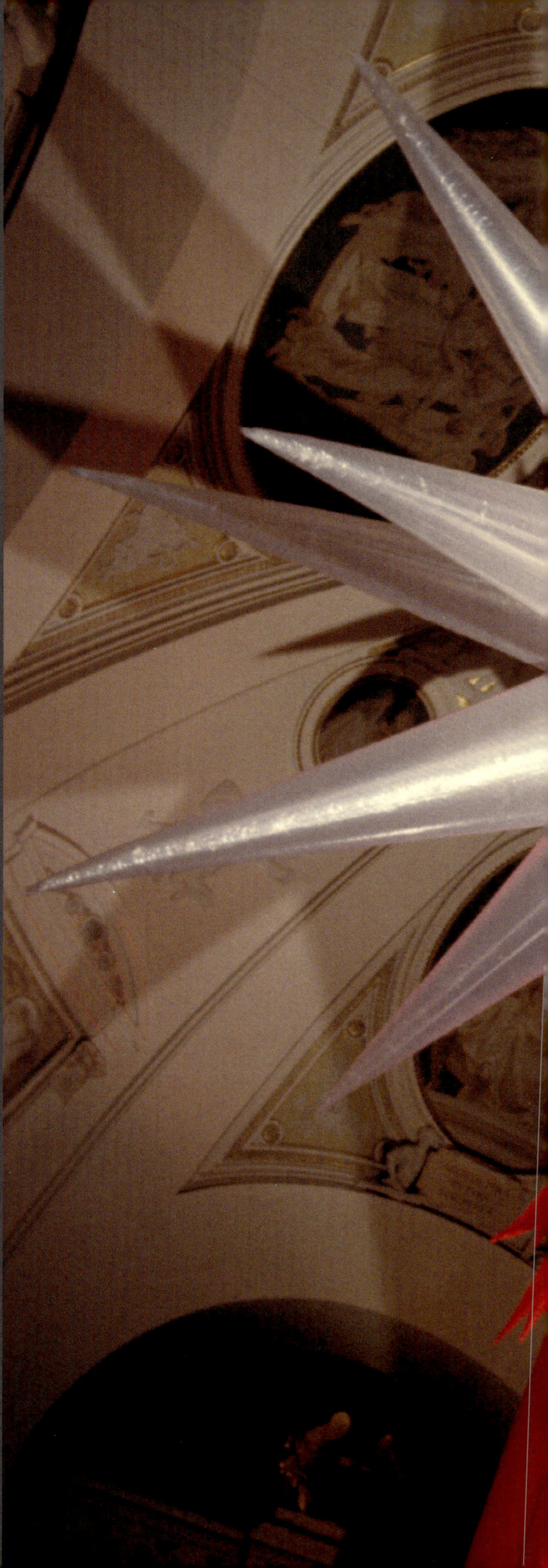

Right: Otto Piene, "Red Rapid Growth," four inflatable flowers soar into the MFA's Sargent Rotunda to reach the murals designed by John Singer Sargent; part of the exhibition *Earth, Air, Fire, Water: Elements of Art* curated by Virginia Gunter and including other CAVS artists, Museum of Fine Arts, Boston, February 4-April 11, 1971
Photo: Otto Piene Collection

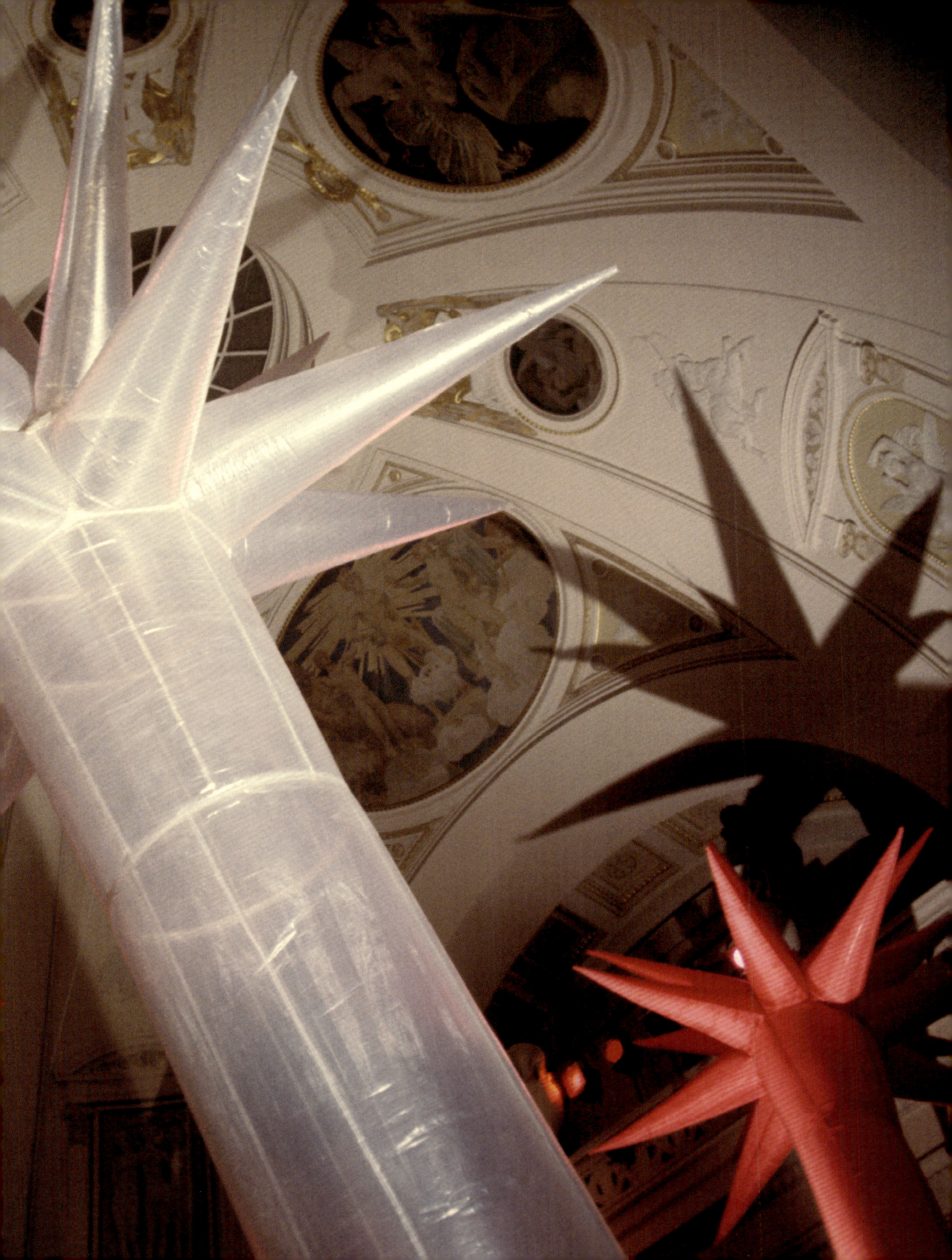

The Artist as Curator

CAVS artist fellows developed exhibitions with kinetic, interactive forms suffused with light as a plastic medium.

> A set program is not envisaged for the Center, but rather a sequence of major collaborative projects. Some involve all the Fellows, others teams of two or three. The creative use of light on a great scale has been designated a major interest. (György Kepes, *The Center for Advanced Visual Studies* inaugural catalog, 1968, 21)

The direct intervention of artists was required for exhibitions that sought to redefine art through new media, technologies, and scientific apparatus. It was not enough to rely on career exhibition curators.

As the media changed, the artists' vision also penetrated beyond the confines of traditional exhibition venues:

> The isolated little cells of space in which we now display works of art are suffocatingly small for the fluid power of light in action. The new, rich intensities of artificial light sources should become part of the big fabric of the night cityscape. The mirroring windows and interpenetrating mobile vistas are the proper background for creative figures that will be shaped by moving contours of lights. (Kepes, *The Center for Advanced Visual Studies* inaugural catalog, 29)

Art at the Center penetrated all aspects of human endeavor, including food, sleep, clothing, celebrations, and the built environment, and would lift off the earth into the sky and space. Art, and artistic events were not confined to museums:

> At its best the urban lightscape is one of the grand sights of our age. . . . This accidental magnificence contains the promise of a great new art: the orchestration of light. (Kepes, Ibid., 24)

CAVS exhibitions reflected its international community, ranging across the US and beyond to Europe, Asia, Mexico, Canada, and South America. Usually democratically conceived and based on voluntary contributions, more than thirty exhibitions were presented between 1968 and 1994. They addressed a wide range of themes that resulted in lively—if not messy—exhibitions with new ways of communicating with the public. "More than ever before the concepts of what art is are broader and more experiential," wrote CAVS artist Virginia Gunter, who produced the 1971 *Earth, Air, Fire, Water: Elements of Art* exhibition at Boston's Museum of Fine Arts. Her title, Producer, was "the museum's word, obliquely acknowledging the show biz nature of the event." (Kramer, *The New York Times*, Feburary 4, 1971, 28)

> This exhibition of necessity moves outside the Museum walls and even within the gallery space the works do not all remain constant and static during the exhibition. The artists represented by this exhibition are aggressive in their assertions that the visual aesthetic experience is not enough. The work may totally surround the viewer; it may necessitate the use of other senses; it may invite physical participation; it may be a demonstration of a system or process. Indeed the work may exist only in descriptive verbal form. All of these possibilities challenge the assumption of the past that the experience of art is essentially contemplative. (Gunter, *Earth, Air, Fire, Water: Elements of Art* catalog, 1971, 5)

Venues ranged from museums to unconventional public exhibition spaces like the BMW Headquarters in Munich or the Massachusetts Avenue entrance of MIT, Lobby 7, where the sun creates a light show of its own twice a year when it aligns with the entrance of the building to illuminate the full length of the Infinite Corridor.

Artist and poet Elizabeth Goldring was CAVS Exhibits and Projects Director from 1979 to 1993,

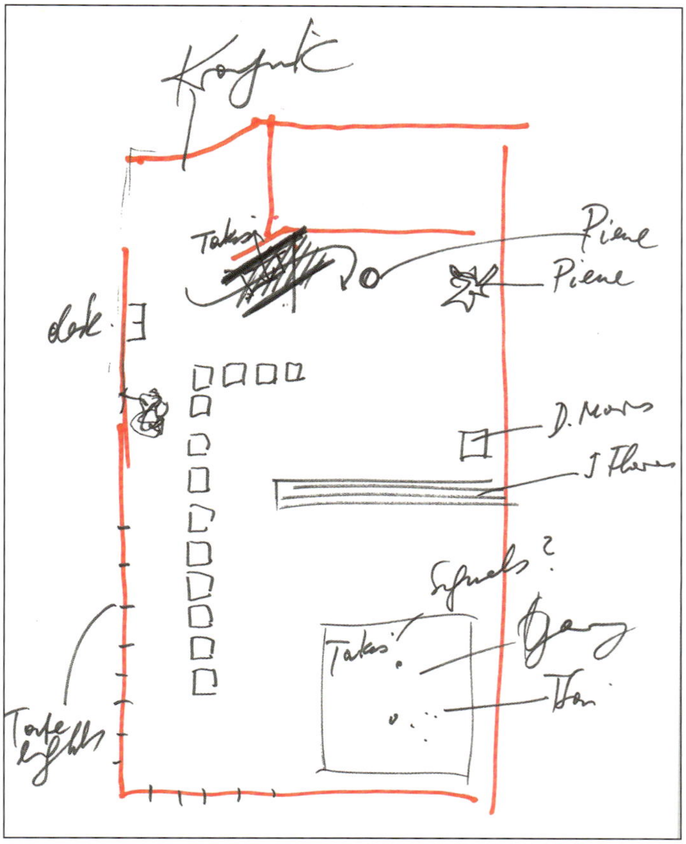

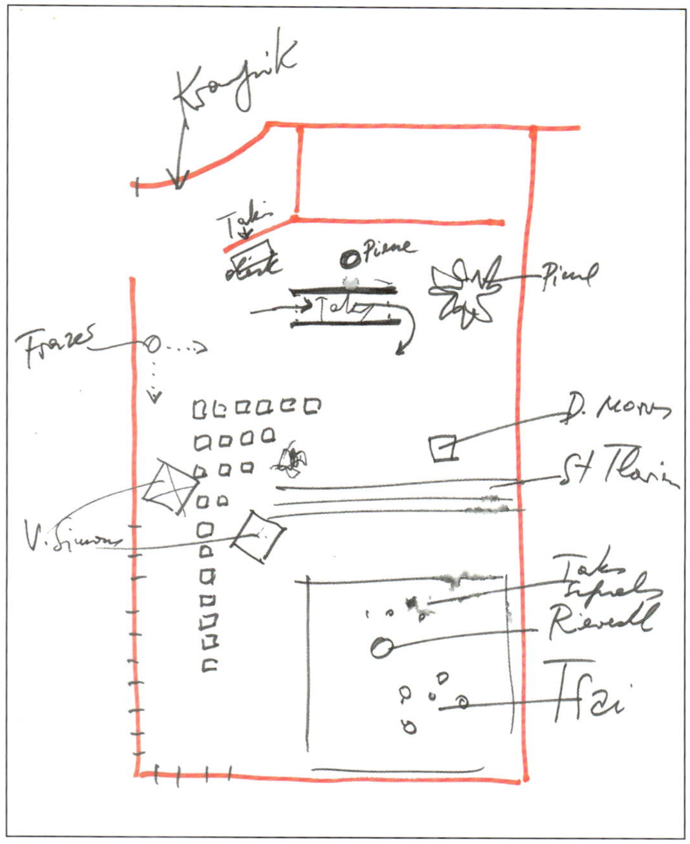

during the heyday of the international group exhibitions that complemented individual exhibitions by CAVS fellows. Countless memos crossed her desk detailing the exhaustive logistics required by technologically innovative art. Steam was plumbed from under the streets; refractive light traversed the Charles River; sensors triggered musical tones in stairways—the physical plant as artist's palette depended on imaginative solutions and unusual relationships. While technology played an important, sometimes obvious role, these unexpected juxtapositions focused on art with a capital A.

Above: György Kepes, design sketches for CAVS exhibition, *Exploration*, Hayden Gallery, MIT, February 28-March 29, 1970
Source: CAVS Special Collection

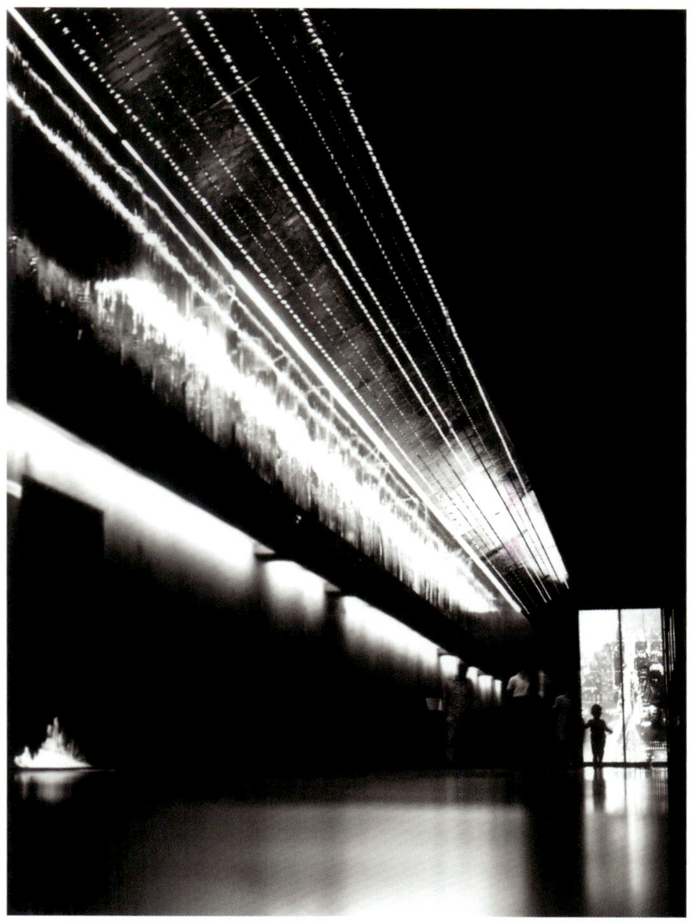

Right: György Kepes, "The Nightscape of the City," T. McNulty and O. Stevens, collaborators, *14th Triennale di Milano*, Milan, 1967-68
Photo: CAVS Special Collection

The *Exploration* Dilemma

The first CAVS group exhibition, *Exploration*, was originally conceived for the controversial tenth *São Paulo Biennial* (1969). Kepes was chosen to "direct the design and construction of the United States entry, working with a number of the Fellows at the Center." (Smithsonian Institution memo, May 1969) The repressive tactics of Brazil's military regime and the issuance of Institutional Act No. 5 (AI-5)—which suspended political and civil rights and imposed heavy censorship—set off an international protest of the *São Paulo Biennial.* The tenth edition (1969) was nicknamed the "Boycott Bienal," with eighty percent of the invited artists refusing to participate.

> Since the end of March, all efforts at the Center have been focused on preparing and designing an exhibition for the tenth Bienal de São Paulo (the major international art exhibition in the world). It was a singular honor for the Center to be asked by the Smithsonian Institution's National Collection of Fine Arts to represent the United States and take over the planning and execution of the United States section. Though there were major obstacles—a limited budget, limited time for preparation, and complex ideological and human problems—the Center accepted the challenge, for it offered an opportunity to express collectively the convictions of the artists working at the Center. Social and environmental interdependence, as fundamental to a livable contemporary life, is the core of this conviction. To symbolize this basic belief, our intention was to form, with a wide range of vital art works, an environmental community, which, like a synergetic system, can exist only through the interconnected workings of its members. To make the exhibition as convincing and as comprehensive as was warranted by its role and theme, the Director invited twelve other artists from outside the Center to contribute to its common objective. To the great disappointment of the Center, more than one third of the participants withdrew from the exhibition (some of them in the last few days before the shipping date) for complex political reasons. They had chosen to boycott the exhibition as their way of protesting against the political situation in Brazil, which had brought serious hardship to fellow artists there. (Kepes, *MIT Report of the President*, 1969)
>
> Generally I got along with Kepes until the spring of 1969. The main issue of my dissent was his insistence that the Center participate in an international exhibition at the Sao Paulo Museum. I guess I was far more in touch with the New York art scene. I tried to convince him he was making a political mistake as far as the New York art world was concerned, which of course was far left of center. At the time Brazil was a military dictatorship, with considerable outcry over its death squads and lack of human rights. To much of the free world, Brazil in 1969 was an anathema. To Kepes what mattered was how it looked to MIT's administrators on the Center's resume. Harold Tovish declined, but he was forgiven because he was a staunch leftist whose credentials were well known. Most of the Center flew down. As far as I know no one was abducted. In retaliation for my disloyalty, Kepes had Sylvania cut off my supply of electroluminescent tapes. It was no hardship since they had been terribly generous. I had enough tapes to mount four more environments. I took the cessation of free tapes as a sign from the Gods that perhaps one should work elsewhere. In retrospect after thirty-five years, Sao Paulo is more like Timbuktu than Venice or Kassel for the New York art world—it could not have cared less in all probability. Still, 1969 was a hyperpolitical time in American history. (Jack Burnham interview)

Kepes ultimately opted to cancel CAVS participation in the *São Paulo Biennial*, writing in a letter to Fellow Vassilakis Takis:

> As you must know by now, I had to make the decision to withdraw from the Sao Paulo Bienal; and to be honest—against my best convictions. The more communication I had with leading Brazilian intellectuals who are living in Brazil, the more inclined I was to believe that an honest exhibition, and the chance to have personal contact with young Brazilian artists, would have been more positive than negative. (Kepes, letter to Takis, July 25, 1969)

Both experienced exhibition designers, Kepes and Piene actively shaped the many exhibitions CAVS would go on to mount, as well as overseeing the critical documentation and accompanying materials.

Beginning with Kepes, exhibitions focused on environmental and civic projects that could not be contained in museums. The same was true under Piene, who emphasized collaborative events and public celebrations as part of exhibitions. Exhibitions mounted on home turf—such as *Exploration* (1970) at MIT's Hayden Gallery (which then traveled to the Smithsonian Institution in Washington DC as an expanded show titled, *Explorations*), *Weather* (1973) and *Food* (1975) in MIT Lobby 7, and *Light/Space/Time* (1994) at the MIT Museum—were labor-intensive, and often required the artists to be on site to help install and maintain their pieces. The poster for *Exploration* at the Hayden Gallery was one of many designed for CAVS events by Jacqueline Casey (1927-1992). Casey worked at the MIT Office of Publications (Design Services Office) for over 30 years and became its director in 1972.

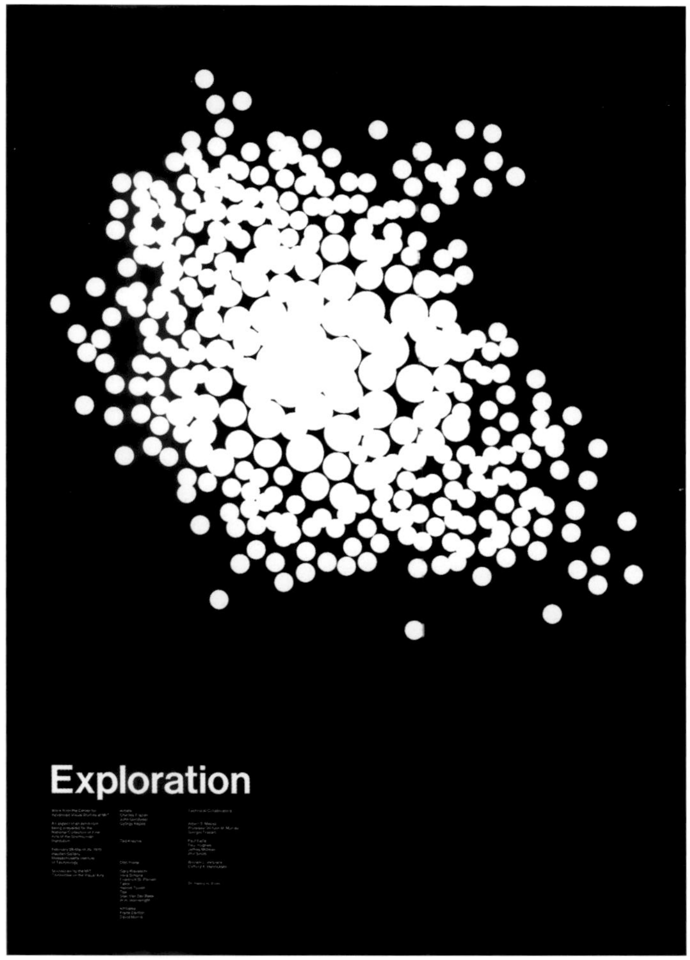

Top: Jacqueline Casey, poster, *Exploration*, CAVS exhibition, Hayden Gallery, MIT, February 28-March 29, 1970, listing:
Artists
Charles Frazier, John Goodyear, György Kepes, Ted Kraynik, Otto Piene, Gary Rieveschl, Vera Simons, Friedrich St. Florian, Vassilakis Takis, Harold Tovish, Wen-Ying Tsai, Stan VanDerBeek, W. H. Wainwright
Affiliates
Frank Carlton, David Morris
Source: CAVS Special Collection

Transparent plastic sheets are placed between crossed polarized screens. Whenever stress occurs, the refractive index of the material is altered and bright color lines will appear which contour the principal stresses. (Kepes/Wainwright, "Photoelastic Walk" description, *Explorations* brochure, Smithsonian Institution, 1970)

Exploration, CAVS exhibition, Hayden Gallery, MIT, February 28-March 29, 1970
Far right: Piene, "Venus of Willendorf"; floor, Kepes and Wainwright, "Photoelastic Walk"; back wall, Kraynik, "Video-Luminar Light Mural"
Photo: Nishan Bichajian

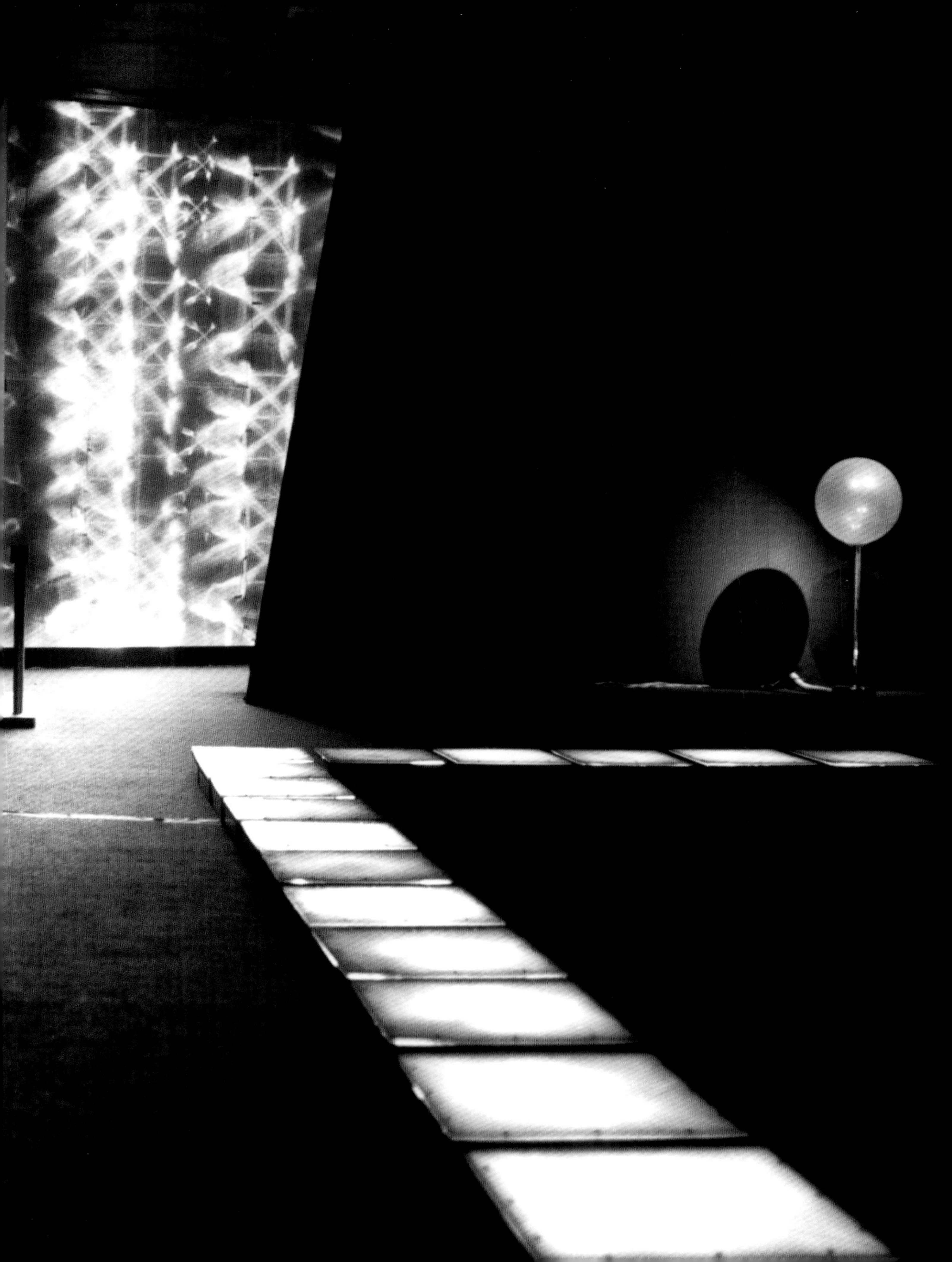

Sensors pick up and translate changing patterns of light. Television picture can be "off the air," closed circuit for audience participation, or videotape. (Kraynik, "Video-Luminar Light Mural" description)

The lights echo what happens on the screen; if your picture appears on the screen, wave your hands in front of your face. (Kraynik, *Explorations* brochure, Smithsonian Institution, 1970)

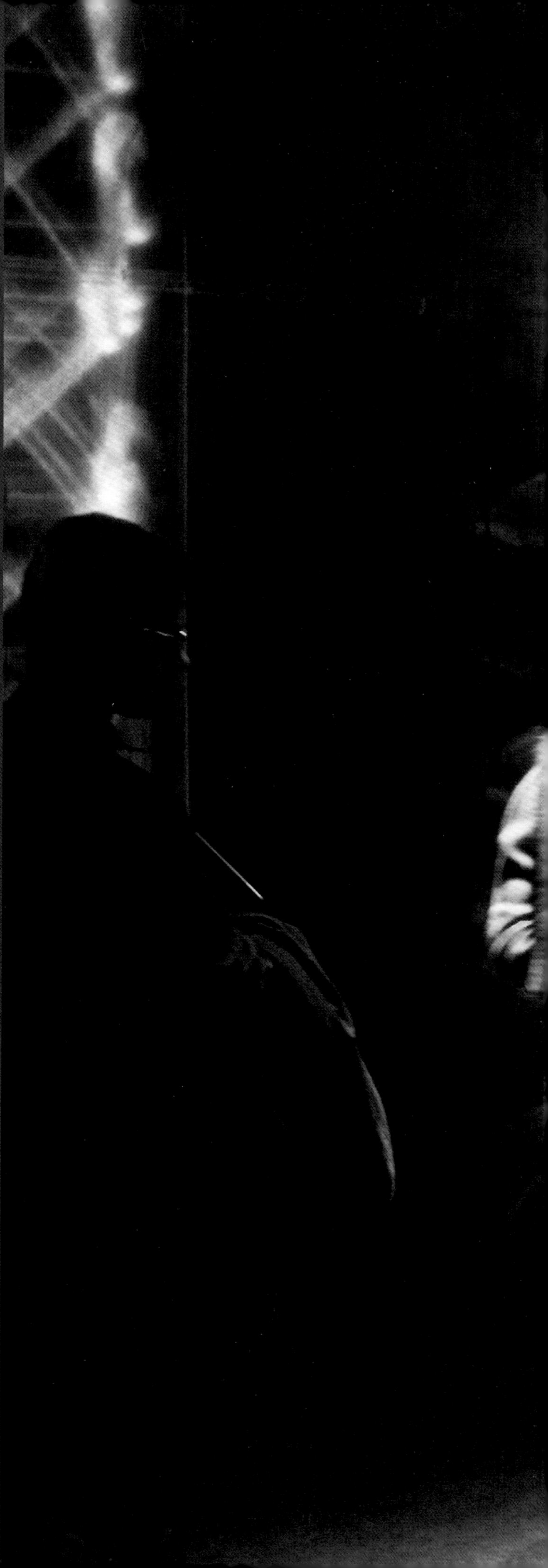

Exploration, CAVS exhibition, Hayden Gallery, MIT, February 28-March 29, 1970
Shown: Kraynik, "Video-Luminar Light Mural"
Photo: Nishan Bichajian

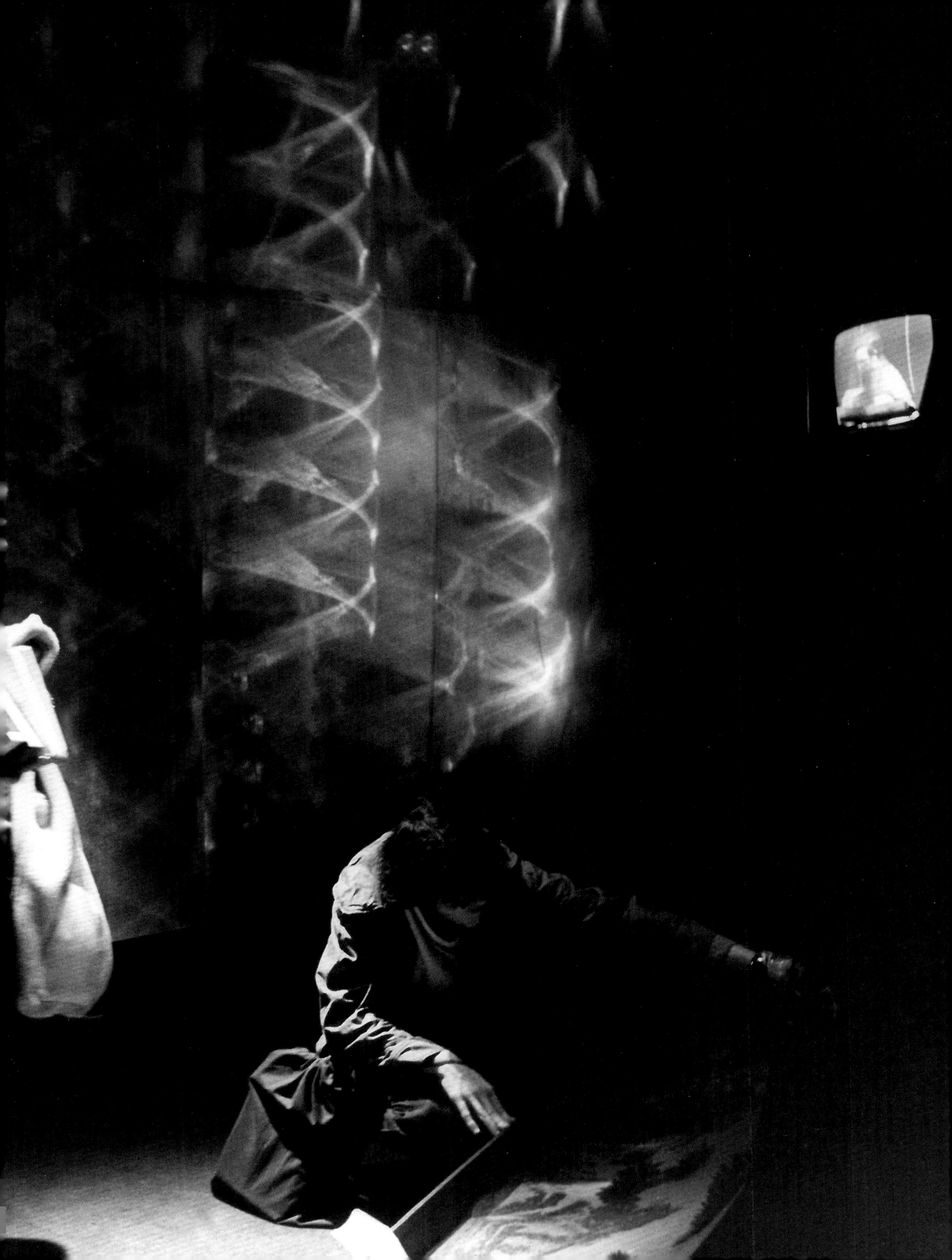

Tethering the sculpture and mounting it on swivels permits it to be responsive to any air movements. It is kinetic without being programmed movement. . . . The material is highly reflective and as the modules move and rotate they reflect and flash color and images.
(Vera Simons, *Explorations* catalog, 1970, 16)

The rods vibrate in a constant harmonic motion, while the pulse rate of the strobe light changes by audio feedback from the viewer. When he whistles or claps his hands, the vibrations of the rods seem to change from gentle undulations to rapid rhythm.
(Wen-Ying Tsai, *Explorations* catalog, 1970, 20)

Clap hands, whistle; microphone is located on ceiling.
(Wen-Ying Tsai, *Explorations* brochure, Smithsonian Institution, 1970)

Look up.
(Vera Simons, *Explorations* brochure, Smithsonian Institution, 1970)

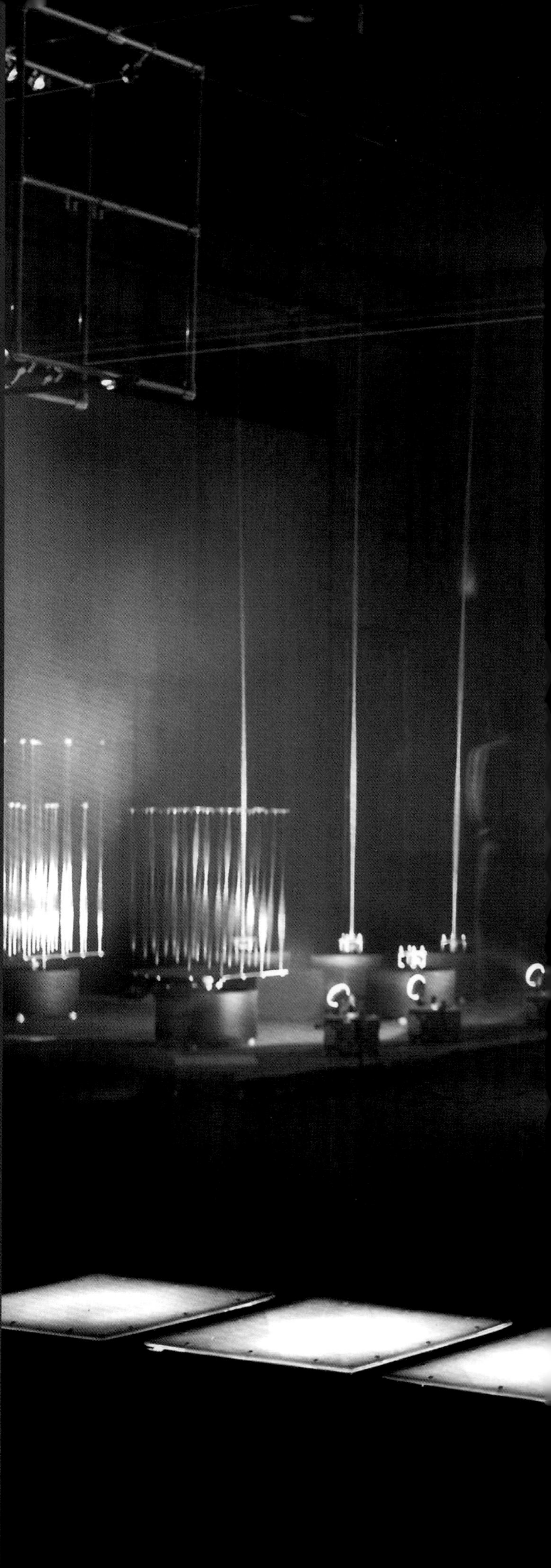

Exploration, CAVS exhibition, Hayden Gallery, MIT, February 28-March 29, 1970
Shown: Kepes and Wainwright, "Photoelastic Walk"; Piene, "Venus of Willendorf"; Vera Simons "Tetrahedrons"; Tsai "Cybernetic Sculptures"; St. Florian, "Imaginary Ceilings"
Photo: Nishan Bichajian

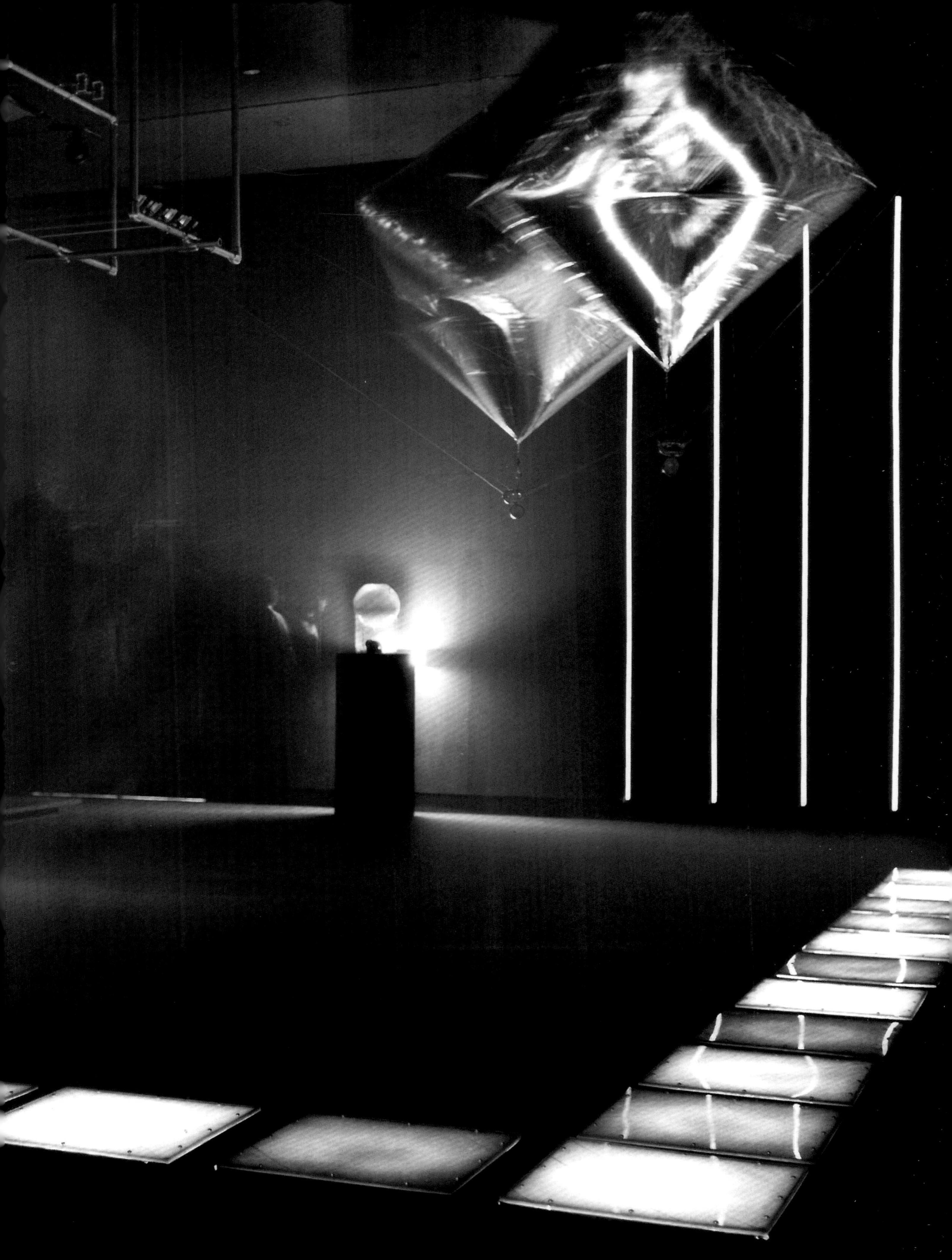

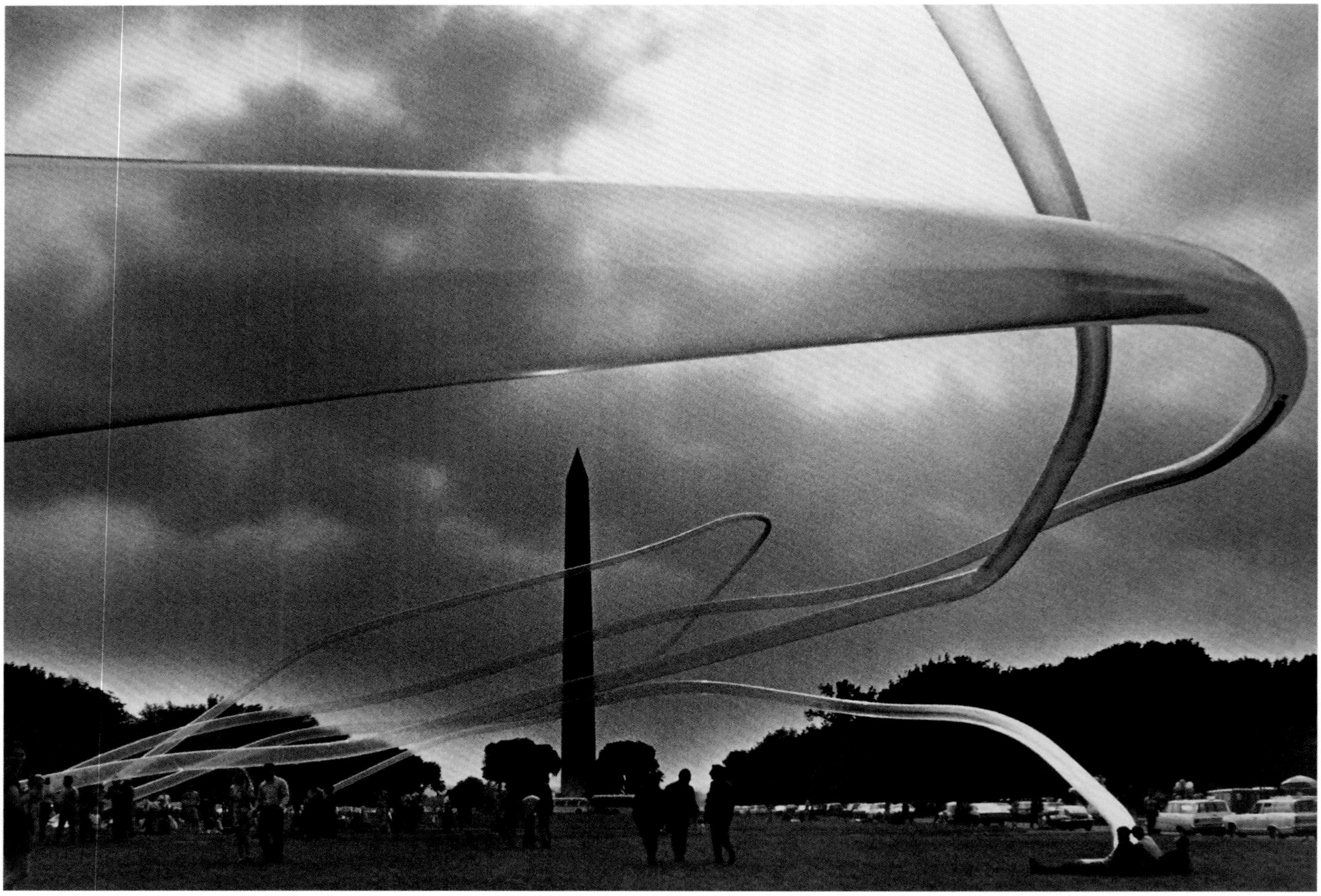

When the show, retitled *Explorations*, moved to the National Collection of Fine Arts at the Smithsonian Institution in Washington, DC (April 4-May 10, 1970), artists were added to the exhibition, along with sky events on the National Mall, including Charles Frazier's "Sky Poem," Otto Piene's "Washington Sky Ballet," and Vera Simons' "Flower Drop."

> You are about to join in a celebration of light, heat, cold, air, electricity, magnetism—forces so omnipresent in our environment that we forget to wonder at their power and beauty. . . . We hope that you will touch, walk upon, walk through, or whistle at the many works that are marked for viewer participation on the map. ("A Guide to *Explorations*," Smithsonian Institution, 1970)

In 1971, exhibitions with CAVS artists included *Earth, Air, Fire, Water: Elements of Art* at the Museum of Fine Arts, Boston; *Art in Civic Scale* at MIT Building W11 during CAVS's May 19 symposium; and *Tsai: Cybernetic Sculpture Environment, New Work Done at the MIT Center for Advanced Visual Studies*, at MIT's Hayden Gallery.

Above: Otto Piene, "Washington Sky Ballet," sky event for *Explorations*, Smithsonian Institution, National Mall, Washington, DC, May 7-8, 1970
Photo: Mike Lien, *The New York Times*

Facing page top: Otto Piene, "Charles River Rainbow," inflatable tube spanning the Charles River, February 13, 1971; part of the exhibition *Earth, Air, Fire, Water: Elements of Art*, Museum of Fine Arts, Boston
Photo: Mira Cantor

Facing page bottom: Otto Piene, "Red Rapid Growth," inflatable sculptures, part of the exhibition *Earth, Air, Fire, Water: Elements of Art*, Museum of Fine Arts, Boston, February 4-April 11, 1971
Photo: Otto Piene Collection

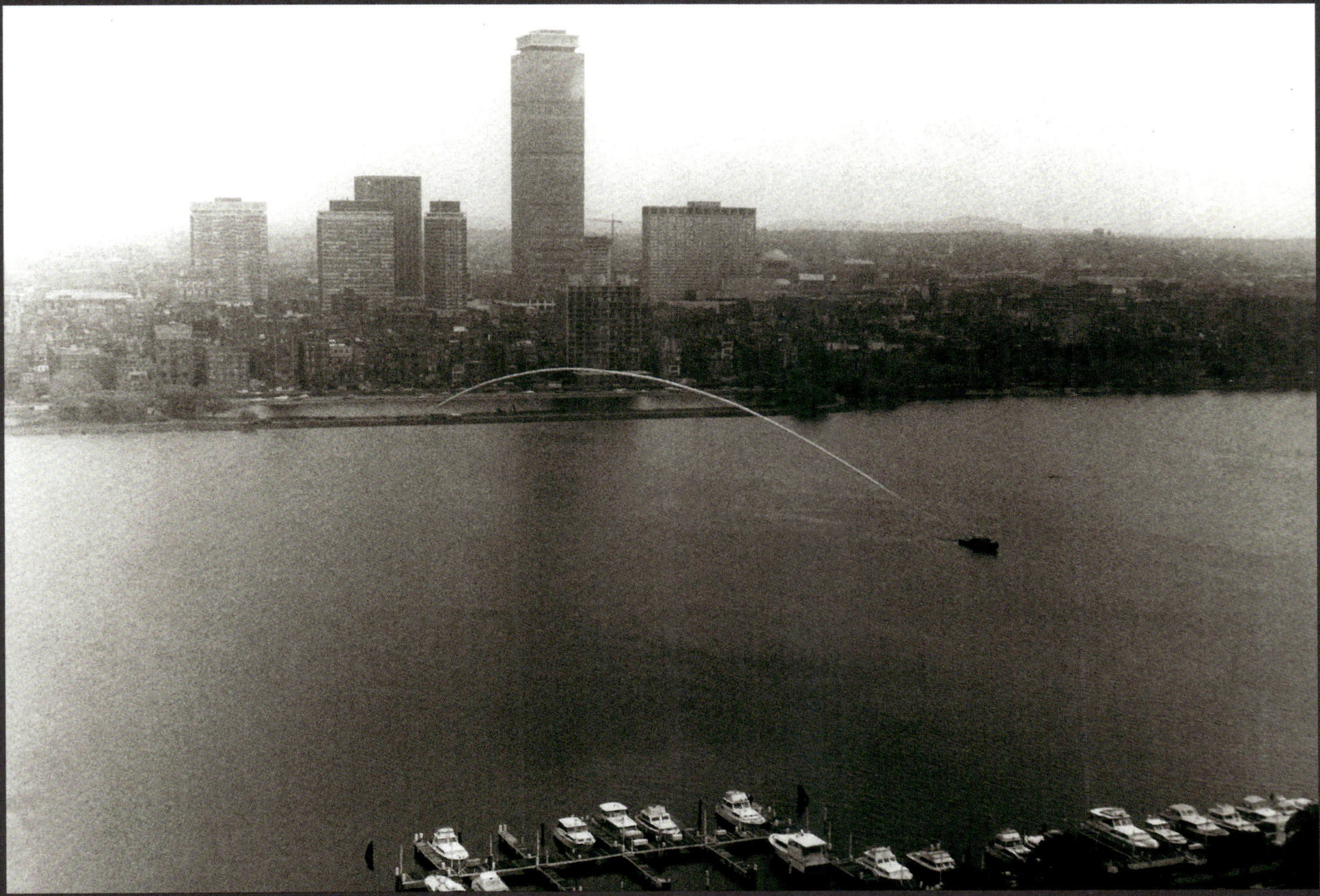

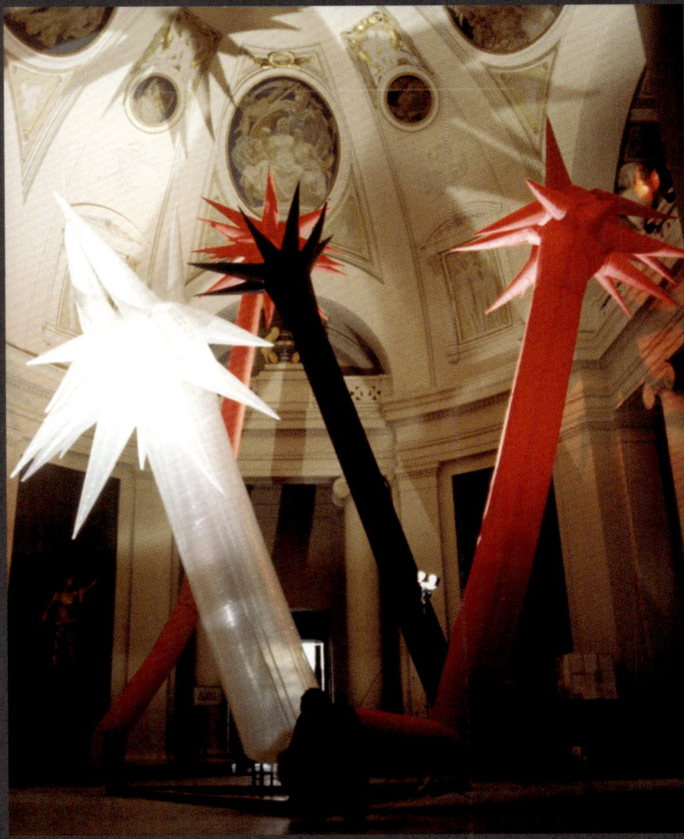

Entitled *Earth, Air, Fire, Water: Elements of Art* and consisting for the most part of very large and highly unorthodox constructions, tableaus and happenings, the exhibition is the final event of the museum's centennial program. . . . Inside the museum however, spectacular sights abound. One of the largest and most beautiful is Otto Piene's "Red Rapid Growth." This brilliant kinetic construction of four flowerlike forms made of polyethylene and silk and reaching 35 feet into the museum rotunda, is a real dazzler, with its sensuous, air-filled red and purple members inflating and subsiding in a programmed rhythm symbolizing the natural cycle of growth and decay. On Feb. 13, Mr. Piene will also stage one of the show's non-museum events: a "Rainbow," made of polyethylene and helium, across the Charles River. (Hilton Kramer, "Boston Museum Mounts an Unusual New Art Show," *The New York Times*, February 4, 1971, 28)

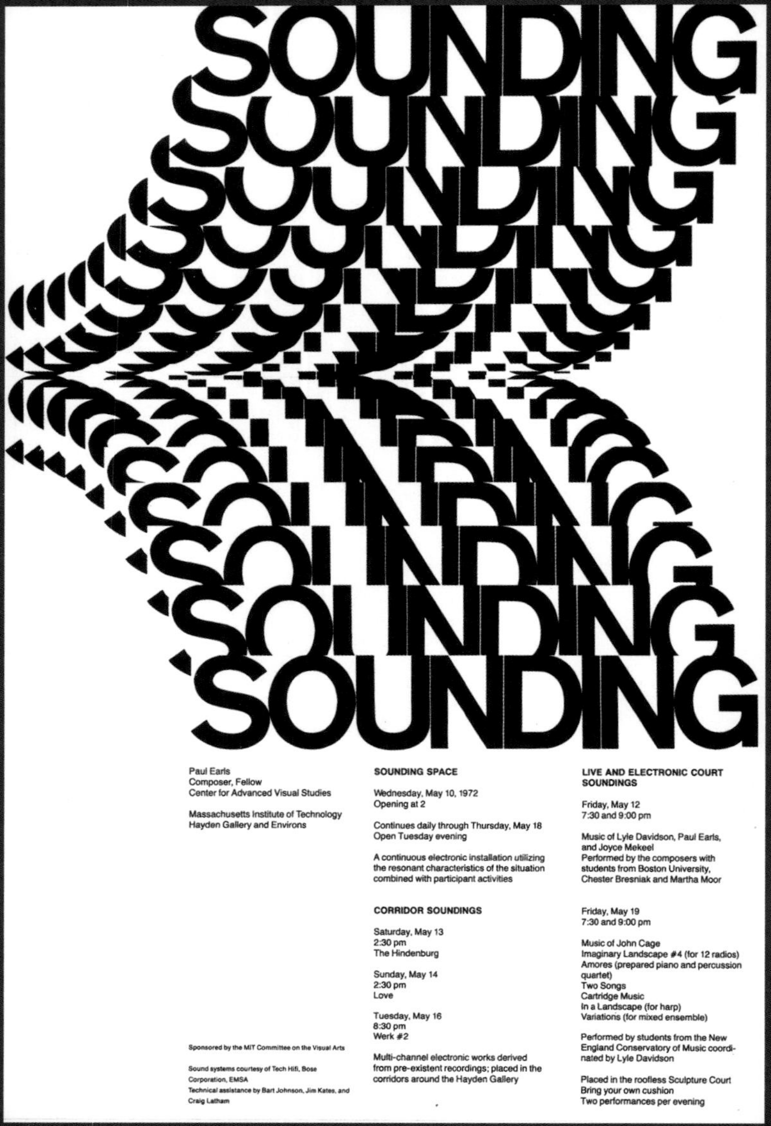
SOUNDING
SOUNDING
SOUNDING
Paul Earls
Composer, Fellow
Center for Advanced Visual Studies
Massachusetts Institute of Technology
Hayden Gallery and Environs
Sponsored by the MIT Committee on the Visual Arts
Sound systems courtesy of Tech Hifi, Bose Corporation, EMSA
Technical assistance by Bart Johnson, Jim Kates, and Craig Latham
SOUNDING SPACE
Wednesday, May 10, 1972
Opening at 2
Continues daily through Thursday, May 18
Open Tuesday evening
A continuous electronic installation utilizing the resonant characteristics of the situation combined with participant activities
CORRIDOR SOUNDINGS
Saturday, May 13
2:30 pm
The Hindenburg
Sunday, May 14
2:30 pm
Love
Tuesday, May 16
8:30 pm
Werk #2
Multi-channel electronic works derived from pre-existent recordings; placed in the corridors around the Hayden Gallery
LIVE AND ELECTRONIC COURT SOUNDINGS
Friday, May 12
7:30 and 9:00 pm
Music of Lyle Davidson, Paul Earls, and Joyce Mekeel
Performed by the composers with students from Boston University, Chester Bresniak and Martha Moor
Friday, May 19
7:30 and 9:00 pm
Music of John Cage
Imaginary Landscape #4 (for 12 radios)
Amores (prepared piano and percussion quartet)
Two Songs
Cartridge Music
In a Landscape (for harp)
Variations (for mixed ensemble)
Performed by students from the New England Conservatory of Music coordinated by Lyle Davidson
Placed in the roofless Sculpture Court
Bring your own cushion
Two performances per evening

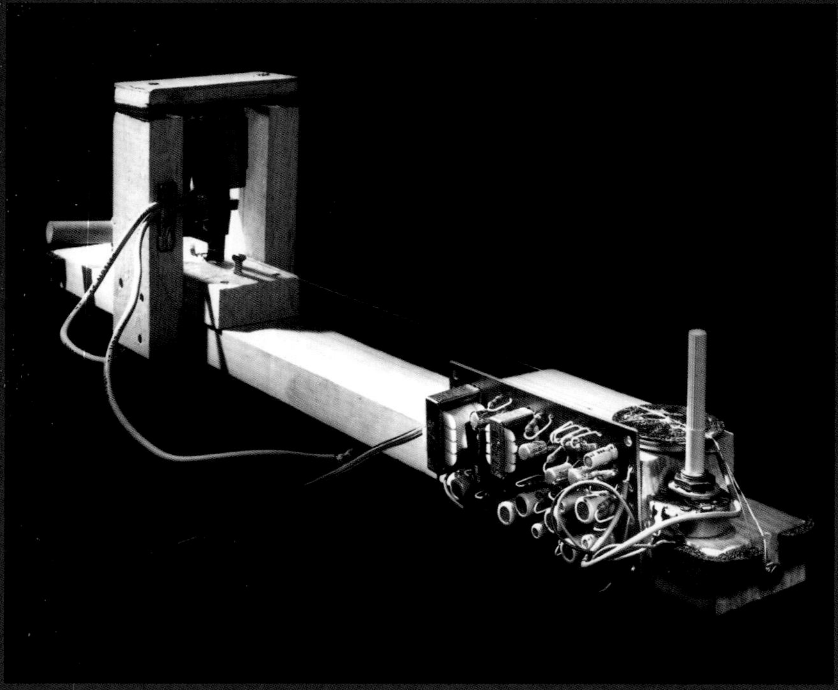

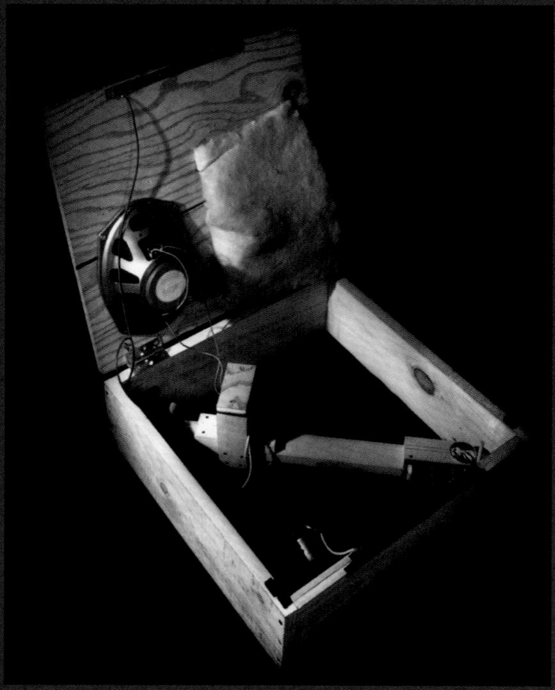

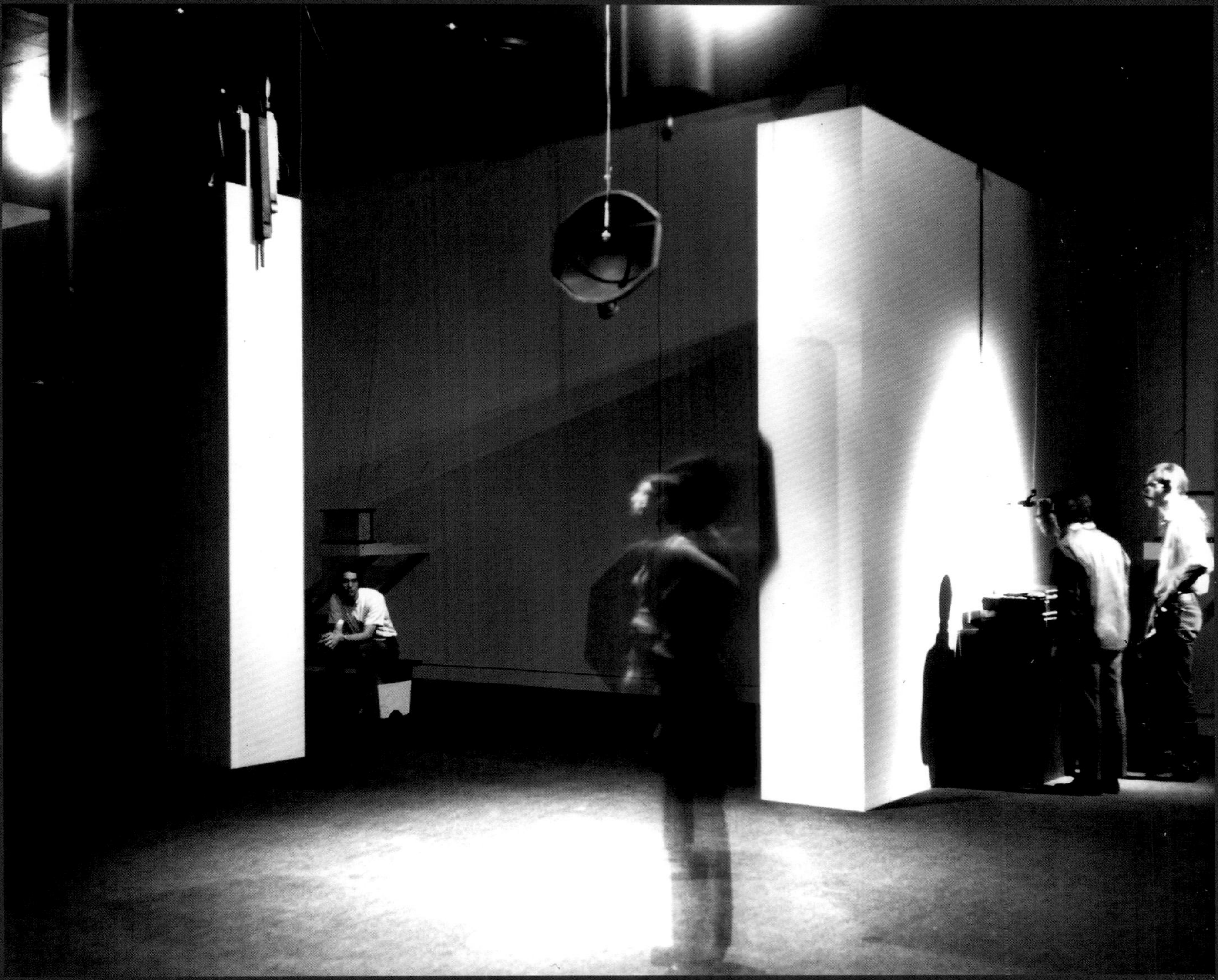

In the early years, MIT's Hayden Gallery regularly exhibited new work by CAVS fellows, for example: Takis, *Phenomenological Sculpture: Evidence of the Unseen*, Hayden Gallery, MIT, November 15 - December 8, 1968; and Tsai, *Cybernetic Sculpture Environment, New Work Done at the MIT Center for Advanced Visual Studies*, Hayden Gallery, MIT, February 6 - March 3, 1971.

Sound, electronic manipulation, and instrument designs were the subject of ambient environments, concerts, and interactive installations by composer Paul Earls. *Sounding Space* —one of three events, along with *Corridor Soundings* and *Live and Electronic Soundings*—was installed at the Hayden Gallery, MIT, May 10-19, 1972.

Above, and facing page top right: Paul Earls, *Sounding Space*, installation views, Hayden Gallery, MIT, May 10-19, 1972
Photos: Nishan Bichajian

Facing page top left: poster for Paul Earls' "Sounding" events, Hayden Gallery, MIT, May 10-19, 1972
Source: CAVS Special Collection

Facing page bottom: Paul Earls, string instruments, Hartford Sound Floor, May, 1972
Photos: Nishan Bichajian

Only a few exhibitions were portable enough to travel. *Multiple Interaction Team* (1972-1974), coordinated by artist fellows Maryanne Amacher and Friedrich St. Florian, traveled to major venues, including the Museum of Science and Industry, Chicago; the Exploratorium, San Francisco; the Cincinnati Contemporary Arts Center; New Orleans Museum of Art; Science Museum of the Franklin Institute, Philadelphia; Museum of Fine Arts, St. Petersburg, Florida; the Hall of Science, New York City; and the California Museum of Science and Industry, Los Angeles.

> For the tasks for which artists are being called upon today are epic tasks, civic and social in character and environmental in dimensions. To live up to this task the artist must make use of the social confrontations of their time, the scientist's conceptual models, and the engineer's new arsenals of technology without giving up primary sense data, the directness and immediacy of personal poetic vision. This exhibition represents works of art that were created in the climate of this belief. (Kepes, *Multiple Interaction Team* catalog, 1971-74, 6-7)

> "M. I. T." is an experiment with art and the sense of community, the sense of place, in which the public and others artists will interact with works of art. (*Multiple Interaction Team, A Traveling Exhibition-Event*, proposal, CAVS, 1972)

Facing page: *Multiple Interaction Team*, opening, CAVS exhibition at the Museum of Science and Industry, Chicago, November 15-January 15, 1972
Photo: "Plans and Perspectives: Budget," portfolio, 1974, CAVS Special Collection

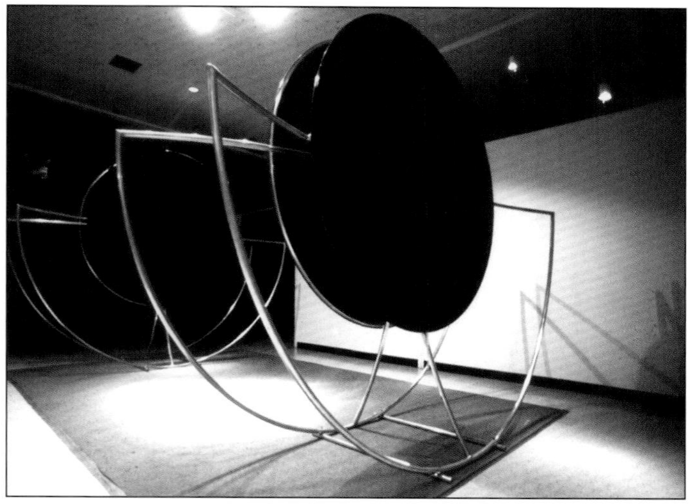

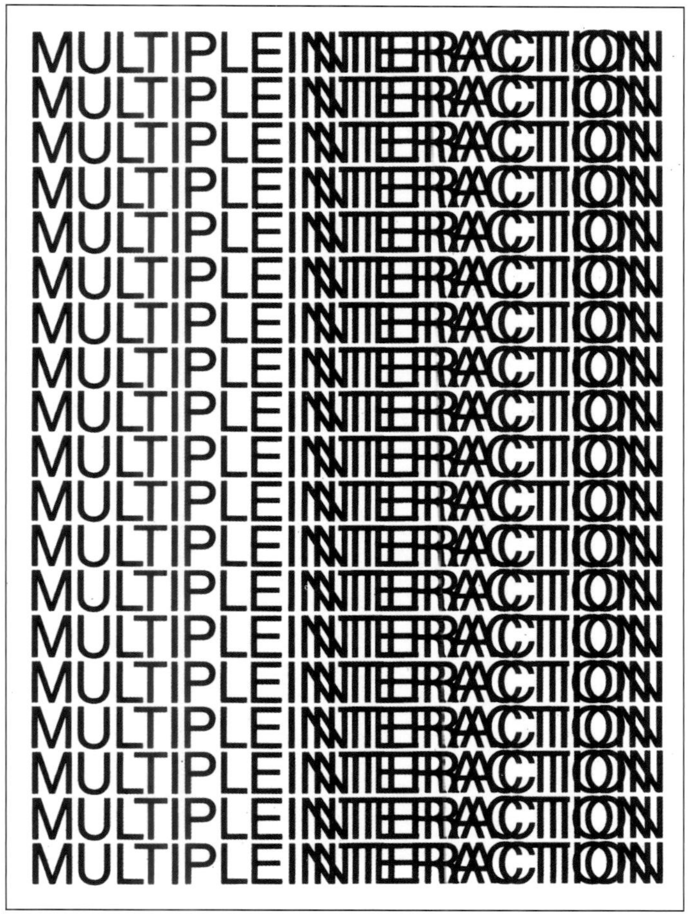

Top: Joe Moss, "Sound Sculpture," *Multiple Interaction Team*, CAVS exhibition, Science Museum of the Franklin Institute, Phildelphia, 1973
Photo: Joe Moss

Bottom: *Multiple Interaction Team*, cover of catalog for CAVS traveling exhibition, MIT, 1972-74
Source: CAVS Special Collection

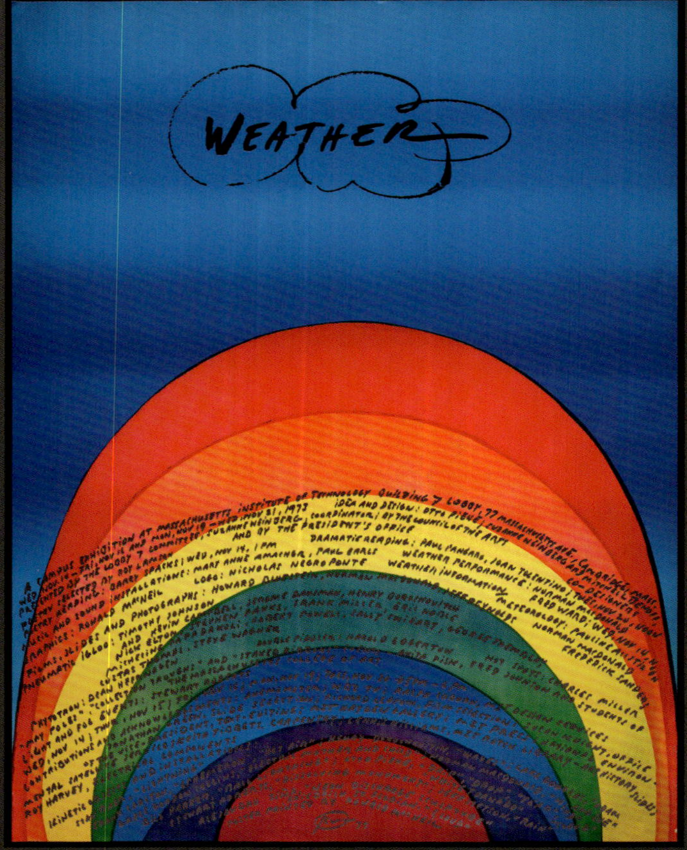

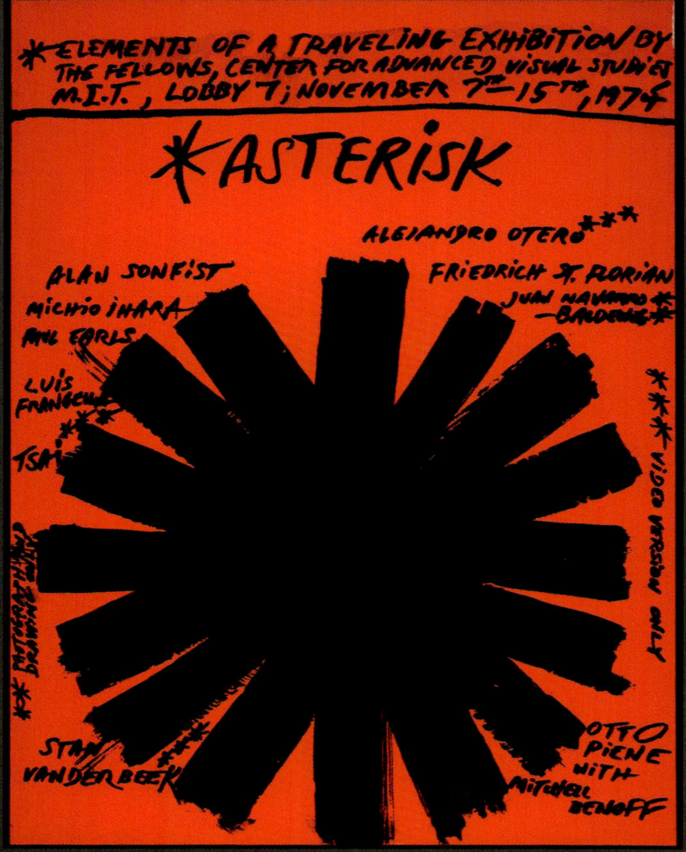

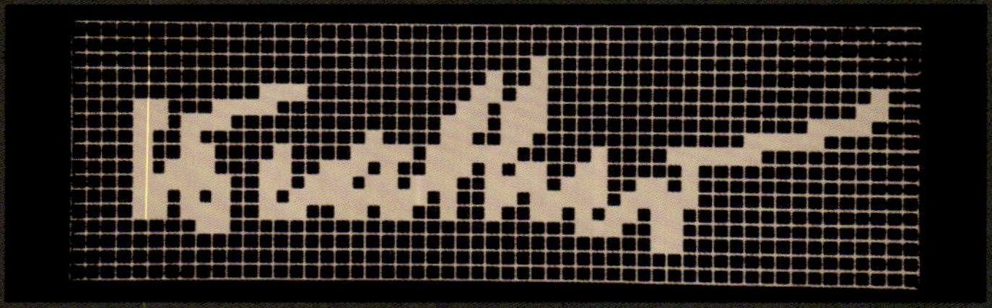

Top left: poster by Otto Piene for the *Weather* show, CAVS fellows and students, Lobby 7, MIT, November 14-16 and 19-21, 1973

Bottom left: "*Weather* logo, Nicholas Negroponte, Associate Professor, Department of Architecture. The logo for the campus exhibition *Weather* has been generated by the pattern recognition facilities of The Architecture Machine." (*Weather* show flyer, 1973)

Top right: **Asterisk, *Elements of a traveling exhibition by the fellows, Center for Advanced Visual Studies*, poster by Otto Piene, MIT, Lobby 7, November 7-15, 1974

Source: CAVS Special Collection

Facing page: Children interact with the CAVS exhibition, *Dialogue for the Senses*, designed by György Kepes with Paul Earls, Friedrich St. Florian, and Michio Ihara, with CAVS fellows, "an exhibition in non-visual perception" (Kepes letter, March 3, 1972), for the new Tactile Gallery, Wadsworth Atheneum, Hartford, Connecticut, May 10 - Thanksgiving day, 1972

Photos (details): Edward Saxe

Kepes' style, in keeping with Bauhaus tradition, was formal with unifying design elements. He worked with Amacher, Earls, St. Florian, and Ihara, the artist-curators for *Dialogue for the Senses* (1972), on an exhibition for the blind at the Wadsworth Athenaeum in Hartford, Connecticut. The pieces moved and responded to visitor presence.

> The exhibition, entitled *Dialogue for the Senses*, aimed at stimulating both the unsighted and sighted into discovering their unexplored range of sense responses and appealed to all non-visual (tactile, thermal, kinesthetic) senses in various forms of interaction. Almost all fellows, as well as research affiliates and a number of students from the Department of Electrical Engineering and the School of Architecture and Planning, worked together to create this exhibition. (Kepes, *MIT Report of the President*, 1972, 4)

Such body-centric bio-art continued at CAVS with works such as Jennifer Hall's *Out of the Body Theatre*, which dealt with epilepsy; Joe Davis' artistic explorations using DNA, beginning with *Microvenus*; and Elizabeth Goldring's *Visual Language for the Blind* and *Seeing Machine Camera*. Goldring, herself visually challenged, was familiar with tactile, interactive installations as an exhibition designer at the Smithsonian Institution and at the Boston Children's Museum, and became acquainted with the Center's work through *Dialogue for the Senses*.

Group shows always ran into problems of space allocation, and conflicting sound and light levels. Artworks were often ephemeral in nature, and video and photographic documentation, along with the catalogs, became the only lasting artifacts.

Video documentation practiced at CAVS by Vin Grabill and Ellen Sebring, and photographic documentation by Nishan Bichajian, Walter Dent, Goldring and others often constituted an art form as well as a record. Many of the invitational group exhibitions connected to conferences and performances were initially designed through creative brainstorming at the Center's fellows' meetings.

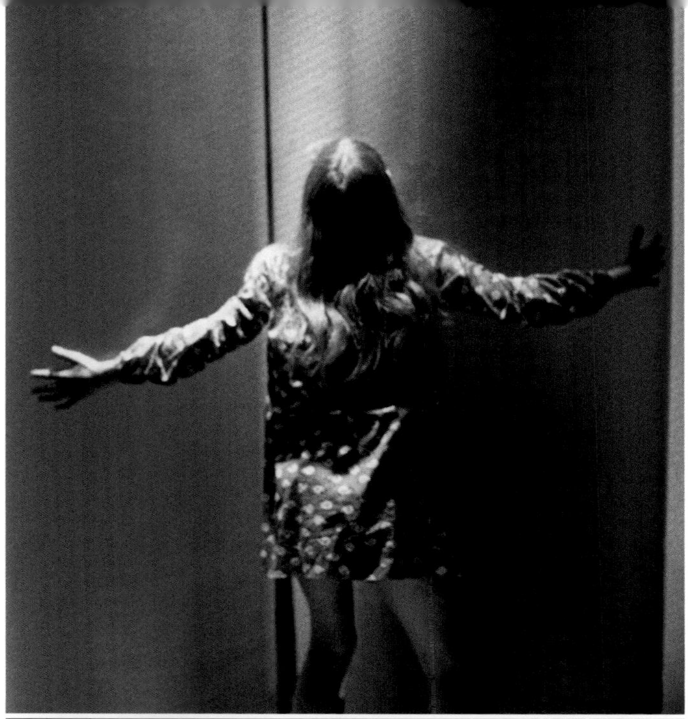

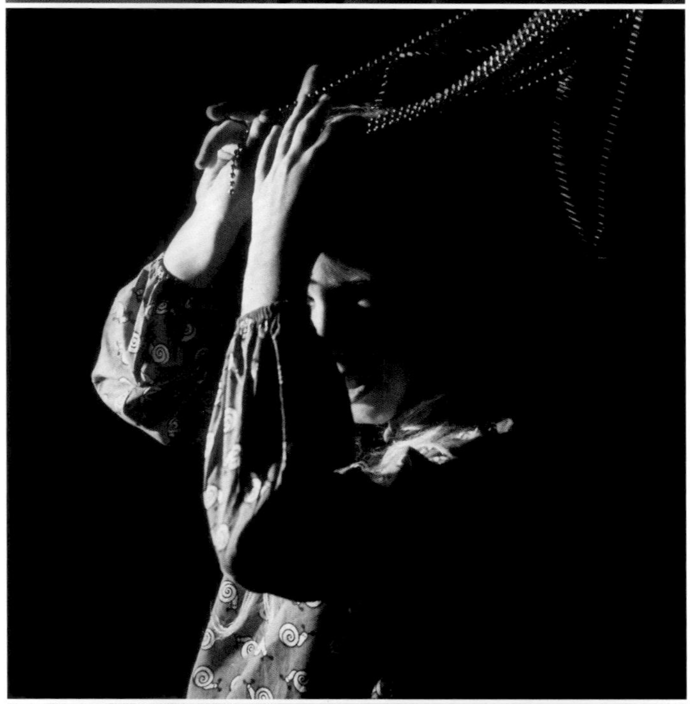

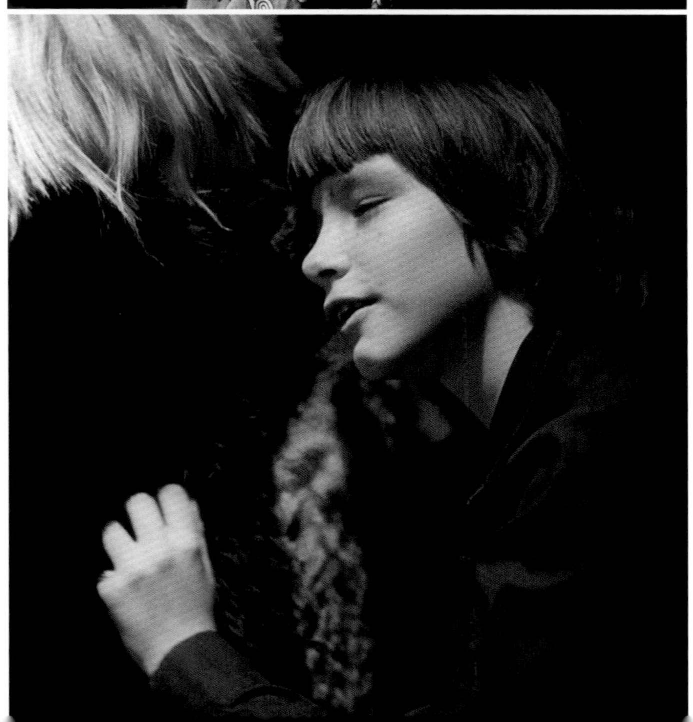

Boston Celebrations

Boston Celebrations Part 1: Jubilee Projects (1975) and *You Are Here, Boston Celebrations Part 2: Long Wharf Projects* (1976) were exhibitions commissioned by the Institute of Contemporary Art, Boston (ICA). The national bicentennial celebration, initially slated for Boston, led to this commission, "With the thought in mind that the Bicentennial is a celebration not only of the past but also of the future" (ICA director Gabriella Jeppson):

> The projects are done in this jubilee spirit and range from the serious to the fantastic, from the possible to the impossible, involving a variety of media. This is the first of two exhibitions. . . . The second show will be dedicated to studies of a more lasting nature. The Fellows intend to improve the environment by means of sculptural and ecological installations on a large scale, that is, the scale of the city of Boston. (Gabriella Jeppson, ICA Director, *Boston Celebrations Part 1: Jubilee Projects* catalog, 1975)

Indeed, celebration was key. Piene's essay in the Part 1 catalog opens with the phrase, "I Celebrate; You Celebrate, He, She, It Celebrates." Both exhibitions presented models, drawings, photographs, and audio descriptions of kinetic ideas based on the movement of wind, weather, and people.

> *Part 1* is composed of Jubilee Projects in the nature of the 1875 jubilee, a temporary feast of parades, concerts and get-togethers. Part 2 will present plans for lasting installations in Greater Boston, artists' contributions to the city which will enhance the environment and the citizens' life in it for a decade, a century, or longer. (Piene, *Boston Celebrations Part 1: Jubilee Projects* catalog, 1975, 8)

The *Jubilee* project yielded conceptual portfolios of events and temporary installations. The catalog lists work from seventeen artists. Lowry Burgess proposed, in collaboration with Paul Earls, "to open the

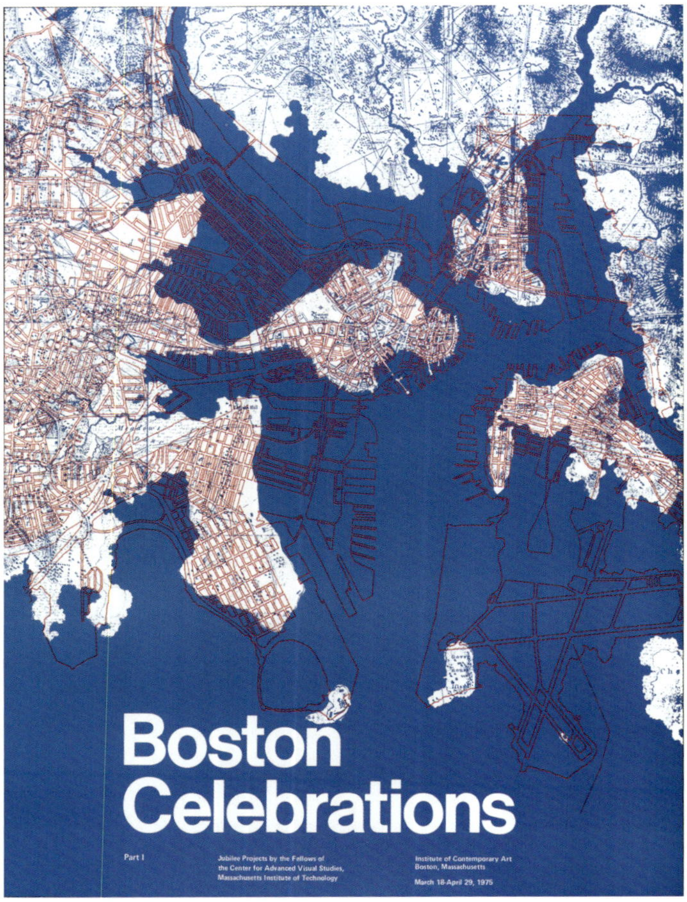

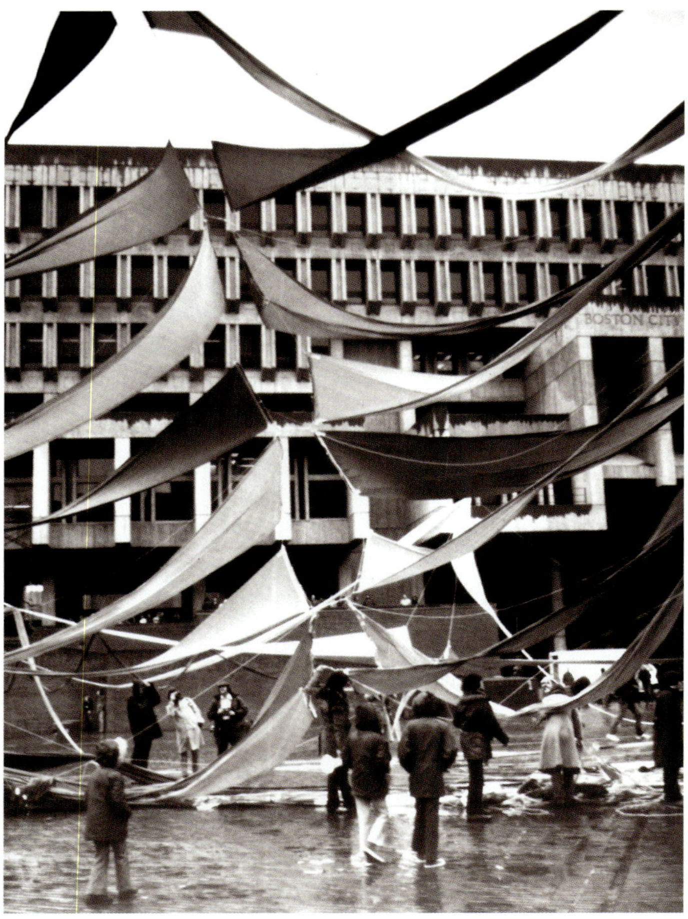

Bicentennial period with one hundred large bells, twenty bands, a canopy, hand bells, carrier pigeons, poetry, parades, and history." (Burgess, *Boston Celebrations Part 1: Jubilee Projects* catalog, 1975, 20) Kepes offered "A proposal for vertical light beams that are sensitive to the surface motion of the sea," and a list of design specifications that included:

> To bring back lost "Sundays" in the sensory experience of urban surroundings;
>
> To introduce a new aesthetic dimension of urban landscape through . . . the luminous accidental richness of the urban nightscape;
>
> To instill in the citizens of Boston and surrounding cities . . . a sense of group identity;
>
> To involve the "play" instinct of individual human beings. (Kepes, *Boston Celebrations Part 1: Jubilee Projects* catalog, 1975, 34, 36)

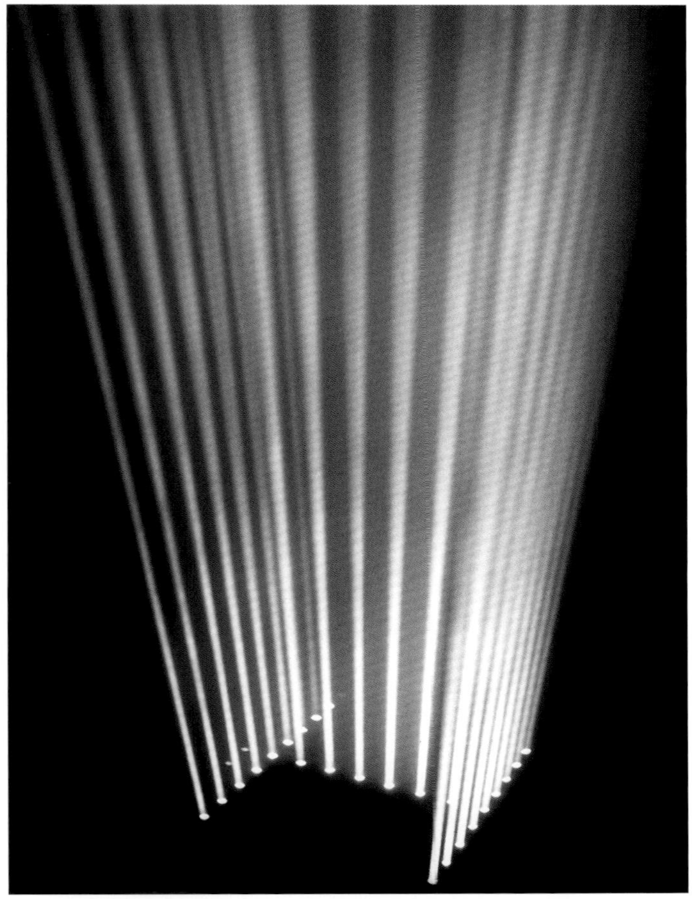

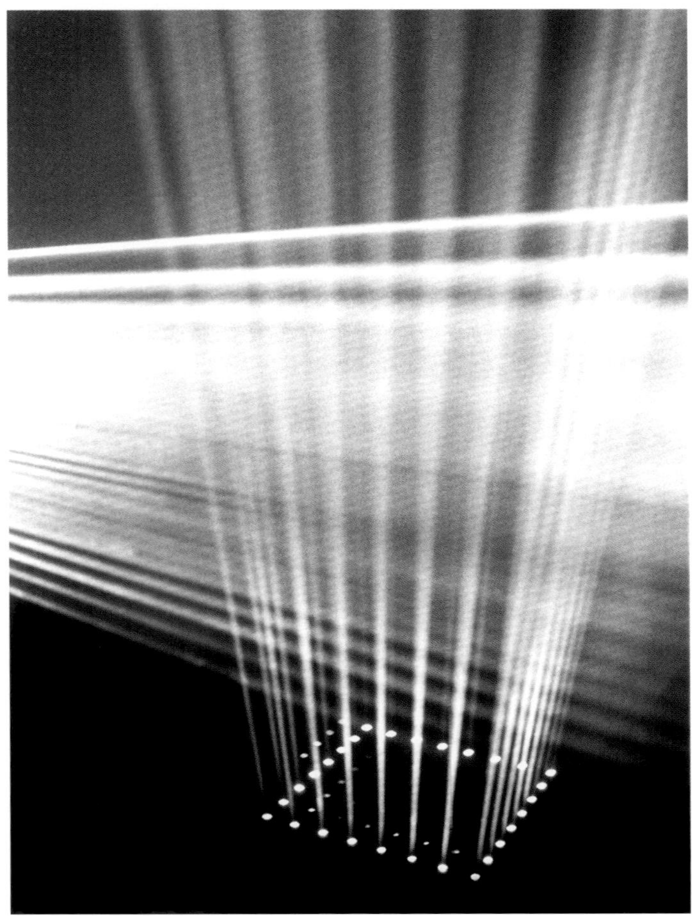

Right: György Kepes, simulated mile-long programmed luminous wall, proposal for the Boston Harbor Bicentennial, originally designed in 1964-65
Photos: Nishan Bichajian

Facing page top: *Boston Celebrations Part 1: Jubilee Projects*, catalog cover by Friedrich St. Florian (on the map, white depicts Boston circa 1775, red depicts Boston circa 1975), Institute of Contemporary Art, Boston, March 18-April 29, 1975
Source: CAVS Special Collection

Facing page bottom: Lowry Burgess with Paul Earls, "City Ring—Band and Bells," dress rehearsal for April 19, 1975, City Hall Plaza event, Boston, February 22, 1974
Photo: Lowry Burgess

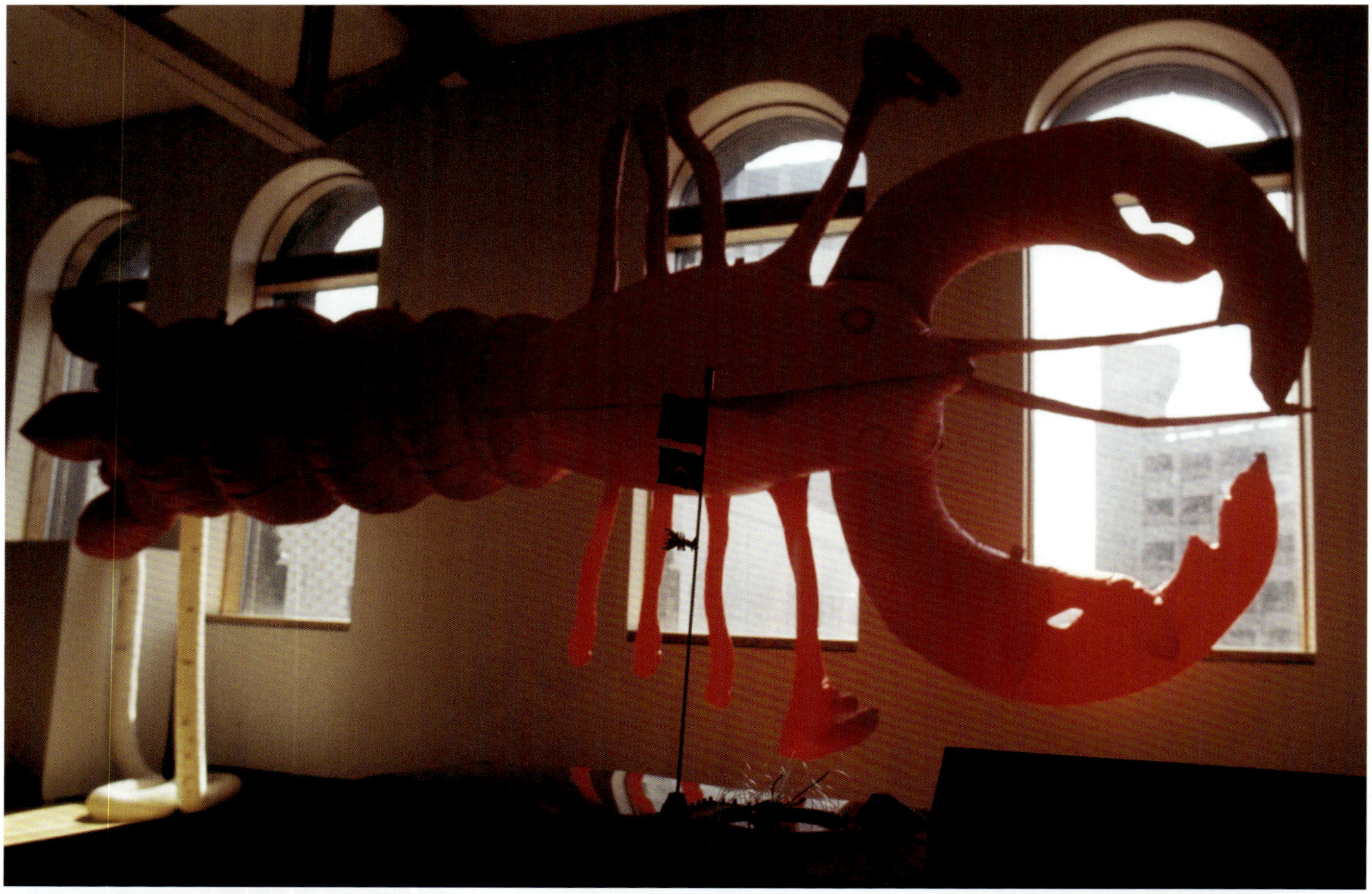

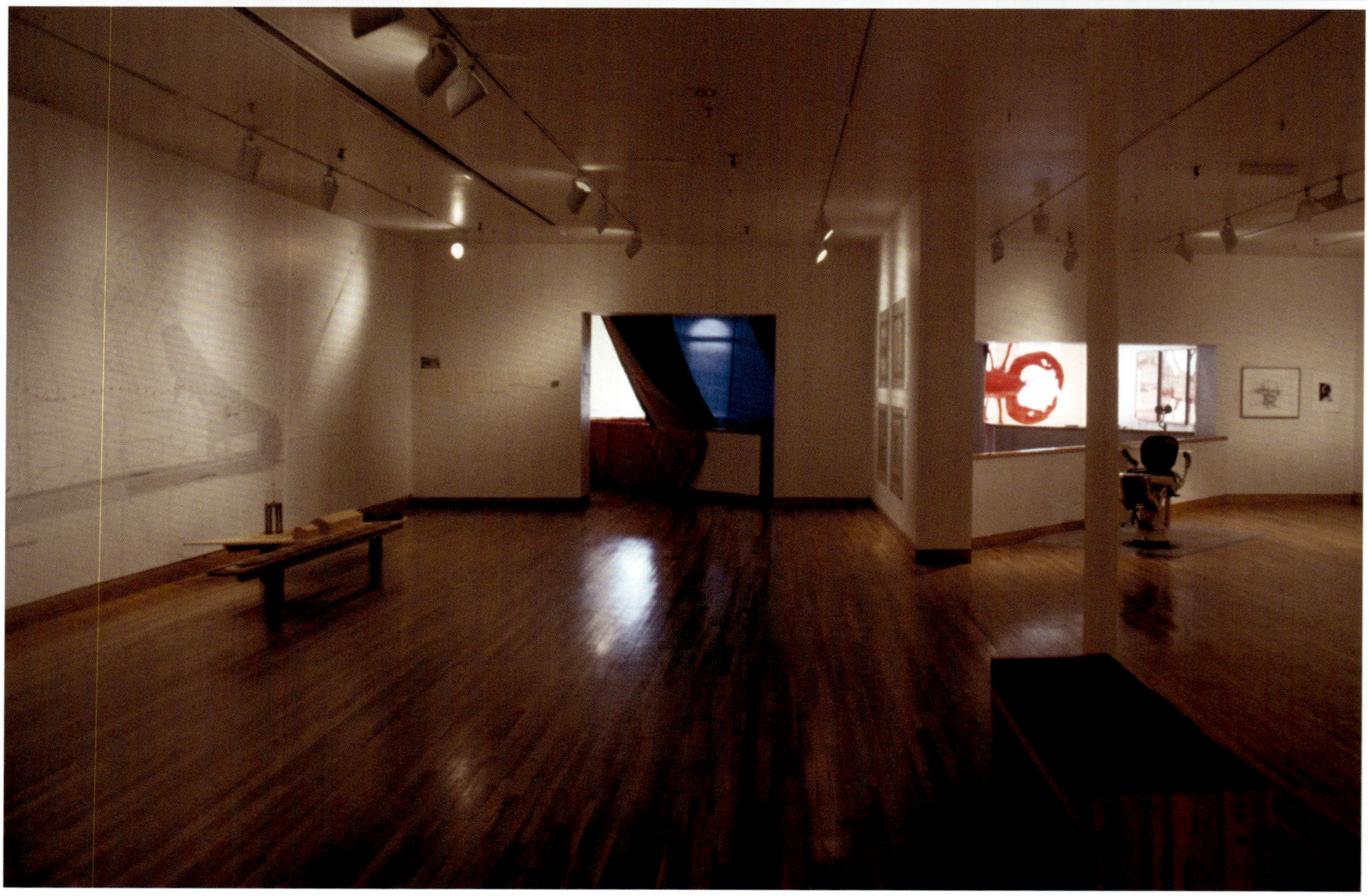

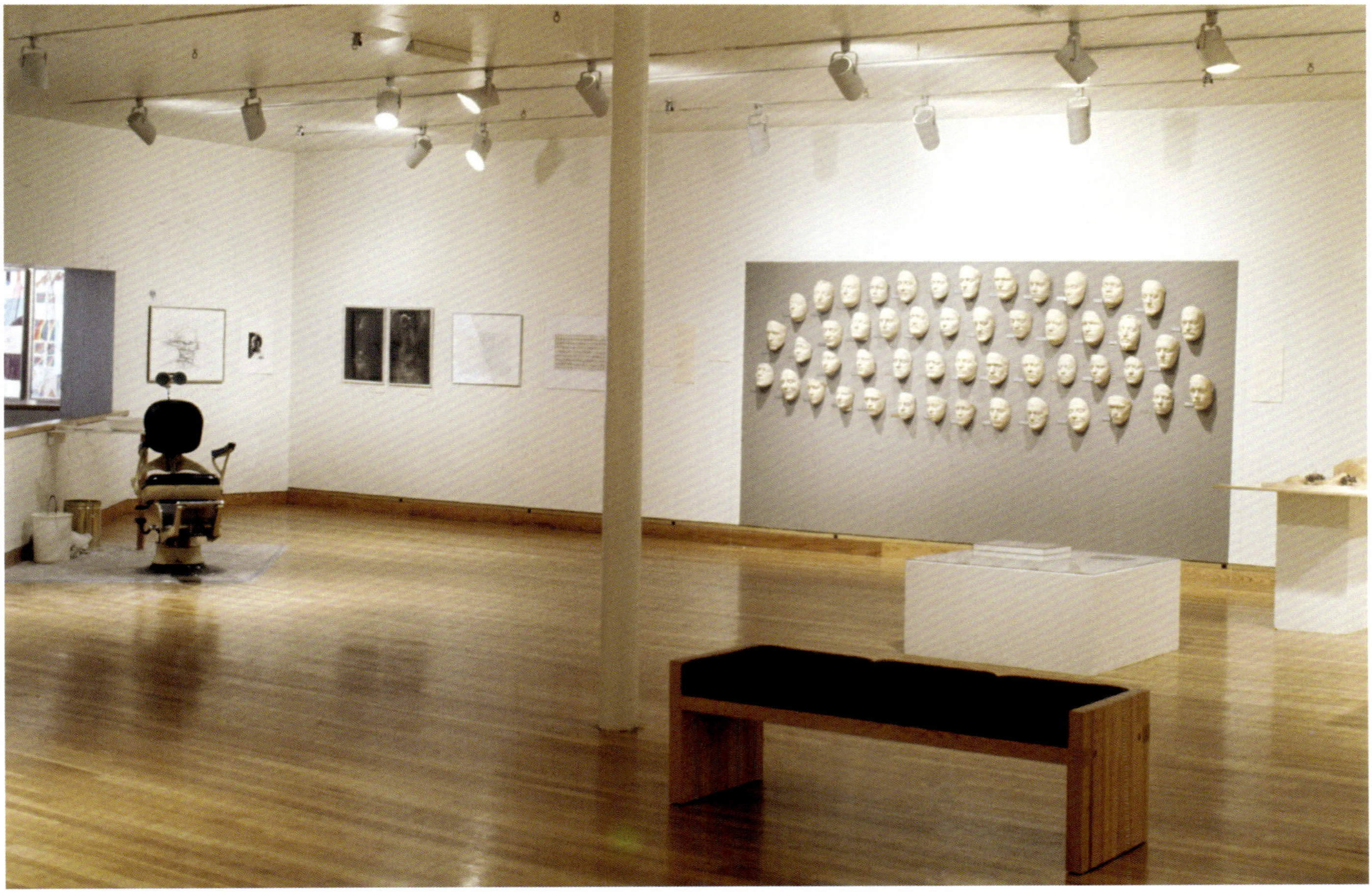

You Are Here: Boston Celebrations Part 2 suggested more permanent monuments designed to improve specific sites through environmental and sculptural means. "The exhibition describes six artists' visions for Long Wharf," Goldring wrote in the catalog. These artists included Burgess, Ihara, Kepes, Carl Nesjar, Piene, and Harold Tovish. Tovish's installation enabled visitors to sit for a mask-making session, and take away a clay mask of their face, described in the catalog:

> . . . a wall which is visible and accessible to the passerby is covered with cast stone lifemasks of people who live in the neighborhood. (*You Are Here: Boston Celebrations Part 2*, catalog, 1976, 82)

Goldring's historical component occupied almost as much exhibition space as the proposed artworks. The exhibition, subtitled *Environmental Art*, was on view at the ICA from March 13-April 12, 1976.

Above: Harold Tovish, "Our People: Neighborhood Project"

Facing page top: Otto Piene, "Lobster Windsock" and "Lobster Fountain" sculptures

Facing page bottom: exhibition view

All images from the exhibition, *You Are Here: Boston Celebrations Part 2, CAVS proposals for Long Wharf,* Institute of Contemporary Art, Boston, March 13-April 12, 1976

Photos: Elizabeth Goldring

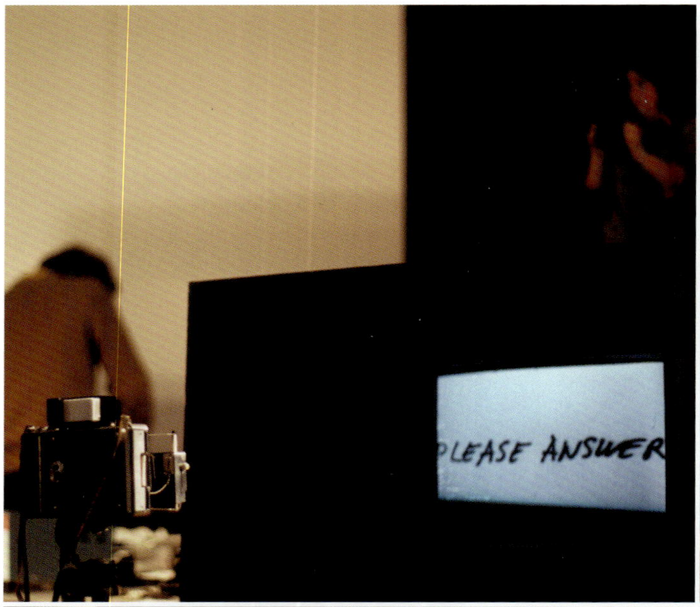
PLEASE ANSWER

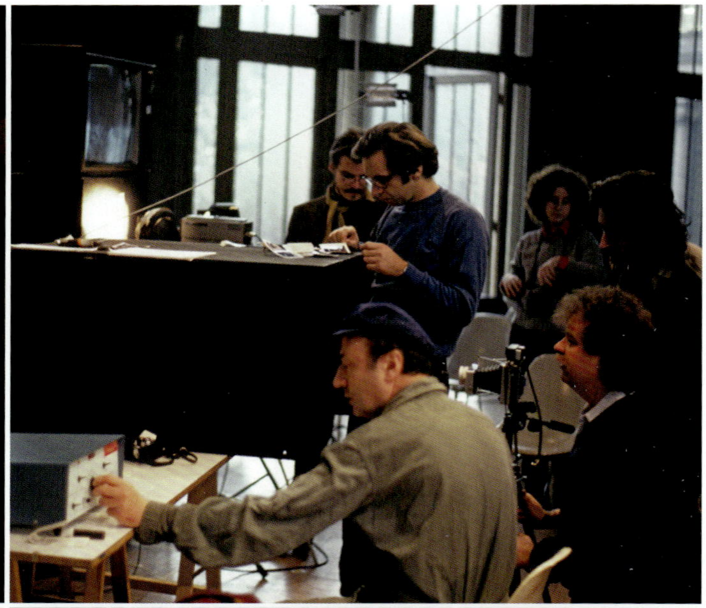

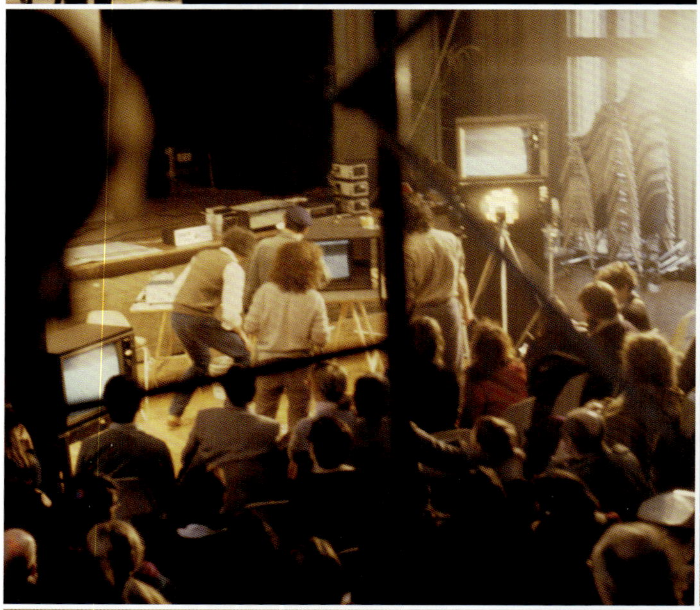

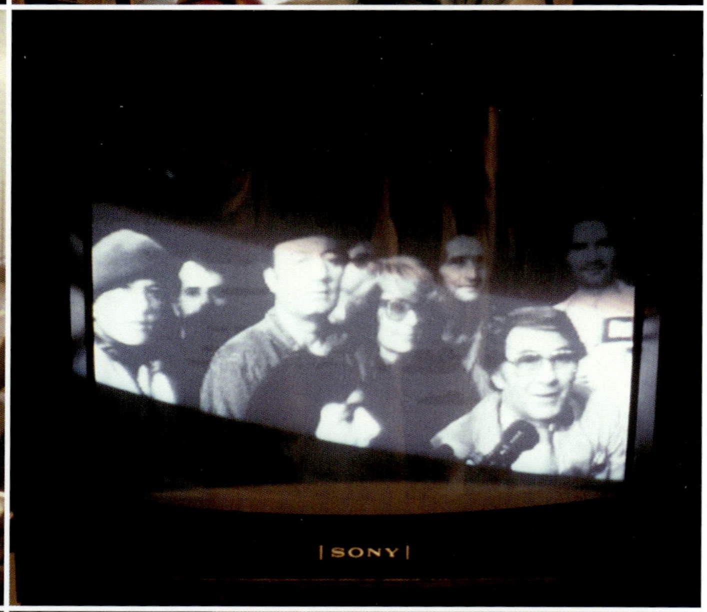
SONY

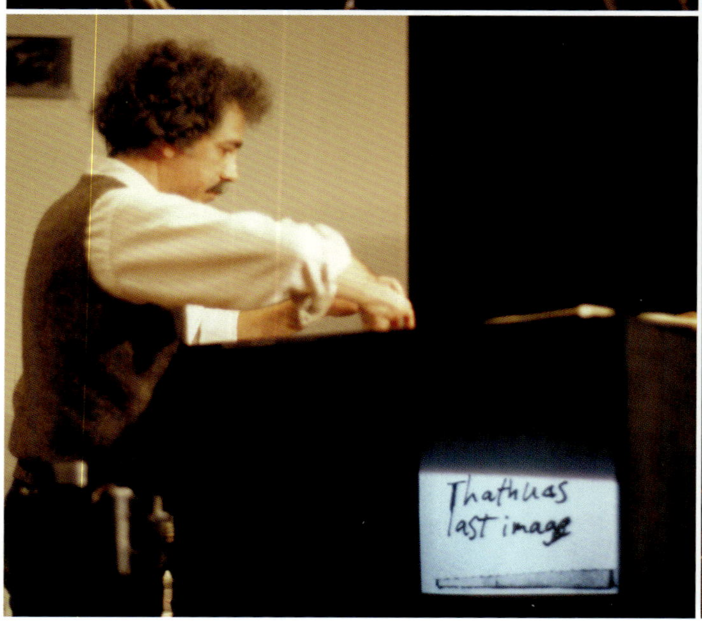
That was
last image

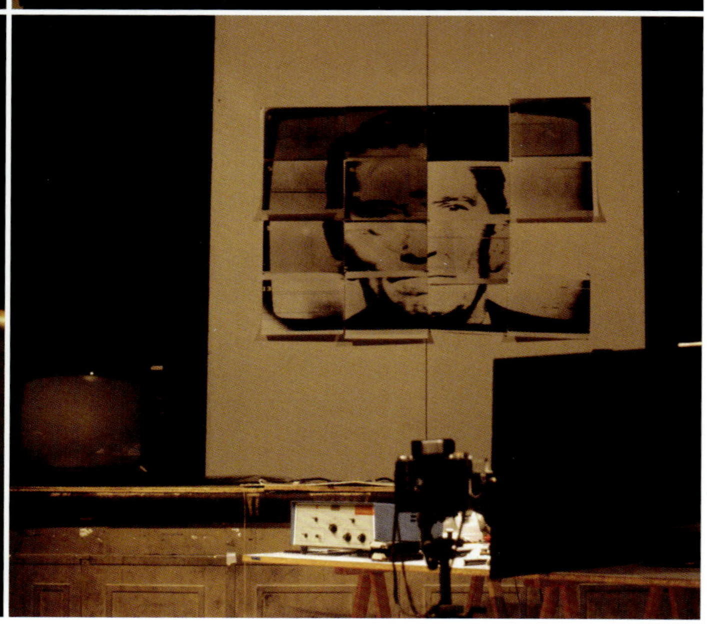

Video and Telecommunications

CAVS engaged in virtual exhibitions that challenged the time-space divide. During the February 1, 1981 telecommunications event, *Interfaces*, Slow Scan Television (SSTV) generated a live link between CAVS in Cambridge and the American Center for Students and Artists in Paris. Slow Scan, or narrowband television, transmitted static pictures at intervals ranging from several seconds to several minutes. Conceptualized by American Center director Don Foresta and CAVS Fellow Aldo Tambellini, *Interfaces* enabled artists to create a performance artwork from the long-distance transmissions and time lapses while images slowly materialized in horizontal scan lines as audiences watched at both locations. The real-time visual link was technically fragile, and reminiscent of early space exploration transmissions, which also used a pre-commercial version of SSTV.

> My idea was to make a mural of Giscard d'Estaing and Ronald Reagan at the American Center in Paris. The group at MIT was receiving d'Estaing and Paris was receiving Reagan. . . . What is Slow Scan? First you have ham radio; Slow Scan is ham television. . . . You have to be synchronized together, set the hour. The camera takes your image and puts in the Slow Scan line by line, in bits. It sounds like a Morse code: beep-beep. (Aldo Tambellini interview)

> I was the first virtual fellow. . . . You realize how primitive the technology was. We had to open the phone receiver and connect alligator clips into the phone. To overcome interference, we wrapped the phone in a scarf! We had the darkroom in the basement. We would take Polaroid negatives off the camera and a guy would hold them up in the air as he ran down to the basement so they would dry. After the photo was developed he would dry it in the same way coming back. (Don Foresta interview)

A portrait of newly inaugurated US President Ronald Reagan was compiled in Paris from Slow Scan images transmitted from CAVS. At the same time, artists at CAVS—Roland Baladi, Peter Codella, Betsy Connors, Sarah Dickinson, Ken Kantor, Bernd Kracke, and Skip Tenczar—received and assembled a corresponding portrait of French President Valéry Giscard d'Estaing.

> The face Paris sent [to CAVS] was of Giscard d'Estaing . . . It was so slow: the image from Paris took hours and hours to form on the CAVS large projection screen. (Luc Courchesne interview)

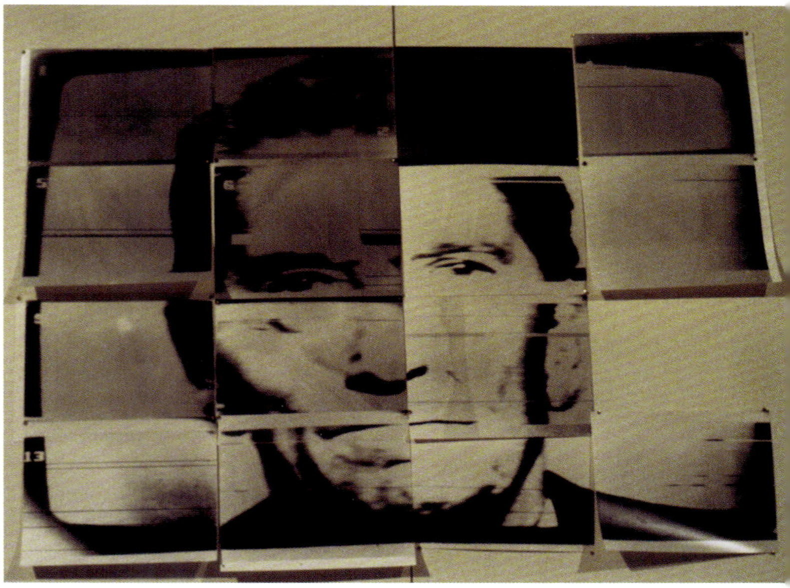

Above: Slow Scan portrait of President Ronald Reagan
Facing page: *Interfaces*, American Center for Students and Artists in Paris exchanges images with CAVS, MIT in Cambridge, via Slow Scan Television, February 1, 1981 (top to bottom, left to right): 1. American Center staff person and Anne-Marie Stein with video camera; 2. Aldo Tambellini with American Center staff and Stein; 3. Don Foresta, with Stein, staff, and audience; 4. (on the TV screen) Elizabeth Goldring, Aldo Tambellini, Vin Grabill, Catherine Ikam, Otto Piene, American Center staff members; 5. Don Foresta; 6. compiled Slow Scan image of President Ronald Reagan.

Photos: Elizabeth Goldring and Otto Piene Collection

Centervideo included the telecommunications event at the American Center, and a CAVS video exhibition, February 1-6, 1981, that traveled to the Kölnischer Kunstverein, Cologne, March 6-10, 1981, and to Bordeaux, France, accompanied by CAVS Fellow Betsy Connors. The catalog reproduced a telex from Don Foresta in Paris to Otto Piene in Cambridge, rife with machine-generated idiosyncrasies that required a "note on typos" from the human TLX Operator.

> TELEVISION IS THE FIRST MEDIUM OF MASS ONE-TO-ONE COMMUNICATION TO BE DEVELOPED. WHY THIS IS SO IS NOT CLEARLY UNDERSTOOD, EVEN BY THE SPECIALISTS THAT MAY EMPLOY IT, ALTHOUGH THEORIES ABOUND. (Foresta telex to Piene (excerpt), *Centervideo* catalog, 1981, 5)

Centervideo included some thirty CAVS artists with varied approaches to the newly accessible time-based video medium, ranging from social-political commentary to media self-reflections and musically inspired experiments in visual aesthetics. Portraits in the catalog were generated by a "Time Machine" designed to manipulate time and image by Piotr Kowalski.

> It is not unreasonable to think that processes involving soft media which harden later like spun pottery, oil painting on canvas, which both employ a vibrating tool or support, may have captured ambient sounds in the process of their fabrication. There may be worlds of words and sounds from people of millenniums past all around us which it would be worth to decipher. "TM II" . . . will be, like "TM I," a public oriented and interactive piece which will capture both the existing sound and the public's image and actions, and restitute them backwards in real time on TV screens and through sound systems. As with "TM I," both forward and backwards-going time events will coexist in the same area. (Piotr Kowalski, *Centervideo* catalog, 1981, 68)

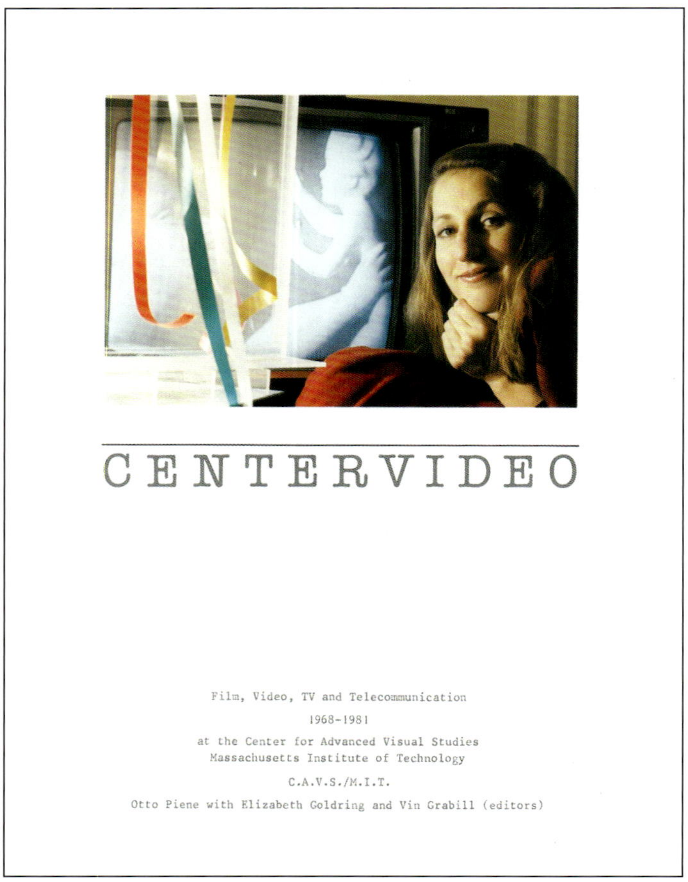

Facing page: Piotr Kowalski, "Objekt mit Rotierenden Spiegel," Ronald Feldman Fine Arts, New York City, 1979
Photo: Eeva Inkeri

Facing page inset top: Piotr Kowalski, portrait of Otto Piene created by "Time Machine II," CAVS, MIT, November 3, 1980
Image: Piotr Kowalski

Facing page inset bottom: Piotr Kowalski, "Time Machine II," 1980
Photo: Piotr Kowalski

Left: *Centervideo* catalog, cover with image by Vin Grabill of video artist Ellen Sebring, editors: Piene, Goldring, Grabill, CAVS, MIT, 1981
Source: CAVS Special Collection

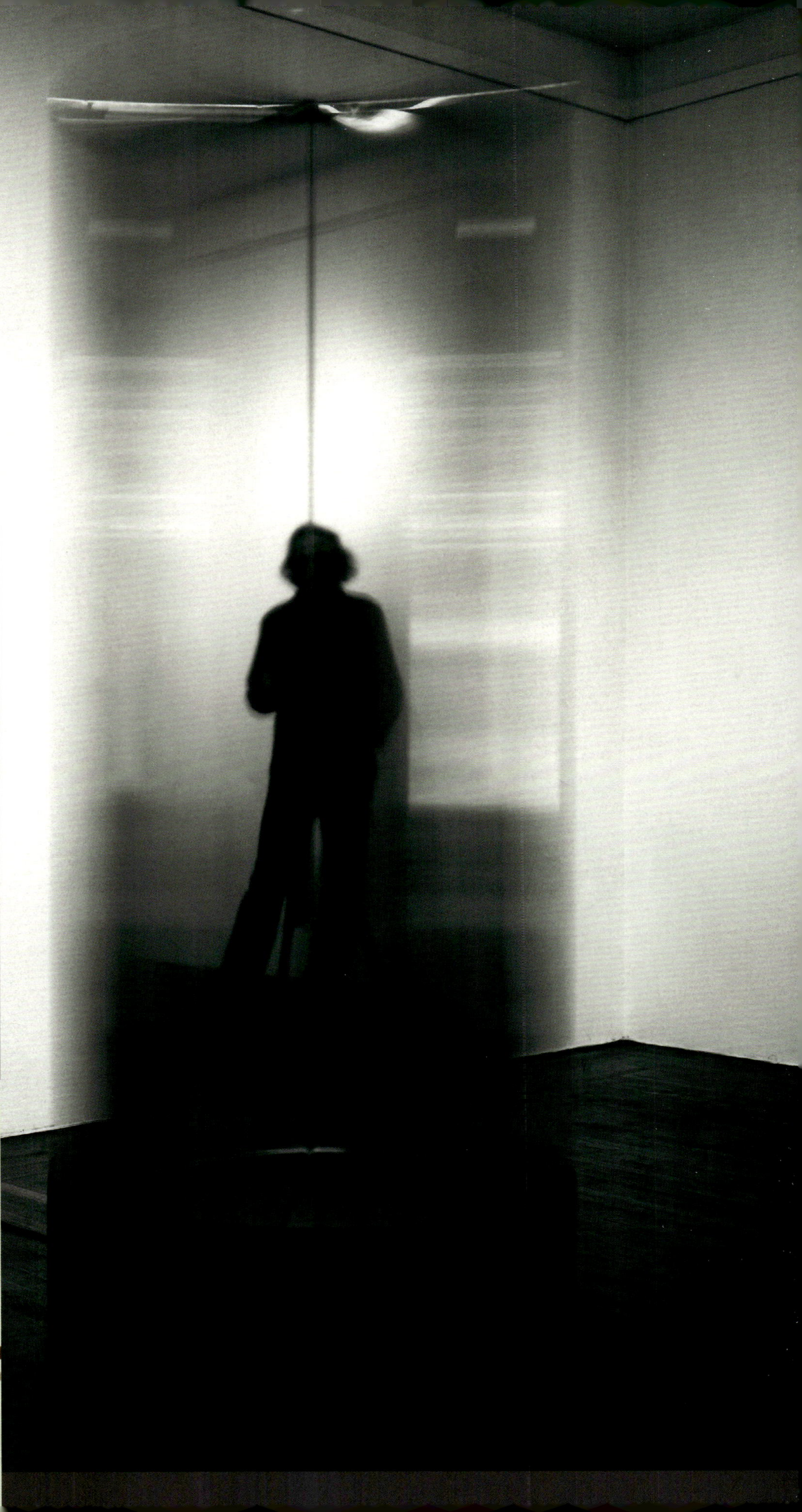

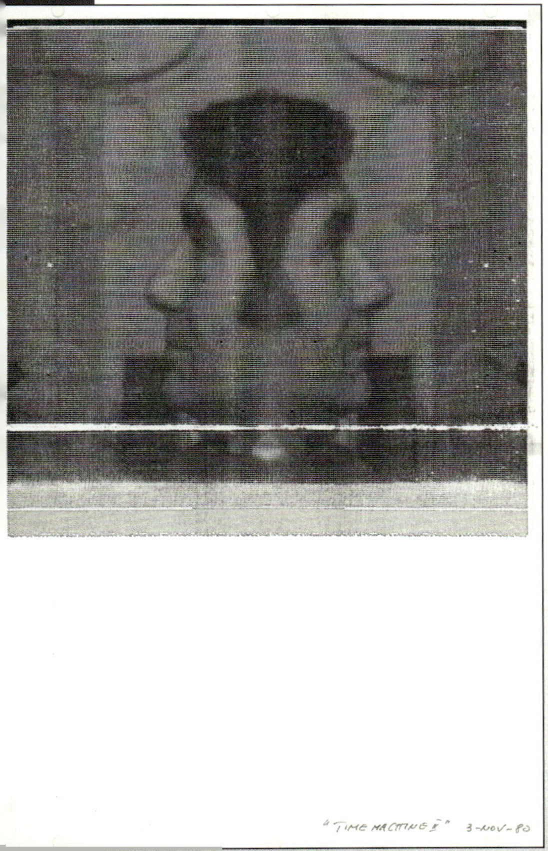
"TIME MACHINE II" 3-NOV-80

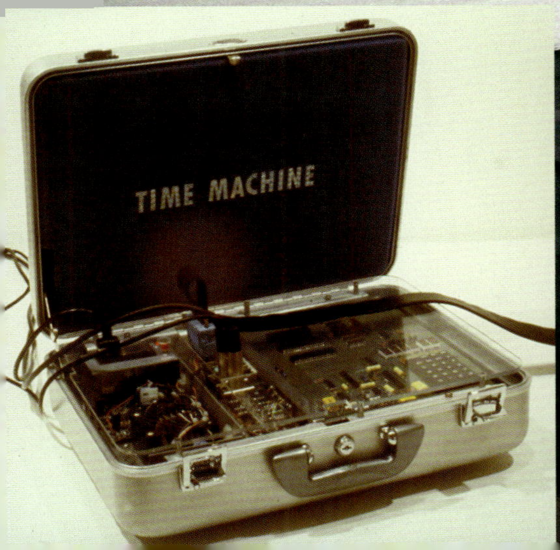
TIME MACHINE

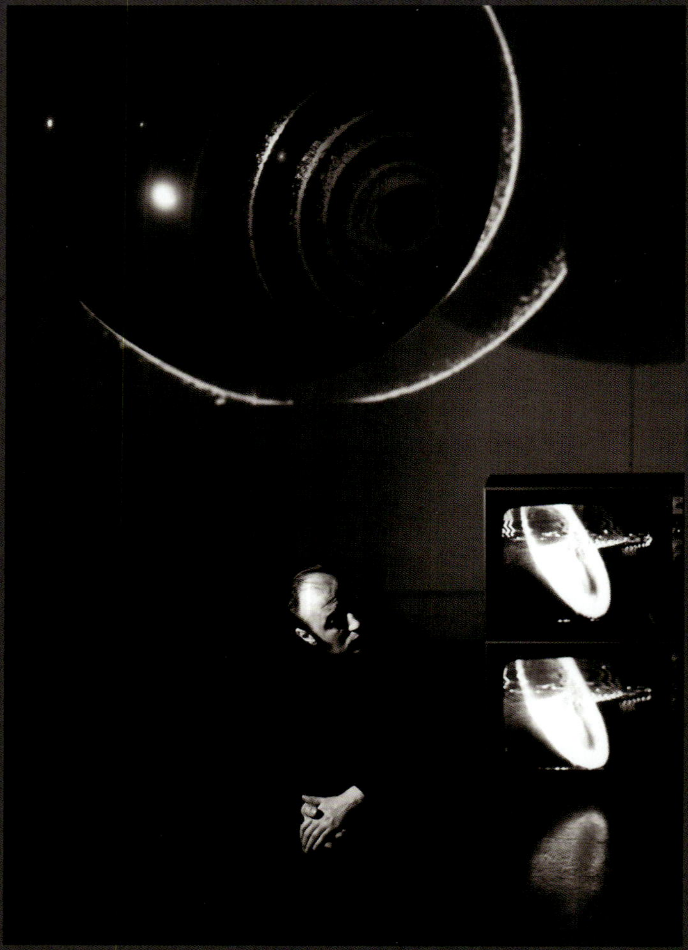

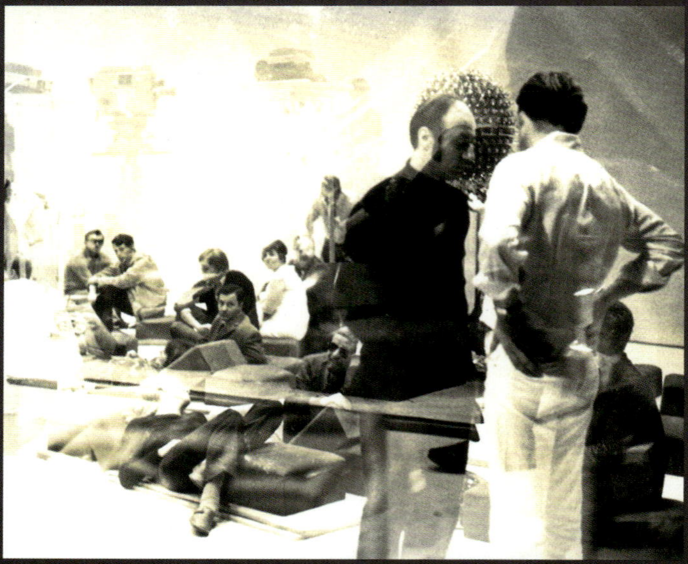

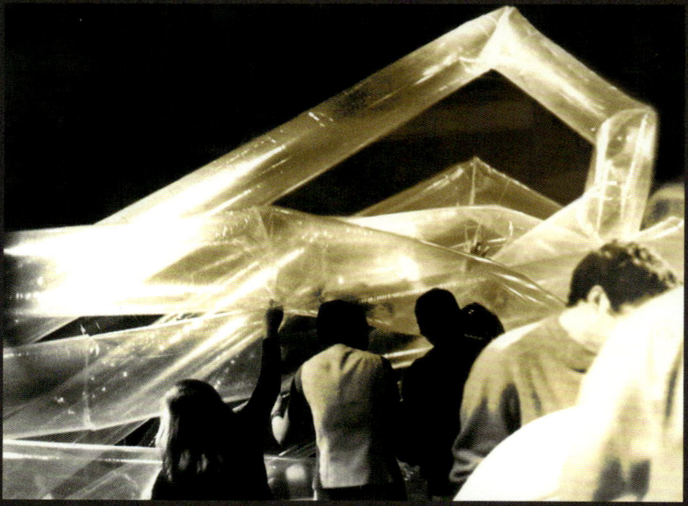

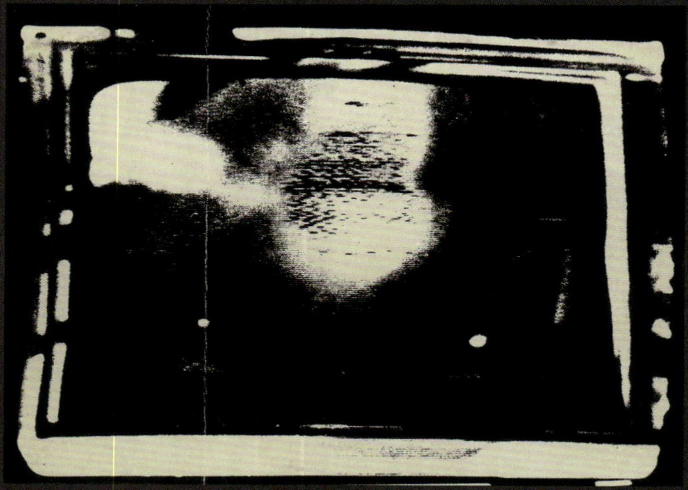

Top left: Aldo Tambellini in the environment of "Black Zero," Black Gate Theatre, New York City, 1967
Photo: Ruspoli-Rodriguez

Bottom left: Aldo Tambellini, image from first videotape, 1967
Photo: Aldo Tambellini

Top and bottom right: Otto Piene and Aldo Tambellini, "Black Gate Cologne," photos made during two 45-minute broadcast recording sessions described as "ein Lichtspiel" with audiences, inflatables, light sculptures, video, and projections blended with real-time image processing, Electronic Studio, WDR, Cologne, 1968
Photos: Peter Kliem

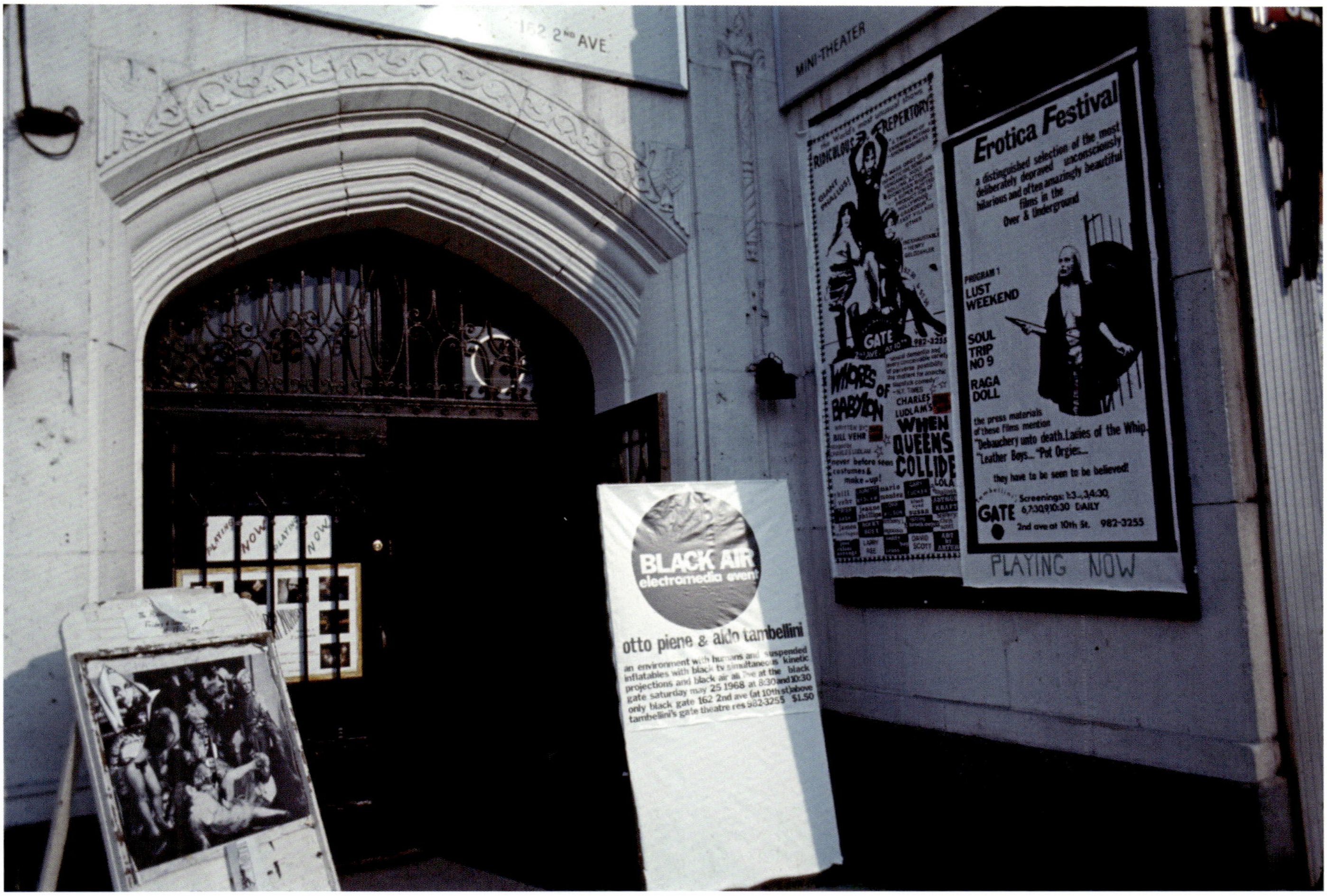

Aldo Tambellini wrote, "We are the primitives of a new era." (*Centervideo* catalog, 1981, 37) He described the political and tactile evolution of his percussive video studies in black that began with glass slides:

> Instinctively I took needles and scratching tools taking away the surface and my forms appeared—the spinning disk, the non-existing moons, the black suns. (Tambellini, *Centervideo* catalog, 1981, 33)

Like Piene, Tambellini was influenced by shattering experiences of bombings while growing up during World War II. The two artists issued the *Black Gate* newspaper in 1967 in which Tambellini wrote, "black gate because the marketable commodity art object is disintegrating under the energy of light," and Piene wrote:

> Painting does not light itself; motionless sculptures are in the way; objects inhibit travelling. . . . The Black Gate is open to ride on a light beam. (*Black Gate* newspaper, Issue 1, New York, April, 1967)

They mounted events, including Piene's *Proliferation of the Sun* in New York City and at WDR in Cologne (Westdeutscher Rundfunk Köln). *Black Gate Cologne* is considered to be the first broadcast television program by artists.

Above: Otto Piene and Aldo Tambellini, "Black Air: electromedia event," poster text: "an environment with humans and suspended inflatables with black tv simultaneous kinetic projections and black air all live at the black gate . . . 162 2nd Avenue above Tambellini's gate theatre," New York City, May 25, 1968
Photo: Otto Piene Collection

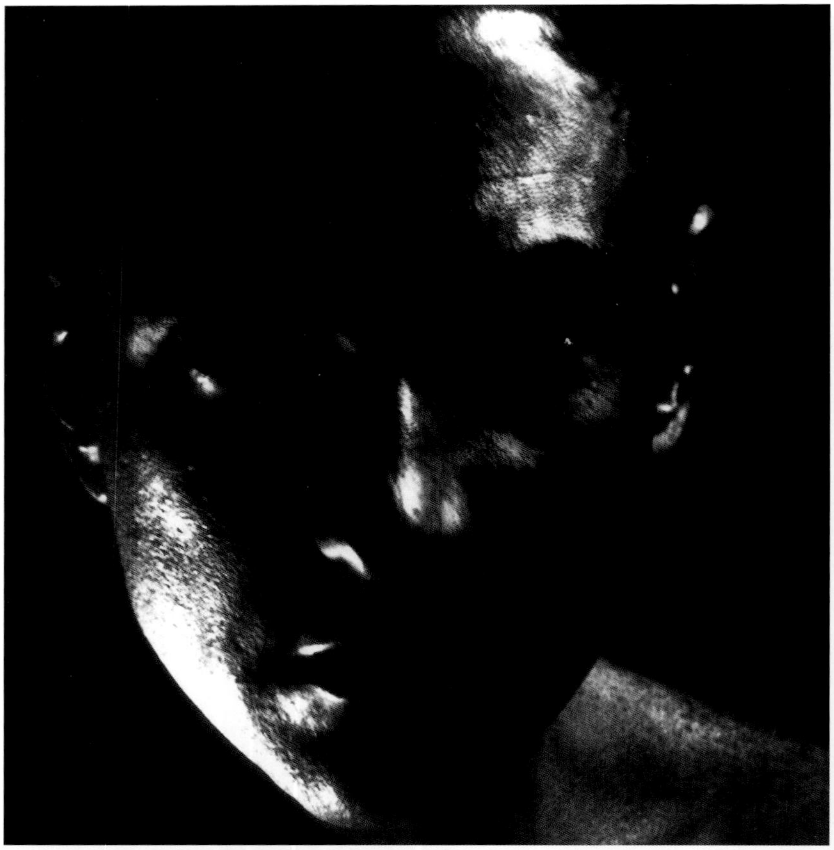

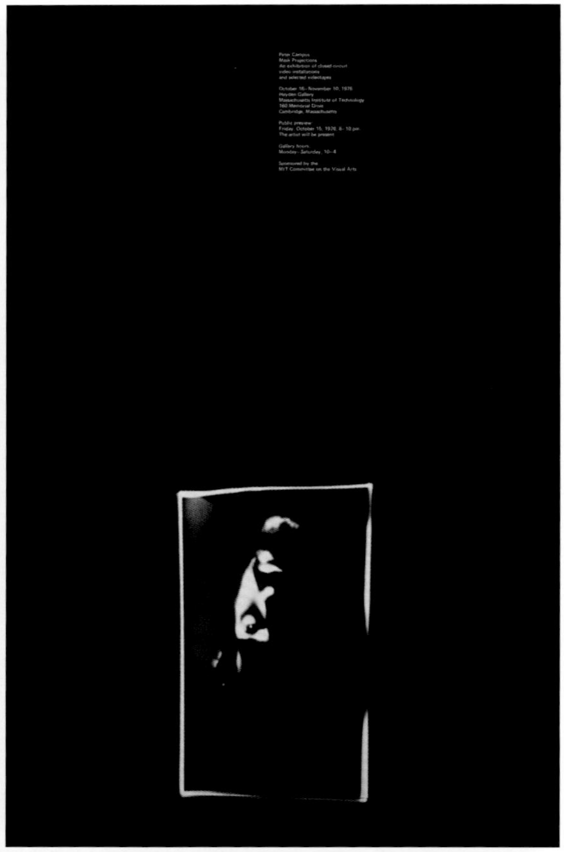
Peter Campus
Mask Projections
An exhibition of closed-circuit
video installations
and selected videotapes
October 16–November 10, 1976
Hayden Gallery
Massachusetts Institute of Technology
160 Memorial Drive
Cambridge, Massachusetts
Public preview
Friday, October 15, 1976 8–10 pm
The artist will be present
Gallery hours
Monday–Saturday, 10–4
Sponsored by the
MIT Committee on the Visual Arts

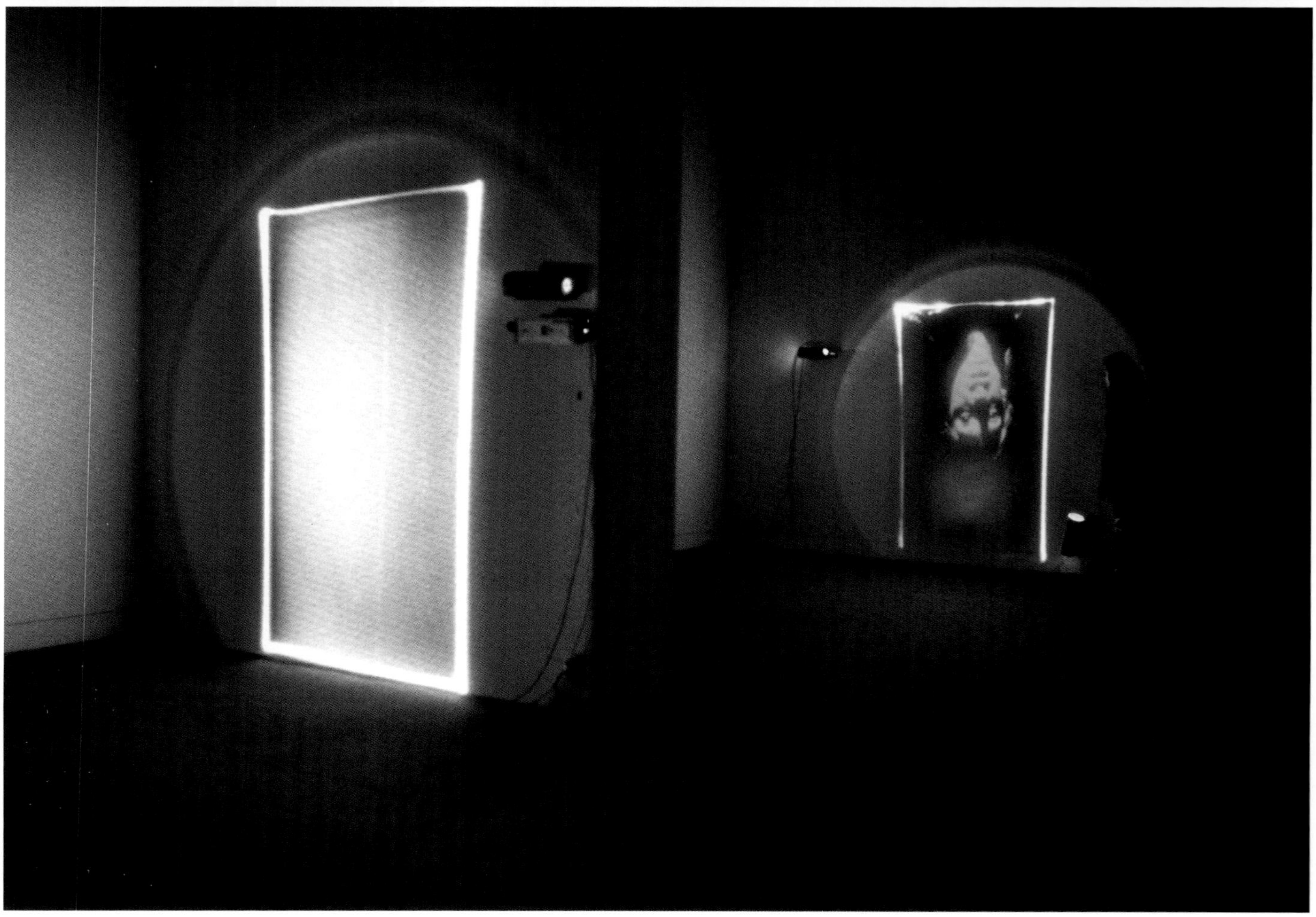

The artist statements in the *Centervideo* catalog showed diverse backgrounds and approaches to the video medium:

> I don't think it's trying to get rid of the totality of self. I'm trying to get rid of the reflection of self, the illusion of self to get down to the inner self. (Peter Campus, 46)
>
> . . . world of toys and small objects . . . elements of humor and surprise . . . the TV monitor as miniature theater for clay models and sets . . . (Betsy Connors, 50)
>
> For me making videotape is analogous to making music: the finished form comes from layering images and sounds along a measured time line. (Vin Grabill, 61)
>
> I have assumed for some time that video and computer animation are the new "painted" images of our time, the changes in these tools have made it possible for the evolution of "cinema" into "optical painting" and illusory visual thinking. (Stan VanDerBeek, 102)
>
> Video is more than a way to preserve/ present/transmit. It can be a tool to degenerate/regenerate/amplify time and vision and language. (Ken Kantor, 66)
>
> I see both terms time and space as determinants of instant/culture. Cultural transgression happens by the confluence of a local culture (indigenous) and a supra-local one fermented by the dynamics of a third one—the "culture of media." (Muntadas, 80)
>
> My intent was to show the black and vibrations, the industrial hell-symbol of a subway ride through the raw and innocent sensuality of babies' faces. (Michael Naimark, 82)
>
> In "Journeys From Berlin/1971" there are five characters: a man, woman, and adolescent girls who "appear" only on the sound track; a female psychoanalytic patient in her early 50s; and a therapist, or analyst, played alternately by a man, woman, and nine-year-old boy. . . . The film utilizes formal strategies that gained prominence in the premises of 60s art- and dance-making, namely the inherent expressiveness of quotidian gesture and imagery and the relativity of meaning resulting from "radical," random, or ambiguous juxtapositions. (Yvonne Rainer, 89)

Facing page top left: Peter Campus, "Man's Head," photoprojection, 10'x10' installation, 1978, Paula Cooper Gallery, New York City, January 16-February 7, 1979
Photo: CAVS Special Collection

Facing page top right: poster for Peter Campus exhibition, *Mask Projections*, "an exhibition of closed-circuit video installations and selected videotapes," Hayden Gallery, MIT, October 16-November 10, 1976
Source: CAVS Special Collection

Facing page bottom: Peter Campus, "num," and "lus," closed circuit video installations, Hayden Gallery, MIT, 1976
Photo: Nishan Bichajian

Boston became a center for video art. WGBH, the local Public Broadcasting Service (PBS) station, invited six artists—Allan Kaprow, Nam June Paik, Otto Piene, James Seawright, Thomas Tadlock, and Aldo Tambellini—to create the first US national broadcast of artists working with television, *The Medium is the Medium* (1969). The program opened with the question, "What happens when artists take over television?" WGBH continued to advocate for experimental video and CAVS artists through the *New Television Workshop* (1974-1993).

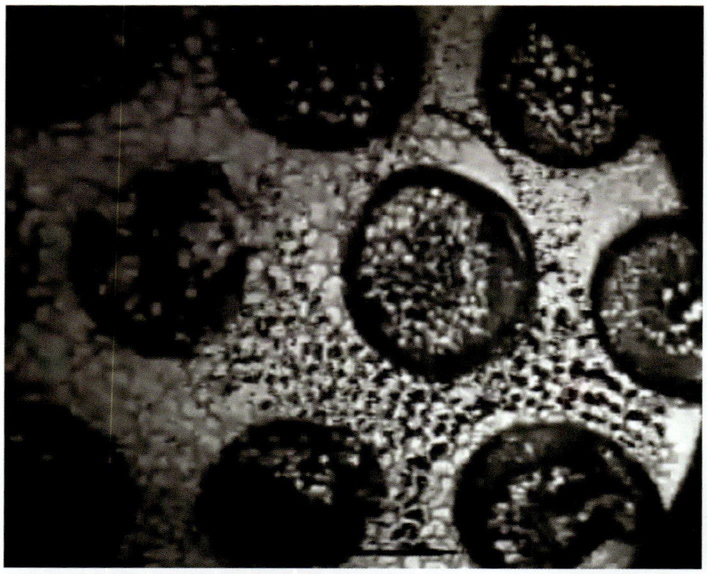

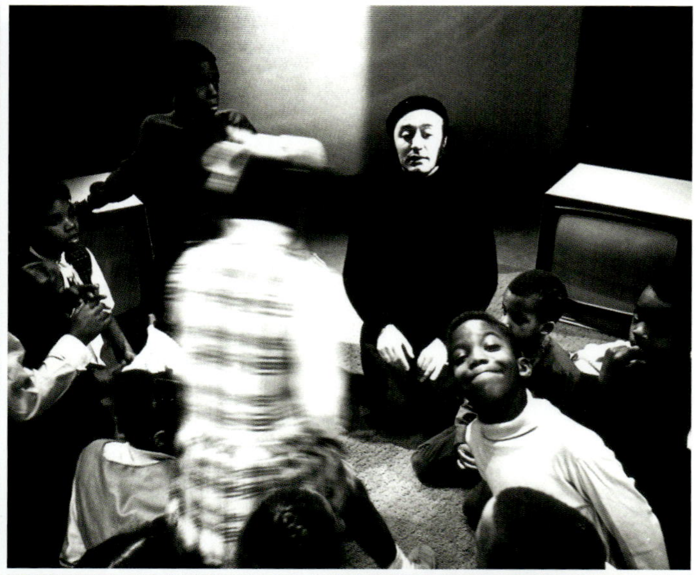

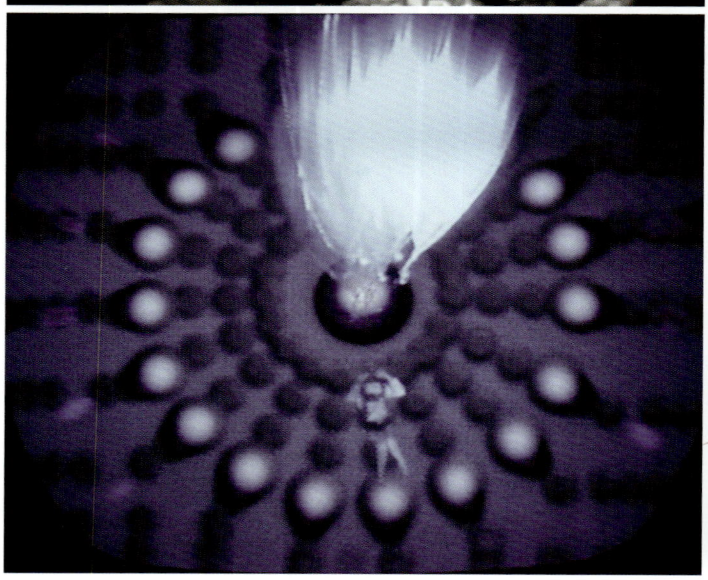

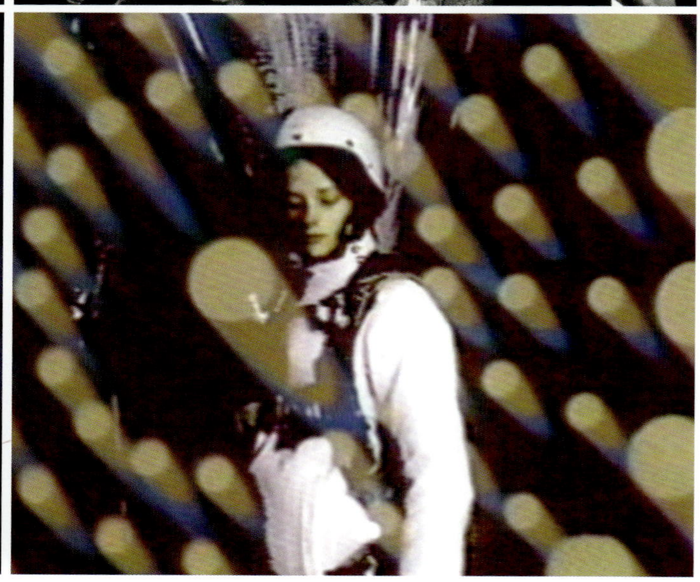

Top left and right: Aldo Tambellini, "Black," with 30 children, slides, film and monitors, *The Medium is the Medium*, WGBH-TV, Boston, 1969
Photos: CAVS Special Collection

Bottom left and right: Otto Piene, "Electronic Light Ballet," with flying performer and light rasters, *The Medium is the Medium*, WGBH-TV, Boston, 1969
Photos: Otto Piene

Facing page top: Betsy Connors, with her animation sets, video shoot, Center for Advanced Engineering Studies, MIT, 1979
Photo: CAVS Special Collection

Facing page bottom: Betsy Connors, "Batteries Not Included," video animation, *New Television Workshop*, WGBH-TV, Boston, 1979
Photo: Betsy Connors

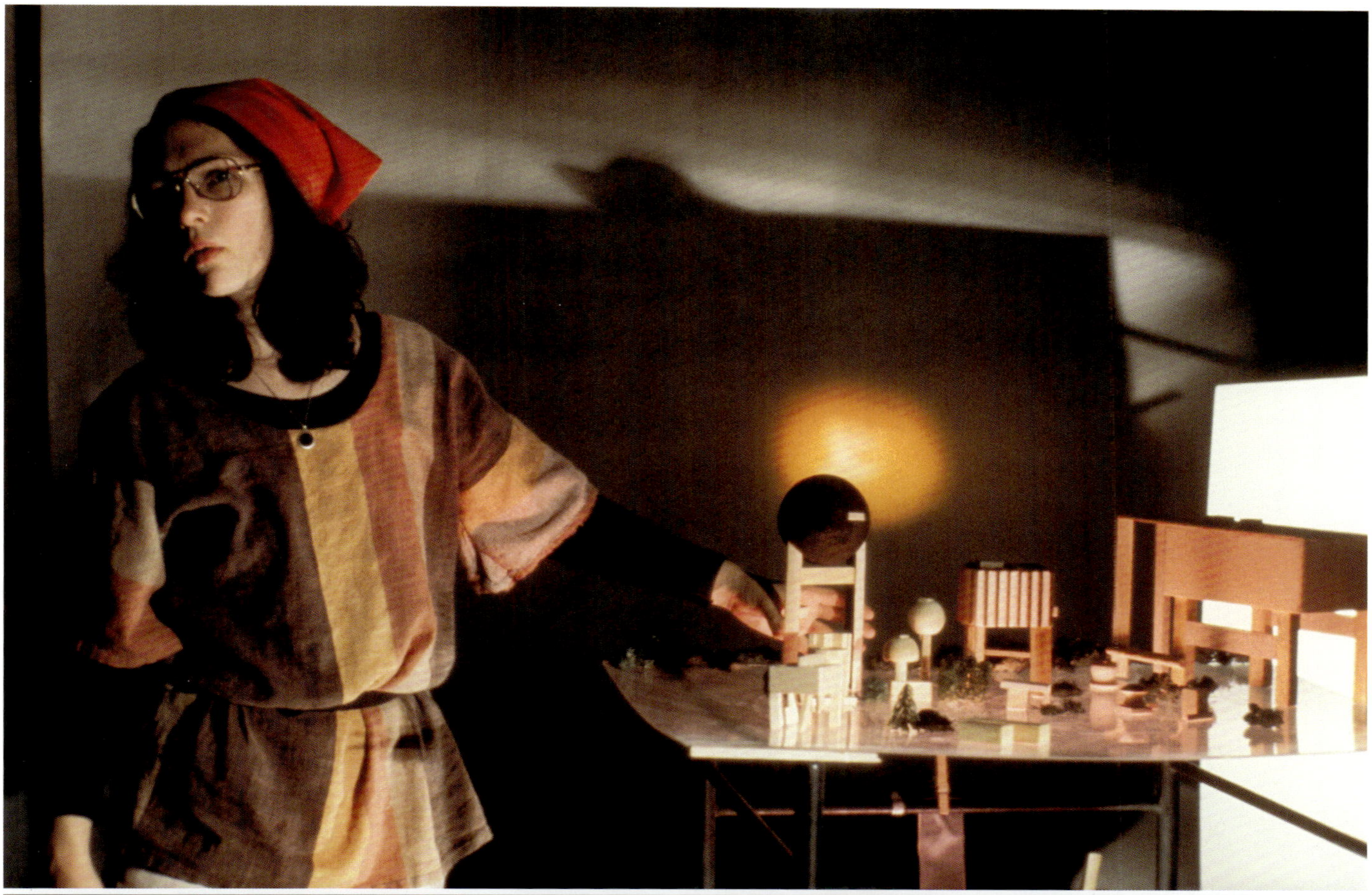

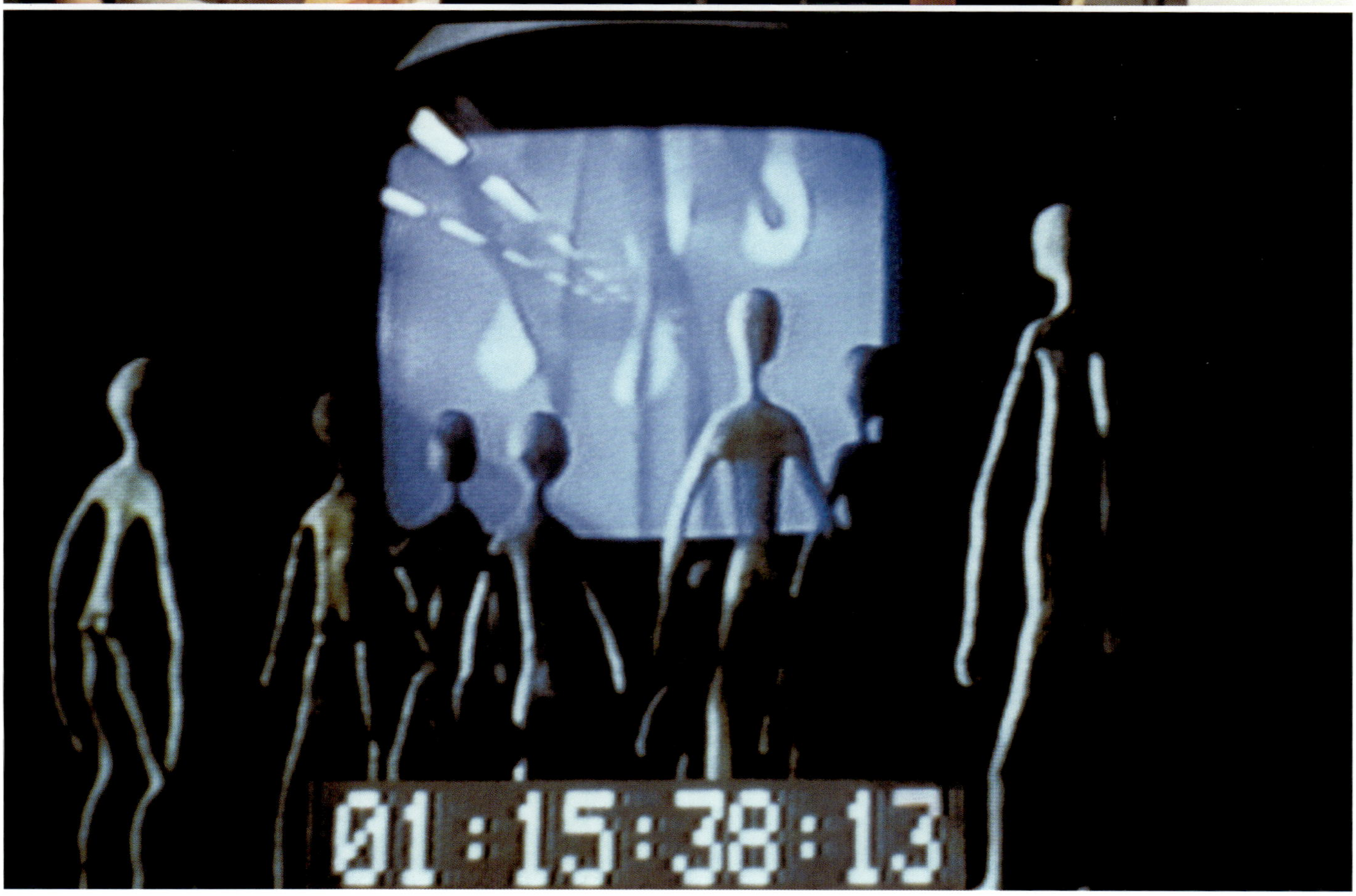
01:15:38:13

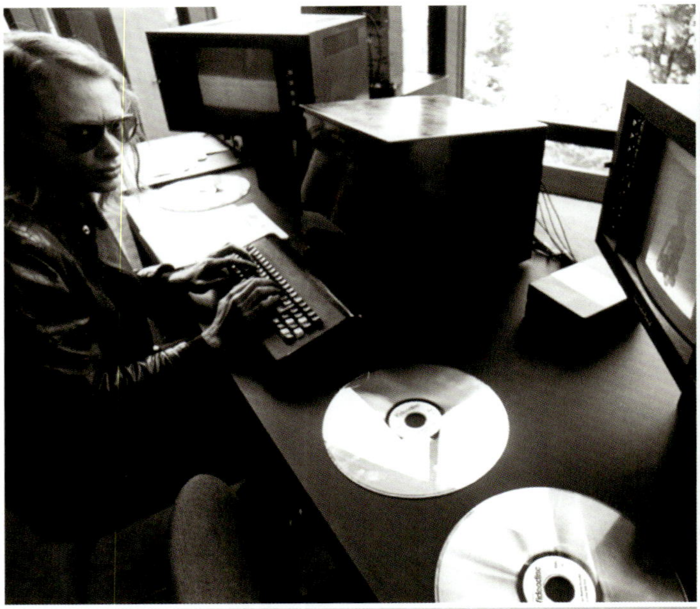
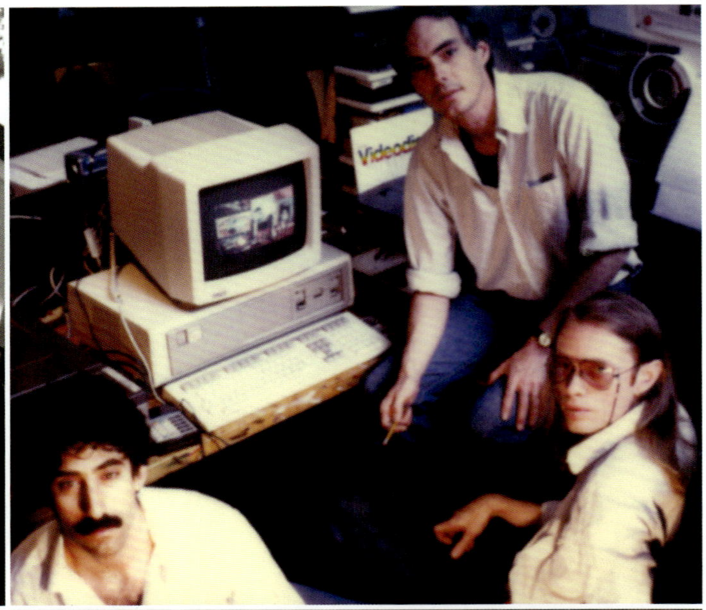
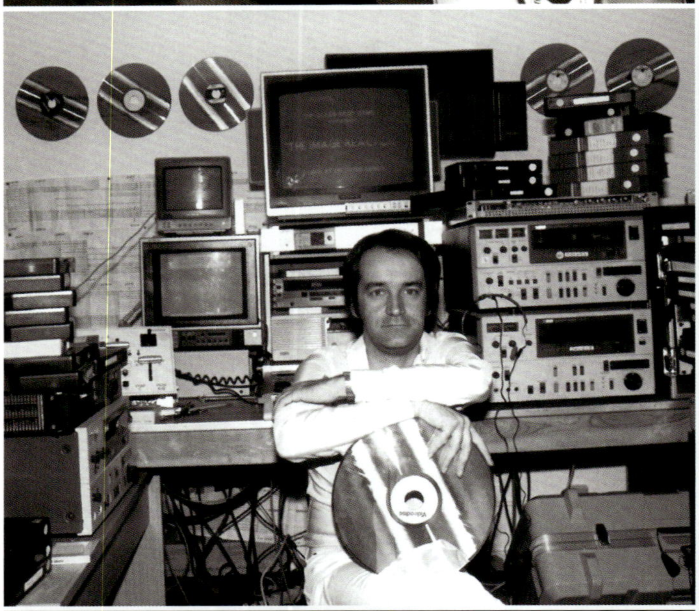
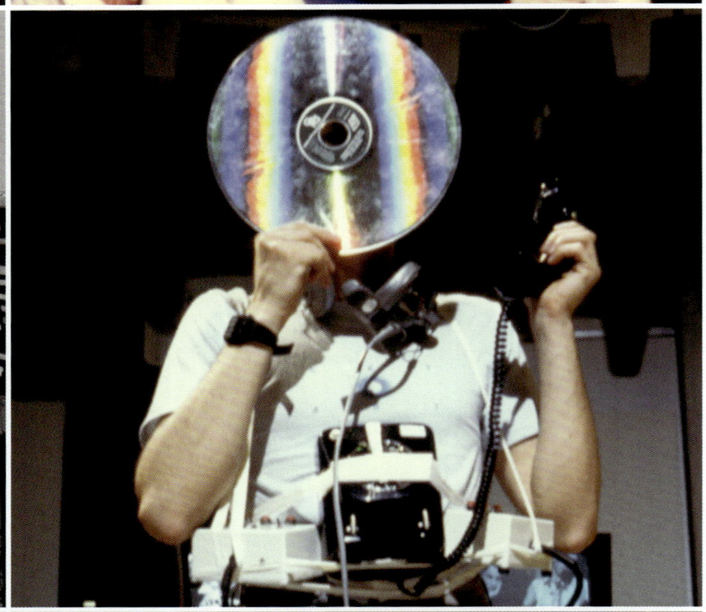
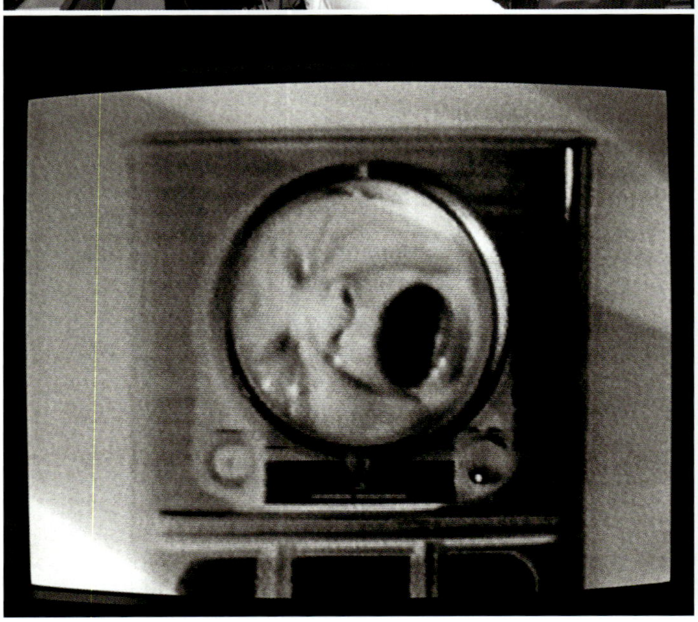
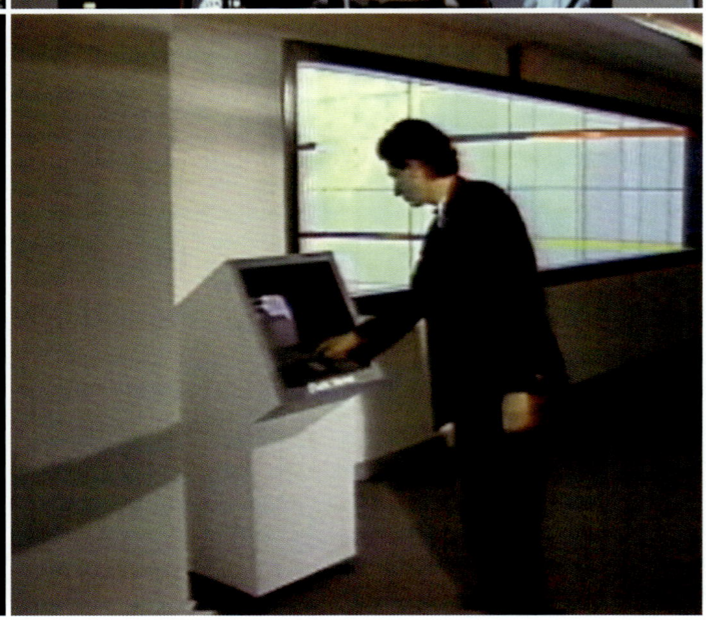

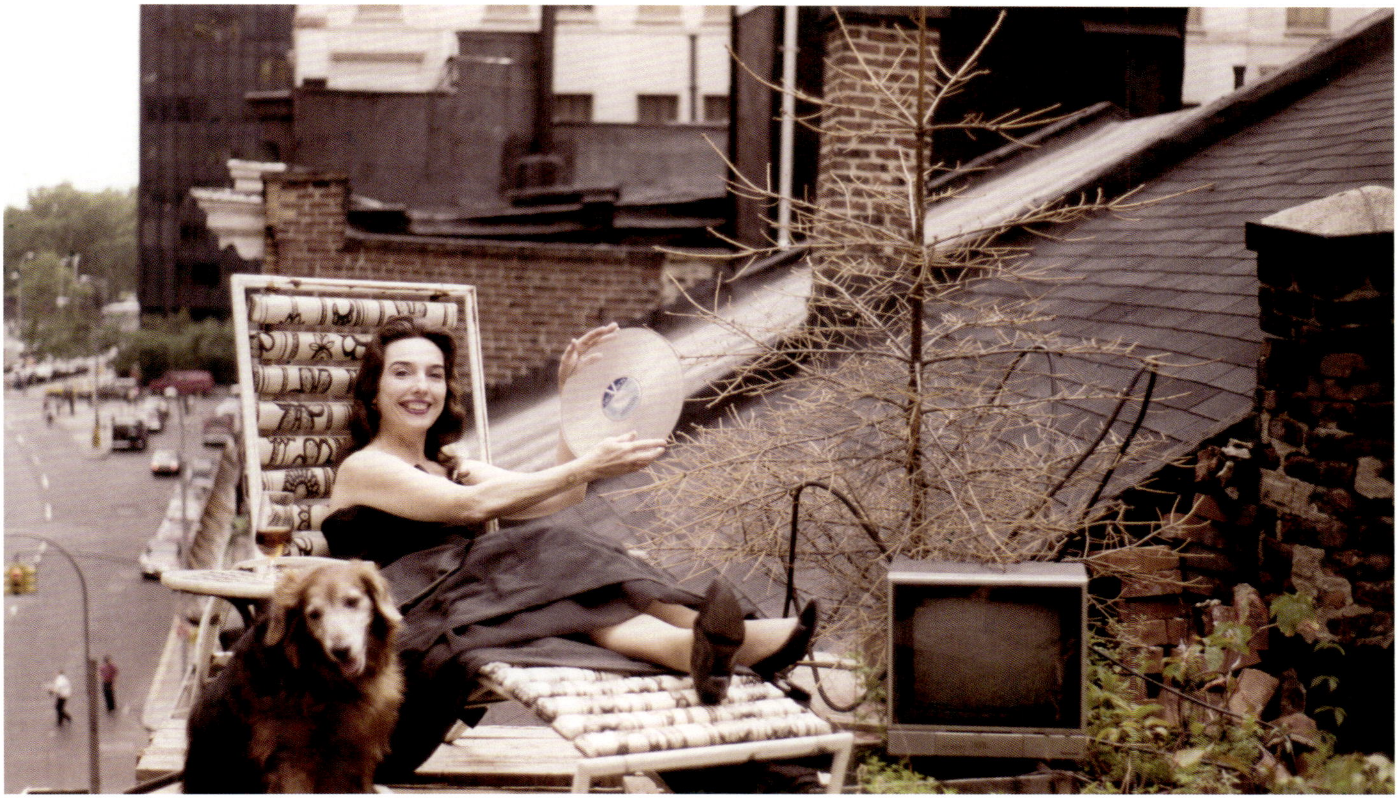

Nascent digital technology offered new opportunities for virtual traveling exhibitions. CAVS produced *Centerdisk* (1982), *Skydisk* (1983), and *The Sky Art Interactive Video Disk Travelling Exhibition* (1984). Fellows Rus Gant, Vin Grabill, and Ben Davis developed the interactive content for these laser discs, specifically designed for this first commercially available optical disc storage medium.

Centerdisk was an attempt to create a visual archive that was reproducible, distributable. We had papers, slides, audio and video tapes, but the videodisc was the first time you could combine all those things in a single format. (Rus Gant interview)

Above: Charlotte Moorman displays *Centerdisk* (1982) on the rooftop of her Manhattan home, New York, 1983
Photo: Vin Grabill

Facing page top left, Rus Gant, "Vision Machine," *Sky Art Conference '82, Ars Electronica*, Linz, 1982
Photo: Vin Grabill

Facing page top right: (left to right) Uriel Levy, Ben Davis, and Rus Gant, videodisc production at MIT, 1982
Photo: Vin Grabill

Facing page middle left: Titus Leber, with his interactive videodisc, "The Image Reactor," CAVS, MIT, 1984-85
Photo: Janina

Facing page middle right: Bernd Kracke, "Data Network," performance, *Sky Art Conference '82, Ars Electronica*, Brucknerhaus, Linz, 1982
Photo: Otto Piene Collection

Facing page bottom left: Peter D'Agostino, "Double You (and X, Y, Z)," interactive videodisc installation, 1985
Photo: Peter D'Agostino

Facing page bottom right: "Elastic Movies," Luc Courchesne operates laserdisc interactive installation including works by CAVS artists Courchesne, Bill Seaman, and Ellen Sebring, produced by Film/Video section, Media Lab, MIT, 1984
Video still frame: Luc Courchesne

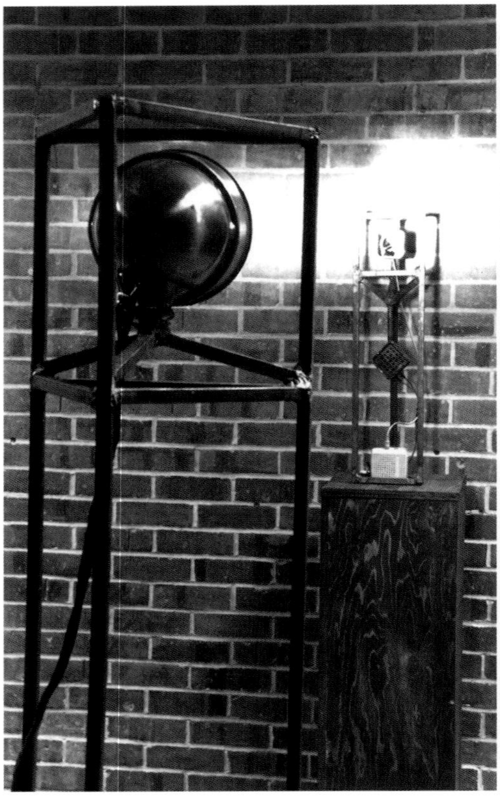
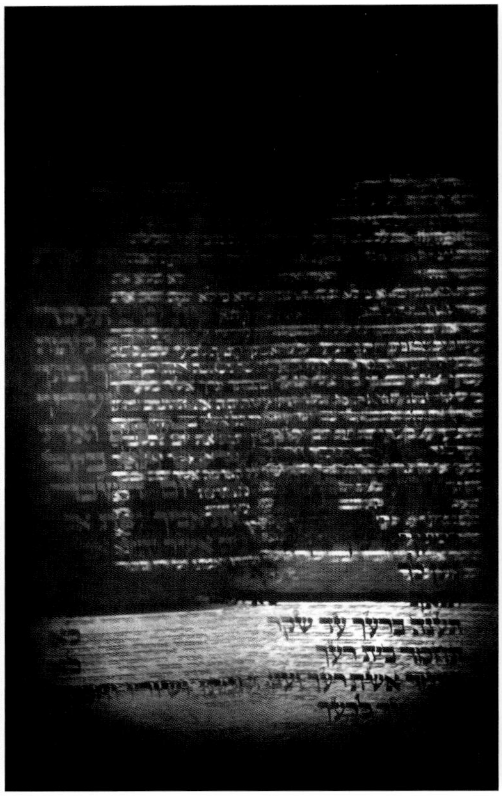
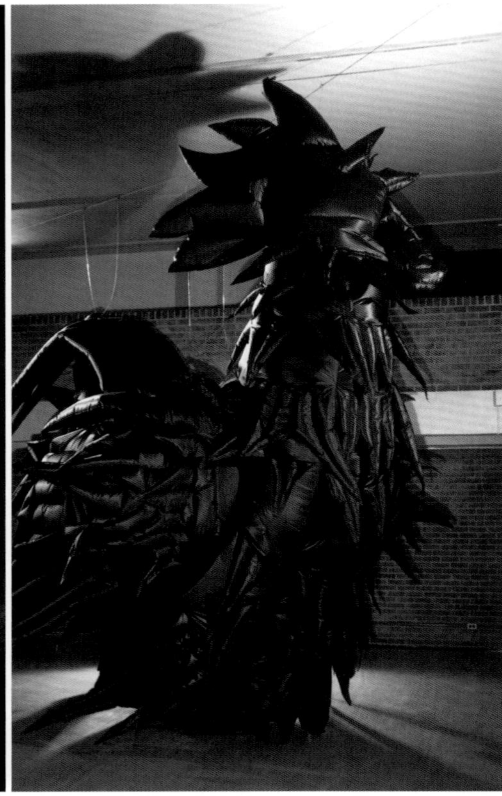

Bottom, left to right across both pages:
Otto Piene, "Prism Star in Light Vessel" and "Light Paths" 1987
Otto Piene "Star of David" 1987
Harold Tovish "Region of Ice" 1984
Elizabeth Goldring and Vin Grabill, "Eye/Sight" video installation, 1987
Joe Moss "Speak, Be Heard" 1973/85/87
Mel Alexenberg, "Torah Spectrogram" 1987

Above, left to right:
Joe Davis, "One Light" 1987
Dieter Jung, "The Ten Commandments: Stone of Light" hologram, with Fred Unterseher and John Perry, 1987
Otto Piene, "Sekhvi Rooster" 1987

All images from *LightsOROT*, CAVS exhibition, Yeshiva University Museum, New York City, 1987
Photos: Walter C. Dent

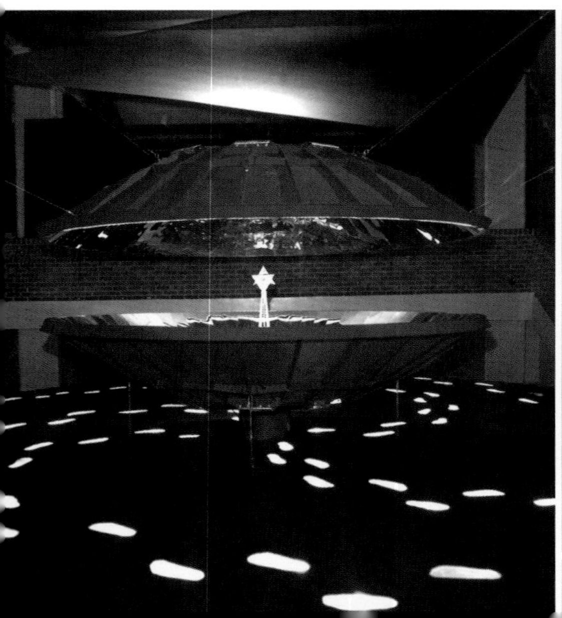
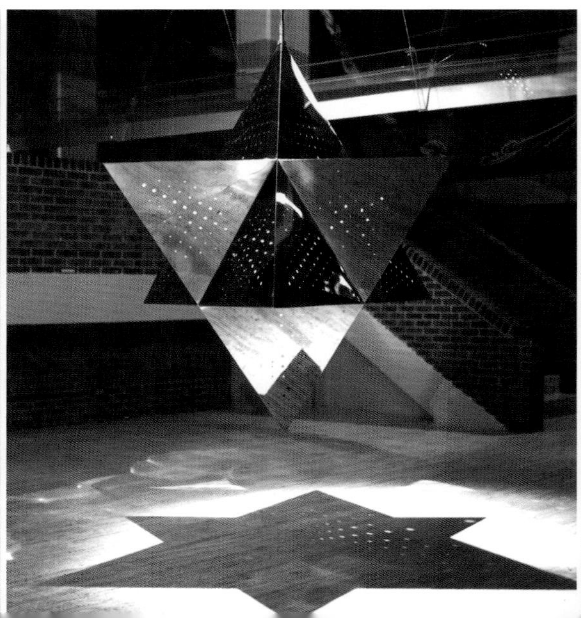
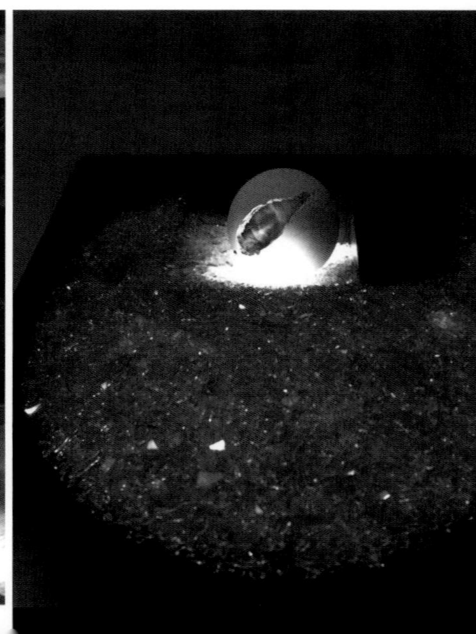

Focus on Light

CAVS exhibitions incorporated light works and thematic considerations of light in large group shows, such as *Celebration of Light*, (*CAVS exhibition for the 350 year Celebration of the City of Savonlinna*, Finland, June 10-August 6, 1989) and *Otto Piene und das CAVS* (Badischer Kunstverein Karlsruhe, 1988). Piene's informal exhibition design relied on the strength and unexpected juxtapositions of individual contributions.

With *LightsOROT* (Yeshiva University Museum, New York City, 1987), although constrained by a specific cultural context, the interplay of works retained freshness through the environmental performances of light and sensory installations. The catalog itself became a memorable artifact, a book drilled through its center to physically let in light.

> *LightsOROT*, the brainchild of CAVS artist Fellow Mel Alexenberg, is an '80s light art show. It is an attempt to reconcile the spirituality of traditional religious beliefs and contemporary art through the medium of light, reflecting new alliances as well as increased familiarity with tools, flexibly adapted. The result is a kinetic light exhibition transporting metaphor and inviting human participation. Human hands, feet, eyes and voices pervade the strobes, lasers, and hosts of microprocessors to spark the light. (Goldring, *LightsOROT* catalog, 43)

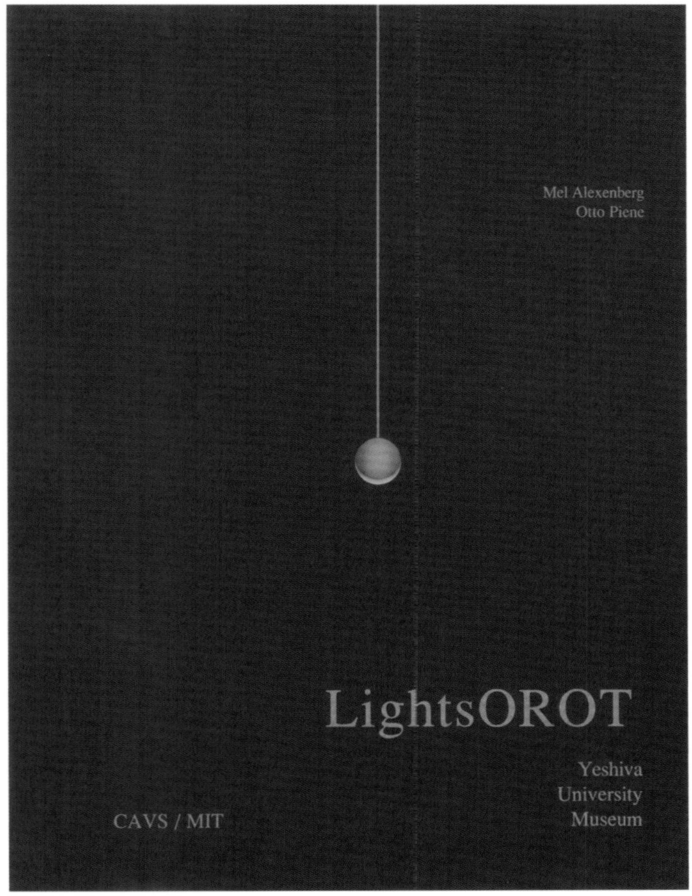

LightsOROT catalog, designed and edited by Mel Alexenberg, Otto Piene, Bill Seaman, and Ellen Sebring, CAVS, MIT, 1987

"The dye-cut is a light frame. It invites "real light" to infiltrate the writing." (Piene, *LightsOROT* catalog, 1988, 3)
Source: CAVS Special Collection

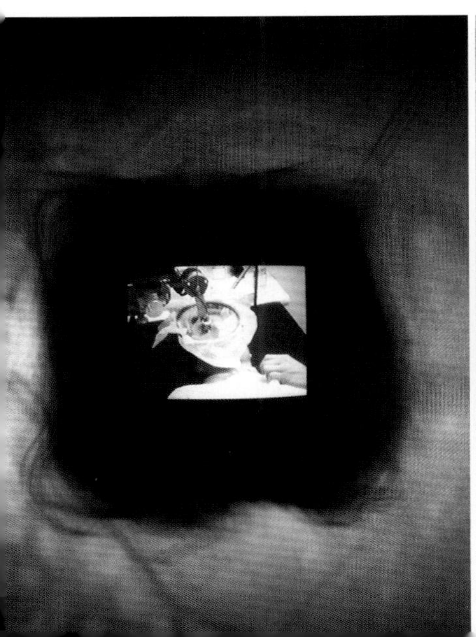

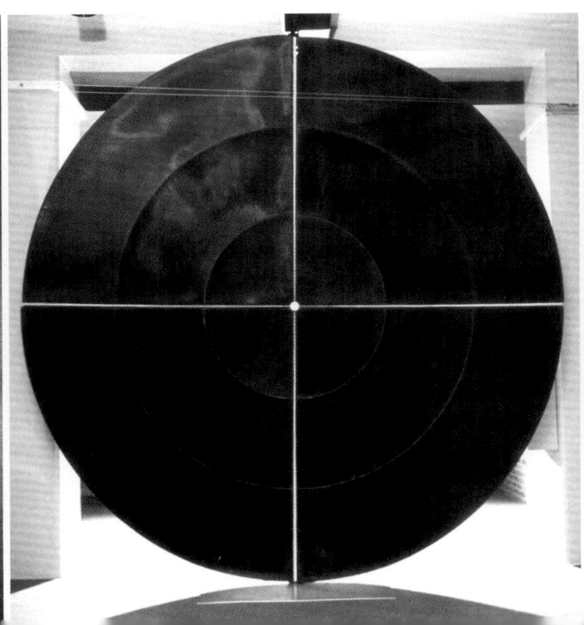

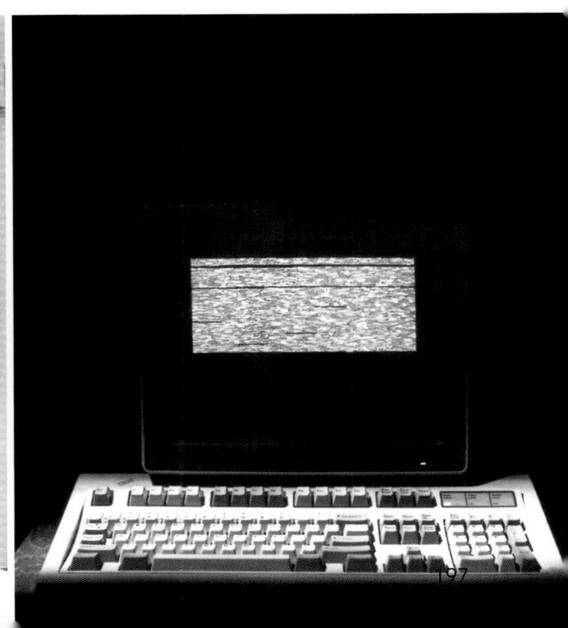

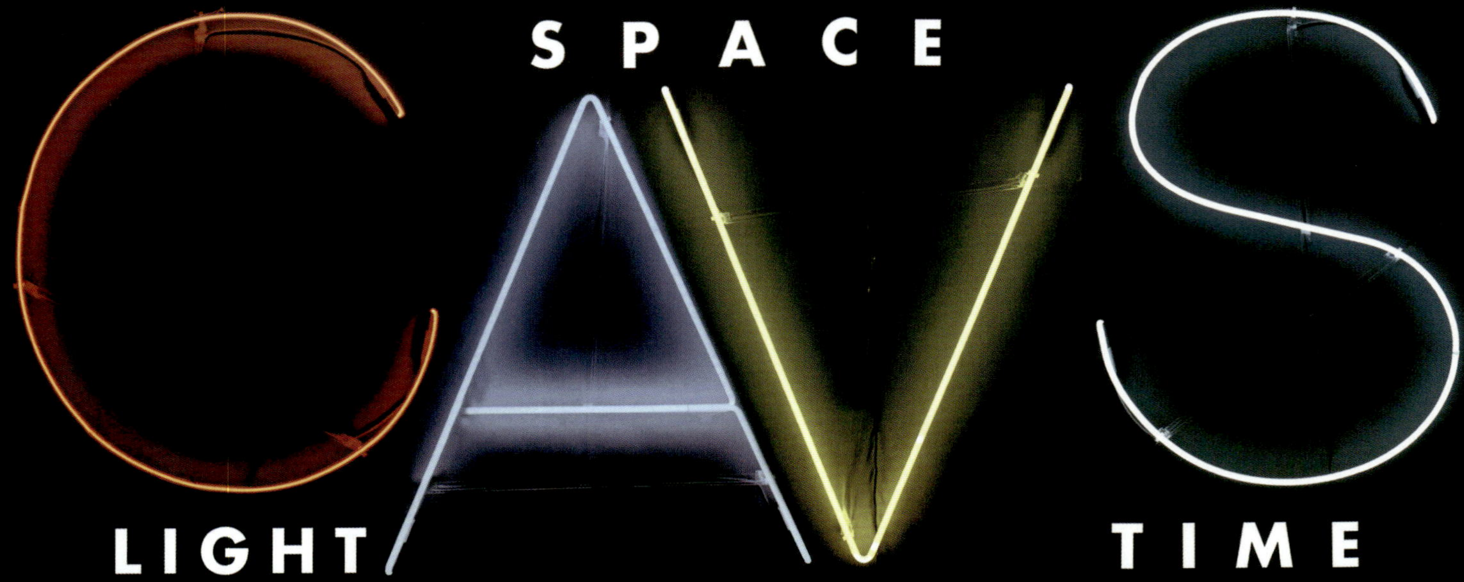

CAVS/MIT - 25 YEARS

a center
for advanced visual studies
retrospective
through 2 october 1994
at the mit museum and cavs

a reception for the artists saturday 4 june 2 to 4 pm

at the mit museum and cavs 265 massachusetts avenue cambridge

CAVS FELLOWS

MARC ADRIAN
WERNER AHRENS
MITSUNOBU AKAHORI
MEL ALEXENBERG
MARYANNE AMACHER
KARIN BACON
KEN BAKER
SARAH BAKER
ROLAND BALADI
JUAN NAVARRO BALDEWEG
ERIC BEGLEITER
MITCH BENOFF
MICHAEL BERNARD
NISHAN BICHAJIAN
JOHN BOESCHE
NANCY IRENEE BOCHENEK
JOAN BRIGHAM
SHAWN BRIXEY
GLORIA BROWN-SIMMONS
MAURICIO BUENO
LOWRY BURGESS
JACK BURNHAM
JAMES LEE BYERS
BILL CADOGAN
PETER CAMPUS
MIRA CANTOR
ERNST CARAMELLE
BOYD FRANK CARLTON
HARRIET CASDIN-SILVER
MARK CHOW
VARDEA CHRYSSA
JURGEN CLAUS
BILL CROSBY
ATTILA CSAJI
ELIZABETH CONNORS
MURIEL COOPER
LUC COURCHESNE
PETER D'AGOSTINO
DAN DAILEY
ROBERT D'ANCONA
DOUGLAS DAVIS
JOE DAVIS
BEN DAVIS
BRUNO DE LARD
ROBERT DELL
WALTER DENT
JUAN DOWNEY
PETER DROEGE
PAUL EARLS
DALE ELDRED
DOV ELATH
MARK FAVERMAN
DIANE FELLOWS
DAVID FINN
ROBERT FISHER
SCOTT FISHER
DONALD FORESTA
LUIS FRANGELLA
NINA FRANKENHEIM
CHARLES FRAZIER
BETH GALSTON
RUS GANT
JOSE WAGNER GARCIA
GREGORY GARVEY
WENDELIN GLATZEL
DERITH GLOVER
ELON GOITEIN
ELIZABETH GOLDRING
CYNTHIA GOODMAN
JOHN GOODYEAR
VIN GRABILL
GEORGE GREENAMYER
VIRGINIA GUNTER
LAWRENCE ALAN HAHN
RON HAYS
NATHAN HEMENWAY
MAREK HOLYNSKI
PIERRE-ALAIN HUBERT
MICHIO IHARA
SETSUKO ISHII
CATHERINE IKAM
TOSHIRO ITAKURA
CHRISTOPHER JANNEY
JON JEIBMANN
BART JOHNSON
DIETER JUNG
KENNETH KANTOR
THOMAS B. KASS
HAREL KEDEM
GYORGY KEPES
LAURA KNOTT
MILTON KOMISAR
PANOS KOUROS
PIOTR KOWALSKI
ELLEN KOZAK
BERND KRACKE
BOB KRAMER
TED KRAYNICK
ROCKNE KREBS
SHELLEY LAKE
LESLIE LARKIN
KAY LARSON
NICHOLAS LASCARIS
THORBJORN LAUSTEN
TITUS LEBER
BILLY LEE
HAROLD LEHR
RICHARD LERMAN
URIEL LEVI
PAUL LEVY
FRANCES MASLAND
PAUL MATISSE
MANDY McINTYRE
THOMAS McNULTY
JOYCE MEKEEL
MARK MENDEL
CLARA MENERES
ANTONIO MIRALDA
MIT MITROPOULOS
CHARLOTTE MOORMAN
AVATAR DA SILVA MOREAS
DAVID CURT MORRIS
MIKE MOSER
JOE MOSS
GROVER MOUTON
ANTONIO MUNTADAS
KEVIN MURPHY
MIKE NAIMARK
CARL NESJAR
JOHN AVERY NEWMAN
JOSE NUNO
ATSUSHI OGATA
ALEJANDRO OTERO
NAM JUNE PAIK
BILL PARKER
DONALD PASQUELLA
CHARLES PATRICK
OTTO PIENE
JOHN POWELL
JOSE de PRADA
ROBERT PREUSSER
KEIKO PRINCE
YVONNE RAINER
STANLEY RESNICOFF
GARY THOMAS RIEVESCHEL
SETH RISKIN
BRYAN ROGERS
ROBERT ROSINSKY
JON RUBIN
LEES RUOFF
PAUL RUTKOVSKY
NARMINE SADAGE
THOMAS SANSONE
BILL SEAMAN
ELLEN SEBRING
FRIEDRICH St. FLORIAN
GUNTHER SCHNEIDER-SIEMSSEN
ALLAN SCHWARZ
BARBARA SHAWCROFT
LEON SHIMAN
TODD SILER
JULIAN SIMON
VERA SIMONS
ALEJANDRO SINA
GEORGES SINGER
FRANK SMULLIN
ALAN SONFIST
STACEY SPIEGEL
STEVE S'SOREFF
IAN STRASFOGEL
TAL STREETER
PETER STRUYCKEN
ATHENA TACHA
VASSILAKIS TAKIS
ALDO TAMBELLINI
DON THORNTON
HAROLD TOVISH
WEN-YING TSAI
MARY TSOUTI
PIERRE VAN BERKEL
STAN VANDERBEEK
PAM VANDIVER
GRETCHEN VAN DYK-DEAN
TOM VAN SANT
SALLY WEBER
JUDITH WECHSLER
GILLIAN WISE-CIOBOTARU
HOWARD ALAN YORK
TAKESHI YOSHIZUMI
JOSE MARIA YTURRALDE
WALTER ZENGERLE

MASTER of SCIENCE in VISUAL STUDIES

DAVID BRIAN ATHERTON
SARA BAKER
ERIC BEGLEITER
SHAWN BRIXEY
GLORIA BROWN-SIMMONS
ELIZABETH MARY CAVICCHI
M.E. CHAN-BERNARD
PETER CODELLA
LUC COURCHESNE
WALTER DENT
SHARON DUNN
EBON FISHER
NINA FRANKENHEIM
BETH GALSTON
LORILEE GARBOWSKI-MALLET
GREGORY GARVEY
SARAH GEITZ
DERITH GLOVER
JONATHAN GOLDMAN
VIN GRABILL
JENNIFER HALL
GREGORY HAUN
NATHAN HEMENWAY
CHRISTOPHER JANNEY
JON JEIBMAN
CARYN JOHNSON
CATHERINE JUDGE
KENNETH KANTOR
HAREL KEDEM
SUE M. KIM
LAURA KNOTT
PANOS KOUROS
ELLEN KOZAK
BERND KRACKE
EDSON LAKE
DAVID MARTIN LARKIN
URIEL LEVI
MARTHA LOZANO GUERRA
GERALD McCARTY
MIKE MOSER
MIKE NAIMARK
GEORGE NUMRICH
ATSUSHI OGATA
IRIS OLIVERAS
MARGARET ORTH
JOHN POWELL
SETH RISKIN
DON RITTER
ROBERT ROSINSKY
LEES RUOFF
TOM SANSONE
BILL SEAMAN
ELLEN SEBRING
TODD SILER
DANIEL SPIKOL
DON THORNTON
MARTA VAHTAR
JARLATH WALDRON
SALLY WEBER
DIANE WILLOW
XAGORARIS ZAFOS

Ph. D. in INTERDISCIPLINARY STUDY

TODD SILER - Psychology and Art

SPECIAL STUDENTS

PETER ANDERSON
BARRY ARONS
CLAUDIA BUCHER
MING CHAN
SARAH DICKINSON
HELGA FRANZ
WENDELIN GLATZEL
WOLFGANG HAHN
DANIEL HILL
CHRIS KIRWAN
ED LE POULIN
TERRY BARBARA MASE
CHEEKONS MOK
NINA PAPA
BOB PENA
ANGELA PERKINS
ALAN RATH
NICOLAS REEVES
RON ROCCO
LORANCE RUSSEL
DULCINEA STARR
ROB STUPAY
MICHAEL TIMPSON
GEORGE TSALIKAS
BETSY WALDRON

The exhibition has been supported in part by the Council for the Arts at MIT, Department of Architecture, and the Eugene McDermott Fund for the Arts.

Electronics products for the exhibit have been provided by Sony, in celebration of 25 years of Trinitron television technology.

Computer equipment has been provided by Apple Computer Inc.

Light/Space/Time (1994) at the MIT Museum celebrated the 25th anniversary of CAVS, and crowned the Piene-directed exhibitions and events. Works included light art in the form of neon, light ballet, interactive video, and holography. Goldring and Grabill's "Eye Room" used a medical device, the Scanning Laser Ophthalmoscope (SLO), to explore visual language through interactive audio-visual eye stations. The exhibition was a feast for the senses from Piene's fire painting, "Sky Writing" to Bill Seaman's recombinant poetry. With his wife, Juliet, György Kepes attended the opening of *Light/Space/Time* in his wheelchair. Goldring recalled that he stayed a long time, and seemed to look with interest and admiration at the wall-sized moving image retina projections in the "Eye Room." In 2017-2018, the MIT Museum marked the 50th anniversary of the Center's founding with several shows: *György Kepes Photographs: From Berlin to Chicago, 1930-1946*, and *György Kepes Photographs: The MIT Years, 1946-1984*. Opening in 2018, *Celebrating the 50th Anniversary of the MIT Center for Advanced Visual Studies*, featured installations by CAVS artists, including holograms from the Museum's holography collection, light and sound art, video art, video documentation, and kinetic and inflatable sculptures. In the spirit of CAVS exhibitions, a small group of former fellows and graduates met over the summer months to consider proposals for a future CAVS exhibition.

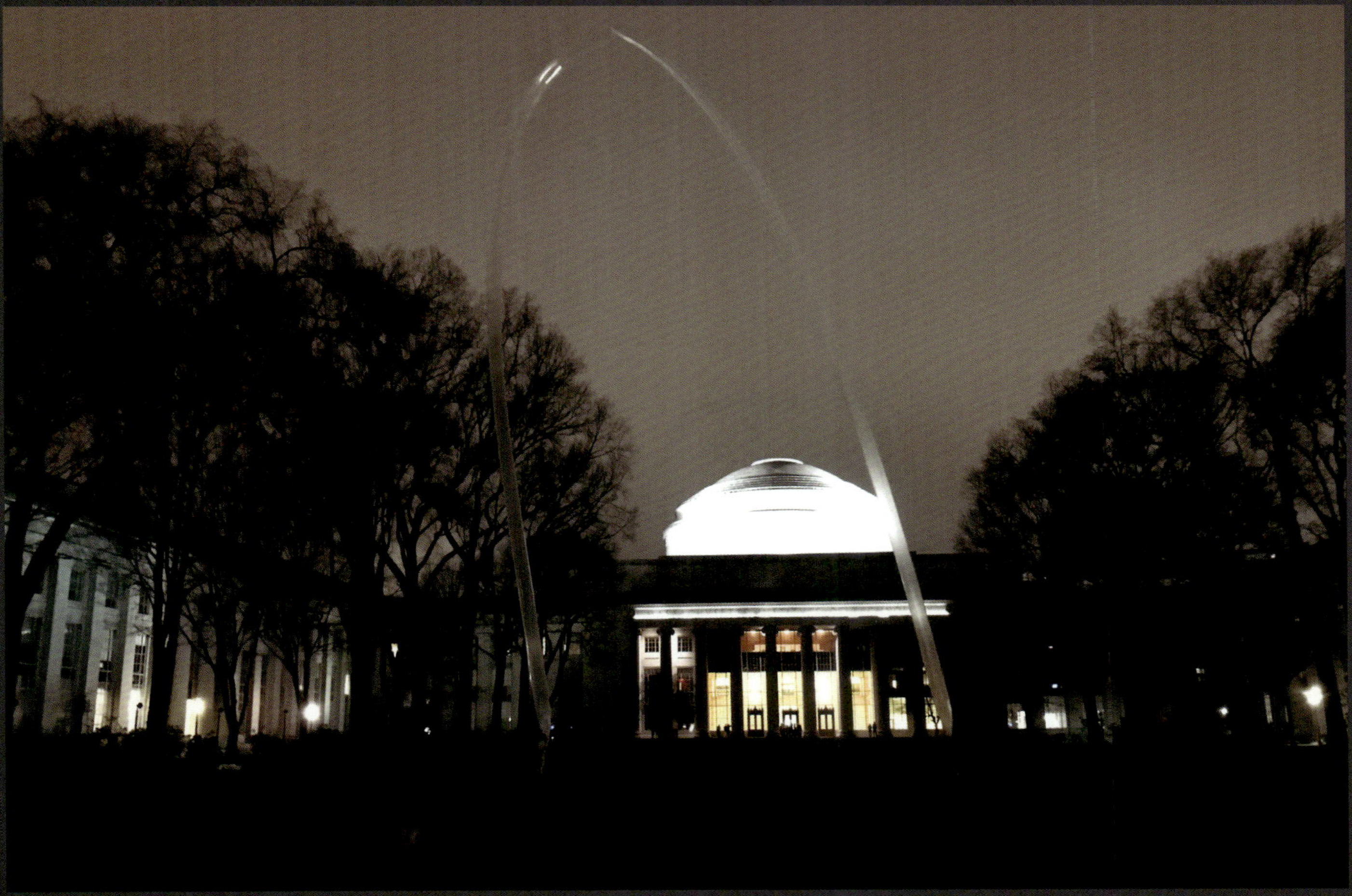

Facing page: *Light/Space/Time: CAVS/MIT - 25 Years*, poster, CAVS retrospective, MIT Museum, 1994
Source: CAVS Special Collection

Above: sky event tribute to Otto Piene in front of the Great Dome in Killian Court at MIT, part of *Celebrating the 50th Anniversary of the MIT Center for Advanced Visual Studies*, MIT Museum, Cambridge, April 27, 2018
Photo: Luc Courchesne

Chapter 6

Artransitions: Symposia and Conferences

CAVS embraced a classical definition of the word "conference," convening artists, scientists, engineers, and public officials around the emerging field of art, science, and technology. Kepes invited luminaries from diverse disciplines for seminars. Piene emphasized art creation through global multimedia celebrations, performances, events, and live transmissions that extended beyond MIT and Boston. CAVS championed art at civic scale, public art, and academic involvement in art and technology. The 1975 *Arttransition* conference advanced the idea of art-as-transition between mind and matter in a manifesto, printed in the catalog, for artists in a technological society. *Arttransition* opens a window on the many conferences that incubated and disseminated ideas and practice at CAVS.

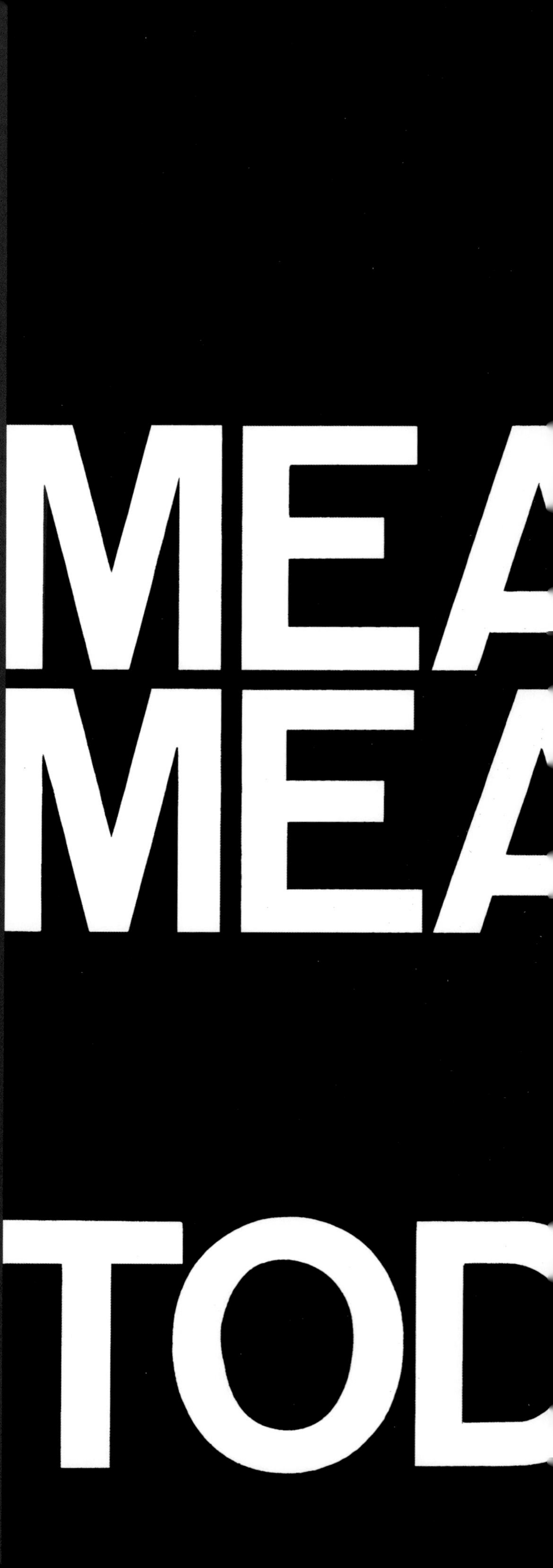

Right: "Means + Meanings in Today's Art," poster (detail) by Jacqueline Casey for a lecture by György Kepes, Kresge Little Theatre, MIT, December 11, 1975
Source: CAVS Special Collection

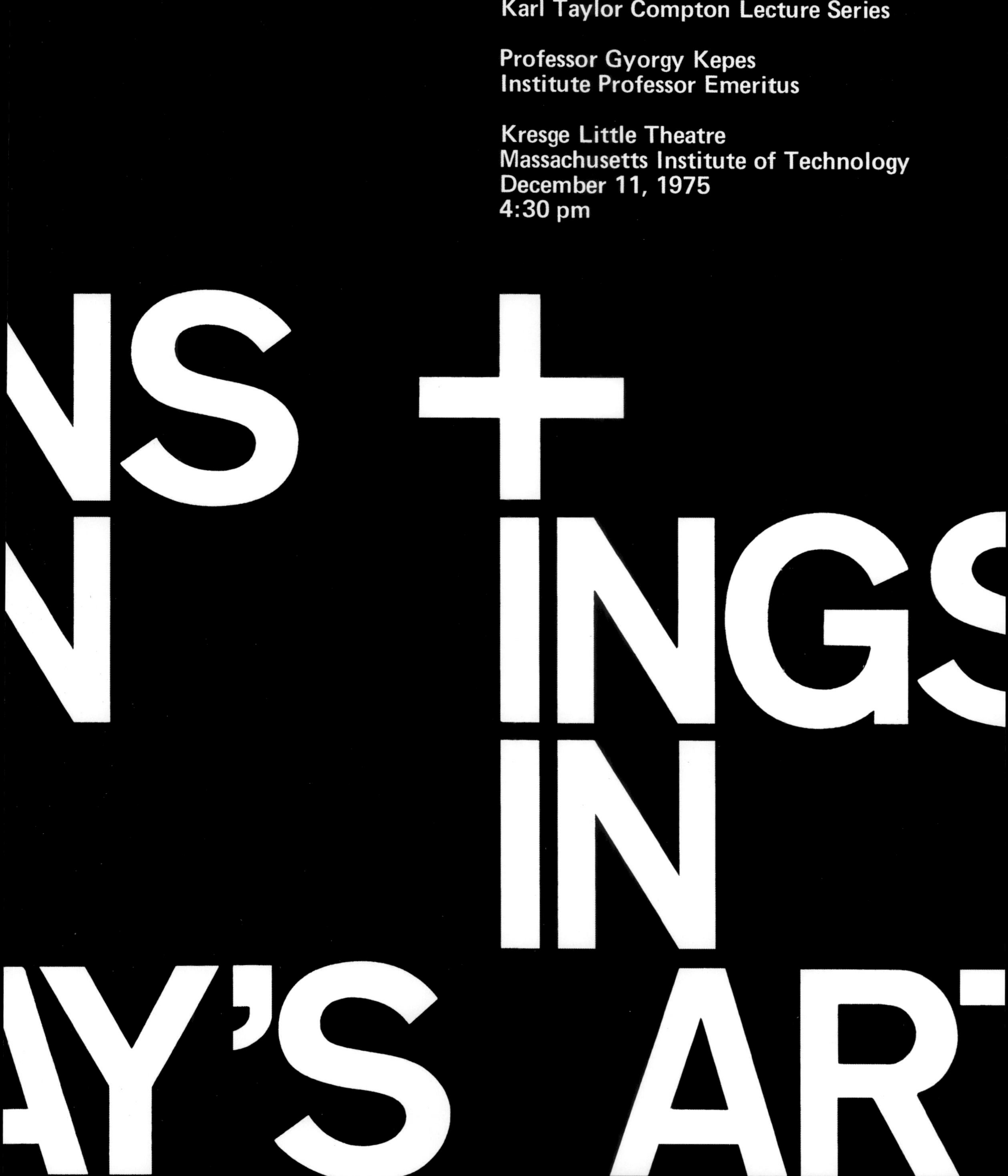
Karl Taylor Compton Lecture Series
Professor Gyorgy Kepes
Institute Professor Emeritus
Kresge Little Theatre
Massachusetts Institute of Technology
December 11, 1975
4:30 pm
NS +
N INGS
IN
AY'S AR

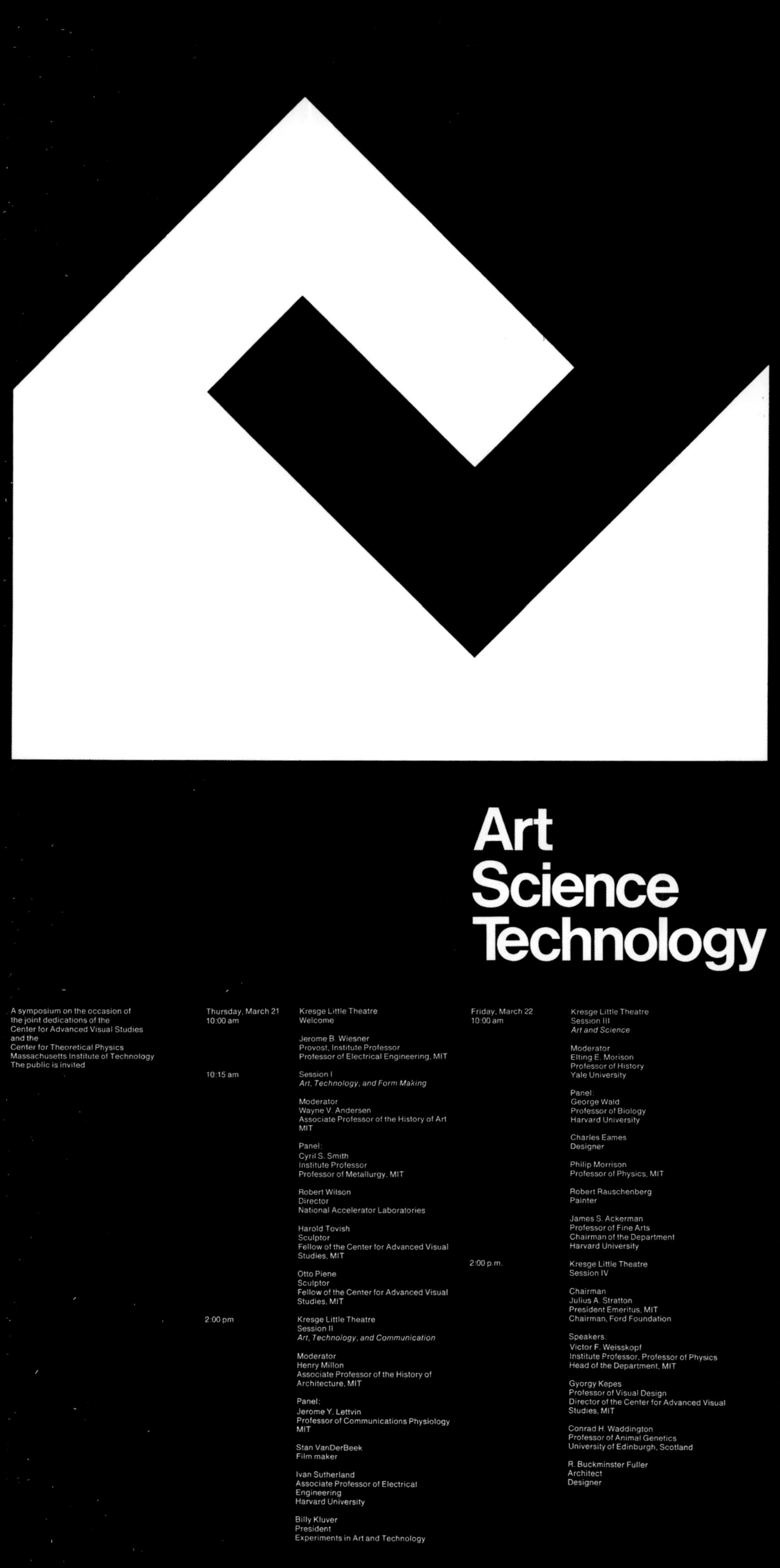

Poster by Jacqueline Casey for *Art Science Technology*, "A symposium on the occasion of the joint dedications of the Center for Advanced Visual Studies and the Center for Theoretical Physics" (poster text), CAVS inaugural symposium, Kresge Little Theater, MIT, March 21-22, 1968 Source: CAVS Special Collection

Kepes Symposia

> And the Center with its six fellows initially—five plus Kepes—was an instant breeder of concepts. (Otto Piene, Matthew Wisnioski, "Arts / Science / Technology" in *A Second Modernism: MIT, Architecture, and the 'Techno-Social' Moment*, 2013, 776)

Under György Kepes, symposia explored the intellectual links between art, science, and technology, and under Piene, expanded multidisciplinary artistic practice to a global stage. CAVS's enduring impact began with these early symposia. Kepes was personally acquainted with many of the leading scientific thinkers and theorists of the day who believed in, or at least respected, his vision of an emerging field of art, science, technology, and environment. Extended and redefined under Piene's directorship, the primary mission nonetheless remained "to inspire and to make possible exceptional artistic work." (*Arttransition*, catalog 1975, vii)

The inaugural *Symposium on Science and Art* (March 20-22, 1968) was jointly sponsored by the Center for Theoretical Physics and the Center for Advanced Visual Studies. Science came first in the title, with proceedings in the main Kresge Auditorium, while art, listed second, convened downstairs in the smaller Kresge Little Theater. This in no way diminished the luster of the panelists, including physicists Philip Morrison and Victor Weisskopf; professor of communications Jerome Lettvin; filmmaker Stan VanDerBeek; president of E.A.T. Billy Klüver; designer Charles Eames; painter Robert Rauschenberg; professor of animal genetics Conrad H. Waddington; electrical engineer Ivan Sutherland; director of the National Accelerator Laboratories Robert Wilson; and architect and designer R. Buckminster Fuller.

> The world as a set of structural systems does not divide into the two distinct territories, scientific knowledge and artistic and poetic vision. Rather, our rational grasp and our grasp of sensibilities both exist within a common structure of motivations, values, and communications. Art and science complement each other; together they constitute the full horizon of creative mental life. (Kepes, *The Center for Advanced Visual Studies*, inaugural catalog, 1968, 3)

CAVS artist fellows also participated in the panels, listed in the program as "Art, Technology, and Form Making," "Art, Technology, and Communication," and "Art and Science." The following year, fellows Ted Kraynik, Otto Piene, Wen-Ying Tsai, and Stan VanDerBeek opened their studios for a series of informal seminars. Jack Burnham organized a forum at MIT, *Artists and the Scientific Community* (January 15, 1969). The Center's radical ideas on art and education were presented at the annual conference of the College Art Association. Kepes presciently targeted environmental issues in symposia such as *Art in Civic Scale* (May 19, 1971), and *Art and Environment* (February 14-15, 1972), the urgency of which he expressed in the brochure:

> The purpose of the Symposium is to bring into focus the role of the artist in the needed revitalization of our man-created environment. The participants are architects, city planners, artists, and others who are not only deeply concerned about our troubled environmental situation but who share a confidence that the imaginative power and sensibilities of the artist are crucially important today both in the development of a new civic consciousness of our environmental "common" as well as in giving a new rich climactic focus to our impoverished environment. (Kepes, *Art in Civic Scale* brochure, CAVS, MIT, 1971)

Education was the focus of *Arts and the University: The Visual Arts at MIT* (February 6-7, 1974). Buckminster Fuller and Robert Motherwell joined Kepes for a closing conversation.

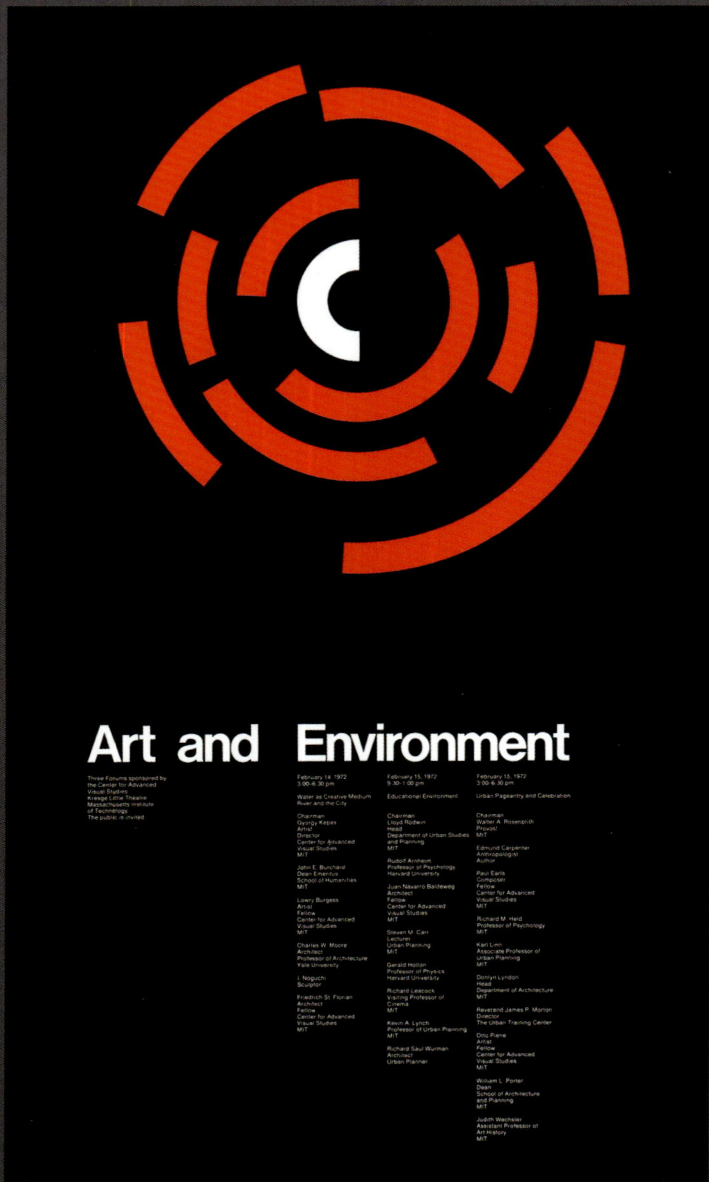

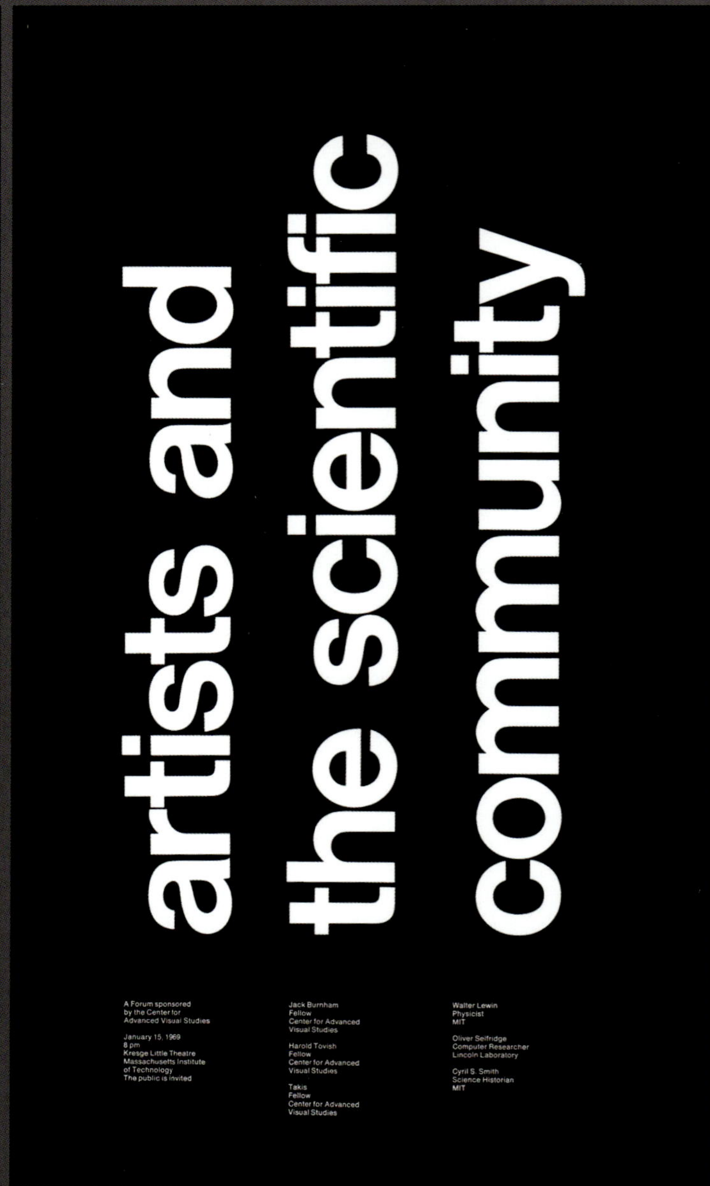

Both pages: Posters designed by Jacqueline Casey for CAVS symposia at MIT: *Art and Environment,* February 14-15, 1972; *Artists and the Scientific Community*, January 15, 1969; *Arts and the University*, February 6-7, 1974
Source: CAVS Special Collection

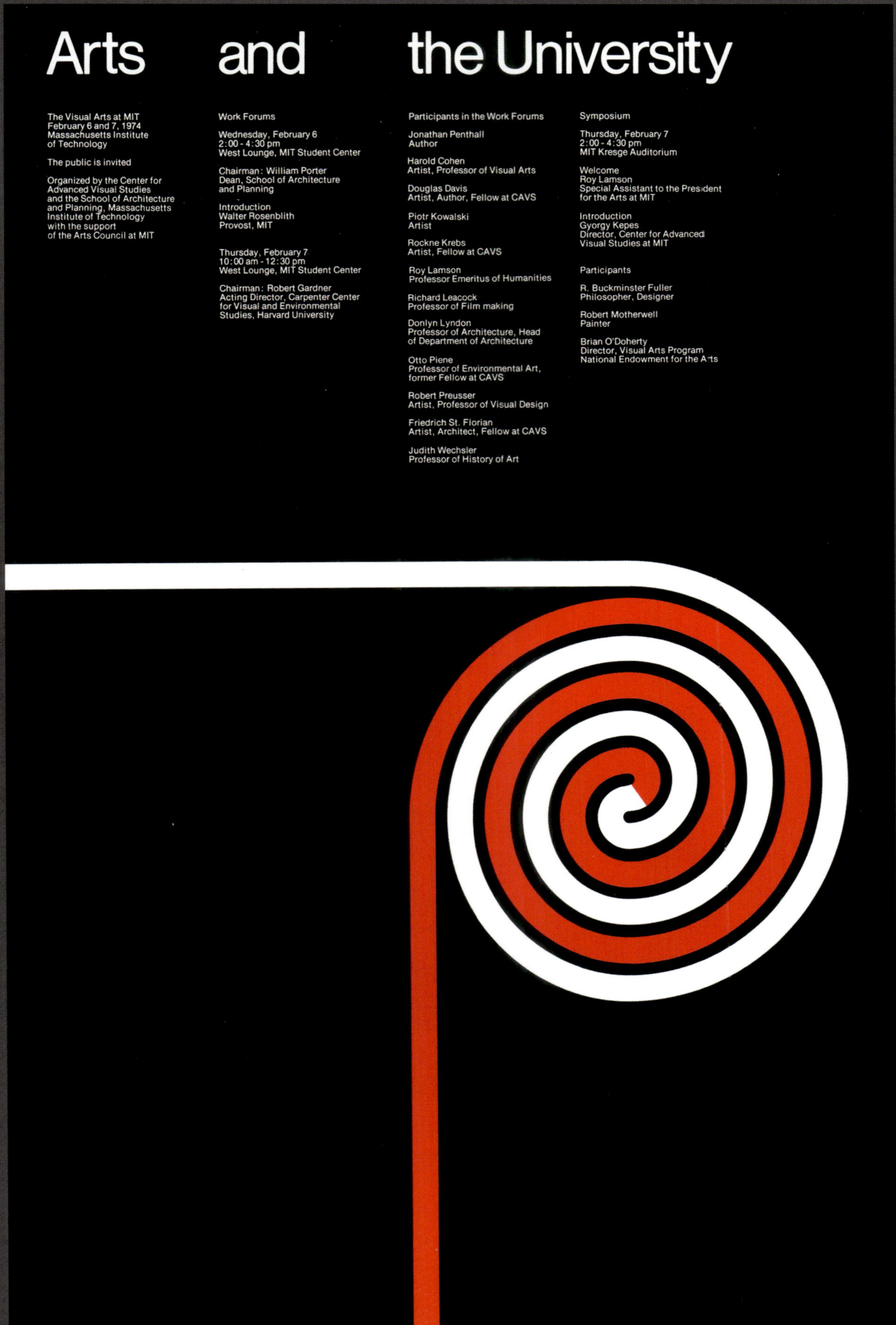
Arts and the University
The Visual Arts at MIT
February 6 and 7, 1974
Massachusetts Institute of Technology
The public is invited
Organized by the Center for Advanced Visual Studies and the School of Architecture and Planning, Massachusetts Institute of Technology with the support of the Arts Council at MIT
Work Forums
Wednesday, February 6
2:00 - 4:30 pm
West Lounge, MIT Student Center
Chairman: William Porter
Dean, School of Architecture and Planning
Introduction
Walter Rosenblith
Provost, MIT
Thursday, February 7
10:00 am - 12:30 pm
West Lounge, MIT Student Center
Chairman: Robert Gardner
Acting Director, Carpenter Center for Visual and Environmental Studies, Harvard University
Participants in the Work Forums
Jonathan Penthall
Author
Harold Cohen
Artist, Professor of Visual Arts
Douglas Davis
Artist, Author, Fellow at CAVS
Piotr Kowalski
Artist
Rockne Krebs
Artist, Fellow at CAVS
Roy Lamson
Professor Emeritus of Humanities
Richard Leacock
Professor of Film making
Donlyn Lyndon
Professor of Architecture, Head of Department of Architecture
Otto Piene
Professor of Environmental Art, former Fellow at CAVS
Robert Preusser
Artist, Professor of Visual Design
Friedrich St. Florian
Artist, Architect, Fellow at CAVS
Judith Wechsler
Professor of History of Art
Symposium
Thursday, February 7
2:00 - 4:30 pm
MIT Kresge Auditorium
Welcome
Roy Lamson
Special Assistant to the President for the Arts at MIT
Introduction
Gyorgy Kepes
Director, Center for Advanced Visual Studies at MIT
Participants
R. Buckminster Fuller
Philosopher, Designer
Robert Motherwell
Painter
Brian O'Doherty
Director, Visual Arts Program
National Endowment for the Arts

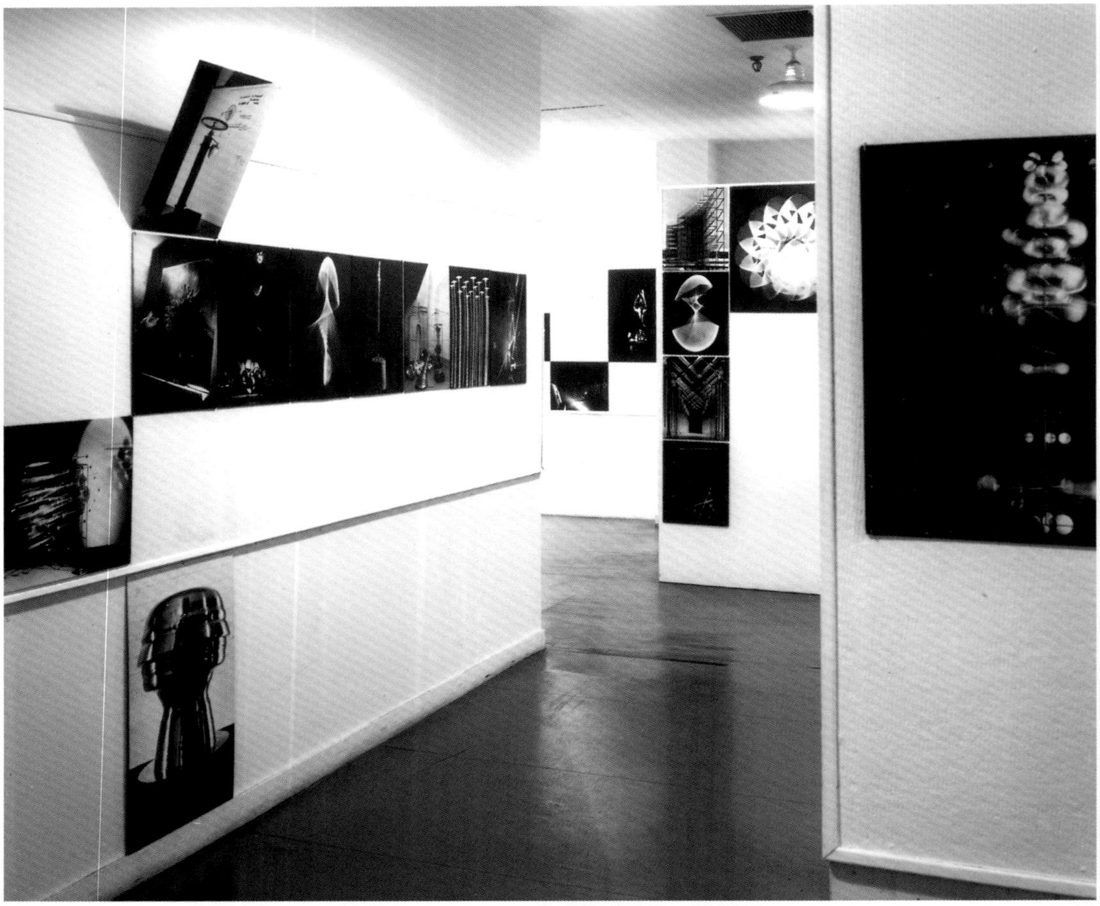
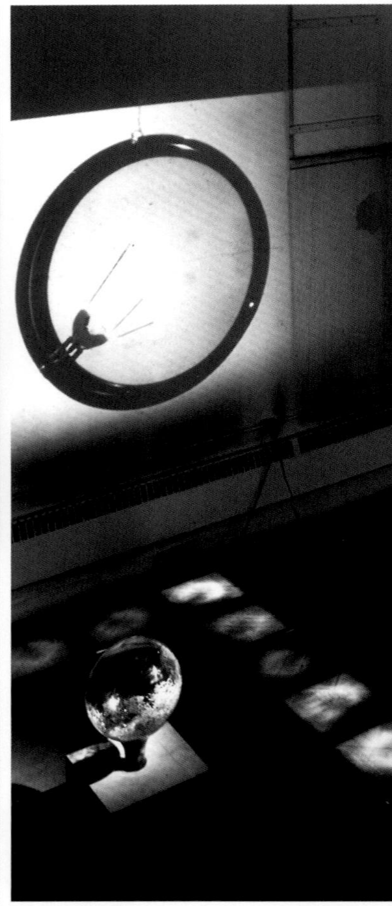
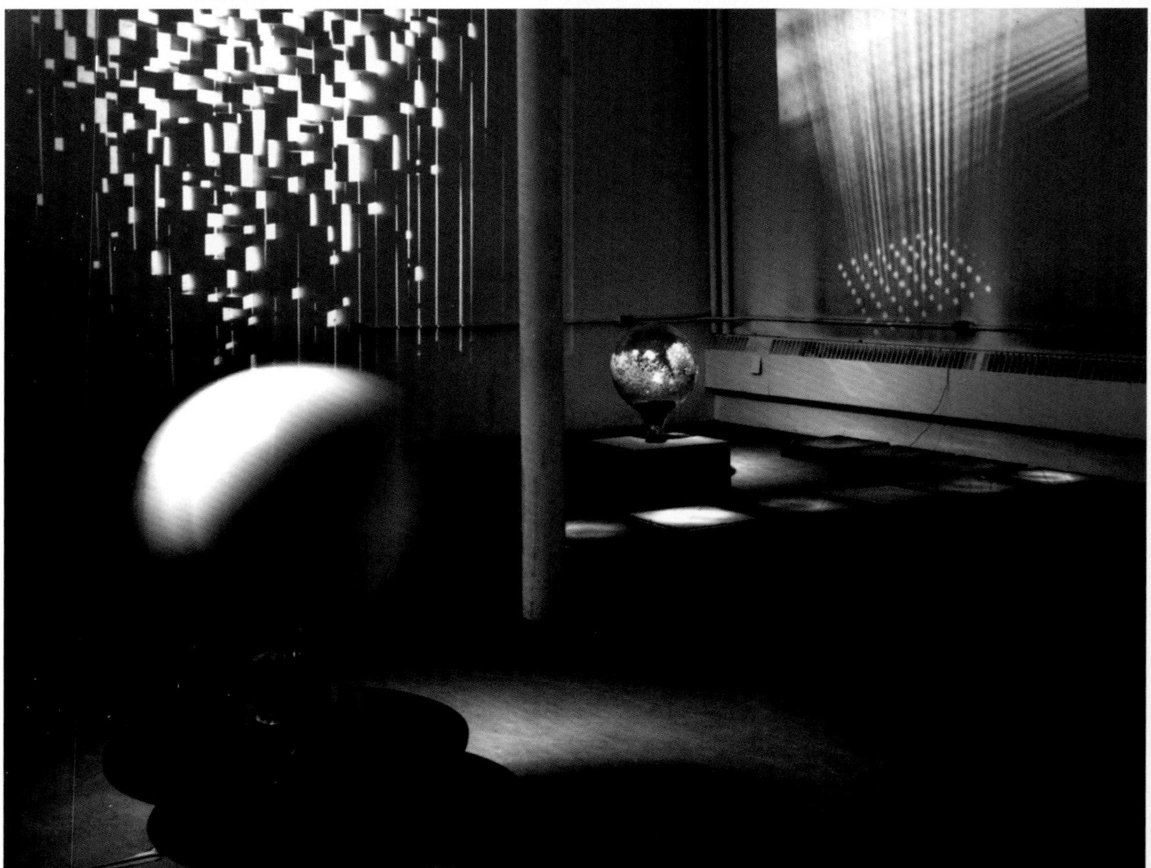
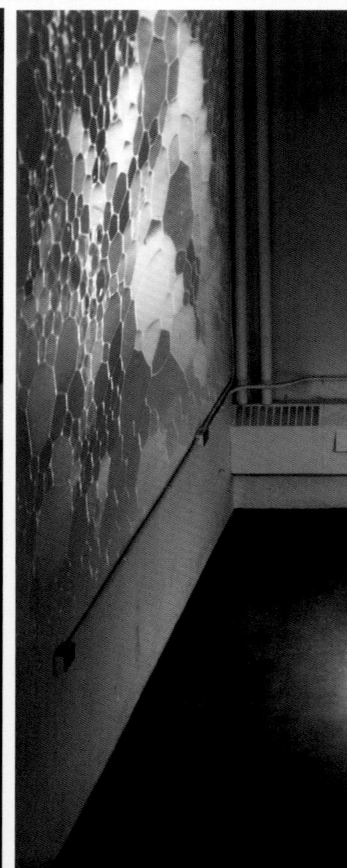

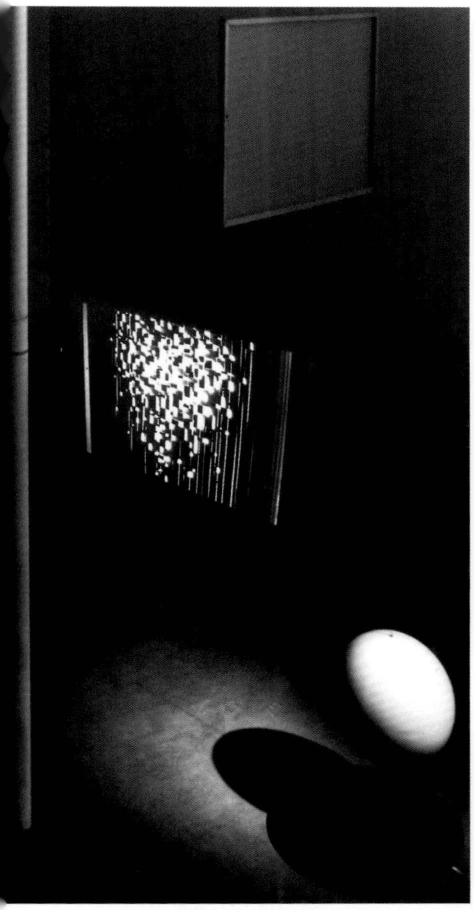

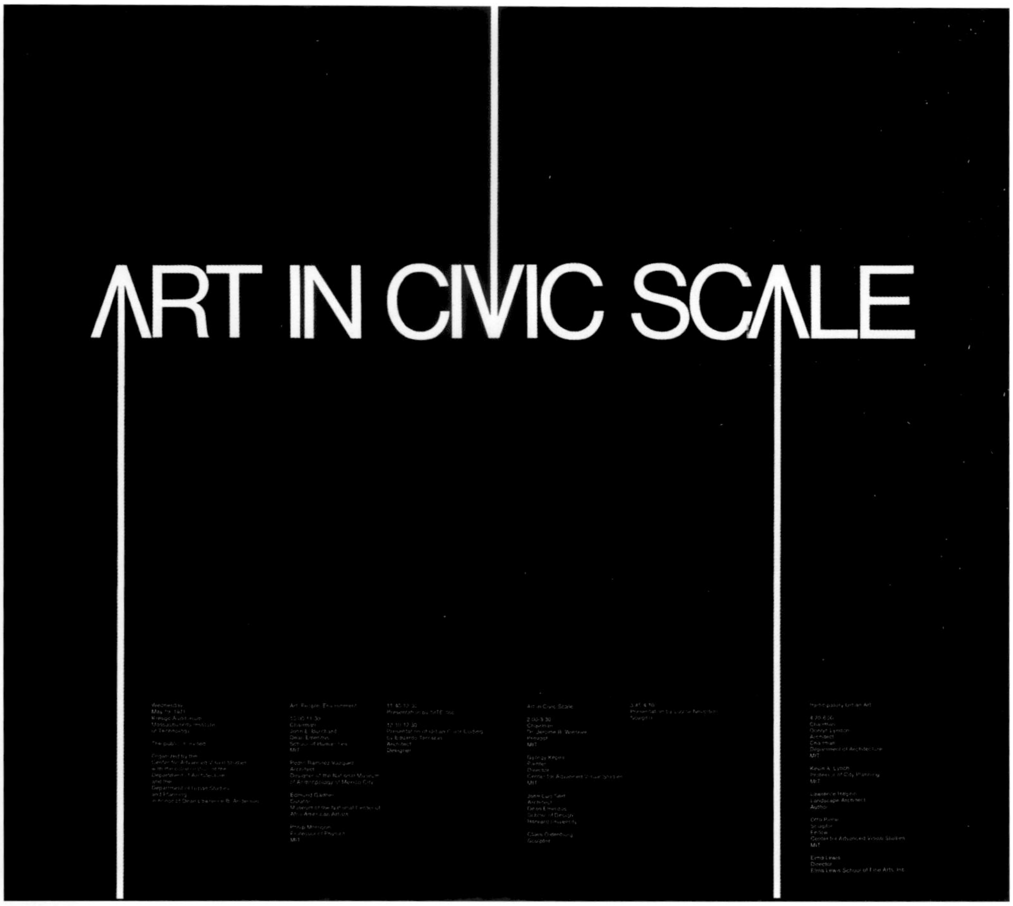

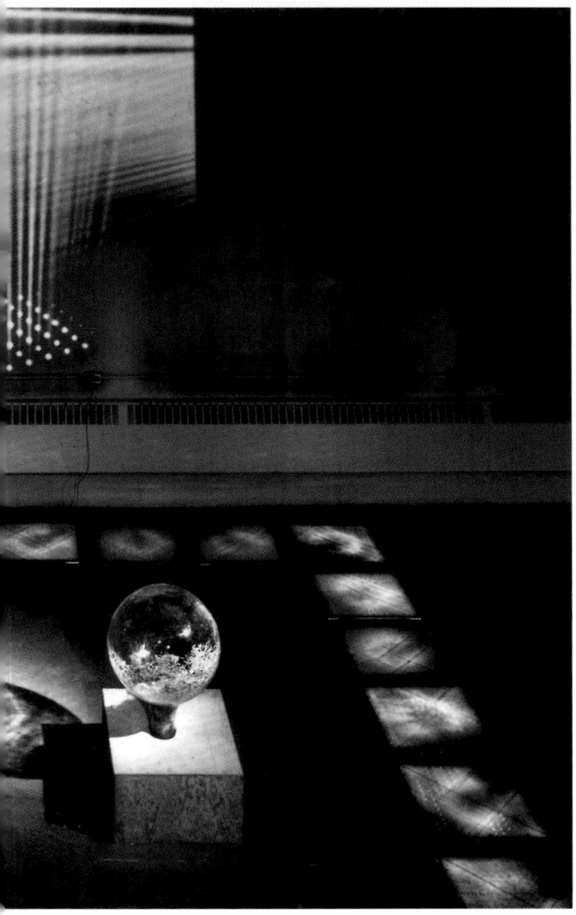

In addition to Kepes and Piene, panelists at the *Art in Civic Scale* symposium included:

John E. Burchard, Dean Emeritus, School of Humanities, MIT
Donlyn Lyndon, Chairman, Department of Architecture, MIT
Jerome B. Wiesner, Provost, MIT
Edmund Gaither, Curator, Museum of the National Center of Afro-American Artists
Lawrence Halprin, Landscape Architect, Author
Elma Lewis, Director, Elma Lewis School of Fine Arts, Inc.
Kevin A. Lynch, Professor of City Planning, MIT
Philip Morrison, Professor of Physics, MIT
Louise Nevelson, Sculptor
Claes Oldenburg, Sculptor
Josep Lluís Sert, Dean Emeritus School of Design, Harvard University

Above right: *Art in Civic Scale* poster by Jacqueline Casey, 1971
Both pages: exhibition by CAVS fellows at Building W11 for the CAVS symposium, *Art in Civic Scale*, Kresge Auditorium, MIT, May 19, 1971
Photos: Nishan Bichajian

Piene Conferences

> I am honored to salute György Kepes, founder of this Center for Advanced Visual Studies for tolerating this follow-up of his accomplishments as an organizer of previous CAVS conferences. (Piene, *Arttransition* catalog, CAVS, MIT, 1975, vi)

New CAVS director Otto Piene acknowledged the vital role of conferences in the 1975 *Arttransition* catalog. *Arttransition* may be seen as the transitional event between the two directorships. Piene's bi-continental art life, with monthly commutes between Cambridge, and Düsseldorf, generated opportunities for CAVS artists to create works for international exhibitions. Embracing the themes of celebration and the canvas of sky and space led to bold interactions with the public, as well as new art forms and methodological approaches that united an inner and outer universe.

> The artists' particular discipline of consequentially investigating their intuitions reflects a cognitive style complementary to logical forms of search. The result—art—is the living link of thought and matter, of the cerebral and the physical, because it is at once spiritual and sensual whether in painting, on television, or in public celebrations. (Piene, *Arttransition*, 1975, xi)

The foundation Kepes built by cultivating critical conversations between leaders in previously divided disciplines served as a platform for Piene's conferences to expand the field of art, science, and technology through art making. Group and individual artworks often stood on precarious technological footing, exposed in a raw, newborn state to public view. CAVS embarked on international exhibitions and performance-driven conferences that tested its resilience in realizing nascent works that in some ways embodied "our biological condition."

> We do not claim that any artist can provide miracles, ultimate solutions to perennial problems, or more than momentary psychic energy at times. But as in our biological condition the artist can be a father, a mother, or a child in our contemporary intellectual family. (Piene, *Arttransition* catalog, 1975, xi)

Piene conveyed his thinking-by-doing approach in his book, *More Sky*, which took an artist's sketchbook approach to theory:

> Elements and technology are the means that permit the revival of large-scale artistic activities. Wind may be considered the siccative, while fire is a new gel. Technology permits the artist to talk to many, design for many, and execute plans for many. It's time now to do more than project: it's time to act. (Piene, *More Sky*, 1970, Preface)

Themes drove the creation of collaborative artworks and exhibitions that demonstrated the tenets of the emerging field through practice. Leading thinkers and artists were invited, but young artists and scientists with new ideas, students, and women were also full participants for the first time at CAVS. The topics that shaped the emergence of art, science, and technology as a field can be seen by reading the programs and catalogs that document CAVS conferences, beginning with the inaugural *Symposium on Science and Art* of 1968 and culminating with *Arttransition 90*.

Artist-curator Virginia Gunter was the organizing director of the first *Arttransition* conference held at MIT, October 15-19, 1975. Piene wrote about the layered meanings of the conference title, "Arttransition," in the catalog.

Facing page: *Arttransition* catalog, designed by Muriel Cooper, CAVS conference, MIT, October 15-19, 1975
Source: CAVS Special Collection

ARTTRANSITION can be understood as a mediating process that leads the traditional and conventional practice of art into greater obligations towards the information and education of an ever-more curious public, via technology.

ARTTRANSITION marks the evolution of responsible artists moving out of solipsistic inner concerns of *l'art pour l'art* to the vast world of human dwellings; the abounding elements of nature; the needs of our electronic information industry in an archaic stage; the increased demands of education for a technology-supported leisure society; the educational and experiential needs of underprivileged countries.

ARTTRANSITION means an opening up of the artist to the challenge of the new media—to communicate in a virtually intimate way with a monumental audience, pushing beyond earthly surface-dwelling.

To engage in ARTTRANSITION does not mean that artists must give up traditional ways and tools of expression, but rather that they extend them beyond the narrow limits of the art-infatuated art world into the art-starved, shrewdly skeptical "real" world which will ideally become a world of art if a balance of peace is maintained by generating creative energies to build, rather than to destroy, a multifaceted, richly nuanced, humane environment for everybody.

To direct oneself to ARTTRANSITION involves the essential democratic effort: to maintain individuality—the artist's identity—while coalescing the artists' effort and those of scholars, scientists, engineers, and politicians beyond gaining philosophical insight and toward building a refined and safe world that reflects human psychic complexity. Growing material needs will become unmatchable if we do not realize that they are the physical equivalent to growing and changing spiritual and intellectual needs.

ARTTRANSITION means acting and creating artworks according to the insight that to humankind all matter is spiritual, a form of energy which we are here to articulate. To render matter into human, peaceful energy no one can work alone anymore. The artist needs the planner, the engineer needs the artist, the sociologist needs the scientist/scholar as Einstein needed Mozart. Collaboration occurs on two planes: general exchange of values on the one hand and intended, concerted, and carefully planned action on the other. We share scientific and artistic information on a basic human level as well as in practical, intricate, and professional cooperation. The latter mode of group work is the most badly needed when it comes to providing shelter and a garden of experience for ourselves and our fellow citizens.

(Otto Piene, from the essay "ARTTRANSITION," *Arttransition* catalog, CAVS, MIT, 1975, xi)

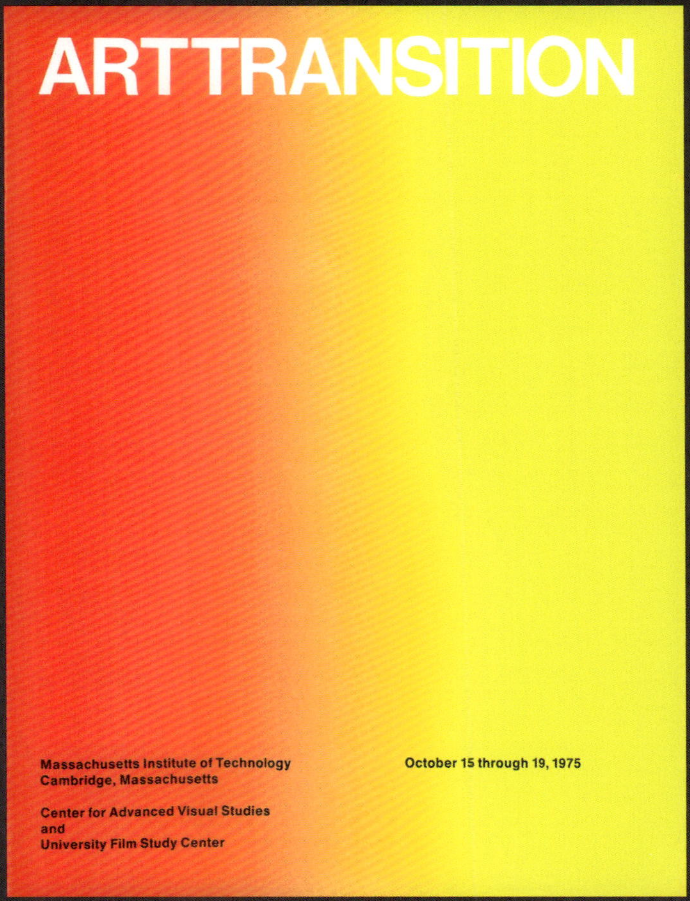

Arttransition

Conference program

CAVS, MIT, Cambridge, Massachusetts, October 15-19, 1975

Panels

New Art/Science/Technology
Harriet Casdin-Silver, artist; John Nolan, President, Massachusetts College of Art; Jasia Reichardt, Director, Whitechapel Art Gallery, London; James Seawright, Director, Visual Arts Program, Princeton University; Sonia Landy Sheridan, Head, Generative Systems Department, Art Institute of Chicago School

New Education/Education Towards New Art
James Ackerman, Professor of Art History, Harvard University; György Kepes, Founder, Center for Advanced Visual Studies, MIT; Roy Lamson, Special Assistant to the President, MIT; John Nolan, President, Massachusetts College of Art; Manfred Schneckenburger, Director, documenta 6, Kassel; Jerome Wiesner, President, MIT

Environmental/Large Scale Art
Christo, artist; Nancy Holt, artist; Rockne Krebs, artist; Otto Piene, Director, Center for Advanced Visual Studies, MIT; Jeffrey Shaw, Eventstructure Research Group, Holland; Jan Van Der Marck, Director, Hopkins Center Art Galleries, Dartmouth College

New Media/Video/Communication
David Cort, artist; Douglas Davis, artist; Juan Downey, artist; Les Levine, artist; Matko Meštrović, Professor, Zagreb University, Yugoslavia; Gerald O'Grady, Director, Center for Media Studies, SUNY at Buffalo; Aldo Tambellini, artist; Howard Wise, President, Electronic Arts Intermix, New York City

Expansion of the Arts
Karin Bacon, celebrationist; Otto Piene, Director, Center for Advanced Visual Studies, MIT; Sonia Landy Sheridan, Head, Generative Systems Department, Art Institute of Chicago School; T. M. Stephens, Coordinator, Art Research Center, Kansas City

Art Support Systems
Chloe Aaron, National Endowment for the Arts; Wayne Anderson, Professor of History of Art, MIT; Renee Levine, Managing Director, Center of the Creative and Performing Arts, SUNY at Buffalo; Matko Meštrović, Professor, Zagreb University, Yugoslavia; Jasia Reichardt, Director, Whitechapel Art Gallery, London; Manfred Schneckenburger, Director, Kassel documenta 6, Kassel; Jan Van Der Marck, Director, Hopkins Center Art Galleries, Dartmouth College; Howard Wise, President, Electronic Arts Intermix

Artist as Educator
Virginia Gunter, Director of Exhibitions and Programs, Massachusetts College of Art; Allan Kaprow, artist; Margaret Mahoney, Vice President, Robert Wood Johnson Foundation; Otto Piene, Director, Center for Advanced Visual Studies, MIT; James Seawright, Director, Visual Arts Program, Princeton University

Film/The Aesthetic of the Technical Image
Peter Feinstein, Director, University Film Study Center; Hollis Frampton, filmmaker; Robert Gardner, Acting Chairman, Department of Visual and Environmental Studies, Harvard University; Richard Leacock, Head, Film Section, MIT; Ed Pincus, filmmaker; Deac Rossell, critic

Categories

Environmental Art
Interaction of Art, Science and Technology
Developmental Media Work
Education/comments
Communication
Video Art
Film
Environmental Concepts
Celebrations

Performances

In the 1976 edition of *Leonardo*, Robert Preusser wrote, "In addition to lectures and panel discussions on each of the five days, there were continuous cinema and video screenings, an exhibition at CAVS of works by Conference participants and by artists associated with CAVS and other special events, such as the following"

"Concerto for T.V. Cello and Video Tapes"—a five-hour performance staged at CAVS (Paik, Moorman)

"Neon Rainbow"—a sky event, consisting of a polyethylene tube and neon rods, launched from the MIT campus. (Piene, Siña)

"Fog, Mist and Dreams"—a collaborative experiment involving the projection of a film of computer generated graphics, video and synthesizer. (Brigham, VanDerBeek)

"A Visualization of an Experience within Music"—a film based on Leonard Bernstein's performance of Wagner's *Prelude to Tristan and Isolde*, produced in 1974 by Ron Hays (filmmaker and video artist), utilizing slit-scan, computer graphics, video and synthesizer.

"U.S. Air Weaves"—a new composition by Paul Earls of CAVS, which he conducted in the Main Lobby of MIT

Conference Credits

Chairman: Dr. Jerome Wiesner, President, MIT
Co-chairmen: Otto Piene, Director, CAVS, MIT, and Peter Feinstein, Director, University Film Study Center
Project Director: Virginia Gunter
Assistant Project Director: Rika Langhorst Olson
Exhibition Coordinator: Elizabeth Goldring
Film Program Coordinator: Jon Rubin
Video Program Coordinator: Ann McIntosh
Catalog Editor: Ann Christensen
Conference Graphics: Jackie Casey
Catalog: Muriel Cooper

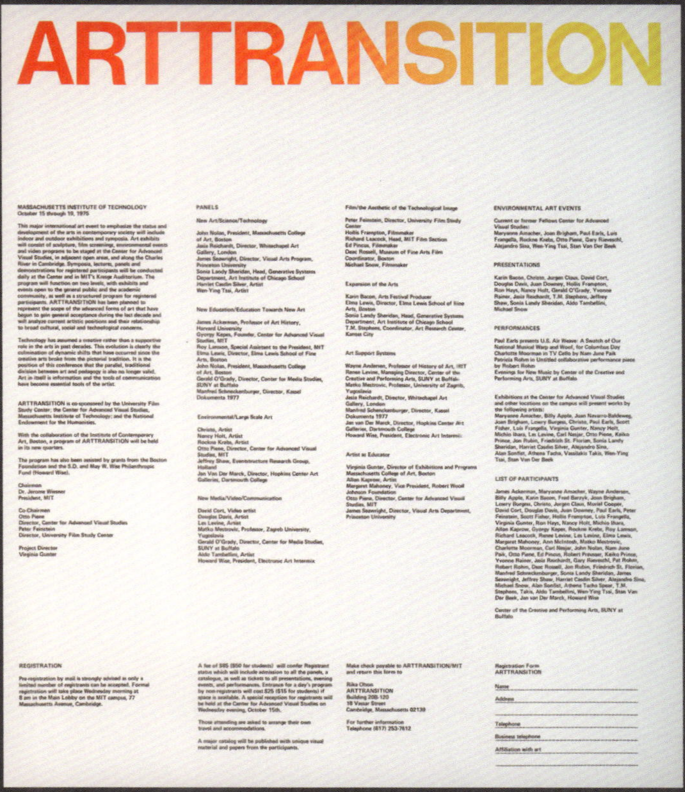

ARTTRANSITION

MASSACHUSETTS INSTITUTE OF TECHNOLOGY
October 15 through 19, 1975

This major international art event to emphasize the status and development of the arts in contemporary society will include indoor and outdoor exhibitions and symposia. Art exhibits will consist of sculpture, film screenings, environmental events and video programs to be staged at the Center for Advanced Visual Studies, in adjacent open areas, and along the Charles River in Cambridge. Symposia, lectures, panels and demonstrations for registered participants will be conducted daily at the Center and in MIT's Kresge Auditorium. The program will function on two levels, with exhibits and events open to the general public and the academic community, as well as a structured program for registered participants. ARTTRANSITION has been planned to represent the scope of the advanced forms of art that have begun to gain general acceptance during the last decade and will analyze current artistic positions and their relationship to broad cultural, social and technological concerns.

Technology has assumed a creative rather than a supportive role in the arts in past decades. This evolution is clearly the culmination of dynamic shifts that have occurred since the creative arts broke from the pictorial tradition. It is the position of this conference that the parallel, traditional division between art and pedagogy is also no longer valid. Art in itself is information and the tools of communication have become essential tools of the artist.

ARTTRANSITION is co-sponsored by the University Film Study Center; the Center for Advanced Visual Studies, Massachusetts Institute of Technology; and the National Endowment for the Humanities.

With the collaboration of the Institute of Contemporary Art, Boston, a program of ARTTRANSITION will be held in its new quarters.

The program has also been assisted by grants from the Boston Foundation and the S.D. and May W. Wise Philanthropic Fund (Howard Wise).

Chairman
Dr. Jerome Wiesner
President, MIT

Co-Chairmen
Otto Piene
Director, Center for Advanced Visual Studies
Peter Feinstein
Director, University Film Study Center

Project Director
Virginia Gunter

PANELS

New Art/Science/Technology

John Nolan, President, Massachusetts College of Art, Boston
Jasia Reichardt, Director, Whitechapel Art Gallery, London
James Seawright, Director, Visual Arts Program, Princeton University
Sonia Landy Sheridan, Head, Generative Systems Department, Art Institute of Chicago School
Harriet Casdin Silver, Artist
Wen-Ying Tsai, Artist

New Education/Education Towards New Art

James Ackerman, Professor of Art History, Harvard University
Gyorgy Kepes, Founder, Center for Advanced Visual Studies, MIT
Roy Lamson, Special Assistant to the President, MIT
Elma Lewis, Director, Elma Lewis School of Fine Arts, Boston
John Nolan, President, Massachusetts College of Art, Boston
Gerald O'Grady, Director, Center for Media Studies, SUNY at Buffalo
Manfred Schneckenburger, Director, Kassel Dokumenta 1977

Environmental/Large Scale Art

Christo, Artist
Nancy Holt, Artist
Rockne Krebs, Artist
Otto Piene, Director, Center for Advanced Visual Studies, MIT
Jeffrey Shaw, Eventstructure Research Group, Holland
Jan Van Der Marck, Director, Hopkins Center Art Galleries, Dartmouth College

New Media/Video/Communication

David Cort, Video artist
Douglas Davis, Artist
Les Levine, Artist
Matko Mestrovic, Professor, Zagreb University, Yugoslavia
Gerald O'Grady, Director, Center for Media Studies, SUNY at Buffalo
Aldo Tambellini, Artist
Howard Wise, President, Electronic Art Intermix

Film/the Aesthetic of the Technological Image

Peter Feinstein, Director, University Film Study Center
Hollis Frampton, Filmmaker
Richard Leacock, Head, MIT Film Section
Ed Pincus, Filmmaker
Deac Rossell, Museum of Fine Arts Film Coordinator, Boston
Michael Snow, Filmmaker

Expansion of the Arts

Karin Bacon, Arts Festival Producer
Elma Lewis, Director, Elma Lewis School of Fine Arts, Boston
Sonia Landy Sheridan, Head, Generative Systems Department, Art Institute of Chicago School
T.M. Stephens, Coordinator, Art Research Center, Kansas City

Art Support Systems

Wayne Andersen, Professor of History of Art, MIT
Renee Levine, Managing Director, Center of the Creative and Performing Arts, SUNY at Buffalo
Matko Mestrovic, Professor, University of Zagreb, Yugoslavia
Jasia Reichardt, Director, Whitechapel Art Gallery, London
Manfred Schneckenburger, Director, Kassel Dokumenta 1977
Jan van Der Marck, Director, Hopkins Center Art Galleries, Dartmouth College
Howard Wise, President, Electronic Art Intermix

Artist as Educator

Virginia Gunter, Director of Exhibitions and Programs Massachusetts College of Art, Boston
Allan Kaprow, Artist
Margaret Mahoney, Vice President, Robert Wood Johnson Foundation
Otto Piene, Director, Center for Advanced Visual Studies, MIT
James Seawright, Director, Visual Arts Department, Princeton University

ENVIRONMENTAL ART EVENTS

Current or former Fellows Center for Advanced Visual Studies:
Maryanne Amacher, Joan Brigham, Paul Earls, Luis Frangella, Rockne Krebs, Otto Piene, Gary Rieveschl, Alejandro Sina, Wen-Ying Tsai, Stan Van Der Beek

PRESENTATIONS

Karin Bacon, Christo, Jurgen Claus, David Cort, Douglas Davis, Juan Downey, Hollis Frampton, Ron Hays, Nancy Holt, Gerald O'Grady, Yvonne Rainer, Jasia Reichardt, T.M. Stephens, Jeffrey Shaw, Sonia Landy Sheridan, Aldo Tambellini, Michael Snow

PERFORMANCES

Paul Earls presents U.S. Air Weave: A Swatch of Our National Musical Warp and Woof, for Columbus Day
Charlotte Moorman in TV Cello by Nam June Paik
Patricia Rohm in Untitled collaborative performance piece by Robert Rohm
Evenings for New Music by Center of the Creative and Performing Arts, SUNY at Buffalo

Exhibitions at the Center for Advanced Visual Studies and other locations on the campus will present works by the following artists:
Maryanne Amacher, Billy Apple, Juan Navarro-Baldeweg, Joan Brigham, Lowry Burgess, Christo, Paul Earls, Scott Fisher, Luis Frangella, Virginia Gunter, Nancy Holt, Michio Ihara, Les Levine, Carl Nesjar, Otto Piene, Keiko Prince, Jon Rubin, Friedrich St. Florian, Sonia Landy Sheridan, Harriet Casdin Silver, Alejandro Sina, Alan Sonfist, Athena Tacha, Vassilakis Takis, Wen-Ying Tsai, Stan Van Der Beek

LIST OF PARTICIPANTS

James Ackerman, Maryanne Amacher, Wayne Andersen, Billy Apple, Karin Bacon, Fred Barzyk, Joan Brigham, Lowry Burgess, Christo, Jurgen Claus, Muriel Cooper, David Cort, Douglas Davis, Juan Downey, Paul Earls, Peter Feinstein, Scott Fisher, Hollis Frampton, Luis Frangella, Virginia Gunter, Ron Hays, Nancy Holt, Michio Ihara, Allan Kaprow, Gyorgy Kepes, Rockne Krebs, Roy Lamson, Richard Leacock, Renee Levine, Les Levine, Elma Lewis, Margaret Mahoney, Ann McIntosh, Matko Mestrovic, Charlotte Moorman, Carl Nesjar, John Nolan, Nam June Paik, Otto Piene, Ed Pincus, Robert Preusser, Keiko Prince, Yvonne Rainer, Jasia Reichardt, Gary Rieveschl, Pat Rohm, Robert Rohm, Deac Rossell, Jon Rubin, Friedrich St. Florian, Manfred Schneckenburger, Sonia Landy Sheridan, James Seawright, Jeffrey Shaw, Harriet Casdin Silver, Alejandro Sina, Michael Snow, Alan Sonfist, Athena Tacha Spear, T.M. Stephens, Takis, Aldo Tambellini, Wen-Ying Tsai, Stan Van Der Beek, Jan van Der Marck, Howard Wise

Center of the Creative and Performing Arts, SUNY at Buffalo

REGISTRATION

Pre-registration by mail is strongly advised as only a limited number of registrants can be accepted. Formal registration will take place Wednesday morning at 8 am in the Main Lobby on the MIT campus, 77 Massachusetts Avenue, Cambridge.

A fee of $85 ($50 for students) will confer Registrant status which will include admission to all the panels, a catalogue, as well as tickets to all presentations, evening events, and performances. Entrance for a day's program by non-registrants will cost $25 ($15 for students) if space is available. A special reception for registrants will be held at the Center for Advanced Visual Studies on Wednesday evening, October 15th.

Those attending are asked to arrange their own travel and accommodations.

A major catalog will be published with unique visual material and papers from the participants.

Make check payable to ARTTRANSITION/MIT and return this form to

Rika Olson
ARTTRANSITION
Building 20B-120
18 Vassar Street
Cambridge, Massachusetts 02139

For further information
Telephone (617) 253-7612

Registration Form
ARTTRANSITION

Name ______
Address ______
______
Telephone ______
Business telephone ______
Affiliation with art ______
______

Right: *Arttransition* flyer, designed by Jacqueline Casey, CAVS, MIT, October 15-19, 1975
Source: CAVS Special Collection

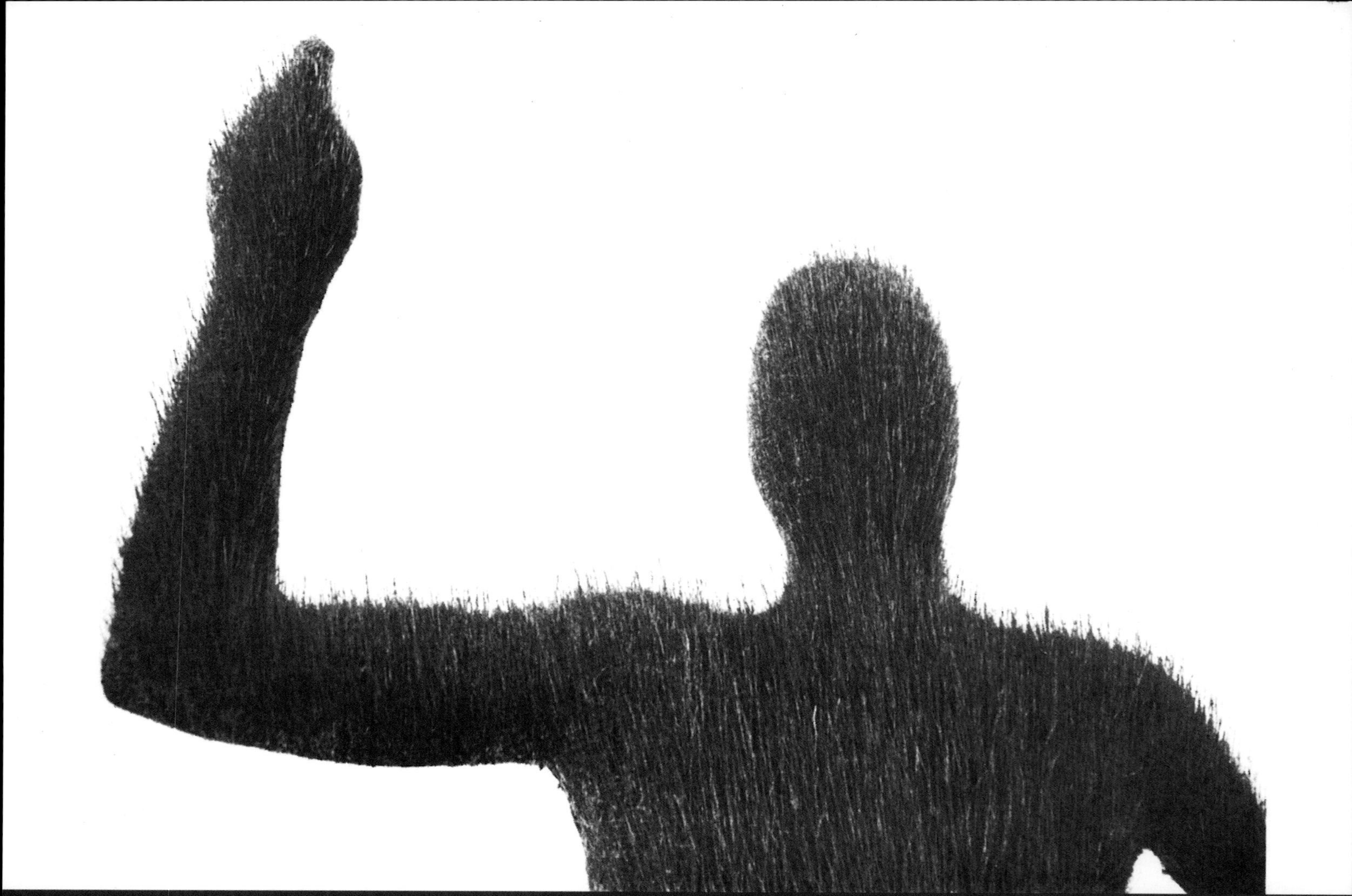

I believe in the transcendence of nature. It is the ultimate source of inspiration. My work is to find forms which signify our connection with the natural order. I do this by planting sculptures called *Lifeforms*. Life takes its form from the elements. We perceive certain combinations of earths, airs, fires, and waters to be alive. And, if art is about life, then it is sensible to realize living ideas with living materials. The contemporary priority is growth. It is the dominant idea of the Western World. As human organization has become more abstracted from the natural order, so has our concept of growth. Now, the word usually denotes a political, technological, or economic condition. Our misunderstanding accelerates our misfortune. Badly conceived and mismanaged "growth" increasingly threatens its host planet and deprives us of participation in ecological reality. *Lifeforms* are not solutions for this dilemma. They are reference points for communications about the ideas they represent. This communication will generate the answers, or not. *Lifeforms* are signs along the way—an attempt to provoke an intimate affair with the earth. For me it is the only way. It is my message. (Gary Rieveschl, *Lifeforms 1973-1976*, Jack Otto & Sons Printers, 1977)

Top: Gary Rieveschl, "Grassman 4," self-portrait, human scale, planted October, 1975, *Arttransition*, CAVS conference, MIT, 1975
Photo: CAVS Special Collection

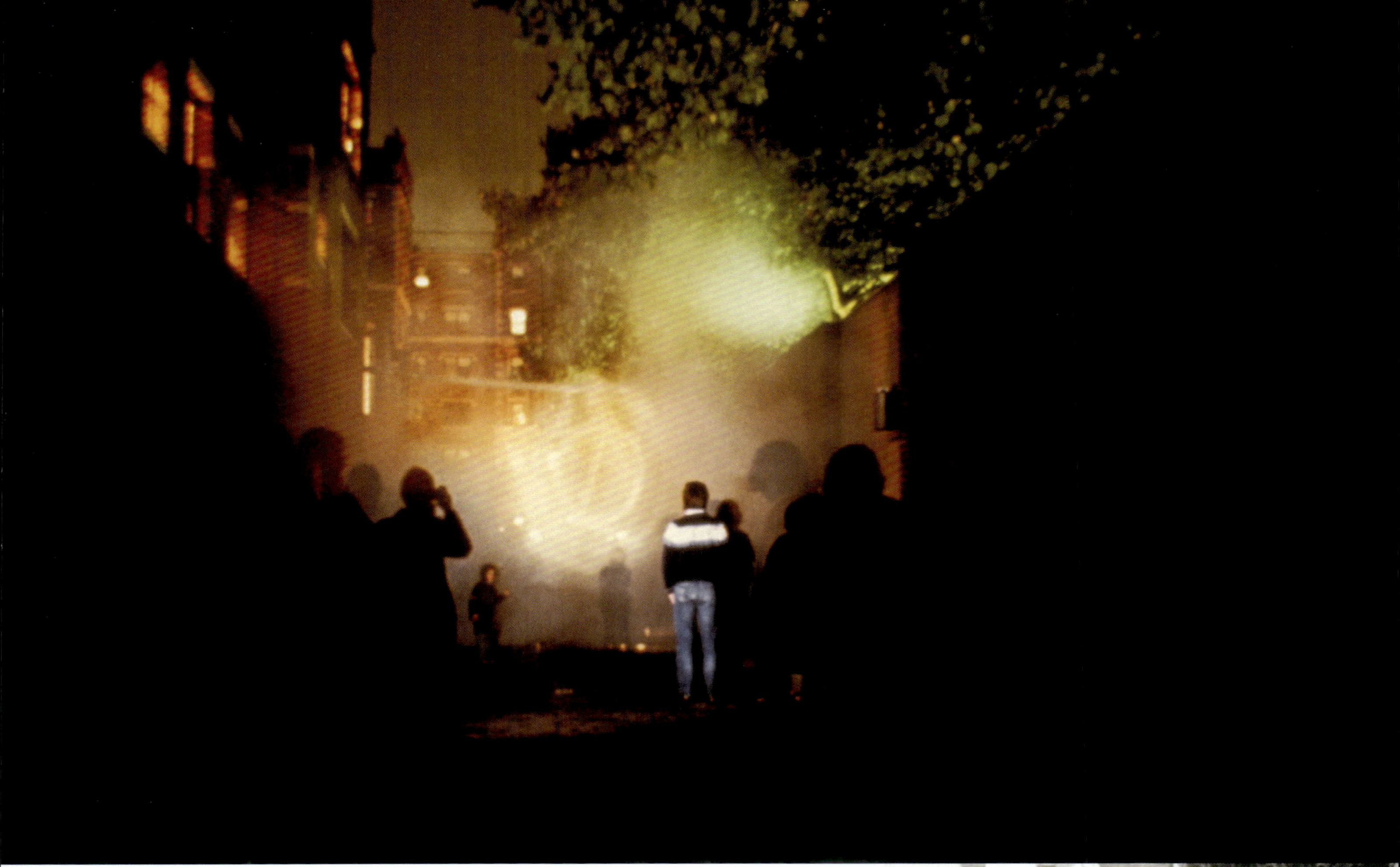

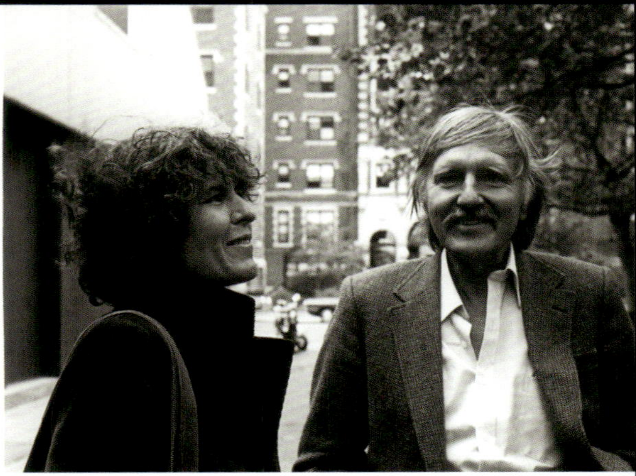

I showed Stan [VanDerBeek]. I just met him. See that photograph? I had met him maybe ten minutes before that was taken. . . . He loved it. He had never projected on steam before, so he was so excited . . . . Stan was just enamored, "Wow, this is great." . . . That's what they called it: "Fog Mist and Dreams." . . . He found a refrigerator in the alley. He put his projector with his film on it, and he tried different things. He tried films that had people, and we said, "Doesn't work. Doesn't work." And then he tried some other film, which was totally color and abstract, and it worked fine. We all said, "Yes. Do that one." . . . Some very proper MIT professors danced through it. And then the next day, which was a Saturday, it was announced that "Steam Screens: Fog Mist and Dreams" would be performed again due to popular demand. And Stan and I were sitting in the audience going, "Oh no," because nobody told us. So we got up and ran down, and got it back together again. It wasn't that hard, but we weren't planning on it. But anyway, that was the beginning of steam at MIT, for me, at least. (Joan Brigham interview)

Top: Joan Brigham and Stan VanDerBeek, "Fog Mist and Dreams," film projections on steam, alley adjacent to Building W11, *Arttransition* conference, CAVS, MIT 1975
Photo: Alejandro Siña

Bottom: Brigham and VanDerBeek, alley adjacent to Building W11, CAVS, MIT
Photo: CAVS Special Collection

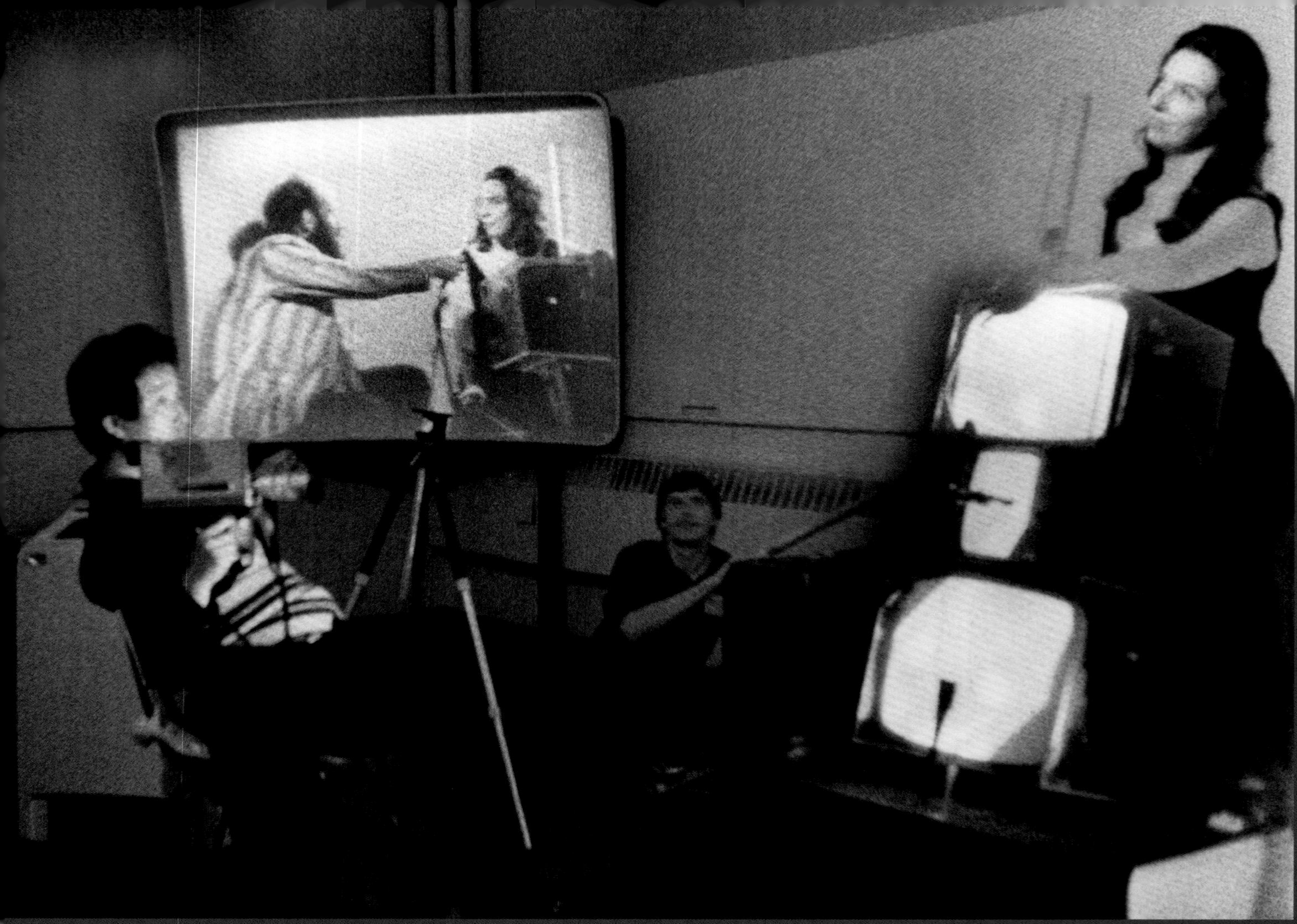

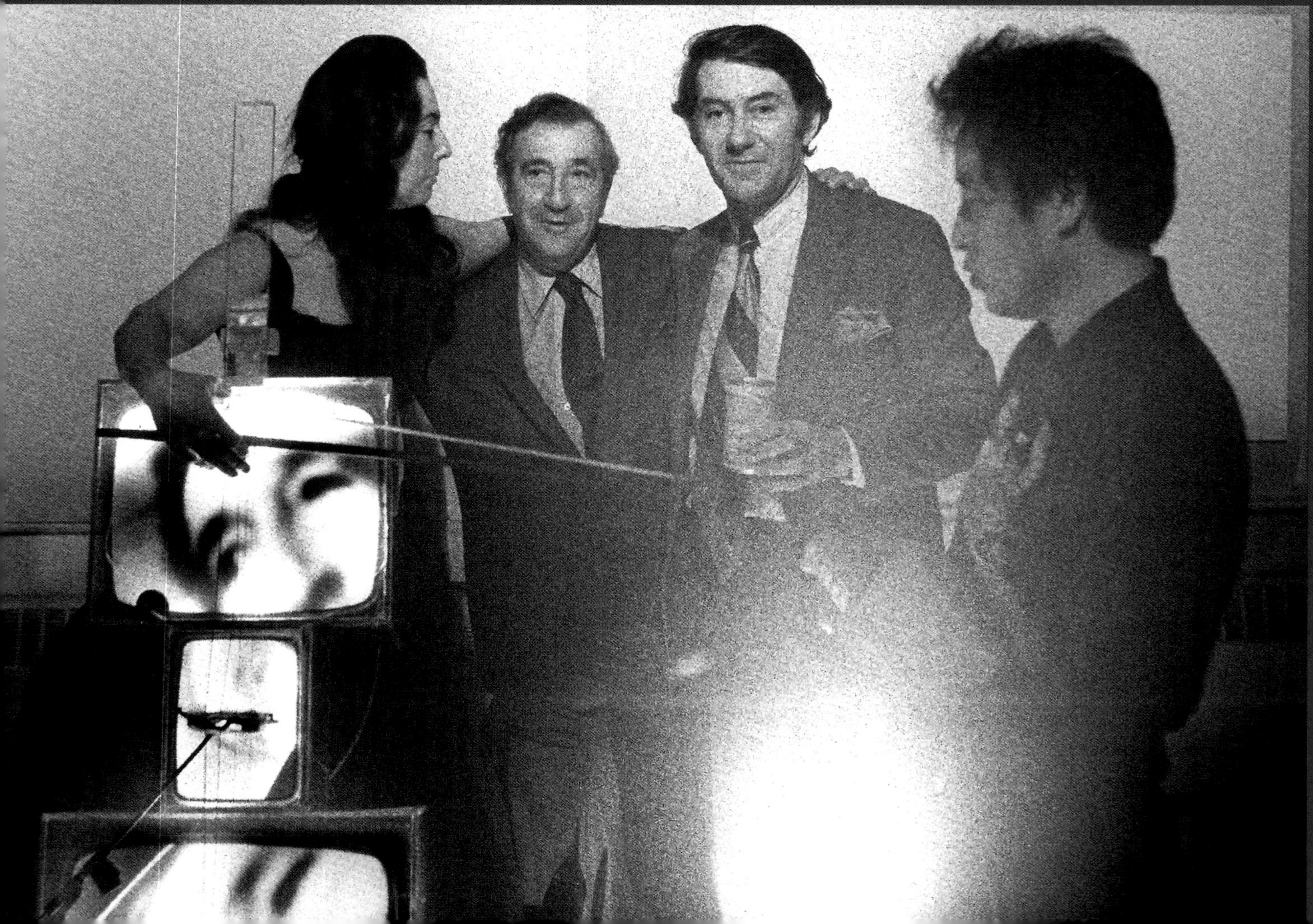

> Beginning with *Arttransition*, each conference had a performance, exhibition, and informal social component together with so many overlapping lectures that one could never take it all in. (Elizabeth Goldring interview)

The first *Ars Electronica* festival was held in Linz, Austria, in 1979. Together with CAVS Exhibits and Projects Director, Elizabeth Goldring, Piene originated five international *Sky Art Conferences* (1981, 1982, 1983, 1986, and 2002). *Arttransition 90* reprised the 1975 conference. The program was conceived by artists in pre-conferences held at the farm of fellow Jürgen Claus in Baelen, Belgium, and later at CAVS.

The four-day *Arttransition 90* symposium and media events were realized together with the participation of The Academy of Media Arts, Cologne; Zentrum für Kunst und Medientechnologie, Karlsruhe; *Leonardo, Journal of the International Society for the Arts, Sciences and Technology*; Carnegie Mellon University/Studio for Creative Inquiry; and CETECH/Jack Lang, Ministère de la Culture, Paris. Contributors included art philosopher and psychologist Rudolf Arnheim; environmental artists Christo and Dale Eldred; media artists Charlotte Moorman and Nam June Paik; and kinetic artist Wen-Ying Tsai. A live telecommunication video connection permitted a two-way dialogue between Laurie Anderson in Cologne and Otto Piene in MIT's Kresge Auditorium.

> The conference has further fueled art-and-technology to spread like brushfire—as is increasingly evident in the forming and expanding of new, respective institutes and schools throughout the world. (Piene, *MIT Report of the President*, 1991)

CAVS would have been unthinkable without the community of mind and body fostered by its conferences. CAVS conferences reflected the social-political themes of the moment, but plunged aggressively into the unrealized and unknown taking art into the sphere of MIT's core mission to explore unfamiliar systems and problem solving.

Charlotte Moorman performed on a 'cello' consisting of electrically amplified strings, and three TV screens while Nam June Paik utilized video processing to record Miss Moorman and the audience on the TV screens. (Robert Preusser, "Report on the 1975 International 'Arttransition' Conference at the Massachusetts Institute of Technology," *Leonardo*, Vol, 228-230, Pergamon Press, 1976)

Facing page top: Charlotte Moorman and Nam June Paik perform Paik's "Concerto for TV Cello and Videotapes," *Arttransition* conference, the Pit at CAVS, MIT, October 15, 1975

Facing page bottom, Moorman and Paik with Jerome Wiesner, President of MIT, and Otto Piene, *Arttransition* conference, the Pit at CAVS, MIT, October 15, 1975

Photos: Roger N. Goldstein

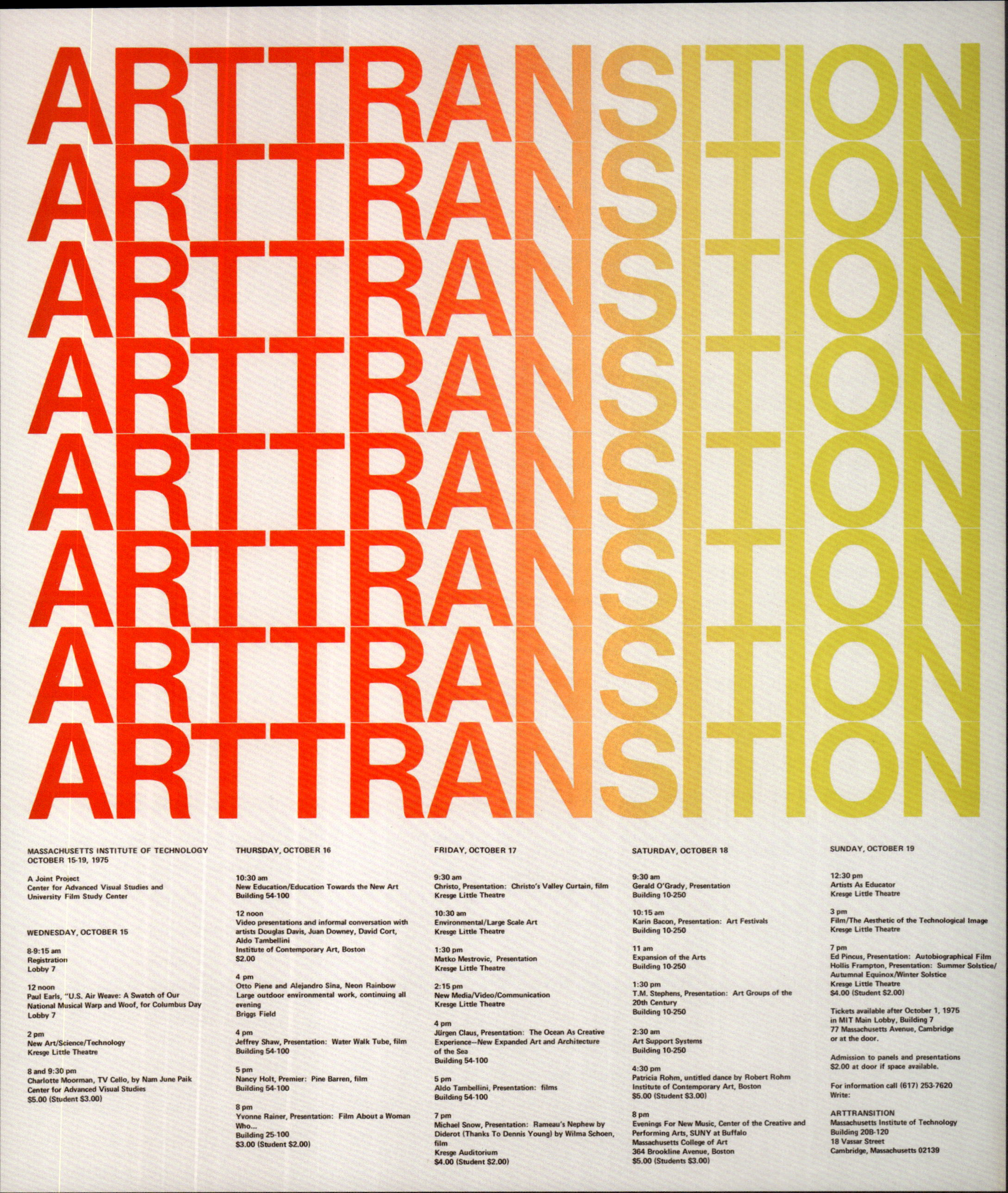

Above: *Arttransition* schedule, designed by Jacqueline Casey, CAVS, MIT, October 15-19, 1975
Source: CAVS Special Collection

Facing page: *Arttransition 90*, poster for the second *Arttransition* conference, CAVS, MIT, October 29 - November 1, 1990 (poster image by Elizabeth Goldring of Piene's "Dialogue de Têtes—Talking Head," with Lees Ruoff, CNAT, Reims, September 22, 1990)
Source: CAVS Special Collection

Massachusetts Institute of Technology
Center for Advanced Visual Studies
40 Massachusetts Avenue, Building WII
Cambridge, Ma 02139
tel.: (617) 253-4415 fax (617)253-1660

ARTTRANSITION

International Conference on Art, Science and Technology **·90** October 29-Nov. 1, 1990

In collaboration with

The Academy of Media Arts Cologne

Zentrum für Kunst und Medientechnologie Karlsruhe

LEONARDO, Journal of the International Society for the Arts, Sciences and Technology

Carnegie Mellon University/Studio of Creative Inquiry, Pittsburgh, PA

and

CETECH, Centre Européen de Technoculture, with the patronage of Jack Lang, Ministère de la Culture, de la Communication, des Grands Travaux et du Bicentenaire, France, and Université de Paris-Dauphine

Sony Headquarters Germany Cologne

Sony Telekonferenz Studio Cologne

Paris Art Center

University of Maryland—Baltimore County

Galerie Löhrl, Mönchengladbach, Germany

With the sponsorship of Frances Balter and Rosalind G. Jacobs

Gyorgy Kepes, Honorary Chairman
Ellen T. Harris, Patron
John De Monchaux, Chairman, Board of Advisors
Otto Piene, Chairman, ARTTRANSITION '90

Lowry Burgess
Elizabeth Goldring — Senior Consultants
Roger Malina

Cynthia J. Goodman, Conference Director

BOARD OF ADVISORS:

Rudolf Arnheim
Steve Benton
David Bermant
Joan Brigham
Annick Bureaud
Jack Burnham
Harriet Casdin-Silver
Jürgen Claus
Betsy Connors
Sarah Dickinson
Douglas Davis
Joe Davis
Paul Earls
Manfred Eisenbeis
Dale Eldred
Ante Glibota
Vin Grabill
Heide Hagebilling
Ellen Harris
Catherine Ikam
Christopher Janney
Catherine Judge
Dieter Jung
Antero Kare
Gyorgy Kepes
Heinrich Klotz
Piotr Kowalski
Richard Lowenberg
Paul Master-Karnik
Paul Matisse
Charlotte Moorman
Philip Morrison
Phyllis Morrison
Atsushi Ogata
Nam June Paik
Frank Pileggi
Frank Popper
John Powell
Bryan Rogers
Manfred Schneckenburger
Lillian Schwartz
Ellen Sebring
Todd Siler
George S. Singer
Tal Streeter
Elie Theofilakis
Harold Tovish
Wen-Ying Tsai
Tom Van Sant
Wibke Von Bonin
June Wayne

CONFERENCE TOPICS

Monday
October 29
- Global House
- The Umbrellas (Christo)
- Sky Art
- Sky Art/Global House (Panel)

Tuesday
October 30
- Electronic Media
- Feminine Perspectives on Art and Technology
- Projects and Publications
- Interactivity in Art (Panel)
- Video Screening
- Artists' Presentations I
- Telecommunications and Networking
- Telecommunications and Networking (Panel)

Wednesday
October 31
- Inner Space/Outer Space
- Public Events and Technological Art in Public Places
- Halloween Breakfast (Carnegie Mellon University)
- Video Artists
- Art and Technology: Centers, Programs and Exhibitions
- Art and Technology: Centers, Programs and Exhibitions (Panel)
- Art and Economics: Toward a Cultural Ecology (Panel)
- Artists' Presentations II

Thursday
November 1
- Tele-Interview: Laurie Anderson at the Academy of Media Arts Cologne, Otto Piene at ARTTRANSITION
- Perspective: New Concepts

EXHIBITIONS

CAVS
- Celebrations by CAVS/MIT
- Gyorgy Kepes — Images
- Harold Edgerton — Images
- Holography — group show
- Bill Bell
- Mark Mendel

MIT Museum
- Paul Earls

De Cordova Museum, Lincoln., MA
- Lowry Burgess

Video Exhibitions
- CAVS
- Video Screening
- Computer Animation
- Interactive Disc Program
- Artists' Video presentations

CAVS
- Biographeteria, Computer Network (Carnegie Mellon University)

PERFORMANCES

CAVS
- Charlotte Moorman
- Seth Riskin

MIT Kresge Oval and Athletic Center
- Drum Fire (Christopher Janney)
- Dialogue de Têtes—Talking Heads (Piene-Earls, Ruoff, Strasfogel)

CCRIB, Boston
- IRO-IRO (Adams, Judge, Ogata)

Tours
- MIT AI Lab (Artificial Intelligence Laboratory)
- MIT Project Athena
- MIT Haystack Observatory, Groton, MA

Staff

Michael Roper	Technical Coordinator
Ingrid Sell	Senior Staff Assistant
John Powell	CAVS Space Coordinator
Martha Werman	Accommodations Coordinator
Beth Wilbur	MIT Office of the Arts
Nat Green	Boston Area Coordinator

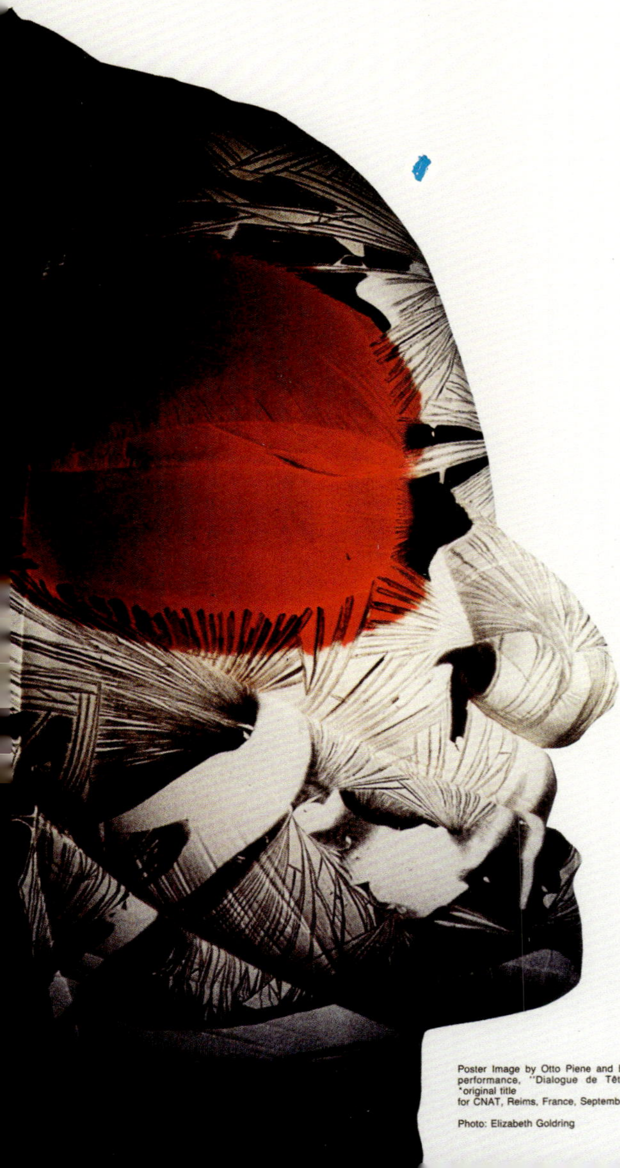

Poster Image by Otto Piene and Lees Ruoff from CAVS performance, "Dialogue de Têtes—Talking Heads"*
*original title
for CNAT, Reims, France, September 22, 1990

Photo: Elizabeth Goldring

Full ticket $250. Single day ticket $75.00
Student tickets with identification $60.00; single day $16.00.
Purchase of five tickets or more—20% off. MIT faculty and students with identification—attend free, tickets from MIT Museum Shop at Stratton Student Center.

Information: 253-8515 or 253-4415

Chapter 7
Sky Art

> Why is there no art in space, why do we have no exhibitions in the sky? Up to now . . . we have left it to war to light up the sky with coloured signs and artificial and induced bursts of flame. Imprisoned mankind does wonders defending itself. When is our freedom so great that we conquer the sky for the fun of it, glide though the universe, play the greatest game in light and space, without being driven by fear and mistrust? (Otto Piene, "Paths to Paradise," *Zero 3*, MIT Press, 1961)

Otto Piene was committed to defining and fostering an art form that propels art into the sky. CAVS hosted five international Sky Art Conferences, convening a collegium of artists and scientists to address and celebrate the peaceful uses of sky and space. Drafted at *Sky Art Conference '86* and presented at UNESCO, the *Sky Art Manifesto* voiced a commitment to transnational artistic explorations of space.

Right: Otto Piene, detail from "The Boston Harbor Project," lithograph, *Sky Art Portfolio*, 1969
Piene sketched and annotated his sky art ideas: 1. the Birds Sculpture, 2. the Flame, 3. the Light Beam, 4. the White Smoke Beam, 5. the Water Beam, 6. the Light Satellite, 7. Light Field in the Water, 8. the Sky Flower on a Mast, 9. the Sky Flower with Hot Air, 10. the Artificial Rainbow, 11. the Nightly Flaming Rainbow, 12. Nightly Display of Artificial Colored Clouds (Artificial Northern Lights), 13. the Fun Vessel, and 14. ∞ [infinite] Combinations Thereof
Source: CAVS Special Collection

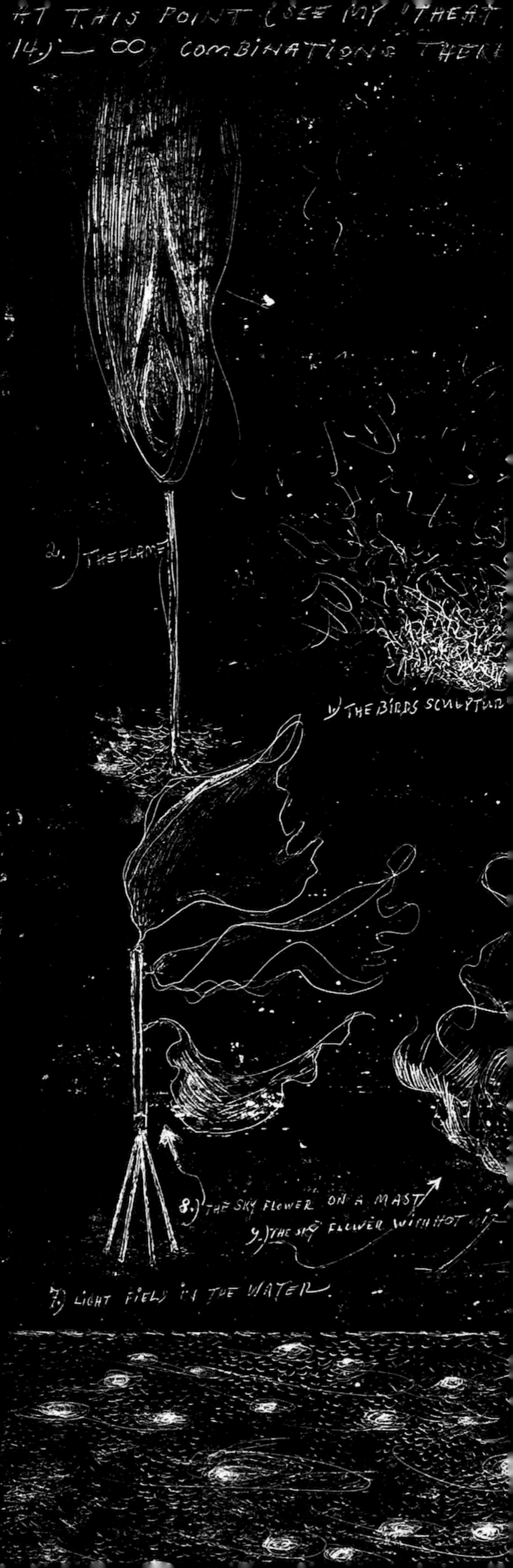

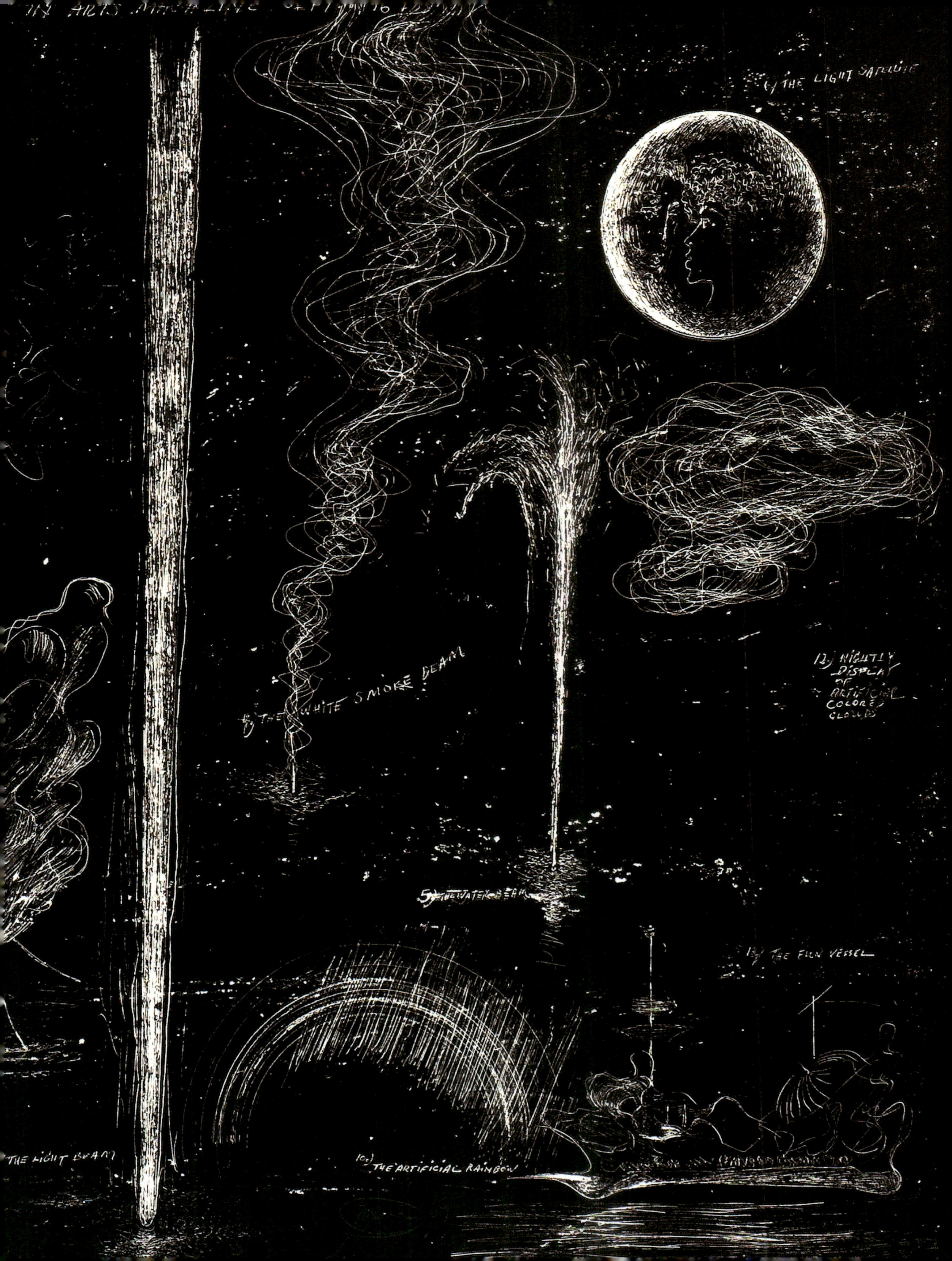

THE LIGHT SATELLITE
12) NIGHTLY DISPLAY OF ARTIFICIAL COLORED CLOUDS
THE WHITE SMOKE BEAM
5) THE WATER BEAM
THE FILM VESSEL
THE LIGHT BEAM
10) THE ARTIFICIAL RAINBOW

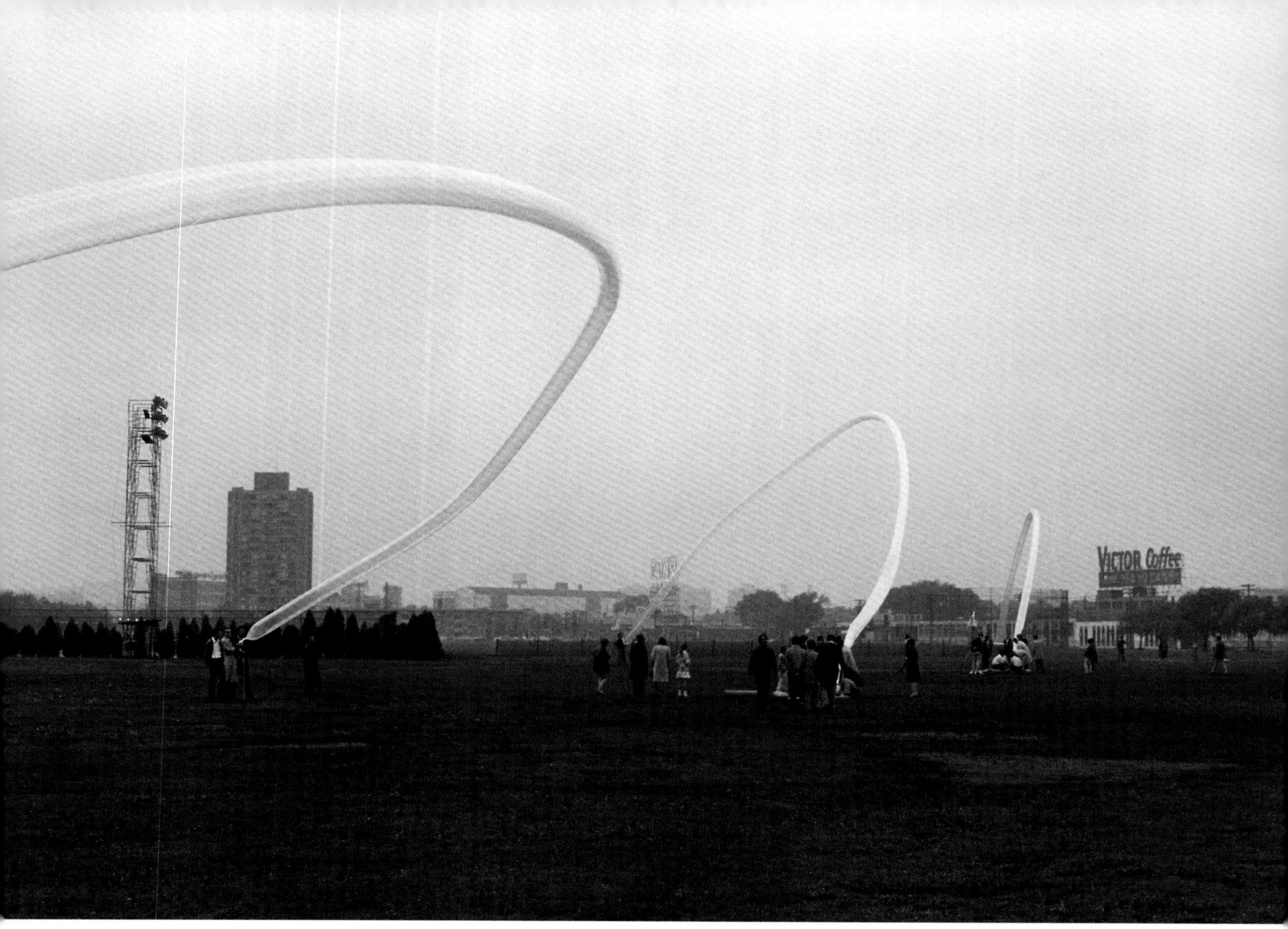

Above: Otto Piene, "Light Line Experiment," Briggs Athletic Field, MIT, May 22, 1968, 7 to 11 pm
Photo: George Claflin

Facing page: Otto Piene, "Manned Helium Sculpture," with performer Susan Peters, television shoot for Piene's segment, "Electronic Light Ballet," of *The Medium is the Medium* broadcast program, WGBH-TV parking lot, Boston, January 5, 1969
Photos: Nan Rosenthal

The 1,000-foot-long, helium-inflated, transparent polyethylene hose, two feet in diameter, composed of four links of hose, each 250 feet long, tied together to form one long line and illuminated in motion in the night sky by a 20 KW arc light; evening, during flight (typed caption on back of "Light Line Experiment" photo)
Photo: CAVS Special Collection

800 feet of 26 inch diameter transparent polyethylene tubing in seven loops, inflated with approximately 4000 cubic feet of helium, attached with ropes and parachute harness to 95 lb. girl for her 30 minute ascent in the air, controlled from the ground by ropes attached to the helium-filled loops and to the girl's harness. (Piene, "Manned Helium Sculpture" description, *Elements* catalog, Howard Wise Gallery, New York City, March 8-April 5, 1969)

Sky as an artistic medium was explored during Otto Piene's tenure as CAVS director at five Sky Art conferences held in the United States and Europe. *Desert Sun/Desert Moon*, a series of installations and events in the Jura Triassic hills outside Lone Pine, California, was part of *Sky Art Conference '86*. Group interactions with the temporal landscape evoked the mythic and immaterial dimensions of sky and space, and, inspired by *Centerbeam*, advanced new forms of collaboration.

During the Group Zero years in Düsseldorf, Piene experimented with balloons and discussed ideas of flying with Yves Klein and other Zero artists. But it was as a fellow at MIT in 1969 that he first coined the term "Sky Art."

> One reason that moved me to go to MIT, to go to the Center, to follow the invitation: I had a faint idea that that would be the place where I might create a Sky Art conference, because that idea had already occurred to me in the mid-60s, no, the early 60s (this was already the mid-60s). "Light Line Experiment" on MIT Briggs Field was quite an interesting thing to happen at that time at MIT, it was something that no one had seen or thought of before, and it was well received. (Piene, *Sky Art Conference 2002* catalog, 2002, 50)

In January 1969, Piene flew his first "Manned Helium Sculpture" for the WGBH Television production *Medium is the Medium*. In 1972, he participated in the closing ceremonies for the Munich Olympics with his 2,000-foot-long *Olympic Rainbow*, and the installation of his projecting diamond sculpture with illuminating spectral flashes day and night.

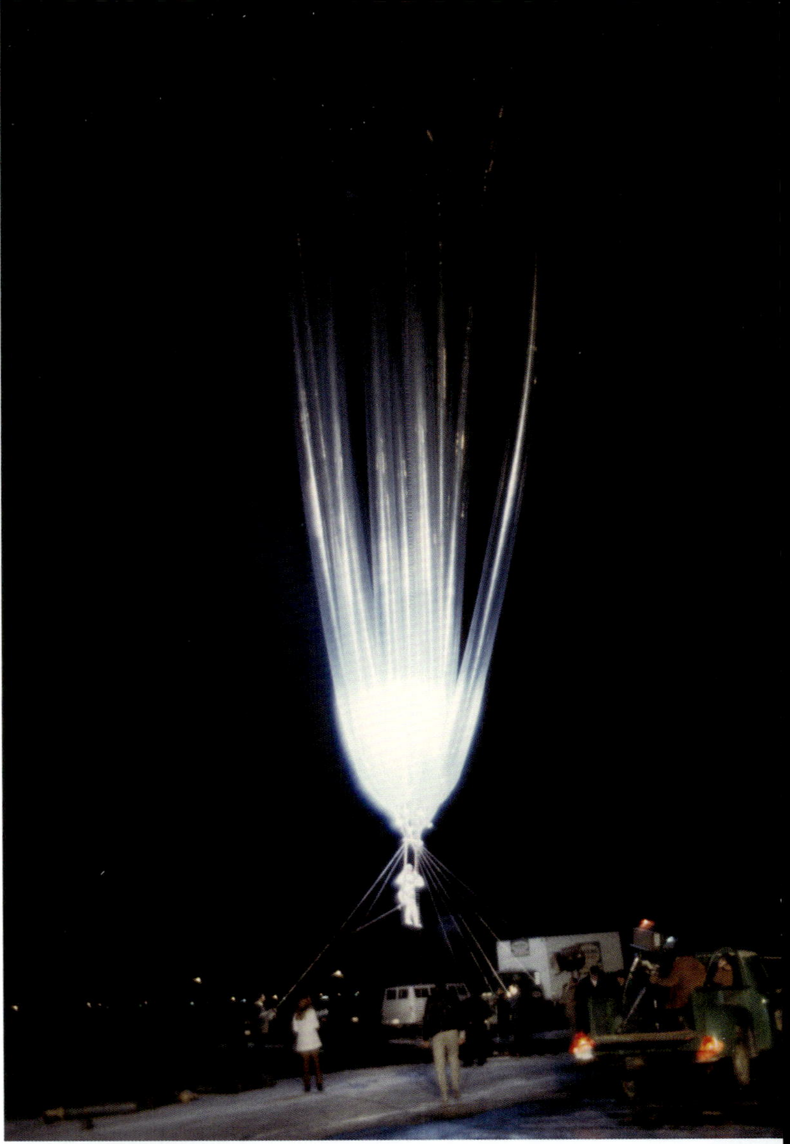

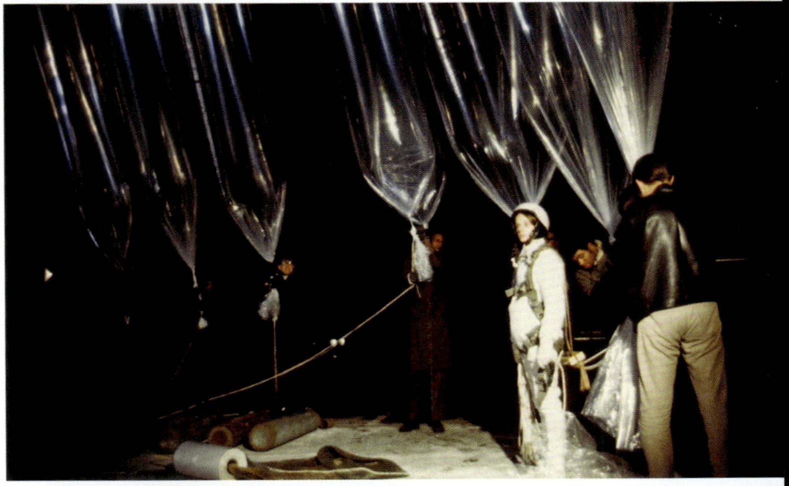

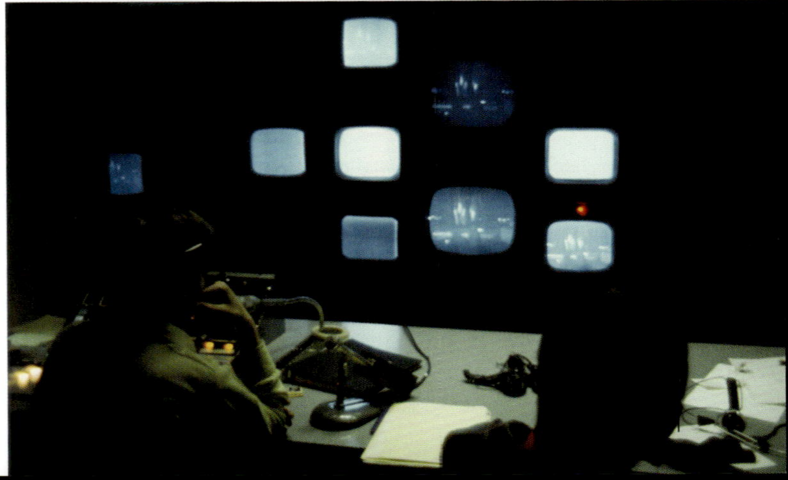

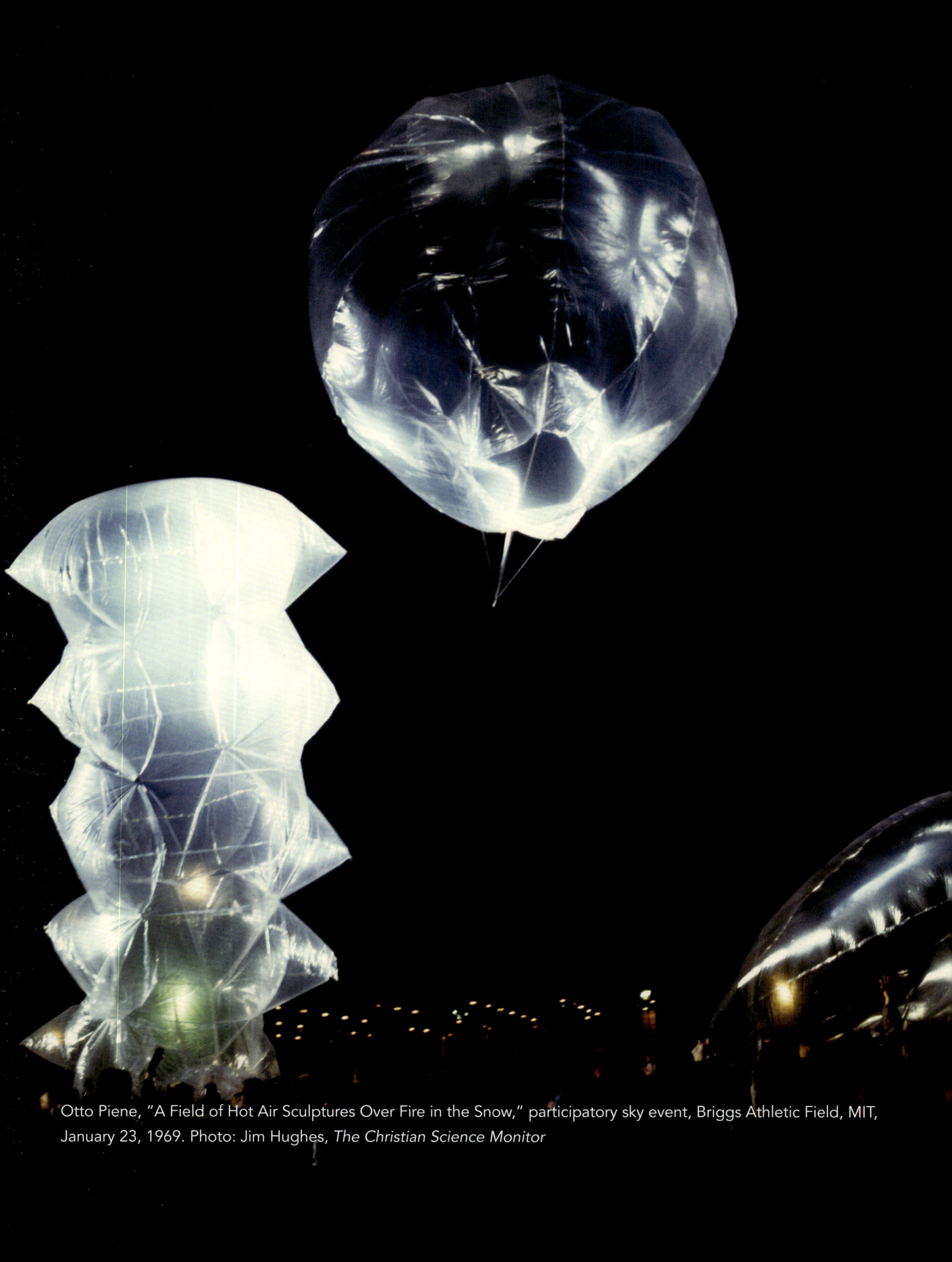

Otto Piene, "A Field of Hot Air Sculptures Over Fire in the Snow," participatory sky event, Briggs Athletic Field, MIT, January 23, 1969. Photo: Jim Hughes, *The Christian Science Monitor*

Piene published twenty-five lithographs as his first *Sky Art Portfolio* (1969), wrote *More Sky* (MIT Press, 1970); and the article, "SKY ART—A Notebook for a Book" (*Arscanada*, 1969). Among the *Sky Art Portfolio* lithographs, two prints, "Flying" and "Looping," were executed after drawings made in 1952.

> Many sky events happen because I draw them first, not all of them but quite a few, and I draw them first in a process of definition through many different stages. Then I turn them into paper sculptures, which lead to patterns and the whole thing turns into reality (which is something that fascinates me). The process is basic work of the artist and the scientist and the engineer. There's this fast translation from your head into visibility and the head/hand relationship still works very well. It's actually faster than any computer I know because you can just keep drawing and define the idea as it flows out of your pen. (Piene, *Sky Art Conference 2002* catalog, 2002, 54)

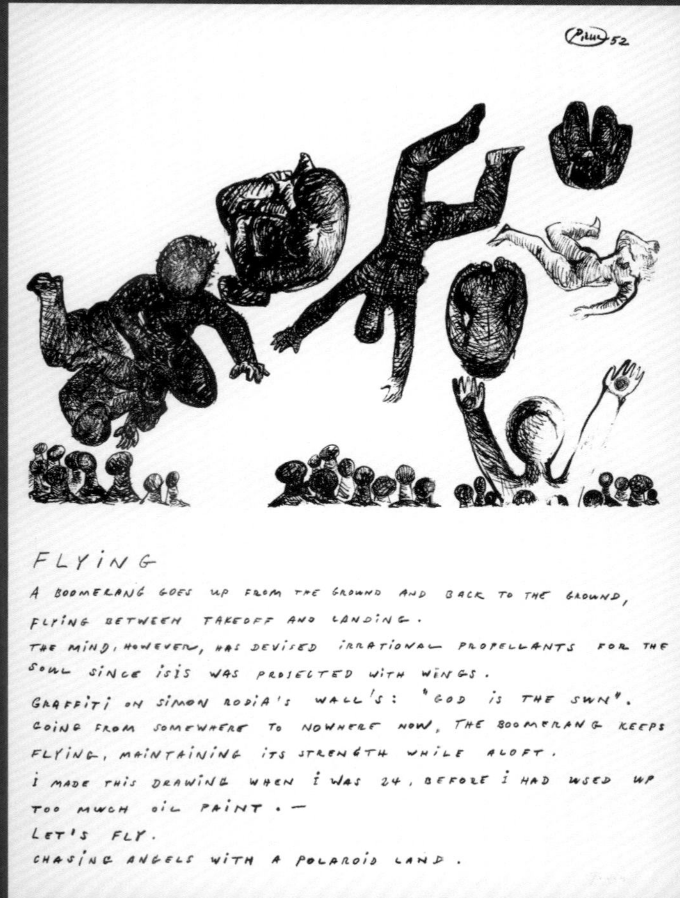

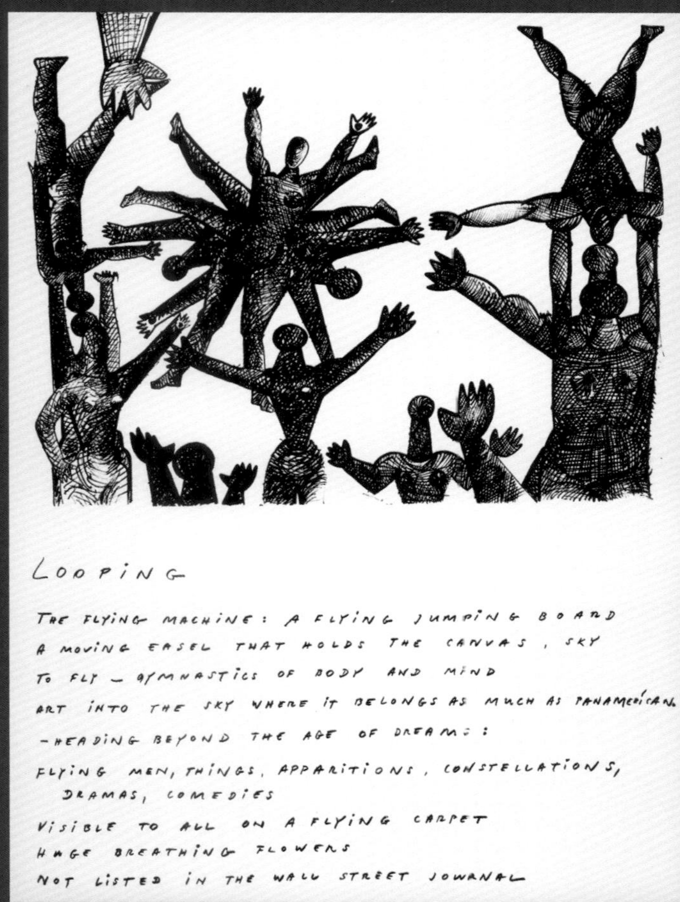

> But as darkness fell, the inflating and releasing of the large unruly balloons showed a new purpose. A truck was nearby holding two large arc lights; these were adjusted to follow its random course when it was released. Under the intense illumination, the balloons seemed to absorb light and emit it afresh. . . . The activity which had seemed aimless before now seemed experimental; the experiment might have been to determine how to give a shape to light. . . . A new humanization of the technological environment may have its beginnings in the spirit of combined (and collective) experimentation and creativity evident in the Piene presentation. (K. B., "Art and hot air and everyone," a story on Otto Piene's "A Field of Hot Air Sculptures Over Fire in the Snow," *The Christian Science Monitor*, January 27, 1969, 7)

Right: Otto Piene, "Flying" and "Looping," lithographs from his *Sky Art Portfolio*, 1969
Photos: Ante Glibota

Sky Art was flying images, light that defined outdoor space and played in the skyscape, and artistic communication that traveled through sky and space via satellite and telecommunications. The material properties of sky art—fluid, insubstantial, and mobile—mimicked the Aurora Borealis and underwater forms, and involved gases, bright lights, and responsive fabrics.

> . . . affordable, non-deteriorating, reusable, recyclable, ultra-lightweight, bright-colored, mirroring cloths, tough as steel and flexible as rubber to become inflatable/collapsible, orbiting image architecture in space. (Piene, "Art-and-Technology: Recent Efforts in Materials and Media," *MRS Bulletin*, 1992).

> With Otto, the whole complex of things shifts. What I felt was a whole set of things coming into place proposed in terms of the Center's direction. The stuff that was driving Otto . . . had this erotic almost decorative base to his work. In a very powerful sense . . . a tremendous force of desire underneath things. On top of that was the glowing presence of the sky. (Lowry Burgess interview)

> . . . the sequence of activities and physical conditions that brings such forms into the air, allowing them to fill each mind present with their buoyant, stately magnitude, and finally the process of deflation, gradual collapse, and descent to earth. The inherent power of such ephemera is that, unlike sculpture as fixed mass, the inflatables project a life cycle through different physical states, each providing a set of natural, non-mechanical changes. (J. W. Burnham, "Ritual Dimension in Otto Piene's Sky Performances," *Otto Piene*, 2010, 379)

Media adapted to the theme of Sky Art by CAVS artists included steam, video art, laser, poetry, and sound. Holography, a light medium that cast three-dimensions into space from a single source, was presented outdoors for the first time in 1977 at *Centerbeam* by holographer Harriet Casdin-Silver, who imagined large holographic images hovering over cities. Casdin-Silver, perhaps counterintuitively, transformed body-centric subjects into light and addressed issues of women's presence, history, equality, and sexuality in feminist subtexts. Physicality, eroticism, and desire were also captured in the life cycles of ephemeral sky art media:

> The artist speaks for her time, we have been living in a technological time but a lot of the art world didn't know it. The Center knew it and supported artists that knew. . . . Paul Matisse in Greece at the Sky Art Conference said he had never seen these projects of mine. . . . One of the contributions I brought to CAVS was a focus on women. . . . I use objects as women's objects, but also create a feminist statement with them. . . . My political agenda was a primary aspect of my creativity long before holography and continued with "Phalli" and "Forks" to "Compton's Buttocks," "A Woman," the male nudes and on. And I believe I was able to impart to the students a recognition of women moving forward, women having a stronger voice, surely an equal voice. (Harriet Casdin-Silver interview)

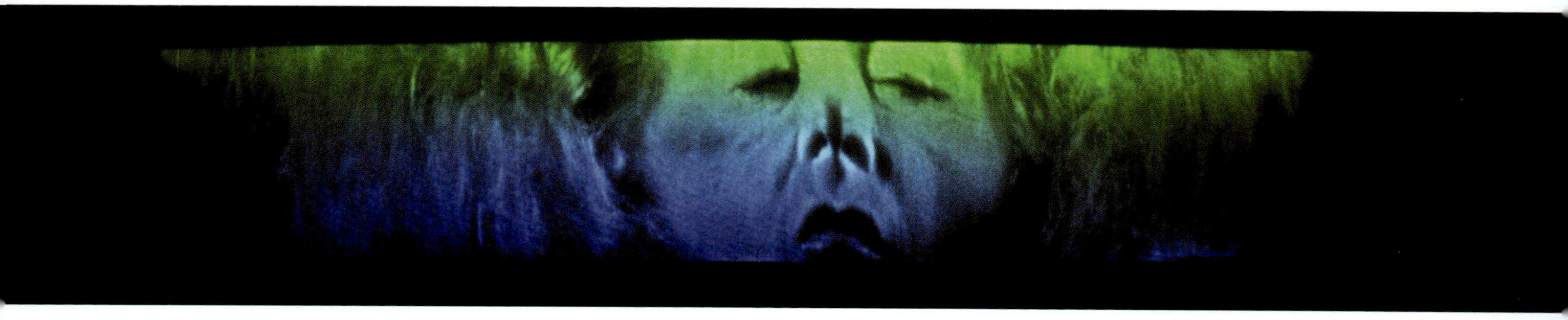

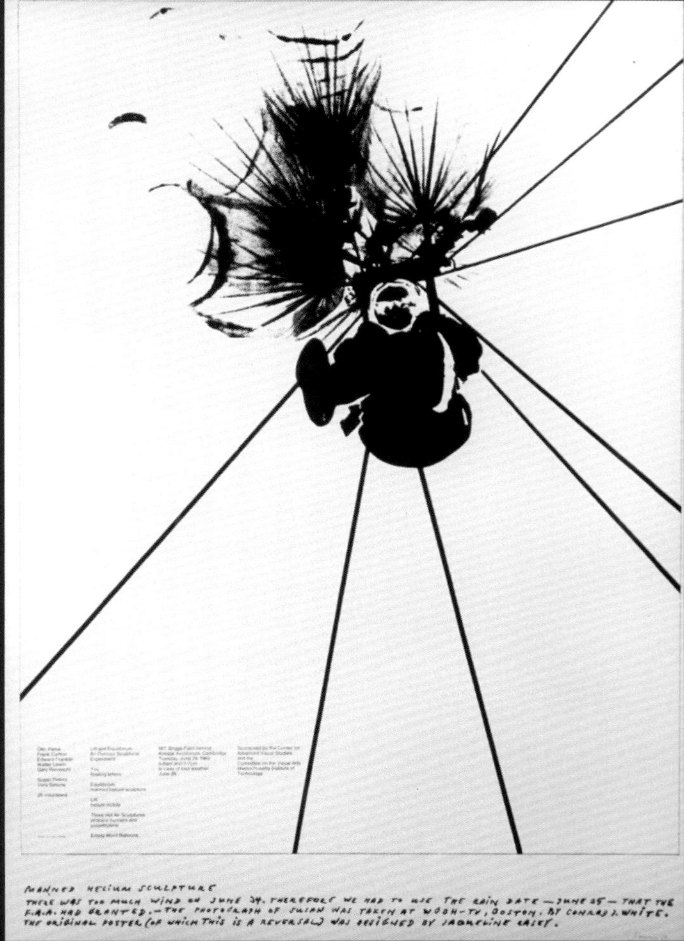

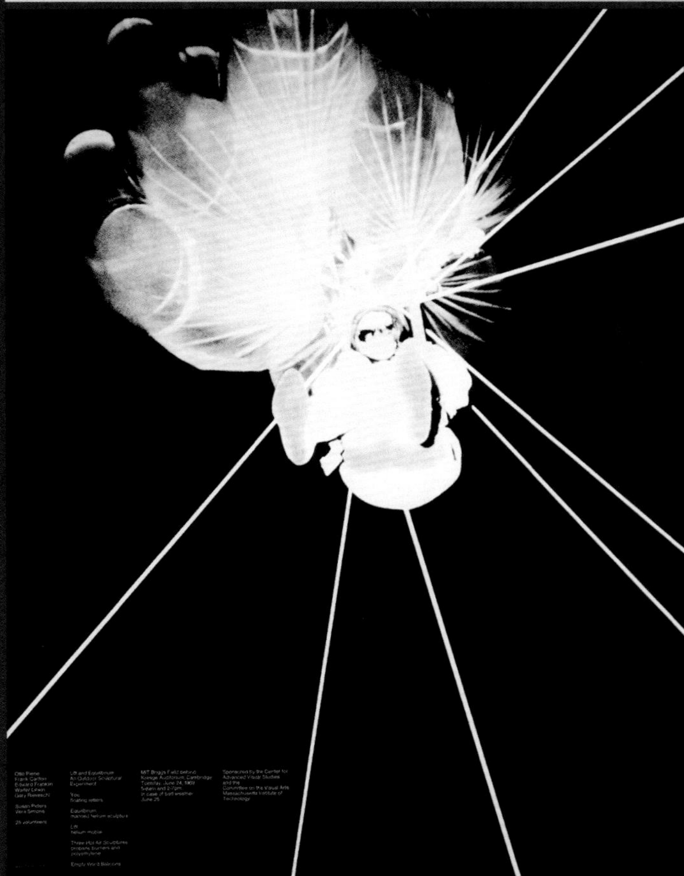

Manned Helium Sculpture
There was too much wind on June 24. Therefore we had to use the rain date—June 25—that the F.A.A. had granted. — The photograph of Susan was taken at WGBH-TV, Boston, by Conrad J. White. The origi poster (of which this is a reversal) was designed by Jaqueline [sic] Casey.
(Piene's handwritten text on lithograph, left, 1969)

Lift and Equilibrium
An Outdoor Sculptural Experiment:
You
  floating letters
Equilibrium
  manned helium sculpture
Lift
  helium mobile
Three Hot Air Sculptures
  propane burners and polyethylene
Empty Word Balloons
Frank Carlton, Edward Franklin, Walter Lewin, Susan Peters, Otto Piene, Gary Rieveschl, Vera Simons, 25 volunteers
(text from *Lift and Equilibrium* poster, left, 1969)

Above left: Otto Piene, *Manned Helium Sculpture*, lithograph from his *Sky Art Portfolio*, 1969
Source: Otto Piene Collection

Left: Jacqueline Casey, poster design for *Lift and Equilibrium*, CAVS all-day event, Briggs Athletic Field MIT, June 25, 1969
Source: CAVS Special Collection

Facing page: Harriet Casdin-Silver, "A Woman," hologram, self-portrait, 1979
Photo: Nishan Bichajian

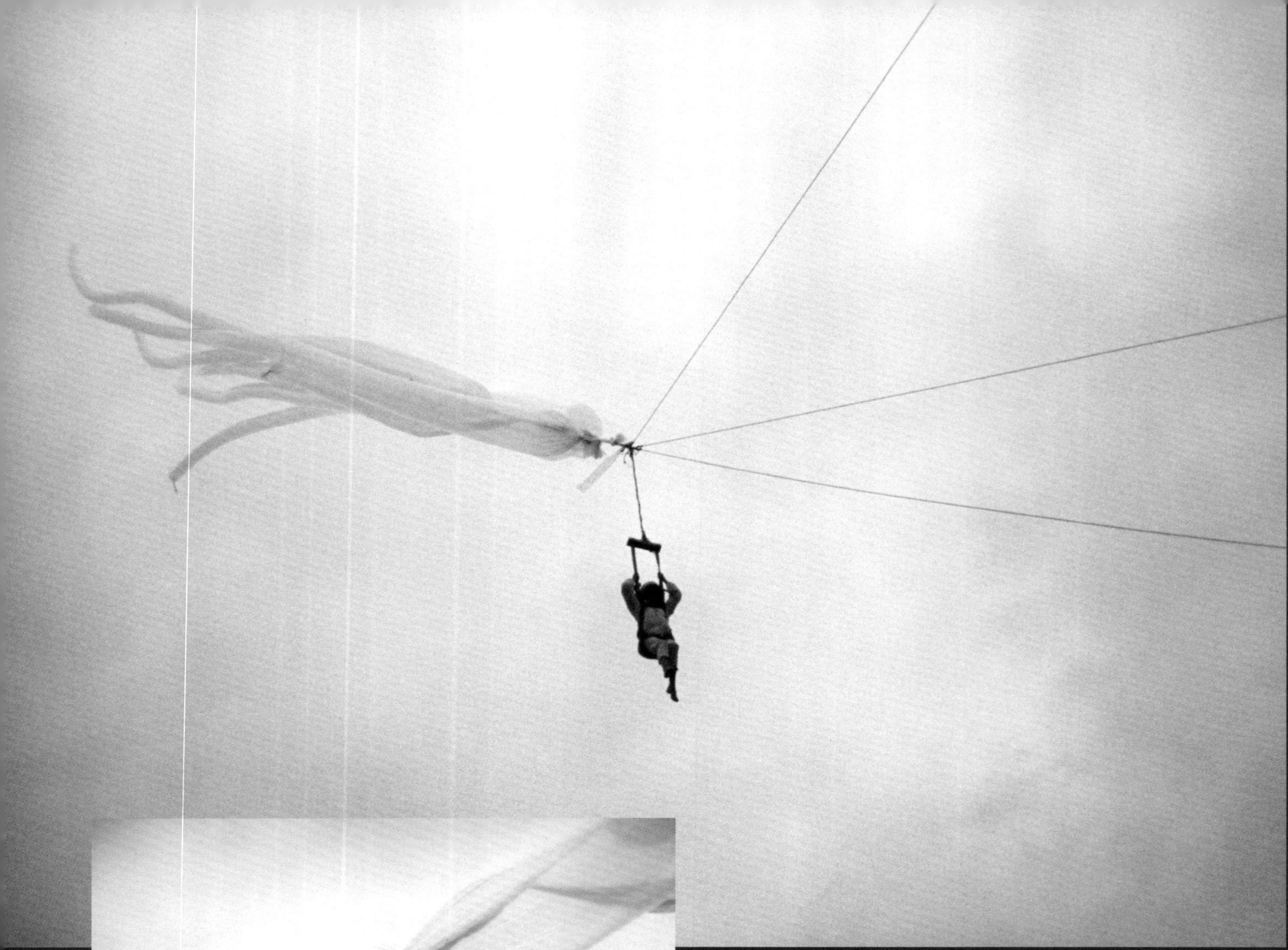

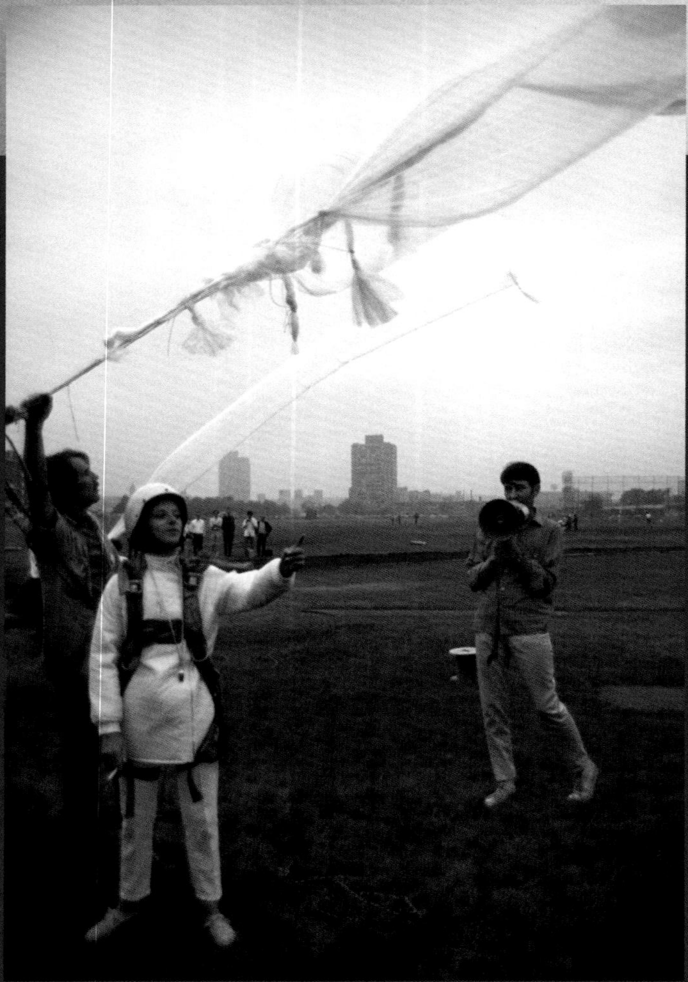

The major event of the day began shortly after 2 pm: "Equilibrium," a manned helium sculpture. An octopus-like assembly of eight helium-filled polyethylene sections, each 41 inches in diameter and 240 feet long, were each joined at one end to a harness in which balloonist Susan Peters rode, and after her, Vera Simons. The other ends of the eight sections were spread in a circle. At first, they were anchored by lines held by a ground crew, then released upward. (CAVS description, *Lift and Equilibrium*, 1969

Above and left: Otto Piene, "Equilibrium: Manned Helium Sculpture," sky event, Piene pictured with megaphone, Susan Peters in harness and flying, *Lift and Equilibrium*, CAVS, Briggs Athletic Field, MIT, June 25, 1969
Photos: Otto Piene Collection

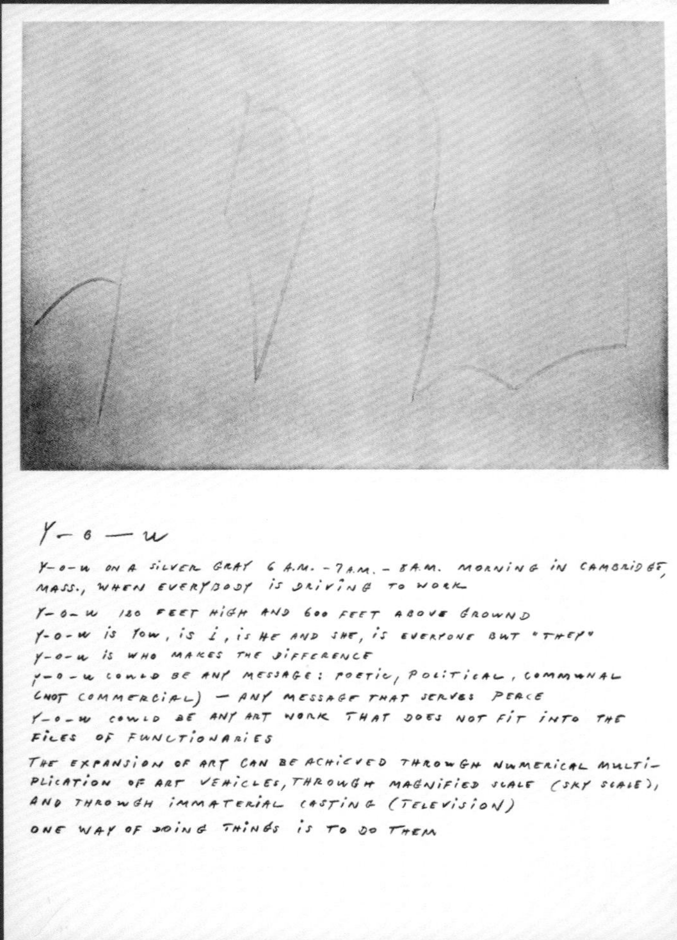

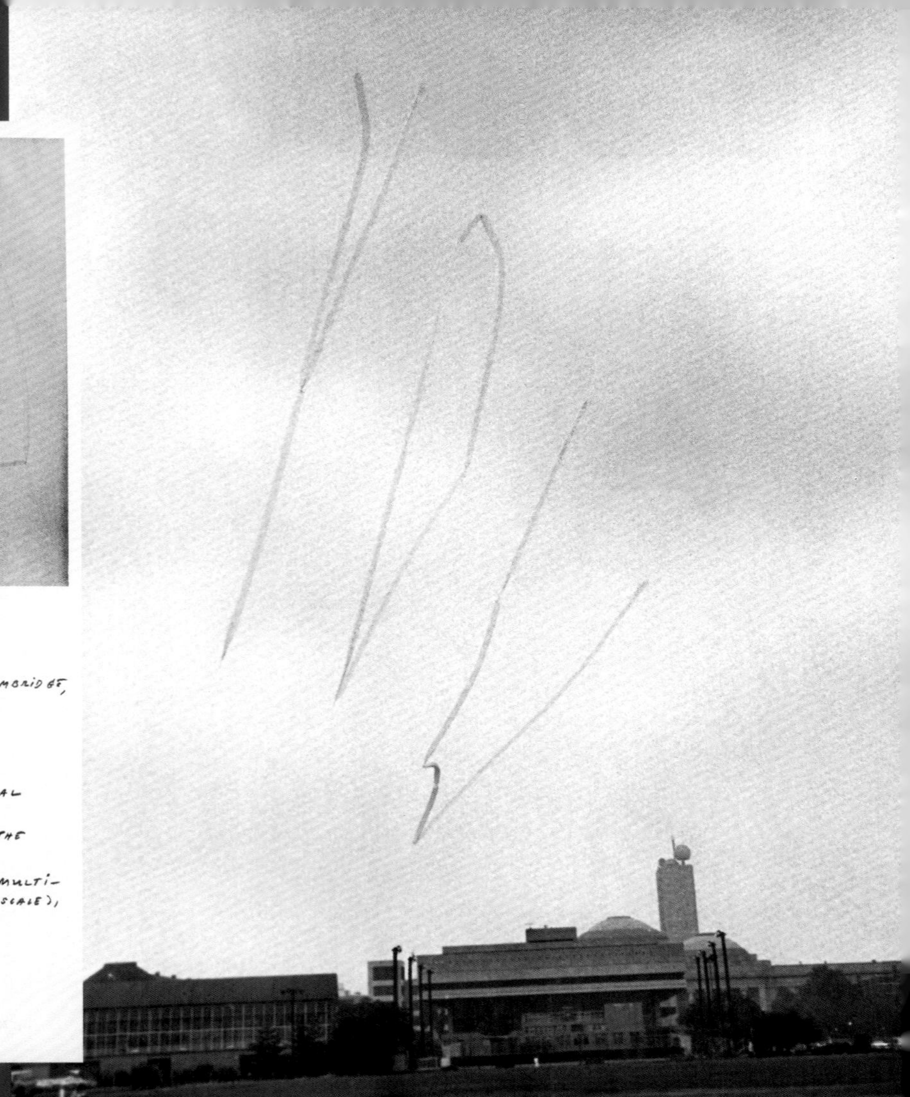

Y-O-U on a silver gray 6 am - 7 am - 8 am morning in Cambridge, Mass., when everybody is driving to work
Y-O-U 120 feet high and 600 feet above ground
Y-O-U is you, is i, is he and she, is everyone but "they"
Y-O-U is who makes the difference
Y-O-U could be any message: poetic, political, communal (not commercial)—any message that serves peace
Y-O-U could be any art work that does not fit into the files of functionaries
The expansion of art can be achieved through numerical multiplication of art vehicles, through magnified scale (sky scale), and through immaterial casting (television)
One way of doing things is to do them
(Piene's handwritten text on "Y-O-U" lithograph from his *Sky Art Portfolio*, above left, 1969)

Above left: Otto Piene, "Y-O-U," lithograph from his *Sky Art Portfolio*, 1969
Photo: Ante Glibota

Above right: Otto Piene, "Y-O-U," sky event, *Lift and Equilibrium*, CAVS, Briggs Athletic Field, MIT, June 25, 1969. Photo: Bob Lyon

July 4th Salute, Boston, June 23-July 4, 1975

CAVS examined celebration as a collaborative art form from a variety of perspectives in the *Workshop on the Theory, History and Practice of Public Celebration*, held June 23-July 4, 1975.

> Probably no group anywhere has sought so intensely in over twenty-three four-hour sessions to explore the virtuosity, importance and necessity of celebrations. Among topics covered by the teaching staff were visual-form inventions; kinetic, sky and environmental art; television; national and religious holidays; funerals and memorial services; the history of celebrations from the earliest to present times; architecture; astrophysics; planning and designing costumes and sound events; light and reflections; and relationships of celebrations to audiences. (Robert Steele, "Report on the 1975 Public-Celebration Workshop at the Center for Advanced Visual Studies, MIT," *Leonardo*, Vol. 9, No. 1, Winter 1976, 58)

Lowry Burgess discussed twelve kinds of memory and the disparate ways they link imagination and celebration. Physicist Walter H. G. Lewin "hammered away on non-astrophysics-minded students to make answers clear to questions about rainbows, artificial and natural color, and light." (Steele, 59) Goldring gave lectures on the history and theory of ritual and public celebrations. The participants became collaborators in the culminating celebration, the "July 4th Salute" mounted on one of Boston's tallest buildings, the Prudential Tower.

> This 'July 4th Salute', conceived by Piene and realized by him and his loyal, hard-working crew, exemplified environmental and sky art. . . . It was a spectacle that Boston will long remember. The 52-story Prudential Tower, which is 750 feet high, was used as a flag pole from which were hung the flag of the U.S.A., the Boston 200 Logo and a rainbow flag. The latter two flags, 30 X 20 ft and weighing about 28 lbs, were made by MIT students. The rainbow design symbolized hope, optimism and peace. The three flags were attached to joined red, white and blue air-inflated polyethylene tubes which were connected by webbing and reinforced by load tapes. The 10 ft-diameter tubes or 'ropes' reached from the ground of the Prudential Plaza to the 48th floor of the Tower. (Steele, 59)

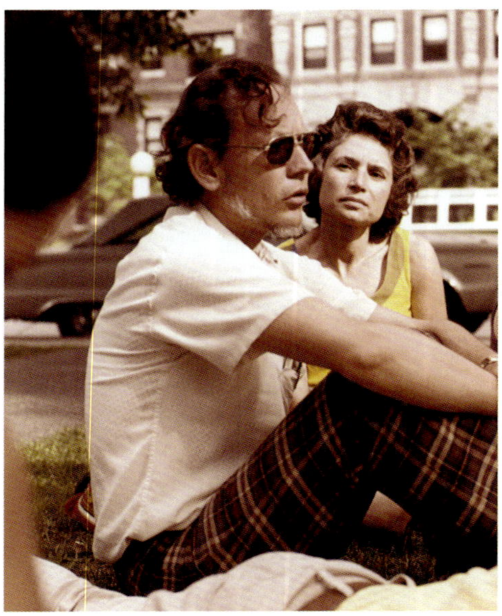
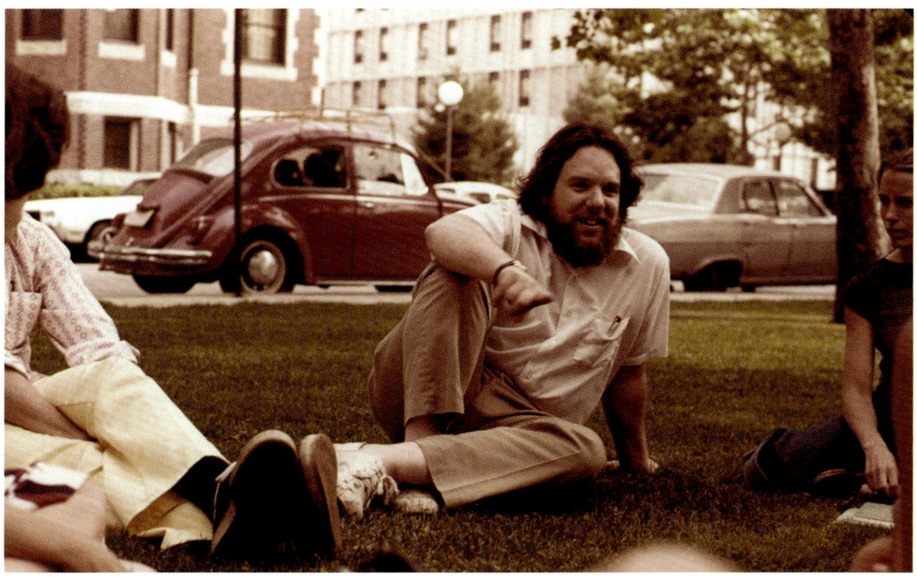

Left: Paul Earls, Zeren Earls; Right: Lowry Burgess, on the MIT Chapel lawn during the *Workshop on the Theory, History and Practice of Public Celebration*, CAVS, MIT, June 23-July 4, 1975. Photos: Brian Dowley

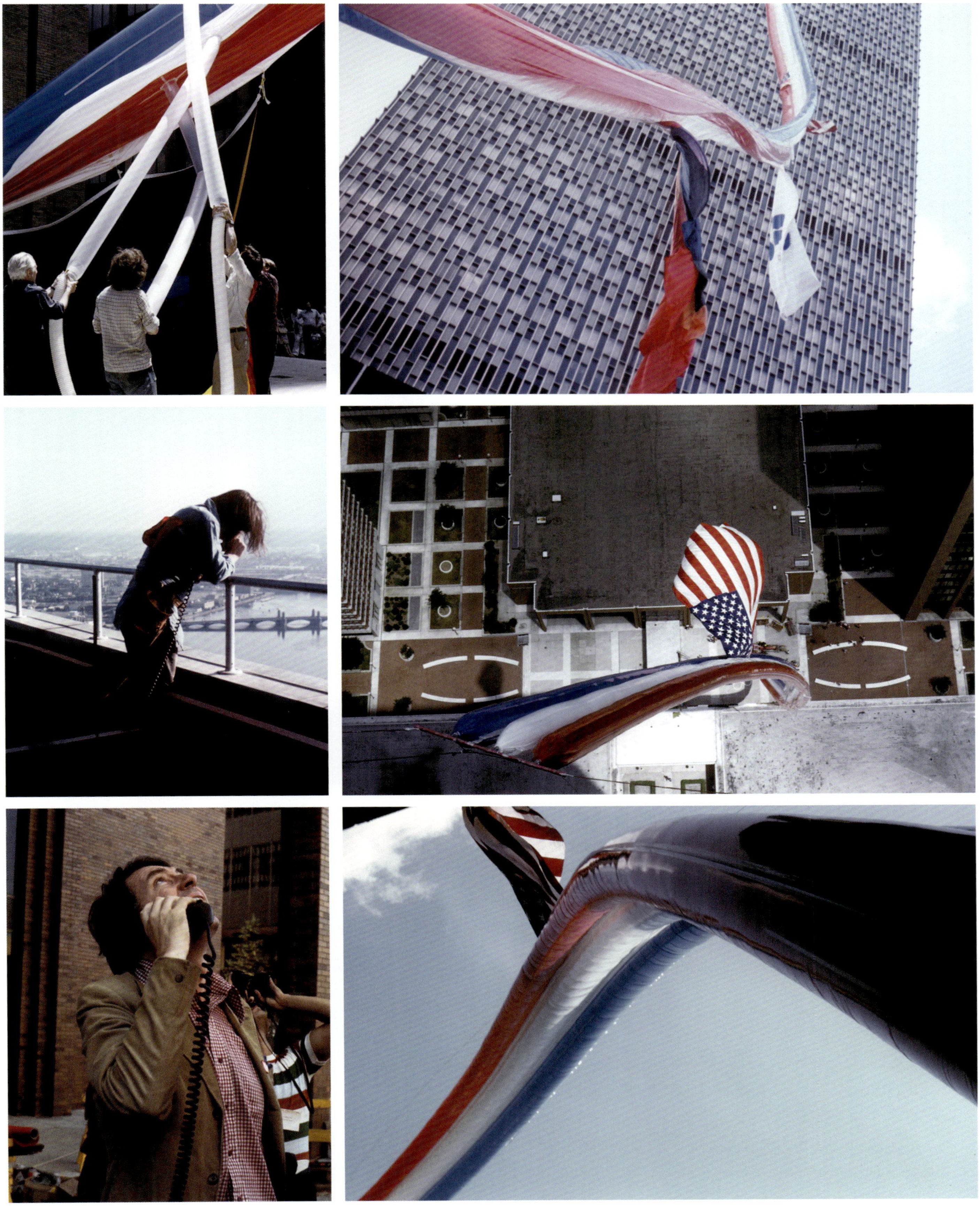

Above: Otto Piene, *July 4th Salute*, sky event, red-white-and-blue inflatable with flags, Prudential Tower, Boston, *Workshop on the Theory, History and Practice of Public Celebration*, CAVS, MIT, July 4, 1975. Photos: Brian Dowley

Sky Art Conference '81, CAVS, MIT, September 25-29, 1981

> It was the beginning of Sky Art. One of the first fellows meetings I ever went to was the Sky Art meeting where Otto presented the idea. I remember people had some strong concerns and questions about it; people had strong opinions. (Sally Weber interview)

Under Piene's direction, CAVS organized the first Sky Art conference in 1981 at MIT. The conference included an exhibition and outdoor sky events, with talks and panel discussions on Sky Art; art, information, and education; life in sky and space; science of the sky; and future perspectives. Kepes offered the keynote lecture titled "Icarus, Daedalus, Antaeus." Bernd Kracke developed "Data Network" to broadcast a continuous Slow Scan link with MIT's Visible Language Workshop where computer graphic collages were made of conference interviews and events. Tom Van Sant laid out an "eye" with mirrors on the desert floor of the California Shadow Mountains. Photographed via Landsat satellite, it captivated artists and scientists alike. This project, *Reflections from Earth*, was reproduced on the *Sky Art Conference '81* catalog cover, and its documentation created ongoing speculation about the possibility of a global event in which the mirror station images by artists would be picked up simultaneously by satellite.

Scientists and engineers also participated in the *Sky Art Conference '81*. Harold Edgerton, inventor of stroboscopic photography, served as honorary chairman. Eugene Larrabee and Paul MacCready spoke on muscle-powered flight. Philip Morrison presented the lecture "1-D, 2-D, 3-D, 4" in which he conveyed a plea that artists and scientists should work together, their collaboration having only just tipped the iceberg of what could be possible, of what could be beautiful. Mark Drela, an MIT graduate student, later MIT AeroAstro professor, flew his handmade "indoor duration model airplane" that at 0.05 grams was one-twentieth the weight of a dollar bill and flew at 2 mph, a bit less than walking speed. Paolo Soleri presented his theories of future eco-space life systems, Frederick Durant spoke on the history of space illustration, and Stephen Crites discussed angelology.

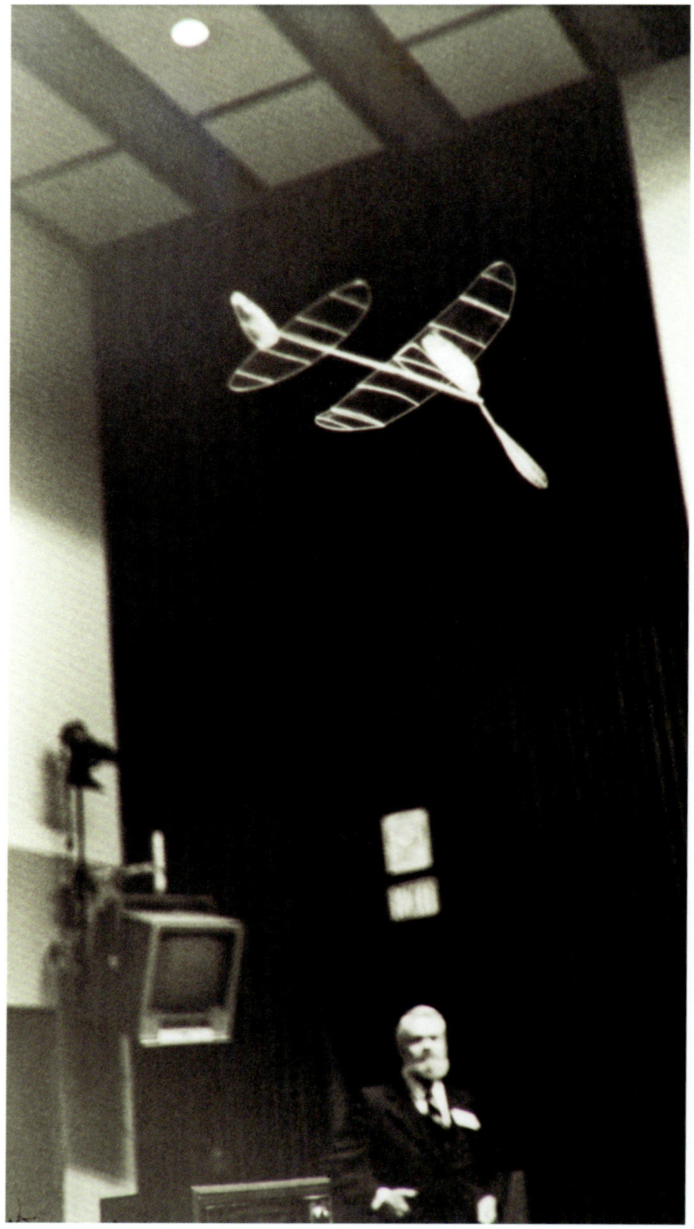

Top: Handmade indoor duration model airplane designed by MIT graduate student Mark Drela; the 0.05 gram plane in flight demonstration was observed in the photo by Eugene Larrabee, MIT Room 10-250, *Sky Art Conference '81*, CAVS, MIT, 1981
Photo: Tal Streeter

This picture was taken from space 600 miles above the earth by the NASA satellite Landsat II on June 11, 1980. The view is of a remote portion of California Desert called the Shadow Mountains. The image represents an eye gazing skyward, measuring 1.4 miles across. The eye is not a land feature, but a picture created by sunlight. Ninety mirrors, each 24 inches square, were arranged on the desert floor and carefully adjusted to reflect sun beams into the satellite's path. The radiant energy from each mirror was recorded by the appropriate sensor at the moment the satellite was scanning its picture of the earth. Though each mirror is small, it saturated a square the size of an acre (about the size of a football field). The entire image is made up of these one acre units, as can be seen in the enlarged pictures. Photos courtesy of NASA, USGS, EROS Data Center. (Tom Van Sant, *Reflections From Earth*, 1980)

As an extension of my concept of 'sculpture as process', which considers the constant flow of information the process that has to be shaped, I shall set up a network of media for the *Sky Art Conference* to process and publish the generated information instantaneously. This "DATA Network" will link the following MIT departments via telecommunication: Visible Language Workshop (image and type processing computer facilities); Information Processing Services (word processing facilities); Graphic Arts (computerized photo typesetting); and Educational Video Resources (MITV-cable network). . . . we will publish daily versions of hard copy news and an 'electronic newspaper' via cable. (Kracke, *Sky Art Conference '81* catalog, 1981, 58)

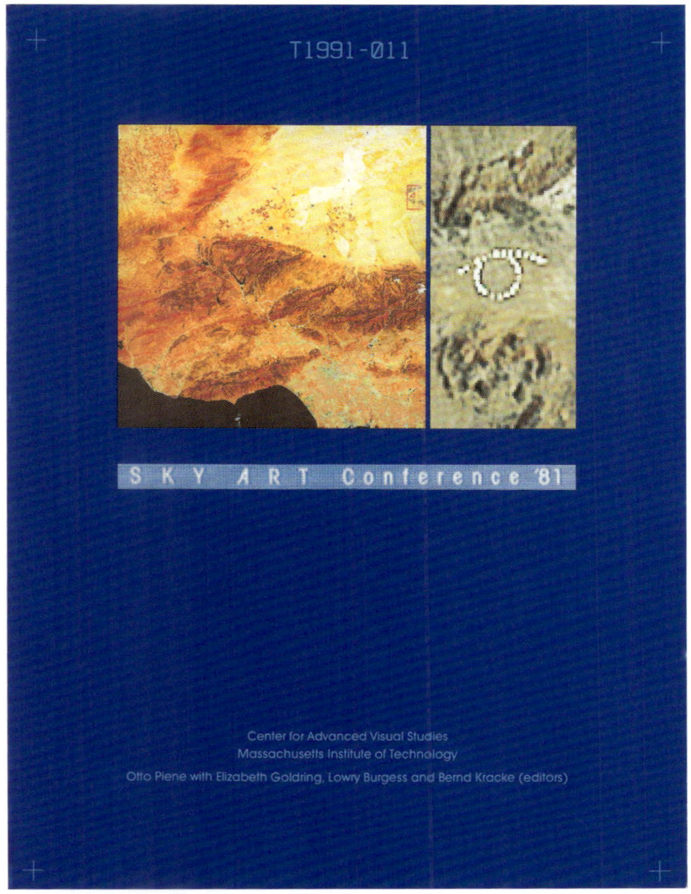

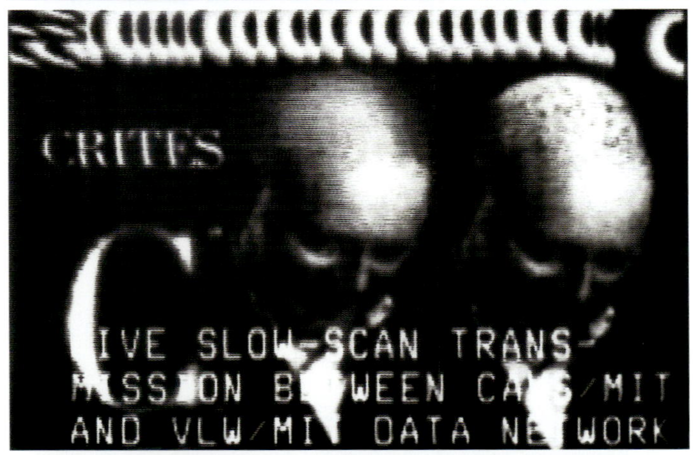

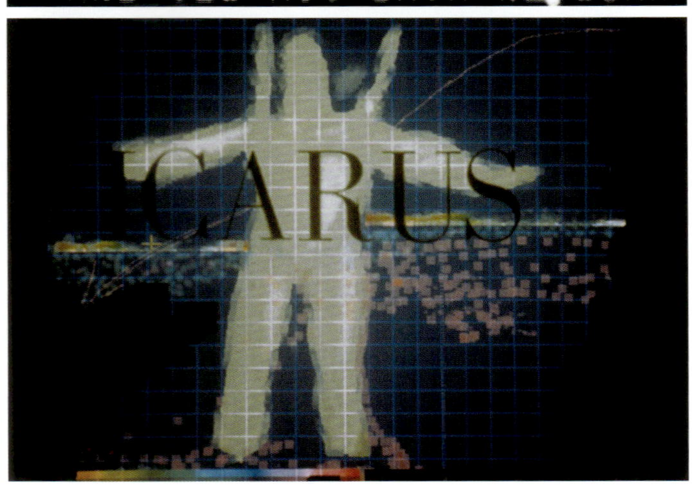

Top: *Sky Art Conference '81* catalog, cover photo taken from NASA Satellite Landsat II, June 11, 1980 for Tom Van Sant's "Reflections From Earth," CAVS, MIT, 1981
Source: CAVS Special Collection

Middle and bottom: Bernd Kracke, "Data Network," live Slow Scan Television electronic conference newspaper, *Sky Art Conference '81*, CAVS, MIT, 1981
Images: Bernd Kracke and Joel Slayton, VLW

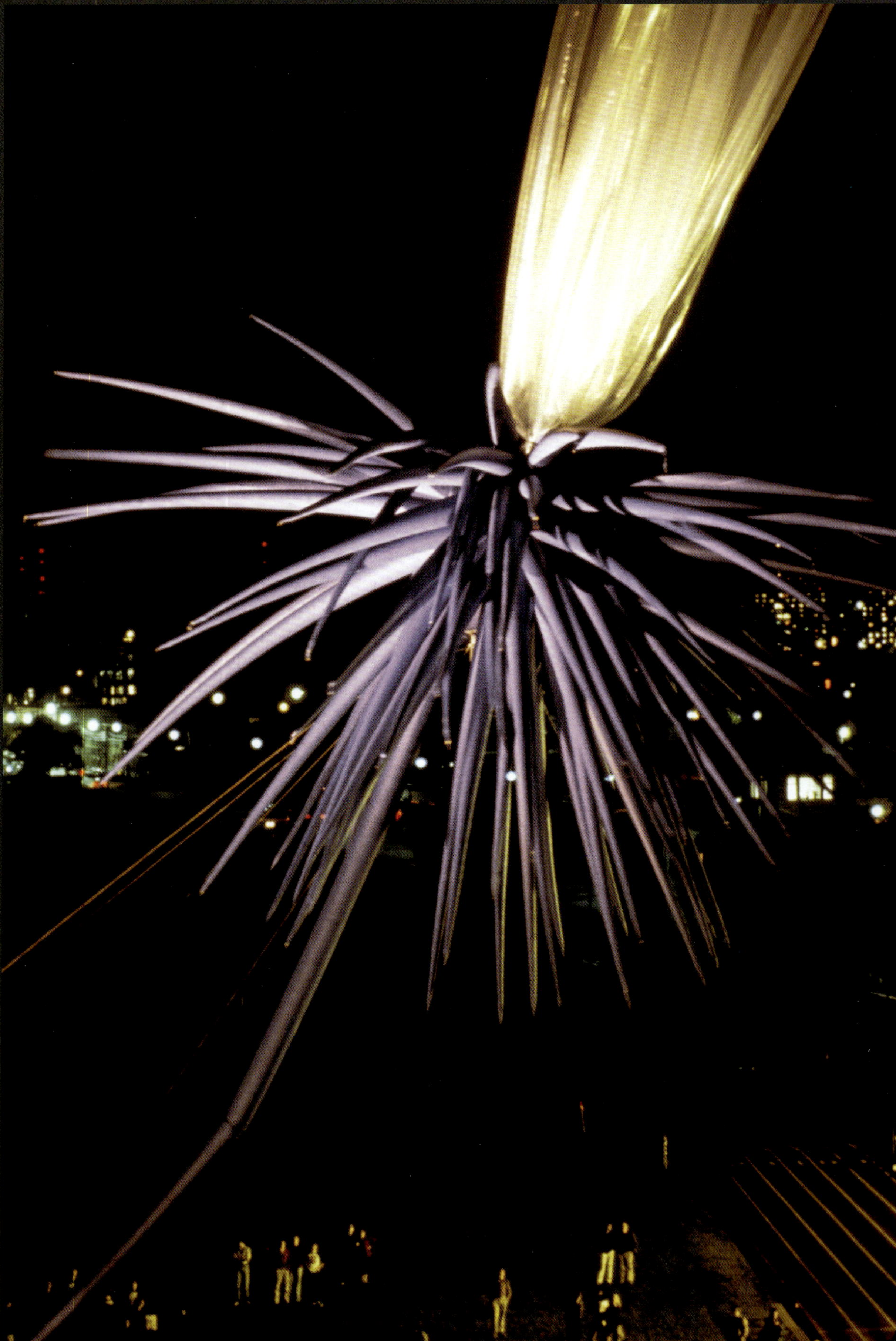

Other speakers included video/telecommunications artist and critic Douglas Davis, who admonished contentless communication; laser artist Rockne Krebs; balloonist/artist Vera Simons; and flier/artist Steve Poleskie. Otto Piene flew his "Blue Star Linz," an air-inflated, helium-lifted sculpture; Dale Eldred installed "Time Incident," creating a solar flash across the Charles River at noon; and Anders Holmquist displayed his "Wind Chapel." Aldo Tambellini staged a telecommunication event involving Australian artists. Conference kite-flying events were created by Van Sant, Jose Maria Yturralde, Howard Woody, and Tal Streeter. Mark Mendel wrote a towed sky poem.

> I don't want to read a poem over and over. One way to approach it is to do work that was temporary. The poem at the Sky Art festival was in the sky for forty-five minutes, and it was gone. (Mark Mendel interview)

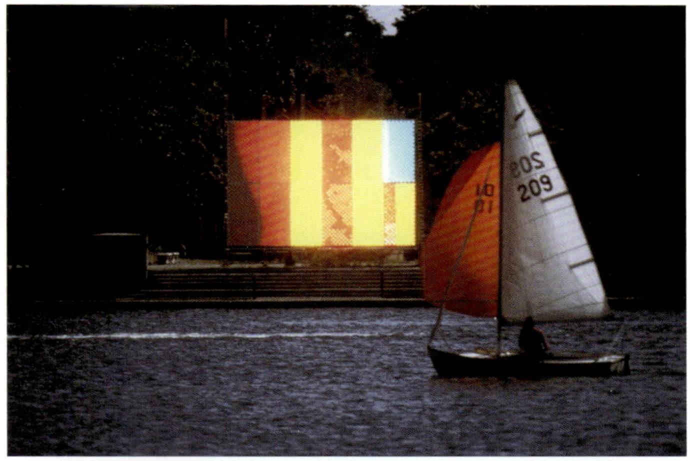

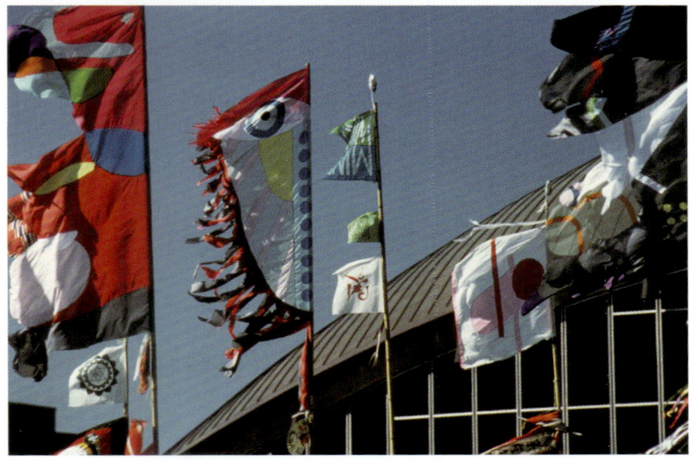

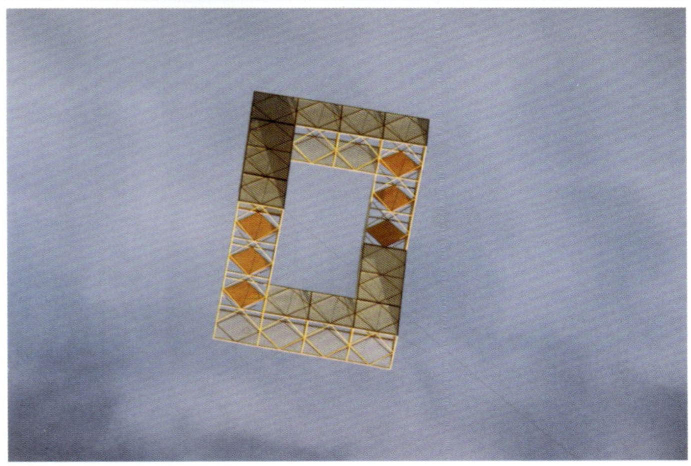

Top right: Dale Eldred, "Time Incident," sun reflective installation on the Charles River, *Sky Art Conference '81*, CAVS, MIT, 1981
Photo: Ted Spagna

Middle right: Anders Holmquist, "Wind Chapel," Kresge Oval, *Sky Art Conference '81*, CAVS, MIT, 1981
Photo: Calvin Campbell, MIT

Bottom right: Jose Maria Yturralde, "Flying Structure," box kite, part of the exhibition, *Sky Art Conference '81*, CAVS, MIT, 1981
Photo: Otto Piene Collection

Facing page: Otto Piene, "Blue Star Linz," inflatable flower (originally designed for Brucknerfest, Linz, 1980), sky event, *Sky Art Conference '81*, Briggs Athletic Field, CAVS, MIT, September 26, 1981
Photo: Calvin Campbell, MIT

Sky Art was a discovery, and I had a series of one-man shows and was also in the l975 Venice Biennial assisted by my friend, Jose de Prada, representing Spain, and with Baldeweg who acted as a conceptual artist. I flew flying structures as kites above the Spanish Pavilion during the day in good weather. The materials were of wood, aluminum, cloth—everything which is natural and fragile on earth and strong in the sky—they were very light structures. (Jose Maria Yturralde interview)

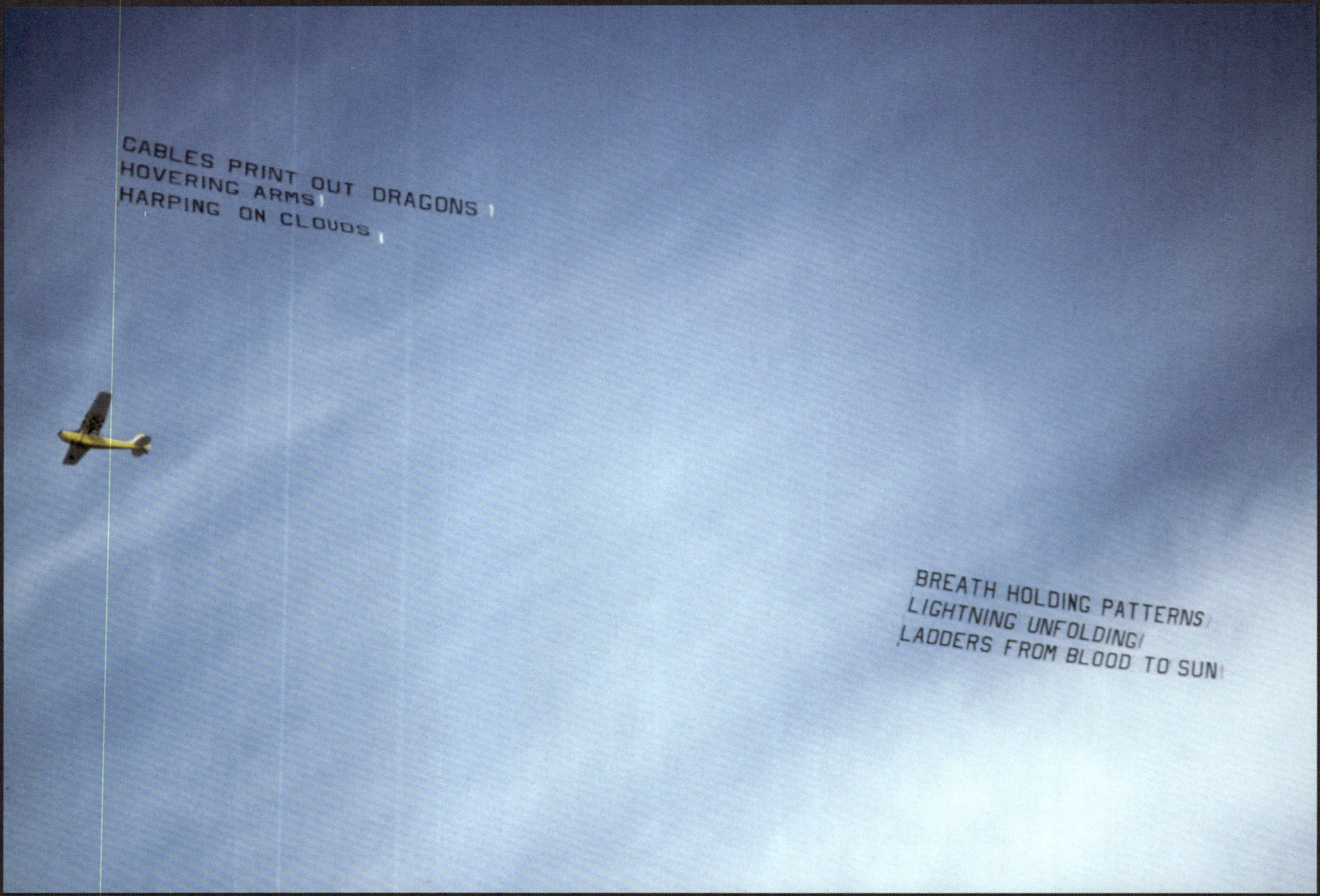

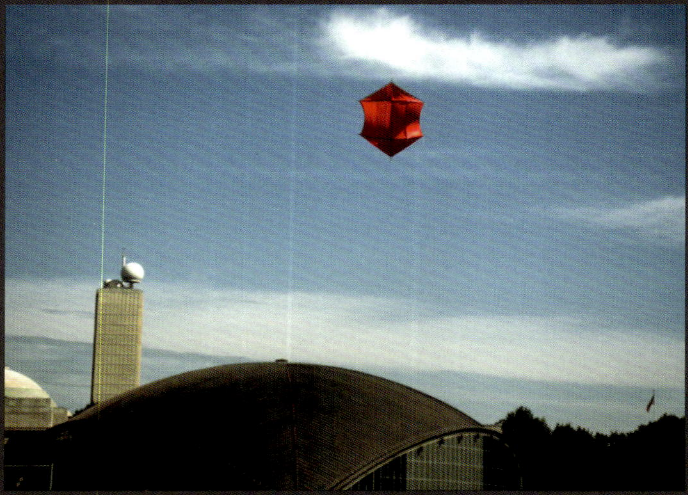

Top: Mark Mendel, "Aerolyric Angelic," poems towed by airplanes, *Sky Art Conference '81*, CAVS, MIT, 1981
Photo: Ted Spagna

Bottom: Tal Streeter, "Kite," flying over Kresge Auditorium, *Sky Art Conference '81*, CAVS, MIT, 1981
Photo: Tal Streeter

The introduction of Sky Art into CAVS was then a problem for some people. I am totally turned on by the sky, the moving out through the sky into the universe, thinking about those things, chin up, a gorgeous thing, looking up beyond the everyday, gravity, heaviness of life on earth. Things open up in the sky on such a grand scale. . . . You set your body and brain as a receptor to all this stuff . . . not immediately evaluating it, not learning what it is, not closing down, just taking it in. . . . All these things are in support of the kite. This is not dissimilar to what Otto is doing. . . . Had NASA been more receptive to artists, had NASA realized a long time ago that they should send a poet out into space, they would be in a much stronger position than they are now.
(Tal Streeter interview)

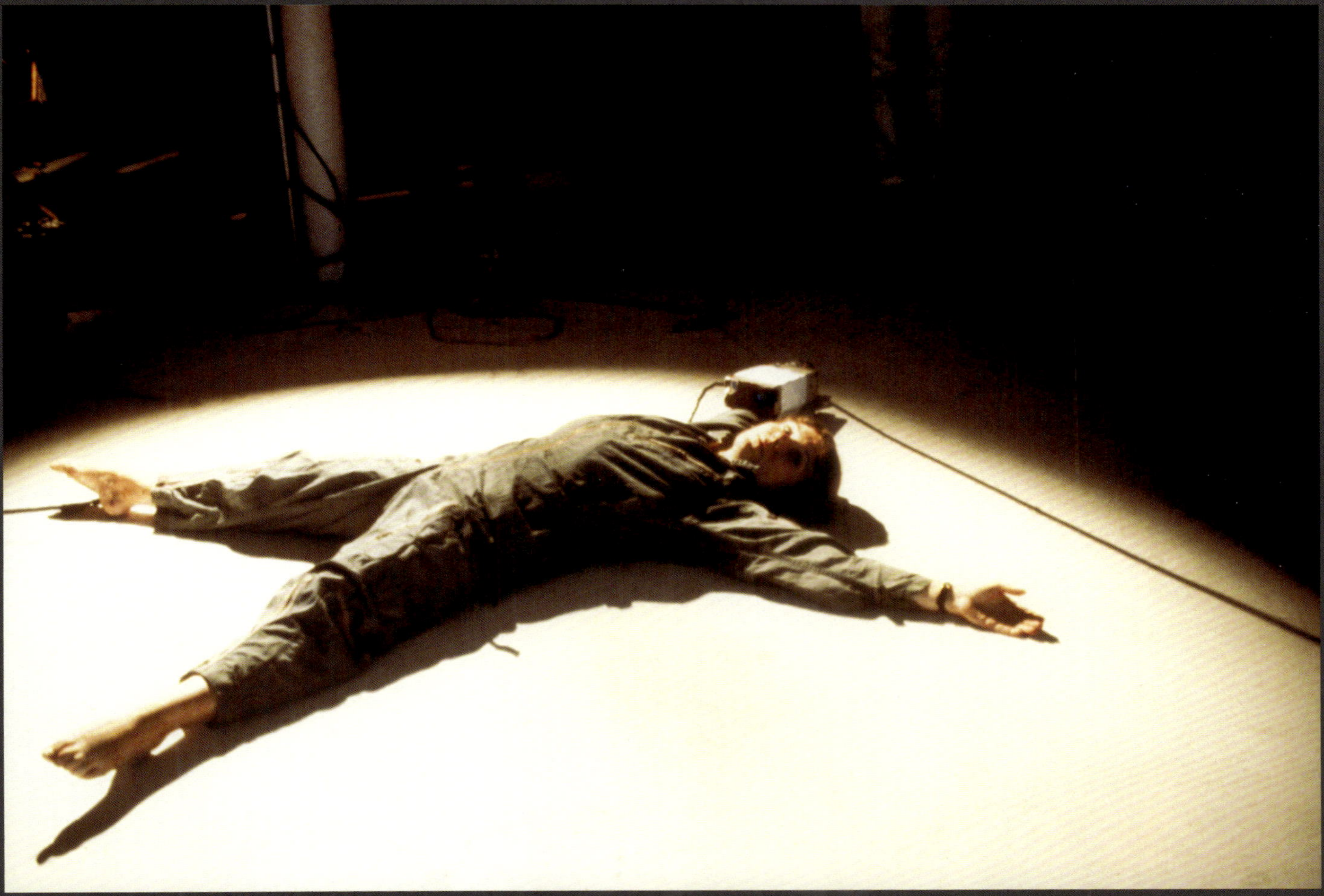

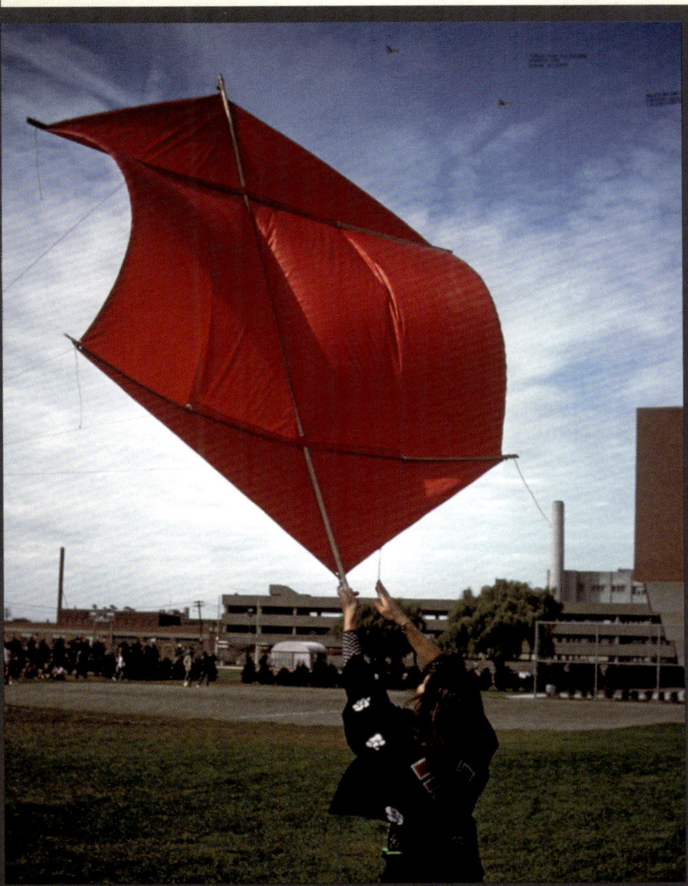

Above and right: Aldo Tambellini/Communicationsphere, "Tele-Sky," performer Sarah Dickinson, SSTV event, CAVS and Metro TV, Sydney, at *Sky Art Conference '81*, CAVS, MIT, September 25-26, 1981. Photos: Aldo Tambellini

Left: Tal Streeter, "Kite," launched beneath Mendel's towed poems, *Sky Art Conference '81*, CAVS, MIT, 1981 Photo: Tal Streeter

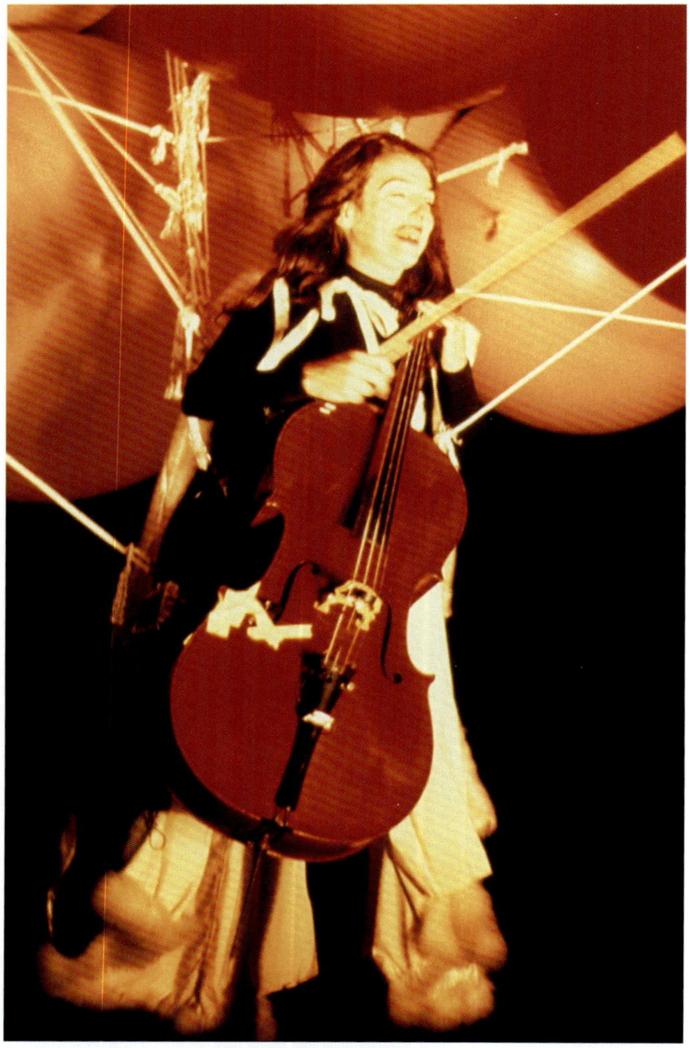

Human flight was addressed by artists and scientists at *Sky Art Conference '81*. Paul MacCready and Eugene Larrabee presented their experiments in human-powered aircraft; Paul Earls flew singer Robert Honeysucker in his staging of the scene, "Flight of Daedalus," from the collaborative sky opera, *Icarus*; and Charlotte Moorman performed Jim McWilliams' version of "Sky Kiss."

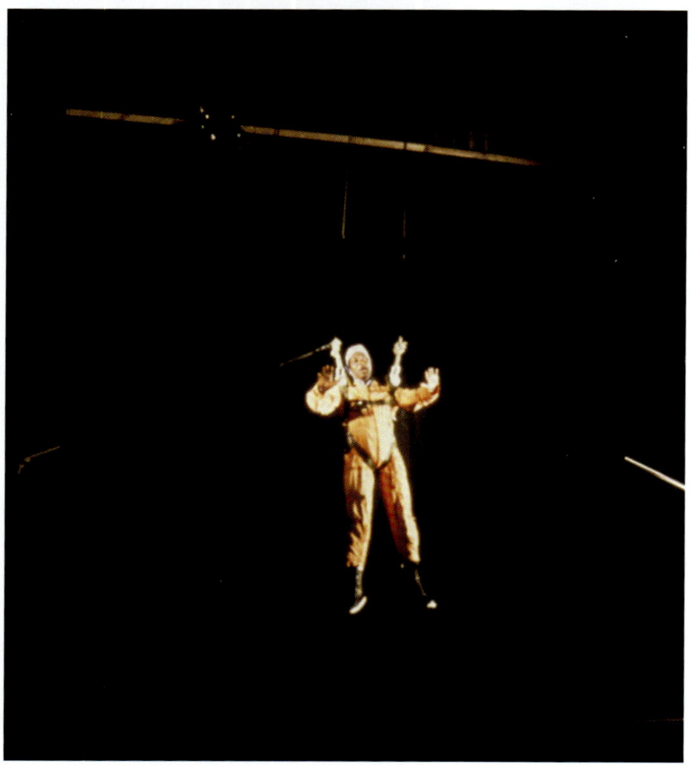

Above: Charlotte Moorman performs Jim McWilliams' version of "Sky Kiss," Kresge Oval, *Sky Art Conference '81*, CAVS, MIT, September 26 and 27, 1981
Photo: Paul Foley

Left: Paul Earls, "Flight of Daedalus," from *Icarus* sky opera, with baritone Robert Honeysucker (pictured), and the MIT Concert Band, Killian Court, MIT, September 25, 1981
Photo: Ted Spagna

Facing page top: Paul MacCready, *Gossamer Penguin*, solar-powered flight piloted by Janice Brown, 1980
Photo: Don Monroe

Facing page bottom: Eugene Larrabee, *Chrysalis*, pedal-driven biplane in flight, 1979
Photo: Steve Finberg

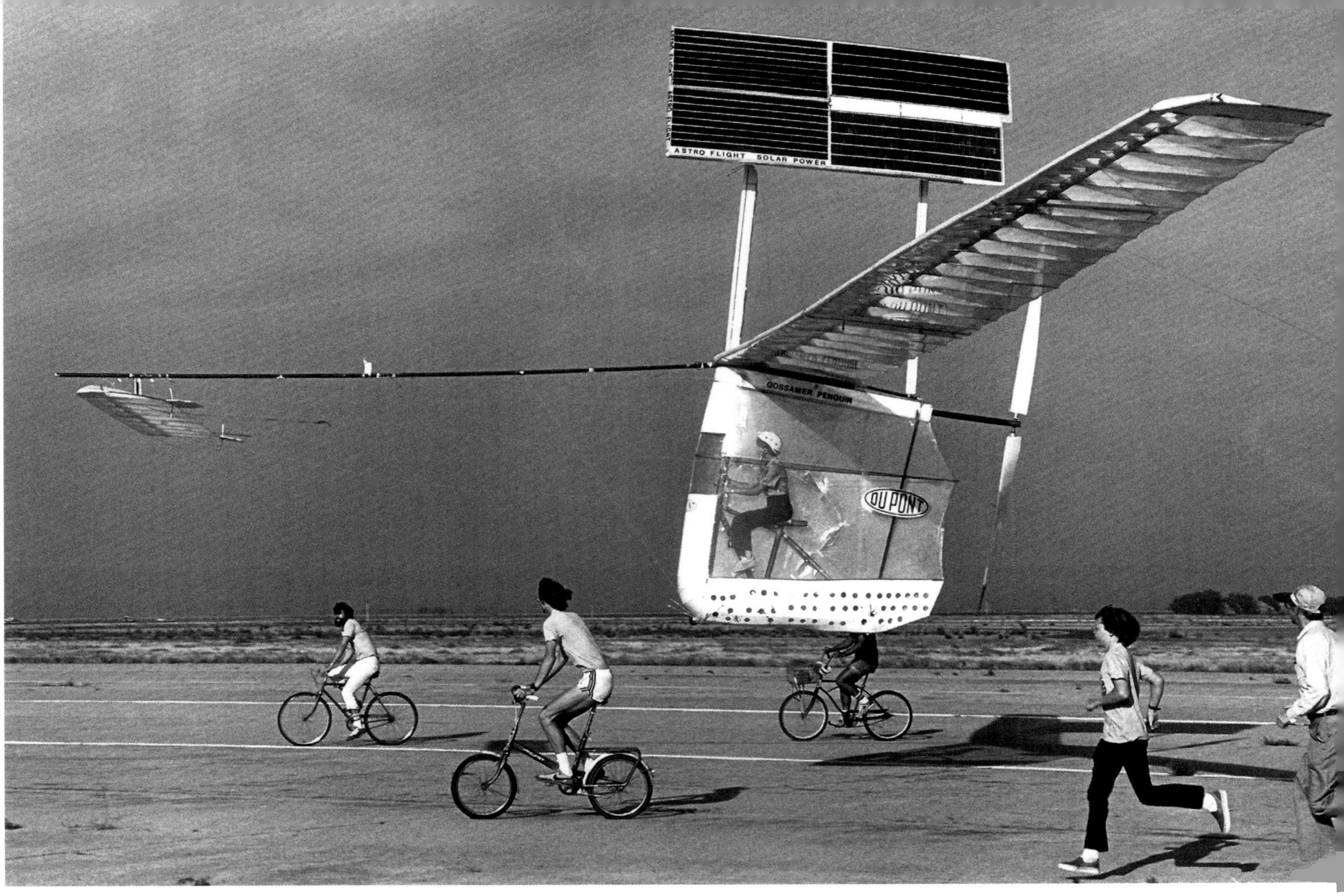
ASTRO FLIGHT SOLAR POWER
GOSSAMER PENGUIN
DUPONT

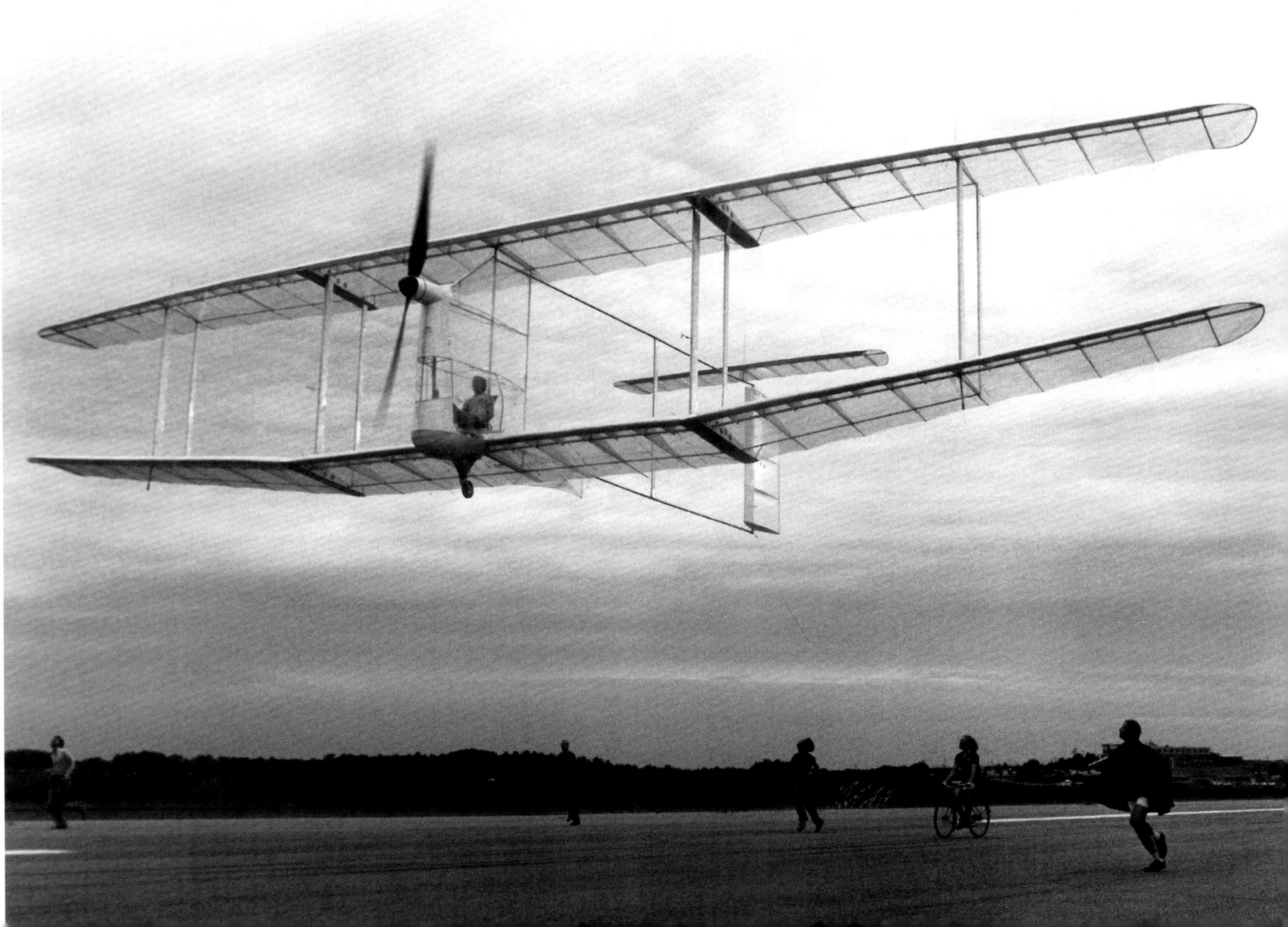

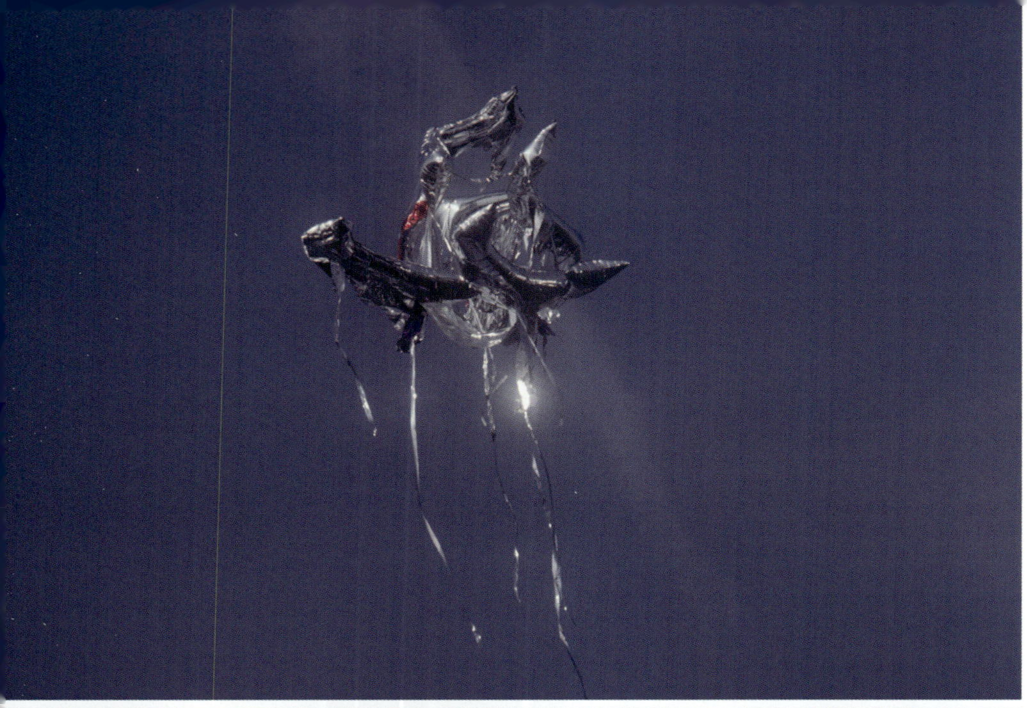

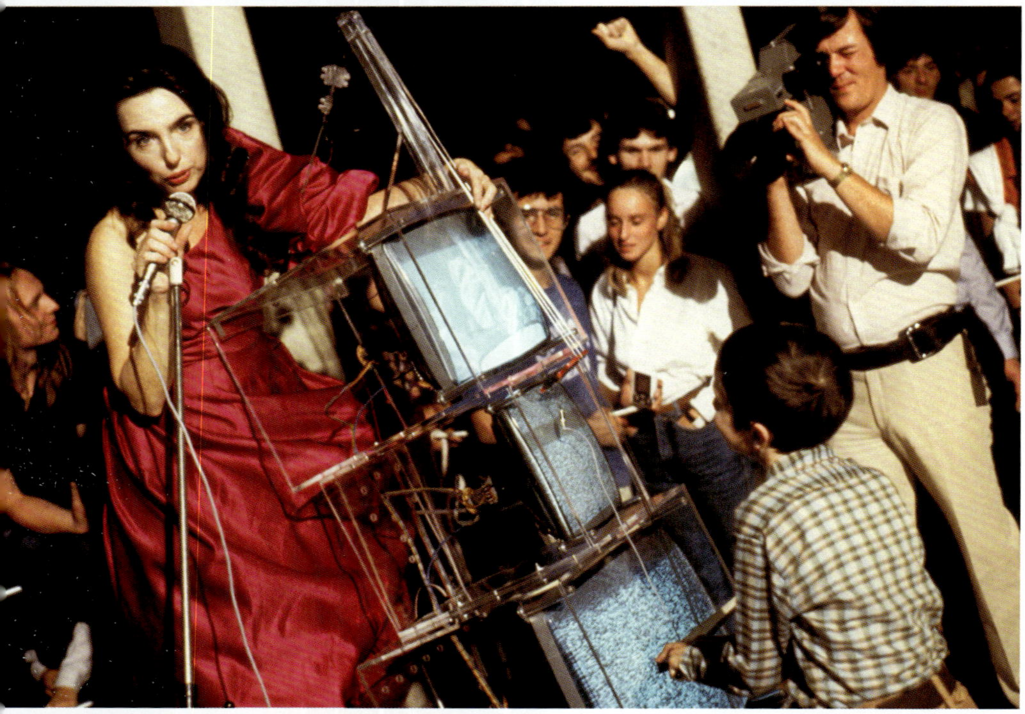

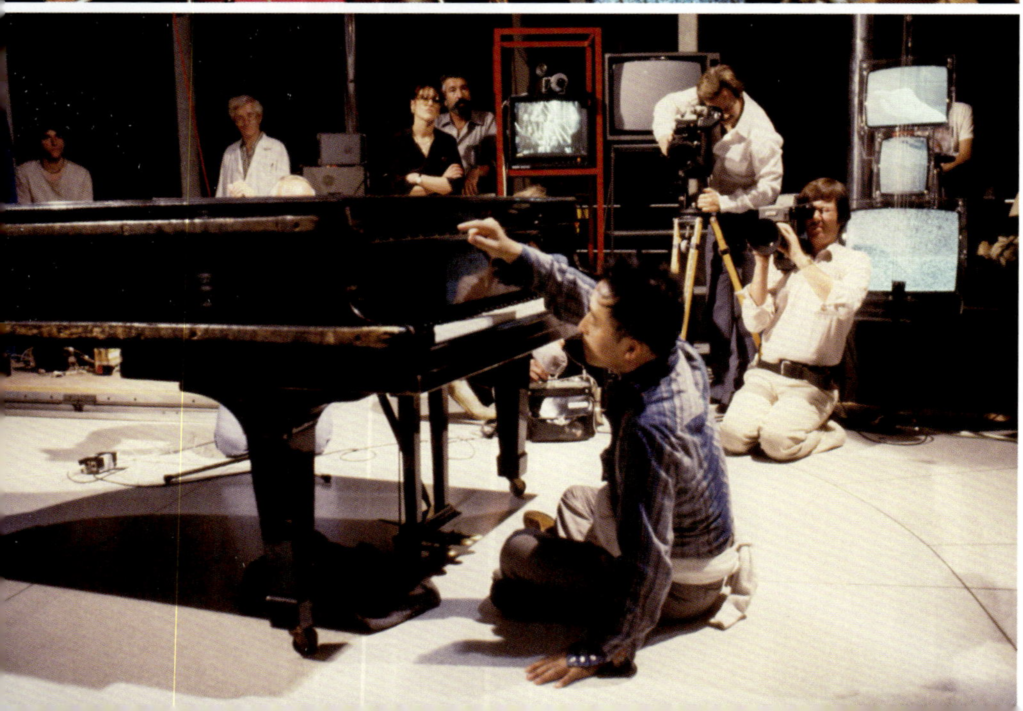

Top: Howard Woody, "Linz Sky Buoy," *Sky Art Conference '82, Ars Electronica*, Linz, 1982
Photo: Howard Woody

Middle and bottom: Charlotte Moorman and Nam June Paik performances, Moorman with Paik's TV Cello, ORF (Österreichischer Rundfunk), *Sky Art Conference '82, Ars Electronica*, Linz, 1982
Photo middle: Vin Grabill
Photo bottom: Otto Piene Collection

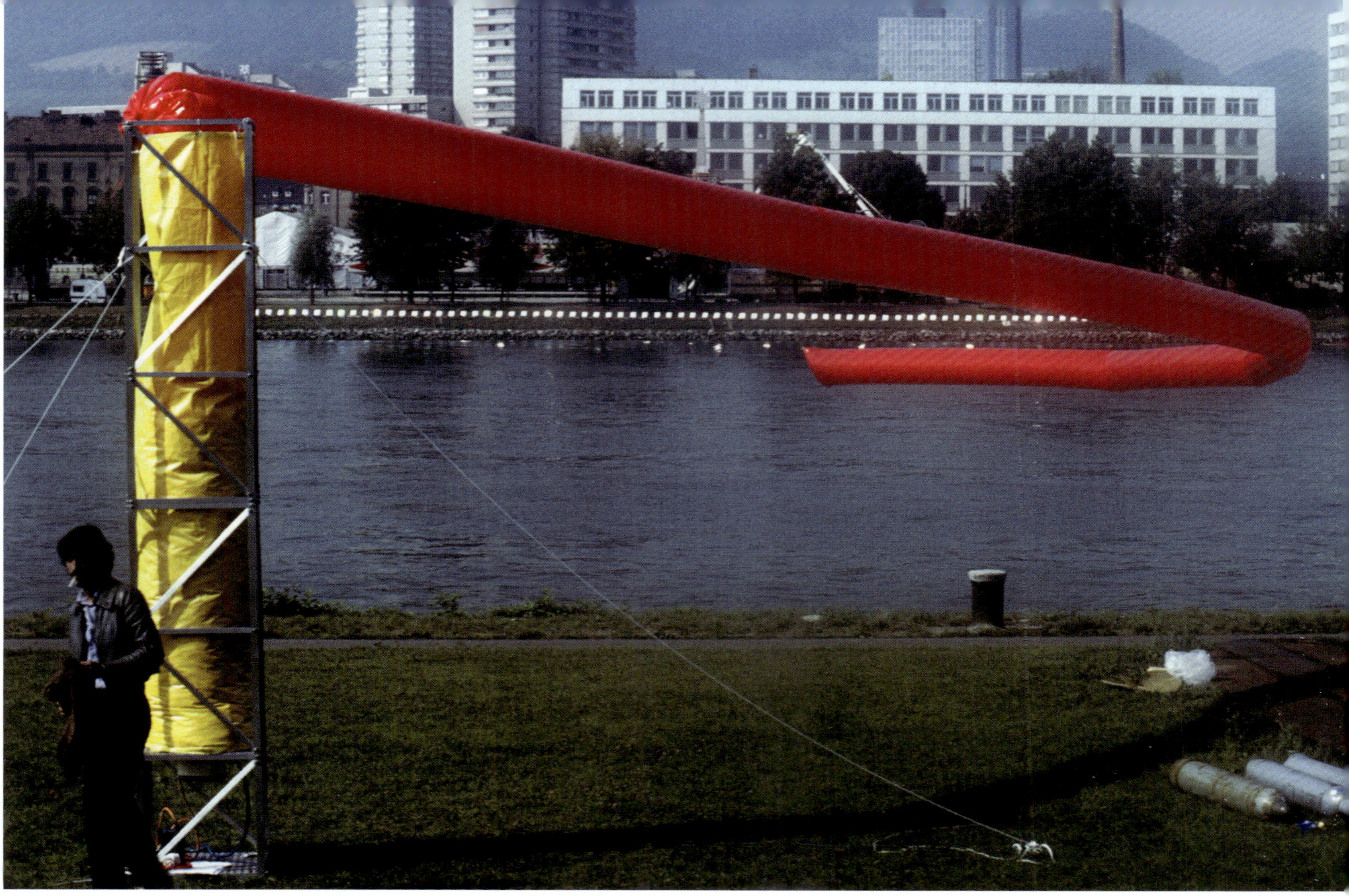

Sky Art Conference '82, Ars Electronica, Linz, September 24-October 1, 1982

The 1982 *Sky Art Conference* in Linz, Austria, featured the German-language premiere of the sky opera *Icarus* by Paul Earls, Otto Piene, and Ian Strasfogel. Outdoor performances included Moorman and Piene's "Sky Kiss"; a mylar angel sculpture flown and tracked by Howard Woody called "Linz Sky Buoy"; and Vera Simons's balloon flight. Yash Pal, Secretary-General of *UNISPACE '82* at the Second United Nations Conference on the Exploration and Peaceful Uses of Outer Space (Vienna, 1982), served as honorary chairman. "International Alarm," was an environmental audio work by Elizabeth Goldring with Edward Le Poulin based on rooster crows.

> Edward Le Poulin and I did a piece called "International Alarm," which played on Walter Haupt's *Linz Sound Cloud*—eight banks of speakers on either side of the Danube and on a barge in the river, which was used in turn by the Bruckner Hall to play Mahler and Bruckner symphonies . . . "International Alarm" was live and processed rooster crows that I'd collected from various parts of the world. At that point, my eyesight was very bad and often as I was recording my poems, rooster crows would get in the way . . . so then I would just listen to the rooster crows and that would become my poem for the day. (Elizabeth Goldring interview)

Above: Tal Streeter, "Linz Line," (foreground), a red inflatable sculpture that hovers over the Danube River; Dale Eldred, "Line of Fire," (background), a bank of diffraction units reflecting, in prismatic colors, the sun's image into the Danube River, *Sky Art Conference '82, Ars Electronica*, Linz, 1982
Photo: Dietmar Löhrl

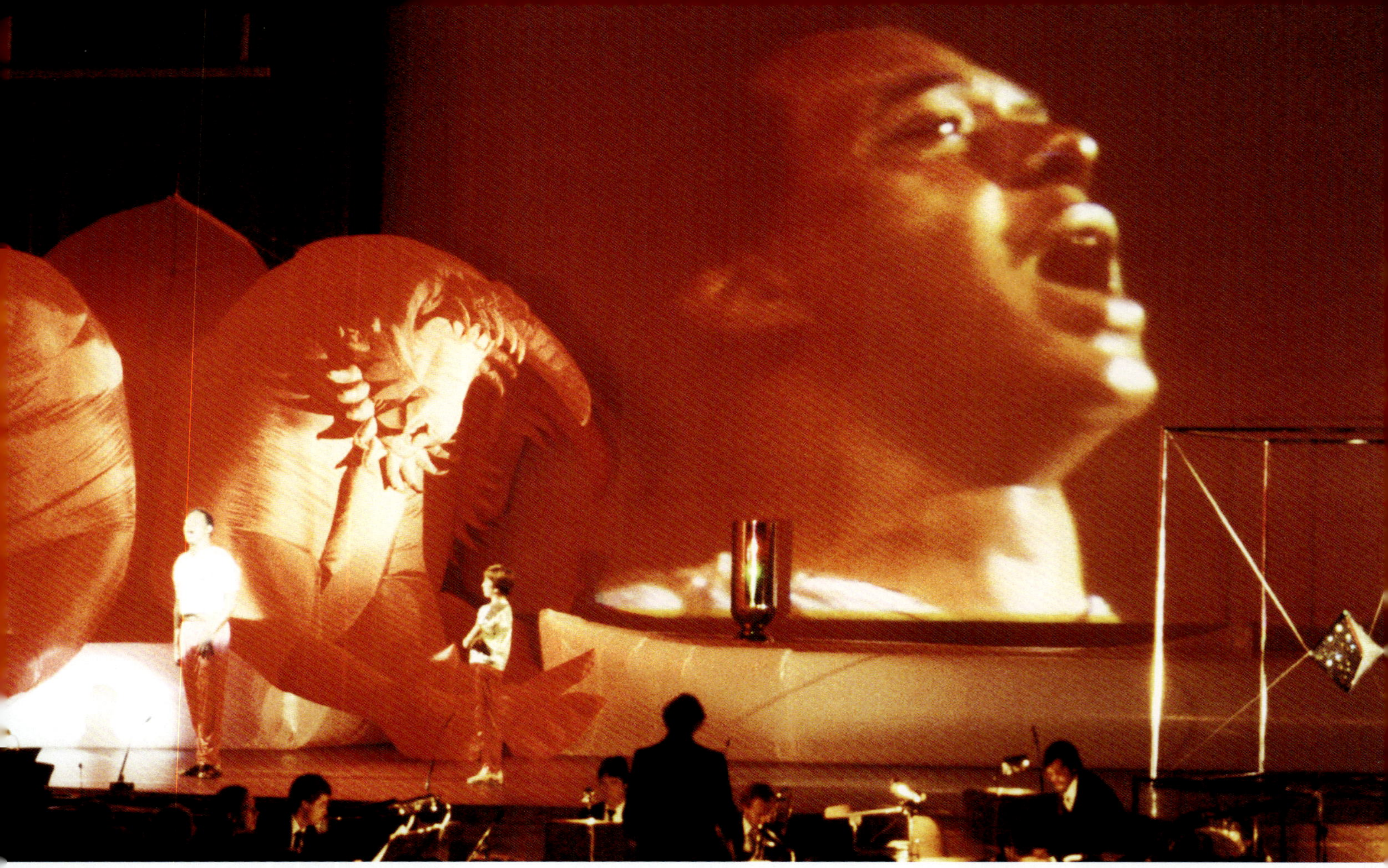

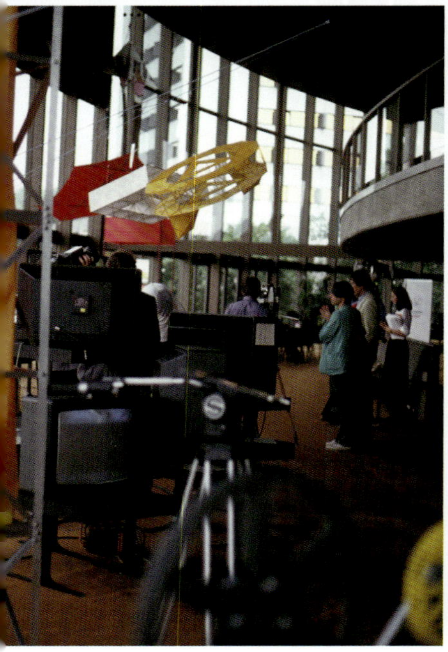

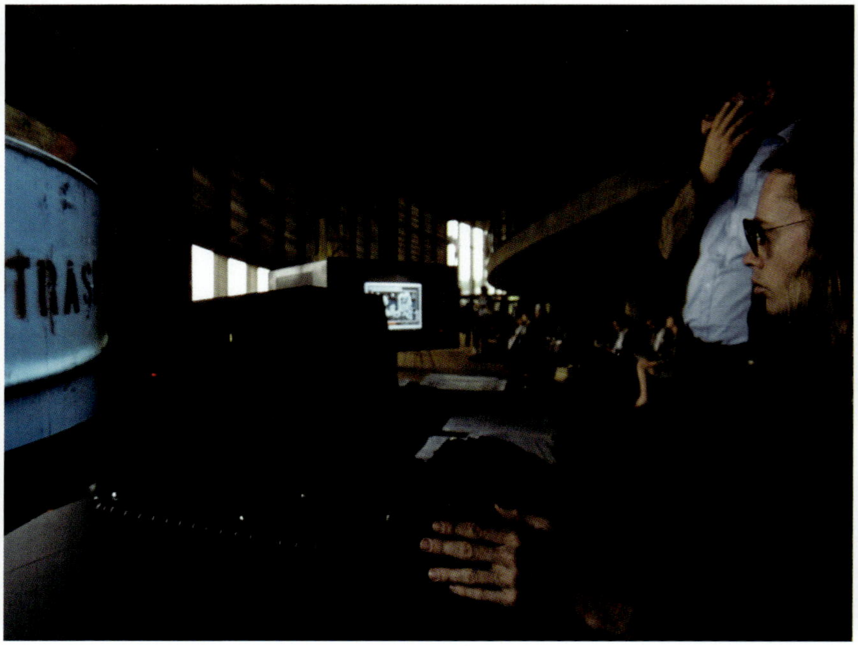

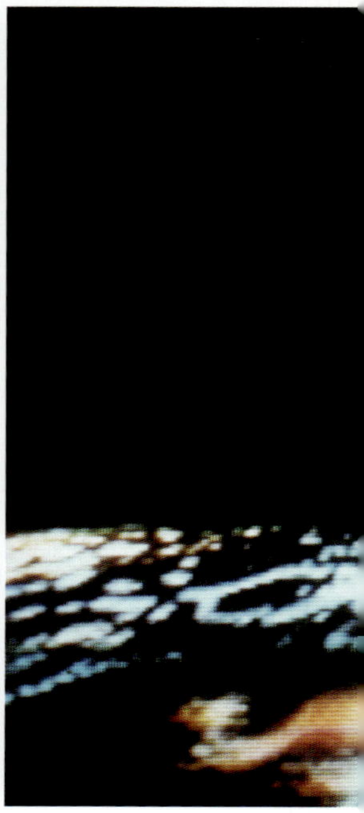

Top: Paul Earls, Otto Piene, Ian Strasfogel, *Icarus—A Sky Opera*, baritone Robert Honeysucker on screen; Boston Musica Viva performs, Richard Pittman, conductor, *Sky Art Conference '82*, *Ars Electronica*, Brucknerhaus, Linz, 1982
Photo: Elizabeth Goldring

Bottom row, left to right across both pages:
CAVS exhibition, with Tal Streeter's "Linz Line"and Jose Maria Yturralde's box kites; Rus Gant, "Vision Machine" laserdisc installation; Ron Hays, video still, sun projections for *Icarus* that visualized Wagner's *Prelude*

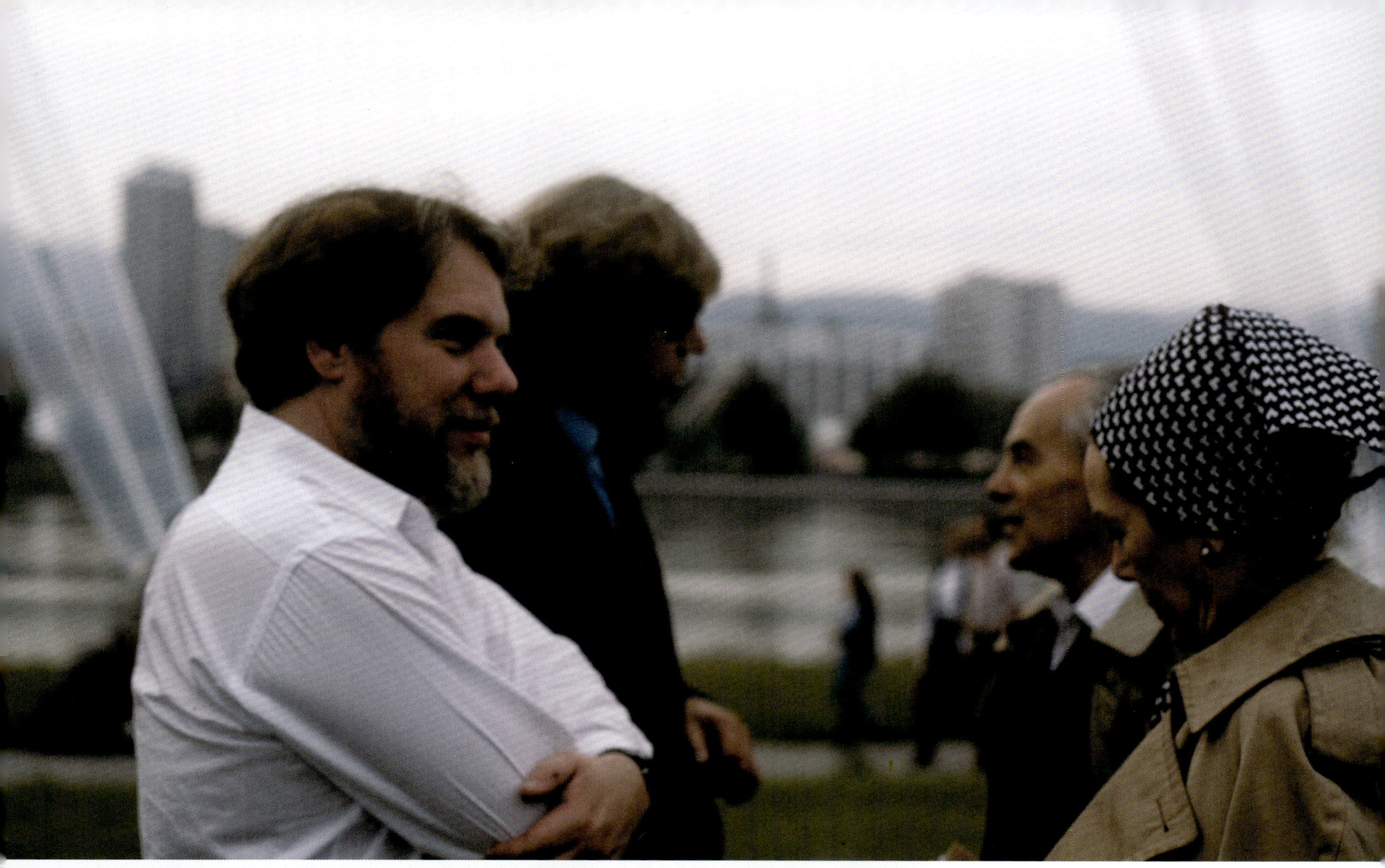

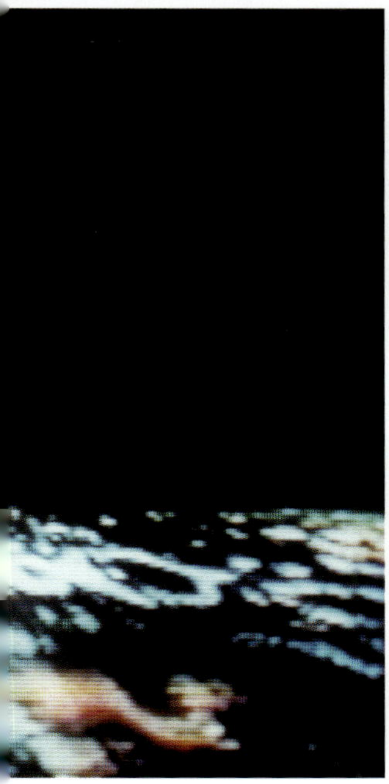

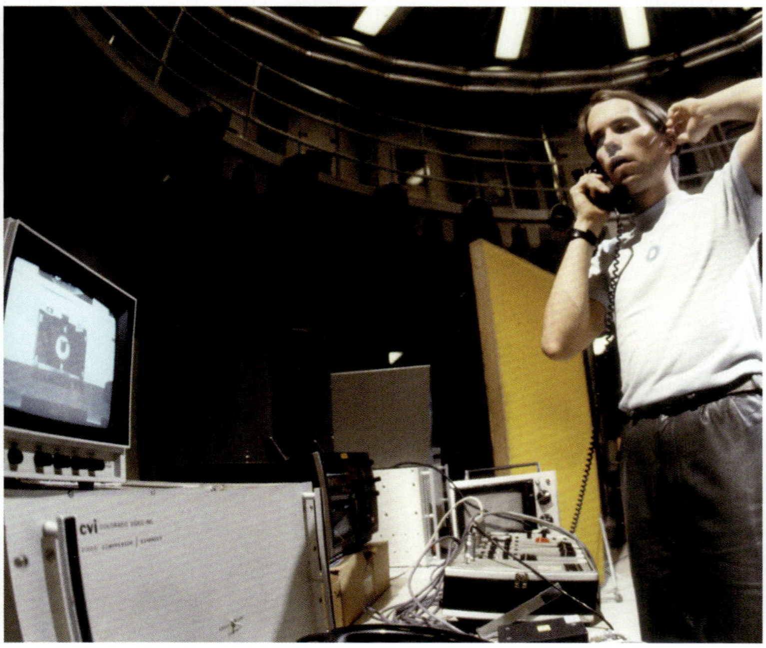

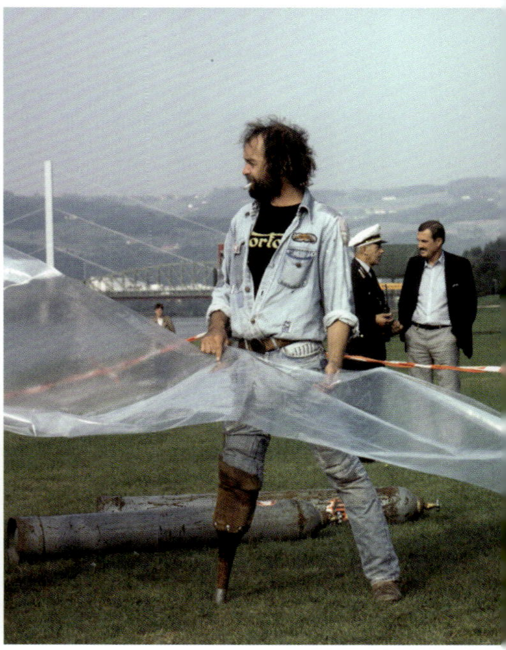

and Love Death using the Paik/Abe Synthesizer (1975); Bernd Kracke, at his "Data Network" telecommunications event, ORF studio; Joe Davis during preparations for sky events, *Sky Art Conference '82*, Brucknerhaus, Linz, 1982
Photos: Otto Piene Collection

Top: Senior Consultant Lowry Burgess with conference Honorary Chairman György Kepes and Juliet Kepes during preparation for sky events, *Sky Art Conference '82*, Brucknerhaus, Linz, 1982
Photo: Elizabeth Goldring

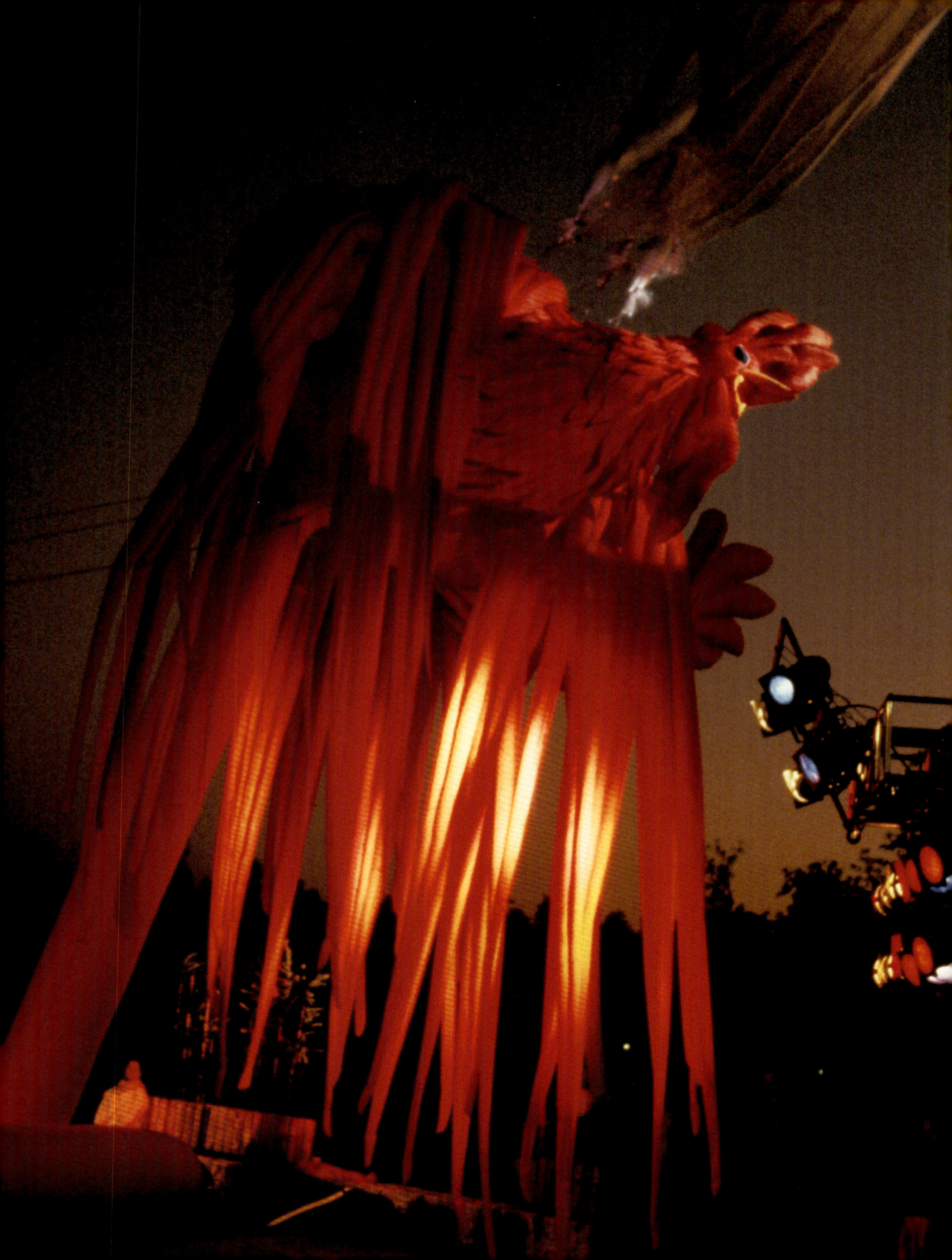

Sky Art Conference '83, BMW Headquarters, Munich, September 24-27, 1983

> The next *Sky Art Conference* for 1983 was being planned for Munich, Germany. I had the opportunity to attend and help put it together. Everyone was busy working with technological instruments, getting the bugs out. It was a high-octane time, a good time to come in....What made it poignant was through my exposure to Moholoy-Nagy and Kepes as a photographer and art history major I had an appreciation for that lineage. (Walter Dent interview)

Sky Art Conference '83, held September 24-27, 1983 at BMW Headquarters in Munich, Germany, brought the sky opera *Icarus* outdoors. Staged at the International Horticultural Exhibition Park (IGA), the opera used floating laser projections and Piene's flying and inflated sculptures as props. Goldring and Piene collaborated on a second version of "International Alarm" featuring Piene's inflatable "Hexagonal Rooster." Japan was represented by a number of participants including the founder of the Kite Museum of Tokyo, sculptor Tsutomu Hiroi, who exhibited miniature kites, and flew a large one made for the conference.

A magical event, "Firecracker Hat" by French pyrotechnics artist Pierre-Alain Hubert, interrupted the scheduled presentations. The exhibition included Christo's drawings for *Surrounded Islands*, Heinz Mack's reflective programmed light sculptures, and Charles Ross's *Solar Burns*. Important contributions of contemporary Sky Art pioneers Yves Klein and Lucio Fontana were also documented.

The Max-Planck-Gesellschaft was represented by a large group of scientists including Reimar Lüst. Lüst gave an opening address on his work with artificial auroras which preceded CAVS Fellow Joe Davis' project for a space shuttle payload. By 1983, attendance and participation from several parts of the world were increasing; examples and documentation of Sky Art projects were proliferating.

Facing page: Elizabeth Goldring and Otto Piene, with sound designer Edward Le Poulin, "International Alarm," Goldring's sound poem featuring rooster crows and "Hexagonal Rooster," one of a series of inflatable sculptures designed by Otto Piene and fabricated by Lees Ruoff, *Sky Art Conference '83*, IGA Park, Munich, September 24-27, 1983
Photo: Walter C. Dent

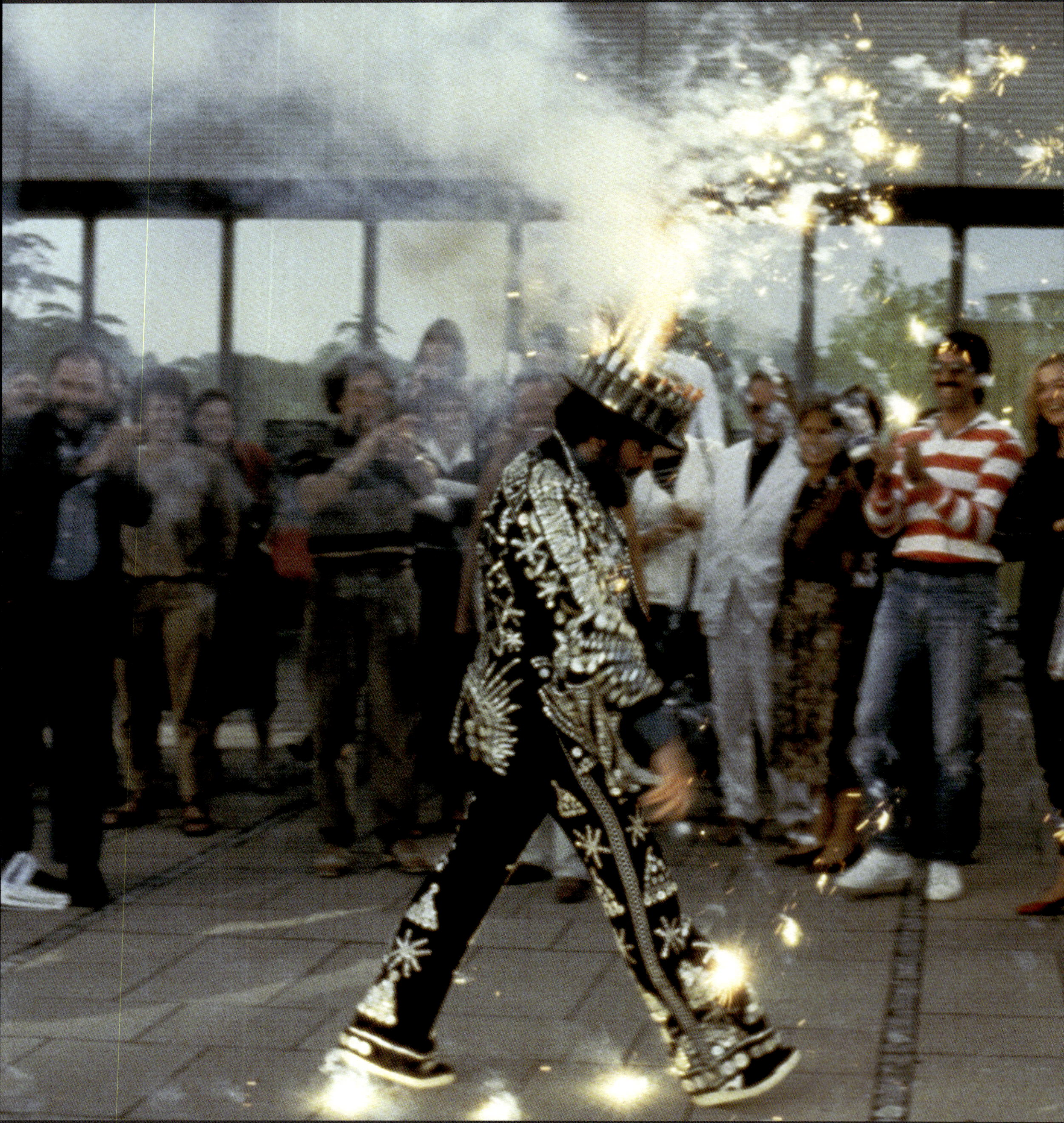

Above: Pierre-Alain Hubert, "Firecracker Hat," *Sky Art Conference '83*, BMW Headquarters, Munich, 1983
Photo: Elizabeth Goldring

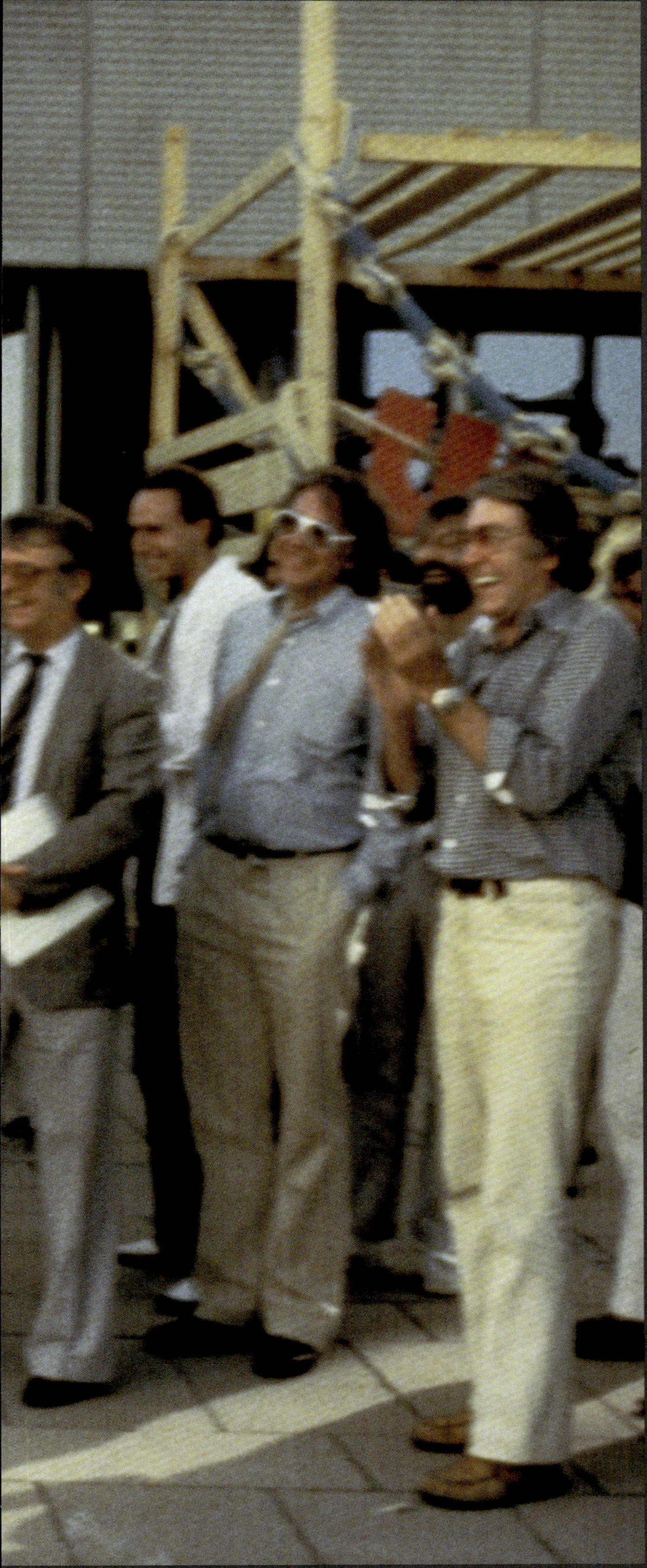

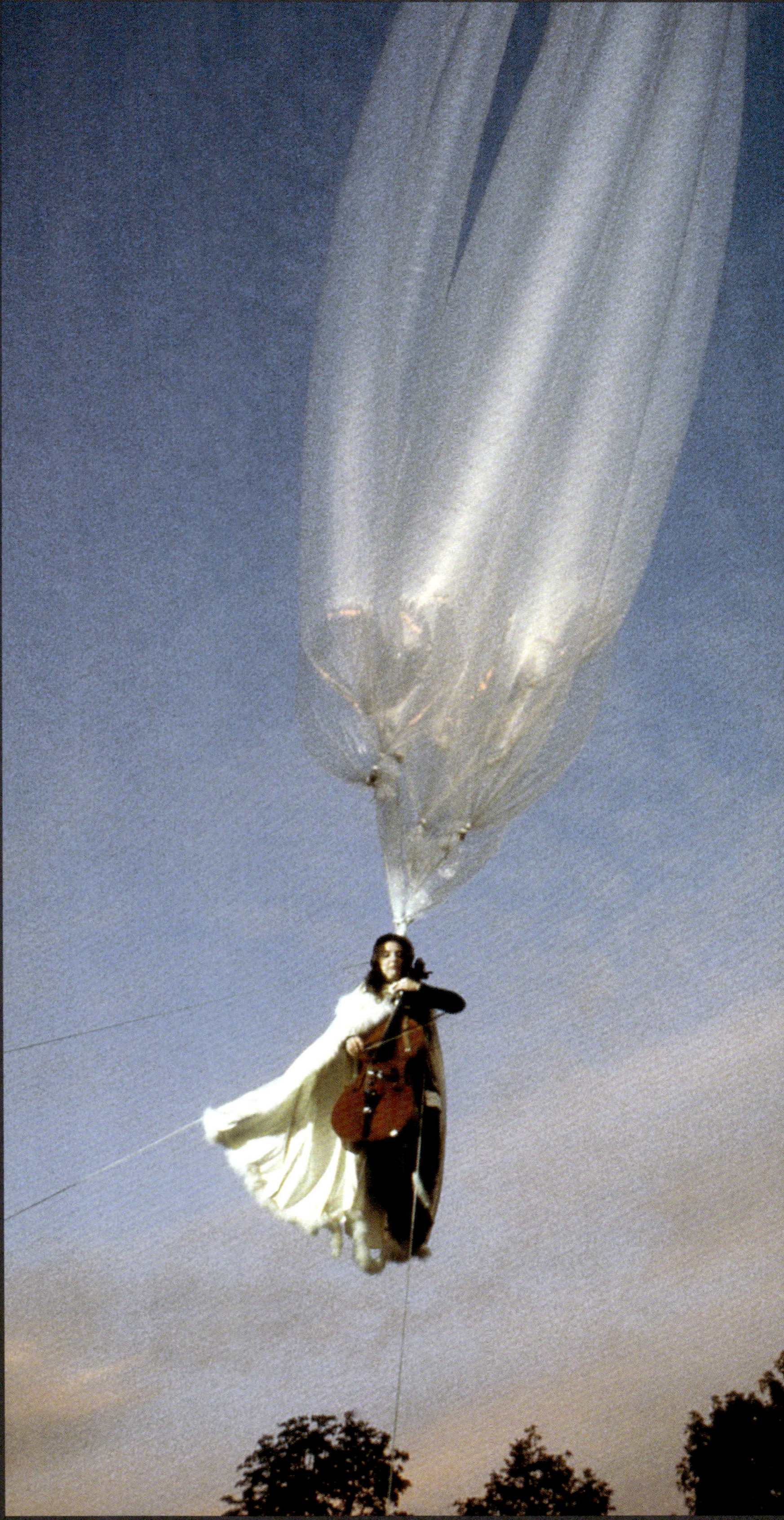

Right: Charlotte Moorman performs Otto Piene's version of "Sky Kiss," *Sky Art Conference '83*, IGA Park, Munich, 1983. Photo: Catherine Ikam

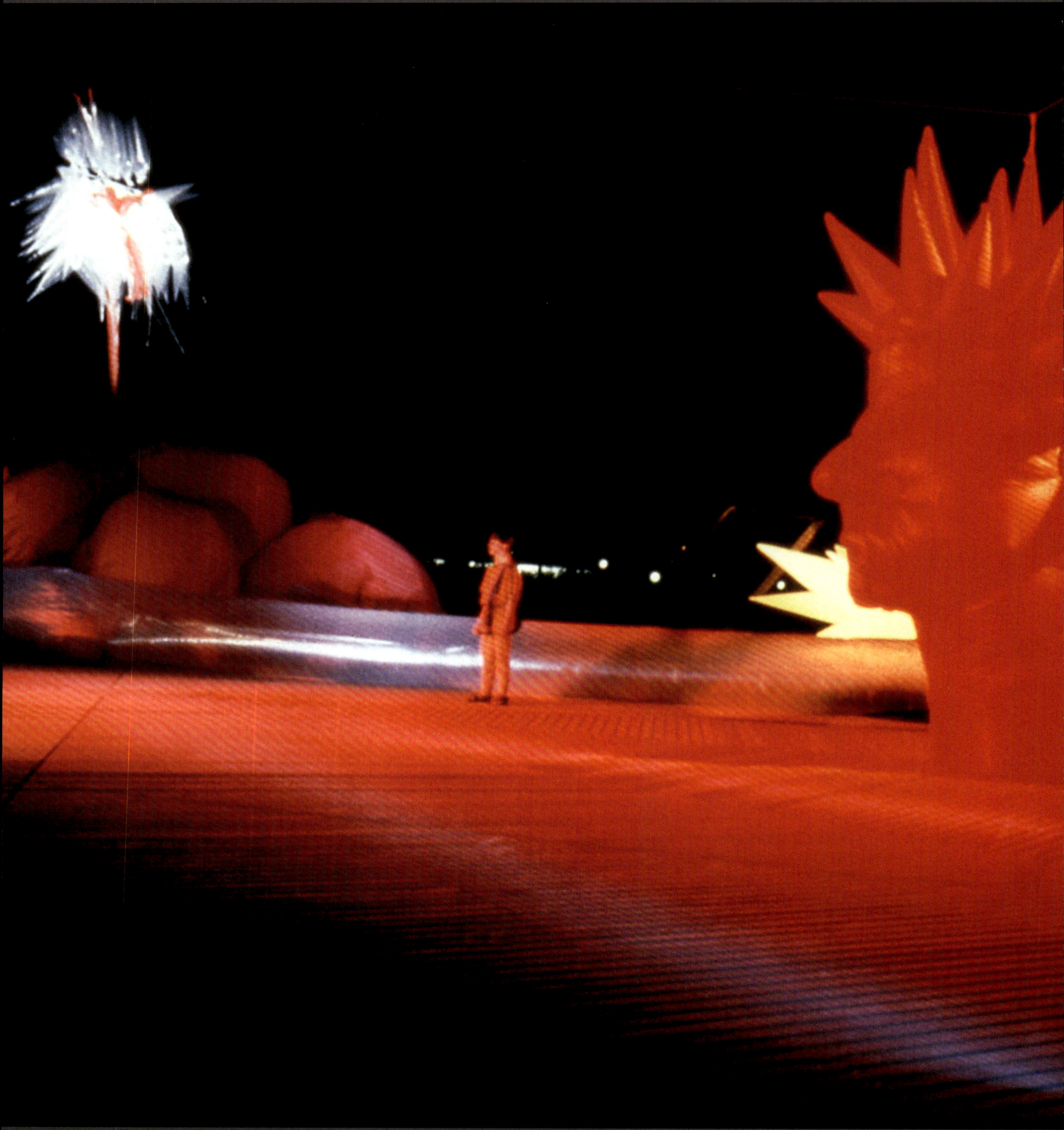

Above: *Icarus—A Sky Opera*, Paul Earls, Otto Piene, Ian Strasfogel, *Sky Art Conference '83*, IGA Park, Seebühne, Munich, 1983. Photo: Walter C. Dent

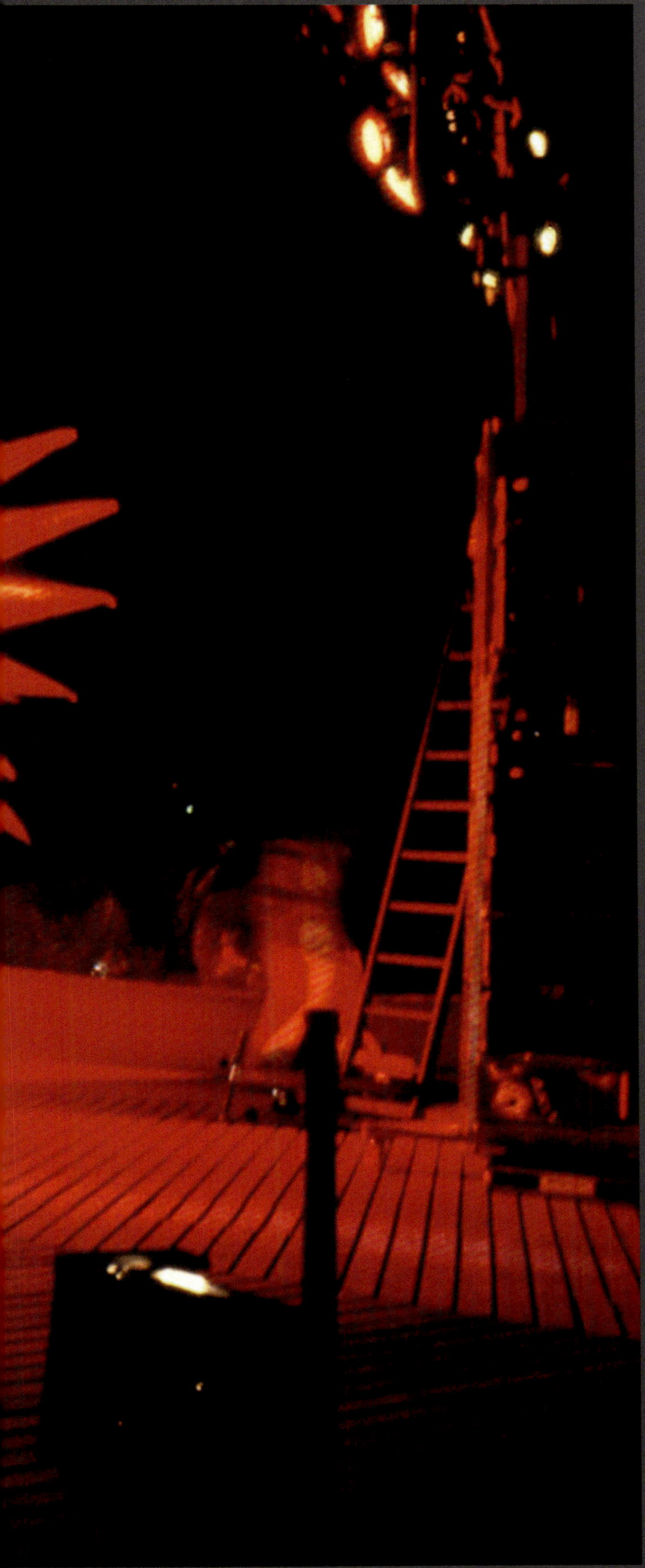

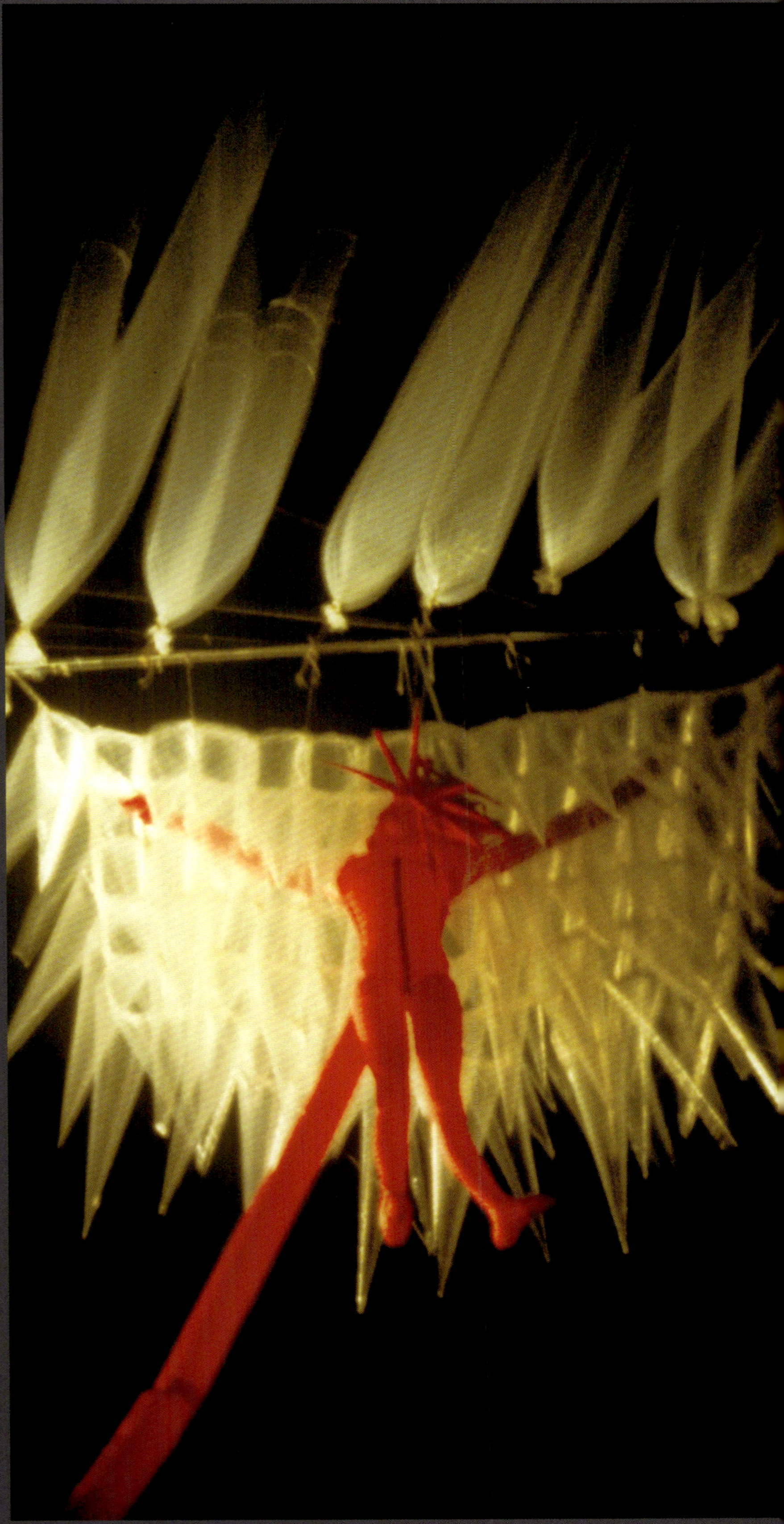

Right: Otto Piene, "Icarus Effigy," inflatable sculpture (1978), flown for *Icarus—A Sky Opera*, *Sky Art Conference '83*, IGA Park, Seebühne, Munich, 1983. Photo: Elizabeth Goldring

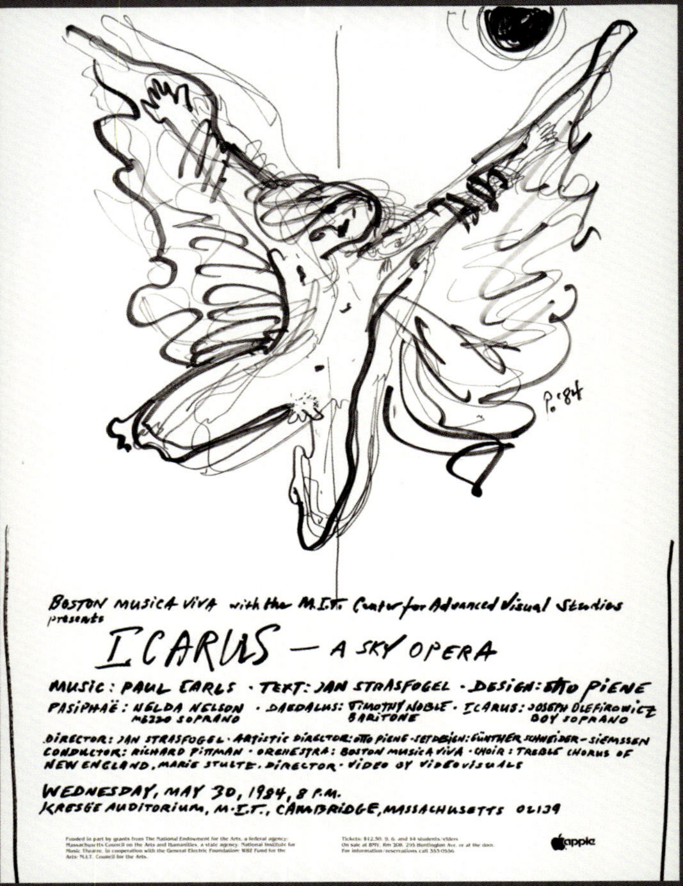

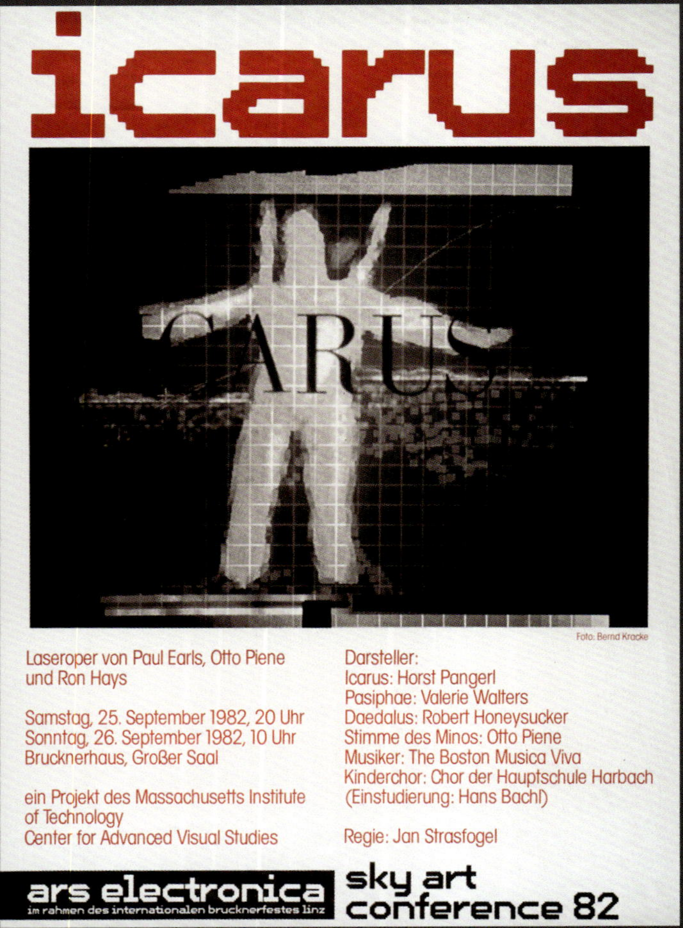

Icarus—A Sky Opera

Another impetus for sky art derives from the age-old desire to fly, as in the legend of Icarus, the tragic hero of Greek mythology. (Goldring, "Sky Art: Art that Flies," *Flying Colors*, 1979)

The Sky Opera, *Icarus*, evolved as a *tour de force* over several years, beginning with the first *Icarus* sketch on the National Mall in Washington, DC, in 1978.

> *Icarus* was first performed in September 1978 on the Mall in Washington, DC as the culminating outdoor event of *Centerbeam*—a performative sculpture and installation by the Center for Advanced Visual Studies, MIT. . . . Otto Piene and I developed the concept further, integrating performers with the visual media and air sculptures we used in other work. We collaborated on the labyrinth scene, combining lasers, music, and inflatable sculptures for performances at the *Biennale für Visuelle Kunst und Graphik* in Vienna in 1979 and in January 1982, at MIT's Hayden Gallery. In September 1981, scenes from *Icarus* were performed for the first *Sky Art Conference* at MIT. Daedalus "flew" through a combination of live performance and superimposed video images. For the *Sky Art Conference '82* in Linz, Ian Strasfogel created a new scenario and text; three new characters (Minos, the Minotaur, and Pasiphaë) were added to the cast. (Paul Earls, *Ars Electronica im Rahmen des Internationalen Brucknerfest*, 1982, 129)

Top: *Icarus—A Sky Opera*, poster drawn by Otto Piene, Kresge Auditorium, MIT, May 30, 1984
Source: CAVS Special Collection

Bottom: *Icarus, Laseroper von Paul Earls, Otto Piene und Ron Hays,* poster with image by Bernd Kracke, *Sky Art Conference '82, Ars Electronica*, Linz, 1982
Source: CAVS Special Collection

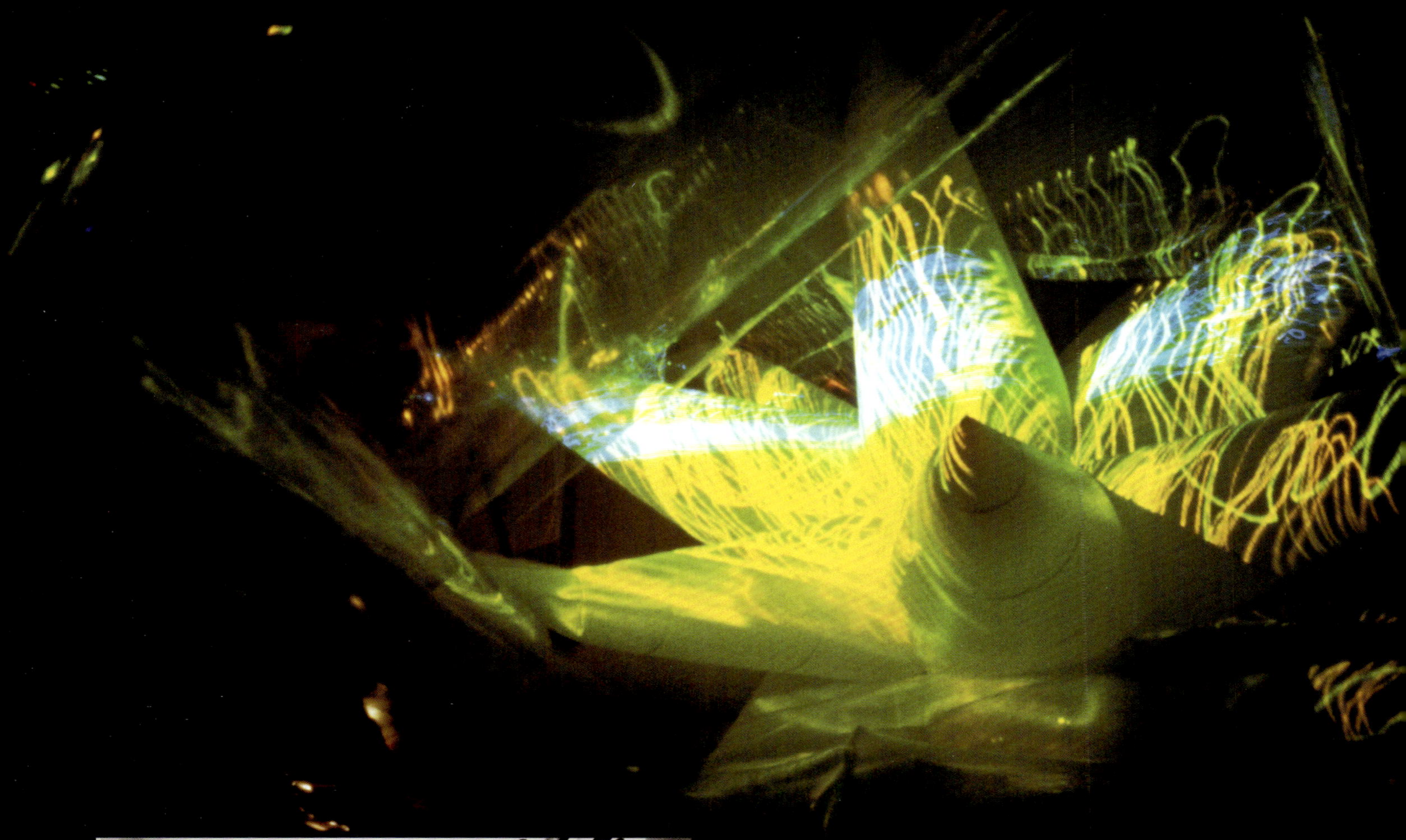

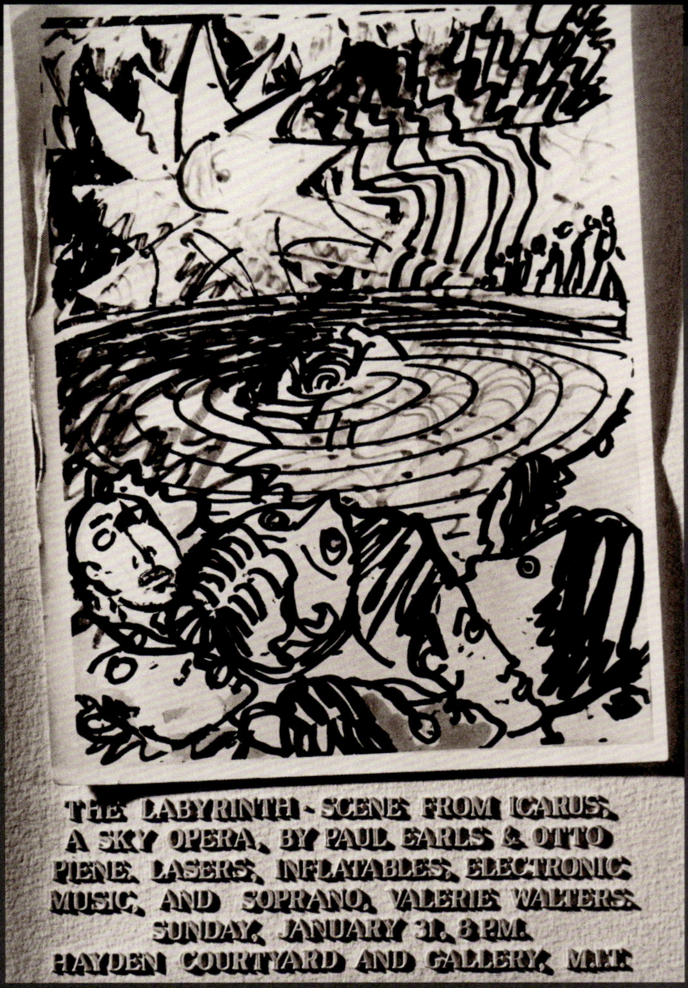

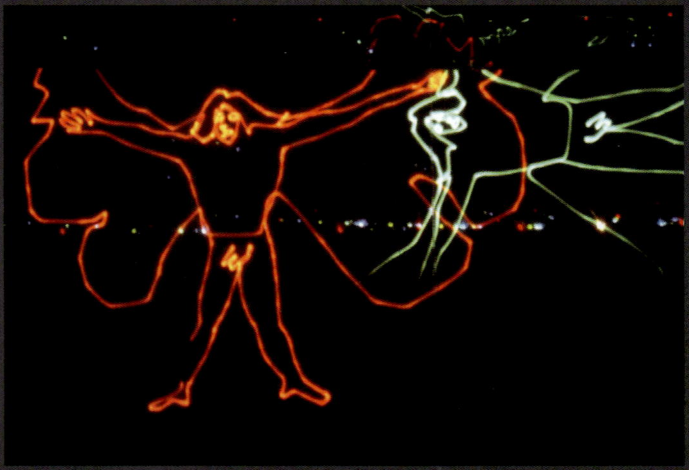

Top and left: Paul Earls and Otto Piene, *The Labyrinth, Scene from Icarus, a Sky Opera*, poster, and inflatables with laser projections, Hayden Courtyard and Gallery, MIT, January 31, 1982
Source: Otto Piene Collection

Above right: laser projections, *Icarus—A Sky Opera*, Paul Earls, Otto Piene, Ian Strasfogel, Kresge Auditorium, MIT, 1984
Photos: Ted Spagna

For Piene, Daedalus, "laden with the problem of problem-solving," represented MIT.

> *Icarus* reveals the correspondence between indoor and outdoor dimensions. What if a sky opera with flying Daedalus and flying Icarus in an outdoor production of environmental dimensions can become an outdoor production, with video substituting for expanses of real space, in limited space—and can become a "laser opera" in an indoor production, with laser projections suggesting sky and sun, in a traditional stage enclosure. Technology takes the broker's role when it comes to trading intensity for breadth and speed of change for cowering reality. Icarus, whom we tend to idolize, is the genius "Jüngling ohne Eigenschaften"—we only know his name; Daedalus, whom we tend to despise, is laden with the problem of problem-solving—a compulsive technocrat with paternal sentiment. They propel each other in an idealistic realm. Their human essence would remain uncommunicable without the mediating passion of Pasiphaë. (Otto Piene, *Ars Electronica im Rahmen des Internationalen Brucknerfest*, 1982, 129)

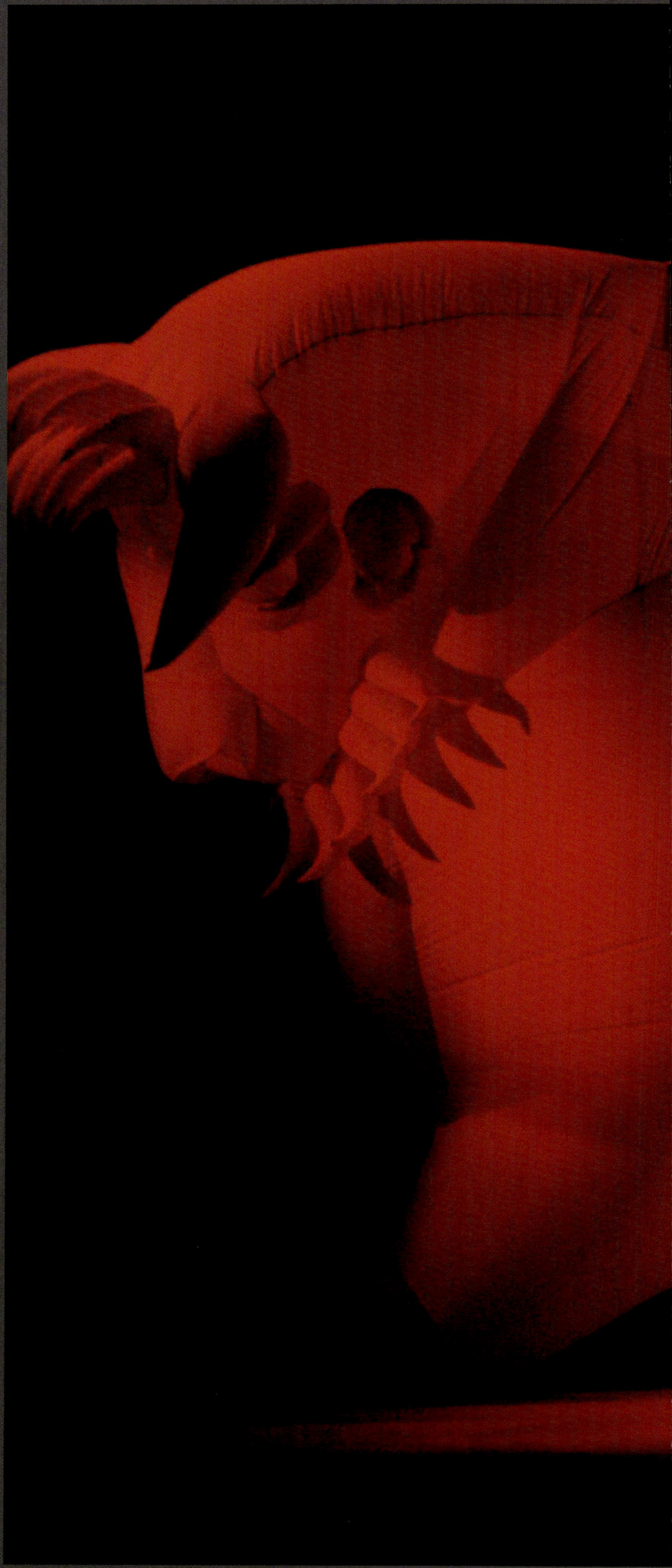

Icarus—A Sky Opera by Paul Earls (music), Otto Piene (design); and Ian Strasfogel (text); with set and costume design by Günther Schneider-Siemssen, Kresge Auditorium, MIT, May 30, 1984
Pictured: mezzo-soprano Nelda Nelson as Pasiphaë; the "Minotaur" inflatable sculpture by Otto Piene
Photo: John Kruse

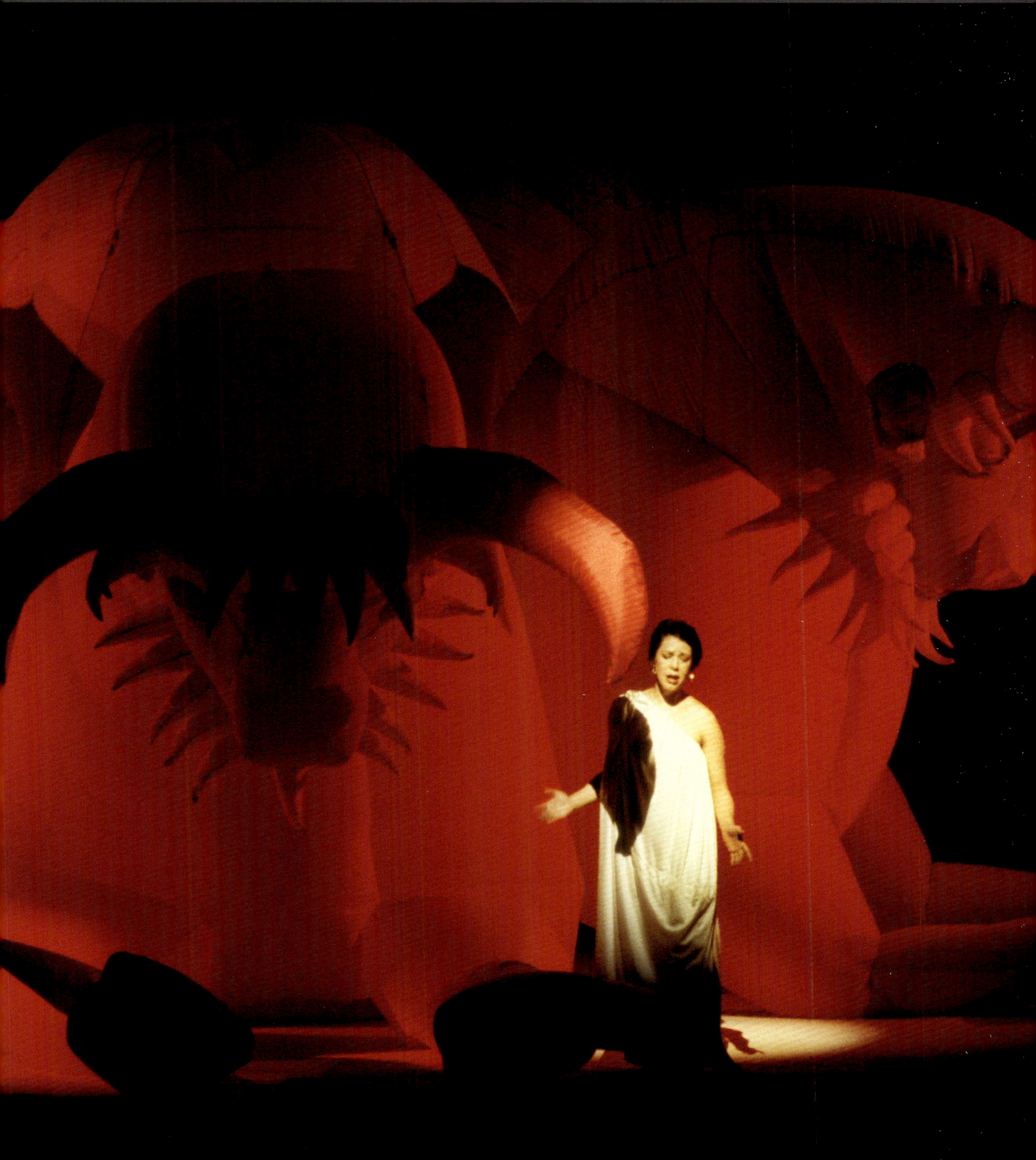

Musician/composer Paul Earls and librettist/director Ian Strasfogel described the tension between passion and technology as it was reflected in the opera and as it constituted a topic of the 1970s and 1980s in the world of art and performance.

> Music Theater constantly seeks new vocabulary, a new syntax. Technology can provide such a new language, expanding the potential of storytelling through sound. For this to happen, all collaborators, musicians, visual artists, singing actor, director, all must agree on the expressive goal. Otherwise it will be mere decorative trimming. The great challenge in our transformation of the Minos-Pasiphaë-Daedalus-Ikarus legend has been to use complex technology for our theatrical purposes. Laser projections, inflated sculptures, still and video projections can be merely delightful toys or ponderous industrial products. We wanted neither. We sought to tame them for our aesthetic and expressive ends. We have used these technological elements for their non-literal, allusive qualities, to expand upon the human and narrative textures found in the Greek legend. We have constantly reintroduced and varied the story's key images of bull, sun, wings, machines. Through this, we have aimed to make the story and characters more powerful than they could ever be in a conventional theatrical format. Through intensive work with our performers, I have sought to balance the vast images of the technology with the intense honesty of the live performer. (Ian Strasfogel, *Ars Electronica im Rahmen des Internationalen Brucknerfest*, 1982, 130)

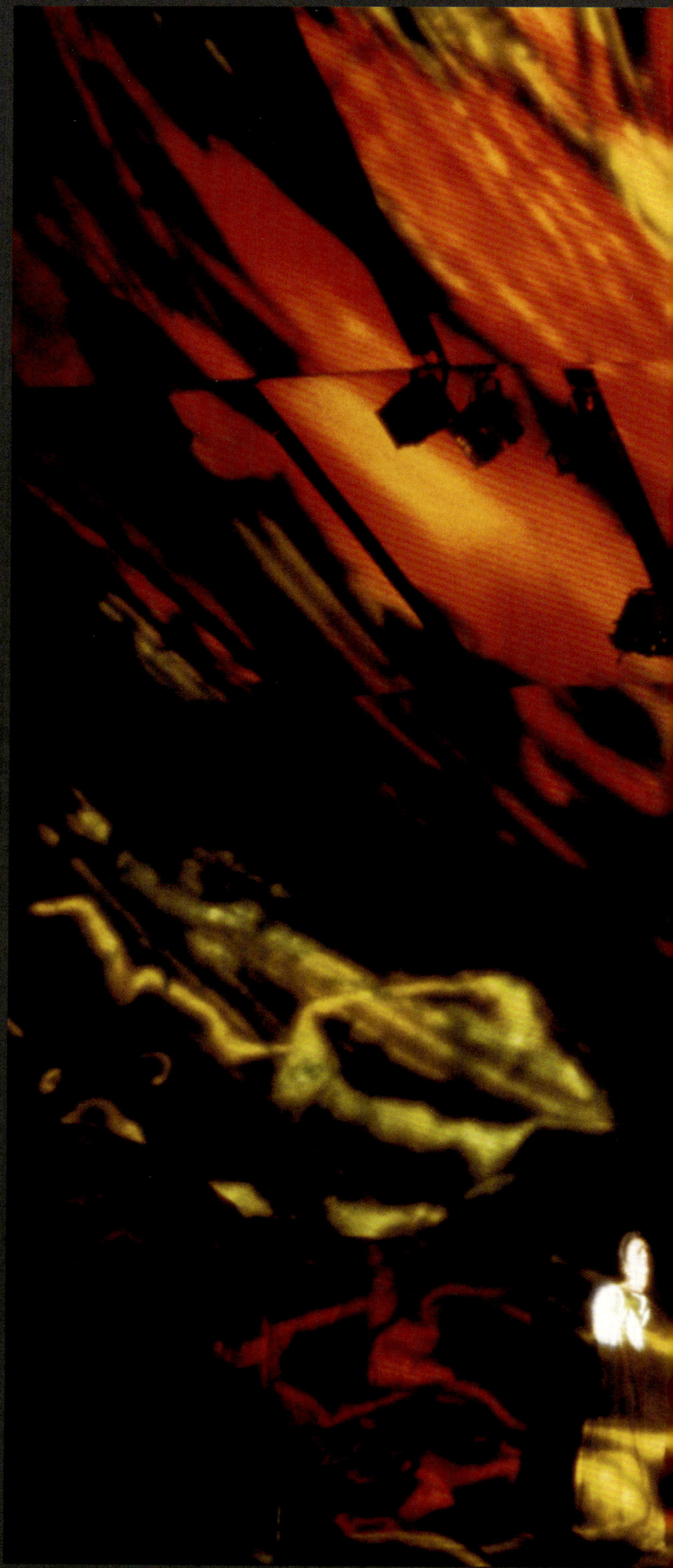

Icarus—A Sky Opera by Paul Earls (music), Otto Piene (design); and Ian Strasfogel (text); with set and costume design by Günther Schneider-Siemssen, Kresge Auditorium, MIT, May 30, 1984
Pictured: mezzo-soprano Nelda Nelson as Pasiphaë; baritone Timothy Noble as Daedalus
Photo: John Kruse

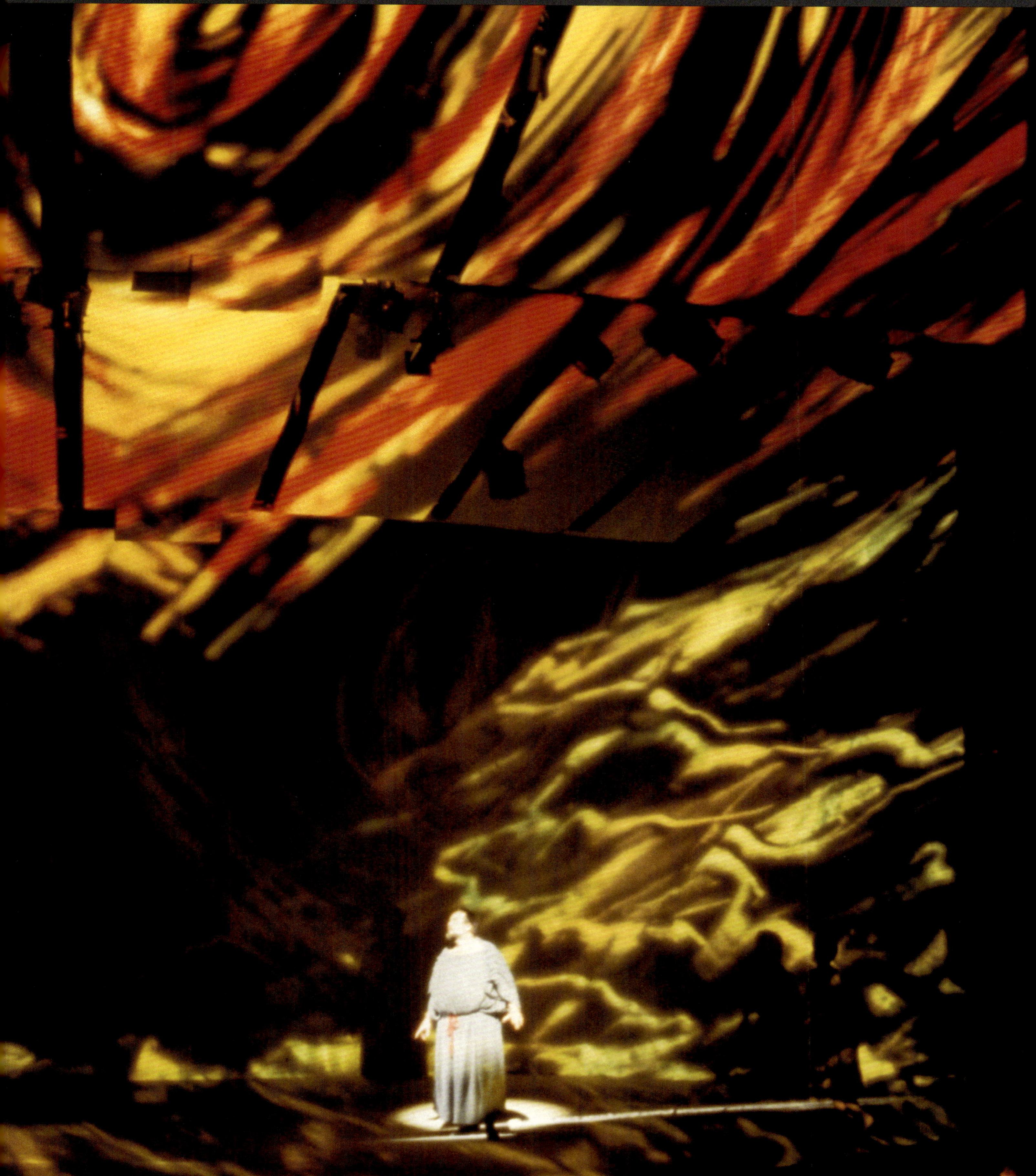

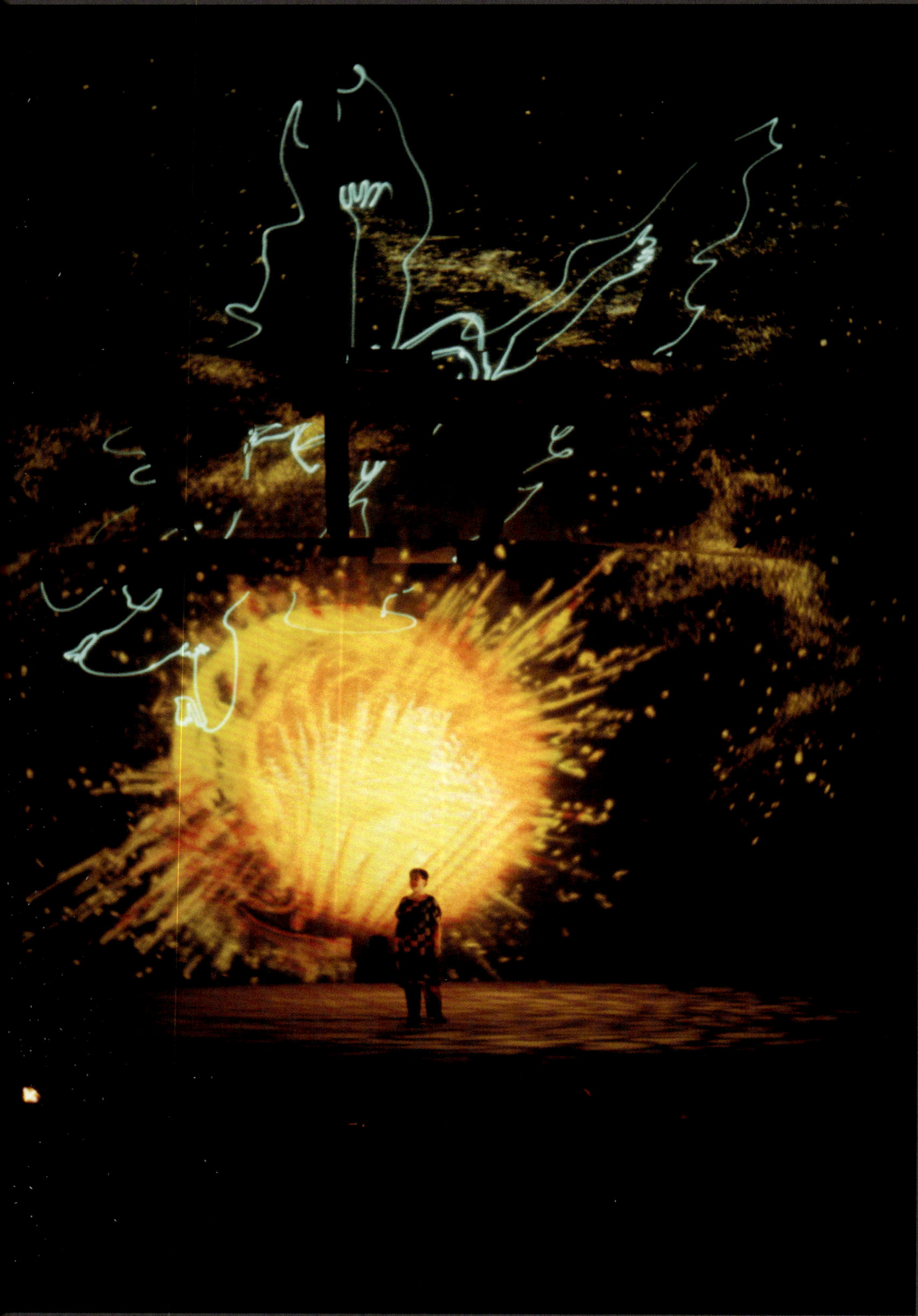

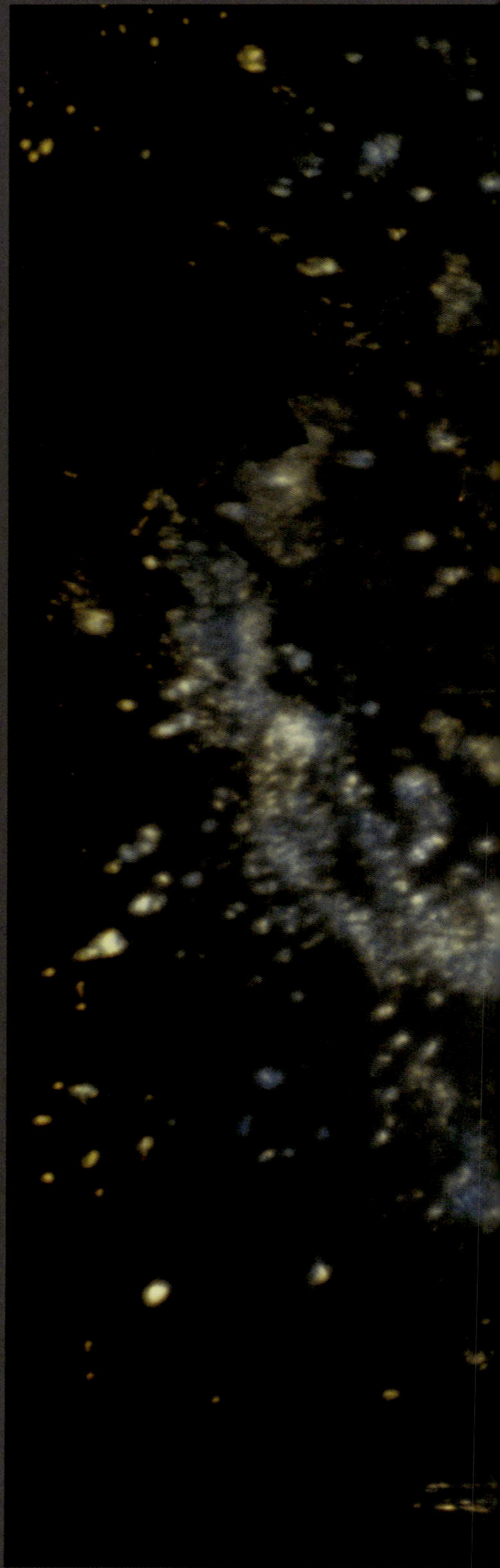

Icarus—A Sky Opera by Paul Earls (music), Otto Piene (design); and Ian Strasfogel (text); with set and costume design by Günther Schneider-Siemssen, Kresge Auditorium, MIT, May 30, 1984

Left: boy soprano Joseph Olefirowicz as Icarus, with Earls' laser projection of Piene's drawing of Icarus' flight

Right: mezzo-soprano Nelda Nelson as Pasiphaë; baritone Timothy Noble as Daedalus; Joseph Olefirowicz as Icarus

Photos: John Kruse

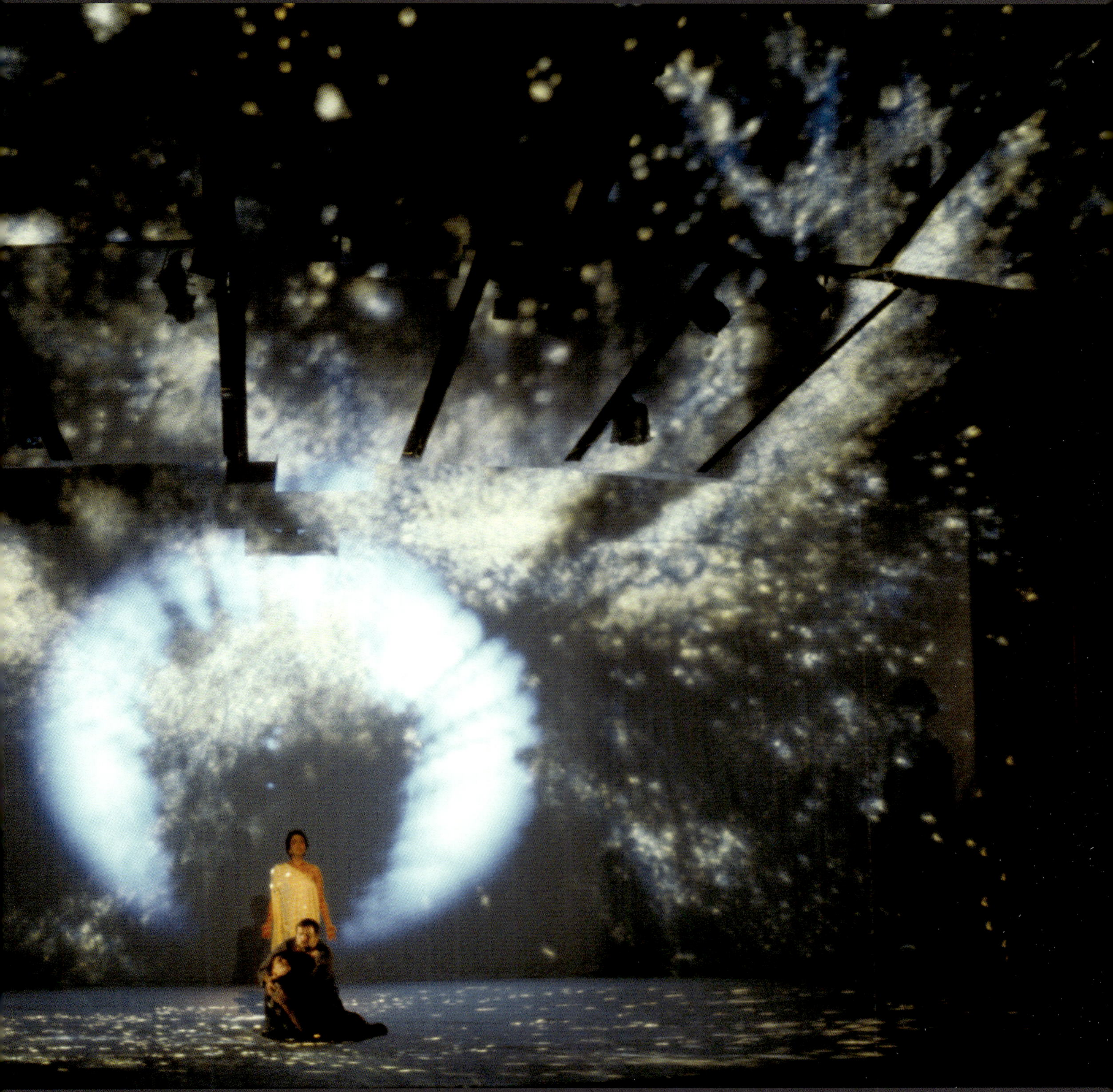

Desert Sun/Desert Moon, Lone Pine, California, June 8-16, 1986

> I have a little place in the high desert in the Owens Valley, inside the Sierra Nevada Mountains and I suggested to Otto that it would be a good place to do a Sky Art Conference . . . . we invited people to come aboard and we met out in Lone Pine. I knew the place, but I scouted the whole area to find the best possible place for us to have access by vehicle but still be where nobody would go. We had the place to ourselves. That's where *Desert Sun/Desert Moon* took place. (Tom Van Sant interview)

The concept for *Desert Sun/Desert Moon* began to crystallize around the spirit of collaboration surrounding Sky Art conference activities. At the 1983 Munich Sky Art Conference, several artists began to talk about creating a collaborative sky artwork on the scale of *Centerbeam* but physically less rigid and dimensionally more expansive. They envisioned a single work burgeoning out of interrelated projects—by artists, for artists, and with artists working directly with each other—which would serve as the major event of the *Sky Art Conference '86*. Much of the text that follows is quoted and paraphrased from a 1987 *Leonardo* article by Elizabeth Goldring, "*Desert Sun/Desert Moon* and the *Sky Art Manifesto*."

Everyone brought their own works and had the freedom to collaborate within the structure of the event. Unlike *Centerbeam*, there was no overarching unifying design. Works happened at different times, and people collaborated and supported each other during these performances and events. Events, installations, and performances radiated out horizontally and vertically like spokes of a wheel about three miles in diameter on the desert floor. The Alabama Hills are a richly twisted aggregate of heaving and sighing boulders. They are made up of meta-volcanic Jura Triassic rock and are among the oldest mountains in North America, sitting in the shadows of the lofty adolescent Sierra Nevada Range and in view of the Inyo Mountains. In a metaphoric sense, the Alabama Hills are the fulcrum for Mt. Whitney and Death Valley, which are 110 miles apart. The area was home for the Cree and Paiute Indians, whose petroglyphs mark rocks in the area. The evanescent colors of sunrises, sunsets, and wild flowers intensify the rock forms. Deep shadows and the pungent odors of sage mollify the vast expanse, as do the Alabama Hills themselves. The Owens Valley below is the dry lake bed of the once brimming Owens Lake, now drained by the Department of Public Works for Los Angeles' water supply. The working dynamics and dialogue with the environment and the ancient magic of the chosen site became even stronger and more evident than anticipated.

Desert Sun/Desert Moon took place from June 8-16, 1986. Most of the artists traveled from Boston to Los Angeles on the same airplane and arrived with great anticipation and a good degree of uncertainty. The participants included Lowry Burgess, Joe Davis, Paul Earls, Dale Eldred, Elizabeth Goldring, Ron Hays, Christopher Janney (also project manager), Charlotte Moorman, Otto Piene, Tal Streeter, and Tom Van Sant. MIT students David Atherton, Shawn Brixey, Laura Knott, and George Numrich were both participating artists and part of a crew of fifteen others. Upon arriving at Lone Pine, the artists surveyed the area to make their final site selections. A mobile home parked next to the power connection served as headquarters. Work lasted from sunrise until late at night, with a three-hour break in the heat of mid-day. A Smithsonian World crew of seven filmed the installations and events, and Vin Grabill documented each work in a video later presented at the *Sky Art Conference '86*.

Facing page top: Otto Piene, "Berlin Star," *Desert Sun/Desert Moon*, Lone Pine, California, 1986

Facing page bottom: Shawn Brixey, Laura Knott and David Atherton, "Photon Voice," Knott dances in front of four fresnel lenses—8", 9", 10", and 11" focal lengths—which, combined, created a camera obscura-type inverted image of her dance in the sky, *Desert Sun/Desert Moon*, Lone Pine, California, 1986

Photos: Otto Piene

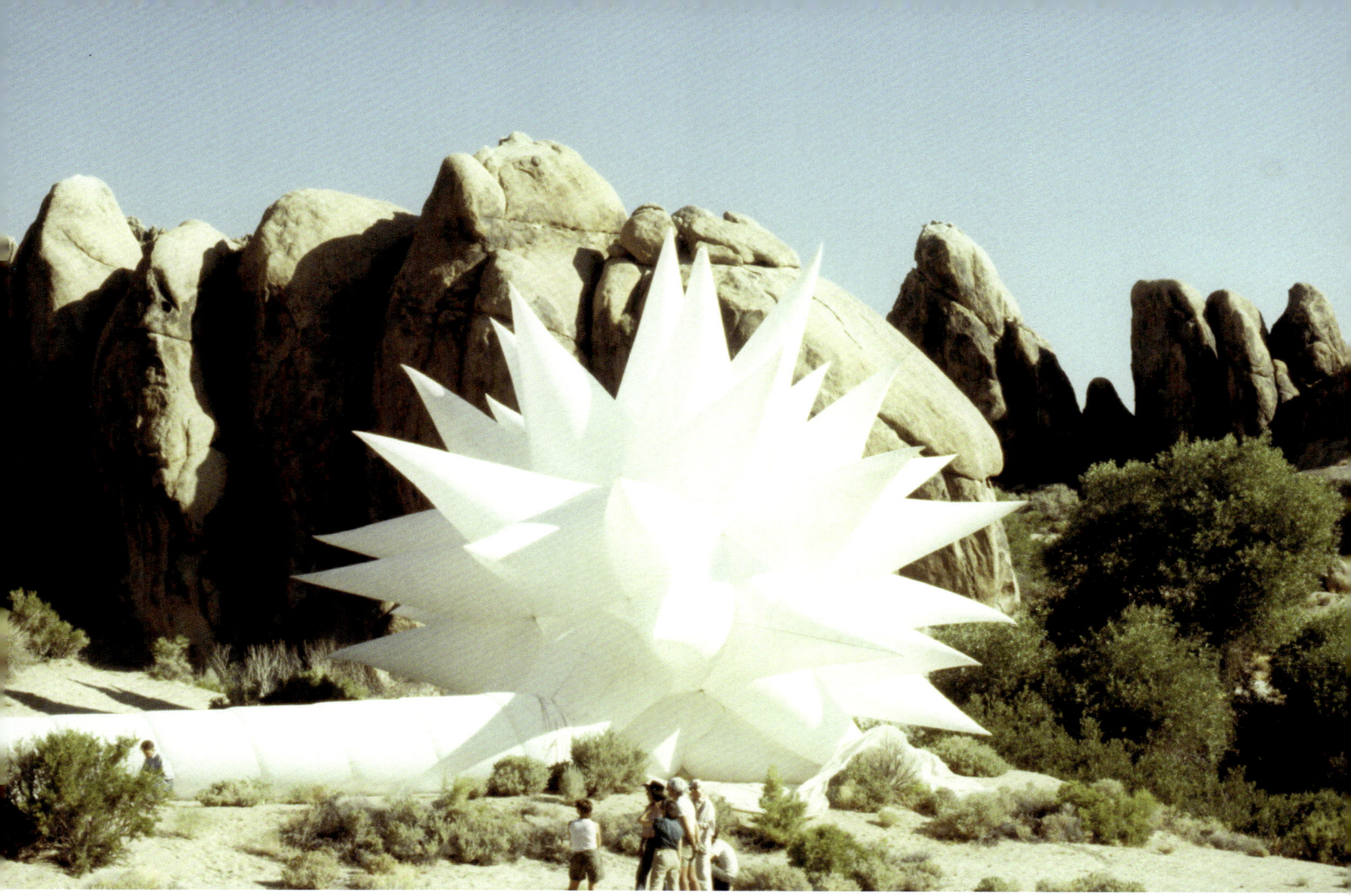
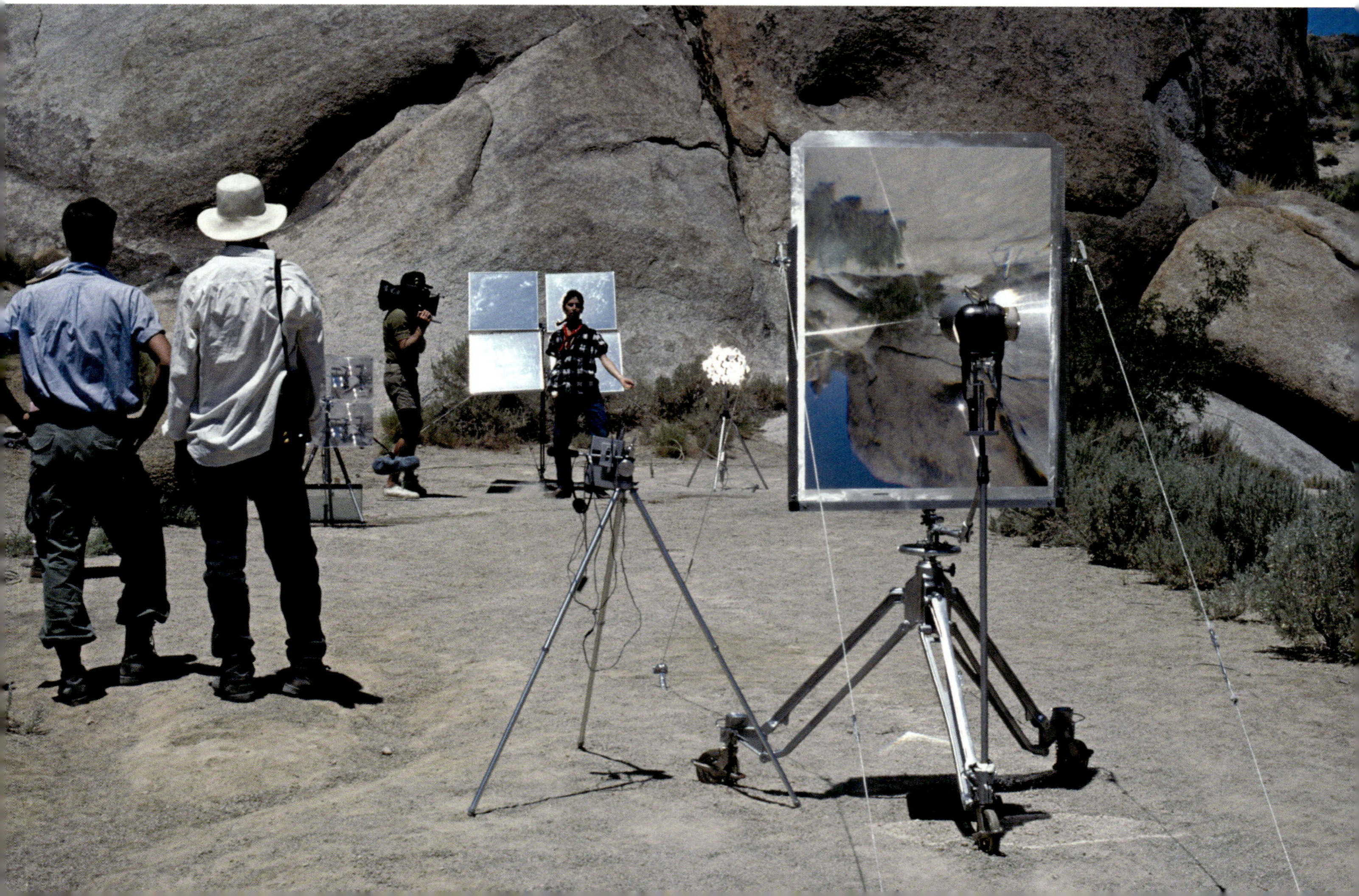

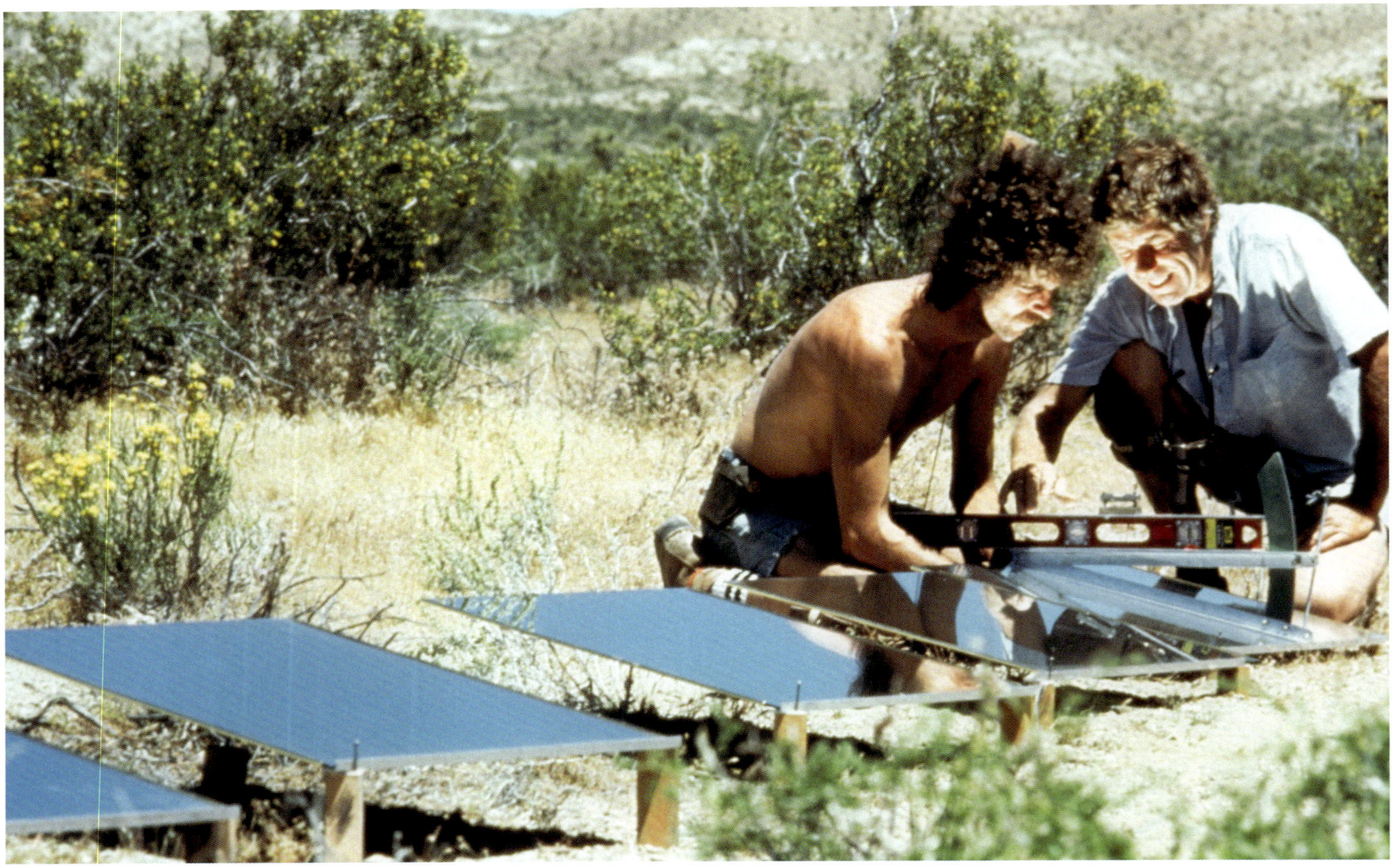

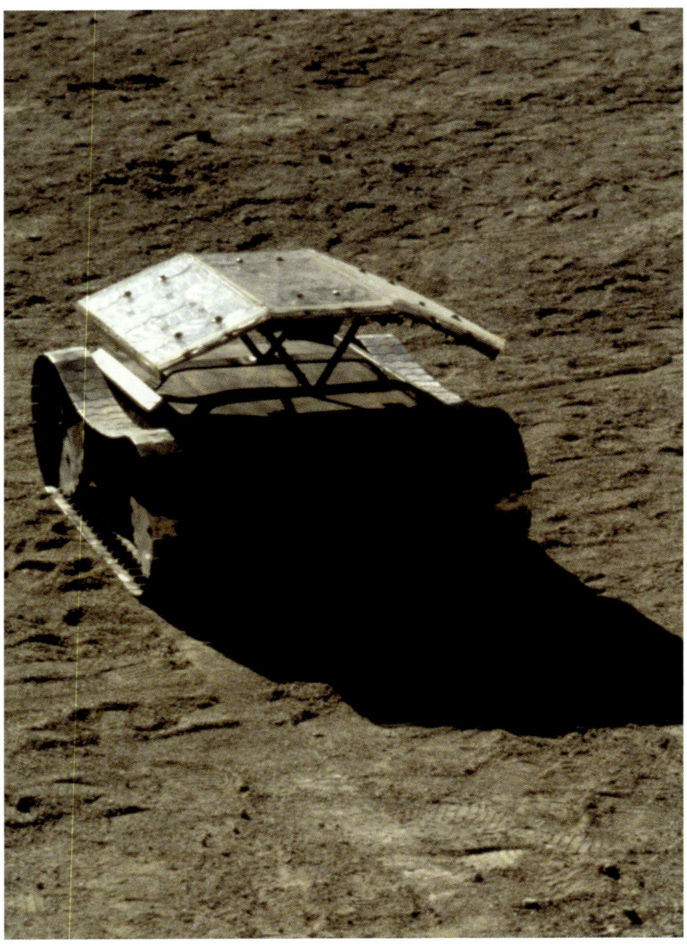

Above: Tom Van Sant, "Desert Sun," 48 mirrors reflected sunlight back to the sensors of a GOES-6 satellite in geostationary orbit at 22,000 miles above the earth. Photo: Shawn Brixey

Left: Joe Davis, "Desert Crawler," recycled steel and aluminum parts, electric motor, solar panels, rechargeable batteries, 1986. Photo: Otto Piene

Facing page top: Tal Streeter, "Red Line," 350 square foot Domina Jalbert parafoil kite. Photo: Tal Streeter

Facing page bottom left: Charlotte Moorman, "Cocoon Concert," with Christopher Janney, inside Piene's "Berlin Star" inflatable. Photo: Elizabeth Goldring

Facing page bottom right: Lowry Burgess, "The Boundless Foundry." Photo: Elizabeth Goldring

All events part of *Desert Sun/Desert Moon*, CAVS in the Alabama Hills, Lone Pine, California, 1986

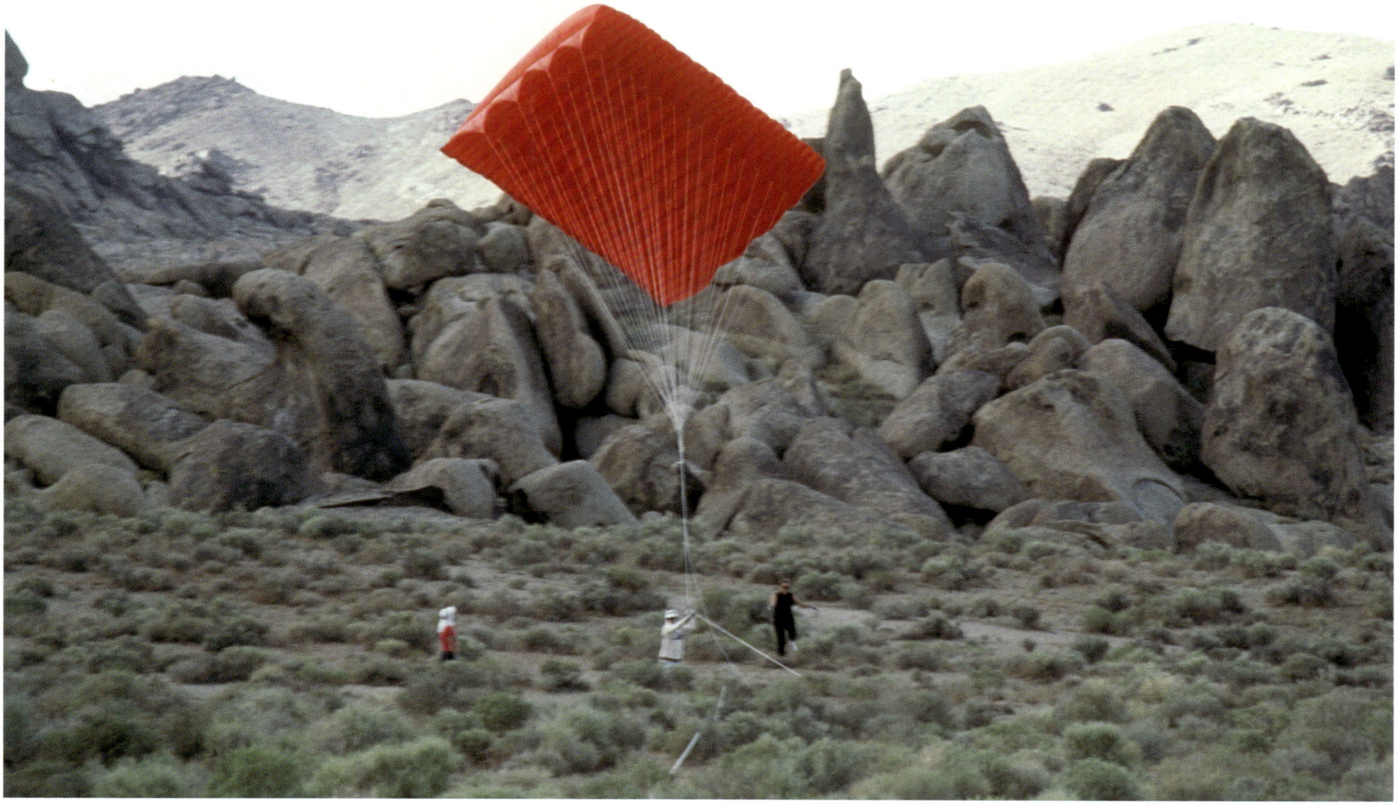

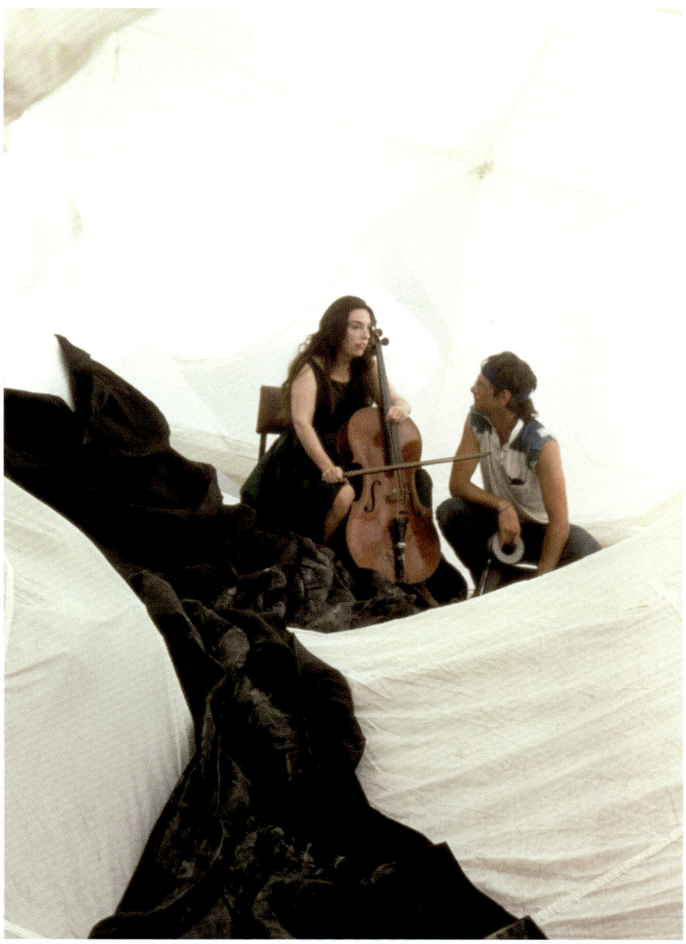

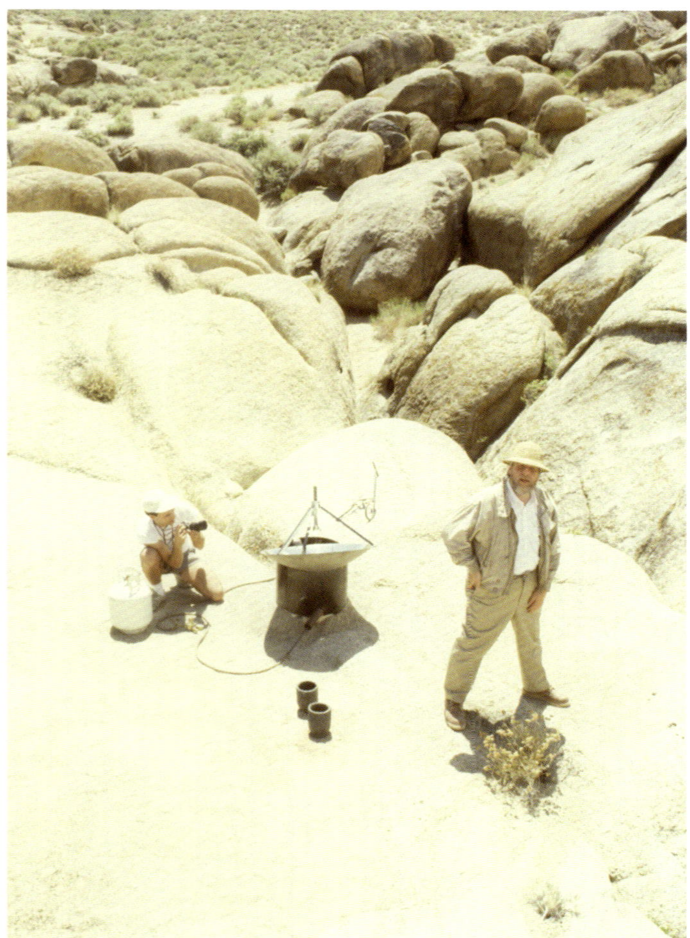

Sky Art Conference '86, CAVS and Other Locations, September 29-October 3, 1986

*Desert Sun/Desert Moo*n was the opening event for the 1986 Sky Art activities stretching from Cambridge, Massachusetts, to California and Alaska, to as far away as Brazil and Paris. The fourth Sky Art conference, which took place at CAVS from September 29 to October 3, 1986, was designed in the spirit of *Desert Sun/Desert Moon* as an intimate gathering of interested artists, scientists, engineers, and public officials. It was dedicated to the memory of former CAVS Fellow Stan VanDerBeek, the pioneer media artist who had been a major figure at previous conferences. By this time, the popularity of the conference as a forum for exchanging ideas and inspiration was evident in the attendance of approximately 150 sky artists from near and far. Participants presented new work, including sky events. At noon on Thursday, Howard Woody launched a "disappearing angel" over the Charles River. Later in the day, Steve Poleskie flew his "Apogee Airway" plane, drawing precise images on the sky canvas above MIT. Joe Moss installed his environmental sound chamber, "Sky Wave," in the Kresge Oval.

Sky Art Manifesto, November 3, 1986 Paris

Perhaps the most unpredicted and valuable part of the *Sky Art Conference '86* was that the panels on the future of Sky Art turned into workshop sessions and produced the initial draft for the *Sky Art Manifesto*. A final version prepared by Lowry Burgess, Otto Piene, and Elizabeth Goldring was sent to artists participating in the conference for comment and signature.

The *Sky Art Manifesto* was also a focal point for the collaborative telecommunication event organized by artists of the University of São Paulo and CAVS fellows, which took place on October 11, 1986. Joe Davis traveled to São Paulo to organize the Slow Scan telecommunication event using six Robot SSTV transceivers. São Paulo participants included multimedia artists Jose Wagner Garcia, Guto Lacaz, and Mario Ramiro; electronic composers Wilson Sukorski and Conrado Silva; poets Augusto de Campos and Julio Plaza; and performance artist Marco do Valle. CAVS Slow Scan projects, which were transmitted to Sao Paulo from CAVS, included Ellen Sebring and David Atherton's "The Happy Man: After the War;" Goldring's "Red Leaves;" and Charlotte Moorman's performance of Nam June Paik's "Sky TV" score, "Slow Scan Study Number One." The evening culminated with the audio transmission from CAVS of Lowry Burgess and Otto Piene reading the *Sky Art Manifesto*. It was endorsed via Slow Scan return with Brazilian artists' signatures.

The preparation of the *Sky Art Manifesto* and its dissemination in São Paulo and at the Paris *Conference on Culture in Space* at UNESCO on November 3, 1986, marked the culmination of the *Sky Art Conference '86*. The *Sky Art Manifesto* commits artists to making art and culture part of international space exploration, evolving new expressions and a synthesis among art, science and technology, humans and space.

The Sky and Space Artists' Manifesto

Photography was invented and popularized by artists: Niépce, Daguerre, Talbot, Morse. Photography as an art form used by scientists and engineers has told us more about deep space than extrapolated data from any other source. During the last decade and a half photography applied to space has advanced space research more than all previous scientific effort.

It is in this tradition that we as artists seek the broadest possible relationship with international space efforts. While photography was developed by artists without foreknowledge of its scientific applications, the cumulative thrust of cultural experience and insight gained from the history of the nineteenth and twentieth centuries leads to our realization that art, science and technology complement each other increasingly as this dramatic century comes to a close.

Interaction among scientists, engineers and artists fosters new media and practice in the arts, and thereby, artists communicate with ever-increasing audiences. To encourage and proliferate art based on integration the Center for Advanced Visual Studies at MIT has for nearly 20 years hosted artists and theorists working individually and collaboratively. Accordingly, a series of international Sky Art Conferences was begun in 1981 and has led to the 1986 Conference, September 29-October 3.

One hundred and fifty artists, scientists, and students convened and decided to address the space agencies through the following statement which we are presenting today at the November 3, 1986, Paris Conference on Culture in Space.

Statement

Our reach into space constitutes an infinite extension of human life, imagination, and creativity. The ascent into the sky is mirrored by the descent into inner space as it reflects the cosmos.

Our release from gravity represents a fundamental shift in human consciousness—flight and release, which open a new dimension of humanity. From the ancient past, artists have formed images and dreams, fired the imagination, built structures of aspiration to give the world wings to fly, and the vision to see new societies in the sky. We live in their cumulative light. Not only here on earth but there in space we must see, touch, feel and think in order to transport soul and spirit. Thus a threshold is crossed where the radiance of art brings expanded awareness into reciprocity with the earth.

"As I stood in the contemplation of the garden of space I had the feeling that I was looking into the ultimate depths, the most secret regions of my own being; and I smiled because it never had occurred to me that I could be so pure, so great, so fair! My heart burst into singing with a song of grace for the universe. All of these constellations are yours, they exist in you; outside your love they have no reality." (Milosz).

We see international implications of our art as fomenting a global consciousness through large-scale display, tele-education/discovery and exploratory play.

Sky artists enthusiastically seek productive alliances with all space agencies. We are asking for the establishment of national and international councils that will advocate specific artistic projects to the appropriate institutions and agencies. Additionally, these councils will assist with the implementation of far-reaching artistic endeavors that will embody subtle and humane purposes.

We pledge our imaginations and skills, our probing spirit and expressive powers in this effort to seek the widest horizon for human insight and experience:

- Interacting with existing vehicles and systems first, then developing special methods, use and implements
- The artist as creator and framer of exemplary phenomena and messages goes into space there to beam signals back to earth
- The artist as explorer of the inner self continues the dialogue with the universe in space
- The artist as frontier poet with the artist's sensory instrumentarium goes into space to widen human perspective on the 'new world'—sky and space
- The artist travels between worlds to harvest tales and images and pass them on to many in places near and far.

Otto Piene, Director
Elizabeth Goldring, Co-Director
Lowry Burgess, Senior Consultant
Paris, November 3, 1986

Sky Art Conference 2002, Delphi and Ikaria, October 15-20, 2002

> Environmental art made way for enhanced understanding of environmental quality and responsibility while dizzying findings in physics and biology exponentially expanded the environment itself. The extent of landscape ("sky") onto which humanity can project its activities has increased by at least several orders of magnitude since 1972. . . . Sky Artists cannot save the world. Instead, they try to save themselves from futility; from the seamless tedium of nature; from every singular monotony and from endlessness itself. Endlessness is our Labyrinth of Daedalus, and there are monsters abiding deep within. It is easy to imagine why we go about building sets of wings. (Joe Davis, Introductory Remarks, *Sky Art Conference 2002* catalog, 2002, 16)

Sky Art Conference 2002 was haunted by the events of September 11, 2001, when the sky turned deadly as the twin towers of the World Trade Center in New York City—once the tallest buildings in the world—were destroyed by two hijacked Boeing 767 jet airplanes. The convocation was held at the site where the mythological character of Icarus flew too close to the sun, melting his wings, and falling into the sea.

> 9/11 became a symbol, and it is a symbol, and it remains a symbol, I think to the world that is yet trying to live in peace and yet trying to redevelop and maintain a sense of harmony. What brings me to referring to the Olympic Rainbow is this: what remained of the closing ceremony of the Munich Olympics happened on 9/11, on September 11th, 1972. So here is what some people call the problematic of Sky Art, the problematic of believing, the problematic of working towards something that is considered or intended positive yet is constantly running into adversity, into conflict, into dramatic events that intend the opposite: disruption and the unsettling of a world ready to become complacent. Well, complacency is something that neither the terrorists nor the Sky Artists ever want. What we want is to work in a place that has a chance or still has a chance to be peaceful, to be untouched, or to return to being untouched, and that's the sky. (Piene, *Sky Art Conference 2002* catalog, 2002, 26)

Piene's commitment to sky art was rooted in his wartime experiences as a conscripted youth soldier in Germany during World War II.

The locations in Greece—Delphi and Ikaria—for the 2002 conference were suggested by poet and CAVS graduate Panos Kouros, on looking out of his corner window over the Acropolis. Visions of sky art, espoused by the assembled sky art practitioners, reached from the depths of inner space to the far reaches of outer space. Members of the Olympics Committee, who attended the conference as guests, invited two CAVS artists to become part of the following year's Olympic Games in Athens: Paul Matisse with his "Olympic Bell," and Mitch Benoff with his "Speed of the Earth."

> . . . let the harmoniously musical sounds of this bell ring out into the Grecian air carrying its special meaning. And so here is the bell that I have brought you, and this is its sound. < a 4' bell sounds musically; repeated several times > . . . I invite you all to go there in the course of our time here at Delphi, and to ring it . . . I hope you will appreciate its harmonious quality, and will consider how differently this world of ours would be if we all were as harmonious together as the tones of this bell. (Paul Matisse, "Sound in Space," *Sky Art Conference 2002* catalog, 2002, 77)

> I not only did the largest sculpture ever done for the Olympics, but it was the Athens Olympics! There were two things you could see at night: the Parthenon and my piece. (Mitch Benoff interview)

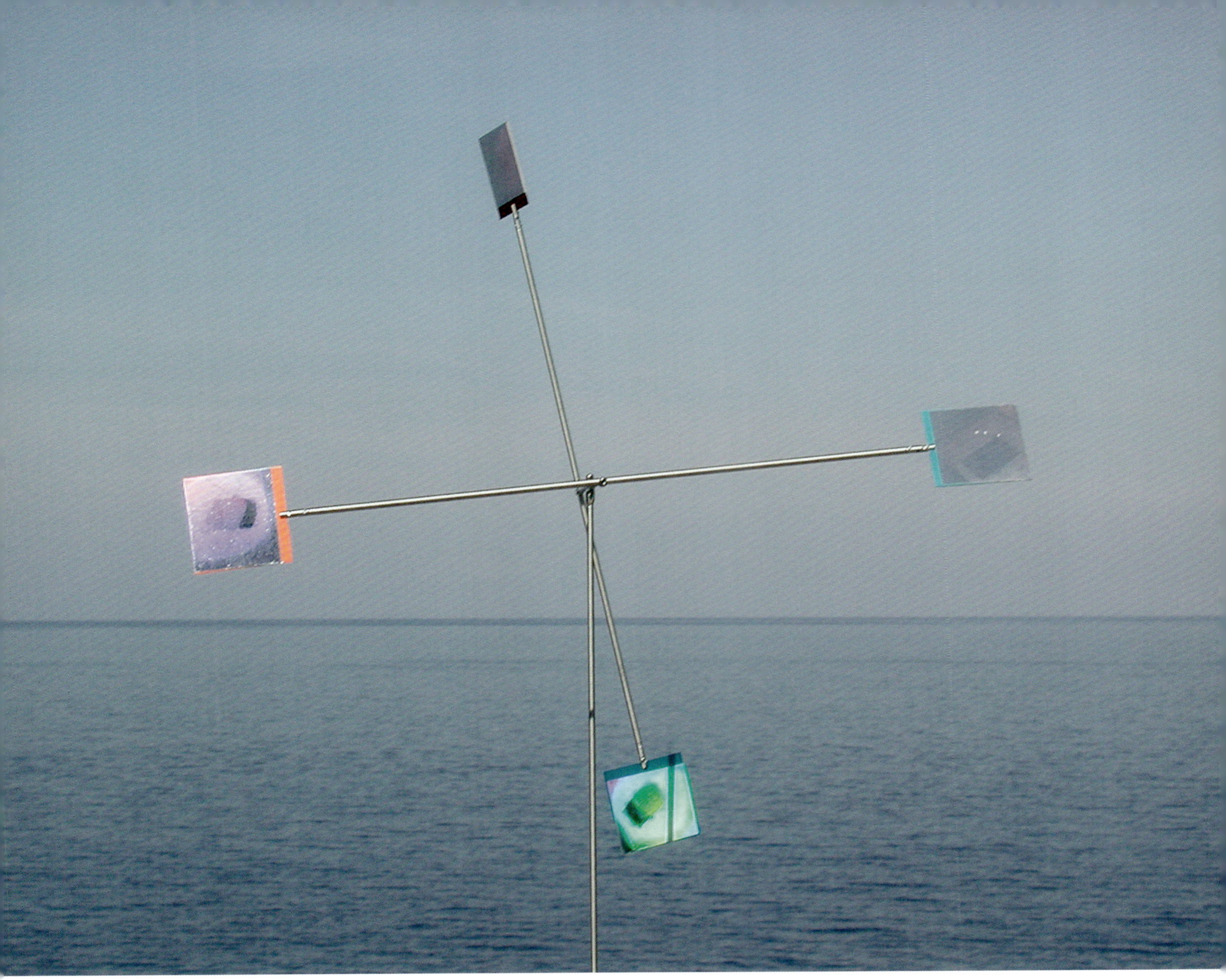

The HoloMobil, composed of stainless steel rods and aluminum panels holding embossed computer generated holograms on each side, turns around x-y-z axes. Air currents activate the sculpture in unpredictable spatial constellations, and the holograms emanate unexpected 3-dimensional images into the air, generated by the incoming light and the vision of the viewer. (Dieter Jung description)

Top: Dieter Jung, "Icarus Eyes," HoloMobile, *Sky Art Conference 2002*, Ikaria, 2002
Photo: Dieter Jung

Epilogue

Dieter Jung's text serves as an elegy and apt conclusion for this chapter dedicated in part to Otto Piene and his celestial vision.

> Otto Piene was intrigued by the challenging diversity and kinetic vitality the skies had to offer, in order to transform intimate concepts and visions into open space, filled with reflections of truth and fiction, fantasy and dreams. The rainbow, containing the visible spectrum of colors, was celebrated as a shadow of the bridge between earth and sky. The *Olympic Rainbow* and Charlotte Moorman's "Sky Kiss" became elementary symbols demanding reconciliation between heaven and earth for a peaceful future.
>
> In this context also scientific papers concerning aeronautics, astronautics and astronomy became a significant enrichment to the scintillating *Sky Art Conferences*, conducted by Otto Piene and poet Elizabeth Goldring. The last conference took place on the Greek island Ikaria, commemorating Ikarus, whose wings out of wax melted in the heat of the sun. Flying objects, lighter than air, are no longer chained to our planet Earth. Sky Art is seen as a continuum of energy integrating multi-cultural, spiritual and religious patterns of the virtual and material world.
>
> I remember Otto Piene's search for the "pure energy" of light, awakening my heartfelt dream to install for Otto a memorial beyond Earth´s atmosphere into deep sky´s unspoiled cosmic regions surrounded by sounds of interstellar space.
>
> Now, after a long time of search and uncertainty the minor planet 359103 is named for Otto. The chosen object, 359103 Ottopiene = 2009 BS, discovered 2009 January 16th by Felix Hormuth (Calar Alto Observatory in Spain), belongs to the inner main astroid belt and is on a relatively stable elliptical orbit. Such being the case, this minor planet behaves to some degree eccentric and shows a special affinity to light. (Dieter Jung, written for *Centerbook*, 2016)

Finally, though much later and technically beyond the scope of this book, the 150th anniversary of MIT was celebrated in 2011 with Piene's sky events flying dramatically over the MIT dome, launched from Killian Court during the *Festival of Art, Science, and Technology*.

Facing page top: Otto Piene, participatory sky event at the *Festival of Art, Science and Technology*, MIT 150th Anniversary celebrations, Killian Court and Great Dome, MIT, May 7, 2011
Photo: Andy Ryan

Facing page bottom: Dieter Jung provided this orbital diagram of the microplanet *Ottopiene*, NASA Jet Propulsion Laboratory, taken April 18, 2018

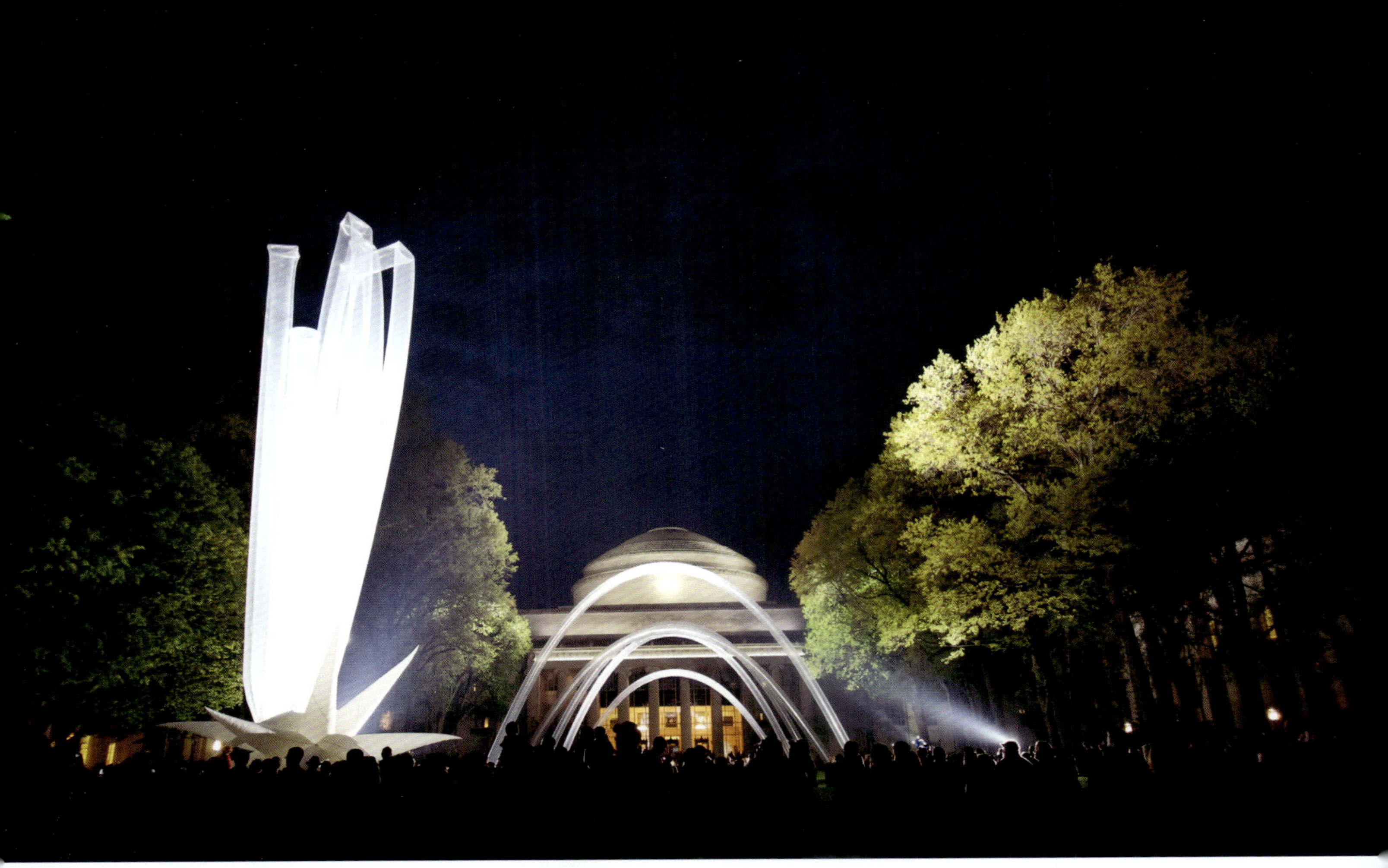

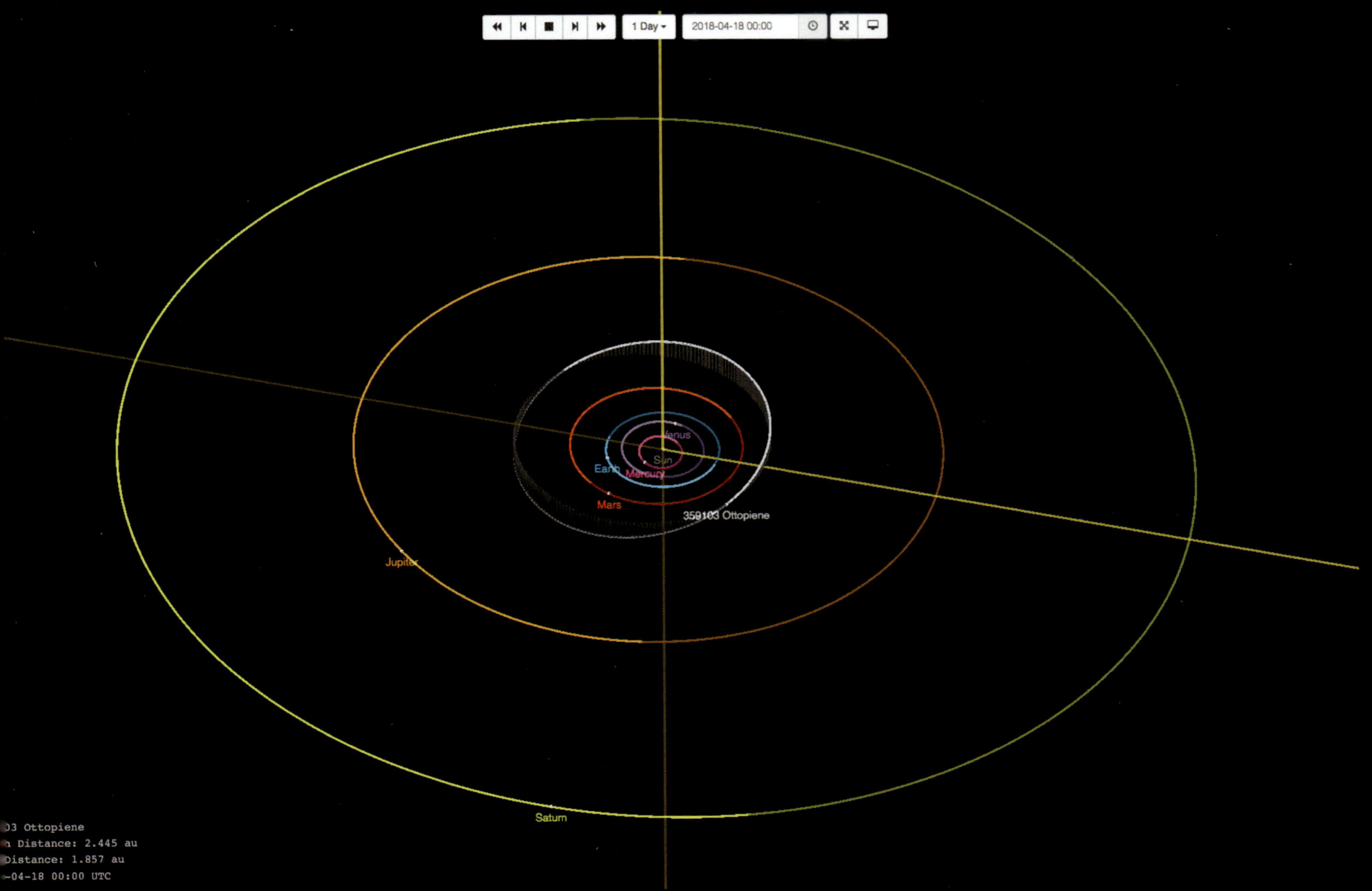
1 Day
2018-04-18 00:00
Venus
Sun
Earth
Mercury
Mars
359103 Ottopiene
Jupiter
Saturn
03 Ottopiene
Distance: 2.445 au
Distance: 1.857 au
-04-18 00:00 UTC

Chapter 8
Genome

The global proliferation of art-science-technology centers can trace its roots to CAVS and the pre-digital era at MIT. The DNA or genome of art-science-technology research, innovation, and artistic communication spread to centers such as the MIT Media Lab, ZKM | Center for Art and Media Karlsruhe, Kunsthochschule für Medien Köln (KHM), and countless others. Indeed, it is commonplace for universities to have media art centers, as once esoteric interactive digital media are now second nature to the children who will confront the biologic postdigital period.

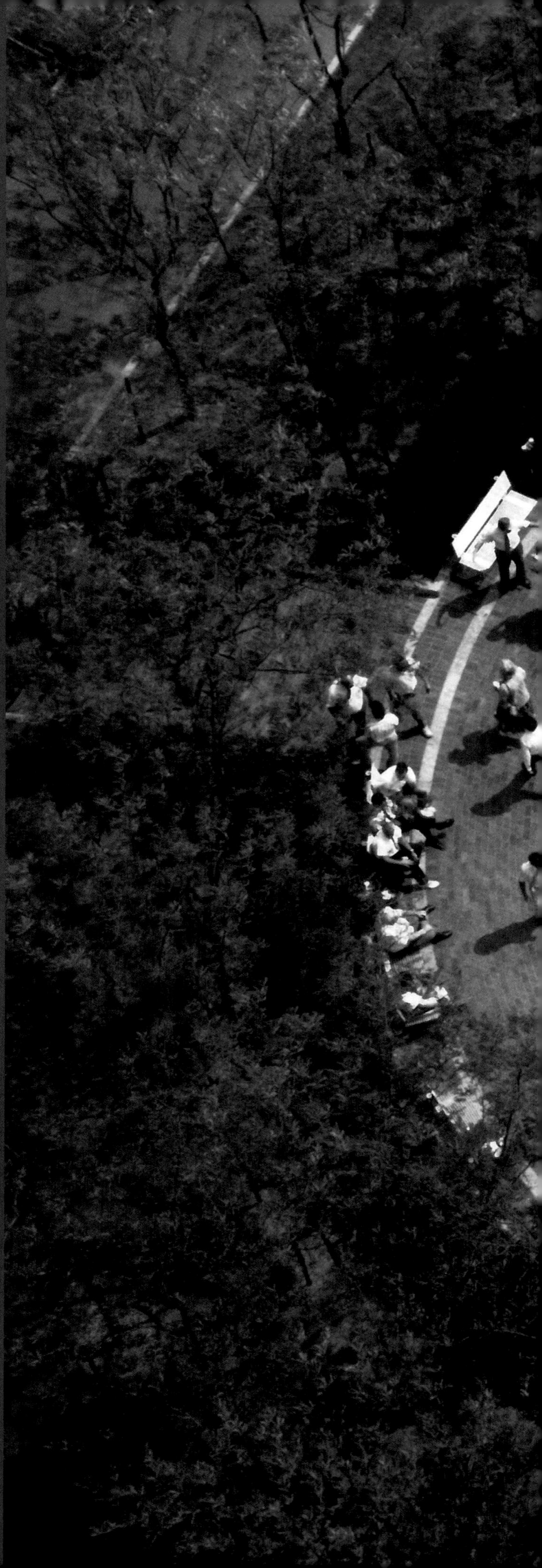

Right: *Galaxy*, fountain with *Earth Sphere* and light sculptures designed and fabricated by CAVS artists Otto Piene, Joe Davis, Joan Brigham, and Allan Schwarz for Thomas Murphy Park, The Halvorsen Company, landscape architects, Kendall Square, Cambridge, Massachusetts, 1988-1990
Photo: Thad Tercyak

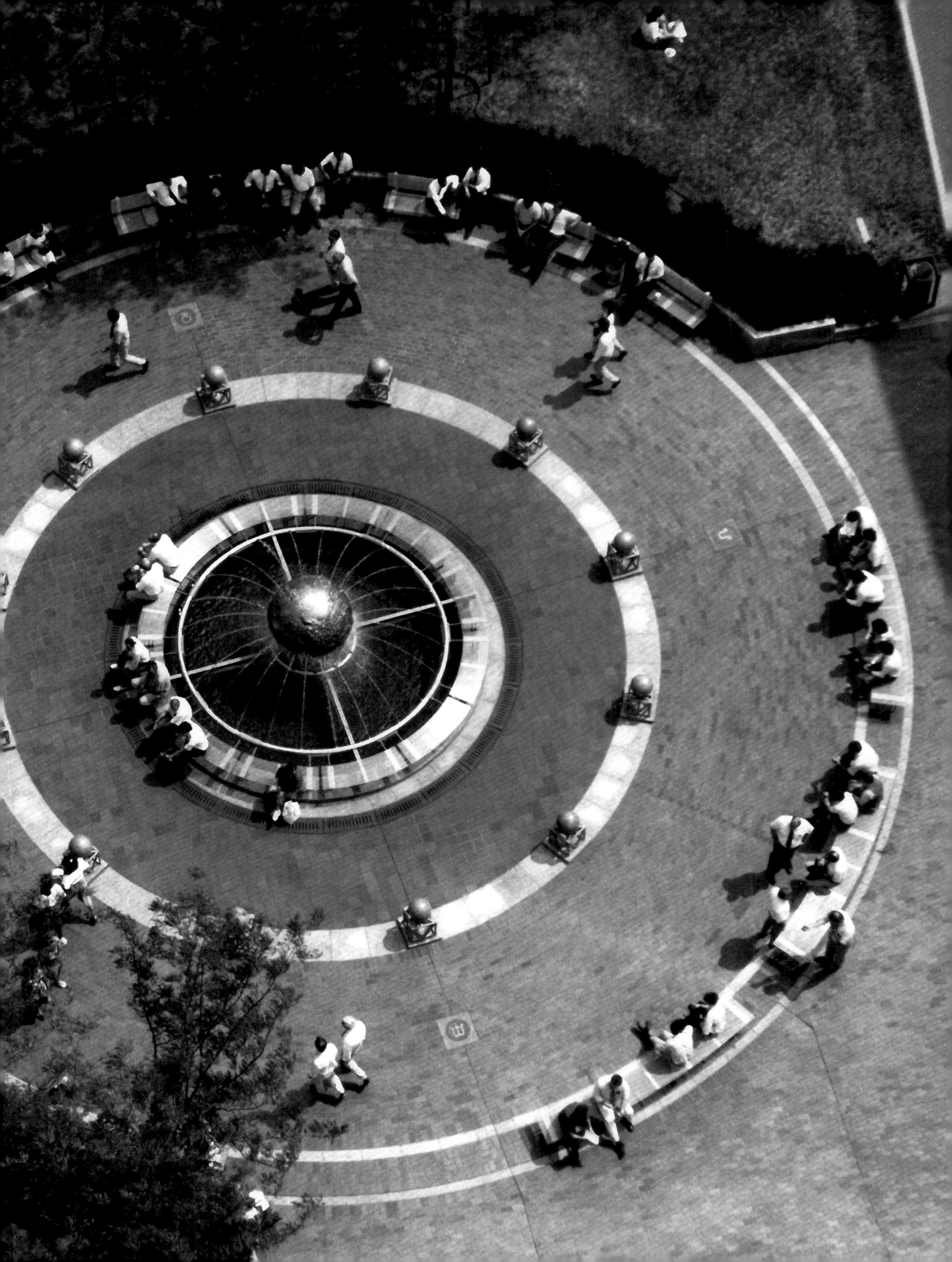

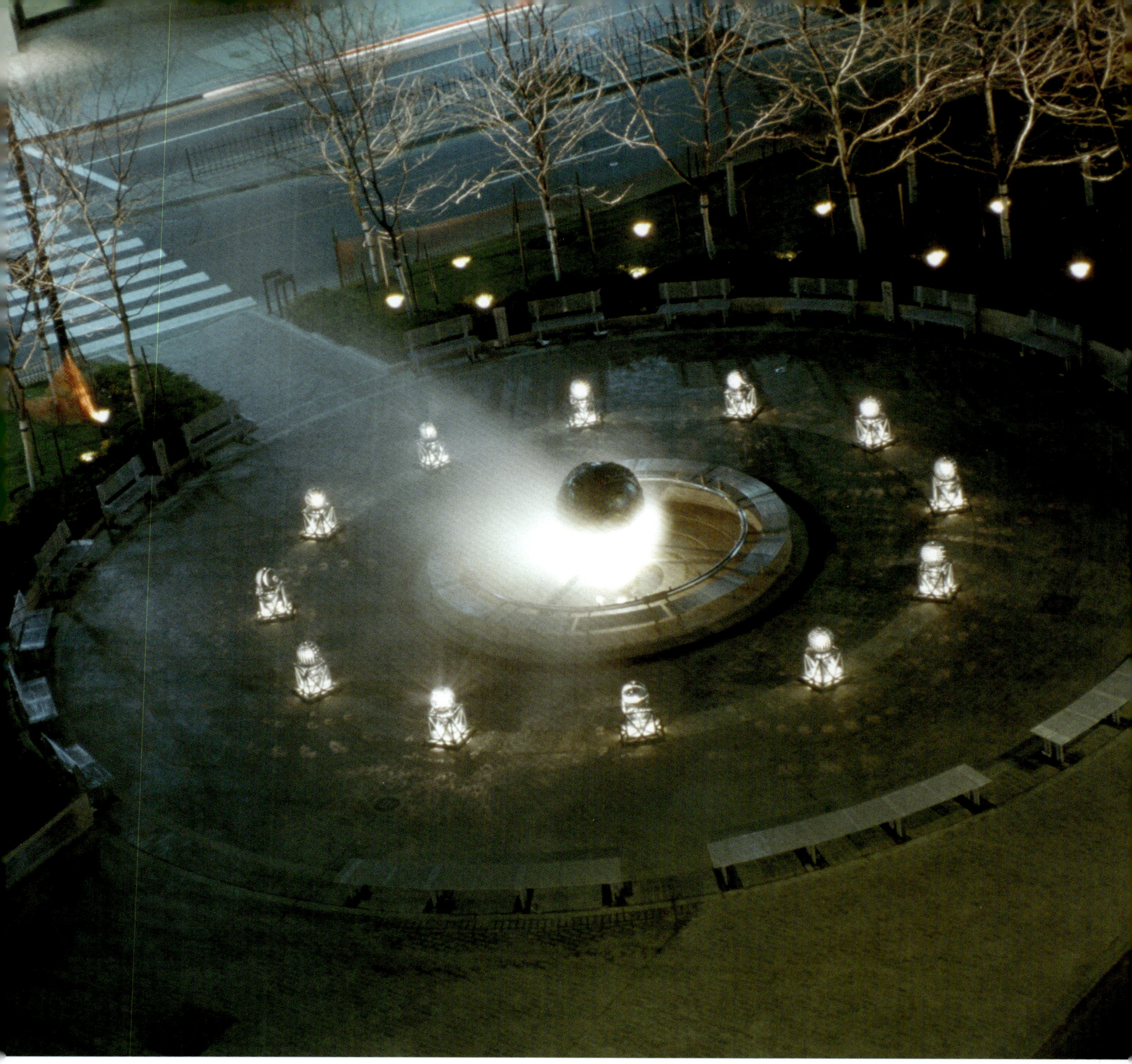

Galaxy fountain is the focal point of a small park designed by Halvorson Company, landscape architects, in collaboration with CAVS director, Otto Piene, and CAVS artists. The "Earth Sphere" appears to float above a cloud of steam in the winter months which in summer turns to water mist. The 15′ wide basin is surrounded by "Moon Lights" suggestive of the phases of the moon. A bronze compass is set within the paving. (Joan Brigham description)

The Spirit of Collaboration

At the founding of the Center for Advanced Visual Studies, there was only one other clearly identifiable art and technology group, E.A.T. (Experiments in Art and Technology). Some years later, a 1990 issue of *Kanal* published by Annick Bureaud, already listed 854 institutions in 31 countries with art, science, and technology programs. Such programs have continued to proliferate, and many former students and fellows of CAVS direct them.

The question of why the Center at MIT was so enduring and irreplicable—that is, what set it apart from other programs?—is often asked. The answer is most likely embedded in a core principle of collaboration as it evolved and was celebrated at the Center. Art historian John R. Blakinger suggested a rather polemic definition of Center collaboration; he identified Kepes' concept of collaboration, instituted at MIT, as on the one hand an example of conversion and on the other an example of complicity. He defined conversion as "the transformation of military research into non-military research" and complicity as "cooperative aims becoming compromised." According to Blakinger, "These two contradictory approaches were at play at the CAVS simultaneously; Kepes was deeply committed to conversion, but he could not escape complicity." (Blakinger, "The Aesthetics of Collaboration: Complicity and Conversion at MIT's Center for Advanced Visual Studies," *Tate Papers*, 2016)

With Piene, CAVS collaborations developed into something else, based on an expanding community of artists living and working together in close physical and human proximity—an interdisciplinary family tree of human participants, which Ben Davis referred to as "both a beacon and a reference point." Collaboration, as it seems to have been so naturally practiced at the Center, remains difficult to reproduce. It was the human factor pitted against the environment of MIT, and yet condoned by it.

Collaboration at CAVS meant not only group, like-minded, and antipodal interactions, but cross-disciplinary actions as well. In a shared synergy, fellows penetrated little-known media, science, and technologies to make CAVS a place for forging languages out of dark matter. The collective impulse to bypass boundaries challenged the MIT deans and administrators faced with budget cuts and institutional change that, in time, emphasized educational goals over independent research as practiced at CAVS.

> Having recently been asked by Dean de Monchaux to Head the Department of Architecture, Jack Myer appropriately opened the Thursday evening lecture series with a talk on change in the Department. . . . Myer explained that these budget cuts, amounting to 8 or 9% over the next three years, will primarily serve to tighten the curricula between the S.M.Arch.S., M.Arch., and S.M.Vis.S. programs. The S.M.Vis.S. and M.Arch. programs will support each other more efficiently, while the Center for Advanced Visual Studies will become attached to the Department of Architecture and its directions reconsidered. (Dorit Fromm, *Architext, Newsletter of the Department of Architecture MIT*, February 28, 1983)

Facing page: *Galaxy*, fountain with *Earth Sphere* and light sculptures designed and fabricated by CAVS artists Otto Piene, Joe Davis, Joan Brigham, Allan Schwarz, for Thomas Murphy Park, The Halvorsen Company, landscape architects, Kendall Square, Cambridge, 1988-1990
Otto Piene, concept and design
Joe Davis, sculpture and fabrication
Joan Brigham, steam
Allan Schwarz, design coordinator
Materials: steel, steam, light, honey locust trees, greenwave shrubs
Photo: Thad Tercyak

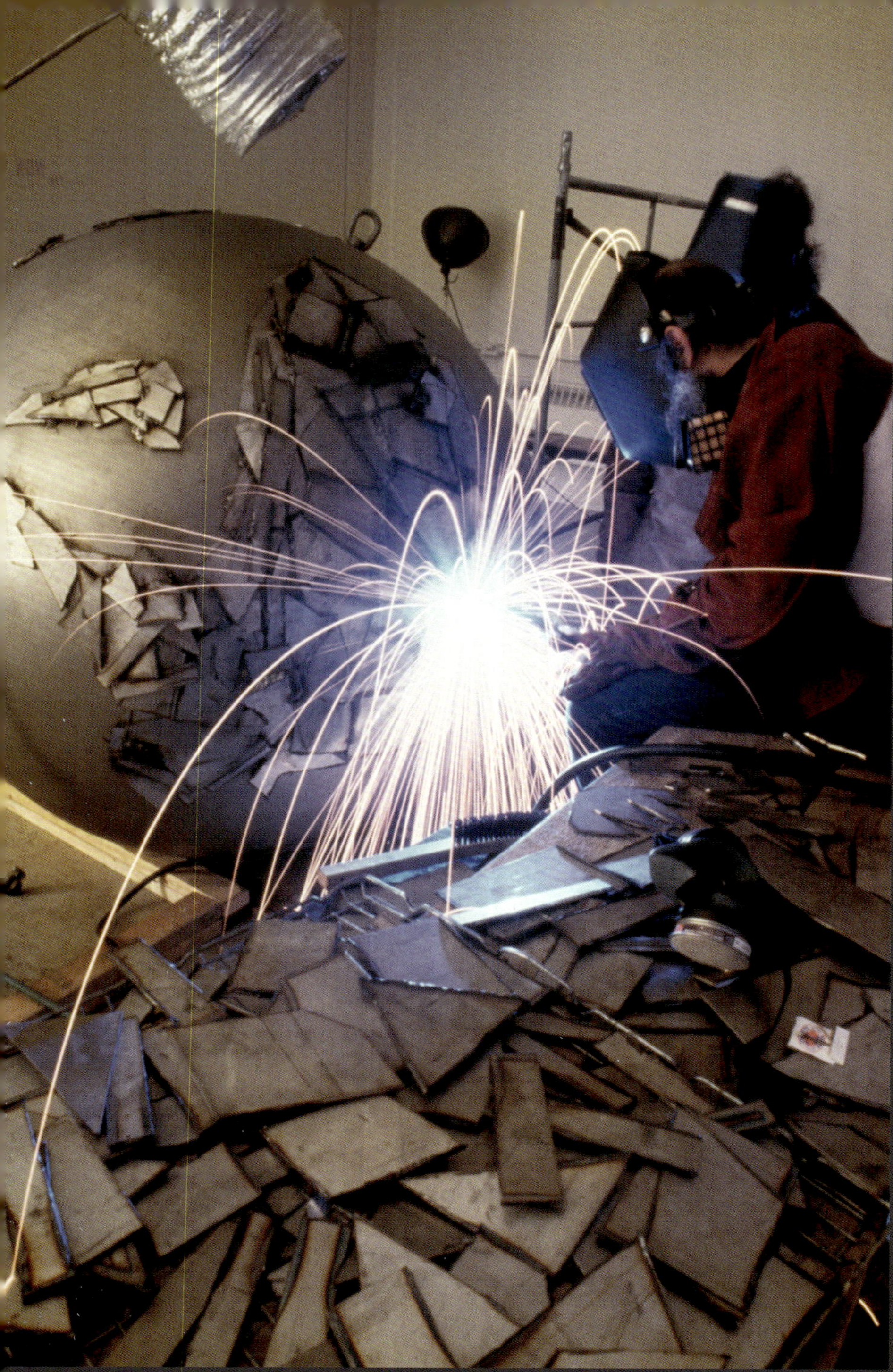

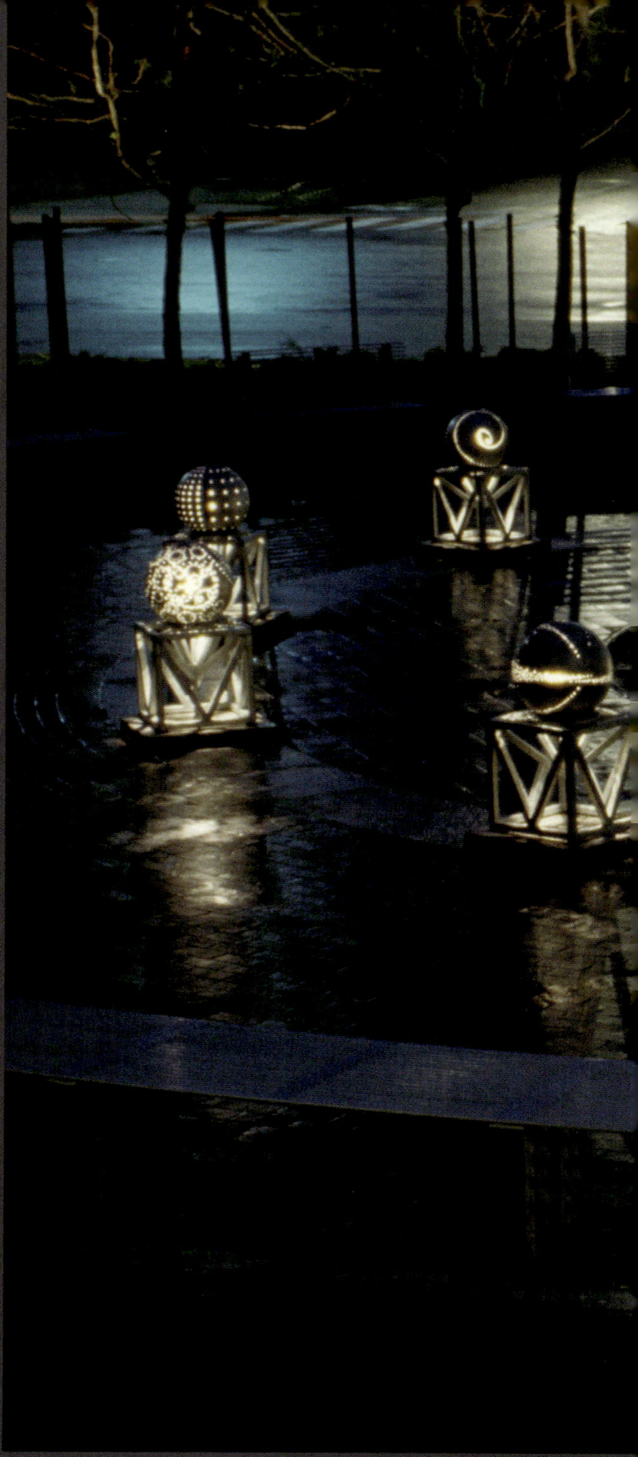

Galaxy is a multi-stage environmental sculpture tightly integrated with park design, so that the traditional boundaries between "park" and "art" are transcended. *Galaxy* celebrates the natural elements of earth, air, fire, water in an urban setting. Its nighttime aspect celebrates light and technology, hinting at its proximity to MIT. . . . pieces of stainless steel were welded onto a 5-foot-diameter stainless steel planetary sphere. Depending on the season, a water spray, mist or steam envelops the sphere, to create its appearance of "floating." (Cambridge Redevelopment Authority, September 14, 1990)

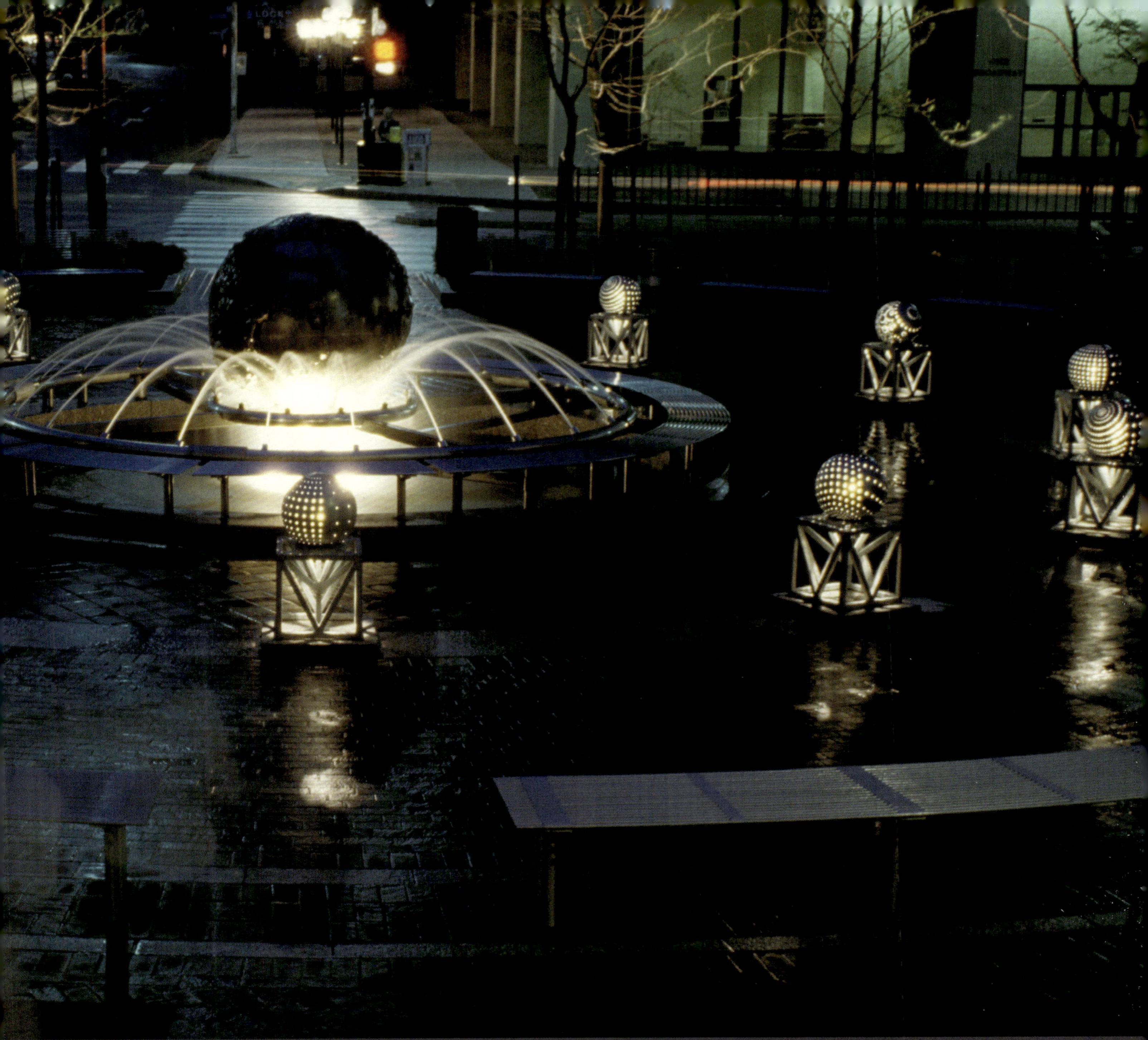

Facing page: Joe Davis fabricating "Earth Sphere," for *Galaxy* fountain, the Pit at CAVS, MIT, 1988-1989
Photo: Otto Piene Collection

Top: *Galaxy*, fountain with *Earth Sphere* and light sculptures designed and fabricated by CAVS artists Otto Piene, Joe Davis, Joan Brigham, Allan Schwarz, for Thomas Murphy Park, Halvorsen Company, landscape architects, Kendall Square, Cambridge, 1988-1990
Photo: Thad Tercyak

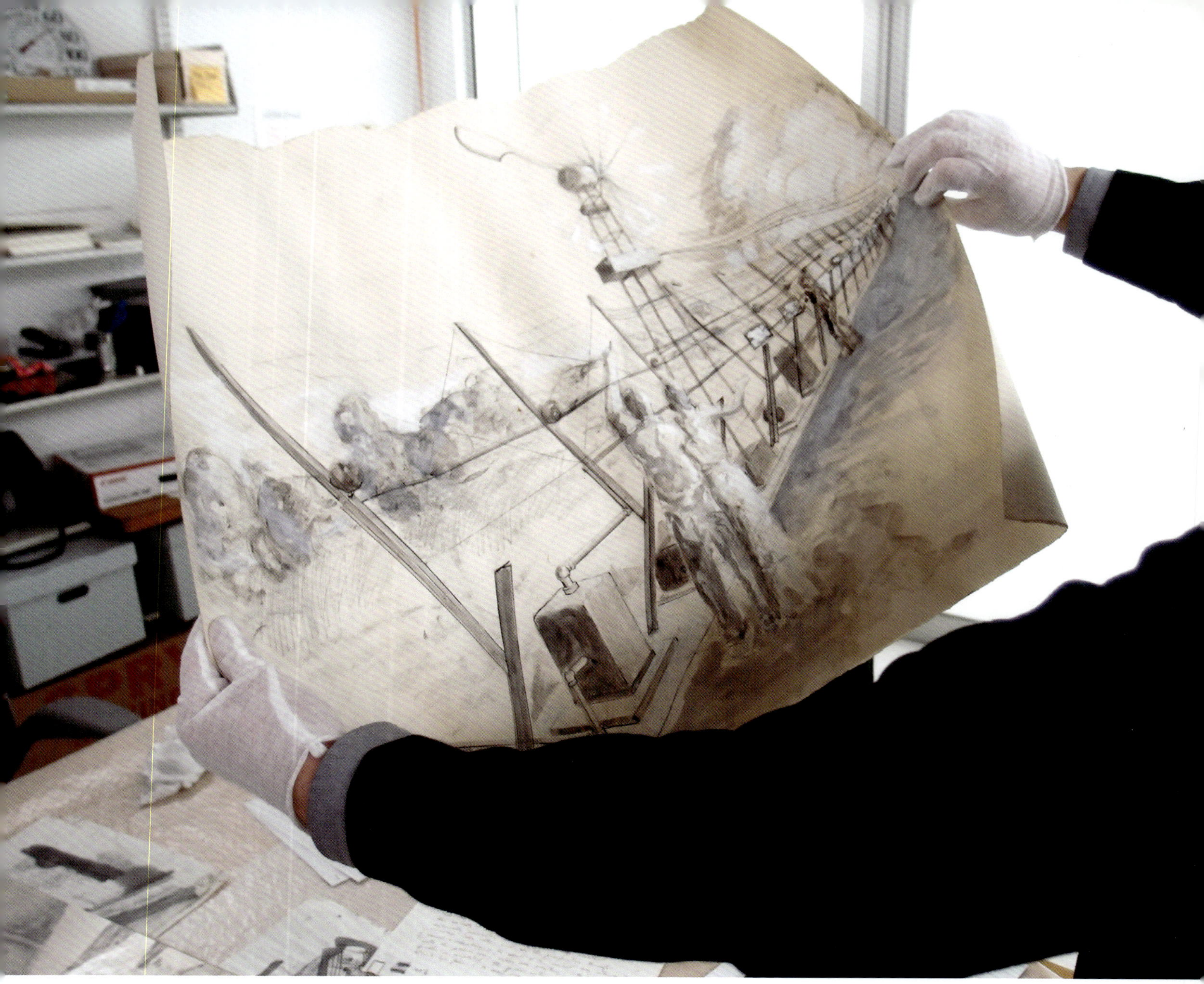

The legacy—the physical, aesthetic, and collaborative "bones"—of CAVS's 1977 project, *Centerbeam*, was examined forty years later in an exhibition mounted by ZKM | Karlsruhe. Peter Weibel describes the project in the text that follows.

Top: Archival view of concept artwork for *Centerbeam* by Lowry Burgess created at CAVS, MIT (1976-77), here, in preparation for the exhibition, *Centerbeam. A Performative Sculpture by CAVS*, ZKM | Karlsruhe, May 24-October 1, 2017
Photo: Tobias Wootton, ZKM | Center for Art and Media Karlsruhe

Centerbeam. A Performative Sculpture by CAVS, ZKM | Karlsruhe, May 24-October 1, 2017

With the exhibition *Centerbeam. A Performative Sculpture by CAVS*, the ZKM | Center for Art and Media Karlsruhe marks its 40th anniversary and, parallel to this year's *documenta 14*, recalls the legendary contribution of the Center for Advanced Visual Studies (CAVS) at the Massachusetts Institute of Technology (MIT) to the *documenta 6* in Kassel in 1977. The spectacular outdoor installation, which was nearly 44 meters long, was a joint project by CAVS Fellows—fourteen artists plus scientists, engineers, and MIT students— based on an idea by the artist Lowry Burgess. The installation utilized the then newest artistic technologies, such as lasers, holography, steam screens, neon, video, and inflatables to create a multimedia "art machine" as an "aqueduct into the 21st century" (Manfred Schneckenburg). By day the main character of the installation, which reminded one of a greenhouse or a pipeline, was participative with prismatic reflections in the water and holographic effects, and by night the *Centerbeam* transformed into a "friendly inferno" (Otto Piene), a power-charged stage for laser projections, concerts, performances, and sky events.

Originally with the goal of completely reconstructing *Centerbeam* the core idea was to create a *Centerbeam 3*, an updated version of the historic installation. The start was in 2013 with a group meeting held at the Piene farmhouse in Groton near Boston, MA, USA. Never documented in its final form, ZKM in collaboration with the MIT Program in Art, Culture and Technology (ACT), the MIT Museum and PAMAL, Preservation and Art-Media Archaeology Lab, Avignon, France, and all the artists and engineers initially involved set themselves the formidable task of reconstructing *Centerbeam*. Months, even years, of meetings, conversations, exchanges and trips to the USA, were needed to determine which form this complex reconstruction should take. All the remnants of the original *Centerbeam* were therefore shipped as loans of the MIT Museum to ZKM. However the reconstruction of *Centerbeam* as an outdoor sculpture in the gardens of Karlsruhe Palace was abandoned for multiple reasons.

In September 2015, a research trip to the USA was organized to meet again with the original artists and engineers involved with the first and second versions of *Centerbeam*. At two meetings a large number of *Centerbeam*'s artists and engineers participated, either on-site or via Skype. Over the years the ZKM has endeavored to produce and overhaul the scientific documentation on *Centerbeam* and its kinetic, performative, and participatory modes of functioning. The current ZKM exhibition, with the aid of historic archive material and surviving remnants of the original *Centerbeam*, explores the problematic of possible methods of such a reconstruction with reference to contemporary issues of media conservation as a case study reflecting on and relevant to the restoration of the media art collection of the ZKM.

The research on the reconstruction of *Centerbeam*, presented by Morgane Stricot as her PhD thesis topic at the Media Art History Conference in Montréal, was recognized in 2015 by the Leonardo Award for Emerging Researchers in Art, Science and Technology. She proposed three different proposals for a reconstruction, ranging from media archaeological reconstruction to total reinterpretation. *Centerbeam* appears not only as an abridged survey of technological advance since the beginning of the first Industrial Revolution. *Centerbeam* introduced with its term "performative sculpture" a new art form, which became dominant in the beginning of 21st century.

— Peter Weibel

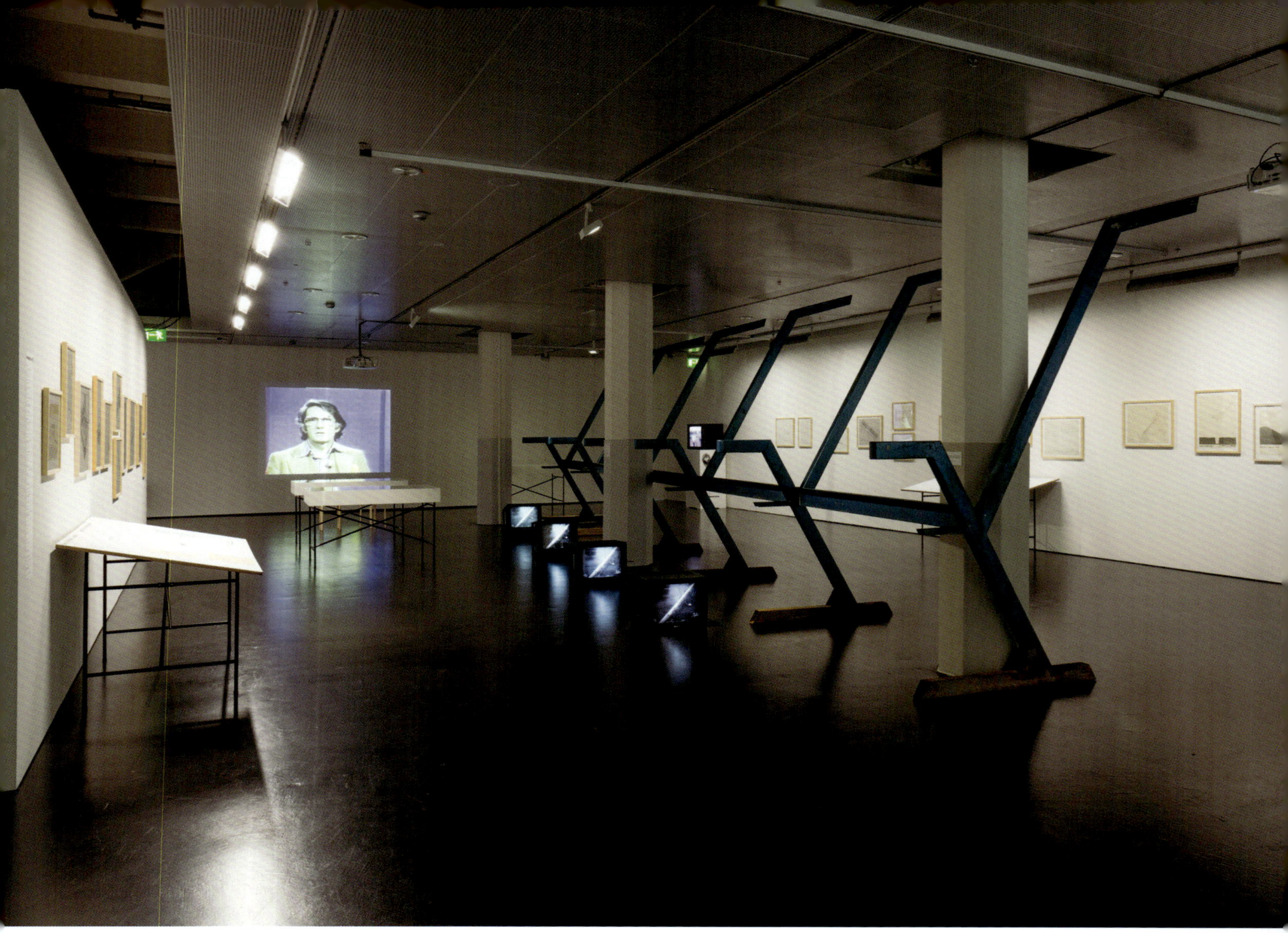

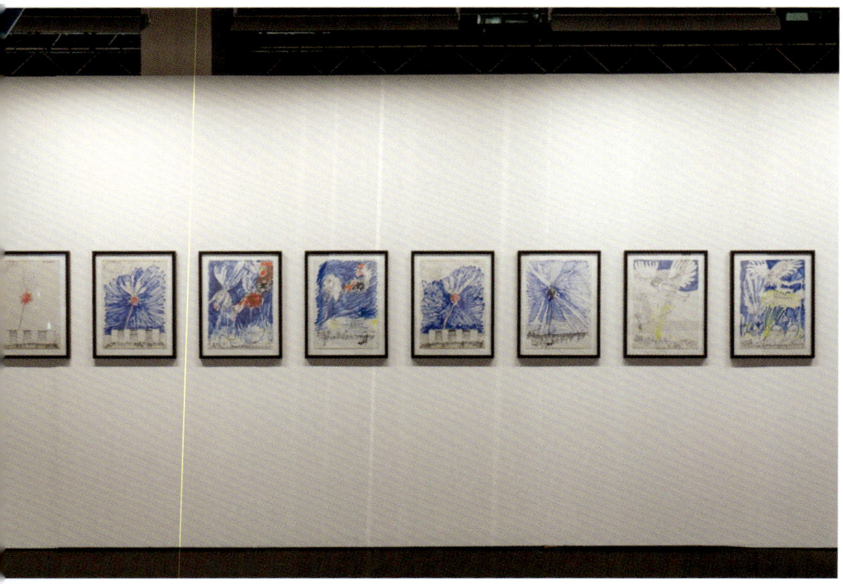

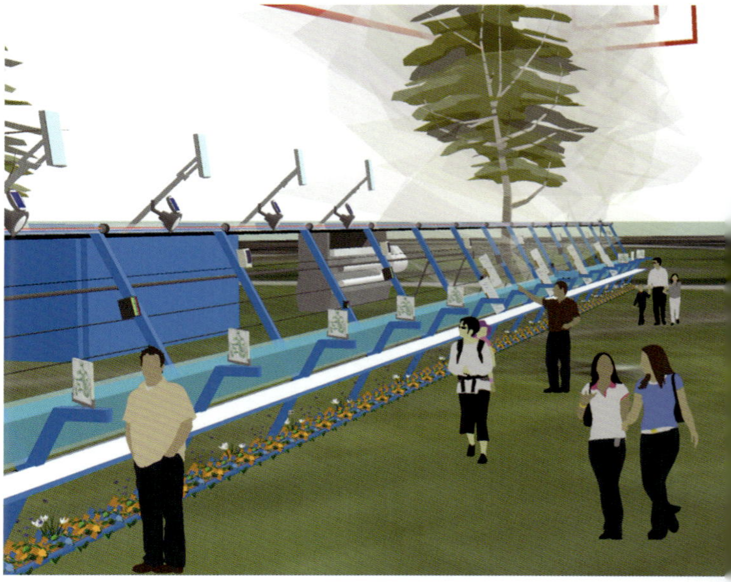

Centerbeam. A Performative Sculpture by CAVS, exhibition views and 3D illustrations of a possible reconstruction of *Centerbeam* at Karlsruhe, ZKM | Karlsruhe, May 24-October 1, 2017
Photos: Tobias Wootton, ZKM | Center for Art and Media Karlsruhe

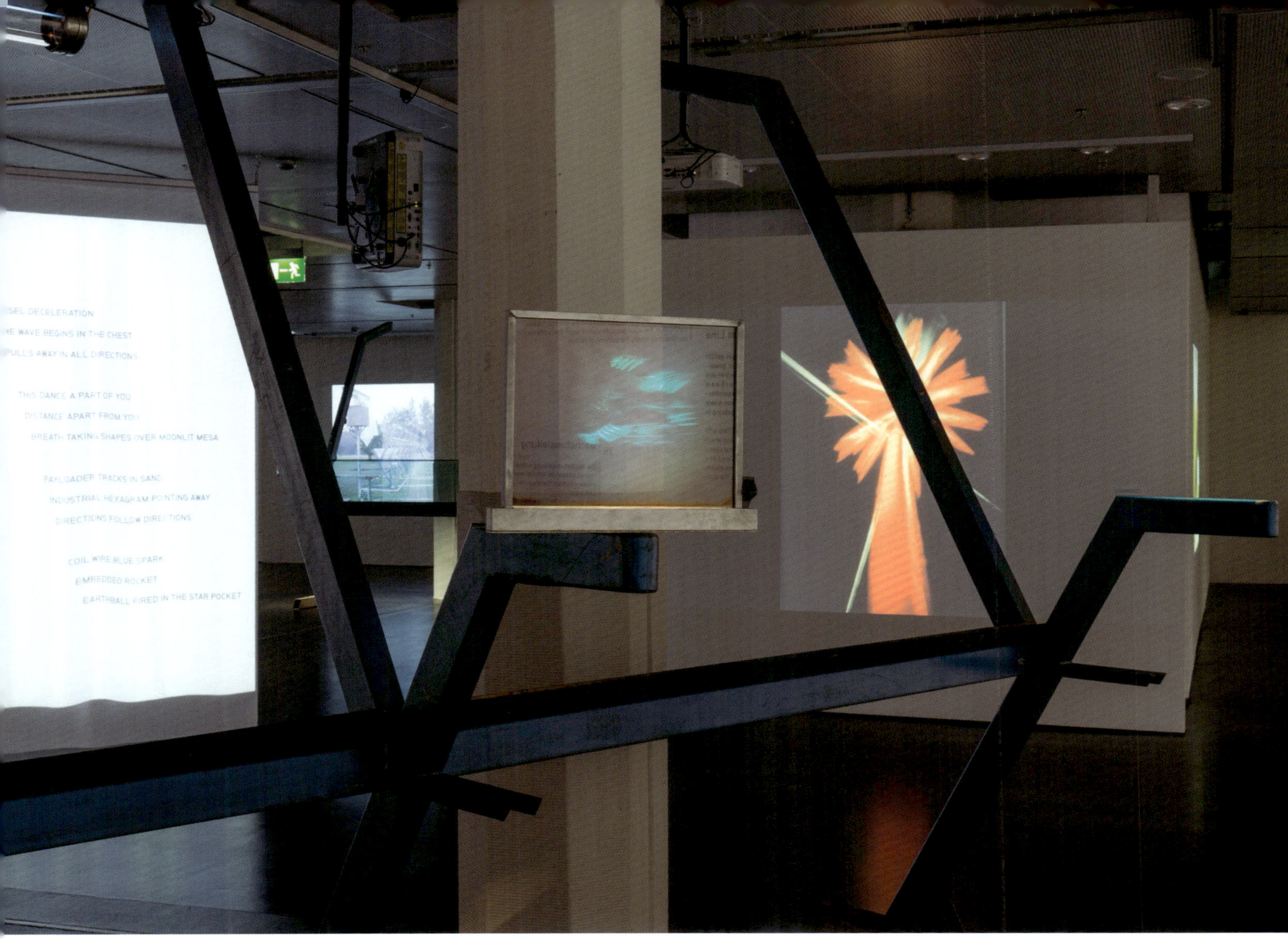
THIS DANCE A PART OF YOU
DISTANCE APART FROM YOU
BREATH TAKING SHAPES OVER MOONLIT MESA
PAYLOADED TRACKS IN SAND
INDUSTRIAL HEXAGRAM POINTING AWAY
DIRECTIONS FOLLOW DIRECTIONS
COIL WIRE BLUE SPARK
EMBEDDED ROCKET
EARTHBALL FIRED IN THE STAR POCKET

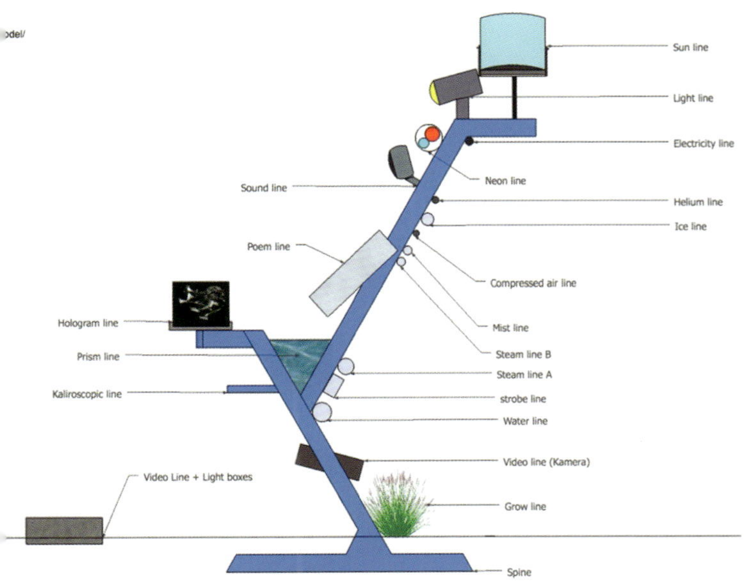
Sun line
Light line
Electricity line
Sound line
Neon line
Helium line
Ice line
Poem line
Compressed air line
Hologram line
Mist line
Prism line
Steam line B
Steam line A
Kaliroscopic line
strobe line
Water line
Video line (Kamera)
Video Line + Light boxes
Grow line
Spine

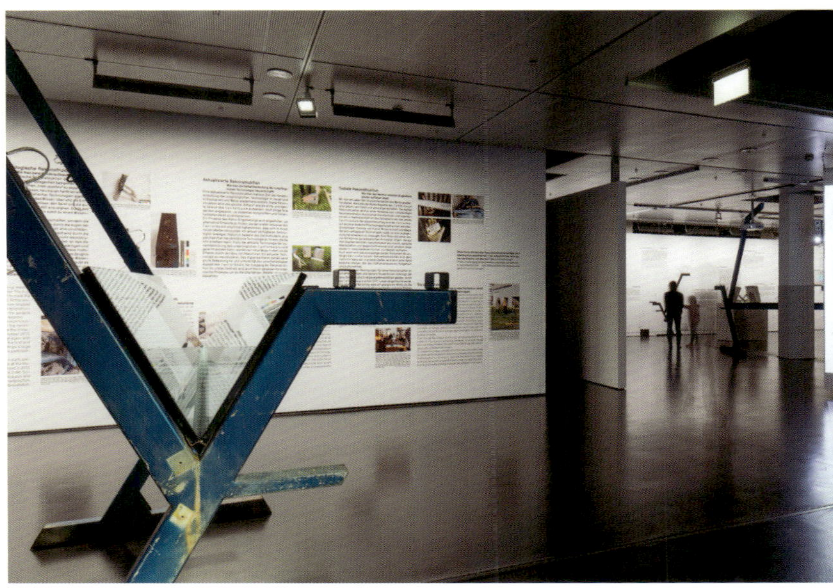

In the following text, Weibel provides background on the "genome" of art-science-technology practice in the twentieth century. "This new alliance of art and technology, from research-based art and art-based research to cross-disciplinary and trans-disciplinary research," he writes, stems from "the DNA of CAVS."

CAVS and the Genome of Art-and-Technology

by Peter Weibel

A striking feature of art education institutions since the eighteenth century is not simply their frequent modification of the profile of their programs to take account of sociopolitical and cultural changes, but that such reorientations of the teaching program were accompanied by giving the institution a new name. Broadly speaking, three main phases can be identified: up to 1900, the academy of fine arts (*Académie des Beaux-Arts*) dominated. In the twentieth century "*fine*," and with it the emphasis on beauty, was banished from art which is why the word "*design*" came to be used increasingly. Finally, one only spoke of "visual research." The word "*art*" disappeared completely.

In the seventieth and eighteenth centuries *Academies of Fine Art* were founded in Europe (Berlin 1696, Vienna 1704, St. Petersburg 1724, Copenhagen 1754, London 1769). The mission of these institutions was not only to foster and propagate a sense of beauty and taste for elegant forms through the arts that were taught there, but also to inculcate love of one's country and respect for everything high and holy.[1] In the mid-nineteenth century, industrialization caused the academies to split off from the trade or vocational schools. After this functional separation into "useful" and "free" art and the distinction between "pure" and "applied" art, it became the remit of the academies to continue working on what was eternally true and pristinely beautiful, while completely disregarding the needs and exigencies of the industrial age.

A School of Painting, Sculpture, and Architecture (MuSChWS) was also founded in Moscow in the mid-nineteenth century; it became the second-largest art school in Russia after the St. Petersburg Academy of Arts. Lenin's decree of 1918 reformed Russia's art schools into Free National Art Workshops (SVOMAS)[2] with the aim of producing highly qualified master artists to work in industry. In the same year the famous avant-garde artist Marc Chagall was appointed Commissar of Arts for the Vitebsk Region. There he founded the Vitebsk Arts College, a revolutionary art institution and one of the most distinguished in Russia. Chagall engaged some of the country's most important artists, including El Lissitzky and Kazimir Malevich, and he envisioned the school would be an independent community of outstanding artists. The old focus on fine arts (painting and sculpture) was extended to include the so-called "production arts," such as architecture, textile design, ceramics, printing, woodwork, and metalwork. In 1920, two Moscow SVOMAS were merged and renamed Vkhutemas[3] (higher art and technical studios), not least due to the engagement of Kazimir Malevich and to El Lissitzky's Proun concept ("project for the affirmation of the new"). At once an embodiment of the ideas of the Russian avant-garde and a revolutionary model of art education, from 1923 to 1926 Vkhutemas saw itself as a center for the study of art, in all its forms and all branches of study, as a universal educational institution and community of artists working in partnership, which included Alexander Rodchenko and Vladimir Tatlin. Similar to the aims of Germany's Bauhaus, it was intended that trades, crafts, and modern technology would lead to art production. However, in practice the differences between the adherents of "pure" and "production" art remained palpable and difficult to reconcile. The programmatic goal, to advance the technicization of art, manifested itself in 1927 with the reform of Vkhutemas as Vkhutein (higher art and technical institute). Instead of a studio environment, the artist-technologists, artist-technical designers, and artist-engineers were to facilitate the rationalization and materialization of artistic work in the environment of an institution. Vkhutein was dissolved in 1930, which spelled the end of interdisciplinary collaboration. Painting and sculpture were transferred to the Leningrad Academy of Arts, and the remainder went to the Architectural-Construction Institute.

It was quite apparent that, from 1900, art had largely abandoned its concept of beauty, which had been essential for the self-image of art academies since the eighteenth century. In the mid-nineteenth century, Karl Rosenkranz, a student of Hegel, published his *Ästhetik des Häßlichen* (1853) [Aesthetics of Ugliness]. Up to that point, beauty as the sensual manifestation of an idea had been the subject of aesthetics; now ugliness and even evil became components of it as well. At around the same time as the *Aesthetics of Ugliness* appeared, Charles Baudelaire's collection of poems *Les fleurs du mal* (1857) [The Flowers of Evil] was published. Both of these authors, Rosenkranz and Baudelaire, whose works are foundational for modern art, welcomed ugliness and evil to the stage of the modern era.[4] The ordinary, the banal, the commonplace, the repellent, the grotesque, the obscene, the evil, the ugly, the diabolic – these all became part of the art canon and cornerstones of modern art when the concept of fine art was enlarged and expanded.[5]

The industrial revolution of the nineteenth century provided the basis for the aesthetic strategies of the twentieth century. This is why Futurism did not sing the praises of the lovely Nike of Samothrace, but eulogized the beauty of automobiles: "We affirm that this wonderful world has been enriched by a new form of beauty, the beauty of speed [...]. A racing car [...] is more beautiful than the *Victory of Samothrace*."[6] This is why Le Corbusier in his books *Vers une architecture* (1923) [Toward an Architecture] and *Le Modulor 2* (1955) [The Modulor 2] made a provocative comparison between modern architecture and the automobile. This is why artists chose to be portrayed as engineers (for example, El Lissitzky's self-portrait *The Constructor*, 1924) and terms such as "*construction*" and "*design*" proliferate in works' titles. This is also why the early volumes of the Bauhaus book series had titles like *Neue Gestaltung* [New design], *Neoplastizismus* [Neoplasticism], *Nieuwe Beelding* [New plastic art] (Piet Mondrian, Vol. 5, 1925), *Grundbegriffe der neuen gestaltenden Kunst* [Basic concepts of the new plastic art] (Theo van Doesburg, Vol. 6, 1925) and journals bore titles like *bauhaus. zeitschrift für gestaltung* [Bauhaus. Journal for design] (1926–1931, edited by Walter Gropius and Laszlo Moholy-Nagy et al.) or *G. Material zur elementaren Gestaltung* [G. Material for elementary design] (1923–1924, edited by Hans Richter).

It was transparently clear: the aesthetic of machines had supplanted the aesthetic based on artisanship. This is the reason why the term "*design*" had superseded the term "*fine art*." Art colleges retained the appellation "academy of fine arts"[7] up to the close of the nineteenth century, but from the mid twentieth century onward they became *Hochschulen für Gestaltung* in German-speaking countries—colleges or universities of design.[8]

As already mentioned above, the famous Bauhaus in Weimar exerted a decisive influence on these developments. Founded in 1919 by Walter Gropius, who was its director until 1928, the aim of the Bauhaus was to create an art that abolished the distinction between artist and craftsman—as formulated by Gropius in the *Bauhaus Manifesto* of 1918. The result was a universal visual language, shaped by exponents of Constructivism like Moholy-Nagy or exponents of de Stijl like Theo van Doesburg as teachers there, which could be applied equally to paintings, sculpture, furniture, buildings, or media such as photography and film.

Even in the 1950s many colleges in Germany continued to invoke the tradition of the Bauhaus and called themselves *colleges of design*, such as the famous *Hochschule für Gestaltung in Ulm* [Ulm School of Design] (HfG), which was founded in 1952 by architect and Constructivist Max Bill, and which in close collaboration with industry brought forth many a design icon. Through Bill, the Ulm School of Design continued the program of the Dessau Bauhaus and expanded it, also with the goal of stabilizing democracy in the young Federal Republic of Germany. Further members of the teaching staff in Ulm were the aesthetician of information theory Max Bense, Abraham Moles, and industrial designer Dieter Rams, whose *10 Good Design Principles* and his designs for Braun consumer goods inspired Steve Jobs and his Apple products.[9]

Many prominent members of the Bauhaus were Hungarian: László Moholy-Nagy, Ernő Kállai, Farkas Molnár, Sándor Bortnyik, Alexander László, and others. The Bauhaus was domiciled in Weimar from 1919 to 1925, subsequently in Dessau from 1926 to 1932, and at the end, from 1932 to 1933, in Berlin. When the Nazis came to power, they closed down the Bauhaus.

Bortnyik returned to Hungary where he founded and directed, from 1928 to 1938, the Műhely School of Commercial Art—also known as Little Bauhaus or Budapest Bauhaus—where, for example, Victor Vasarely studied, one of the founders of Op Art.

László Moholy-Nagy emigrated to the USA, where in Chicago he founded and directed the *New Bauhaus*, later renamed *School of Design*, then *Institute of Design*, and finally the *Illinois Institute of Design* (IIT). Moholy-Nagy was succeeded as director by his student György Kepes, a painter, designer, and art theorist born in Hungary, who in 1967 was commissioned by the Massachusetts Institute of Technology (MIT) in Boston to found a new center, where Kepes taught until his retirement in 1974. On the basis of the alliance between art and machines, the goal of the *Center for Advanced Visual Studies* (CAVS) founded by Kepes was a new cross-linkage of art and science.[10]

The term *Visual Studies* has a European genealogy. In 1960, the Centre de Recherche d'Art Visuel [Visual Art Research Center] was founded in Paris and officially renamed Groupe de Recherche d'Art Visuel [Visual Art Research Group] (GRAV) in July 1961. We notice that in the name of the group the term "research" appears alongside the term "visual art." The members of the Zagreb group *New Tendencies*[11] were inclined to delete the term "art" altogether and replace it with "visual research" to "give rise to a greater awareness of what is in the process of coming into being in the visual arts."[12] This methodological approach to art, which could be considered as being "scientific" from the outset, had been prepared by the practices of Constructivist and Concrete art.

Within the *Nouvelle Tendance*[13] movement the idea arose of abandoning the term "art" altogether, and speaking in the future only of "visual research." Thus, half-way into the twentieth century, art became a field of visual research. Whence the name Center for Advanced Visual Studies (CAVS). After Kepes, Otto Piene, who had once been part of the New Tendencies, was director of CAVS from 1974 to 1994. In this period not only advanced art trends such as Kinetic Art, Light Art, and Sound Art were incorporated, but also new art forms such as Television Art, Video Art, Holography, Performance Art, Telecommunication and Network Art, Digital Art, and BioArt. In other words, CAVS focused for the first time on media. And thus assumed a central, pioneering role in the 1970s. CAVS served as a starting point and model for many institutions founded in the 1980s that engaged with new media in research, teaching, and practice. First and foremost among them is the MIT Media Lab, founded in 1985 by Nicolas Negroponte. At around the same time, in 1984, as professor of visual media design at the University of Applied Arts Vienna, I established the first media art class at a European university.

In the same year a proposal was tabled in Karlsruhe to found an "interdisciplinary center for research, education, and artistic activities,"[14] which culminated in the foundation of two unique institutions: the ZKM | Center for Art and Media Karlsruhe in 1989, and the Staatliche Hochschule für Gestaltung—Karlsruhe University of Arts and Design—in 1992. The ZKM, of which I have been Chairman and CEO since 1999, owes its foundation to a great degree to the inspiration provided by CAVS.[15] Moreover, Otto Piene was one of the ZKM's most important and supportive founding fathers. The ZKM calls itself the "electronic Bauhaus"[16] (Jürgen Claus, who was a Fellow and Research Affiliate at CAVS from 1983 to 1988) and also the "digital Bauhaus" (Peter Weibel) to make it clear that the ZKM, like CAVS, is a further development of the Bauhaus in the age of electronic and digital media.

As we have seen, at the end of the twentieth century not only the term "*fine art*" was deleted, but the term "*art*" itself; "*design*" will also disappear in due course. Their places will be taken by the terms *visual research* or *visual studies* or *visual culture and media*. CAVS played a pivotal role in this paradigm shift.

An important offspring of CAVS, therefore, is the Academy of Media Arts Cologne, whose Founding Director was Manfred Eisenbeis. KHM Director Siegfried Zielinski (1994-2000) succeeded Heinrich Klotz and Peter Sloterdijk at the Karlsruhe University of Arts and Design as Rector from 2016 to 2018. The Center for Media Study of the State University of New York at Buffalo, founded in 1972 by Gerald O'Grady, where the avant-garde filmmakers Hollis Frampton, Paul Sharits, and Tony Conrad as well as the video artists Steina Vasulka, Woody Vasulka, and Peter Weibel taught, was also seminal and pioneering.

Before the advent of CAVS, virtually no art institution had a media art center. Today, however, from Walt Disney's CalArts (California Institute of the Arts) in Los Angeles to CAFA in Beijing, from Nanyang Technological University Singapore to Karlsruhe there is no university without media art—and from Ernst Caramelle to Ute Meta Bauer many former Fellows of CAVS or MIT teach at these institutions. This is another success story of CAVS.

This new alliance of art and technology, from research-based art and art-based research to cross-disciplinary and trans-disciplinary research, will prevail in the twenty-first century, and the DNA of CAVS, beyond any doubt, is in the genome of this new alliance.

1 See Wolfgang Ruppert, Der moderne Künstler. Zur Sozial- und Kulturgeschichte der kreativen Individualität in der kulturellen Moderne im 19. und frühen 20. Jahrhundert, Frankfurt am Main: Suhrkamp 1998, p. 481.

2 Russian acronym for *Svobodnye gosudarstvennye khudozhestvennye masterskiye.*

3 This Russian acronym for Vysshiye Khudozhestvenno Tekhnicheskiye Masterskiye also appears in the literature in the allographic variants VKhUTEMAS, VChutemas and WChUTEMAS. See Hubertus Gassner and Eckhart Gillen (eds.), Zwischen Revolutionskunst und Sozialistischem Realismus. Dokumente und Kommentare. Kunstdebatten in der Sowjetunion von 1917–1934, Cologne: DuMont, 1979; as well as Irina Tschepkunowa (ed.), WChUTEMAS. Ein russisches Labor der Moderne. Architekturentwürfe 1920–1930, WChUTEMAS, exhib. cat., Martin-Gropius-Bau, 05.12.2014–06.4.2015, Berlin: Berliner Festspiele 2015.

4 The ground for the aesthetic revolution of modernism around 1900 had been prepared by the industrial revolution. Previously, the arts had been dedicated to beauty and the sublime—Romanticism, for example; then, with the extension of the arts to embrace ugliness and evil, the foundations of Modern art were laid. Thus, fine arts in the proper sense ceased to exist after the middle of the nineteenth century. Since that time there has been the hegemony of the unfine arts. The fine arts have been deposed, although some museums still call themselves Palais des Beaux Arts as in Brussels. Beauty is now something for fashion, not for art anymore.

5 An ugly object, a urinal (Marcel Duchamp, Fontaine), was presented in 1917 under the title ready-made as art.

6 F. T. Marinetti "Fondanzione e Manifesto del Futurismo (Pubblicato dal Figaro di Parigi il 20 febbraio 1909)," trans. in: Futurism. An Anthology, ed. Lawrence Rainey Christine Poggi, and Laura Wittman (New Haven, CT: Yale University Press, 2009), p. 51.

7 For example, Académie des Beaux-Arts in Paris, Accademia delle Arti del Disegno di Firenze, and Bayerische Akademie der Schönen Künste in Munich.

8 In Germany there are—or were—colleges of design in Berlin (BTK—Hochschule für Gestaltung), in Ulm (1953-1968), in Offenbach am Main (1970-), in Schwäbisch Gmünd (1971-), in Pforzheim (college of design founded in 1971, a merger of the United Goldsmiths and crafts school; expanded in 1992 to include engineering and business school), in Karlsruhe (Staatliche Hochschule für Gestaltung—Karlsruhe University of Arts and Design since 1992), and in Schwäbisch Hall (2000–2013). In Switzerland, the Hochschule für Gestaltung und Kunst Zürich (HGKZ) – Zurich Canton University of Design and Art – evolved from the Kunstgewerbeschule Zürich – Arts and Crafts School of Zurich City, founded in 1878; the Hochschule für Gestaltung und Kunst in Basel – FHNW Academy of Art and Design – emerged from the Basel Zeichnungs- und Modellierschule (school of drawing and modelling) which was founded in 1796.

9 See Jaroslaw: "How Dieter Rams Inspired Some Apple Products": https://theultralinx.com/2014/03/dieter-rams-inspired-apple-products/ (last accessed: 20.3.2018).

10 For example, György Kepes (ed.), *The New Landscape in Art and Science*, Chicago: Paul Theobold 1956; or György Kepes (ed.), *Structure in Art and in Science*, New York: George Braziller, 1965.

11 The term *nove tendencije* [New Tendencies] was also the title of the exhibition in Zagreb in 1961.

12 GRAV published the brochure Groupe de Recherche d'Art Visuel, Paris, 1962, in collaboration with the Galerie Denise René. The brochure contains a text on the New Tendencies, April 1962, n.p.

13 See Margit Rosen (ed.), *A Little-Known Story about a Movement, a Magazine, and the Computer's Arrival in Art: New Tendencies and Bit International*, 1961–1973, ZKM and The MIT Press: Cambridge, MA, 2011.

14 Michael Heck "Die Akteure und ihre Konzepte I: Der Weg von den Projektskizzen zum Basispapier", in: Rolf Funck, Michael Heck, and Peter Weibel (eds.), Das ZKM | Karlsruhe. Die Anfänge der Zukunft, Paderborn: Fink 2014, pp. 65–74, here p. 68.

15 See Zentrum für Kunst und Medientechnologie Karlsruhe: Konzept '88, published by the City of Karlsruhe, pp. 5–8.

16 Jürgen Claus, *Das elektronische Bauhaus. Gestaltung mit Umwelt*, Edition Interfrom, Osnabrück: 1987.

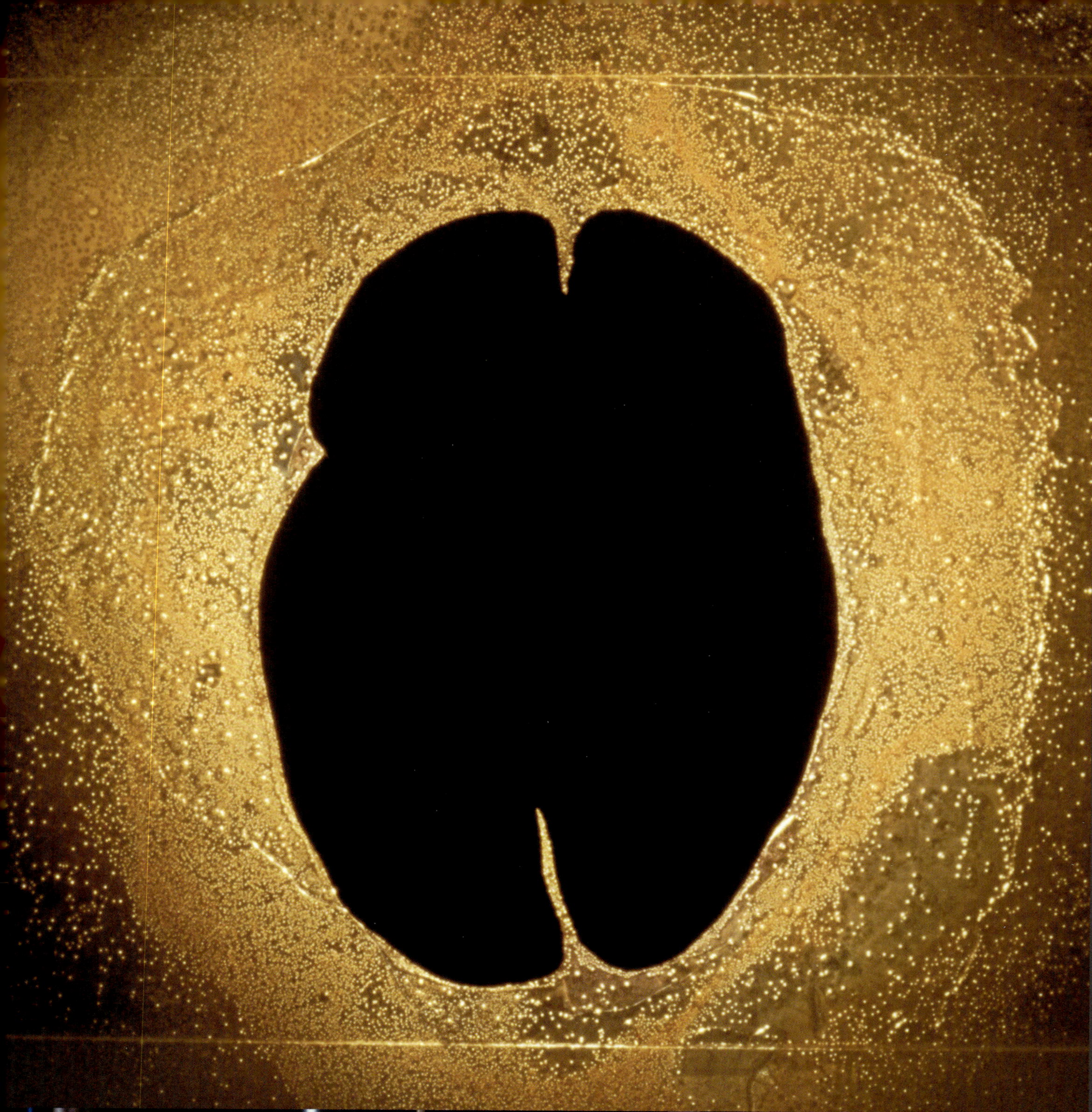
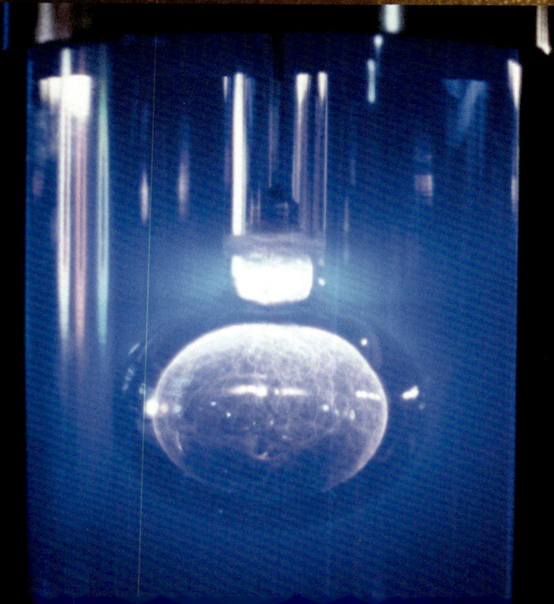
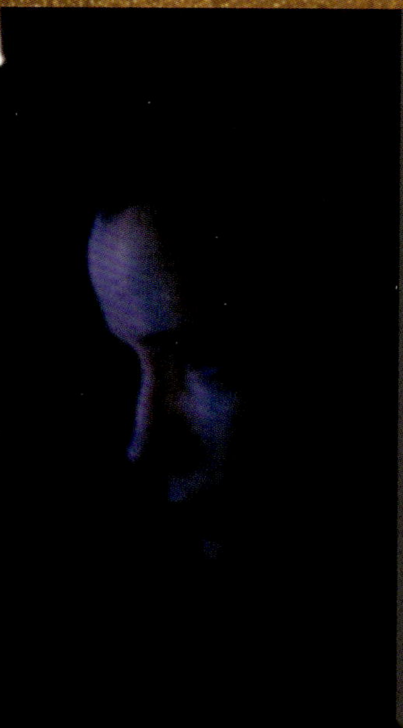
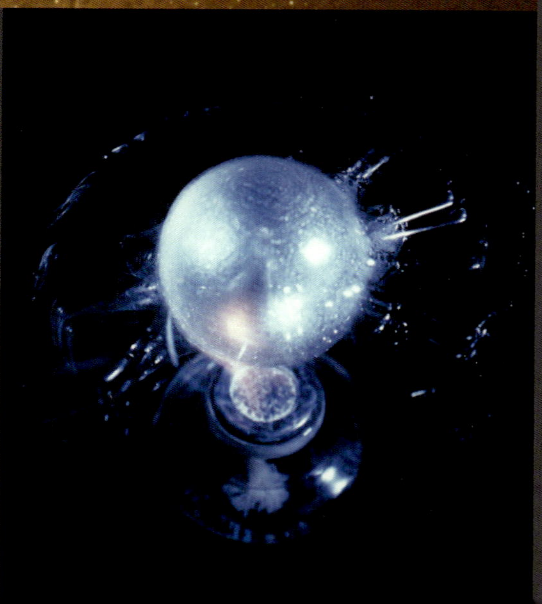

True to a genetic code rooted in the 1960s revolution, CAVS resisted absorption by various entities at MIT and likewise remained ambivalent toward typical definitions of art world success. Increasingly outcast in later years, CAVS advanced the validity of artistic vision, experimentation, and community independent of the corporate partnerships that would have sustained it. Just as it was born of revolution, the Center died in the same spirit of resistance even as new media and the field of art, science, and technology grew up to become mainstream.

Dean of Architecture William Mitchell enthusiastically supported CAVS in the 1990s with a base at the Media Lab, but the move to several new homes at MIT seemed to shift artistic interactions toward collective, individual rather than collaborative works. Already in 1978, in reference to a proposed arts complex, Piene wrote:

> My thoughts for a new art building begin—in Bauhaus tradition—with the building: its shape, function, expression. The structure itself should express art, poetry, human inner movement. (The Eiffel Tower, the Statue of Liberty, space flight were not there before somebody started believing they could be.) The art buildings complex (ABC) is therefore not an architect's or designer's job. It ought to be the creation of a team of involved environmental artists, architects and engineers—an integrated group work (as opposed to committee design) leading to a functional, sculptural edifice whose very shape is vital, organic and represents major human concerns. It could become a human shape; or a rose; or a "star." The building can be made of glass or concrete according to practical needs. The structure should not represent a style, a fashion or a habit. It ought to be an "all spirit" expression of the community of artists, scientists and engineers working in the arts and in industries catering to the arts. (Piene, "The A.B.C. (Art Building Complex), or Arts Building Complex," memo, December 12, 1978)

The response to MIT's criticism of Center activities lies in the collaborative works and discoveries presented in the final chapter of this book, "Work at CAVS." If it was the art world that MIT needed, Kepes was an artist with museum and gallery shows under his belt, and Piene was continually listed as one of the leading artists internationally. While MIT routinely hired leaders in their fields, like Kepes and Piene, the collective focus of the Center placed experimentation over established directions.

The genome of Center collaborations is embodied in many of the fellows' artistic DNA inspiring new generations and new disciplines. The CAVS Special Collection at ACT remains a rich treasure for the analysis of the environment and social-political climate that encouraged both interdisciplinary work and all hands working toward larger goals—the kind of collaboration that forced the development of new fields and new media. Work at CAVS thrived in an unlikely combination of independence and interdependence, tough pragmatism and explosive optimism.

Facing page top: Todd Siler, "Book Form: Neural Nebula," mixed media on plexiglass, 1982
"This Mind Icon sculpture envisions the human brain as a dynamic web of interrelated events, its study expanded to a new discipline: the ArtScience of Neurocosmology, the brain-universe relation." (Todd Siler, *Breaking the Mind Barrier,* Simon & Schuster, 1990)
Photo: Otto Piene Collection

Facing page bottom left: Alejandro Siña, "Corona Discharge," Moira Siña observes the air/water sculpture, 1976
Photo: Nishan Bichajian

Facing page bottom right: Joan Brigham, "Twister," kinetic steam-driven "aeolipile" glass sculpture, 1988
Photo: Walter C. Dent

Chapter 9
Work at CAVS

The story of CAVS lies in the artwork produced there and the narrative advanced through the extraordinary visual output of the Center's diverse international community of professional artist fellows and graduate students. The works themselves convey discoveries within emerging artistic terrains. From the feminist reincarnation of the ancient Venus von Willendorf in Casdin-Silver's life-sized hologram, to painting the Boston skyline with floating flowers in Piene's sky art celebrations, artistic works re-envisioned the physical environment, including outer space, the body, society, nature, and the emergent digital realm. The images reveal the roots of art forms advanced at CAVS, including bio art, celebration, environmental art, holography, laser and light art, music and sound, performance, poetry, sky art, telecommunications, and video art. The visual history evokes the spirit and atmosphere of CAVS, highlighting work that predated much of today's media art by several decades.

Right: Mark Mendel (right) assisted by Wendelin Glatzel, "Slo-Scan Poem," the Slow Scan performance built a wall using bricks painted with words by Mendel; images of the wall were transmitted via Slow Scan at about one image per minute to Limberg, Germany, gradually forming a poem, the Pit, CAVS, MIT, 1979
Photo: Otto Piene Collection

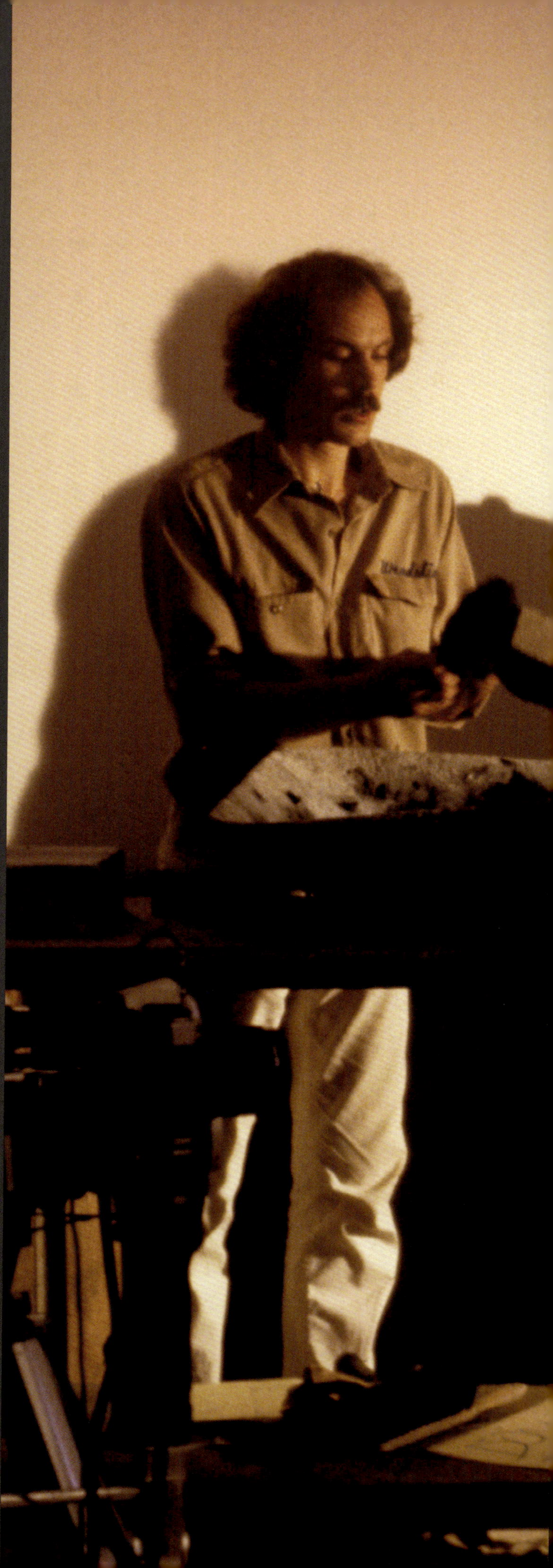

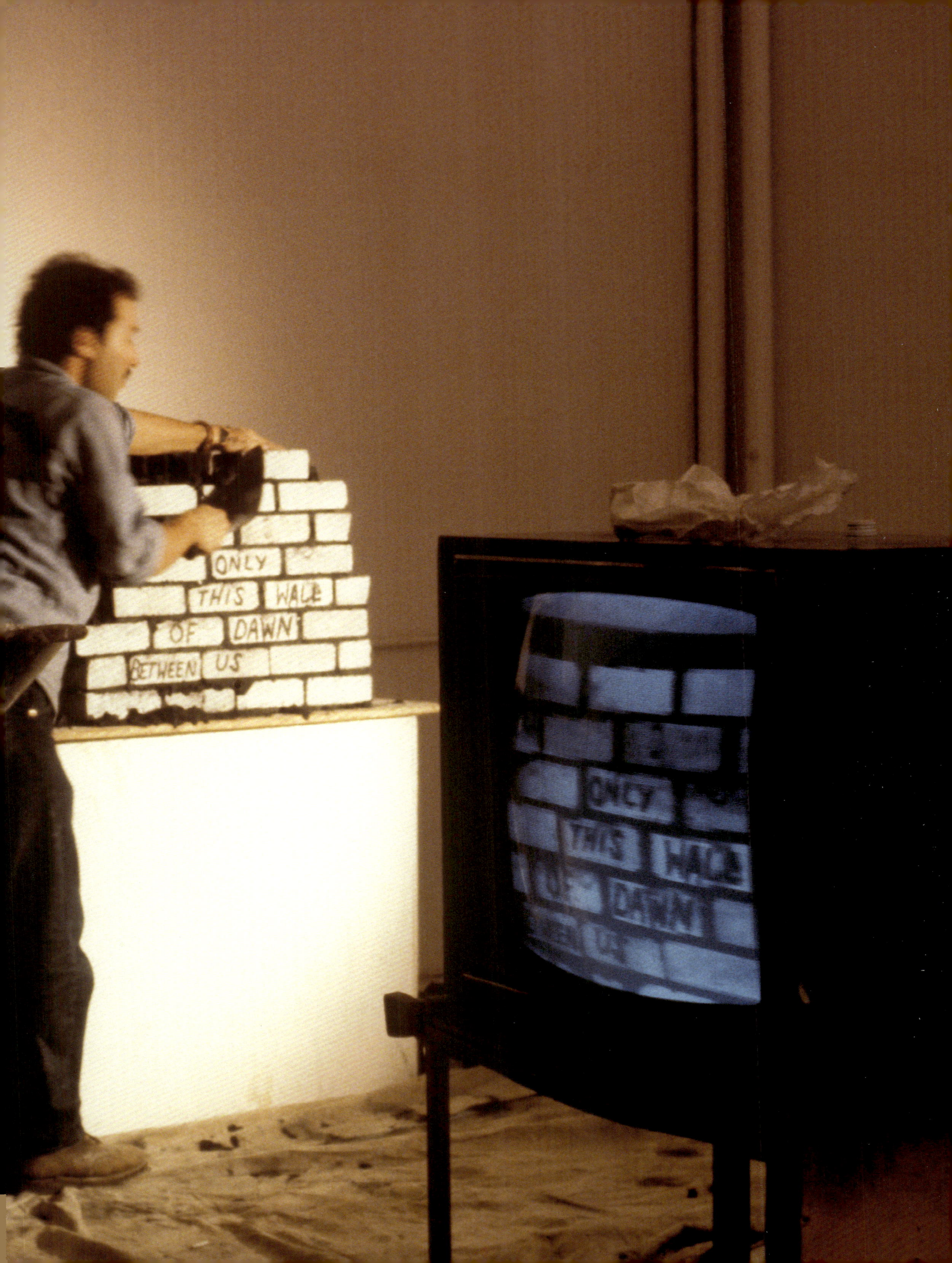
ONLY
THIS WALL
OF DAWN
BETWEEN US

This chapter presents a sampling of work explored by CAVS artists while at MIT. The selections—far from comprehensive—complement other images and descriptions appearing earlier in the book. *Centerbook* covers CAVS as a whole, its methods, and collaborative output rather than individuals. That being said, the diverse trajectories of individual artists energized CAVS as a collective. Alphabetically arranged, this view of the work is impressionistic, rather than literal, as artworks transcend categorization:

Art and Architecture
BioArt | Brain | Body
Celebration
Environment
Food
Holography
Light
Performance
Photography
Public Art | Steam
Sculpture
Sky Art
Telecommunications
Video | Film

Capturing the oeuvre of CAVS is like catching lightning in a bottle. The Center was as much about thought, and interactions that challenged the status quo, as tangible artworks. Jack Burnham's *Beyond Modern Sculpture* (1968), completed while he was a fellow at CAVS, "posited technology as contemporary sculpture's 'controlling force' and predicted that sculpture would 'eventually simulate living systems'." (*Artforum*, February 25, 2019)

> I might have given him [Piene] a syllabus of a course I was teaching called "Systems and Art" based on modern systems theory, very much an anti-object, environmentalist approach that regarded so-called art works as quasi-organic and disposable. (Jack Burnham interview)

True to its roots in the fractious 1960s, CAVS hosted artists whose work challenged the intended uses of technology, scientific protocol, and art world traditions, often playing the role of artist-as-provocateur. The interviews conducted by Joan Brigham reveal the restless urge of CAVS artists to subvert the expected.

> When I came to the Center, video was no big deal for me anymore. I was not interested in doing so-called "art video."... it became a popular thing, and it didn't interest me. When something gets popular, I do something else. (Aldo Tambellini interview)

Like György Kepes, Otto Piene believed that the inspiration for new art forms lay in the environment itself.

> Nature was there to provide inspiration, to provide the scale, to provide the sense of energy, to provide energy itself, to provide the energy of ideas. . . . So, this whole business about energy as an art form, or as the force of art, became very important at the Center. Hence, the addiction to light. The answers to why the Center for Advanced Visual Studies was there, in one form or another in many different formulations, was time and again the great integrating force—light. (Otto Piene, interview by Bill Seaman, 2012)

Facing page top: Gillian Wise Ciobotaru, self-portrait, London, 1982

Facing page bottom left: Gillian Wise Ciobotaru, mural, Barbicon Arts Centre, *Art Into Architecture* exhibition, London, 1981

Photos: CAVS Special Collection

Facing page bottom right: Gillian Wise Ciobotaru, photo (from brochure) of full scale model of three walls for "Art into Architecture" exhibition, the Pit, CAVS, MIT, 1981
Photo: Sarah Dickinson

Art and Architecture

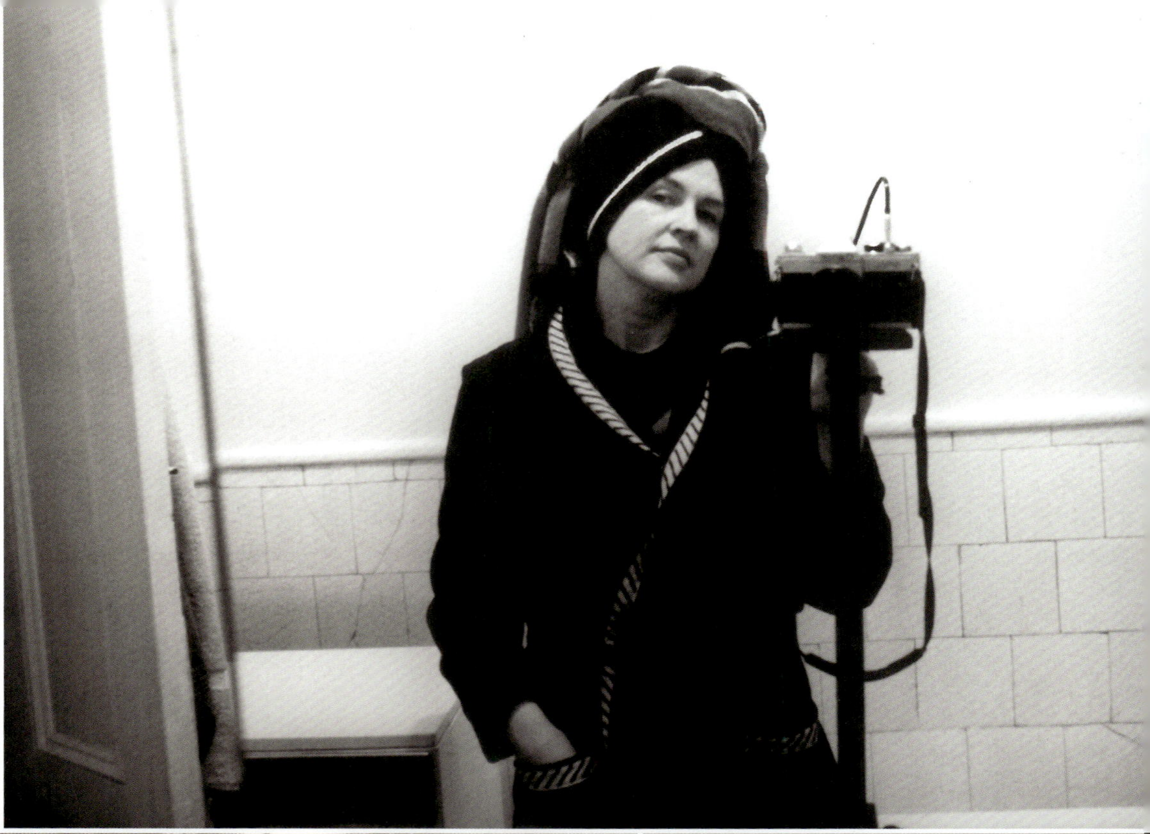

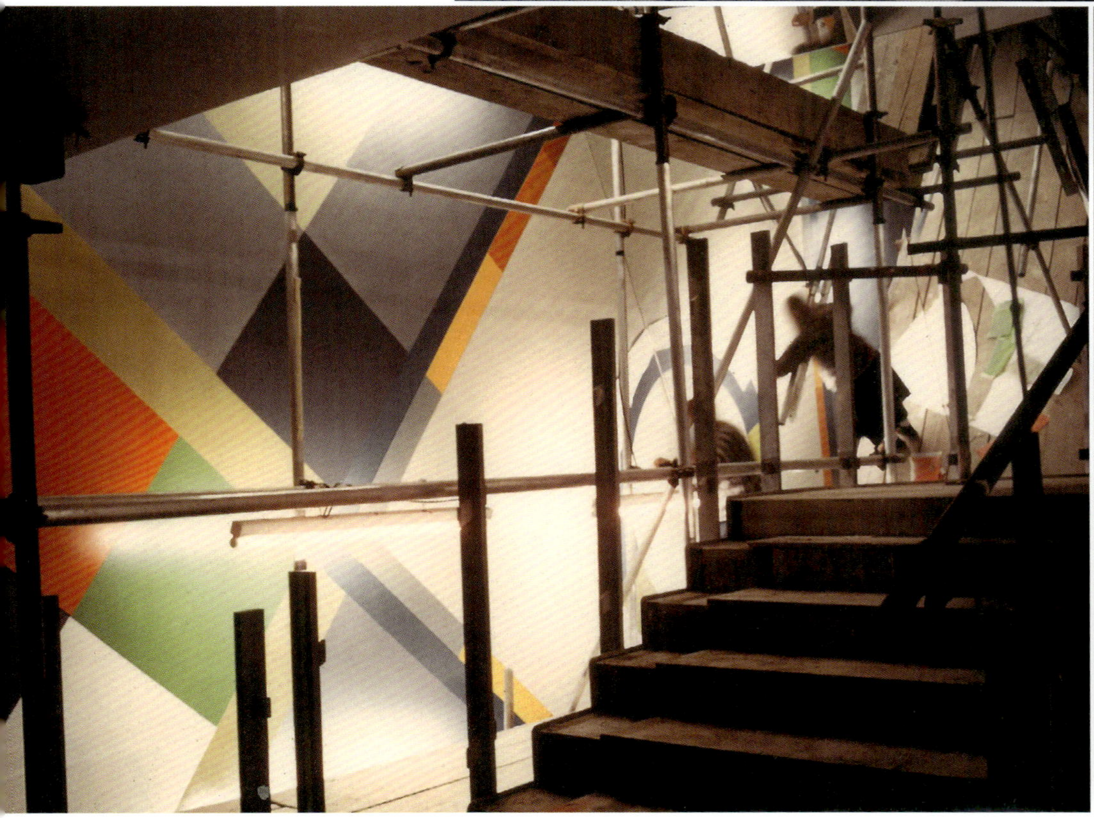

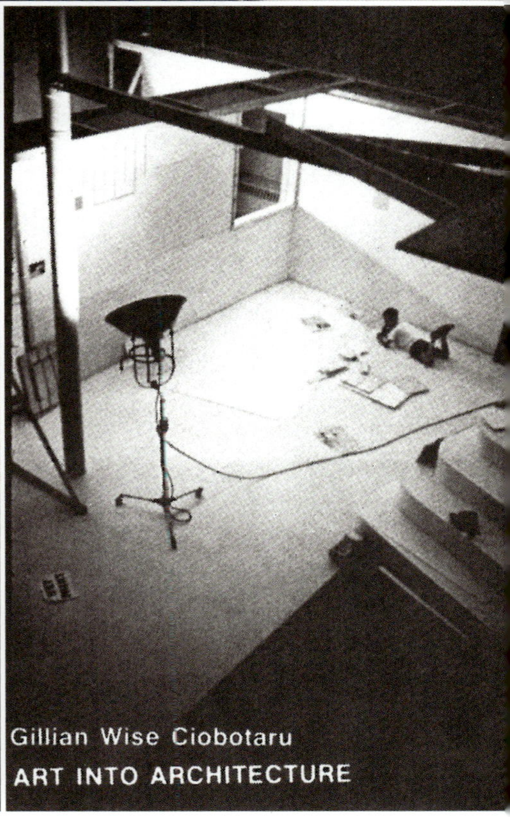

As an artist I have an eye. You know it if you have an eye. . . . It's a phenomenon. You get a signal that it is working. Say you are walking through a room of pictures. You are seeing them through your peripheral vision, and you get a signal that it's working, that something is good. This phenomenon needs research like a formula no matter if the work is classical or abstract. People who have this should be interesting to scientists. (Gillian Wise Ciobotaru interview)

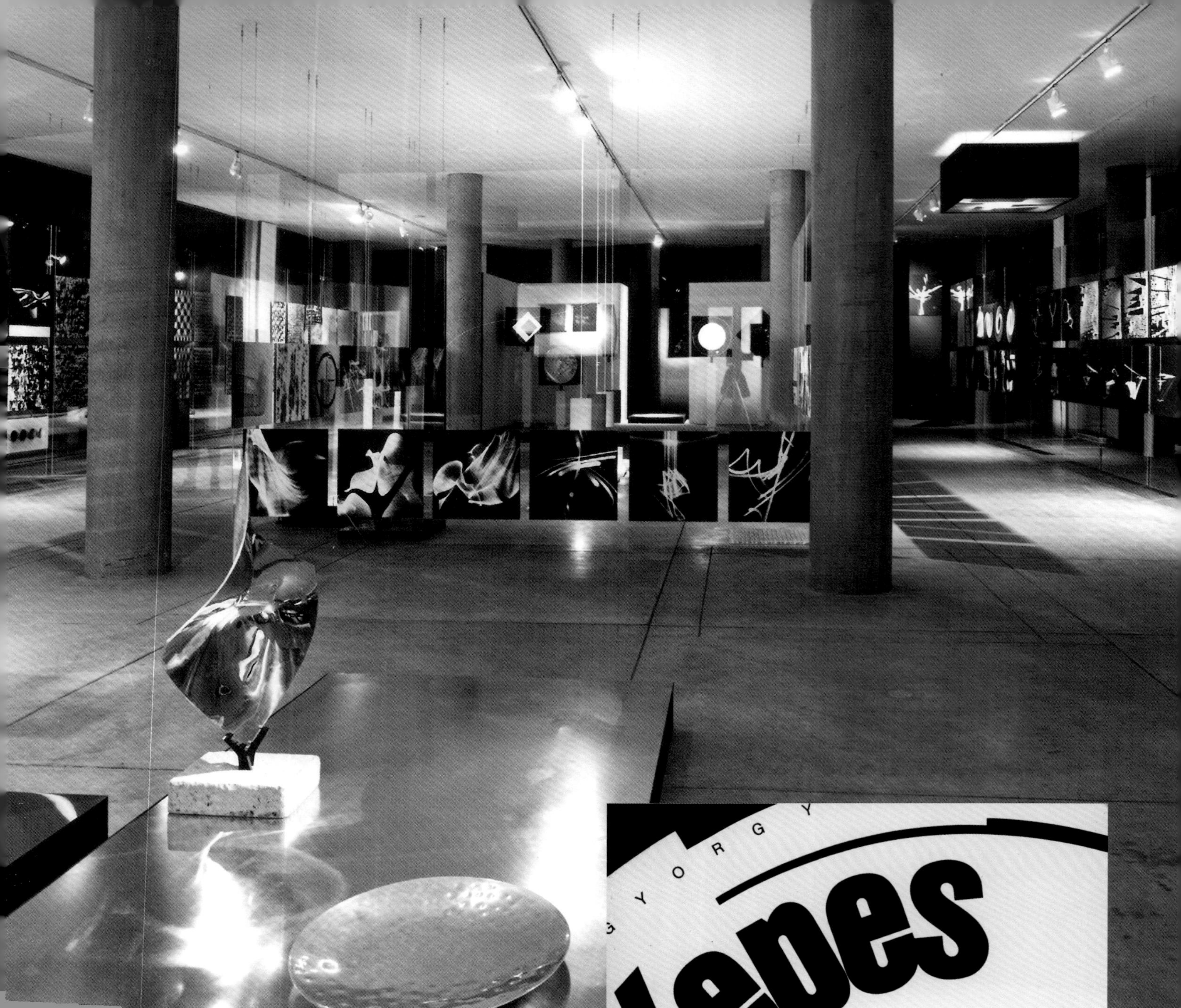

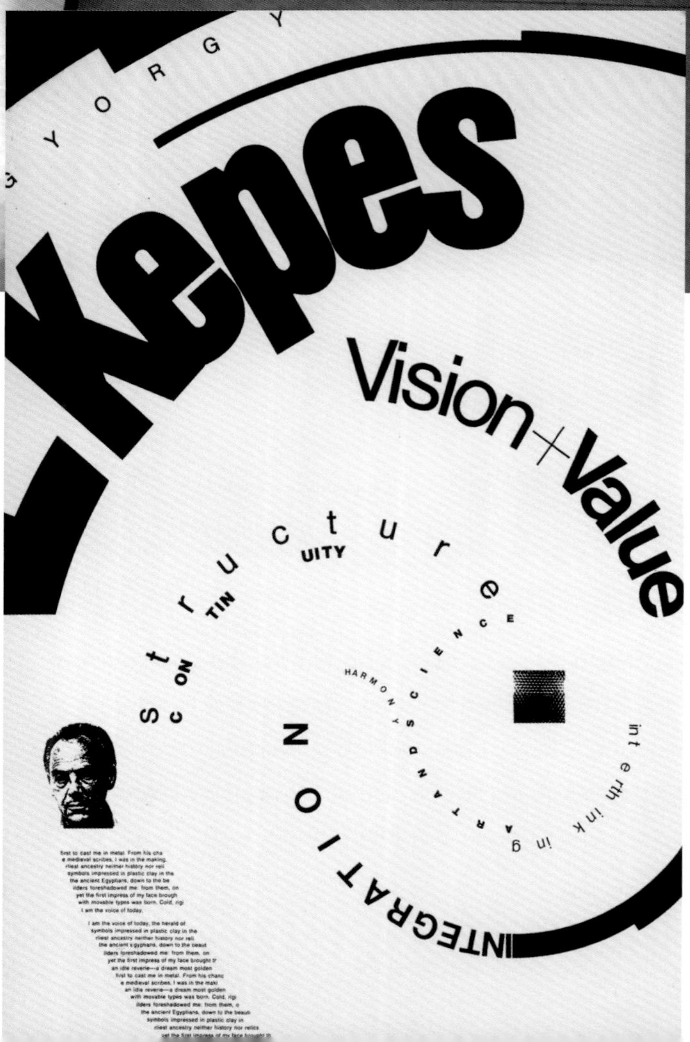

Above: György Kepes, "Light as a Creative Medium," exhibition design, Carpenter Center, Harvard University, 1965-66
Photo: David Van Piper

Right: poster, György Kepes, *Vision + Value* series
Source: CAVS Special Collection

Facing page inset: poster for "György Kepes: The MIT Years, 1945-1977," with image of Kepes' Programmed "Glow Column," 1973, Hayden Gallery, MIT, 1978
Source: CAVS Special Collection

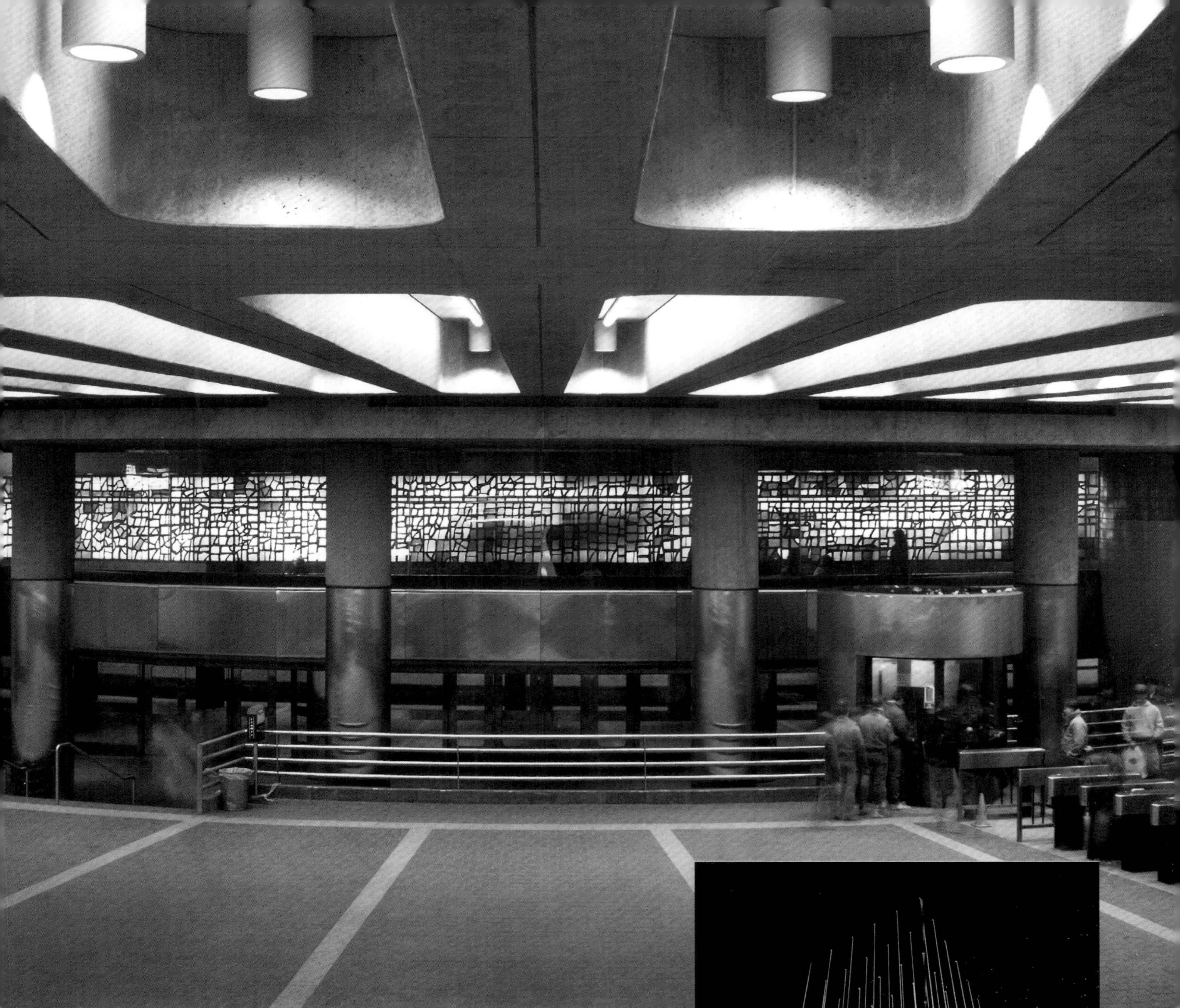

Arriving down at the main lobby, one glimpses a large wall of blue faceted glass, illuminated through a diffusing screen from the back. The rhythmical quality of this luminous stained glass brings forth a substitute sky. Approaching buses light up the slender red glass band which runs the length of the wall. (*Arts on the Line*, Cambridge Arts Council, 2002, 10)

Top: György Kepes, "Blue Sky on the Red Line," stained glass window lit by the headlights of underground buses, Harvard Square MBTA Station, Cambridge, 1985
Photo: Cymie Payne

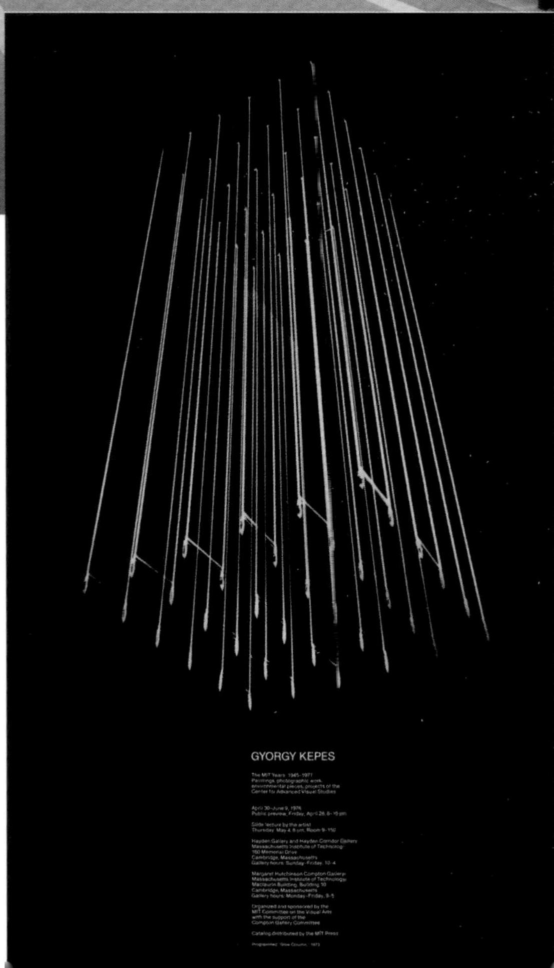

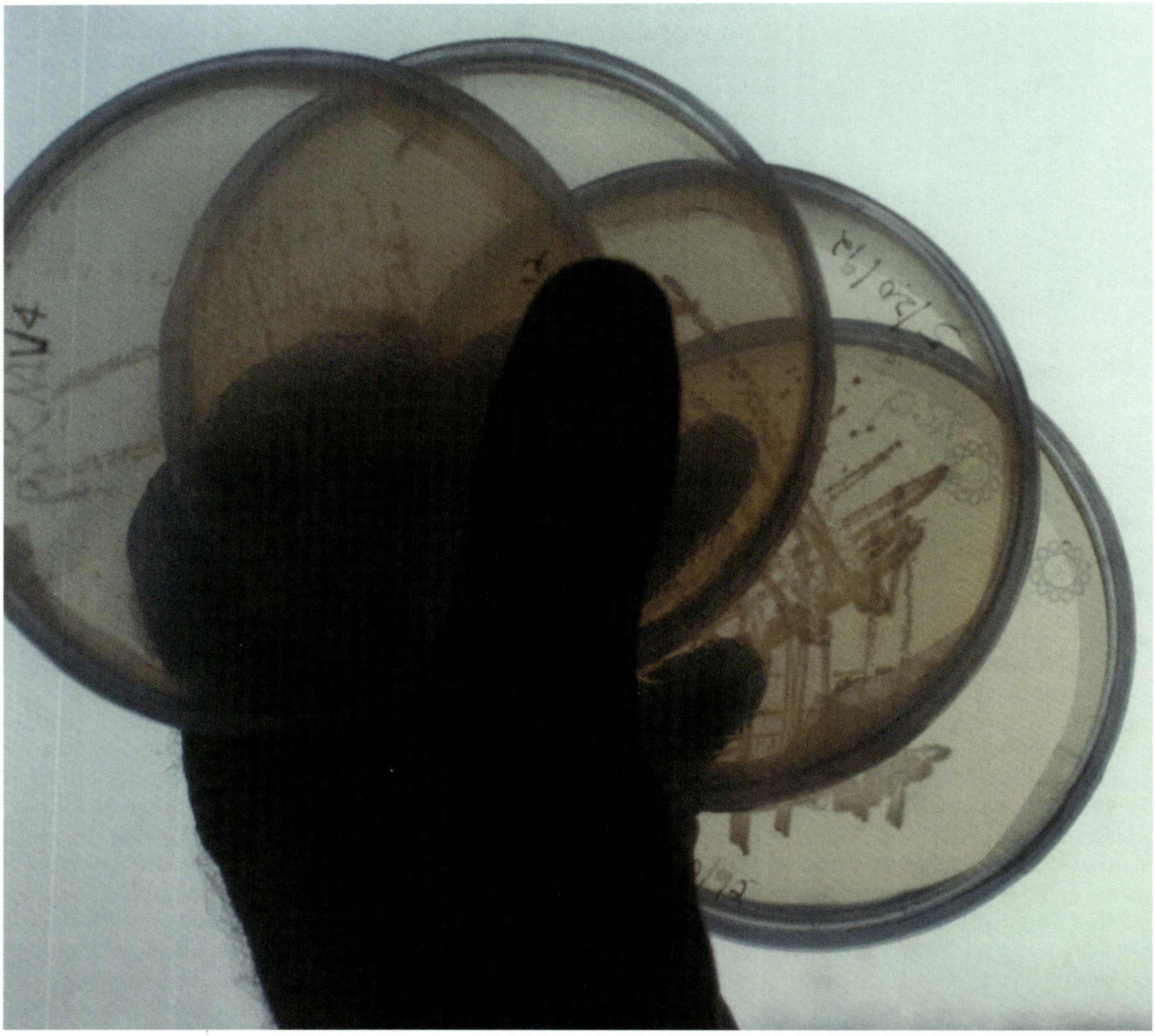

Above: Joe Davis, "*Microvenus* Bacteria on Nutrient Plates," visible colonies, 4" plates, 1986
Photo: Joe Davis

Facing page: Joe Davis
Source: CAVS Special Collection

Working with Dana Boyd, a geneticist at Harvard Medical School, and members of the Rich Laboratory at MIT, Davis has created Microvenus, a synthetic molecule containing graphic information within DNA, and spliced into bacterium that could last for nearly a billion years if released into space. When decoded, a Y-shaped Germanic symbol representing life appears. (Wendy Martens, "Joe Davis and the Art of Science," *Cambridge Chronicle*, April 1, 1993, 21)

Astronomers have spent years searching the heavens for radio messages from extraterrestrials. But now Joe Davis, an artist at MIT's Center for Advanced Visual Studies, says he has a better way. His project, called Micro-Venus, involves putting encoded messages in the DNA of a hardy strain of bacteria, which could duplicate and spread throughout the galaxy—hopefully for clever aliens to decipher. (Steve Nadis, "Intergalactic Bugs," *OMNI*, January 1991, 74)

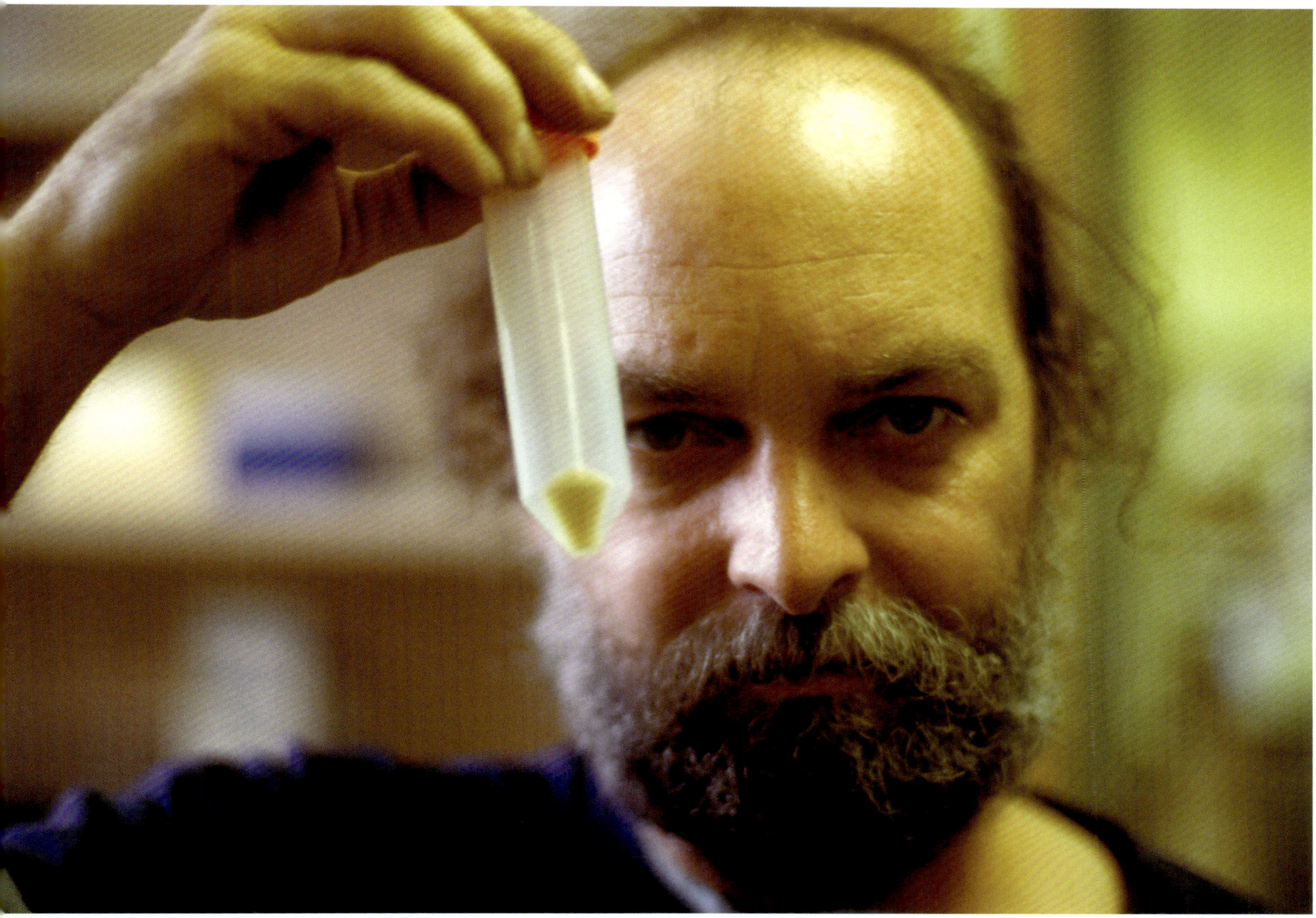

I wanted to learn how DNA holds information and how I might put arbitrary information into such a molecule myself. I learned how to synthesize DNA and then transform bacteria with it so that it would be inherited from generation to generation. The first information I encoded in this way was a line graphic of female genitalia, which also happens to be an ancient Germanic rune, and the symbol radio engineers use to indicate an antenna. I called the project *Microvenus*. It turned out to be the first genetically engineered organism to include digital information and the first one created for artistic purposes. . . . *Microvenus* is smaller than the shortest wavelength of visible light. DNA molecules can't be directly resolved with optical microscopes or even with the most advanced digital imaging equipment. . . . When you want to look at something that small you always have to look at or through something else, like viewing the world through a mirror. Sequencing technology is very precise however. Although it is invisible, it can nevertheless be resolved with amazing accuracy. You can read back the ladder-like DNA molecule rung by molecular rung. There are now batches of *Microvenus* bacteria and *Microvenus* DNA in laboratories all over the world. It has been exhibited in major museums worldwide. It has frequently been cited in the scientific literature, and can now even be found in certain scholarly works of art history. There are architectural and procedural problems for curators who seek to exhibit genetically engineered art in public spaces. Battles went on for several years before such an exhibition was ever realized. (Joe Davis interview)

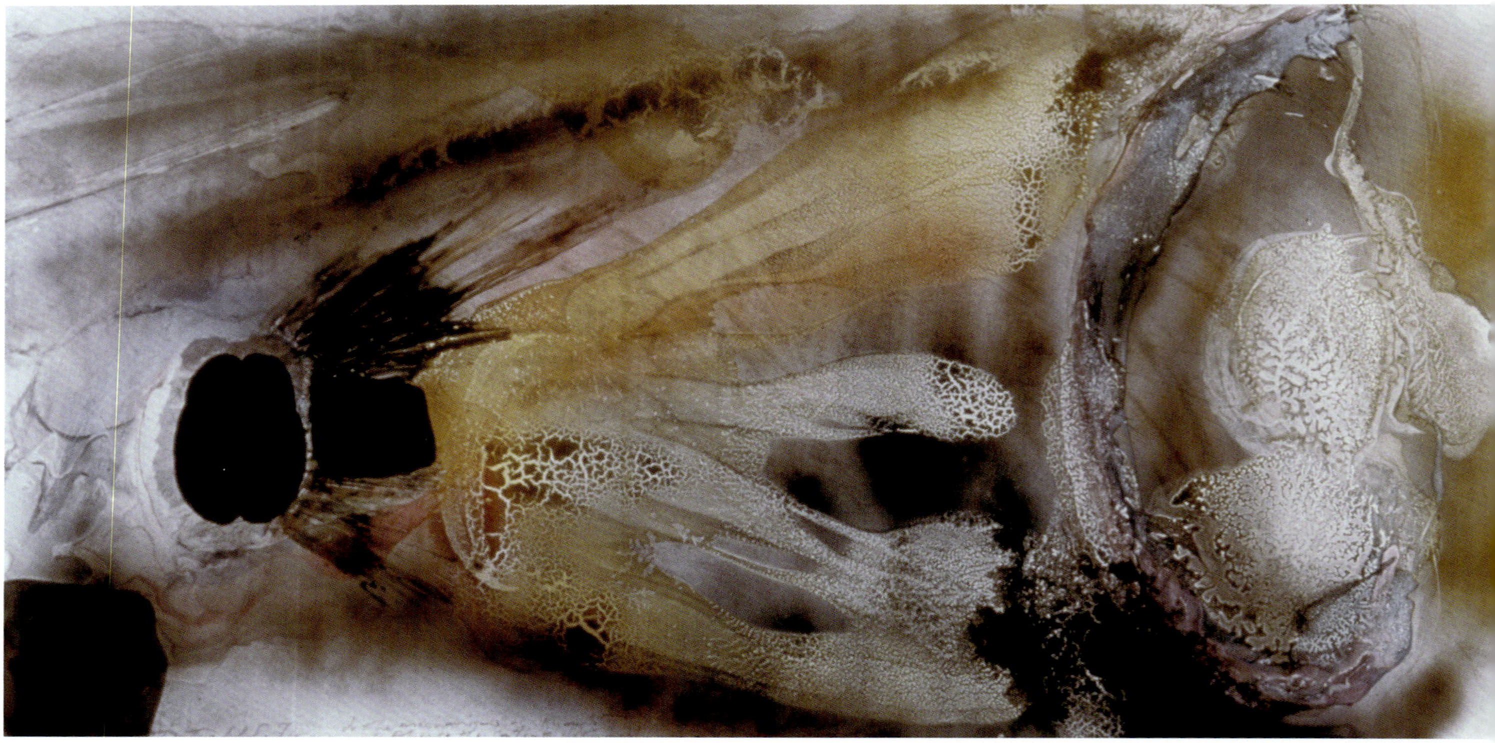

Above: Todd Siler, "The Biomirror," horizontal section of the human brain "illuminating how the bio-electric-magnetic-chemical mirror connects the mind to its creations in an instant of intuition," (Siler description) mixed media on plexiglass, 27"x 55", 1982
Photo: Otto Piene Collection

It was a time of total freedom. I dreamed up and drew out concepts that bridged the divided worlds of nuclear and neural sciences, using art to explore the deep connections between fusion & fission processes that form the universe . . . I freely blurred the boundaries between disciplines. . . . And so the Center remained my virtual research lab where I could dig deeper into the how and why people unconsciously erect those annoying arbitrary mental barriers whenever they fear understanding something or communicating what they think they understand. . . . The Center remains on the fringe, marginalized by the rest of the art world. In fact, the work that was going on there turns out to be very much in the forefront of where that whole world was moving, but they didn't recognize it or see it. . . . Sometimes the "strange attractors" seem invisible to the mainstream of everyday thinking. It's a term from astronomy describing a small and highly concentrated form of energy, like a quasar, a very powerful little force. I think the Center is a strange attractor. It doesn't have to be large in volume to have a large impact. (Todd Siler interview)

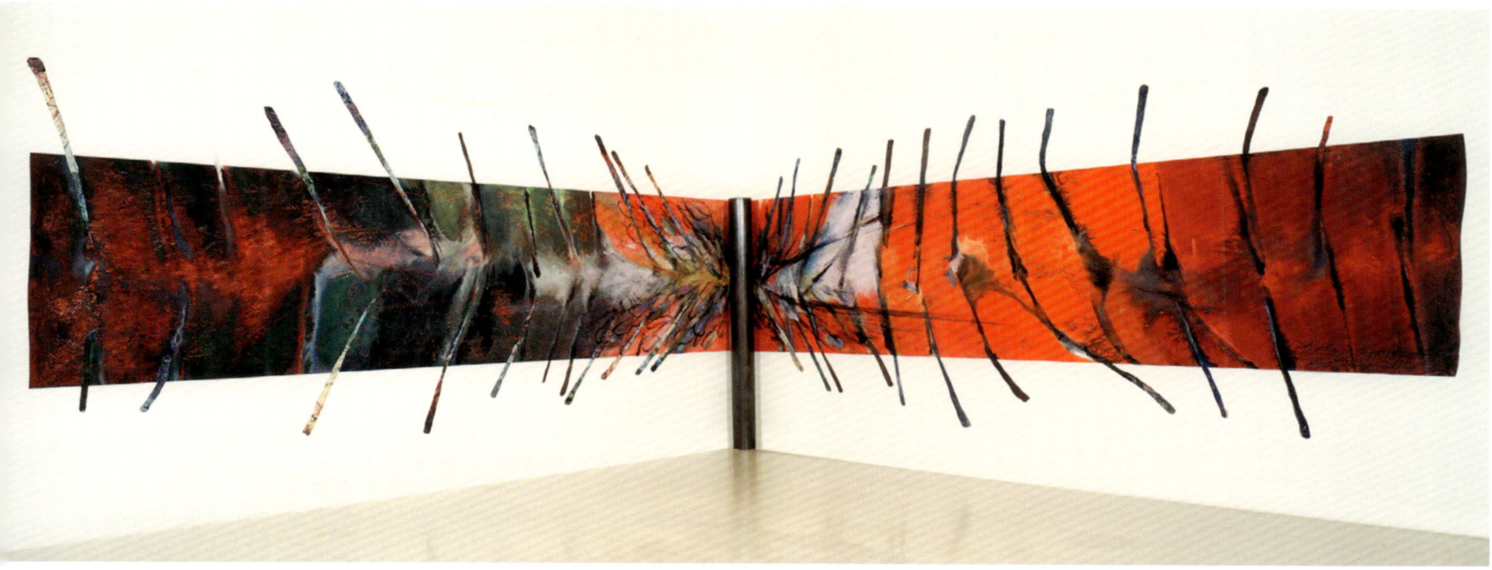

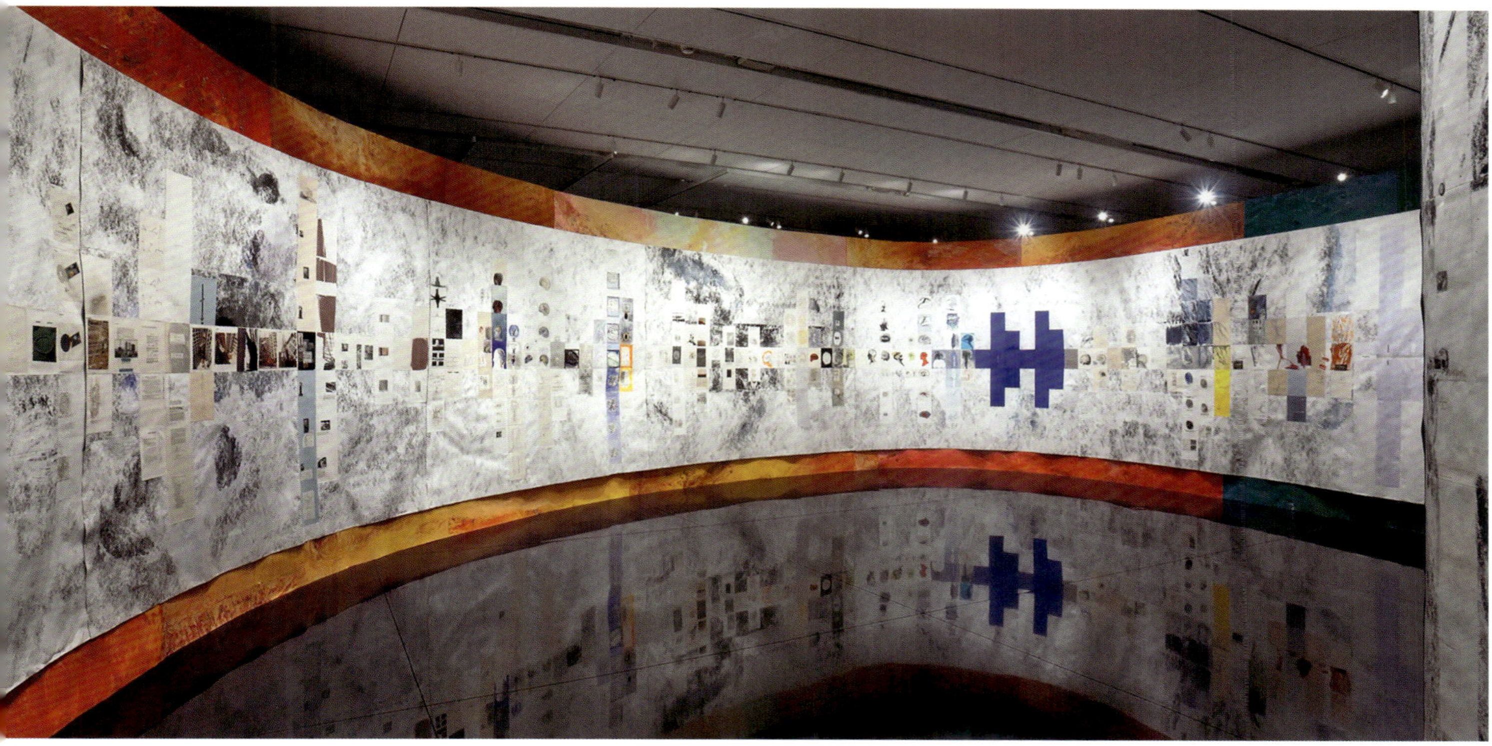

In *Breaking the Mind Barrier: The Symmetries of Nature*, 1986, an 8-foot steel column was placed in a corner of the gallery and flanked by two mixed-media paintings. The paintings represent patterns of thought that go into producing the technical apparatus for a powerful military projectile. Once every three minutes the hollow column emitted a brilliant flash of white light which, like the splitting of an atom, signified the absolute convergence of creation and destruction. (Patricia C. Phillips, "Todd Siler at Ronald Feldman Fine Arts," *Artforum*, May 1987, 147)

Top: Todd Siler, "Breaking the Mind Barrier: The Symmetries of Nature," mixed media on spunbonded synthetic canvas, steel column, photo strobe, sound track, 1986 (Courtesy of Saidye Bronfman Centre, Montreal, Canada and Ronald Feldman Fine Arts, NYC)
Photo: Todd Siler

Bottom: Todd Siler, "The Brain Theater with Thought Assemblies emerging from The Limbic Brain," CU Art Museum, Boulder, 1979-2014
Photo: Jeffrey Wells

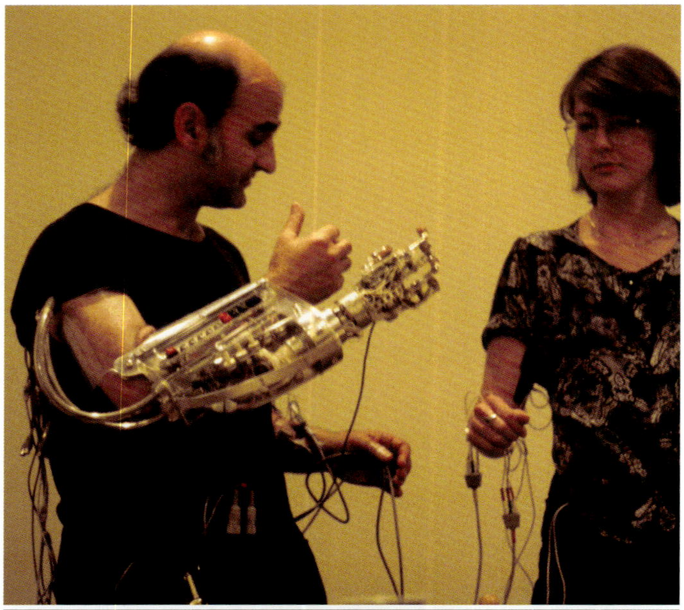

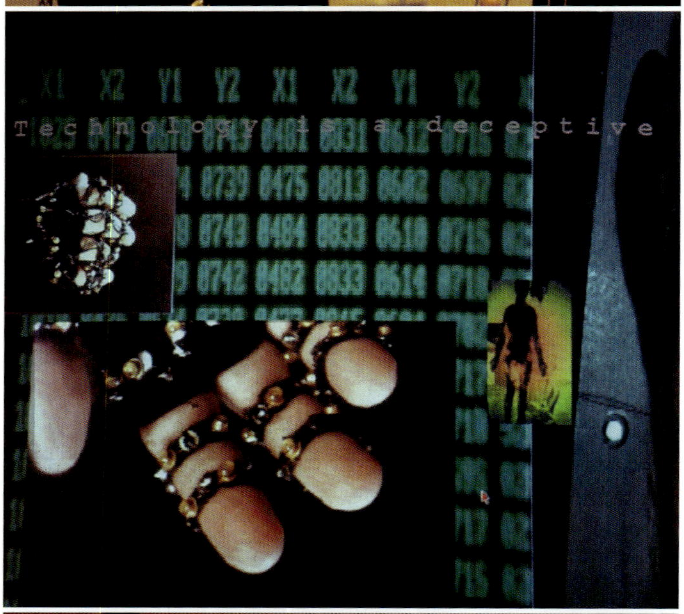

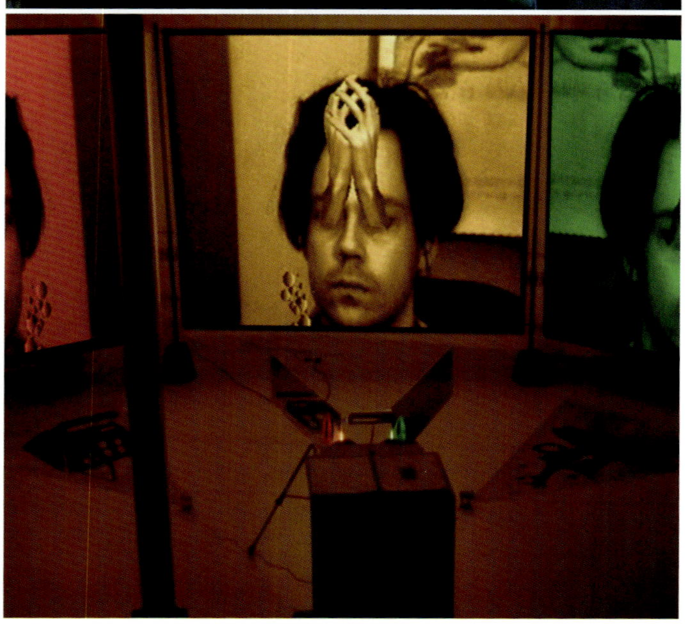

At AMG [Architecture Machine Group], I was building the first LED suit for 3-D optical tracking. This is very interesting to me. For the last eight years people had been doing thesis work on this proposition, but it was all theoretical. It turns out, the reason nobody made a suit before was because none of the engineers knew how to sew an LED onto a sweater. Such a simple and direct phenomenon, they didn't realize at the time in all this heady discourse—or how to glue two 2x4s together, how to make things. . . . NHK-TV in Japan took it on as a big project: sign language, gestures with hand tracking gloves, a lot of money came into AMG but none of it filtered over to CAVS. . . . So, for me going between CAVS and AMG, that paradigm became bigger and bigger. I was noticing why certain things were invested in, and why others were not. This had to do with DARPA research dollars beginning to fade out. I was watching them take the technology of AMG and hand it off on a silver platter to commercialism: Photoshop, development of computers that were affordable. They didn't go through the traditional development schemes, but slid by the content-creators right into mass population and mass production. (Jennifer Hall interview)

The video projector had three beams coming out of it. I placed long mirrors in front reflecting the red, the blue, and the green beam. With one projector I could get a 30′ long wall of video. . . . using an Amiga computer. I would put things on tape and play them through this system, so I was doing computer-based sound and video installations in 1986. . . . MIT was the place where I combined my interest in electronics, interface design, and art, especially as I was surrounded by people who wanted to do the same thing. . . . The Center was exploring works which didn't fit into a category. They were truly avant-garde, and totally unselfconscious. (Don Ritter interview)

The media of my art are crystals, micro-organisms, plants and insects. Micro-organisms reveal the invisible content of the air; the living things that we internalise through breathing, the bacteria that live in us and that we project back into the environment. The crystal enclosures show constant pattern change in the natural genetic structure of inorganic material. The instincts of insects make a choreography of their movements. (Alan Sonfist, catalog, Institute of Contemporary Arts, London, 1971)

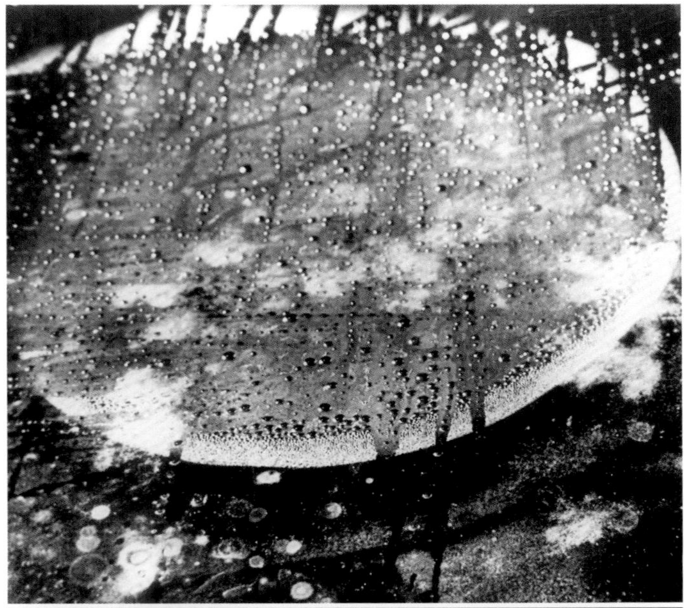

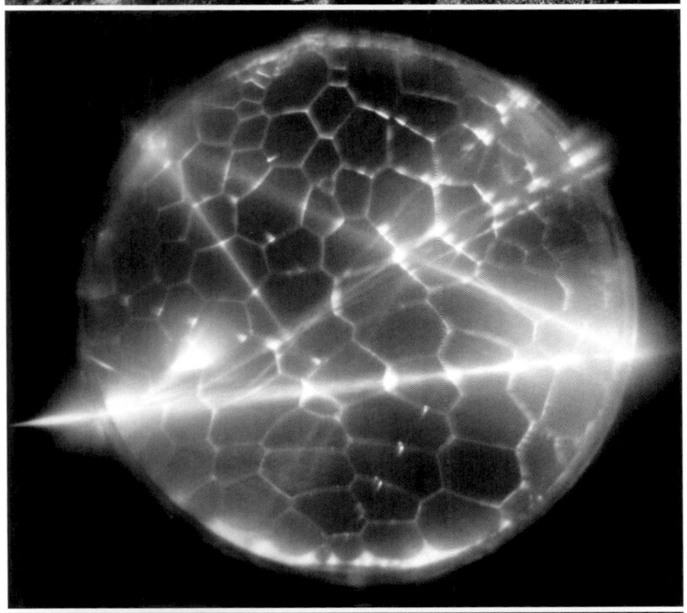

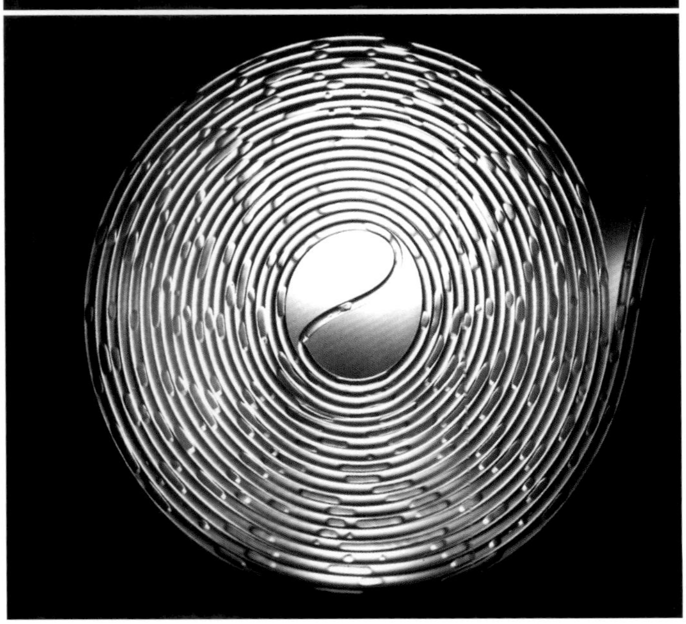

Top right: Alan Sonfist, "Micro-Organism" (detail), 1968
Photo: CAVS Special Collection

Middle right: Mauricio Bueno, "Liquid Prisms," 1971
Photo: Nishan Bichajian

Bottom right: Mauricio Bueno, untitled, 1972
Photo: Nishan Bichajian

Facing page top: Stelarc, the artist demonstrates the "Third Hand," an EMG controlled mechanism, at CAVS, MIT, late 1980s
Photo: Otto Piene Collection

Facing page middle: Jennifer Hall, digital collage (detail), including hand tracking glove (in development as part of "Computer-aided Body Tracking Techniques for Animation," for NHK Television, Osaka), Architecture Machine Group, MIT, 1986
Photo: Jennifer Hall

Facing page bottom: Don Ritter, "Stithy," video-sound installation, created at CAVS, 1988
Photo: Don Ritter

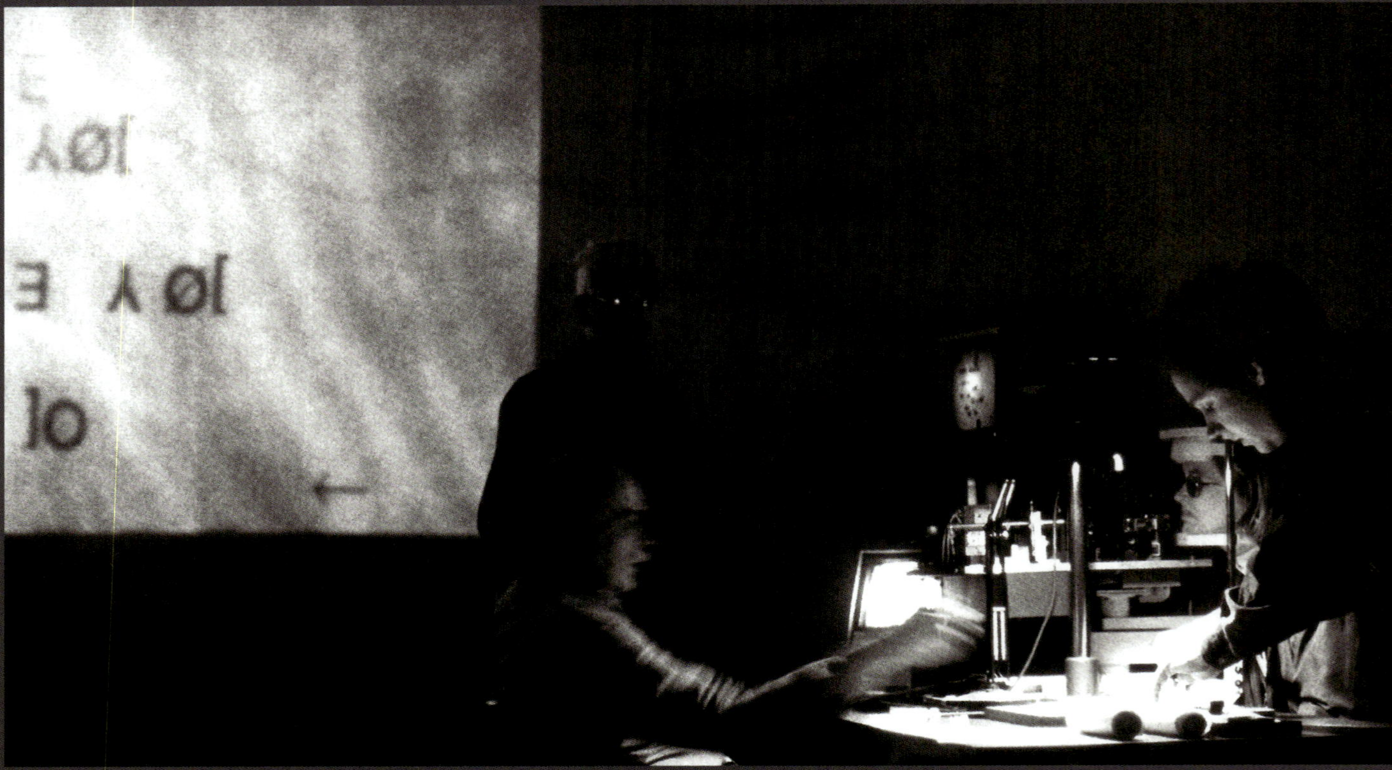

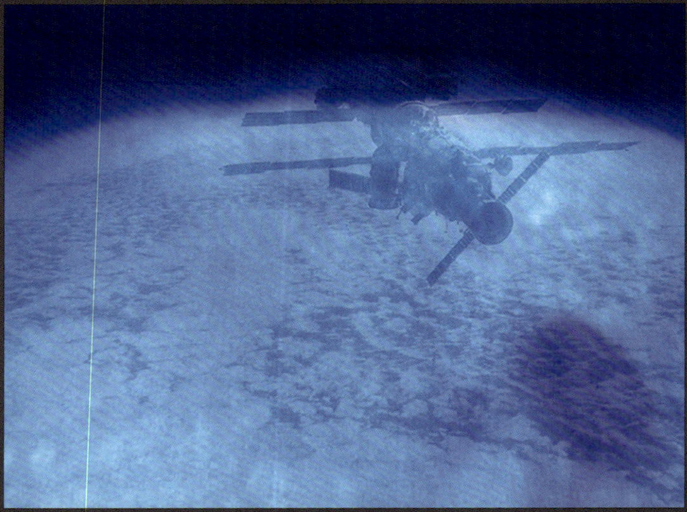

I worked on seeing machines for people who had little or no sight. I did a lot of work with Rob Webb who . . . invented a machine called a Scanning Laser Ophthalmoscope (SLO). My physician sent me over to have my eyes tested when I had no sight, to see if it was worth it to try and do surgery, because they didn't know if I had any retina. Well, they projected stick figures onto my retinas, and I could see some of them, and I was just amazed, and said, "Well, can you write a word?" and they wrote the word "sun," and I could see that word, too. Needless to say, it was the first word I'd seen for a long time. I said, "I've got to meet the man who is responsible for this machine," and the technician said, "He's not accessible really." Well, I did get a hold of him and we began a collaboration which lasted for the entire time I was at MIT. . . . instead of projecting stick figures on people's retinas, I would project words and images, like a face, picked up with a camera, and people who were blind could see these images and words. (Elizabeth Goldring interview)

Top: Elizabeth Goldring, "Retinal Poetry: Experiments in Visual Language seen through the SLO," Goldring at the Scanning Laser Ophthalmoscope, with fellow poet, Panos Kouros (right), and SLO inventor, Rob Webb (seated), CAVS, MIT, April 10, 1991
Photo: Otto Piene Collection

Bottom: Elizabeth Goldring, retina print generated with the SLO, "Lunar Station," 2003
Photos: Elizabeth Goldring

Facing page top: Elizabeth Goldring, retina print, "Wandering Monk, Kyoto, on Ananda's Retina," 2002
Photo: Elizabeth Goldring

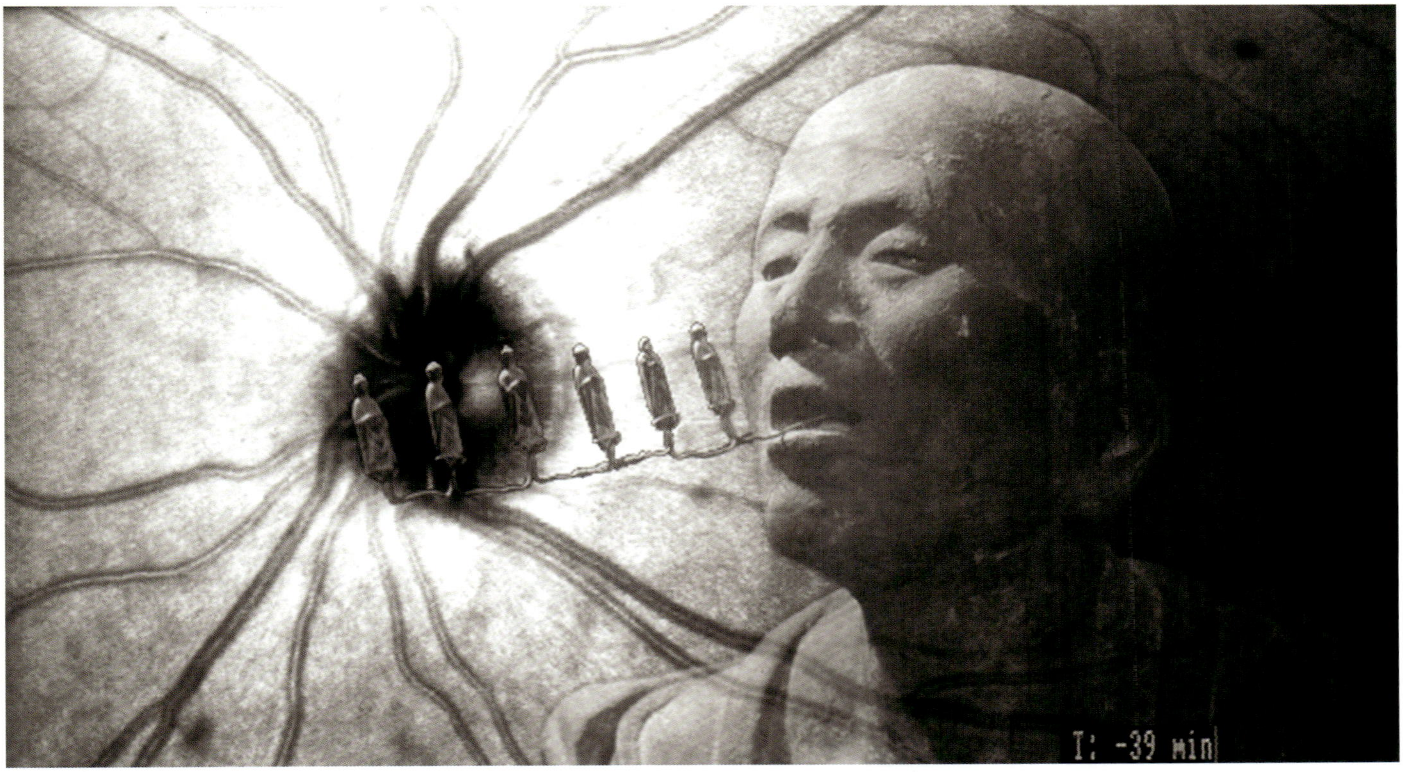

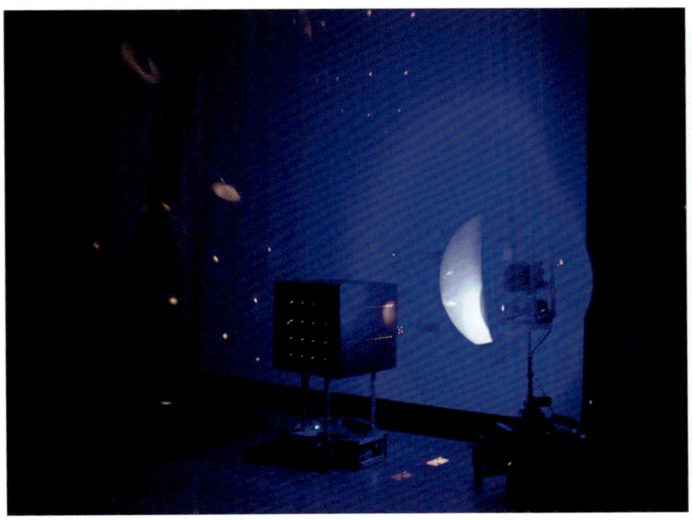

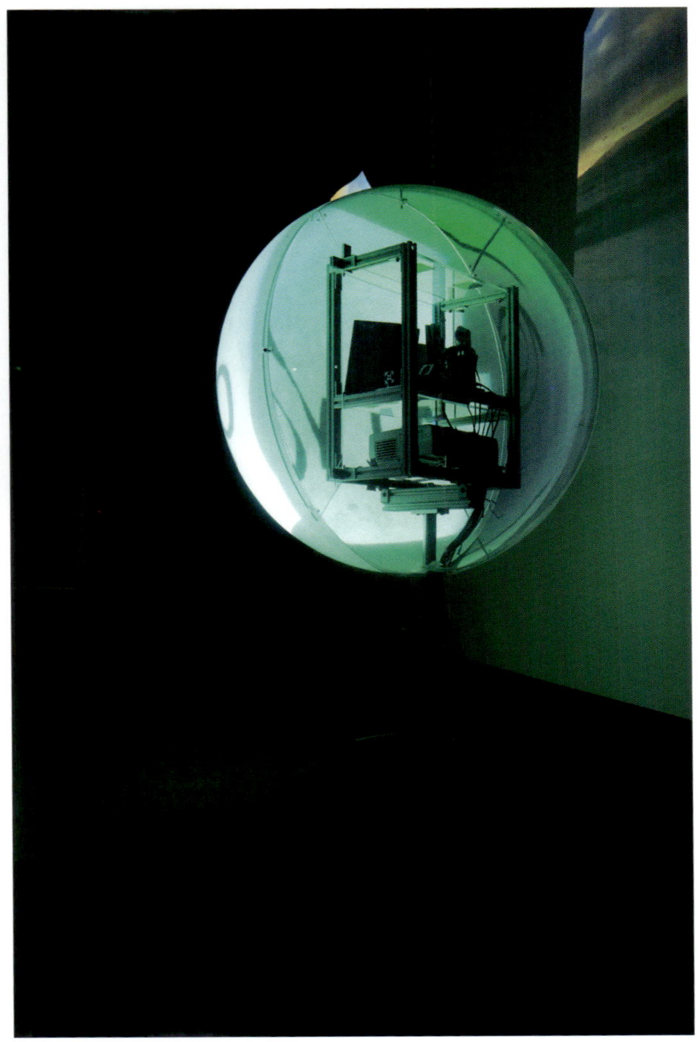

Goldring and her interdisciplinary team at MIT collaborated with avant-garde director, Robert Wilson, to develop robotics that address impaired vision. They rehearsed scenes for "My New Friend SU: The Moon's Other Side," in which Goldring's "Eye Robot" played a role. (Goldring website)

Above left and right: Elizabeth Goldring, "Eye Robot," performance with Otto Piene's "Lightroom," Art, Culture and Technology (ACT), the Cube, MIT, 2012
Photos: Courtesty of the Artist

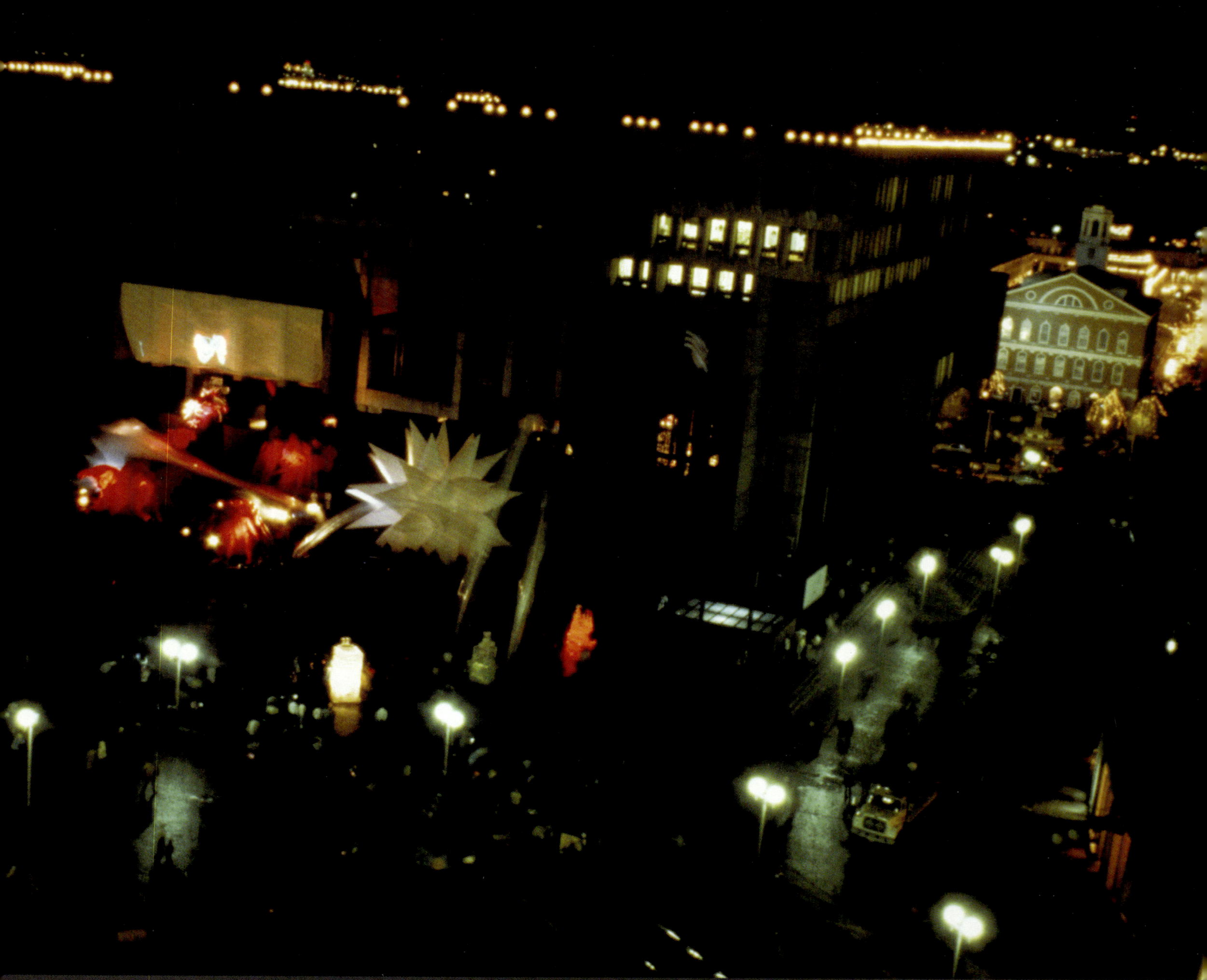

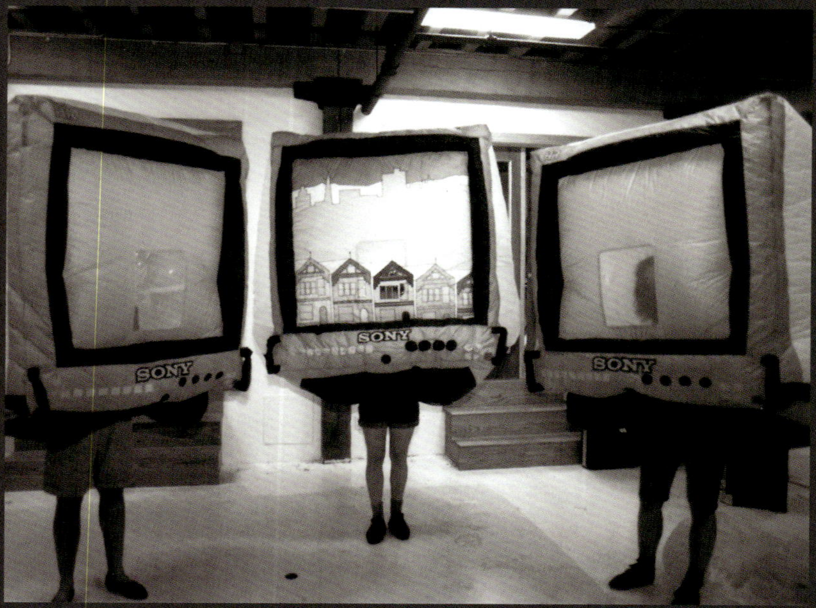

Above: Paul Earls, Otto Piene, CAVS/MIT, "Star Tales," a multi-media performance for lasers, inflatable sculptures, and musicians, First Night, Boston City Hall, New Year's Eve, 1982
Photo: Otto Piene Collection

Left: Jonathan Goldman, "Inflatable Costumes: Sony Monitors," nylon, fan, battery, 1987
Photo: CAVS Special Collection

Celebration

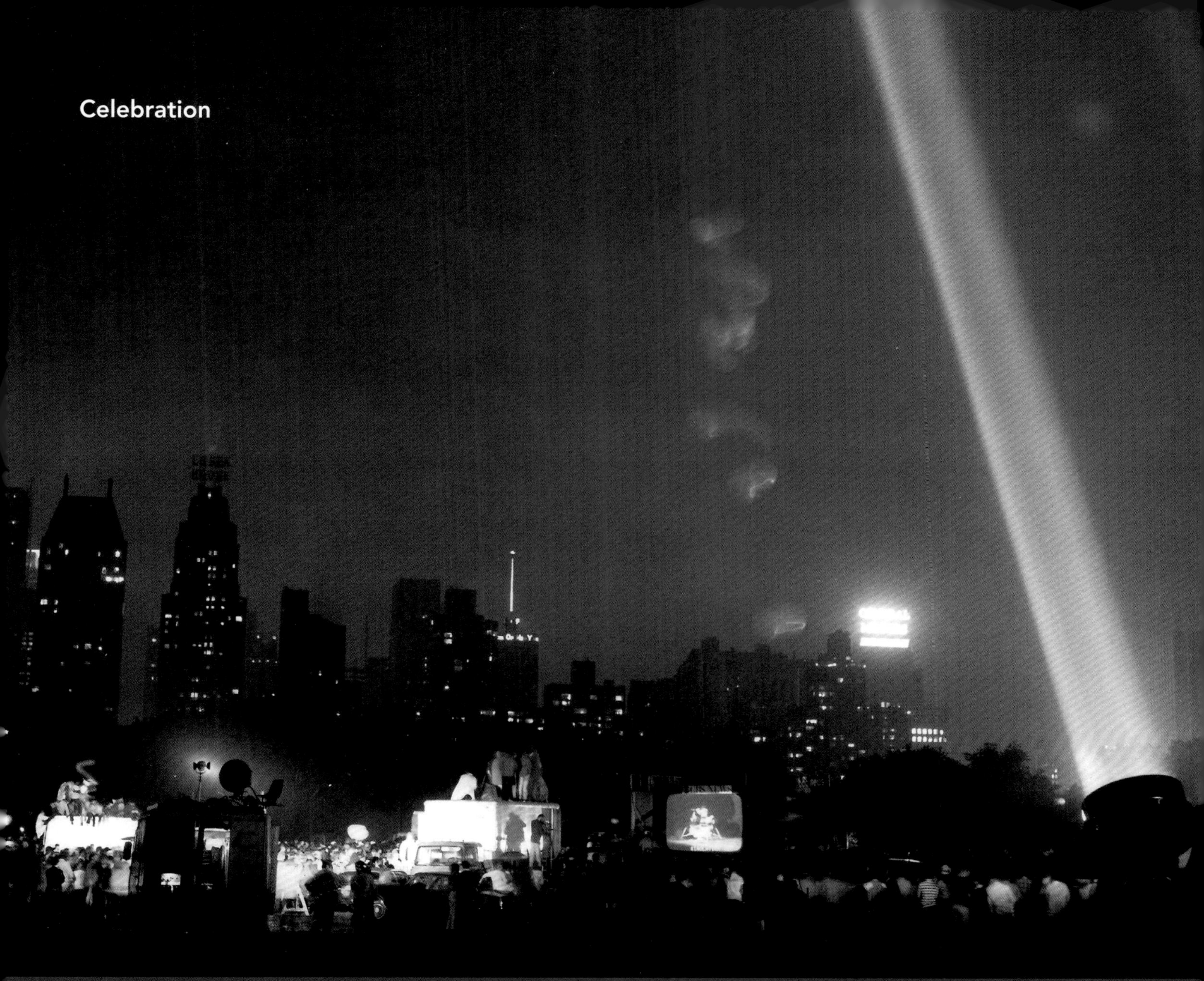

I don't actually remember how I got to the Center, but I do know that Otto was the only person in the art world who acknowledged my work towards creating festivals as an art form. It was an amazing experience to be able to communicate with him, and be a part of a group of people who were thinking in terms of outdoor art, about the relationship between the sky and the weather, the water—artists who cared about participation, who thought about making events. (Karin Bacon interview)

Above: Karin Bacon, "Moonsite," a celebration of the first moonwalk; first look at the whole earth from the sky, Central Park, New York City, 1969
Photo: Daniel McPartlin

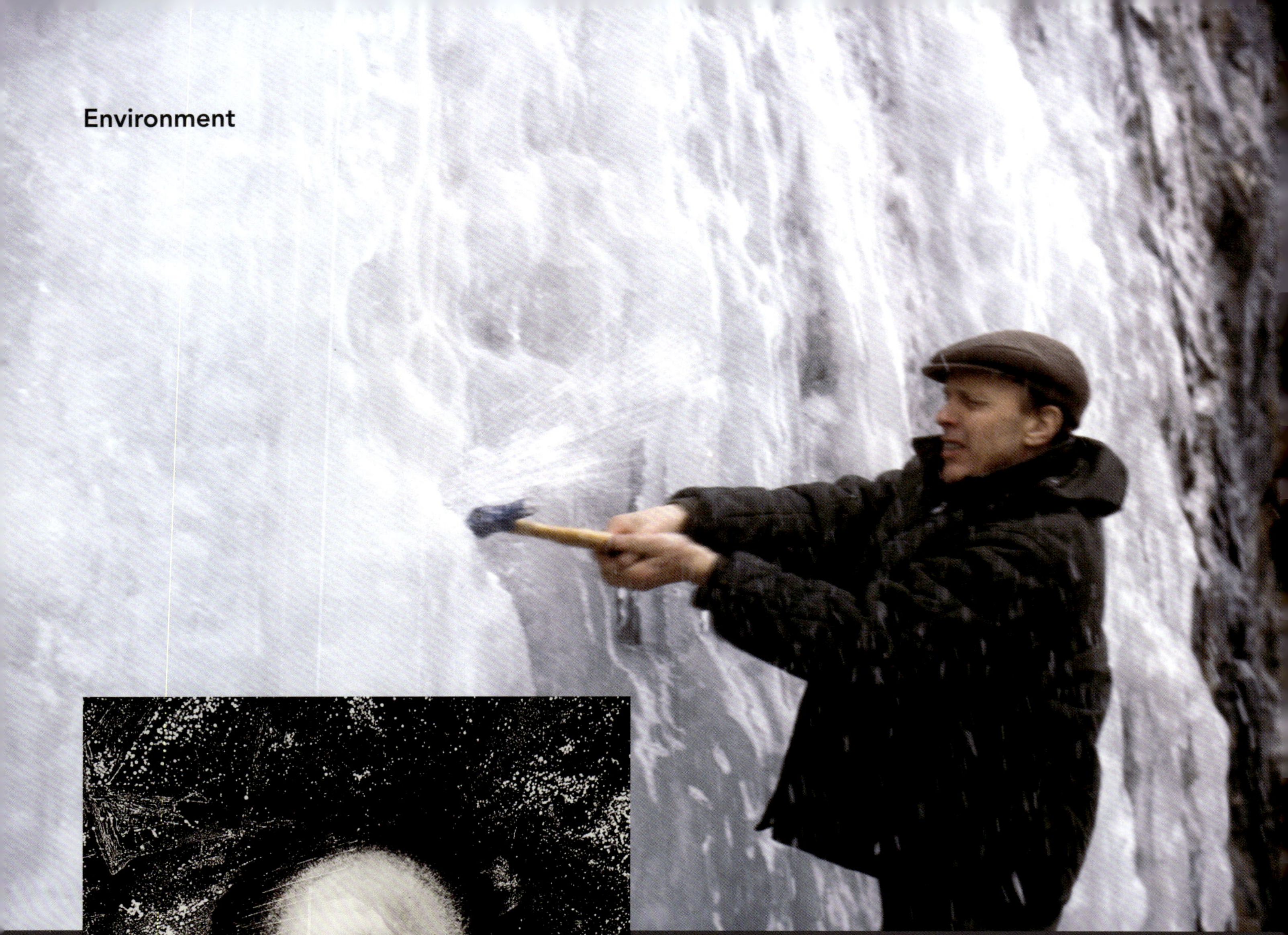

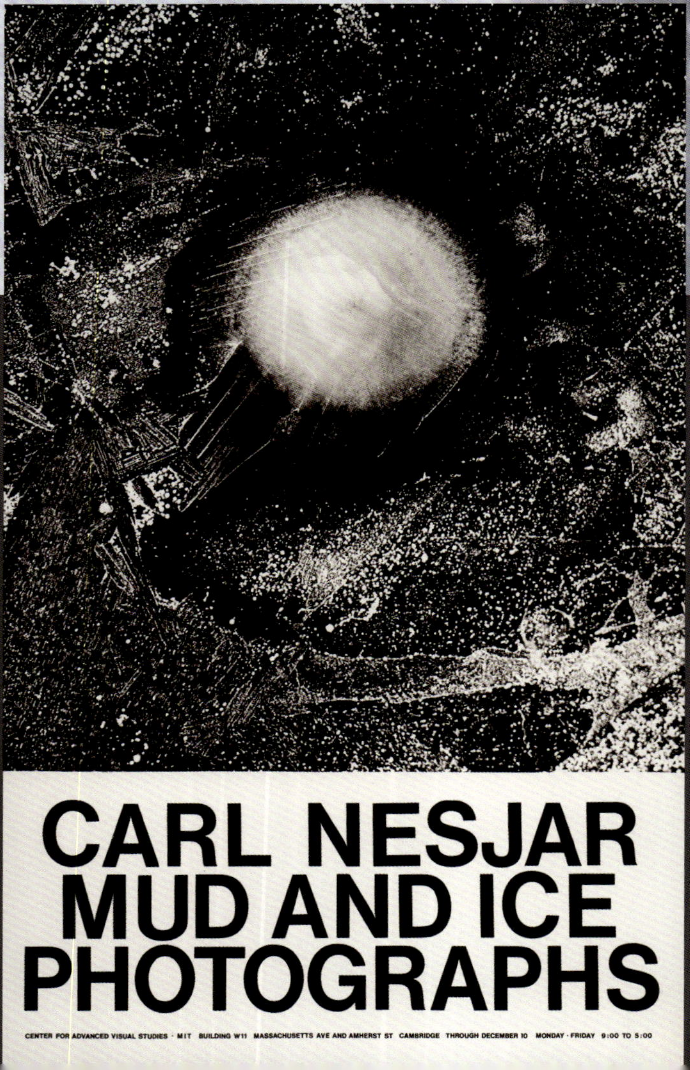

I liked both Kepes and Otto . . . very different people. They were like a family to me. It was one of the most pleasant periods of my life—Lowry was there, John Newman, Harriet Casdin-Silver. Getting this close to so-called technology was new to me. My father was an engineer, and I think I've got a lot of genes from him. Technical problems appealed to me, that is, within a certain sector. I worked in metals and film while there, photographing ice crystals, which I exhibited in the CAVS hall. It opened up new avenues, new possibilities. I was too shy to look up Minor White. I borrowed someone's darkroom once to make prints. (Carl Nesjar interview)

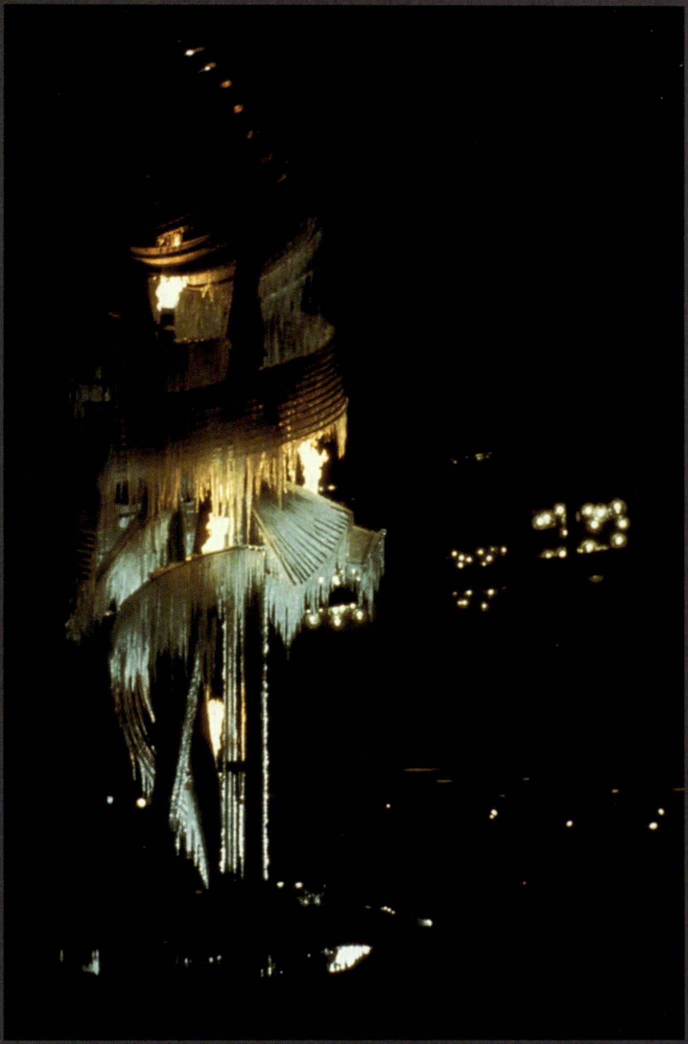

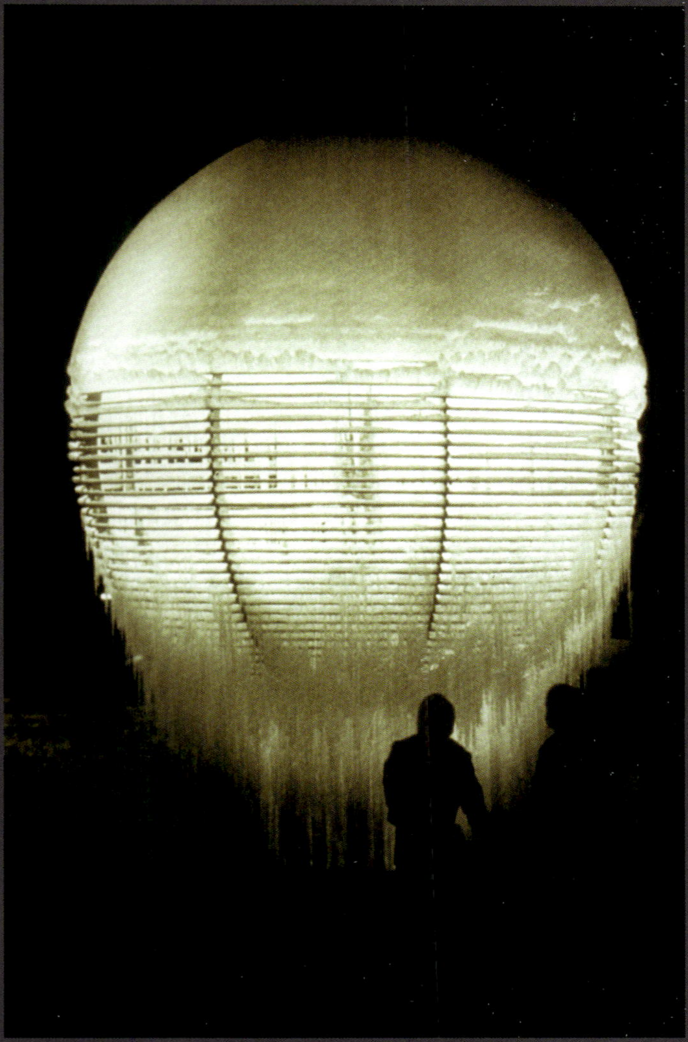

Above left: Carl Nesjar, "Ice Fountain," Norwegian Agricultural University, Ås in Akershus, near Oslo, 1972
Photo: Carl Nesjar

Above right: Carl Nesjar, "All Year Fountain," a globe consisting of a double helix; in summer, water running down the helixes made it look as if the globe is turning, Moderna Museet, Stockholm, 1972
Photo: Carl Nesjar

Facing page top: Carl Nesjar and ice, date unknown
Photo: Otto Piene Collection

Facing page inset: Carl Nesjar, "Mud and Ice Photographs," poster for exhibition at CAVS, MIT, 1977
Source: CAVS Special Collection

I think the most meaningful things I have done are the ice fountains. It's nice to have had one original idea in your life. The idea came out of the Barcelona fountains in 1961. I like Spanish fountains because they emphasize water, not sculptural forms as in Italy. This was probably from the influence of Arabs who came to Spain, the Moorish tradition by people who saved water. They made very modest low-key fountains where the water was the important thing. When I came back to Oslo, I was all excited about trying to make modern fountains. Through this I came upon the idea of fountains which could function all year. (Carl Nesjar interview)

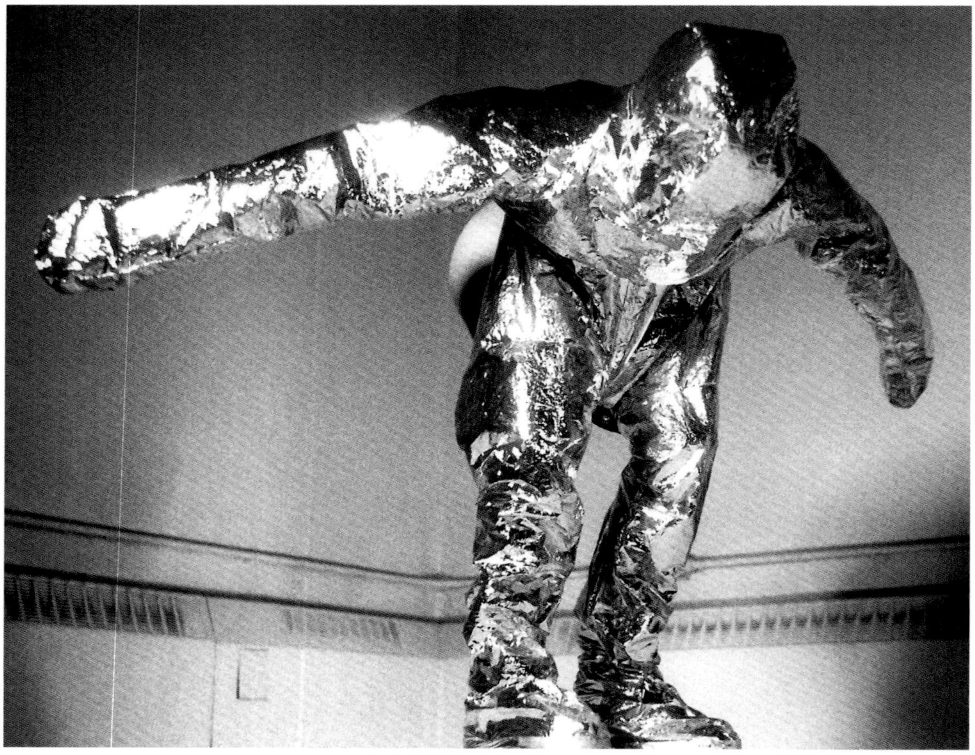

The "Lightproof Suit" is made out of mylar and black satin lining. It is designed to protect the human body from the hazards of excessive exposure to light and images. (Luc Courchesne, photo caption)

Left: Luc Courchesne, "Lightproof Suit," photograph no. 1, CAVS, MIT, 1982
Photo: Robert Rosinsky

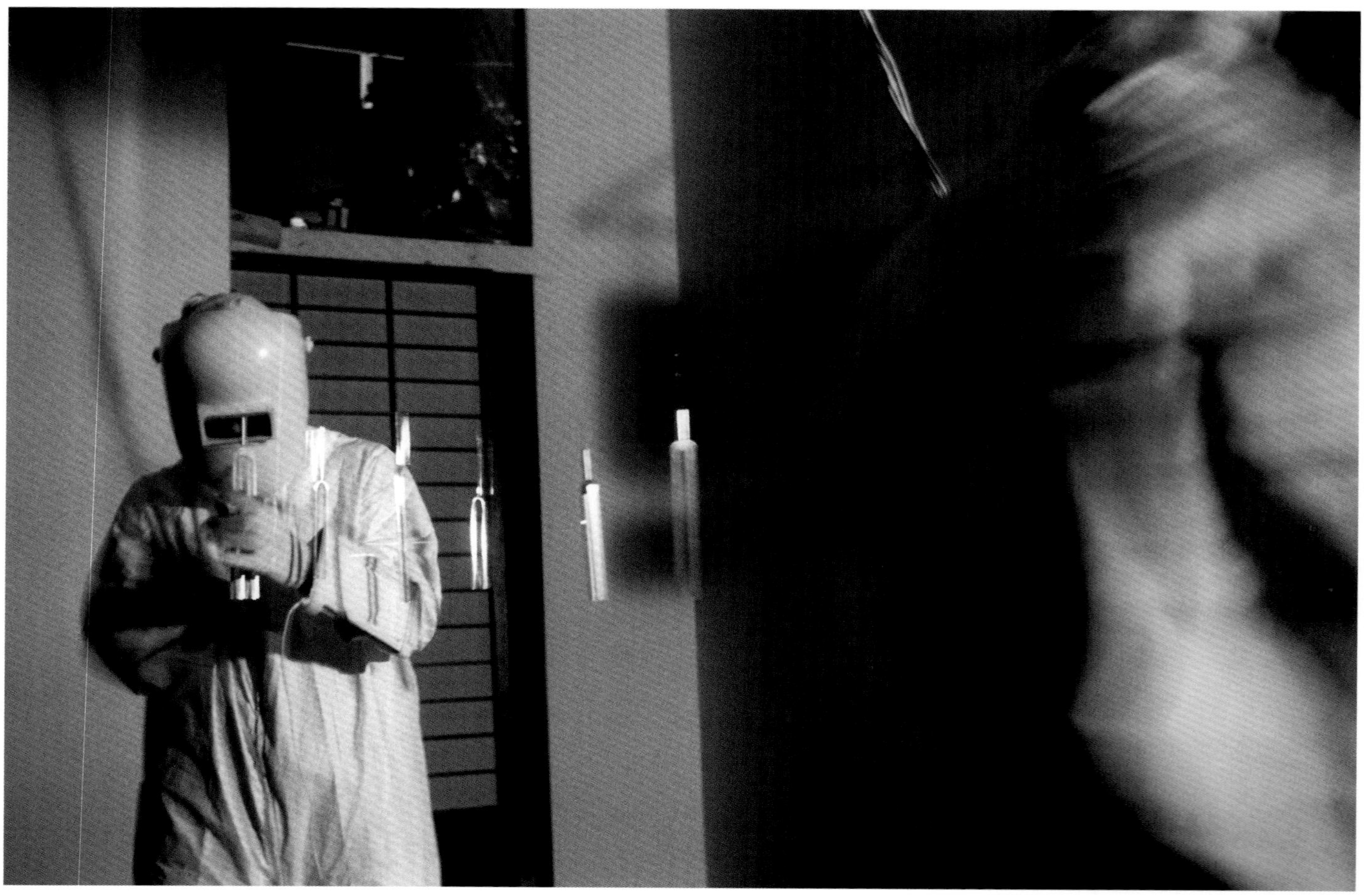

Above: Richard Lerman, performing "Incident at 3 Mile Island—perhaps an Elegy for Karen Silkwood," performers, amplified tuning forks and laser light, World Music Days, Brussels, 1981; John F. Kennedy Center, Washington, DC, 1981; The Milky Way, Amsterdam, 1982. Photo: CAVS Special Collection

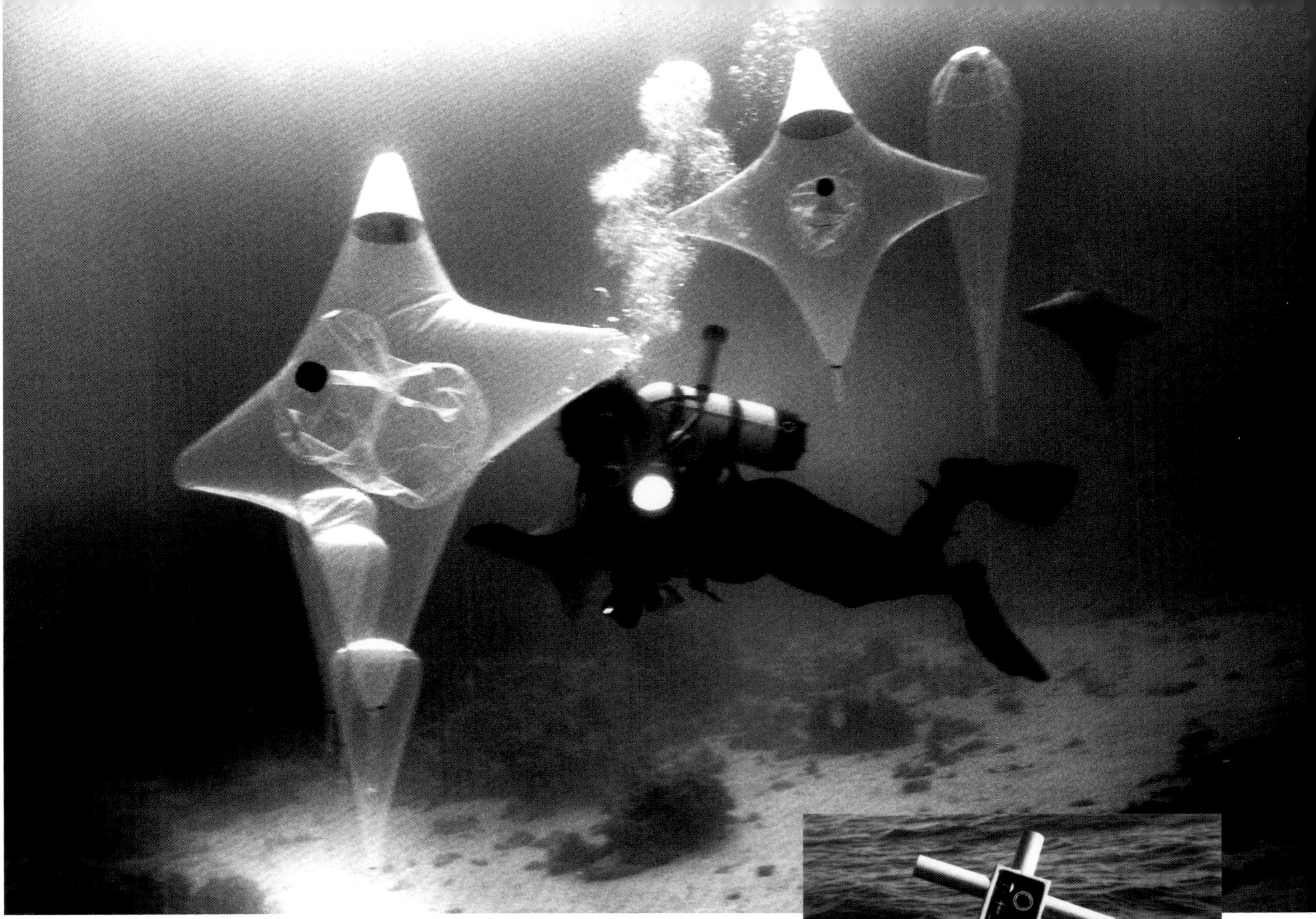

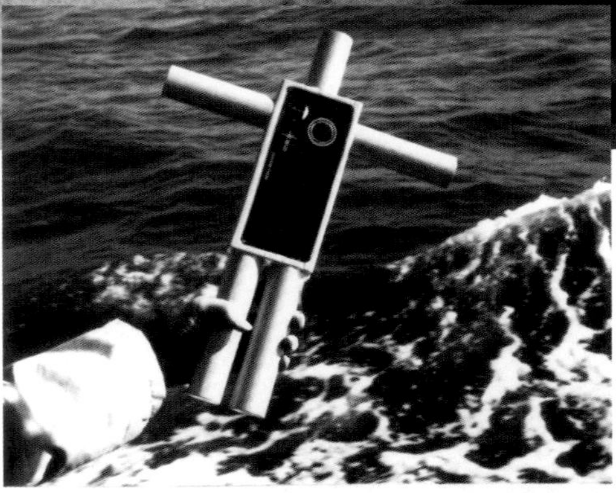

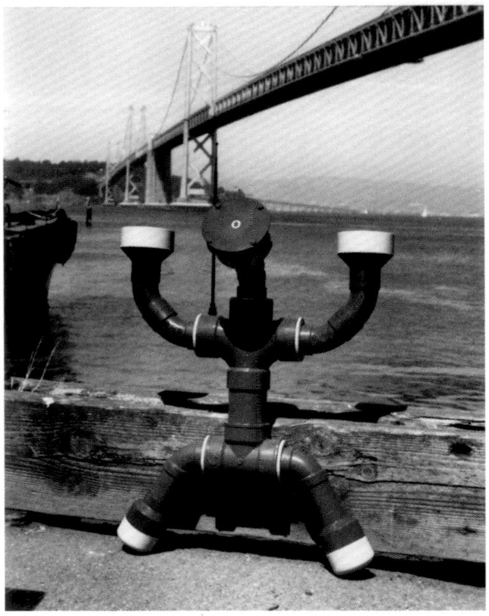

I liked Jürgen Claus a lot, again that European sensibility. He was working with Uri [Levi] doing underwater stuff. There was this whole history of underwater stuff at MIT most people didn't know about . . . Doc Edgerton and Jacques Cousteau . . . they would be over in the pool at MIT doing these bizarre experiments—people trying to come up with a common agenda. There was no communality at all at the beginning, yet over a period of months things would gradually connect, and finally, something that had all these various pieces and components felt like it was one thing. (Rus Gant interview)

Above: Jürgen Claus, "Die Gärten von Sharm" (The Gardens of Sharm), underwater spaces, 1978
Photo: André Foulon

Right inset: Bryan Rogers, "Odyssetron I" (upper image) and "Odyssetron II" (bottom image), "A Cybernaturical Metamodel . . . serial launching of marine space-time capsules evolving towards the first unmanned, marine circumnavigation of the planet earth," (Rogers, *Odyssetron*), San Francisco, 1980
Photos: CAVS Special Collection

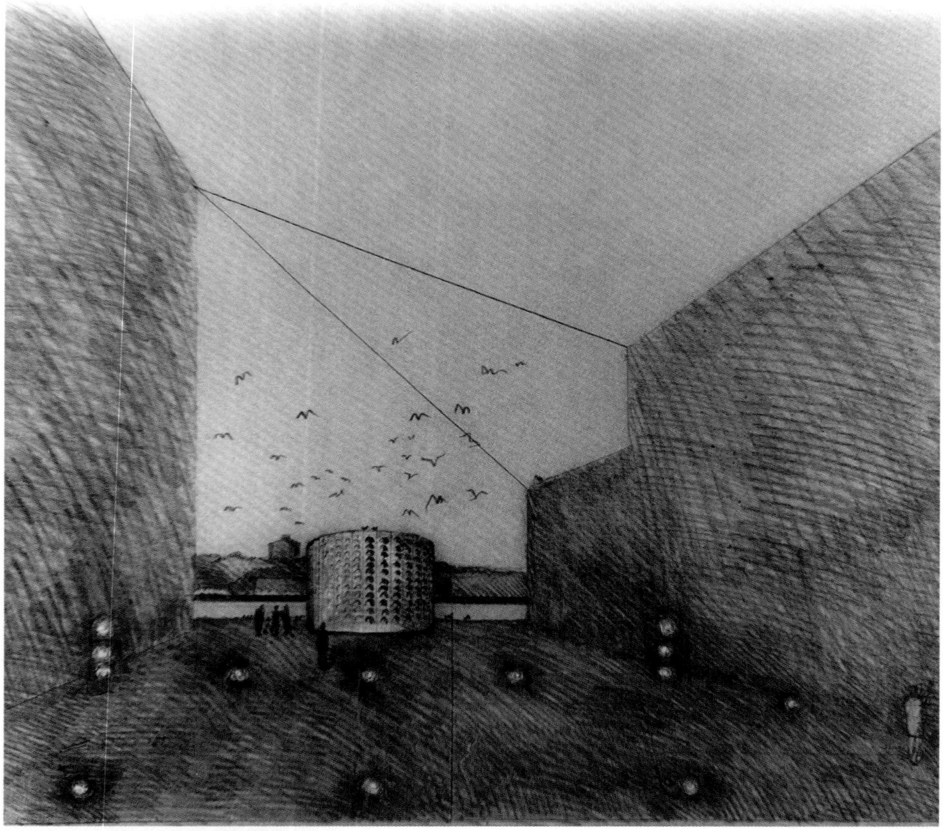

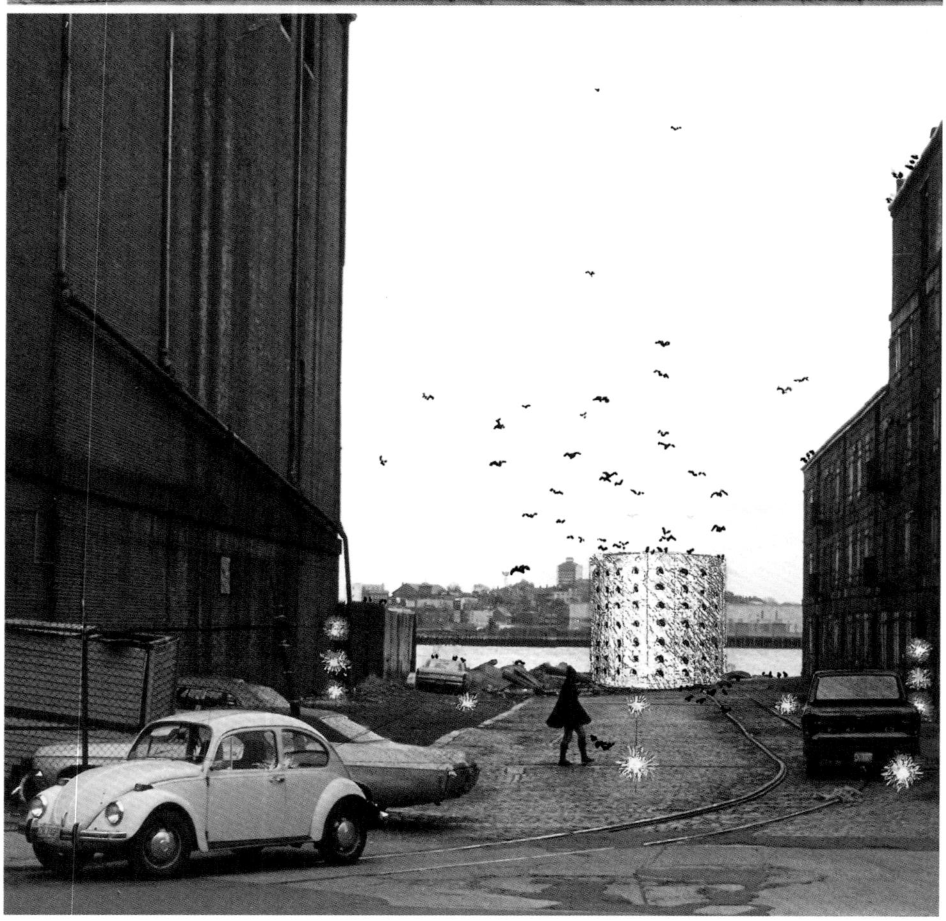

When you stand there, the sounds of the city are abnormal. It's true even near the Charles River on Magazine Beach, you hear the other side of Memorial Drive. Traffic noise is very very steady, not one single noise here and there, but one single texture, white noise. I remember Maryann [Amacher] talking: "we can't just do projects utilizing the noise for the project!" The noise is just overwhelming. . . . Neurological, biological perception seems to be very different, maybe it could be some sort of mechanism, I'm not sure. As artists we all had this perception, it was so keen; you can't wait for it to happen, you just have to work, follow your perception, chase around, what is this all about? . . . It was like exploding into air. It was the same situation as my postpartum depression. At that time I didn't know what to name it—losing senses. . . . I welcomed that chance, artists engineering their own ideas, how to make it visible, in and out. (Keiko Prince interview)

Left: Maryanne Amacher, Luis Frangella, Keiko Prince, "Pigeon House," concept drawing and photo mock up (details), 1967-71
Source: CAVS Special Collection

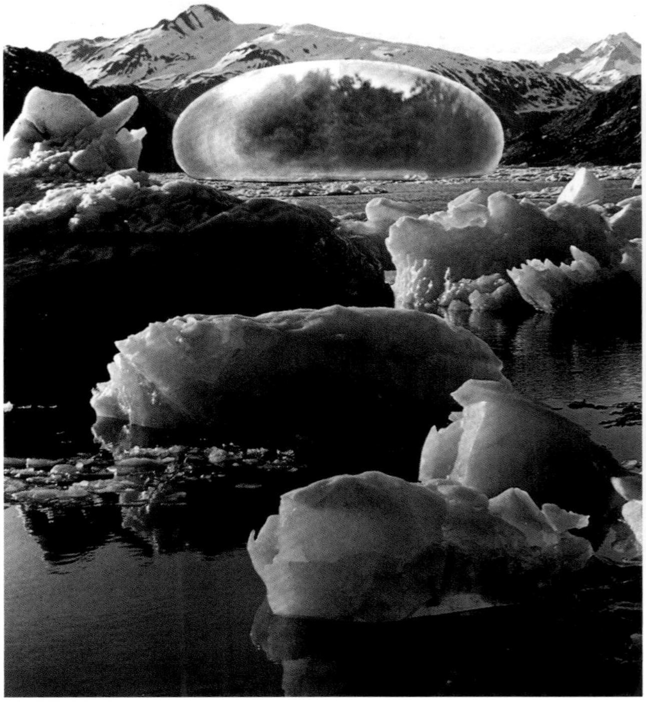 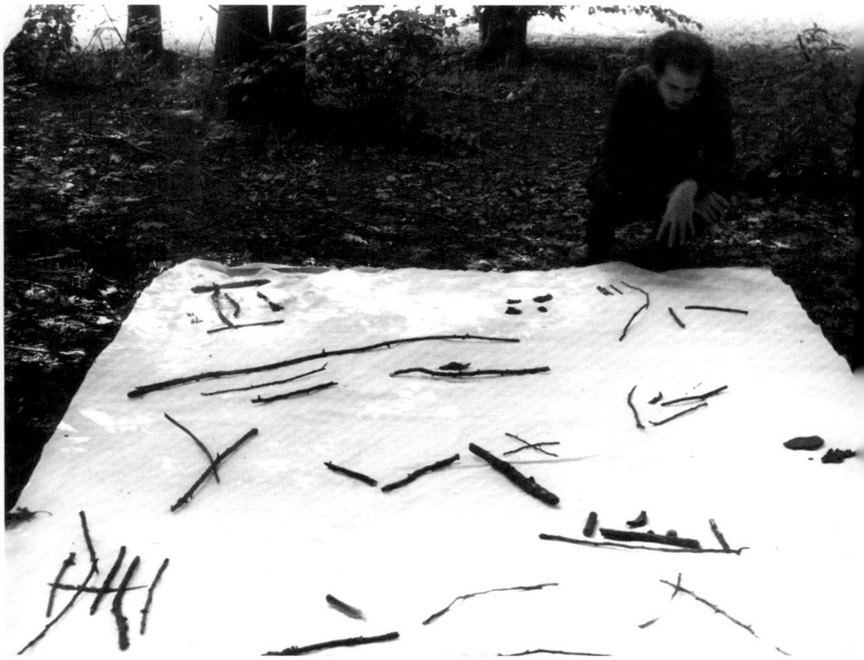

. . . Bucky Fuller's lenses concentrated light, which I connected to a satellite and some buoys. Light goes through the optical fibers, and then you see through another lens, a big sun. . . . Suddenly, I realized that to me, as an artist, if I am talking about light, you make it very simple. You make a drop of oil on a piece of cloth—the light becomes visible. (Juan Navarro Baldeweg interview)

Top left: Juan Navarro Baldeweg, photo, 1971

Top right: Alan Sonfist, "Landscape," element selection, Cologne, 1974
Photos: CAVS Special Collection

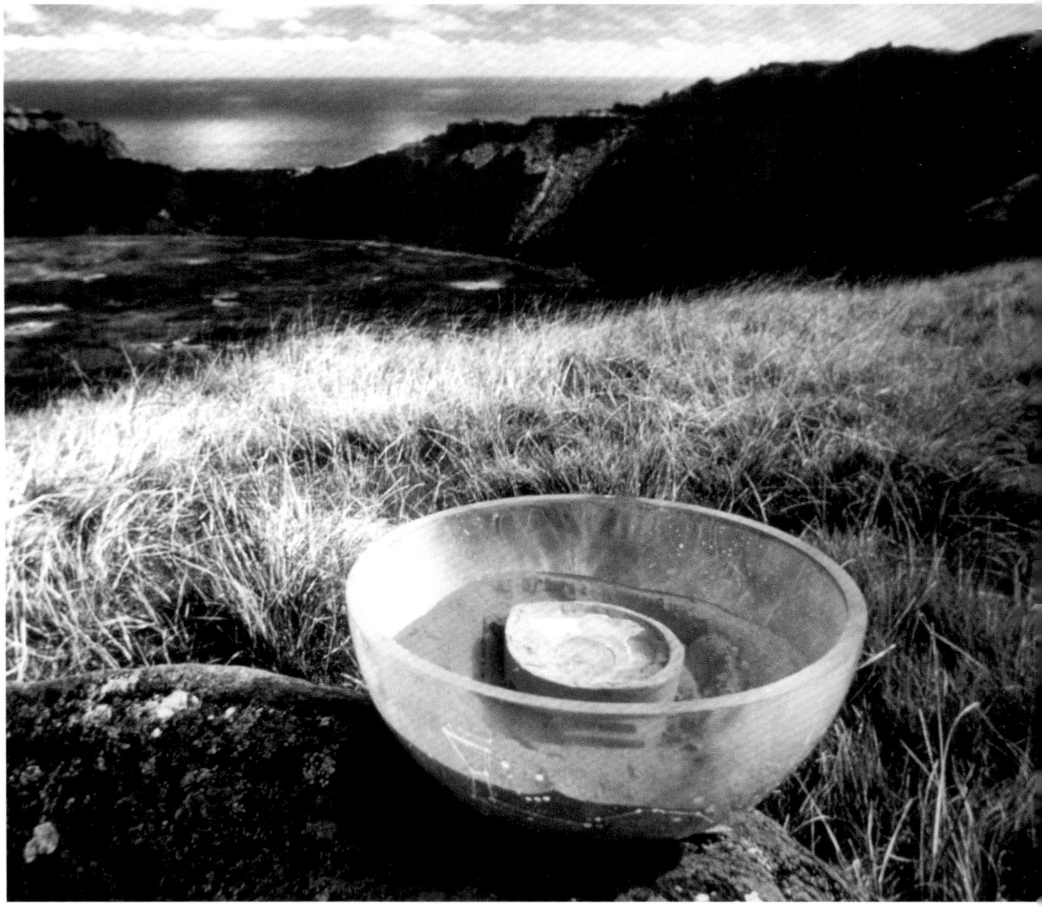

Right: Lowry Burgess, "The Quiet Axis - Part II, The Utopic Vessel," conceptual earthwork sculpture; pictured: assembly on the edge of Rano Kao, Easter Island, 1967-71
Photo: Janet Burgess

The lower half of the Utopic Vessel showing constellations, the vermillion utopic substance, the N'ave, the beeswax collar—footing for the Inner Vessel pointing toward the ocean horizon where the Utopic Vessel is now placed. (Lowry Burgess caption)

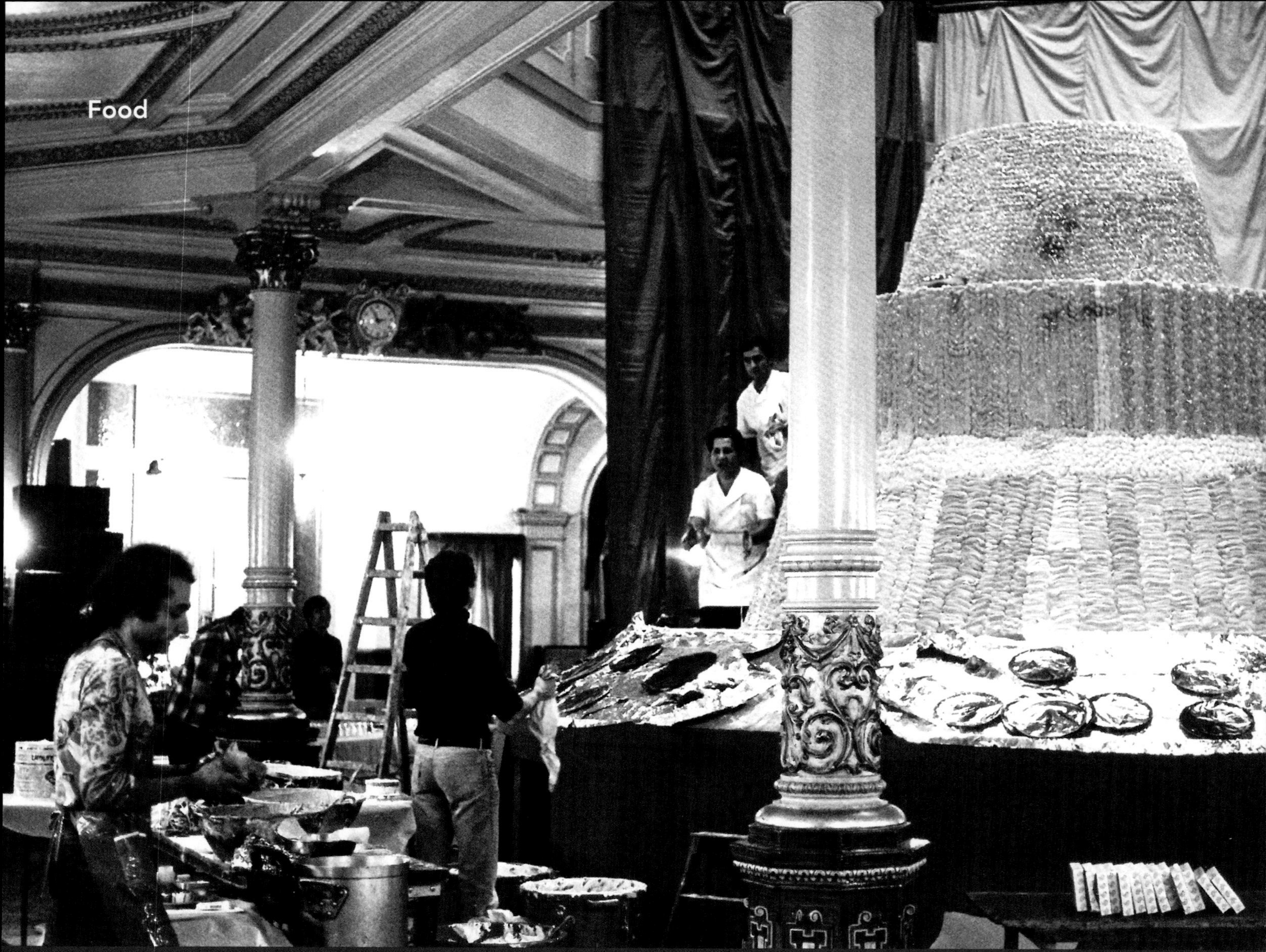

Above: Antoni Miralda (left) working with a team of pastry chefs on his project, "Merenga Hotel Orient," Barcelona, October, 1976
Photo: CAVS Special Collection

My interest [in coming to CAVS] was to figure out how an institution like MIT could match my interests in ritual, celebration, food, starting in the mid-60s. For that, I was spending hours and hours in corridors of big buildings. I find a door that I couldn't get into; the door has a NASA connection. A locked door was for me a dream, thinking how can I get through to that level of higher experimentation. Of course, it never happened. I just did much more common sense things: food and power, air, smells. (Miralda interview)

Facing page top: Otto Piene and food design by Keiko Prince, holographic chocolates (edible holograms) by Eric Begleiter, *80th Birthday Celebration for Otto Piene*, MIT Museum, Cambridge, April 18, 2008
Photo: Ellen Sebring

Facing page bottom left: Otto Piene, "Ice Cream," page from *More Sky*, MIT Press, 1970/73
Source: CAVS Special Collection

Facing page bottom right: Otto Piene, poster for the CAVS "Food Show," Lobby 7, MIT, April 7-20, 1975
Source: CAVS Special Collection

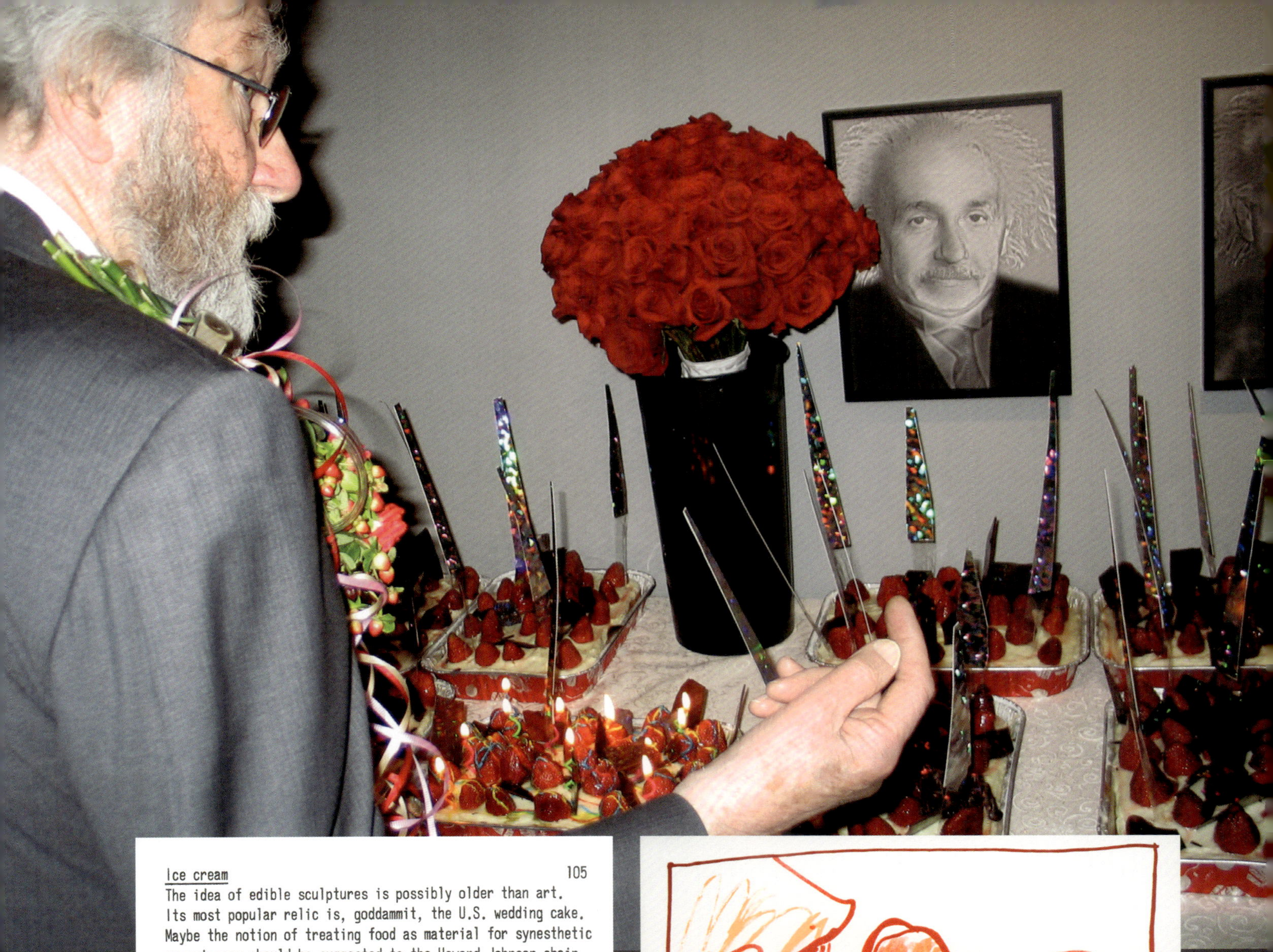

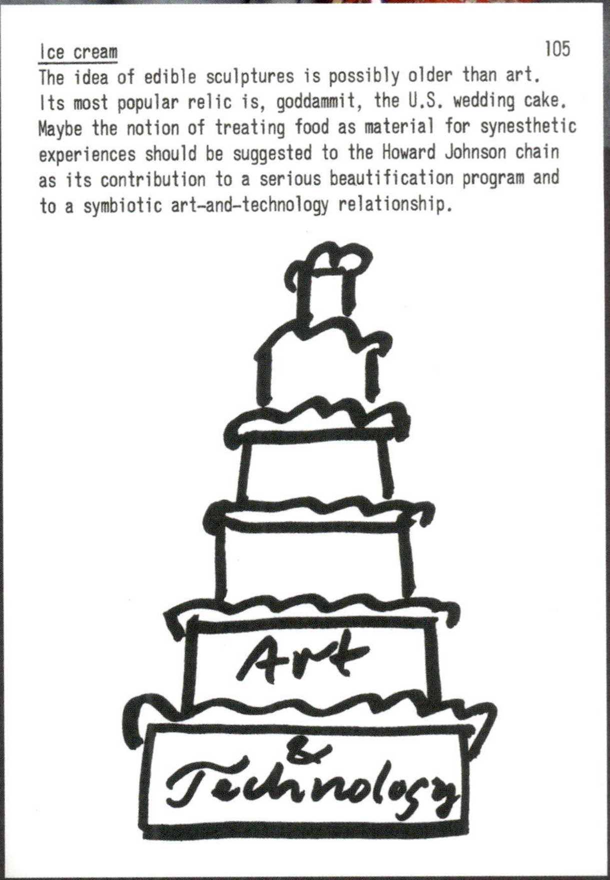

Ice cream 105

The idea of edible sculptures is possibly older than art. Its most popular relic is, goddammit, the U.S. wedding cake. Maybe the notion of treating food as material for synesthetic experiences should be suggested to the Howard Johnson chain as its contribution to a serious beautification program and to a symbiotic art-and-technology relationship.

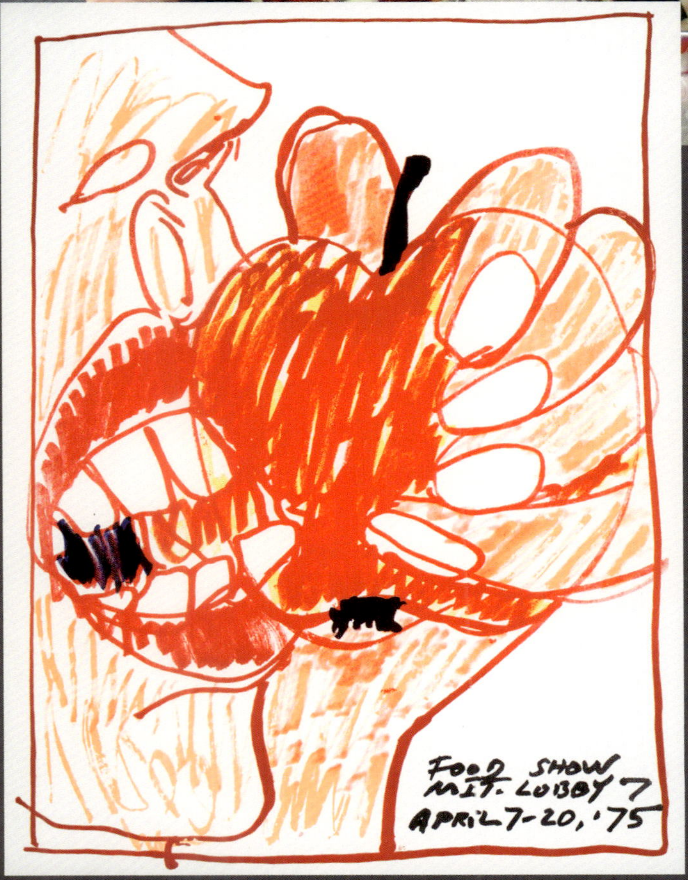

Holography

Light as a medium was shaped into three dimensions by artists who collaborated with scientists to learn and advance holography as a visual art form, beginning with Friedrich St. Florian, Lowry Burgess, and Harriet Casdin-Silver, who collaborated with scientists Steve Benton and Walter Lewin for *Centerbeam*, followed by (in alphabetical order) Csáji Attila, Eric Begleiter, Betsy Connors, Paula Dawson, Susan Gamble, Dieter Jung, Ioannis Michaloudis, John Powell, Don Thorton, and Sally Weber.

Stephen Benton, scientist and director of CAVS from 1996 to 2003, invented rainbow holography in 1968, the year CAVS was inaugurated. The new imaging method allowed holograms to be viewed in ordinary daylight. First at the Polaroid Corporation in Cambridge, and then at MIT where he founded the Spatial Imaging Group in 1982 and became a founding professor at the Media Lab in 1984, Benton was instrumental in connecting CAVS artists to holography and providing facilities, equipment and technical support.

Holography was used to structure light and visualize immaterial architectures and geometries of the imagination. Friedrich St. Florian was the first holographer at CAVS.

> I came up with this idea of an imaginary cloud. Clouds are, in a sense, non-physical. You can put your hands through them . . . When I was a child I had a canopy bed where the inside of the canopy was painted to look like heaven. There was an angel looking out between the clouds. . . .
> We used the largest projection holographic plates available at the time, 24-inch x 72-inch. This project was done when I knew I was coming to the end of my stay at the Center so I called it the MIT Himmelbett. This was my monument to MIT, my "thank you gift." (Friedrich St. Florian interview)

Initially using laboratory and domestic objects, Harriet Casdin-Silver later focused on the human body, transforming holography into an instrument for expressing feminist/humanist perspectives.

> [Holography] is like sculpture made of light. . . . I believe that the art of this time will emanate from the laboratory. (Harriet Casdin-Silver, *Centerdisk*, 1982)

I see my work with aging not so much out of feminism as out of ageism. My old people are proclaiming they are still here and human and contributing. Feminism becomes humanism. This is about where should CAVS go: our time is technological but it is also biological. These two are obvious directions for exploration by new people at the Center: replacing an old lens with a new lens. Artists can help with mind-body integration. I feel that my mind seems to work along the same waves as the very young. Do we want to help mind and body to evolve? Help all species? Or just humans, which ones should we push forward? It is fantastic as well as frightening. . . .

In 1992, I was at the University of Ghent. I used two grown-up children of Professor Pierre Boone (who ran the lab) as models. I made them look unhappy, cringing, trapped—we are all trapped. I showed the Ghent holograms for the 25th Anniversary of CAVS. In Ghent, I also made a hologram of René, a hermaphrodite . . . I toured Ghent and saw the Adam and Eve on the altarpiece of the Ghent Cathedral and shot them. Back at the lab at CAVS, I switched their genitalia and made a new triptych, and in the middle between them, a vertical row of asses. (Harriet Casdin-Silver interview)

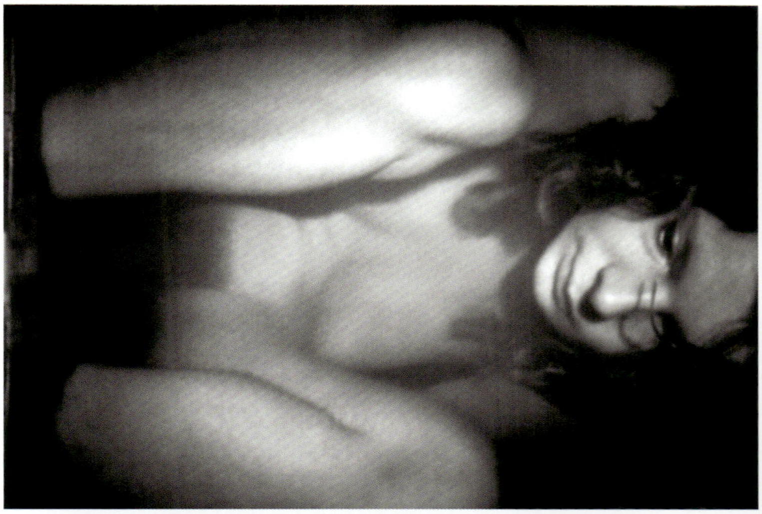

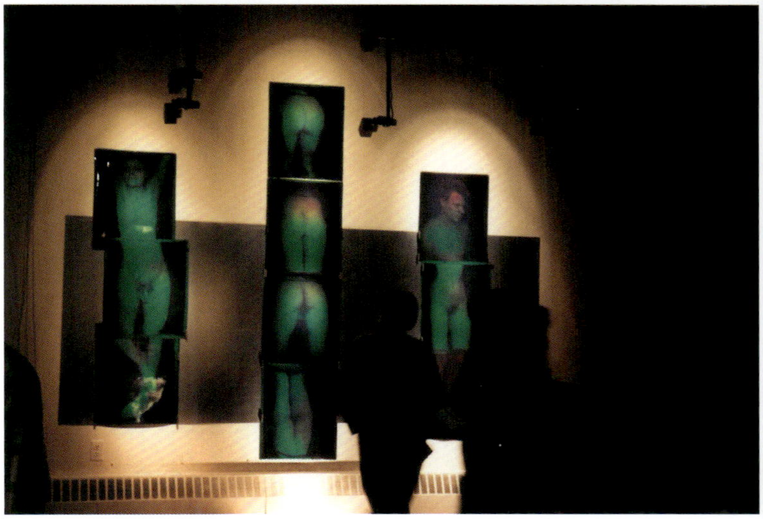

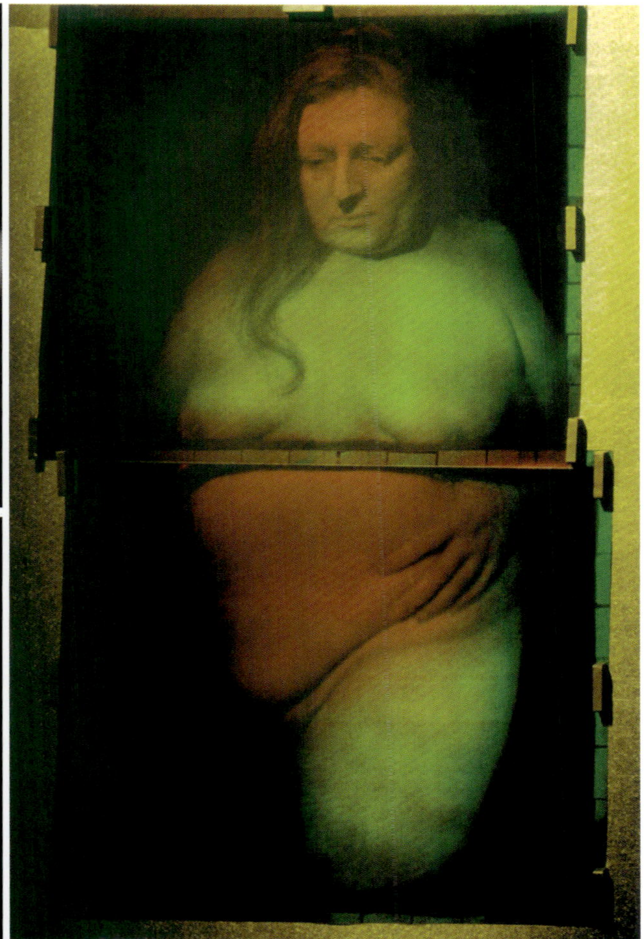

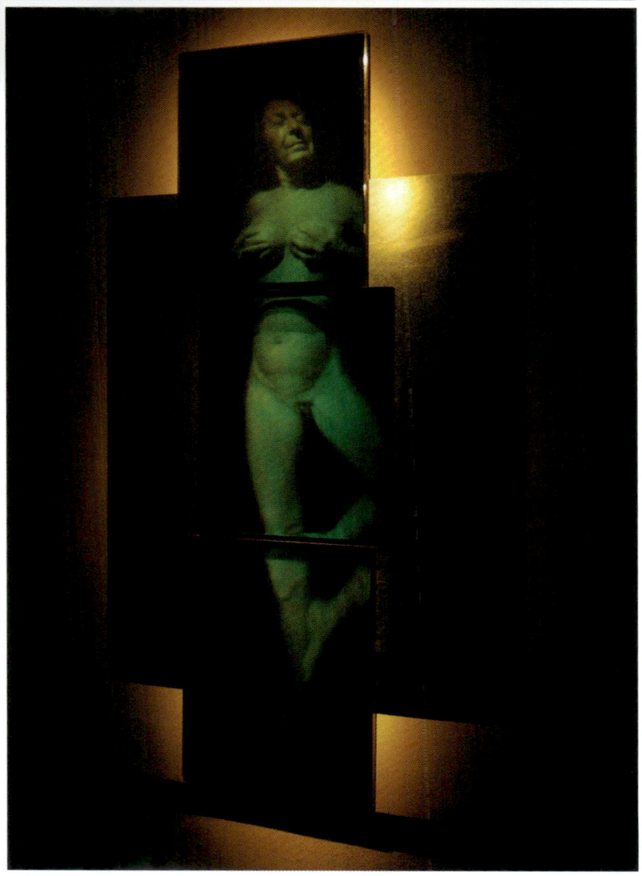

Top left: Harriet Casdin-Silver, "AN I," pulsed laser reflection hologram, shot at Dr. Pierre Boone Lab, Rijksuniversiteit Gent, Ghent, Belgium, 1992

Bottom left: Harriet Casdin-Silver, "To Van Eyck and Bosch," three reflection holograms on metal, 8 x 10 feet, exhibition room, CAVS, MIT, 1994

Top right: Harriet Casdin-Silver, "The René Wall," detail, reflection holograms on metal, 1993

Bottom right: Harriet Casdin-Silver, "70 + 1 + 2," self-portrait detail, three reflection holograms on metal, camera, Todd Gieg, 1998

Photos: Harriet Casdin-Silver

According to Sally Weber, her holograms constitute experiences of being "inside light" through architectural-scale works.

> CAVS Fellow Jose de Prada helped me with my first hologram installation, "Lightscape." Nobody had done anything that big outdoors in holography. It was composed of four arcs of acrylic which held the holographic optical elements made on film. He helped me determine the angles that plastic needed to be in order to be perfectly flat on the ground. I had to build a structure to hold the film. At the time he was doing research on a concept with dimensional pieces of wood which created fabulous forms that all interlocked together. . . . With respect to holography, there was no hierarchy. It was the way the medium was used, the intention: the content needed to come through the technology. I fought tooth and nail about that in holography, because content was sometimes superseded by technology. I found that problematic. What I liked about holography was the distortion. We don't know what a bad lens looks like. Usually everything must be correct and perfect. But a bad lens creates fabulous effects. Holography is a bad lens. At the Center working off the idea that there was no "good" or "bad," we could fail and it was fine and we did. (Sally Weber interview)

While Paula Dawson pioneered Phantombrush software to draw freehand in virtual 3D space, Betsy Connors' holograms reflect her background in film and photography. Using subtle movements, shadows and highlights, she stimulates what she refers to as "peripheral perceptions of nature."

> I am interested in creating a sense of light, as when it plays across water, or interacts with wind as through the leaves on a tree, an almost peripheral awareness, of landscape holography. (Betsy Connors interview)

For Dieter Jung, Lowry Burgess, Eric Begleiter, Setsuko Ishii and others, holography synthesized visionary and physical objects and spaces, giving physical form to psychological and spiritual energy. Burgess made his first holograms at MIT in 1971. His *Moonbounce* was a ritual pouring of the Earth's waters captured on video. Converted to shortwave radio, the video data was projected to the moon, reflected back, and received by the Millstone-Haystack radio telescope in Westford, Massachusetts. The data was then turned into a digital hologram by the MIT Spatial Imaging Group.

> Holography shows the individual how perceptible images are formed, and are being formed, revealing a fundamental creativity at work within everyone's mind. . . . To me, [holograms] were how I see in my mind, pieces of my mind in space and light. They were like visions put in the world. (Lowry Burgess interview)

Facing page top left: Sally Weber, "Focalpoint," a water fountain installation using holographic diffraction grating to project the colors of sunlight, Museum of Science, Boston, April-June, 1983
Photo: Sally Weber

Facing page top right: Sally Weber working on "Focalpoint," November, 1982
Photo: Bruno de Lard

Facing page bottom: Sally Weber, "Lightscape," installation described as, "four segments . . . composed of holographic film made into a series of diffraction gratings in MIT's George R. Harrison Spectroscopy Laboratory, then sandwiched between two layers of plastic." (*MIT Tech Talk*, May 26, 1982), Kresge Oval, MIT, May 1982
Photo: Sally Weber

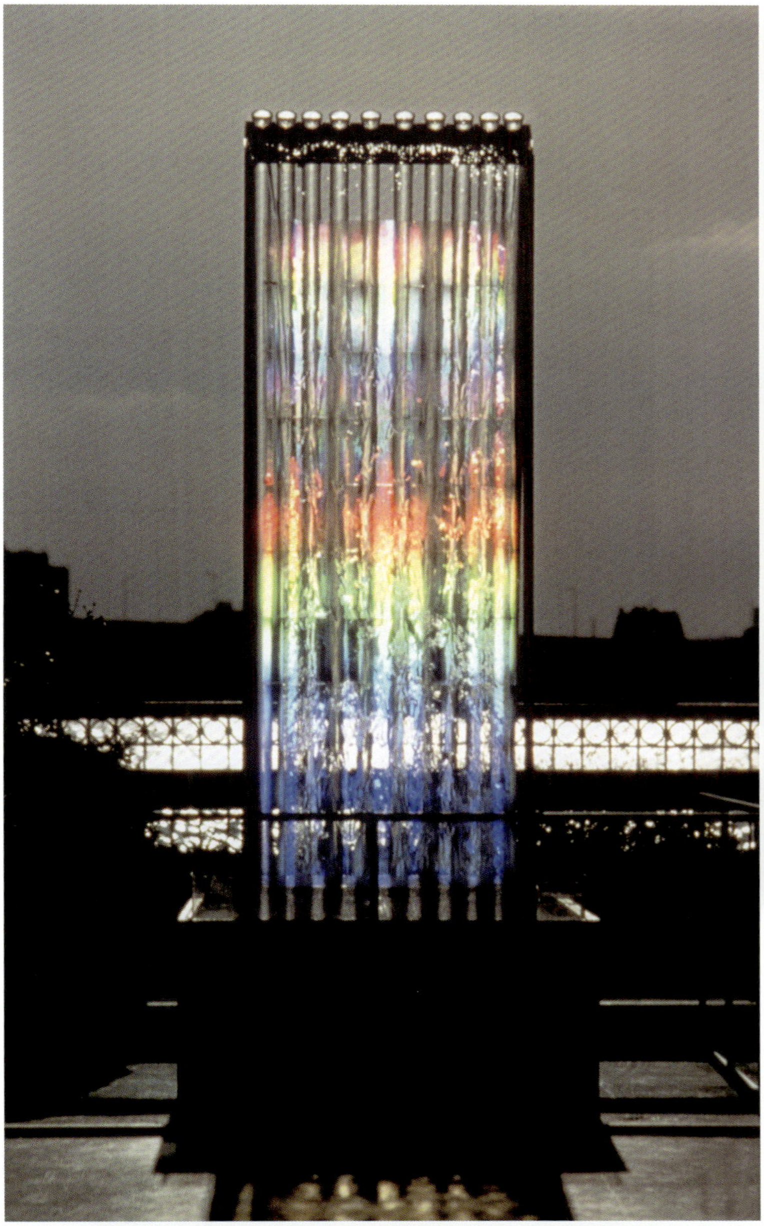

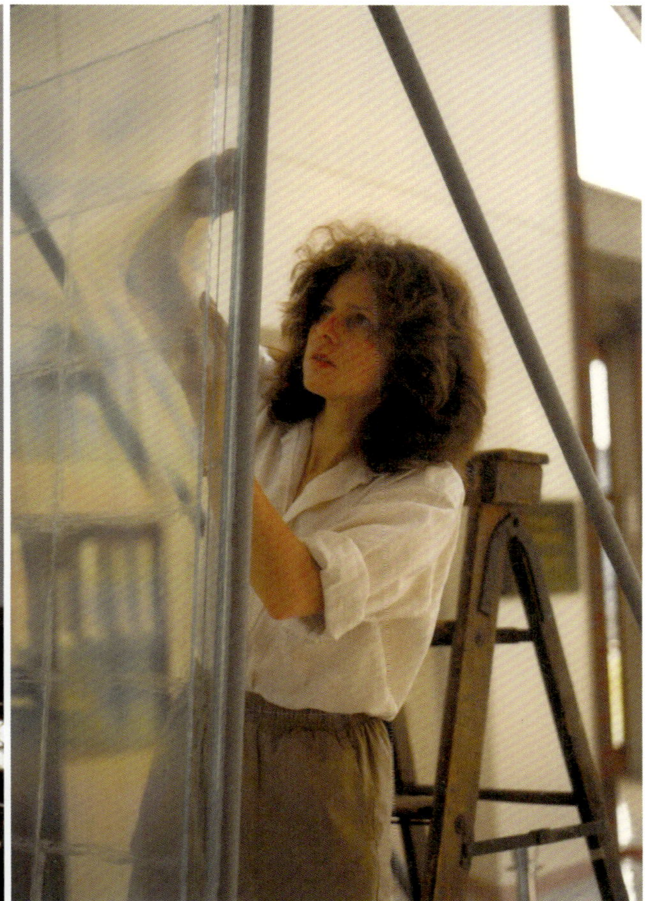

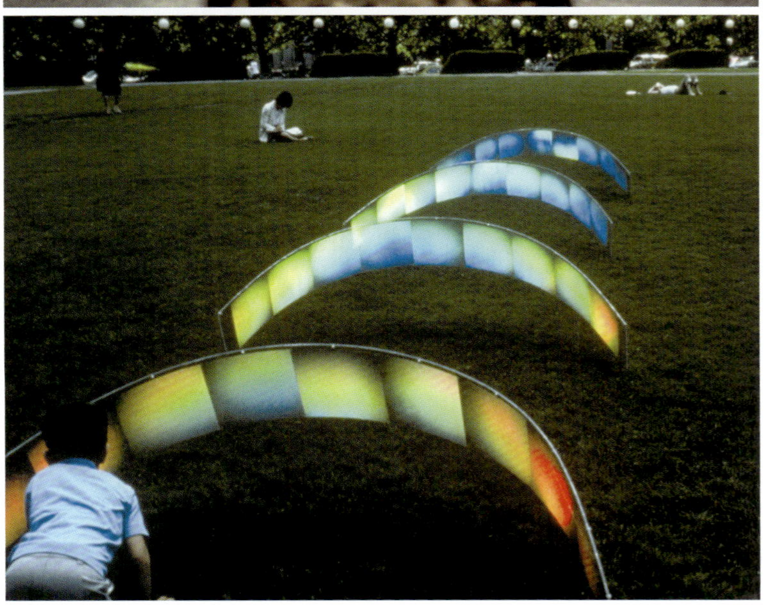

The kind of holography I was doing used very simple gratings which required two beams of laser light. I did a lot of tests and not being of mathematical inclination, testing allowed me to see what the sun angles would do. So, I made holograms, took them outside to see how they projected and then made graphs of what all these different gratings would do. . . . Someone suggested that I go over to Polaroid and see Steve Benton. I got all my notes together, made an appointment, knocked at the door and said, "Hi, I've done all this. This is what I want to do. The sunlight is coming in at these angles. I need to see green at twenty feet. I knew what the other colors would do." Then he said "Oh," worked on his calculator and said, "that's what you need to make to make the holograms." A miracle! That's how I came to know Steve. (Sally Weber interview)

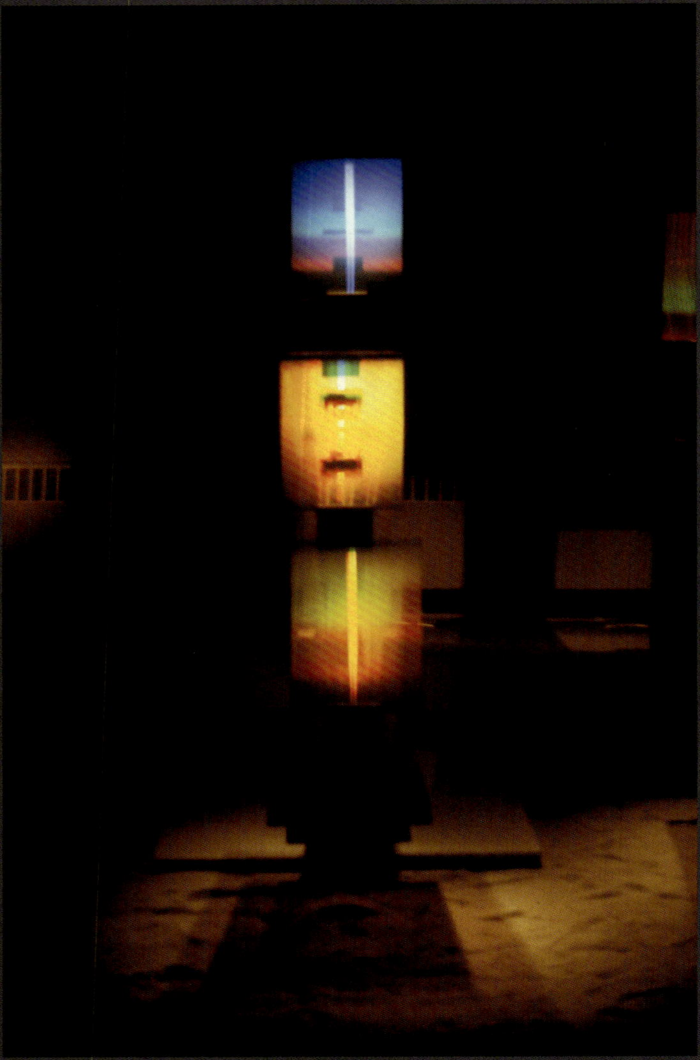

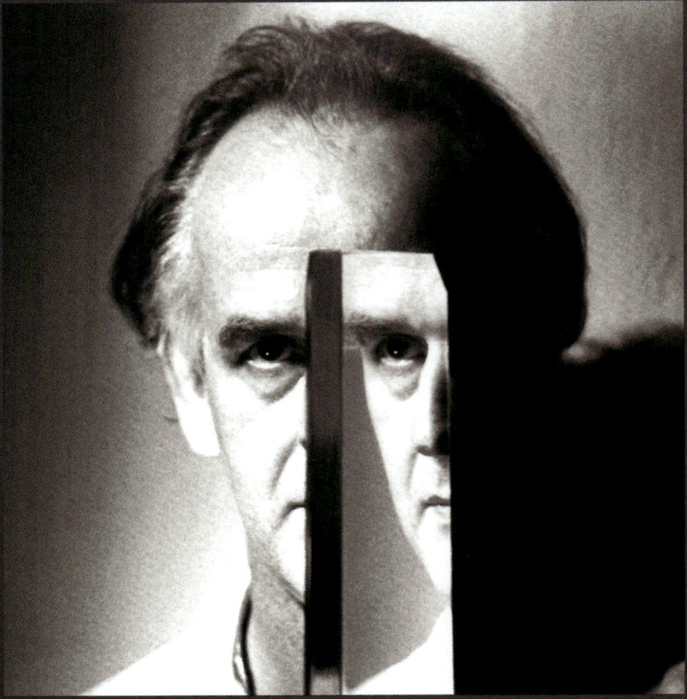

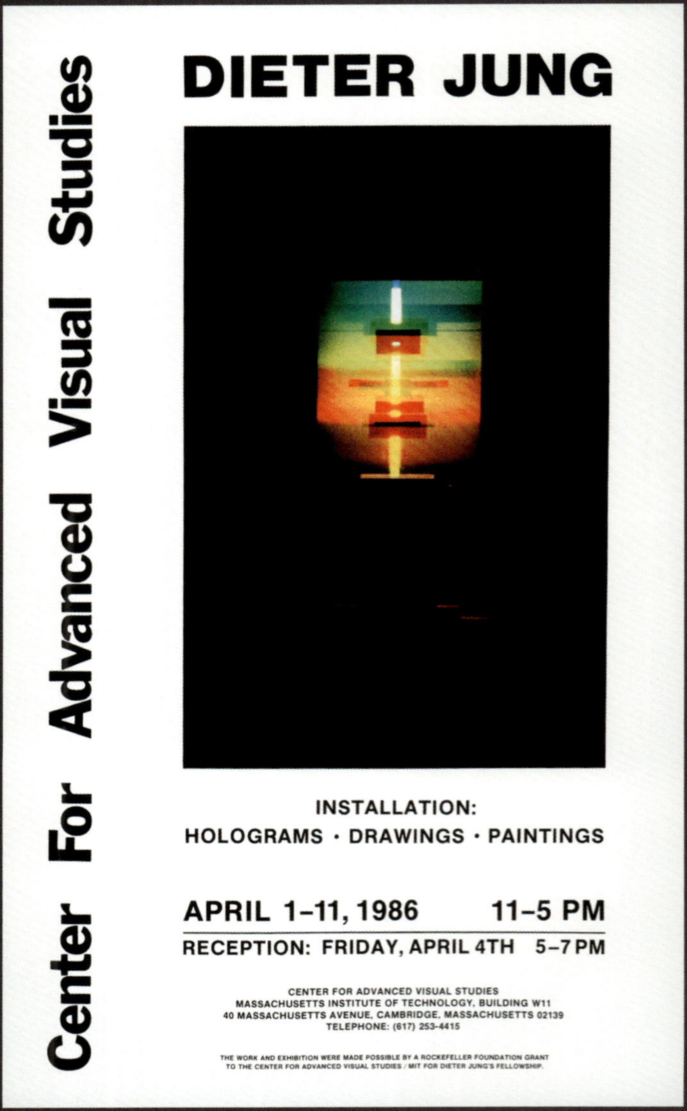

Holography incorporates infinite aesthetic capacities for communication and a continuum of energy, integrating multicultural, spiritual, and religious patterns of the virtual and material worlds. . . . If you break a hologram in many pieces, each part contains the information of the whole! . . . Holograms stimulate the visual senses and apparently reveal fascinating dimensions of thoughts and emotions, awakening dormant layers in the cortex system. (Dieter Jung interview)

Top left and bottom right: Dieter Jung, "Installation: Holograms, Drawings, Paintings," exhibition view and poster, CAVS, MIT, April 1-11, 1986
Source: CAVS Special Collection

Top right: Dieter Jung, artist's portrait
Photo: Dandart

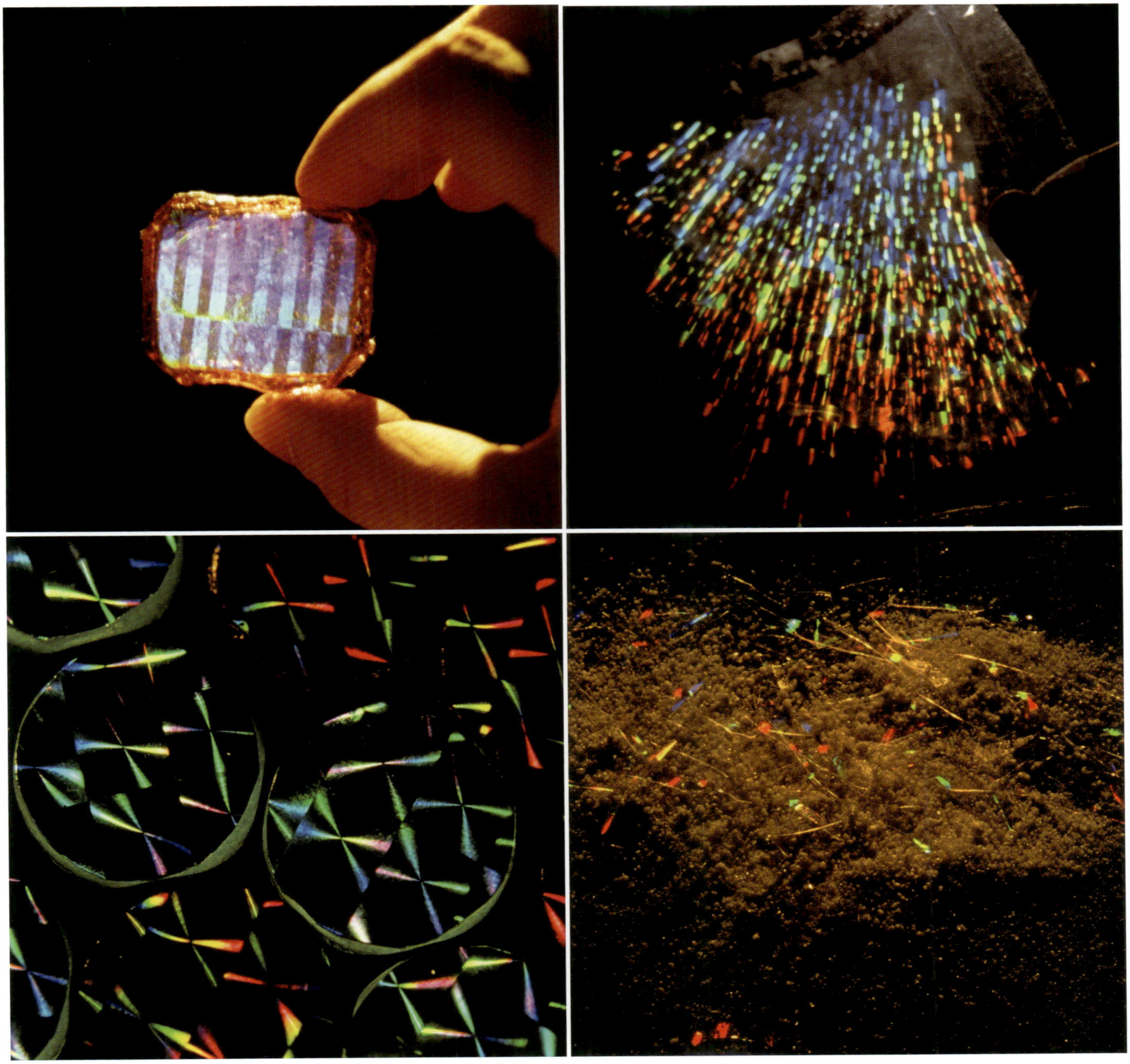

As a graduate student and fellow at CAVS in the early 1980s, Eric Begleiter studied human vision and 3D imaging methods. He invented and patented numerous holographic techniques including edible microstructures to diffract light in controlled ways for consumable products, e.g., edible holographic hard candy, chocolate, and pharmaceutical tablets.

Holography is like a memory condenser, a window into the space of mind and culture, creating products of a hyper-dimensional reality . . . . Chocolate, candy, pharmaceuticals and surface illusions—creating food colors with no chemicals, consuming pure light is a form of alchemy. (Eric Begleiter interview)

Above: Eric Begleiter, edible holographic candy, tablets and granulated sugar, 1988
Photos: Eric Begleiter

Light

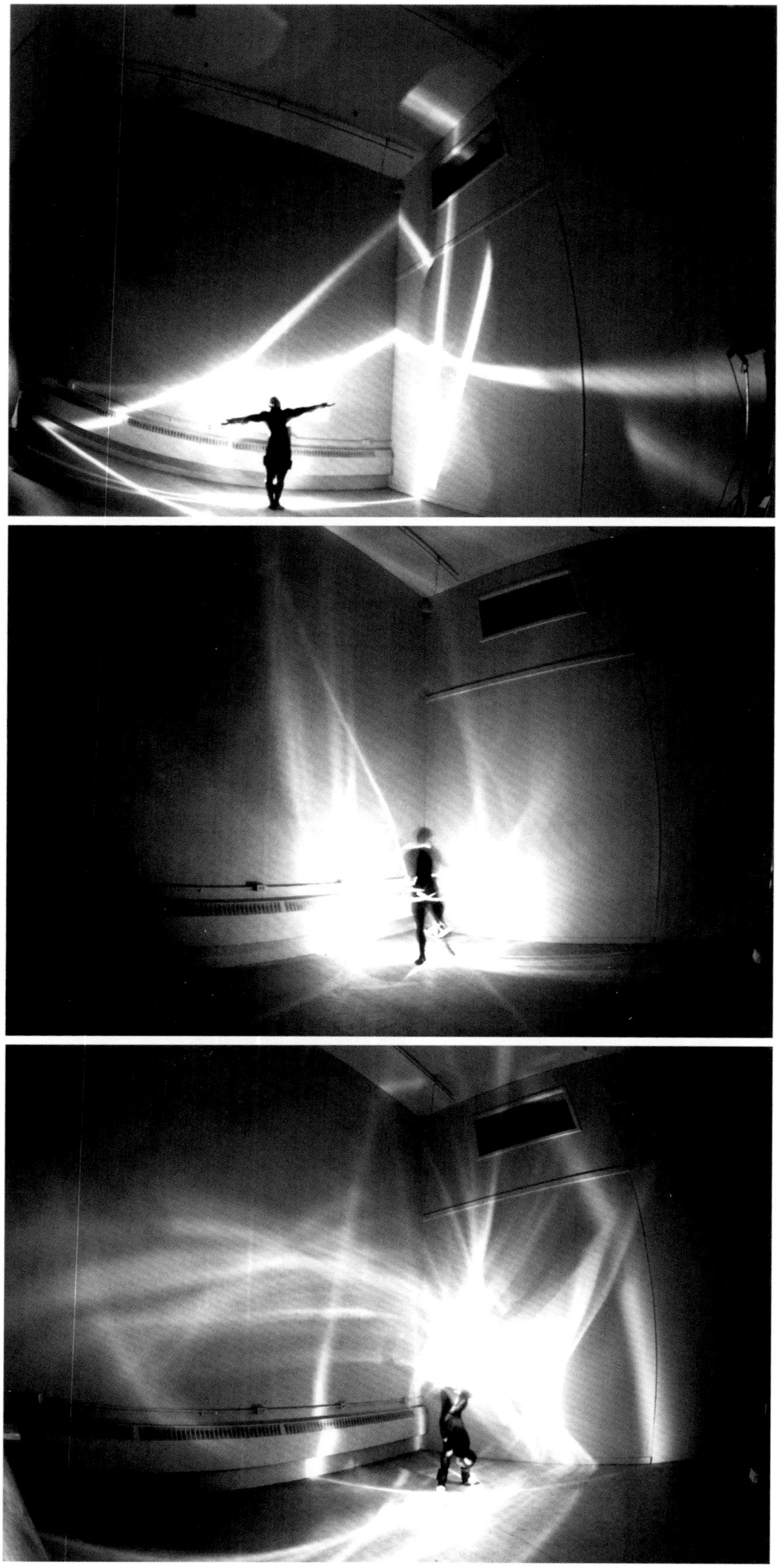

At the outset I drew images of what this project would be, to give it initial form: the human figure circumscribed by a ring of light. . . . I developed light instrumentation that was carried on my back. I made the Pit totally dark . . . The light transposed my body to the boundaries of the room. I was able to "move" the space by articulating the light phenomena through my gymnastic ability. . . . I was just starting out. I later learned what I was up to had a place in the history with the Bauhaus School, Oskar Schlemmer and more—Otto's *Light Ballet*. (Seth Riskin interview)

The light seems to be coming from all directions. It plays across one wall and wanders onto the next, morphing in scale and clarity as it moves. The patterns come and go according to rhythms of their own, but they also work together in unison. At the piece's zenith, as it were, the room is filled with light from multiple sources, creating the sense that one has been thrust into the center of a boxed-in Milky Way . . . (Sebastian Smee, "Otto Piene's 'Light Ballet' dazzles at MIT," *The Boston Globe*, November 20, 2011)

Left: Seth Riskin, "Light Dance," the artist performs with body-mounted light instruments, the Pit, CAVS, MIT, 1989
Photos: Noah Riskin

Facing page: Otto Piene, "Lichtballett," MIT List Visual Arts Center, October 21-December 31, 2011
Photos: Günter Thorn

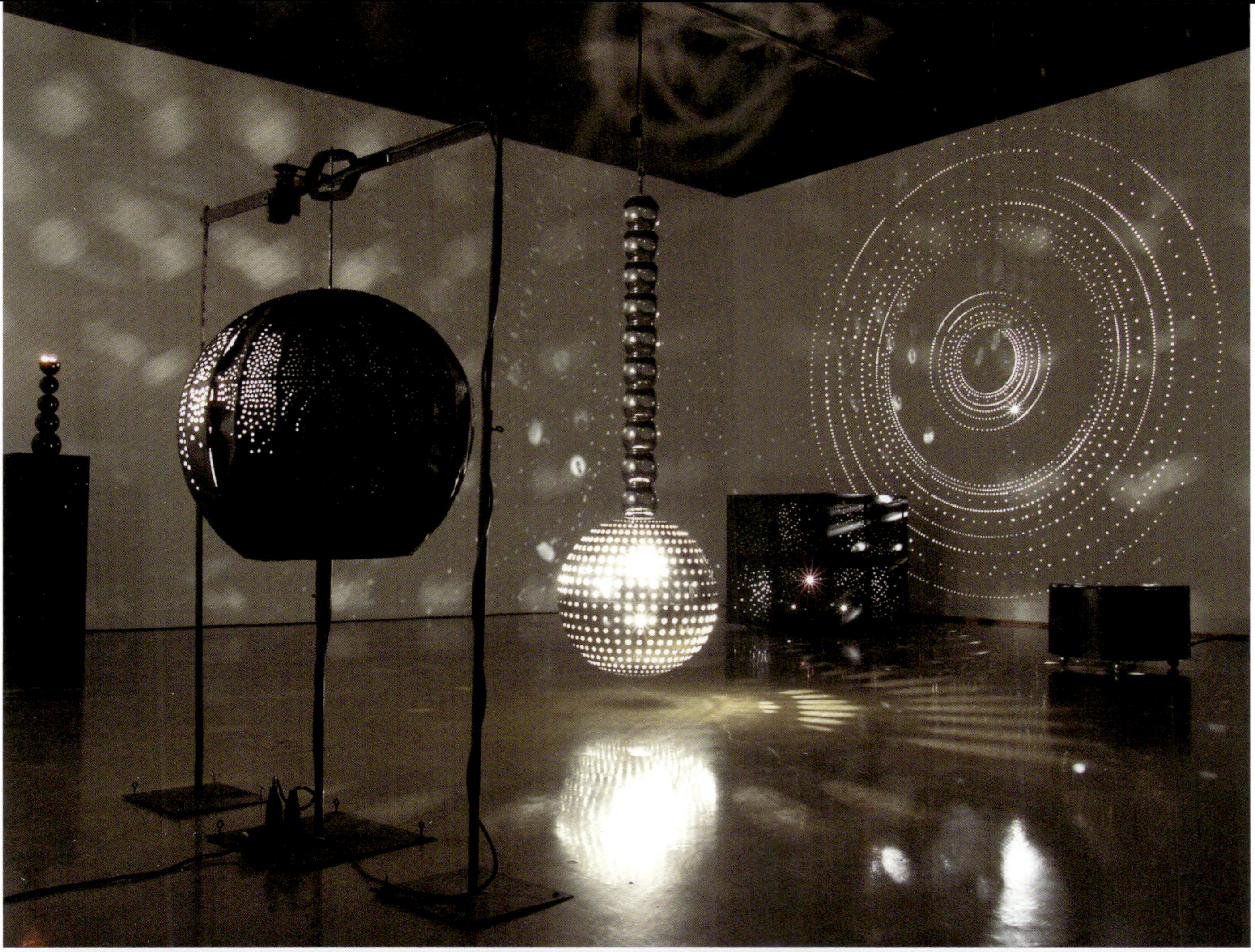

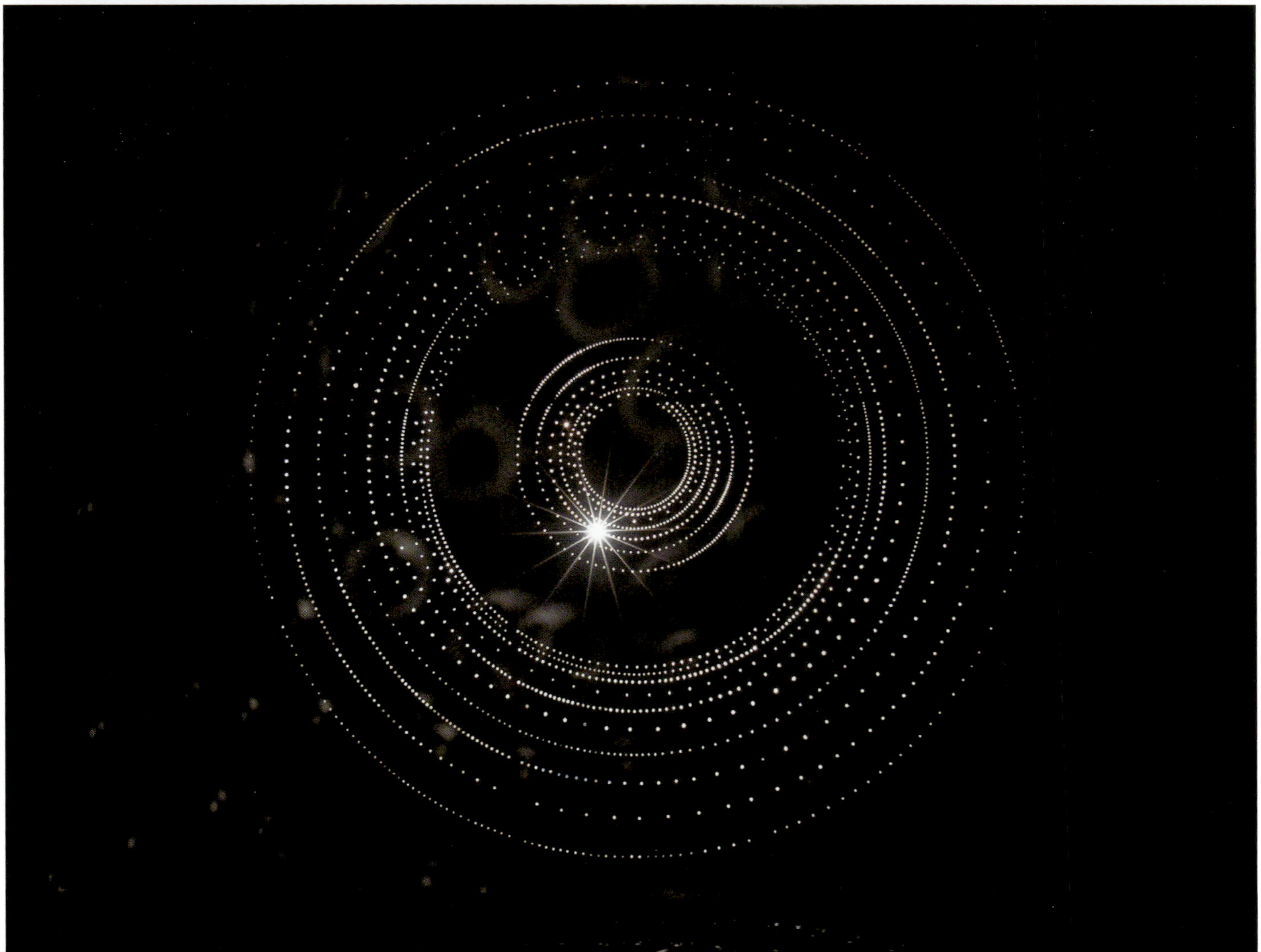

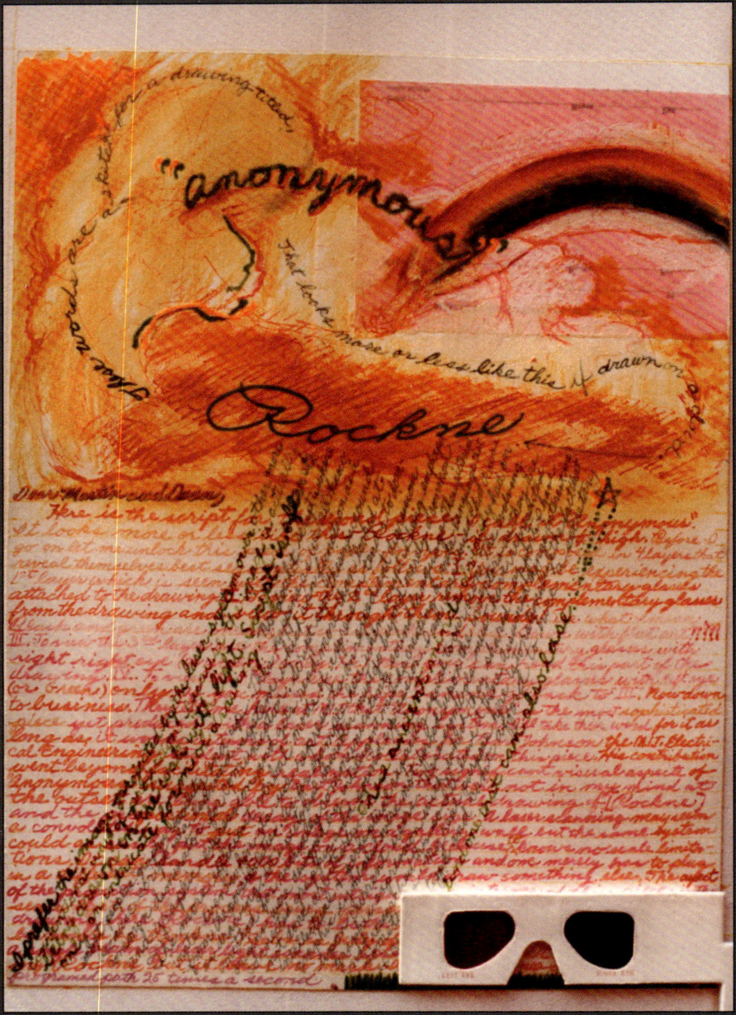

In the drawing *Anonymous*, Krebs examines cursive writing as a combination of anonymous origin and individual expression. This graphic work has four visually distinct layers and can be totally understood only with 3D glasses. One part can be seen with normal vision, one with the glasses' red lens, one with the green lens, and one using both lenses at once. The sculpture "Anonymous," that results from this drawing made use of a scanning device (which Krebs developed with Bart Johnson, an engineer) that created a memory for a laser beam. The piece spelled Krebs' signature in laser light. . . . Krebs strongly feels that sunlight will be a much more important source for art in the future, and that his drawings are partly for those artists who will continue the exploration of this area. (Jay Belloli, *Rockne Krebs: A Retrospective of Drawings, 1965-1982*, 1983, 13, 17, 23)

Left: Rockne Krebs, "Anonymous," collage with 3D glasses, 1974; also labeled, "'Rockne' a stereo drawing/design for a laser scanning projection" (typed caption, CAVS Portfolio, 1974)
Source: CAVS Special Collection

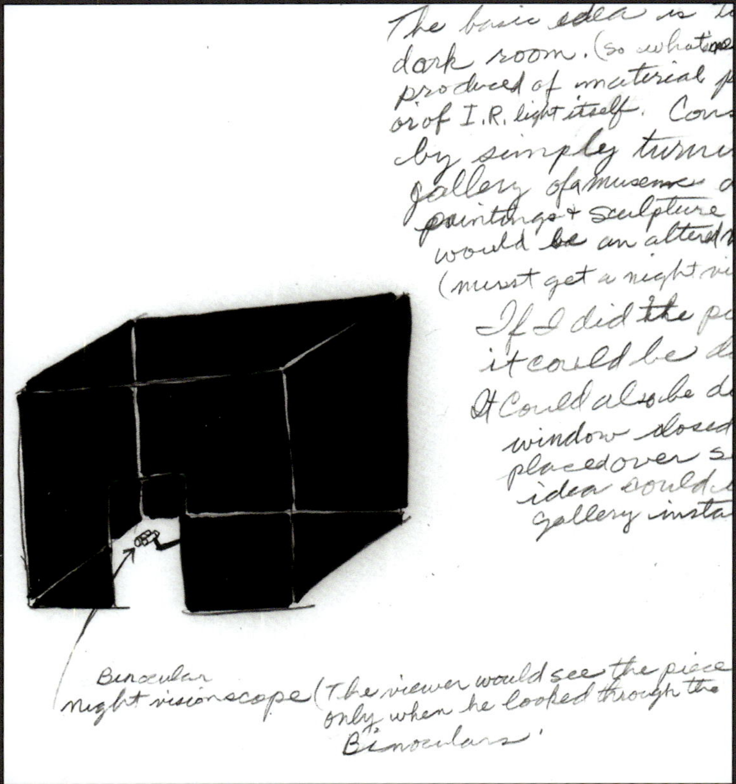

The basic idea is to do a piece in a completely dark room. (So whatever) the piece would be produced of material particularly responsive to infrared light or of I.R. light itself. Consider doing the piece by simply turning off lights in a gallery or a museum already installed with paintings + sculpture. In this case my piece would be an altered vision of existing work . . . (Krebs, handwritten text on "April Fools or Dark Art" drawing, 1973)

Left: Rockne Krebs, "April Fools or Dark Art," detail, drawing with text, April 1, 1973
Source: CAVS Special Collection

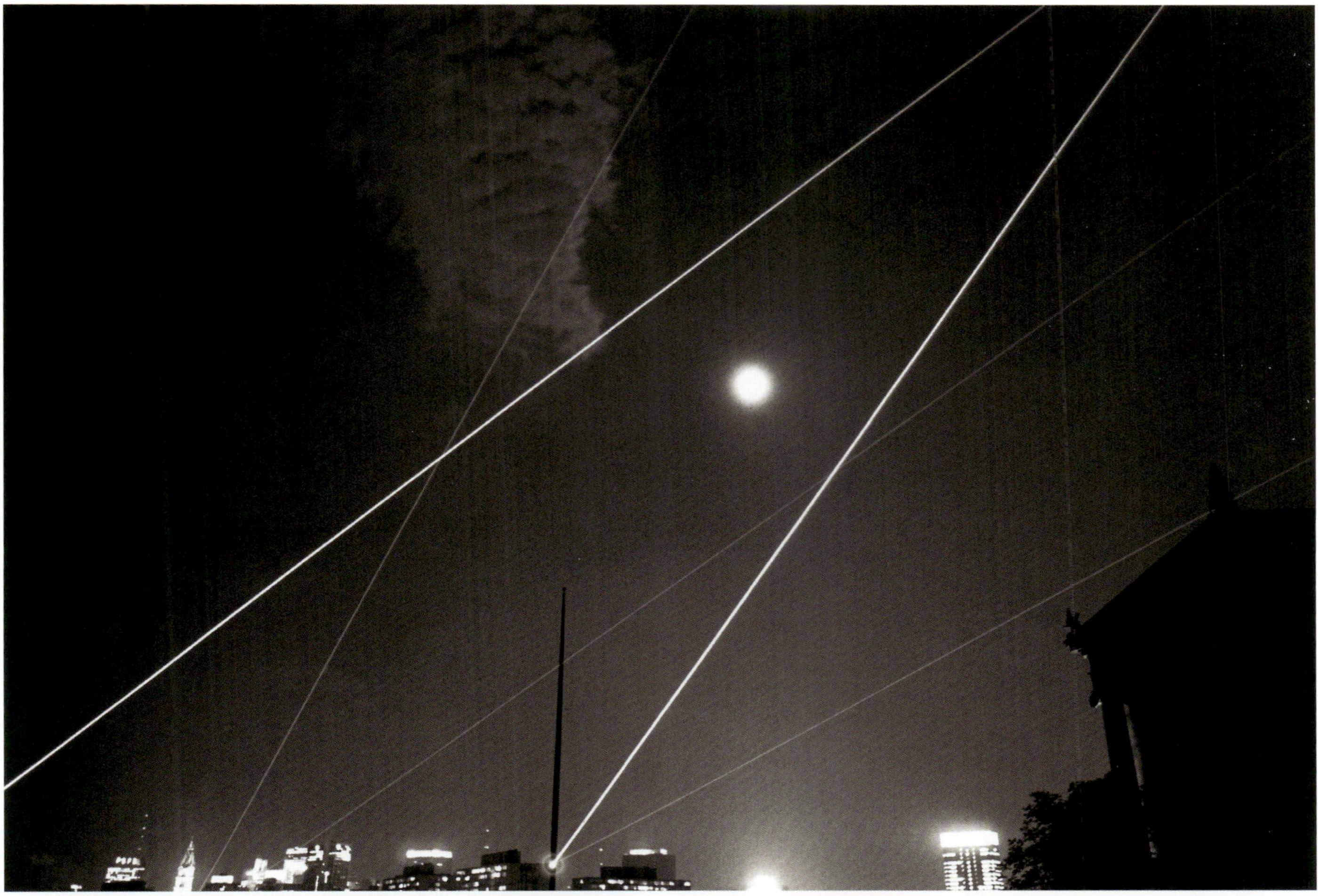

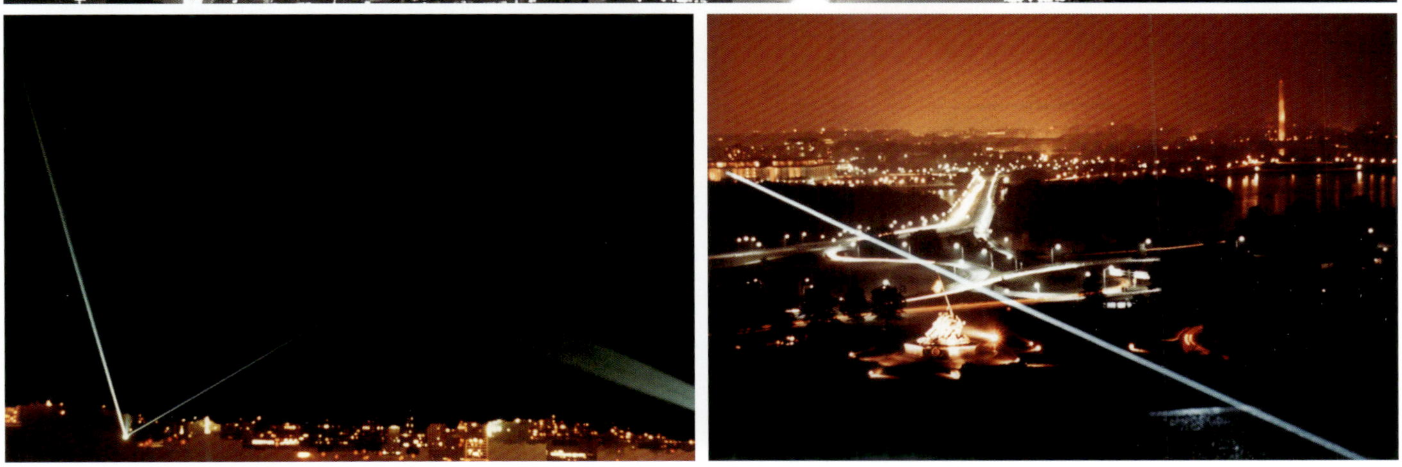

"Irish Light[s]" is made of light, of the night lights of the city, the reflections of the water, the night lights of the sky, and laser beams of brilliant, glowing green. . . . In a trial run on Tuesday night, Krebs, employing mirrors, split each beam in two, producing as he did so a floating "w" of light. Last night he changed one light-line into a spreading cone, what he called "a fat swath of green light." (Paul Richard, "Of Lasers and City Lights," *Washington Post*, May 30, 1974)

Top: Rockne Krebs, "Sky Bridge Green," laser installation, Philadelphia, Pennsylvania, 1974
Photo: Patrick Radebaugh

Bottom: Rockne Krebs, "Irish Lights," part of "Art Now 74," Kennedy Center, Washington, DC, 1974
Photos: CAVS Special Collection

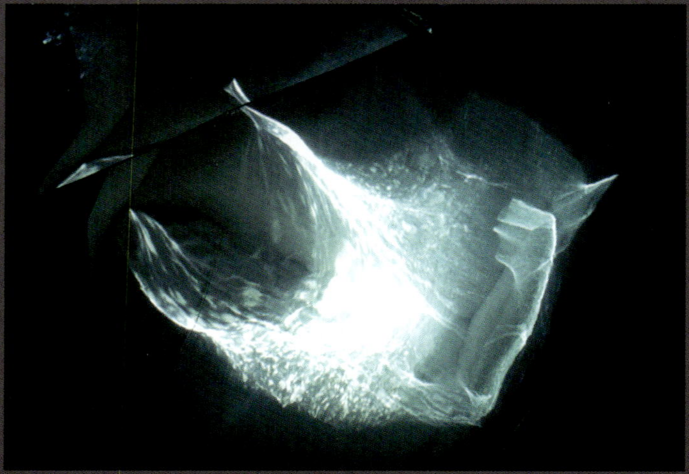

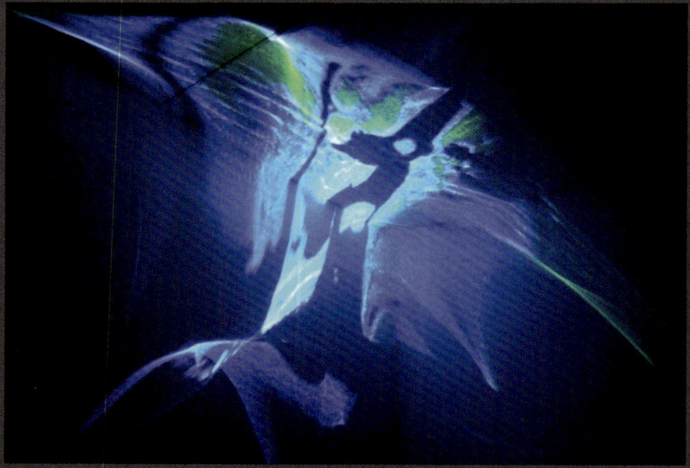

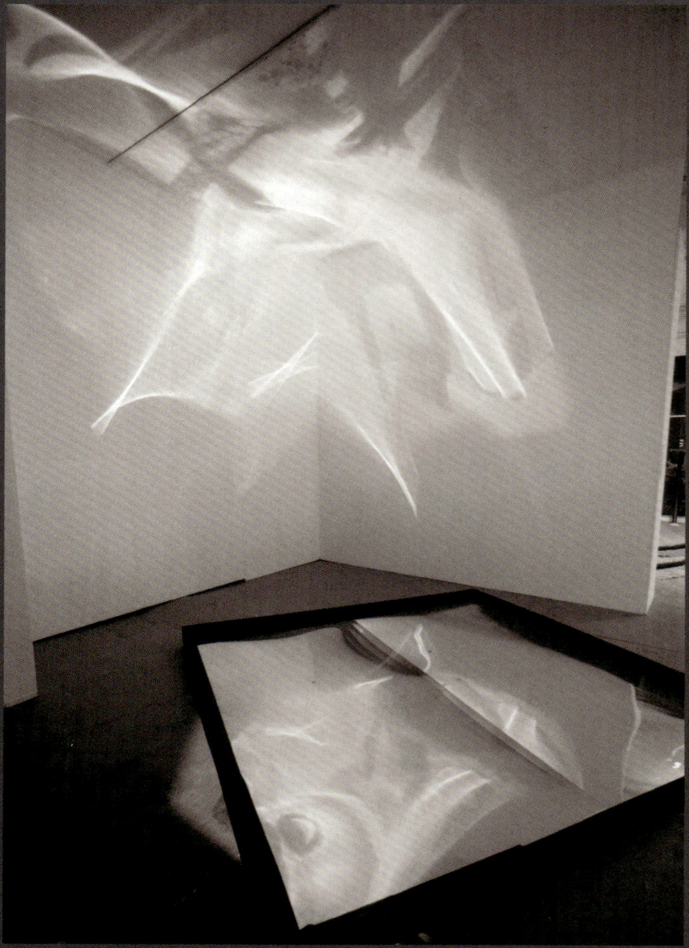

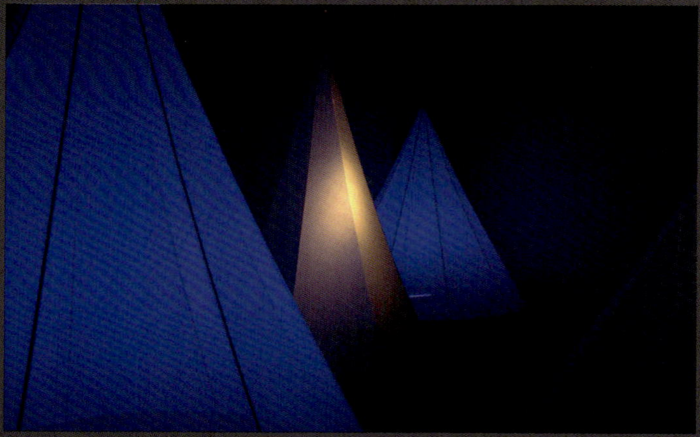

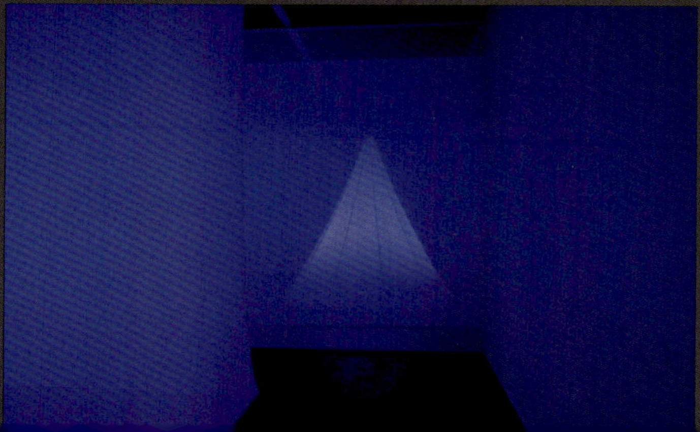

It is about making a place of light: an artificial, a fictional enclosure that physically situates—a sort of self-theater: a walk-in-film. With in it, a drama: a subjective time-space sentience—the *Homo sapien* perceptive holdings of the uncanny (which resides in us, side by side in our everyday activities)—probed, profiled, projected. (Keiko Prince, "Topos," 1978)

Top left and right: Keiko Prince, "Dear Mr. Kukai," installation with water, mirror, poem, 15 x 15 x 15 feet, Mills Gallery, Boston Center for the Arts, Boston, 1993
Photos: Toru Nakanishi

Right: Beth Galston, scrim and light environments, top, "Tepee," Kingston Gallery, Boston, 1986
bottom, "Geometries," Montserrat College of Art, Beverly, Massachusetts, 1987
Photos: Beth Galston

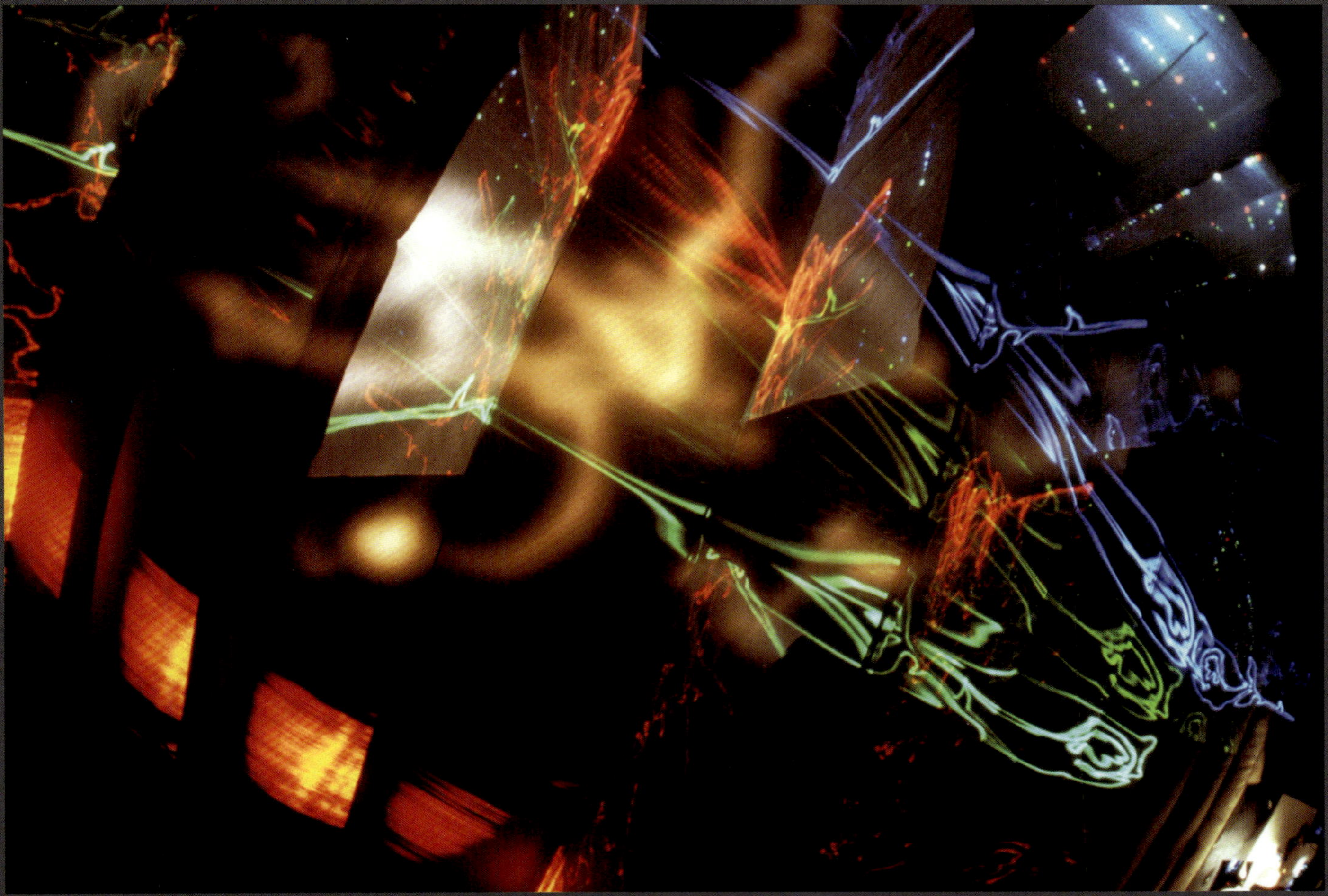

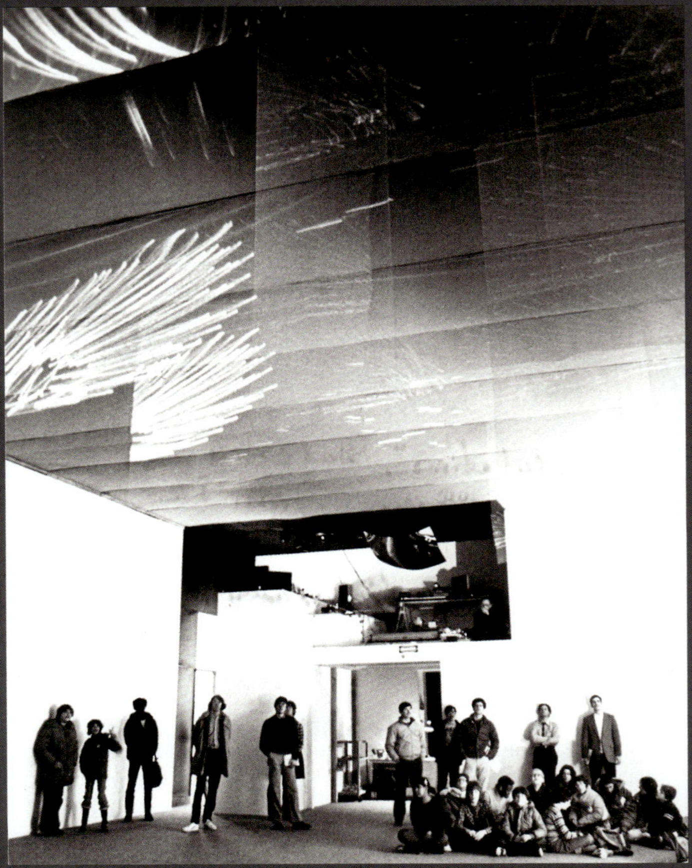

Top: Paul Earls, Otto Piene, "Modulations: Lasers and Music in Space," Hayden Gallery, MIT, 1982
Photo: CAVS Special Collection

Bottom left: Newspaper photo with caption, "Laser show. Students from the Perkins School for the Blind in Watertown are treated to a music and laser show at the Hayden Gallery of the Mass. Institute of Technology. Partially blind students were able to describe the brilliant laser images to their classmates." (*News-Tribune*, Waltham, MA, February 3, 1982)
Photo: UPI

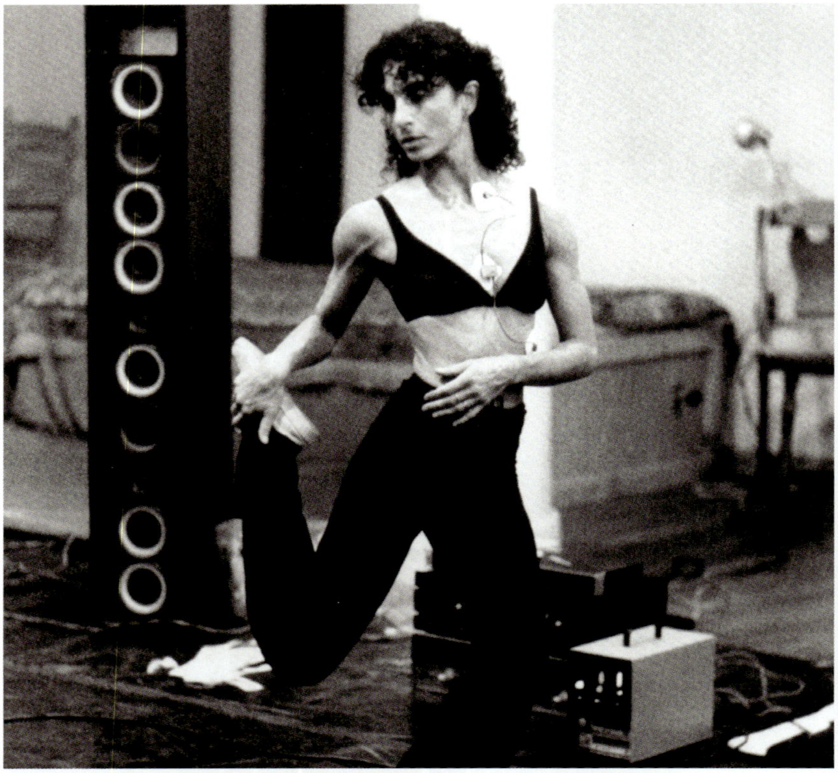

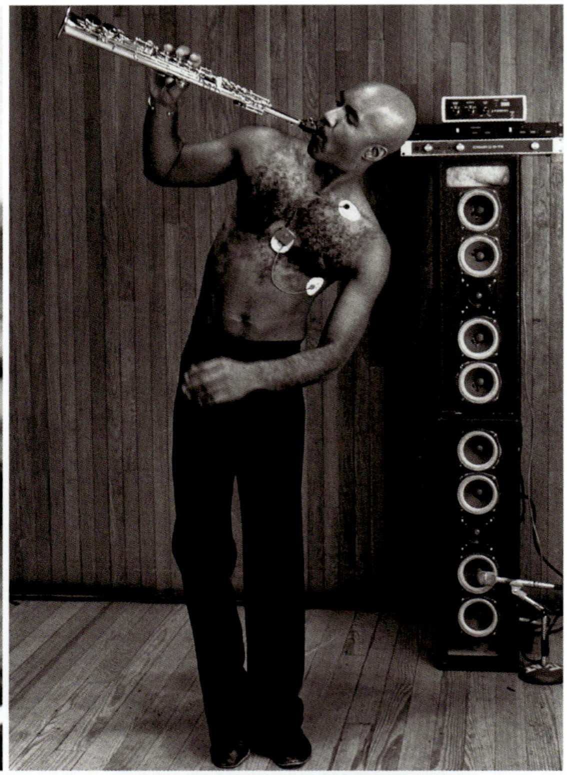

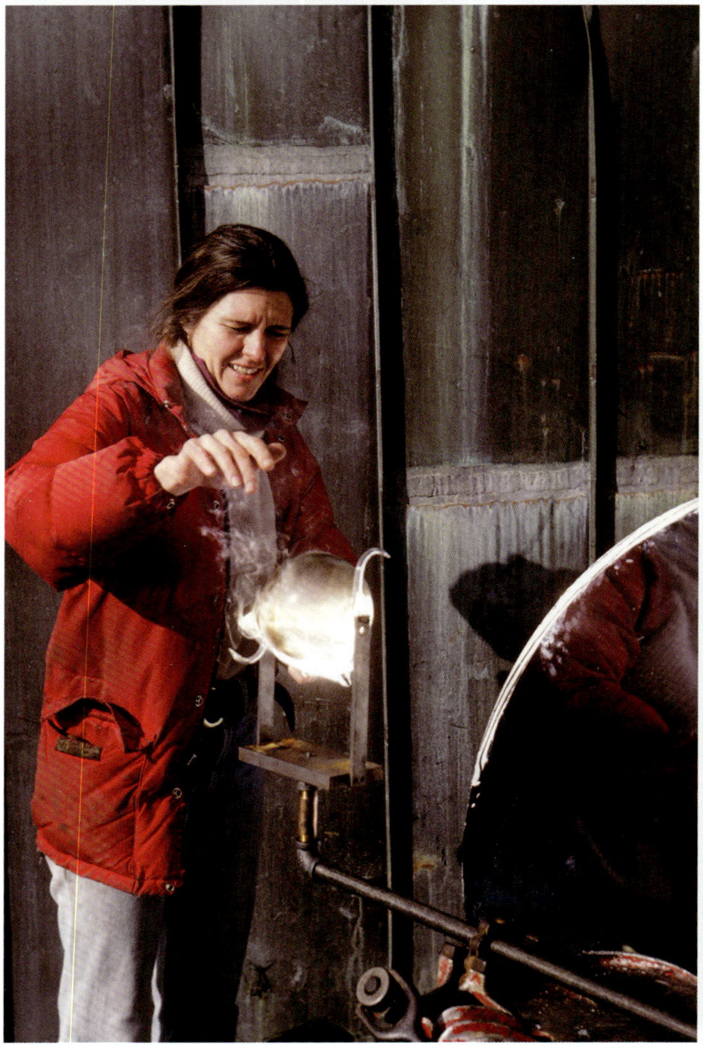

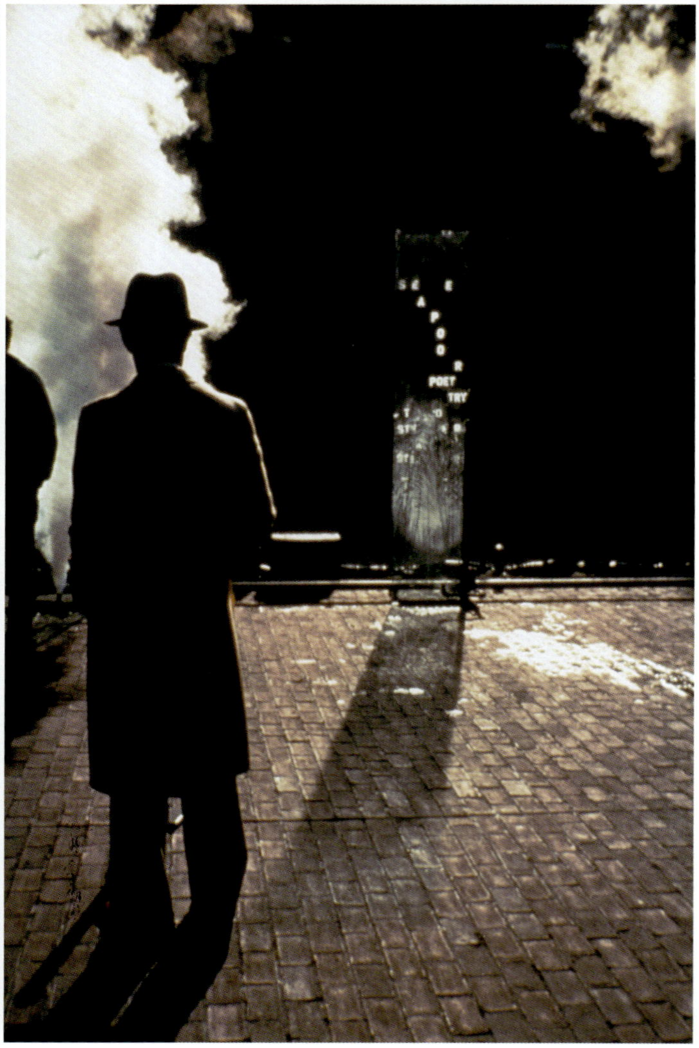
POET
TRY

Performance

Facing page top left and right: Bates/Christopher Janney, "Heartbeats," dancer Sara Rudner (left) and musician Stan Strickland (right), 1983 "*Heartbeats* uses a bio-engineering device which monitors the electrical impulses to a patient's heart and surrounding muscles. Placed on a dancer's chest and amplified through filters and a sound system, this machine provides an unusual percussion track. Layered over sound is live music." (Janney)

Photography

Top right: Robert Rosinsky, "Camera Obscura for Viewing Objects of Desire," interactive sculpture, "proposes that longings and dreams are projections within boundaries of perception" (artist's caption), 1984; and middle right: a room-sized camera obscura study for the artist's SMVisS thesis, *A Lexicon for Camera Obscura*, CAVS, MIT, 1984

Bottom right: Peter Campus, exhibition announcement with photograph "Decay," with text, "Campus' photographs of natural elements, such as rocks, tree trunks and leaves, are scanned into a computer, altered in various ways, then made into black and white or color prints. Photographed backgrounds are replaced by computer-generated patterns. The images . . . blur the standard distinctions between the natural and the cultural or man-made," MIT List Visual Arts Center, 1992

Public Art | Steam

Facing page bottom left: Joan Brigham works on her steam-powered glass "aeolipile" sculpture, 1986

Facing page bottom right: Joan Brigham and Christopher Janney, "Steamshuffle," with poetry by Emmett Williams etched on glass panels illuminated by steam, captioned, "As the steam condensates on the glass surface, words, stenciled in clear acrylic, appear as if traced by a finger on a misty window," Government Center Plaza, Boston, 1981
Photos: CAVS Special Collection

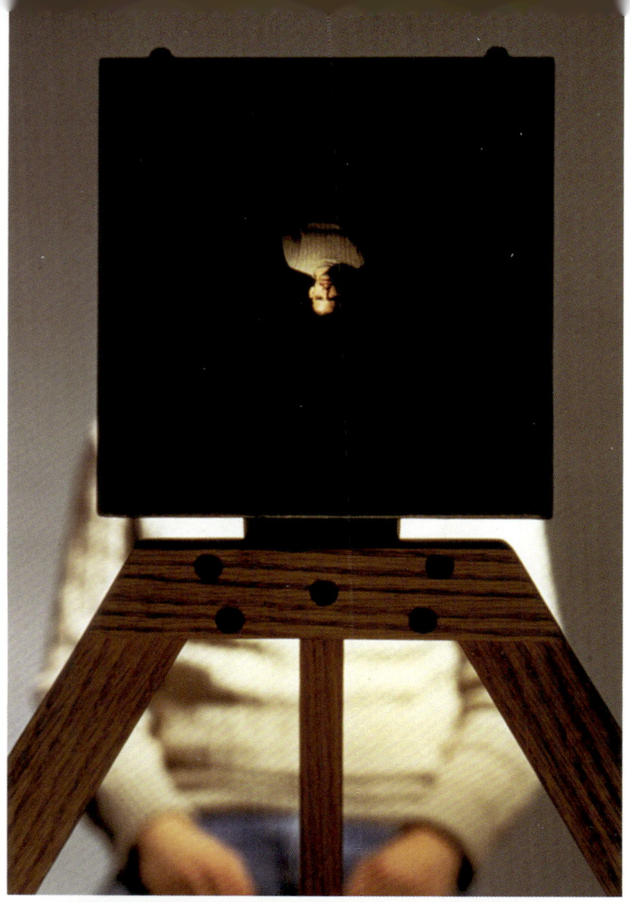

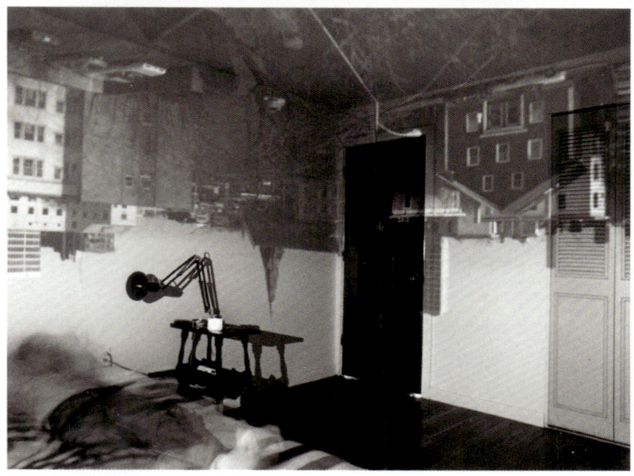

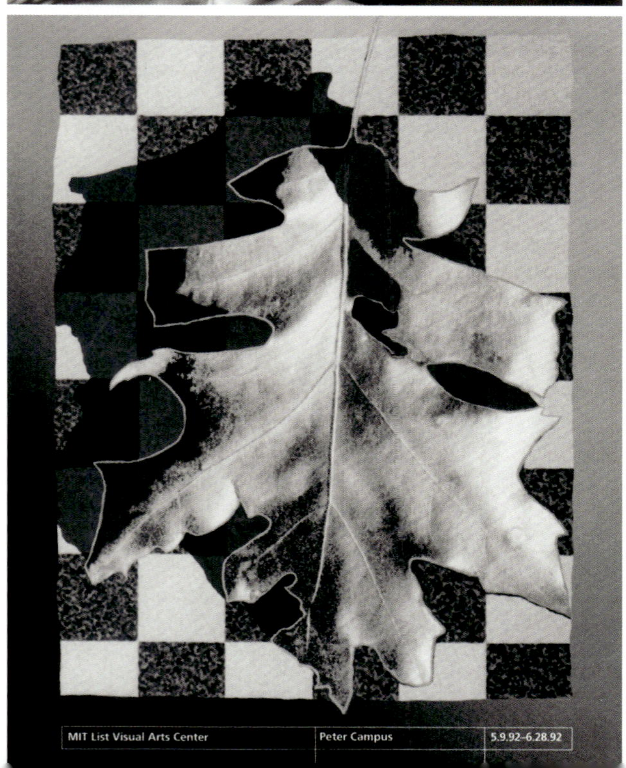

Sculpture

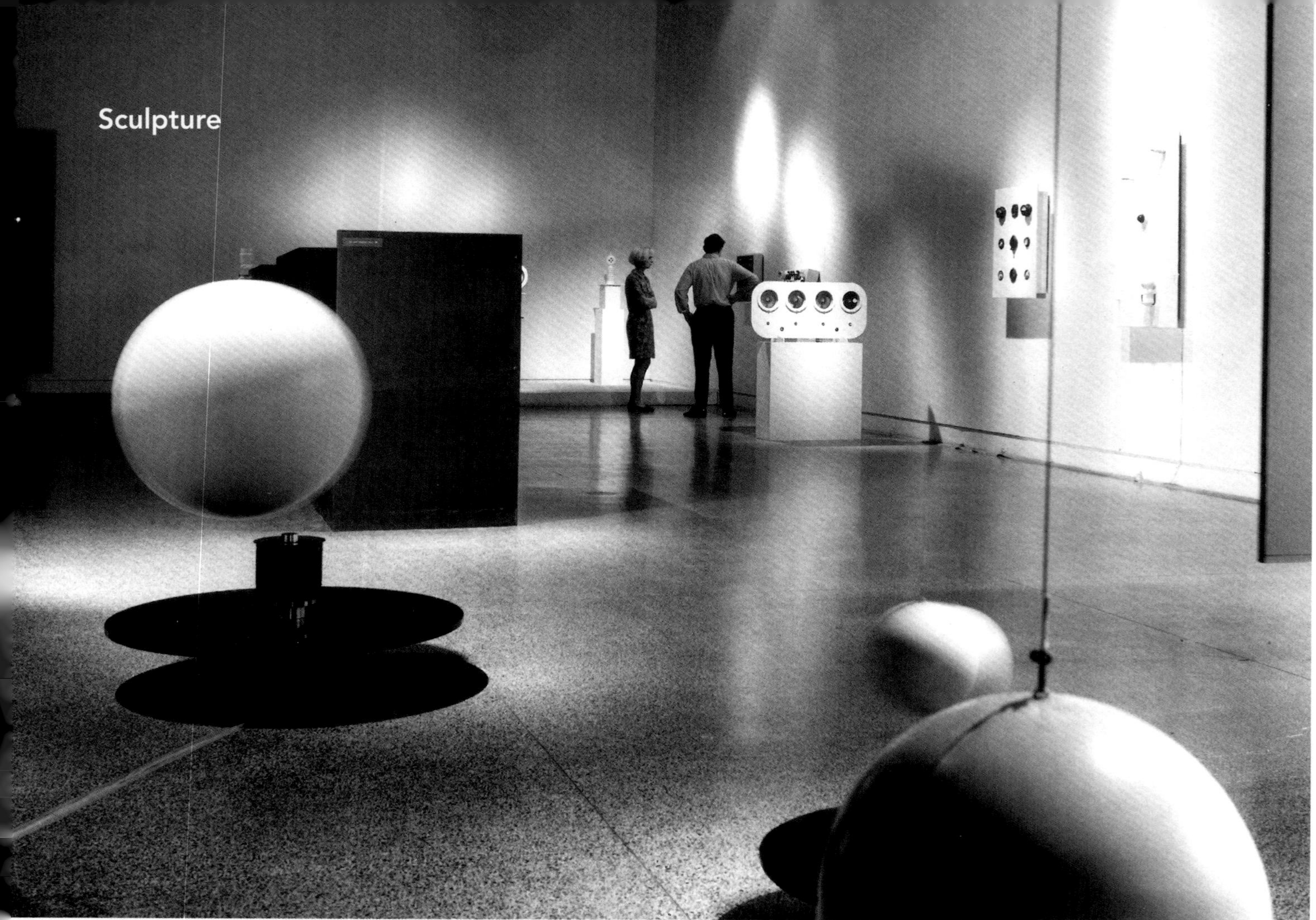

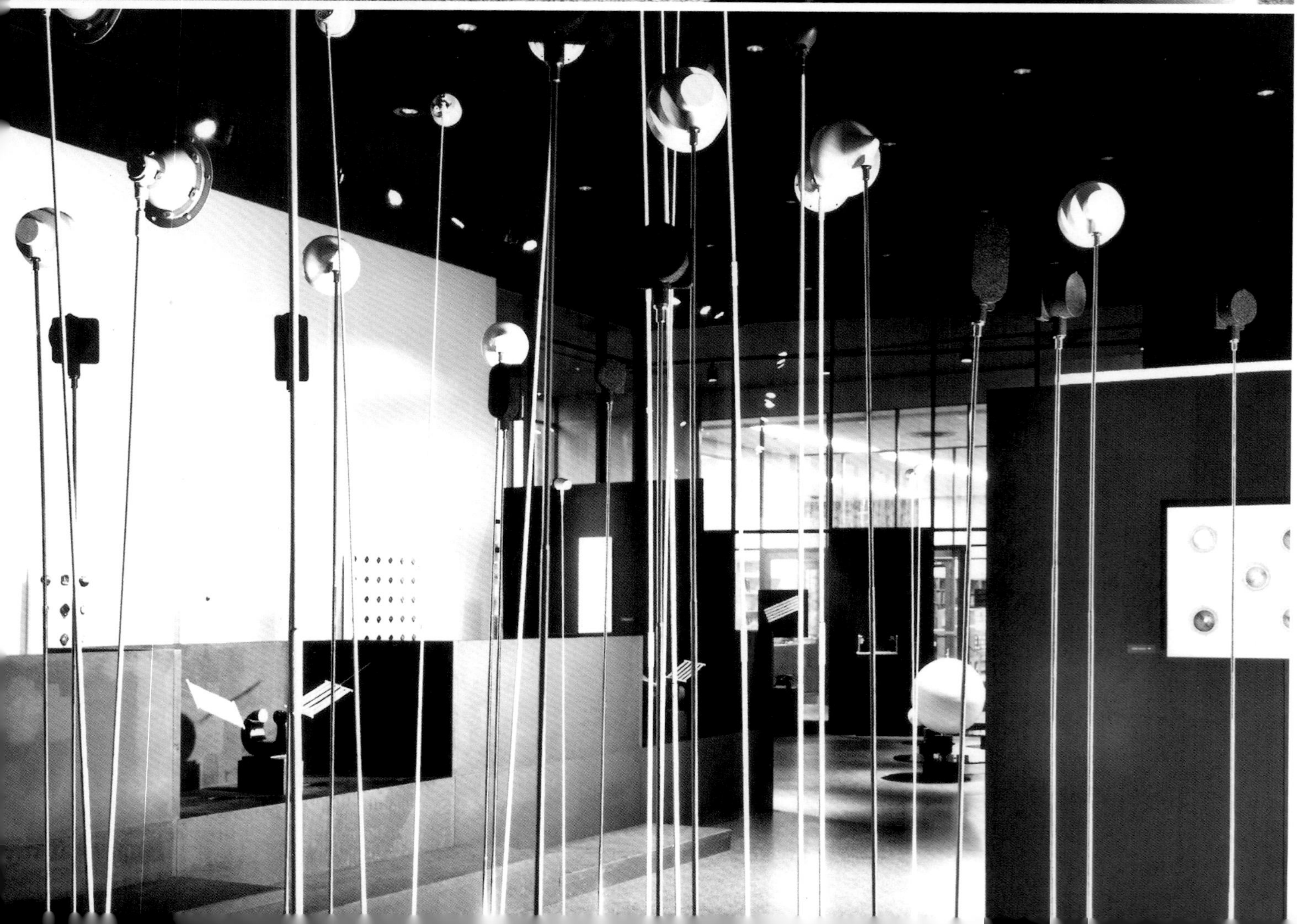

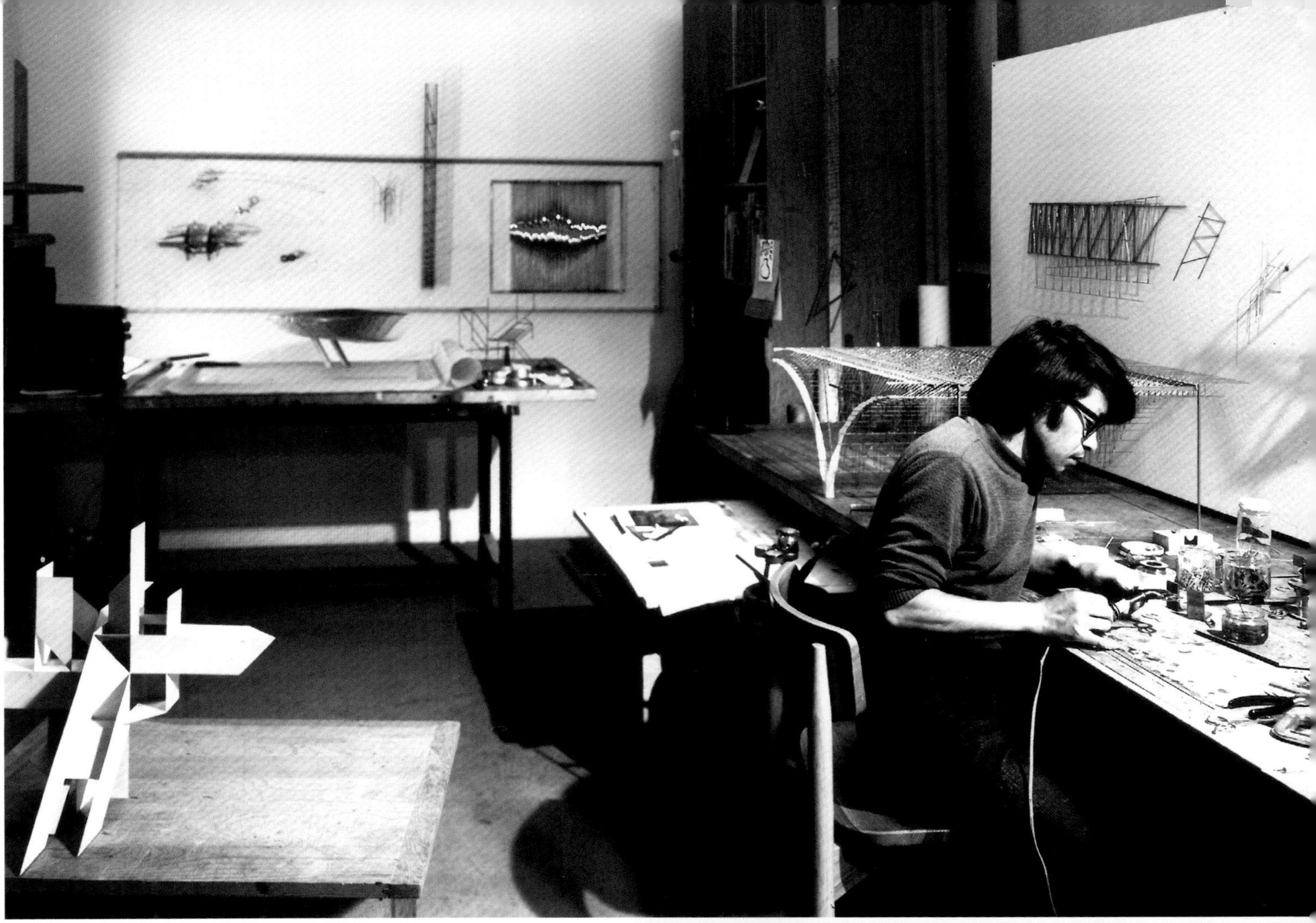

Facing page top and bottom: Vassilakis Takis, exhibition, including top left, "Electromagnetic Sculpture," Hayden Gallery, MIT, November 15-December 8, 1968
Photos: Nishan Bichajian

Top: Michio Ihara in his studio at CAVS, April 10, 1972
Photo: Nishan Bichajian

Right: Michio Ihara, "Suspended Sculpture," AT&T Long Lines Building, Atlanta, 1982
Photo: Michio Ihara

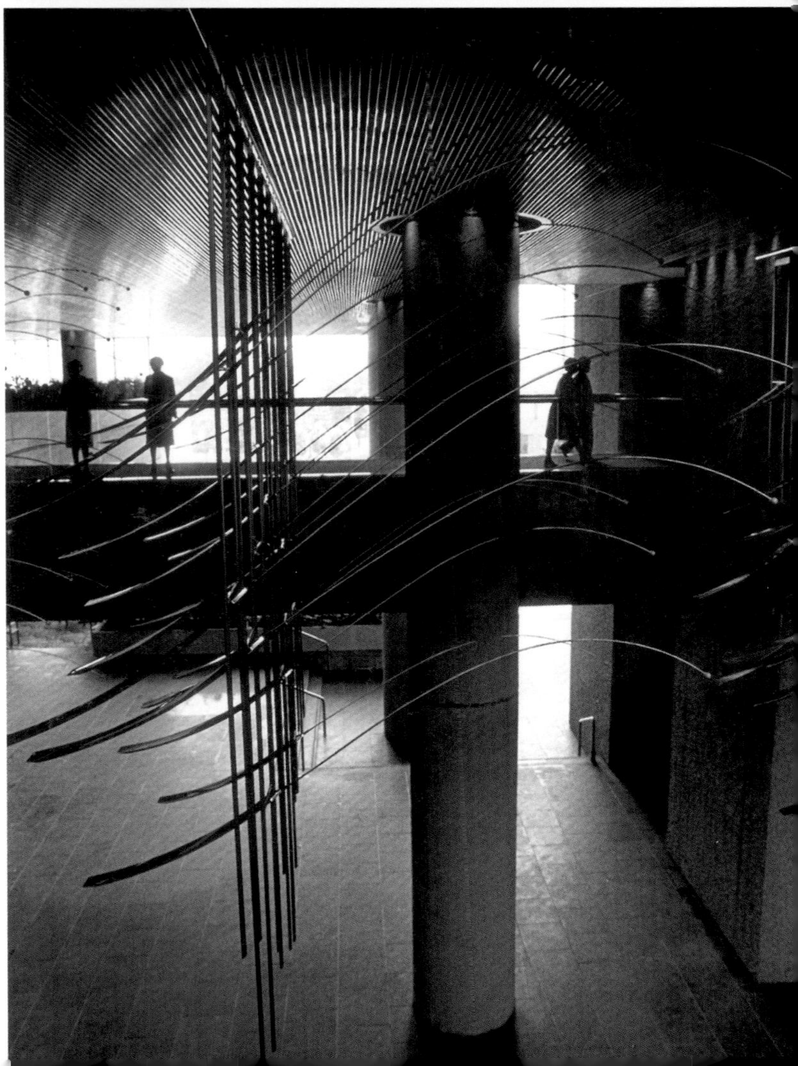

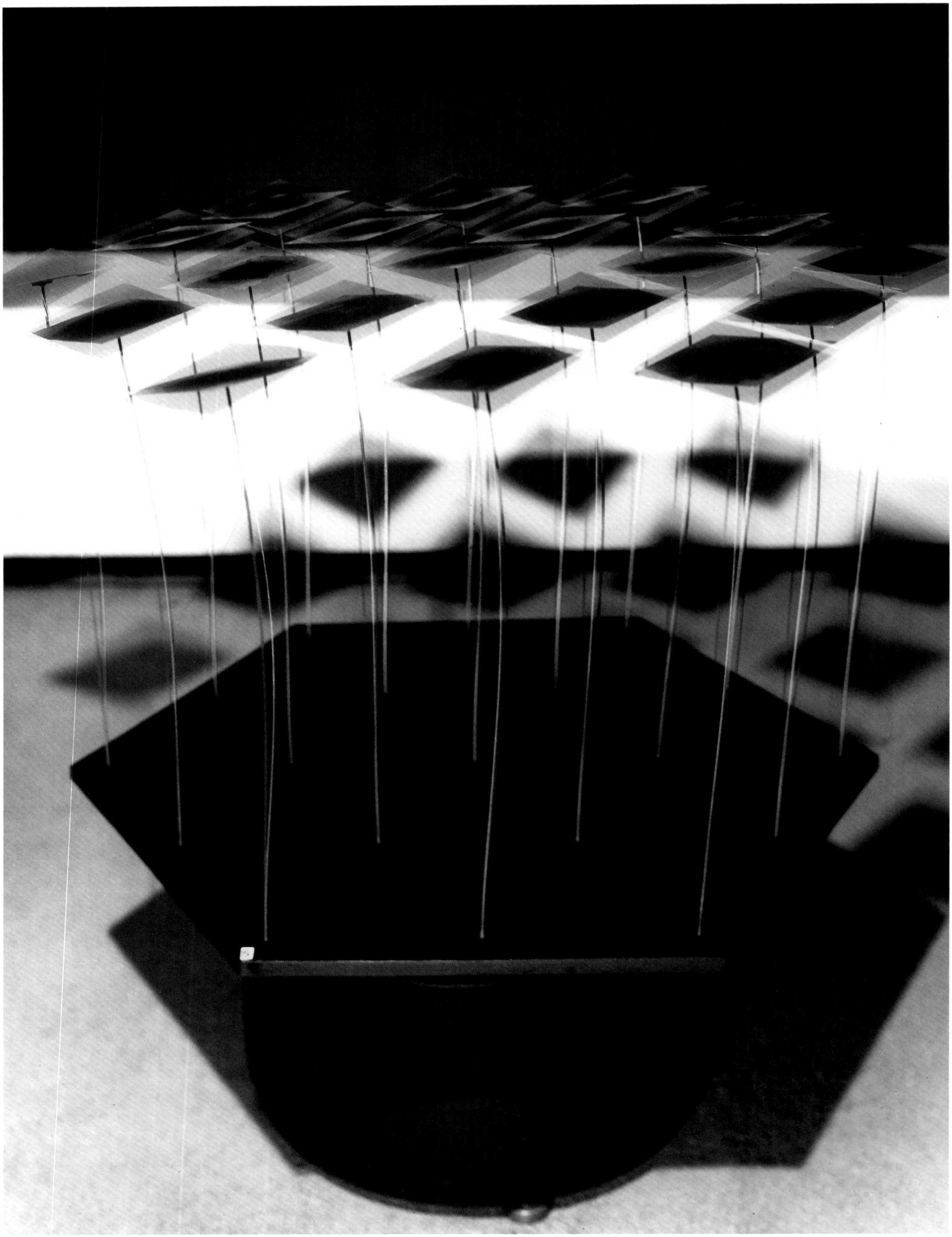

Our culture has made a concerted effort to remove us from nature. . . . We have an artificial sun, artificial stone, artificial weather in my office. We leave this self-contained environment, get into a car, and go to another self-contained environment. I have been a public official on land use boards for many years. I am continually disturbed [that] people would try to destroy our own life-support system—the earth. I tried to make something that was completely transportable. You can remove them from a sensitive environment without causing any damage or changes. I created what I call a "visual ballad" that gently sings the song of the earth. Many environmental artists cause great destruction to the landscape. When it is done in the name of art it seems to be justifiable, but to me it is vandalism. . . . My first project at CAVS was to develop a geothermal simulator. . . . The geothermal hot water circulated through the works like blood so that the copper areas were warm. There were areas of rock crystals, which are regarded by many cultures as embodiments of the earth's energy. . . . I developed a system with the aid of infrared thermometers that was extremely responsive to the micro-climate, and these new sculptures became part of the Center's twenty-fifth anniversary exhibition. (Robert Dell interview)

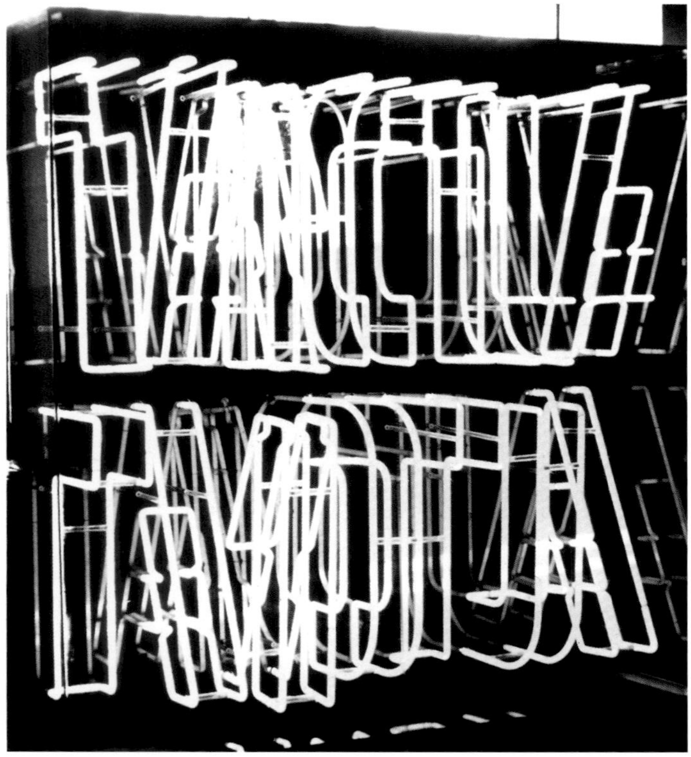

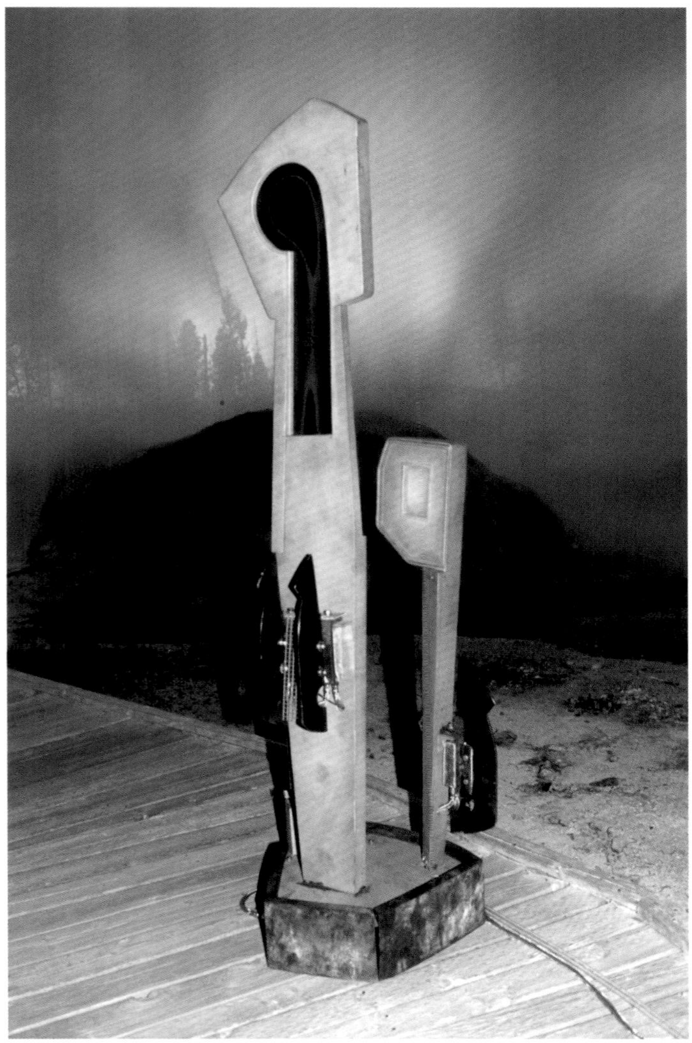

Top right: Chryssa, "Automat," neon sculpture of overlaid letters, 1971
Photo: CAVS Special Collection

Bottom right: Robert Dell, "Hitavaettur" (meaning "guardian of geothermal hot water" in Icelandic), geothermal sculpture powered by hot springs, permanent display, Perlan, Reykjavik, 1988
Photo: Robert Dell

Facing page: Wen-Ying Tsai, "Harmonic Sculpture #5," stainless steel rods, 1968
Photo: Eric Pollitzer

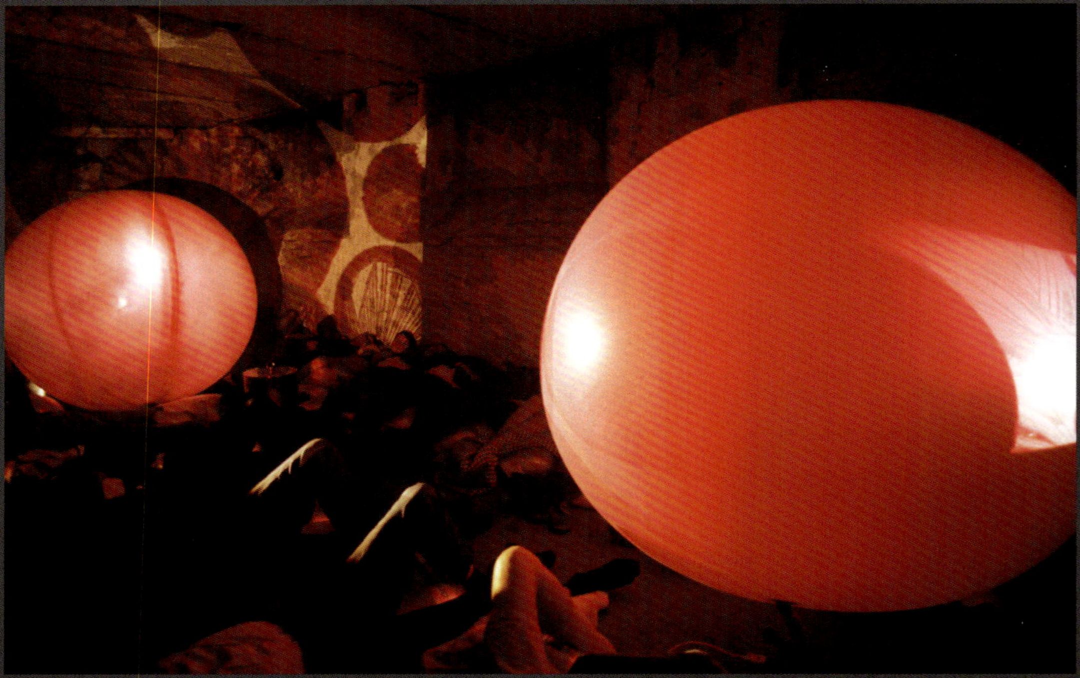

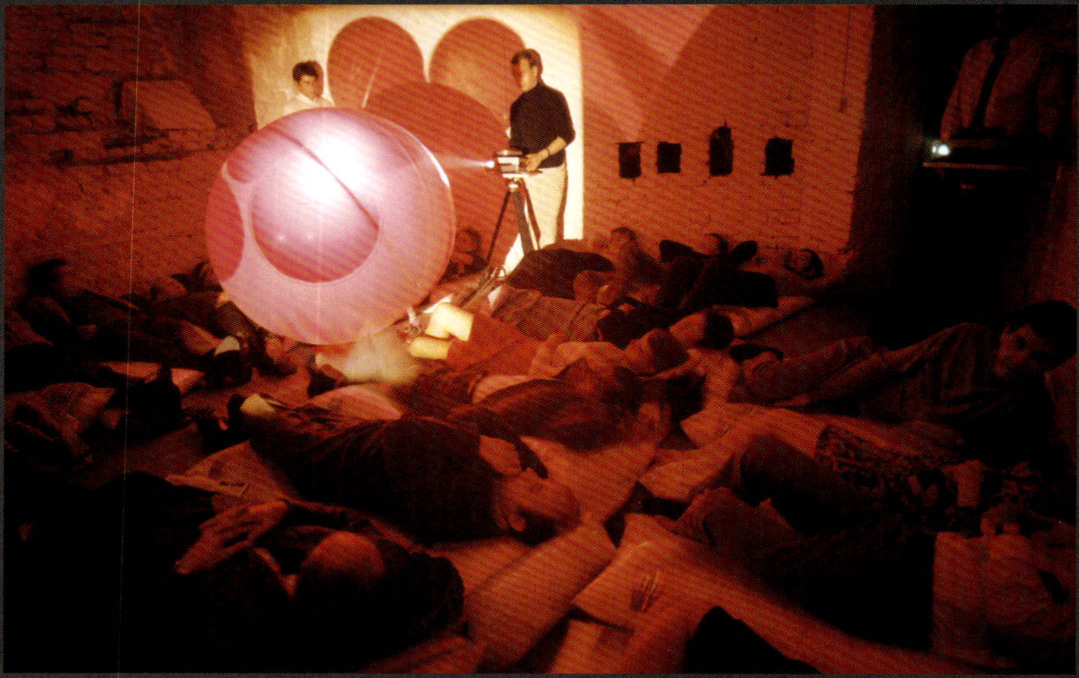

On the one hand, the love of clouds, and the love of light, which is how and why we all live, and the love of other phenomena such as noises, such as music, such as the world of animals, the world of growing and decaying, such as the world of agriculture, such as the world of food and drink—all of these are factors that directly or indirectly come out of the sky. They are elements of celebration that are worth paying attention to, and also interpreting and offering and pointing at, for instance, to teaching at the Center for Advanced Visual Studies, and the minds that met there . . . and that are also part of what we have to prepare ourselves for, as a part of communal feeling and communal celebration. (Otto Piene interview by Bill Seaman, March 6-7, 2012)

Above: Otto Piene, "Proliferation of the Sun," performance, painted glass slide projections, inflatables, and sound, Galerie Art Intermedia, Köln, 1967
Photos: Walter Vogel

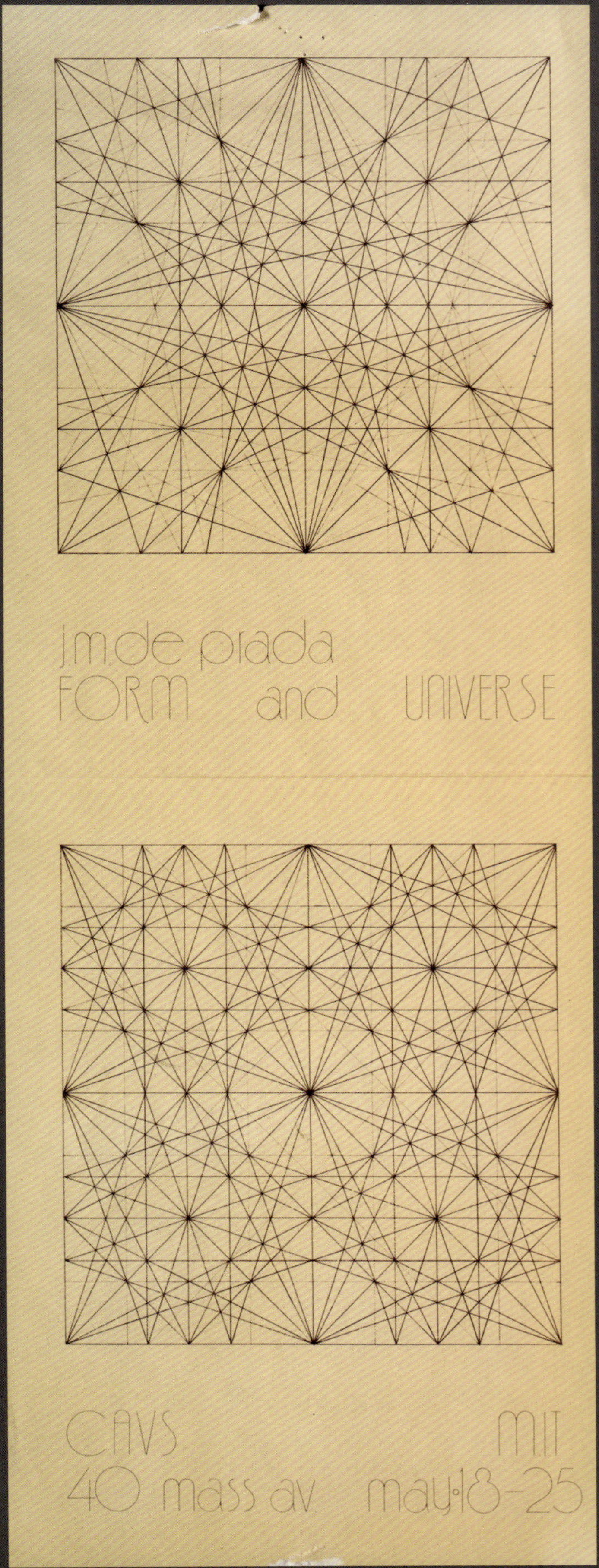

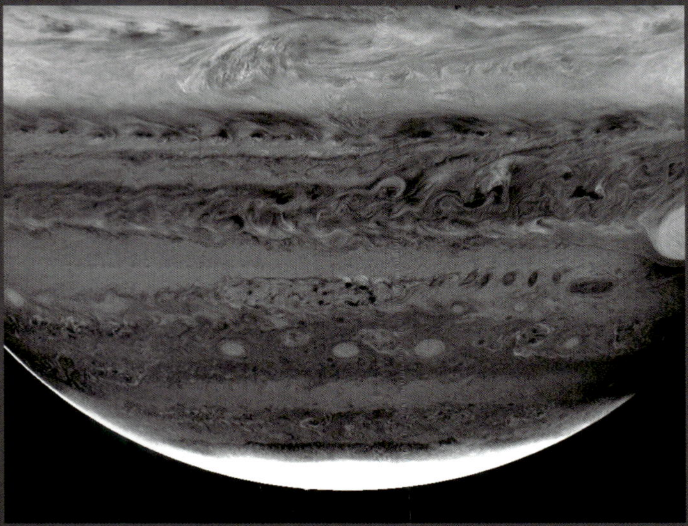

I used the Center to work in 1980 on something related to a conceptual quest connected to art and architecture. There is an old question: why has symmetry been so useful for artists and architects? If symmetry belongs to the natural world as cosmologists would say, the universe should be homogeneous, uniform and isotropic. This property should be important for any scientific and artistic purpose. Or, it is not the case? The terms used by the cosmologists mean that the universe should be identically symmetrical, simultaneously, for every point in space, and in any direction. Thinking in that way I tried to make a three-dimensional model of the universe. The physical universe must be one, not plenty of mathematical points without dimensions, but something real, with bodies and voids. (Jose de Prada interview)

Left: Jose de Prada, "Form and Universe," exhibition, CAVS, MIT, May 18 - 25, 1982
Source: CAVS Special Collection

Top right: R. West and Gloria Brown-Simmons, Planetary Visualization, "Image produced by ratios of methane and non-methane filters for the study of cloud stratigraphy, Cassini Jupiter Encounter, released December 18, 2000, NASA / JPL / University of Arizona," (Gloria Brown-Simmons, Project Description)
Source: CAVS Special Collection

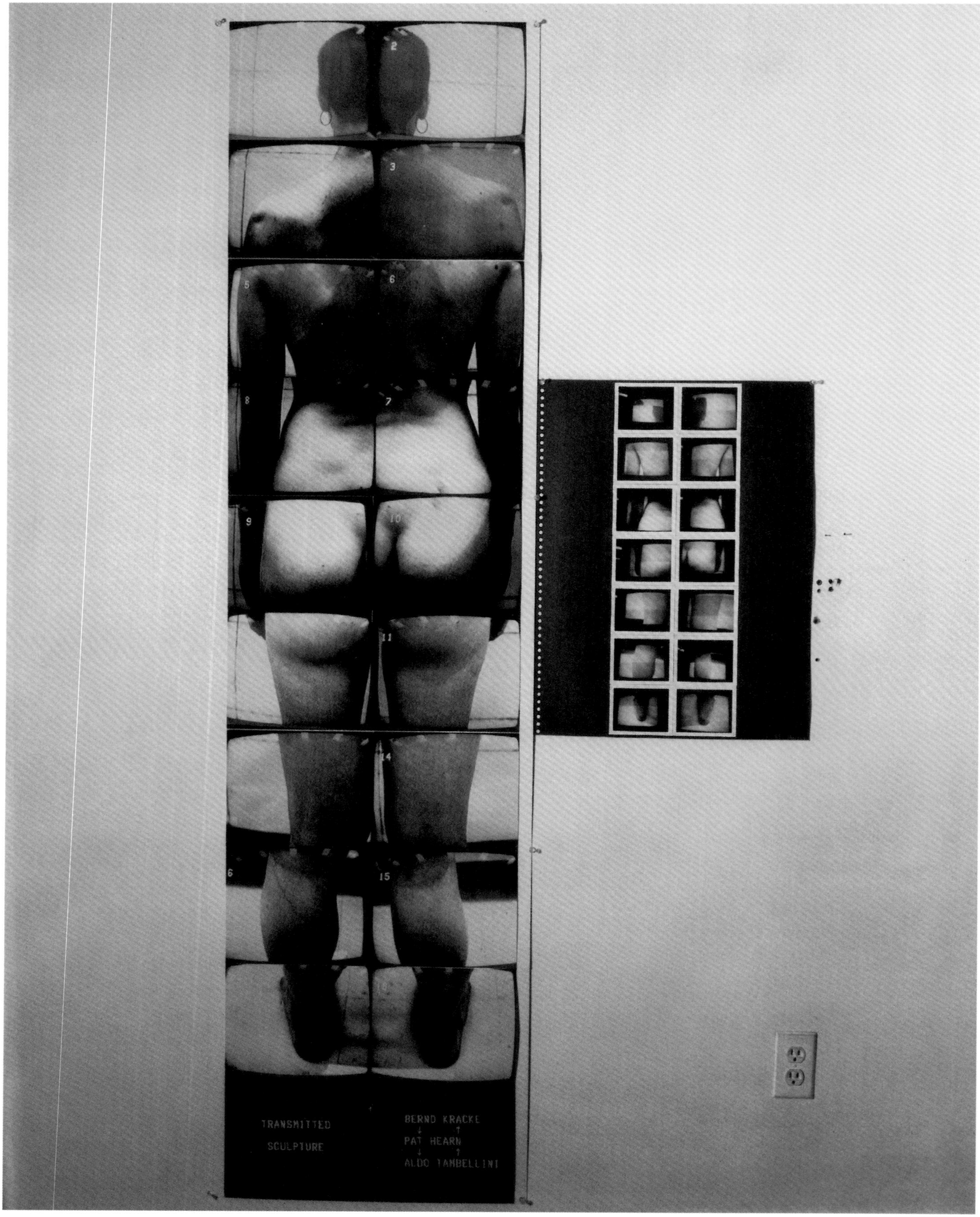
TRANSMITTED
SCULPTURE
BERND KRACKE
PAT HEARN
ALDO TAMBELLINI

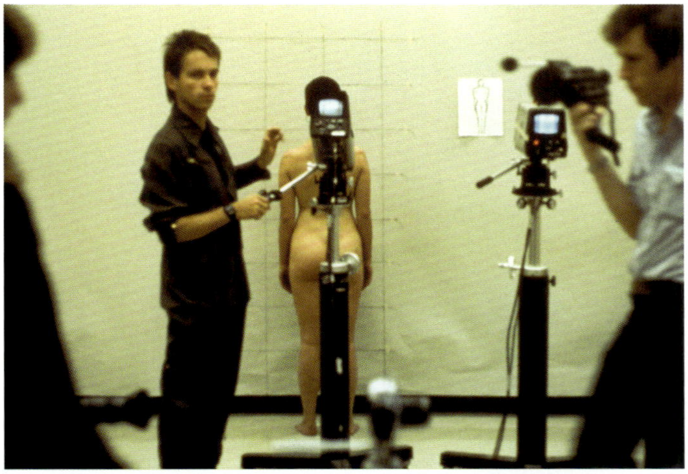

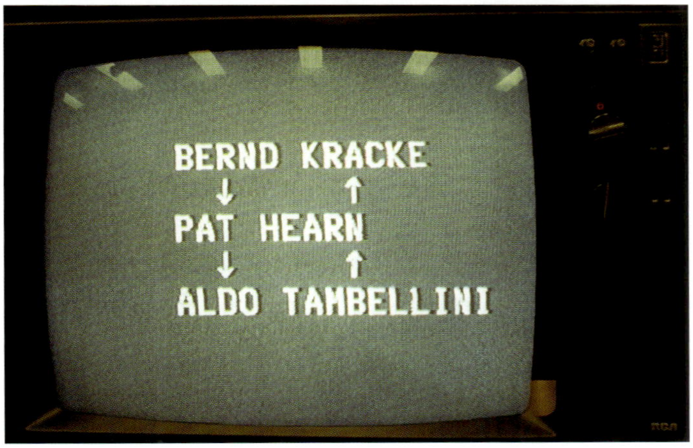

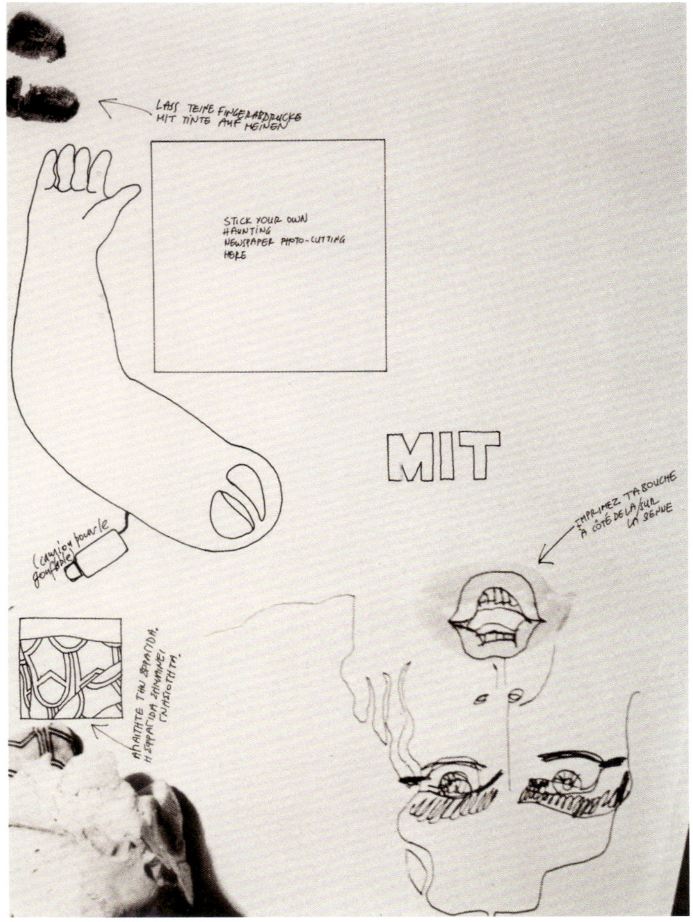

These examples of 'transmitted sculptures' which utilized telecommunication to create models of reality emphasize the dynamic scale of sculpture-as-process. . . . The electronic process of information exchange dissolves past, present, and future in a continuous sequence of timeless images. Image sequences and their representation of time and space can be endlessly and indistinguishably manipulated. Every media image confronts us with its ambivalent qualities of being both real and unreal, present and absent, live and recorded, big and small. "The ITSELF," as Otto Piene mentioned, "changes its self because a 'media-self' is added." This 'media self' or what can be called a 'media reality' is juxtaposed with our self and 'reality'. Forward, backward, fast forward, fast backward, slow motion, single frame, random access, edit, erase, and 'do it all over again' are some of the options for the production of the multi-faceted 'media reality' which influences its distribution patterns as well. (Bernd Kracke, *Sculpture as Process*, thesis, CAVS, MIT, 1981, 66-67)

Facing page and top left: Bernd Kracke, "Transmitted Sculpture I," photomontage and views of Slow Scan event transmitting from MIT to Atlanta, Georgia, with Pat Hearn (future art dealer), performer, in collaboration with Aldo Tambellini's group Communicationsphere, 1980
Photo facing page: Bernd Kracke
Photos top left: Otto Piene Collection

Top right: Mit Mitropoulos, postcard (detail) for "mixed media activities in Brussels" (artist's note to Piene on the postcard) including mail art, June, 1980
Source: CAVS Special Collection

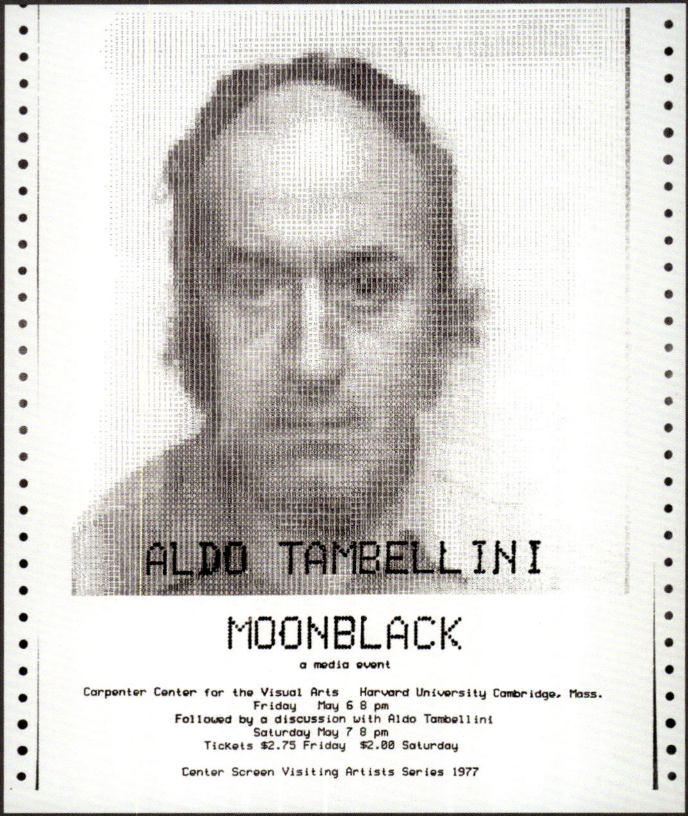

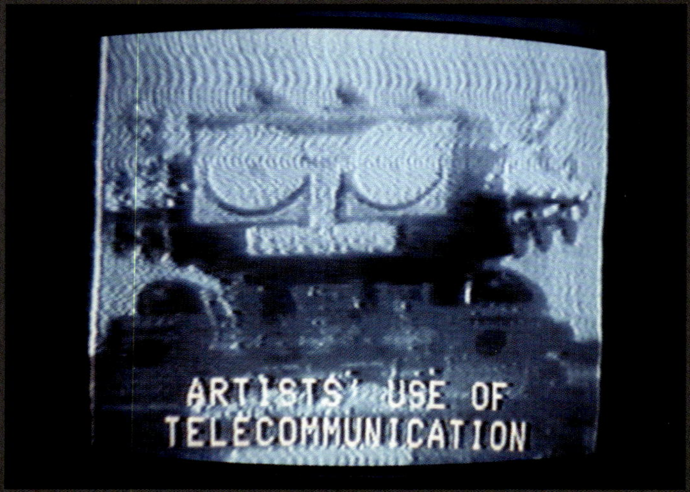

Above: Aldo Tambellini, "Moonblack," poster, media event, Carpenter Center, Harvard University, 1977
Source: CAVS Special Collection

Bottom: Aldo Tambellini, ROBOT from "Artists Use of Telecommunications," Slow Scan television linking MIT, San Francisco Museum of Art, and points in Austria, Japan, Canada, and New York City, Communicationsphere group, February 16, 1980
Photo: Aldo Tambellini

I arrived there at CAVS, and I was new. The first thing we did at the Carpenter Center, Harvard, was a program called "Moonblack"—excellent, in a big room. The concept was arts and science. A "Leonardo" figure was lying on the floor in a parachute outfit, two video cameras strapped to her arms, video projector, overhead camera on Sarah Dickinson . . . We projected film from the '60s riots. . . . "TV to TV" was in Lobby 7, with a relationship to cable people and the Architecture Machine Group. We hooked up a series of monitors at the Student Center so there were two spaces going through cable. It appeared to be two dancers, one in each space dancing simultaneously. I liked the whole idea of interactive programming in two spaces and bringing in the community. . . . Then, I had a show at the Institute of Contemporary Art, Boston . . . called "Communicationsphere." I did an installation of things I'd done before. It was a big show documenting that CAVS period. I had Slow Scan on the street; whenever people would pass, their images would be frozen inside the gallery. I had metal boxes that we bought from *The Herald* or *The Globe*, and I made a generic newspaper called, *The Generic News*. I reproduced things in the news. On the cover was a picture of an anchorwoman. She was a regular in some parts of the country, and she said she was depressed and took a gun and shot herself in front of her TV audience. I was trying to show the effect of television. . . . When we did "Artists Use of Telecommunications" we started with a toy robot. It was a good comical videotape. Toronto is sending . . . scanning takes about 5 seconds. You hear beep-beep-beep and, line by line, it creates a kind of anxiety about what it is going to be like. If you sent the whole image right away, it's not so exciting. . . . I am a strong believer in groups, artists interconnected with each other. I don't believe in the individual artist isolated from society. Art is dependent on society and contributes works to society, and the artist doesn't separate himself. He can do his individualistic work, but I think he should interact with the rest of society, which I ended up doing when I was at CAVS with Slow Scan. (Aldo Tambellini interview)

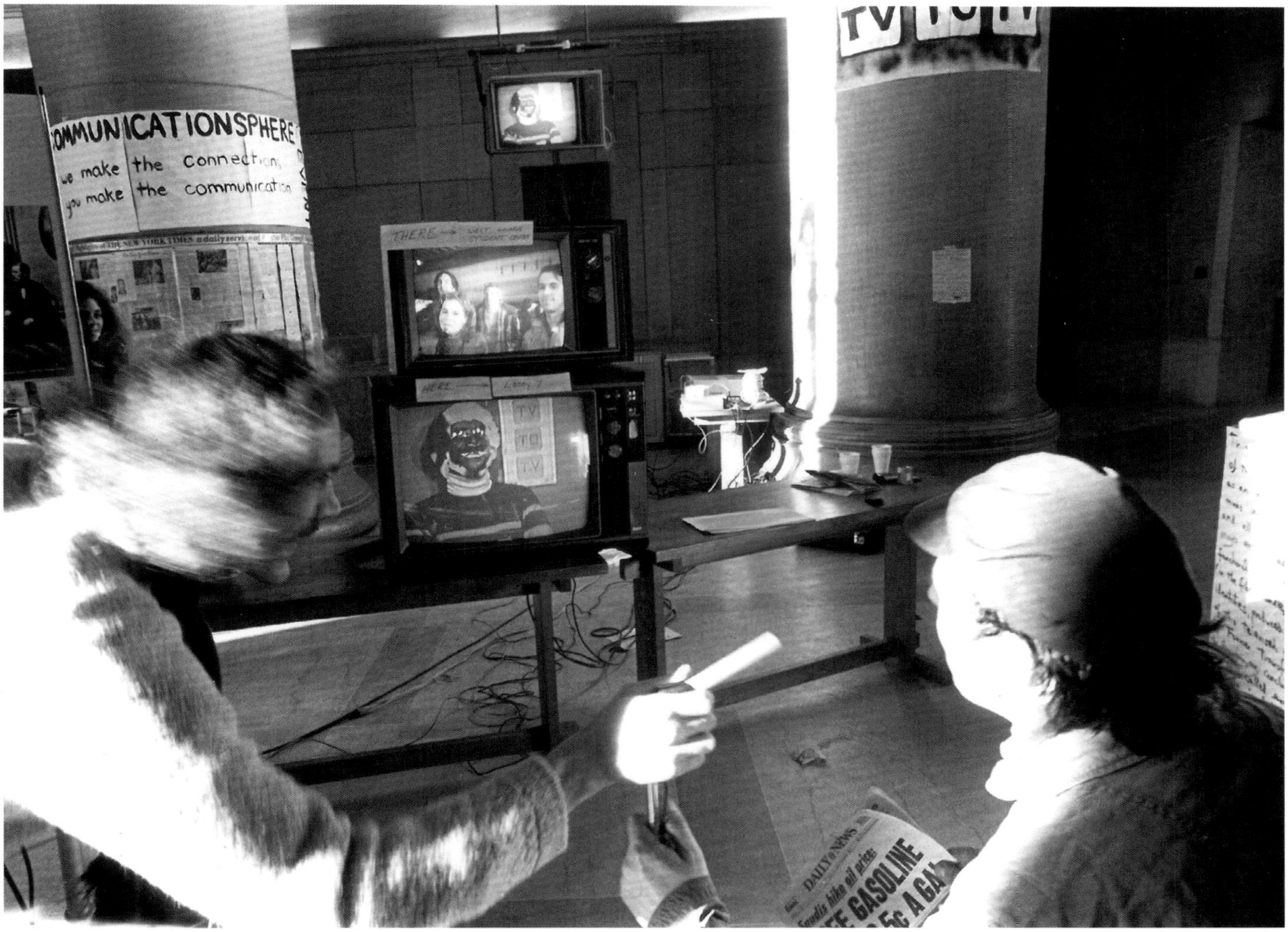

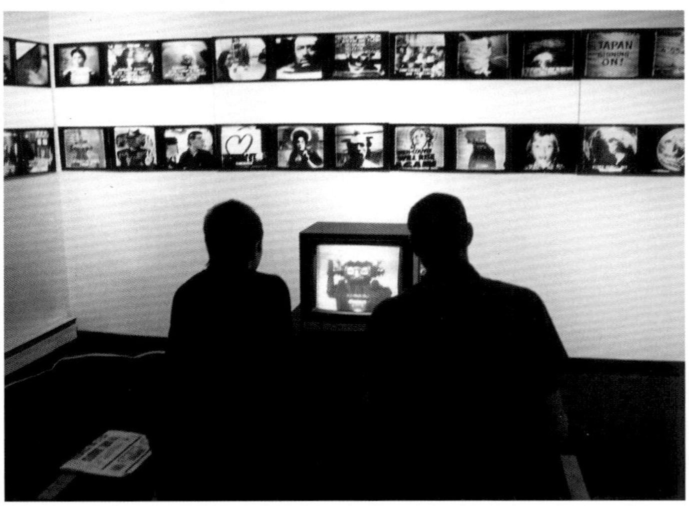

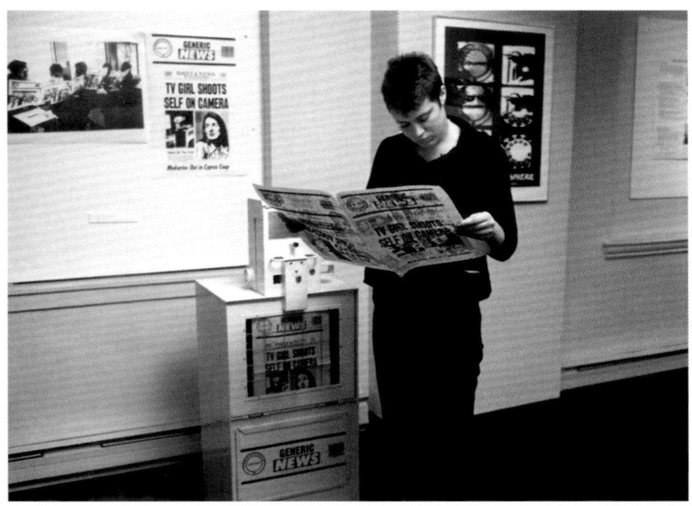

Top: Aldo Tambellini, "TV to TV," pictured, fellow Wendelin Glatzel (left) and Tambellini (right), Communicationsphere group, Lobby 7 and Student Center, MIT, 1979
Photo: Marc Palumbo

Bottom left and right: Aldo Tambellini, "Communicationsphere," exhibition, "Boston Now '83," Institute of Contemporary Art, Boston, 1983
Photos: Aldo Tambellini

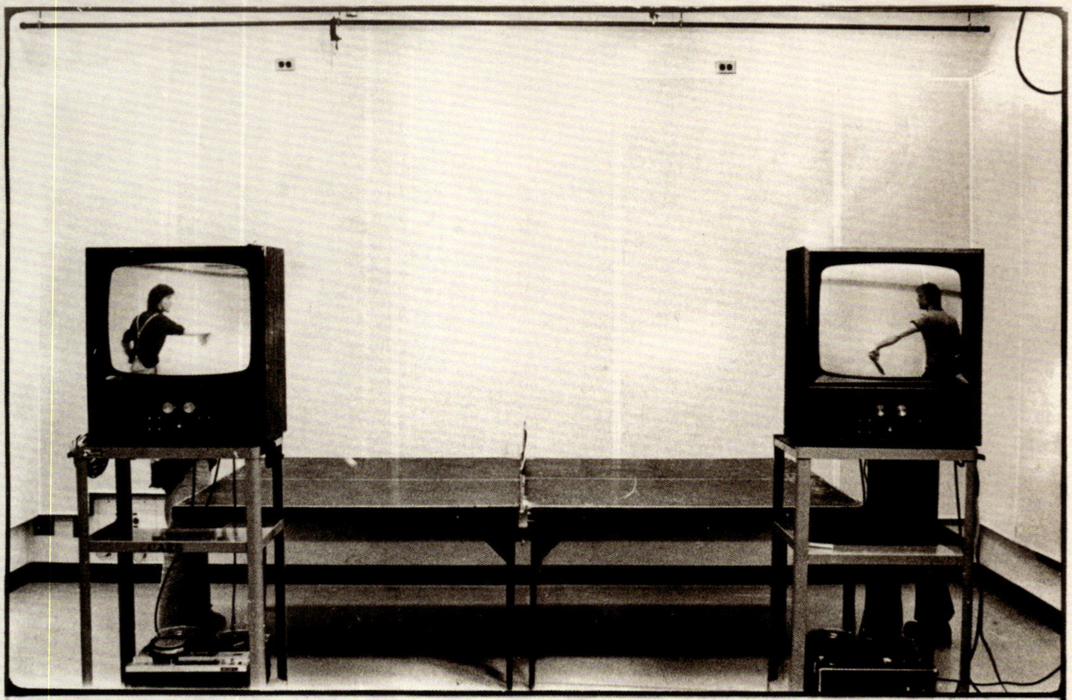

Video-Ping-Pong / Ernst Caramelle 1974
recorded at the M.I.T., Cambridge, Mass.

first presentation: Hayden Gallery, M.I.T.
Cambridge, Massachusetts, January 1975

video-tapes
2x 30 min. (sound)
film: Peter & Dan

Thanks to:
Keiko Prince
Jeff Stetson
Peter Dourmashkin
Dan Krynytzky
Ann Mcintosh
Juan Navarro Baldeweg
Georges Singer
Norman Lang
Mr Adler
Mr Roberts
Kathy Arnold
Friedrich St. Florian
Morris Smith
Otto Piene

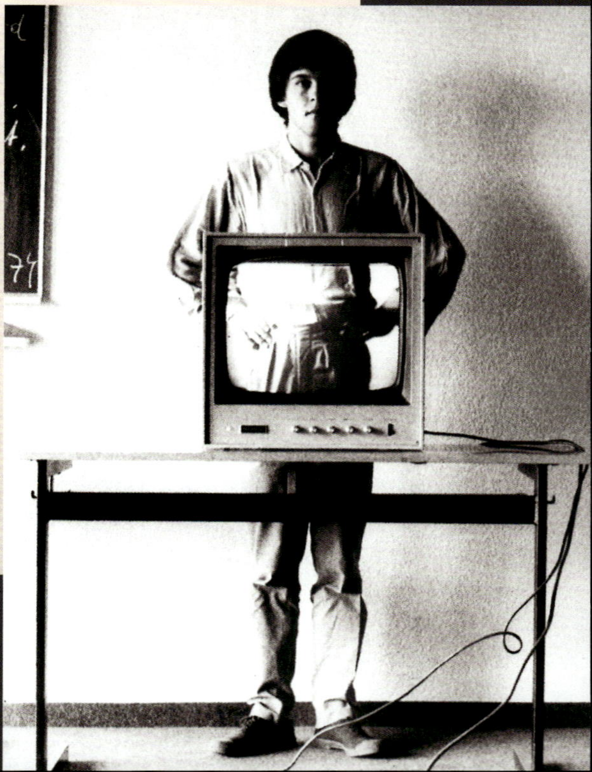

I was totally immersed in editing this film, and it either took two months or may have taken a little longer. . . . I had a sense CAVS was important to a lot of people. It was outside of the mainstream art world, a place for research. I was there in a way under false pretenses. I was working with a technological medium, but not in the sense that everyone else at the Center was. No one there was involved with narrative. (Yvonne Rainer interview)

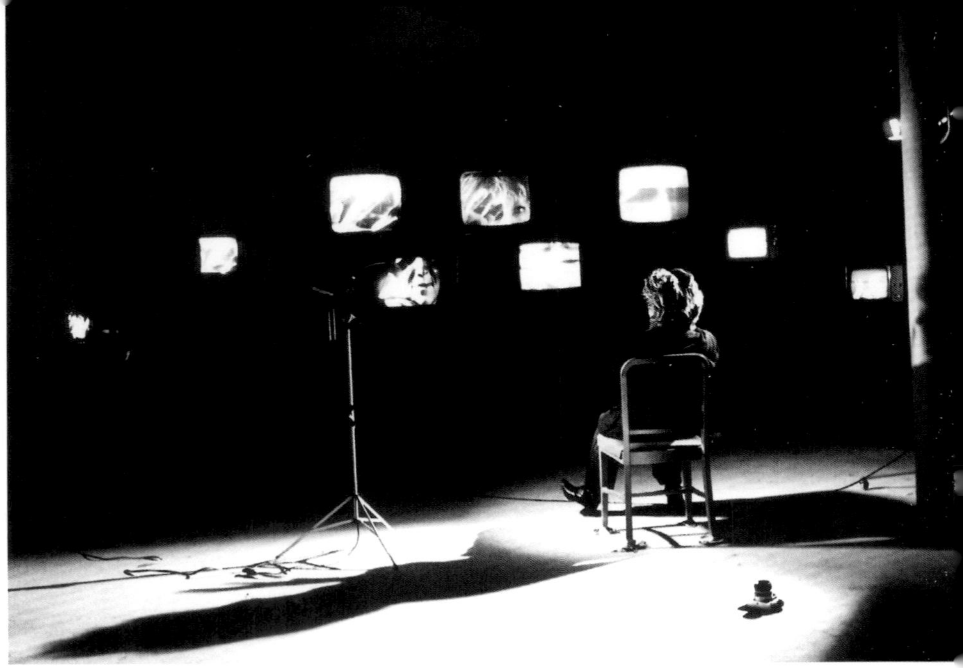

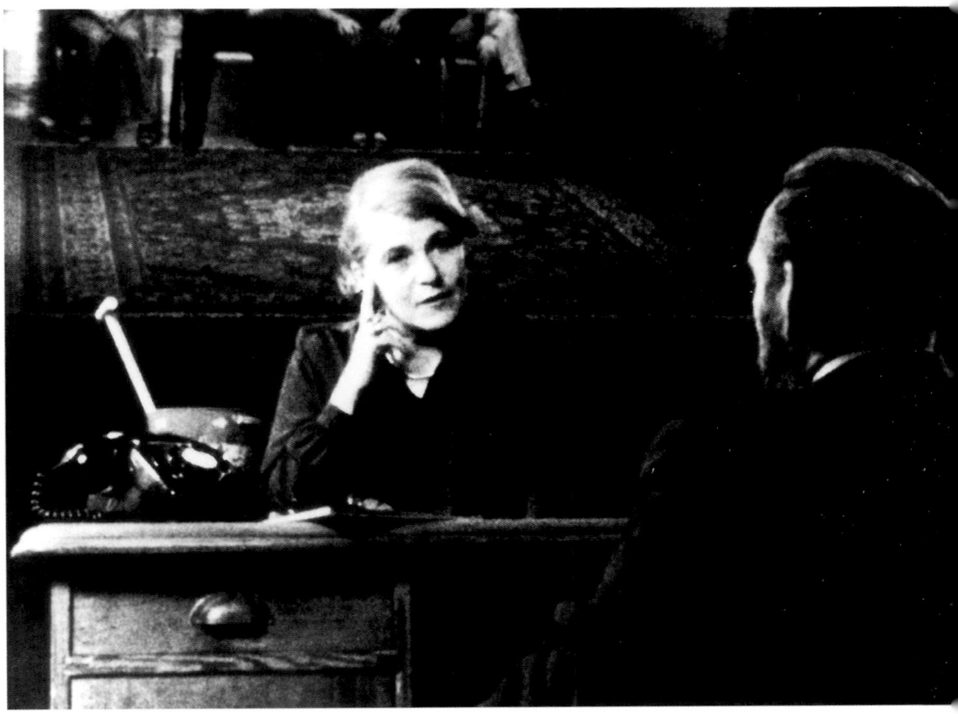

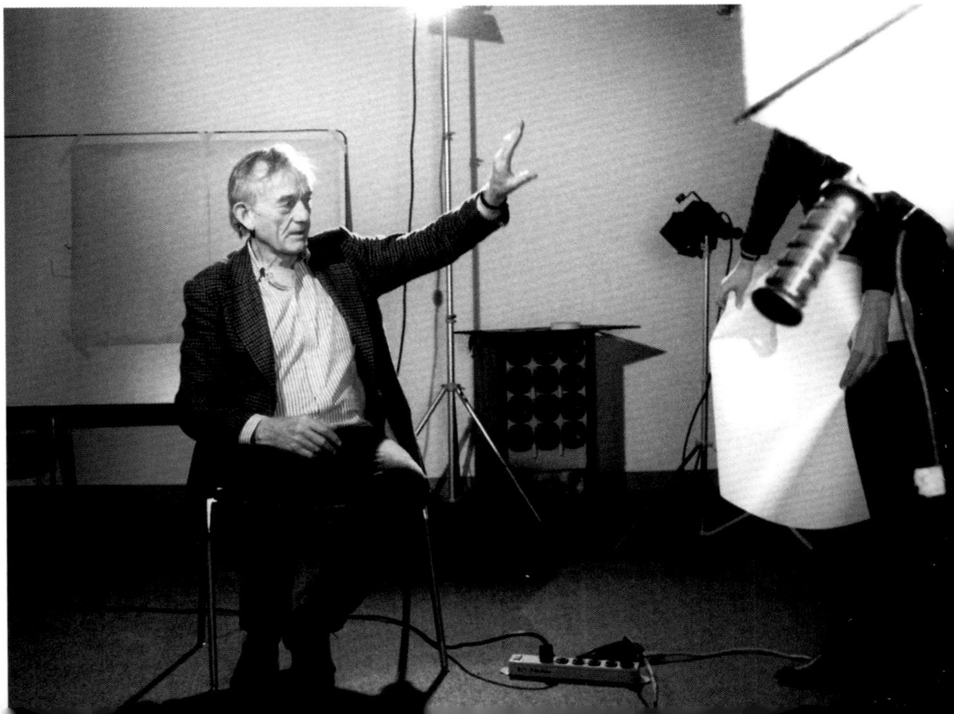

Above right: Catherine Ikam, "Identity III Mars," the artist interacts with her video laser installation, the Pit, CAVS, MIT, 1985
Photo: Veronica Falcão

Middle right: Yvonne Rainer, "Journeys From Berlin/1971," 125 minute film directed by Rainer, with five characters (pictured, actors Annette Michelson and Gabor Vernon), 1980
Photo: CAVS Special Collection

Bottom right: Richard Leacock, Director of MIT Film/Video section, 1985
Photo: Veronica Falcão

Facing page left: Ernst Caramelle, "Video-Ping-Pong" (1974), poster for presentation at Hayden Gallery, MIT, January 1975
Source: CAVS Special Collection

Facing page inset: Ernst Caramelle, "Video-Landschaften," photographed with the artist (detail), 1974
Photo: CAVS Special Collection

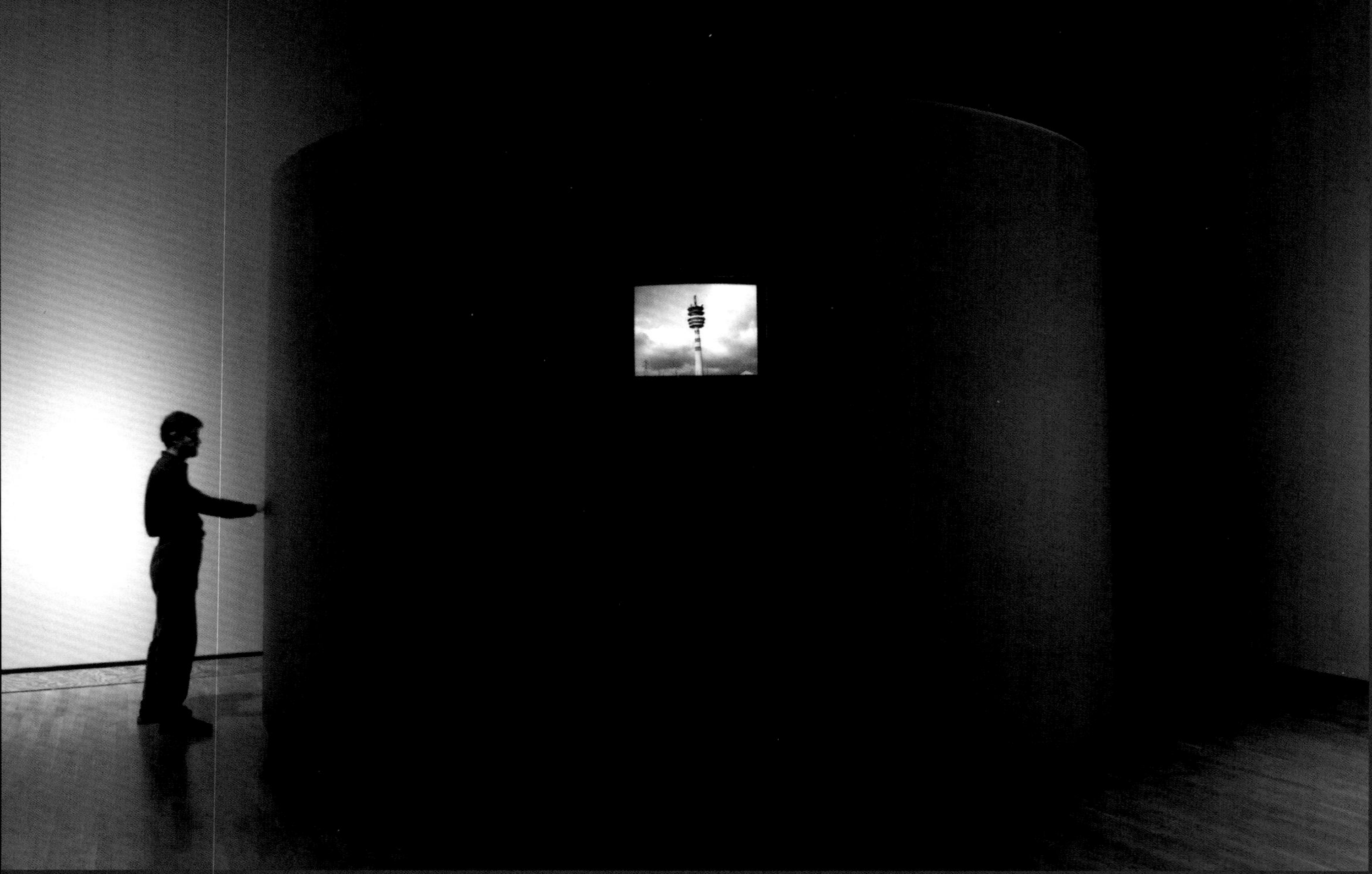

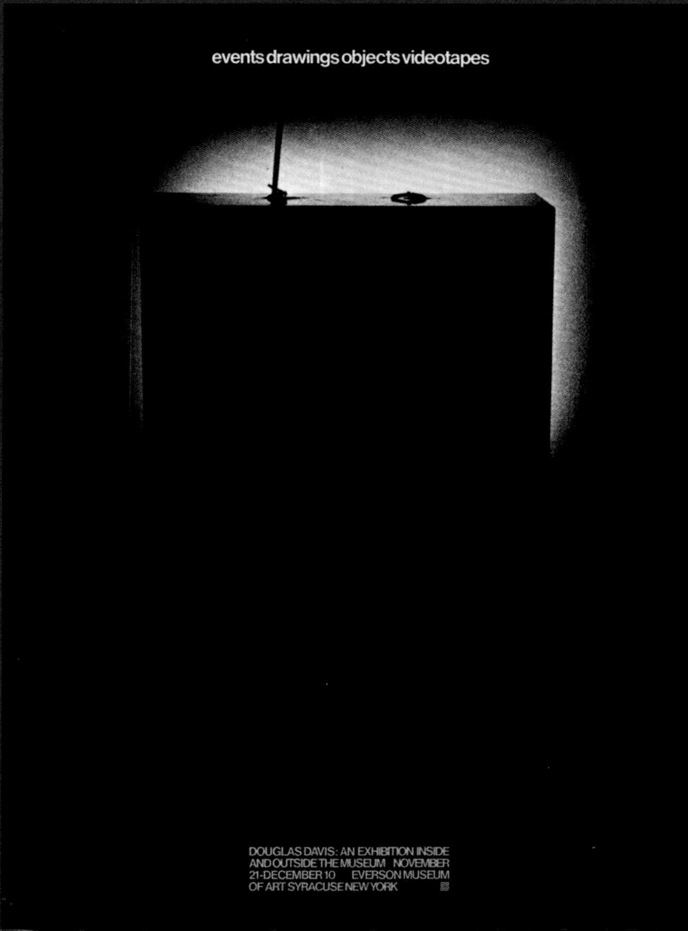
events drawings objects videotapes
DOUGLAS DAVIS: AN EXHIBITION INSIDE
AND OUTSIDE THE MUSEUM NOVEMBER
21-DECEMBER 10 EVERSON MUSEUM
OF ART SYRACUSE NEW YORK

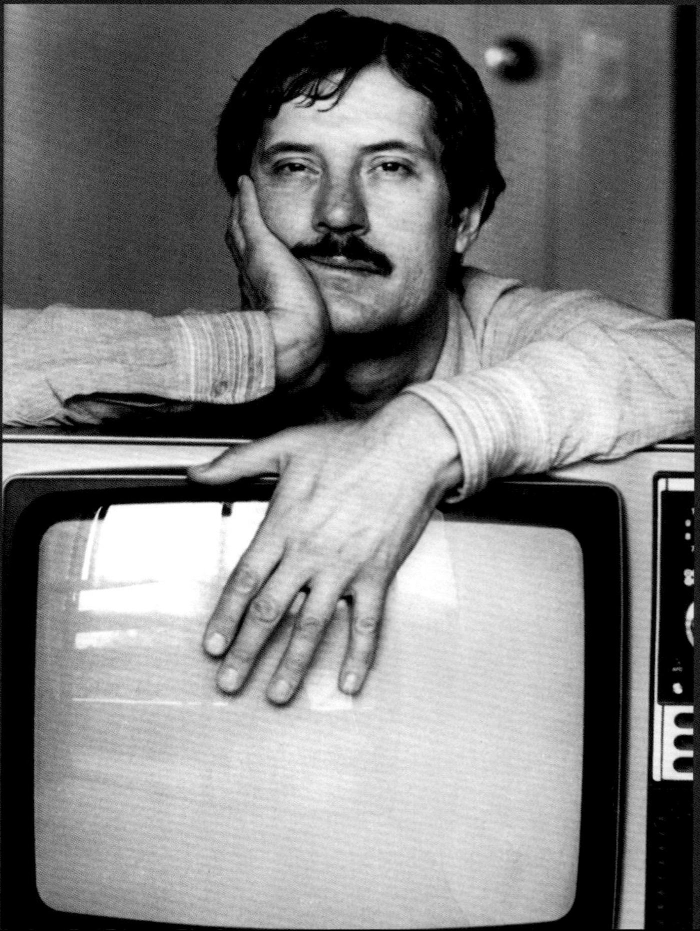

I did the project "On Subjectivity" . . . . The work was thought here [CAVS], developed here, produced here, presented here—all the different steps. I took 50 photos from *Life* magazine, and sent them to people to write captions for an interpretation. Can you recall the day I gave you the image, and asked for your participation, and when you got the book with the final project? For me, it was finished when everybody received the book. Then, the participants could see the scope of the book. When you received the photograph, you had only seen what I gave you. I divided the 50 photos among 250 people so each group of five people had the same photo to write a caption for. I was interested to see what others thought about the same image, and also the relation of one photo to others in *Life* magazine—a kind of American history. (Muntadas interview)

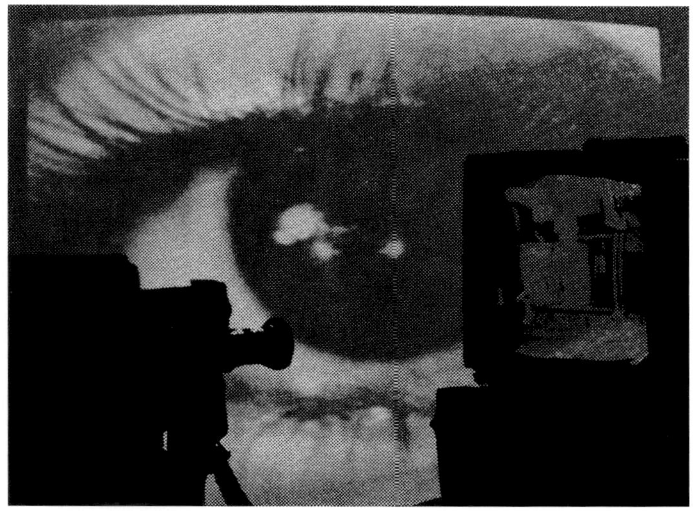

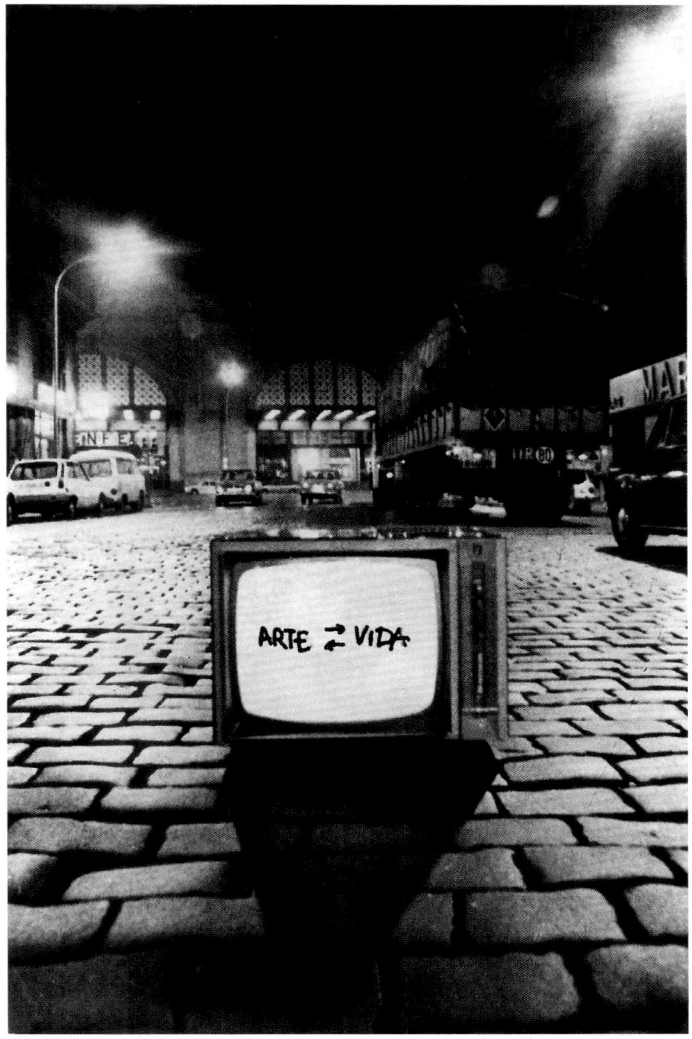

Above right: Muntadas, "On Subjectivity," (About TV), book and 50 minute videotape, 1978
Photo: Muntadas

Bottom right: Muntadas, "Acción Calle Comercio," video installation, Barcelona, 1974
Photo: Muntadas

Facing page top: Peter D'Agostino, "TransmissionS: In the WELL," interactive videodisc installation, Philadelphia, 1990
Photo: Gary McKinnis

Facing page bottom left: poster for "Douglas Davis: Events Drawing Objects Videotapes," with image of Davis' 1971 video installation, "Images from the Present Tense I," Everson Museum of Art, Syracuse, New York, 1972
Photo: Gwenn Thomas

Facing page bottom right: Douglas Davis, artist's portrait
Photo: *The Australian*

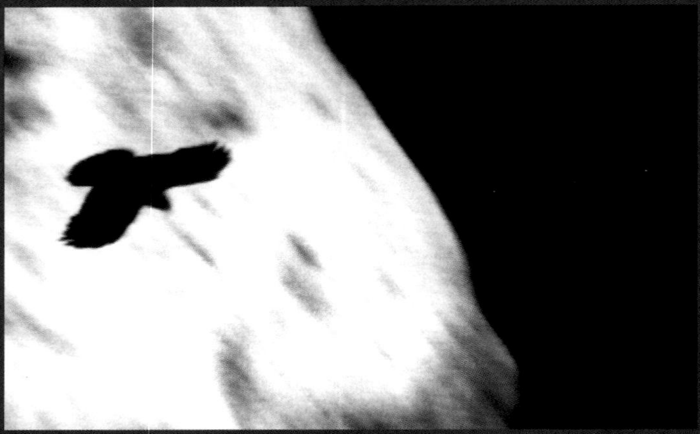
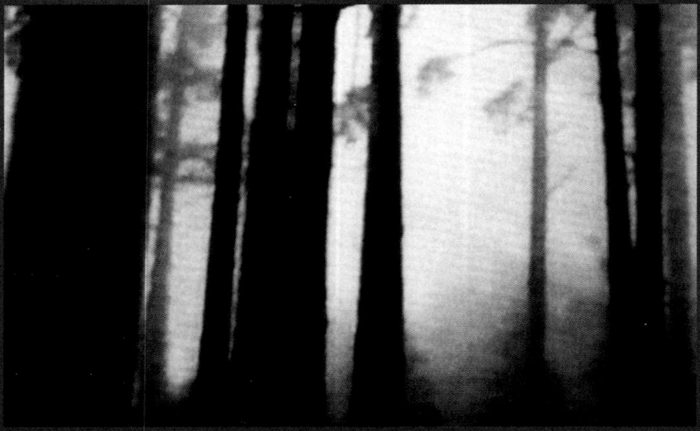
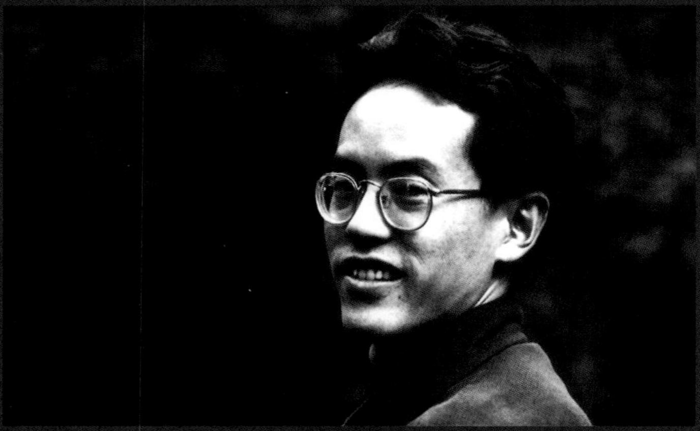

Ogata, a media artist from Japan, said the piece "is a meditative celebration of life and energy . . . *kagen*, a Japanese word, has a double meaning, literally, to increase and decrease; figuratively, to seek the means." His work [is] "intended to awaken our senses . . . to provide a universal sensory experience that transcends our differences in culture, age, race, or gender." (*MIT Tech Talk*, October 25, 1989, 11)

Atsushi Ogata, top,"kagen," videotape, 1988; middle "kiri no uta," videotape, 1988, bottom, artist's portrait, 1988. Photos: CAVS Special Collection

Say there are three objects: images of movies, text, music. As a user of the system, you can position something in the virtual world and you can navigate through the world, and look at it or listen to it from different angles. In other words, I can position any object, put a text in relation to that, I can experience how the meaning shifts as I build the context. The content takes the metaphors out of our linguistic presuppositions, and materializes or virtualizes them. You can actually experience it and point to it. For me, it points to itself, but in a larger way, points to how these media elements become an expanded linguistics, in that they can become operative in a computer-based environment. (Bill Seaman interview)

Facing page: Bill Seaman, "S.HE," still frames from 11 minute videotape, 1983
Photo grid: Bill Seaman

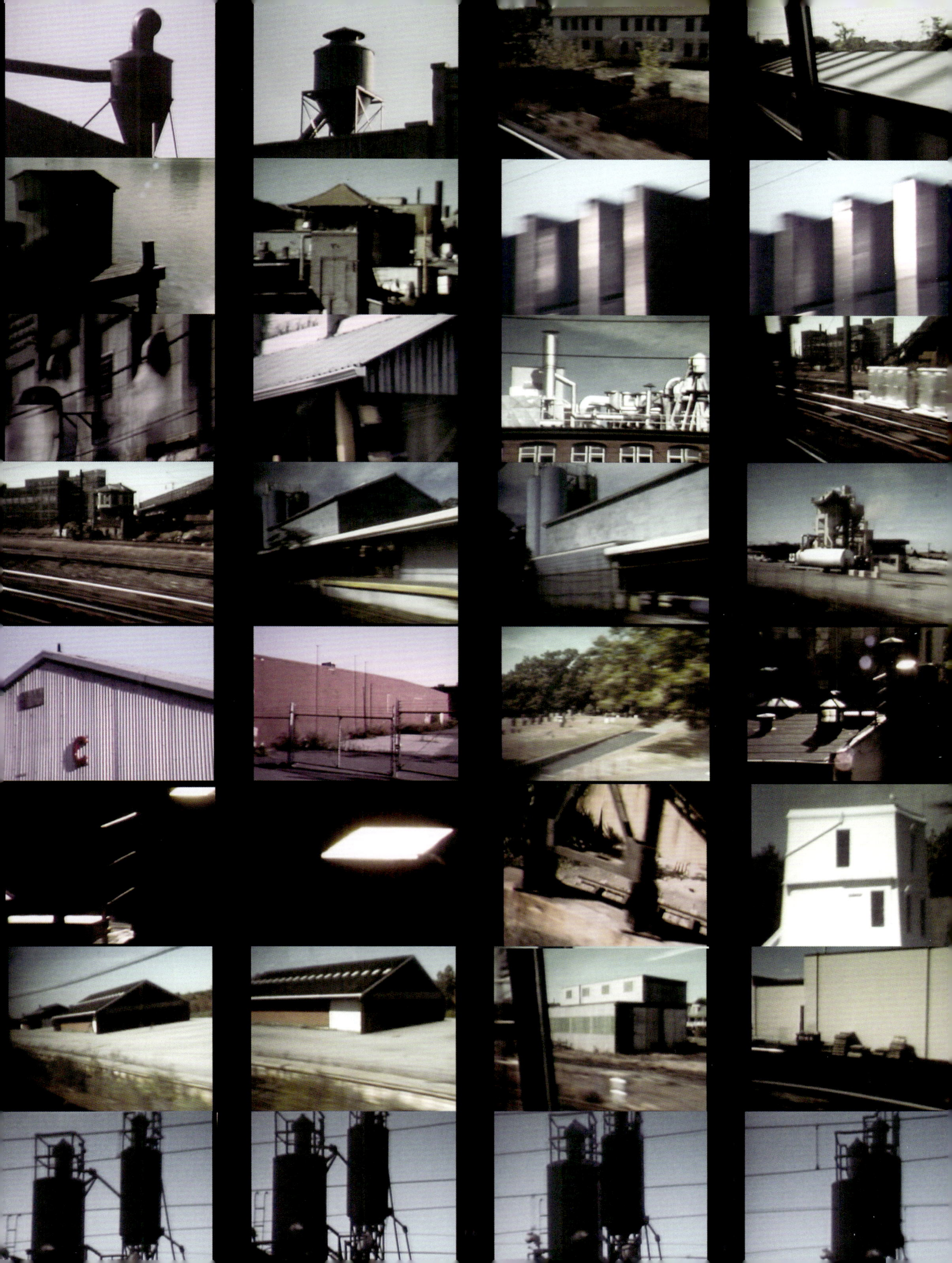

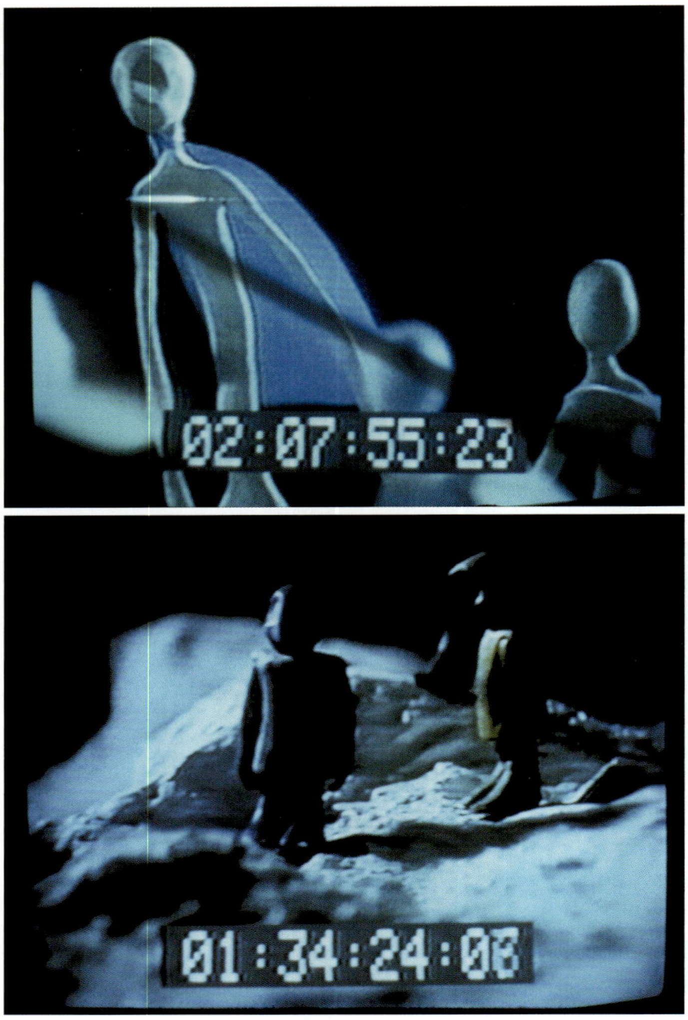

One of the lunarscapes was in a sandbox, probably 4 by 5 feet. I decided to make another lunarscape from plaster of Paris, because when one of the astronauts looked out at the moon, he said it looked like plaster of Paris. . . . I figured "Batteries Not Included" was my big shot at TV, so I decided to reflect something in the past, and did the lunar models. It was different from everything else I had done, since my previous work was related to home and other real life situations that you never see in animation although they are common, like bedrooms and bathrooms. (Randall Conrad, "Animation and Video Art: An Interview with Betsy Connors," *Visions*, Boston Film/Video Foundation, March 1980)

Above left: Betsy Connors, "Batteries Not Included," stills from videotape, a science-fiction animation with futuristic earth sets and lunar landscape, WGBH-TV, Boston, 1979. Photos: Betsy Connors

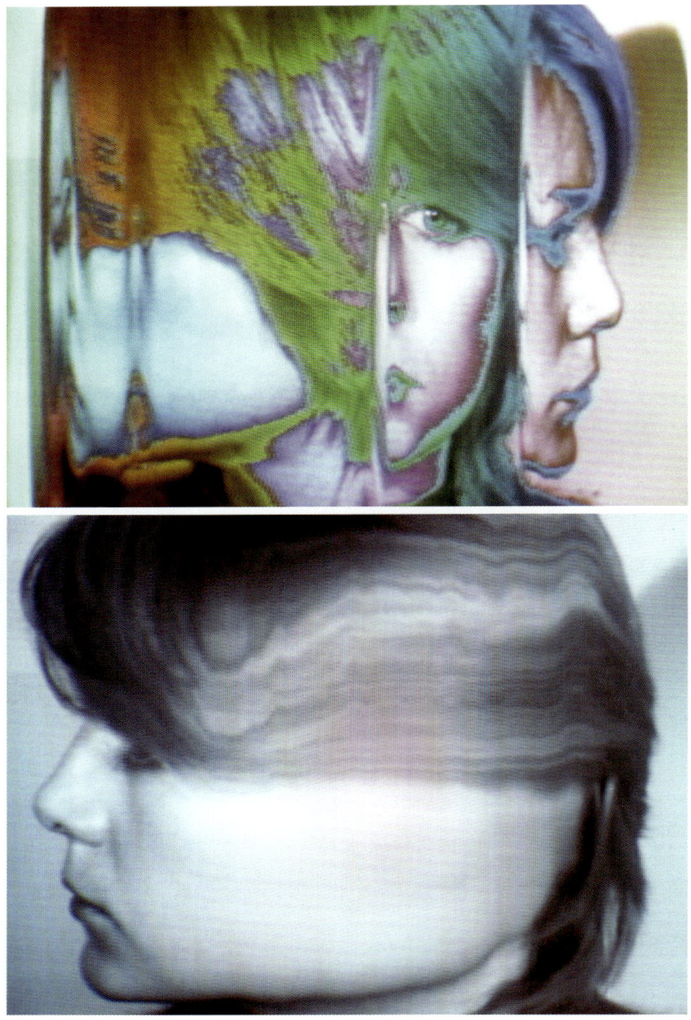

Digital portraits were computer generated using the Ramtek 9300 raster display frame buffers at the Architecture Machine Group at MIT. . . . Facial distortion is the result of movements made by the poseur during the 20 second sweep of the flying spot scanner. (Shelley Lake, *Ars Electronica* catalog, 1980)

Above right: Shelley Lake, "Digital Portraits," 1980 Photos: Shelley Lake

Video could be structured like music, images played like notes over time. I borrowed compositional modes from classical music to structure visual space. But, the camera was also seductive, a sensuous way of seeing and re-seeing the world in its elements, slow, close up, tactile, like entering dreams. . . . MIT was, is, a place where people enjoy confronting the unknown. . . . a place of doing, and the Center got a lot done. This heady mix of freedom and confidence bred results. As a musician who wanted to work with images, I embraced video. I remember thinking that I could do whatever I wanted. There was no form that predated my work to determine how it should look. It felt unlimited. (Ellen Sebring interview)

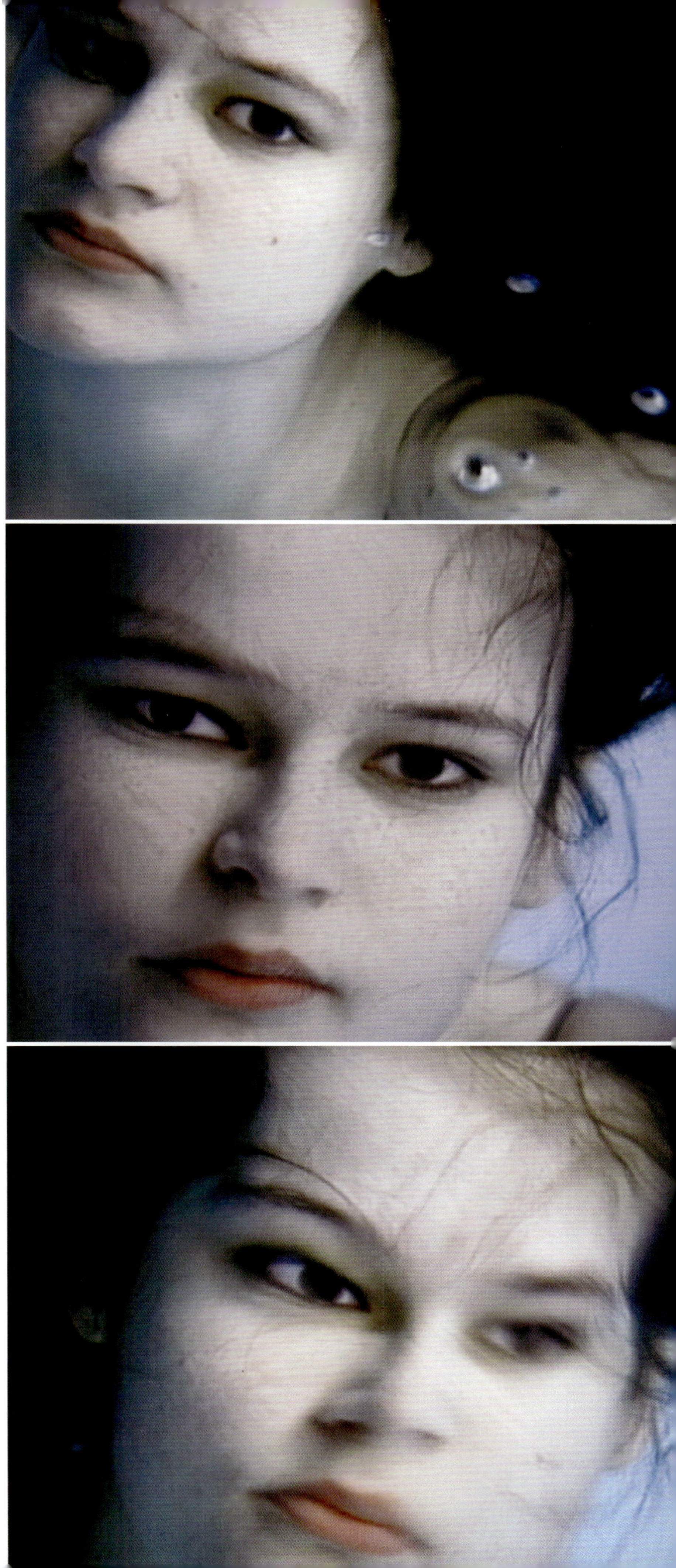

Right: Ellen Sebring, "Aviary," still frames from 28 minute videotape (pictured, Nadja Kutz), commissioned and broadcast by PBS series, *New Television*, 1989; collaborative performance at the CUBE, Media Lab, MIT, 1988
Photos: Ellen Sebring

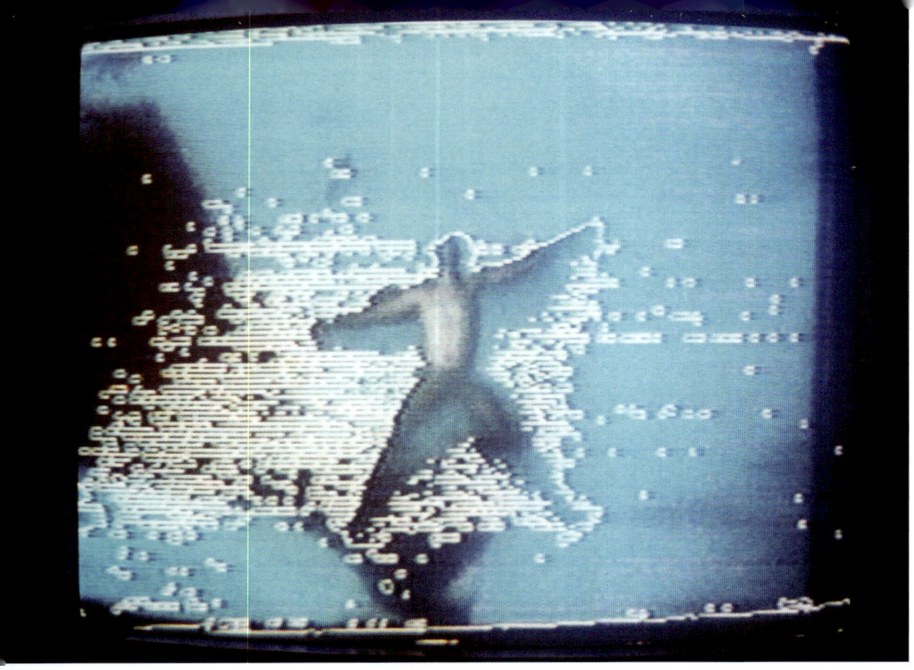

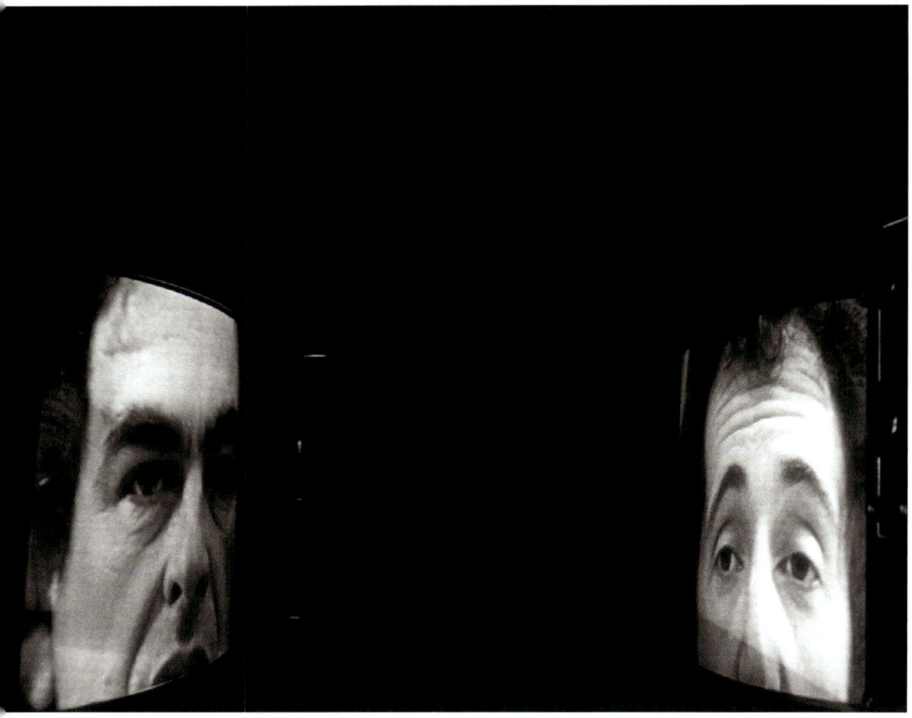

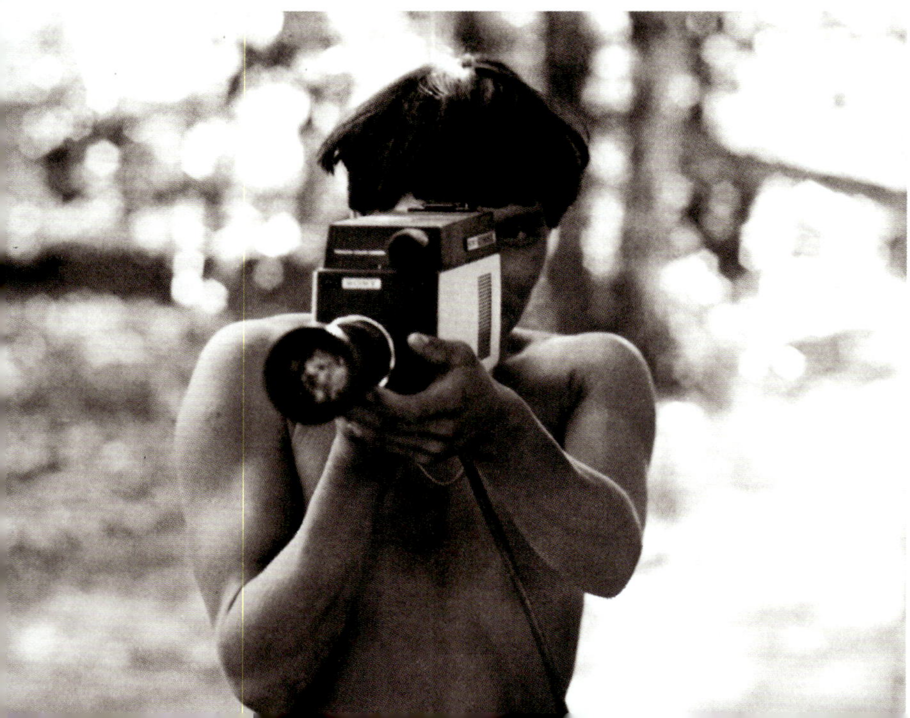

Did CAVS change art? Art is usually produced in isolation, with little recognition by critics and galleries. Art produced at CAVS is largely outside the mainstream of US art. . . . Art is becoming more globalized with virtual programming, web, special effects, the legacy of tech art. I remember Charlotte Moorman's piece inside a tank of water playing Saint-Saëns' "Swan." I like that kind of complexity: old/new technology in combination. Paik as baroque! I do like Muntadas, who makes art inside the medium. For me now, I am interested in the idea of emptiness, nothingness. (Jose Maria Ytturalde interview)

Top: Vin Grabill, video still, "Sketch for Icarus", 1982
Photo: Vin Grabill

Middle: Mit Mitropoulos, "Face to Face," the artist and Richard Lerman on television monitors, video installation, CAVS, 1984
Photo: CAVS Special Collection

Bottom: Juan Downey, location photo, "Video Trans Americas" project, 1973-1977
Photo: CAVS Special Collection

The fact that [Nam June] Paik was coming around with all these gadgets and conglomerations of video things . . . he was associated with the Center. When someone is a pioneer, they tend not to be incorporated into the mainstream. As with Harriet [Casdin-Silver]'s work, it took a long time to say this is an art form, and we will look at it alongside painting and sculpture, and it can contain an expressive meaning. (Beth Galston interview)

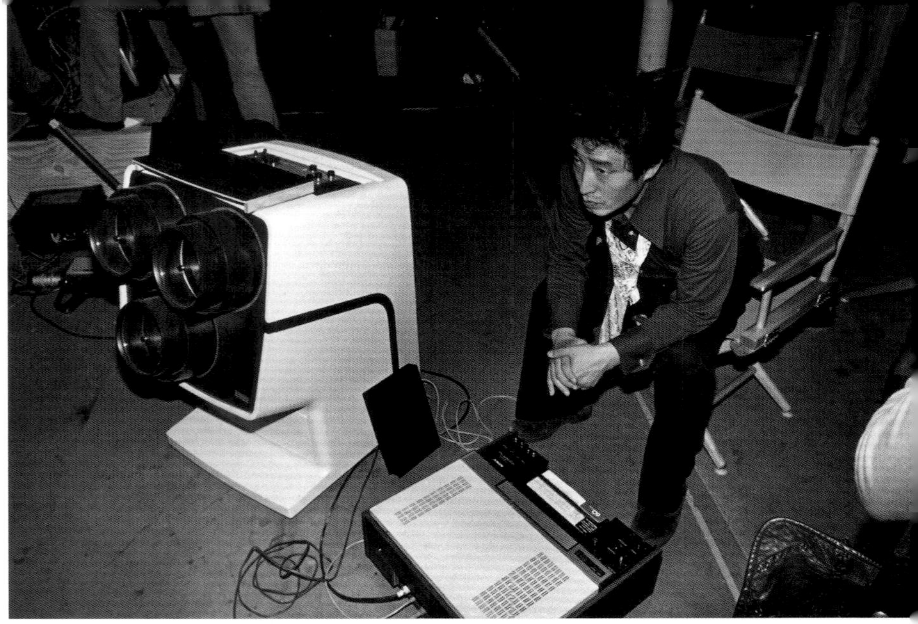

For a long period at the end of the '80s, beginning of the '90s, most of my students hated Nam June Paik. For them, he was dated and old-fashioned, totally irrelevant to their concerns. Now, they are totally fascinated. . . . There's a possibility that this new generation with an interest in what went on in the '70s will pick up some of that same energy again. I do think we need a radical period of experimentation in the arts just to digest everything and integrate it into society. Playing is still an important element. Find out what it can do, what happens if I cut this little thing? I hope it's not just nostalgia. I think it's real. (Don Foresta interview)

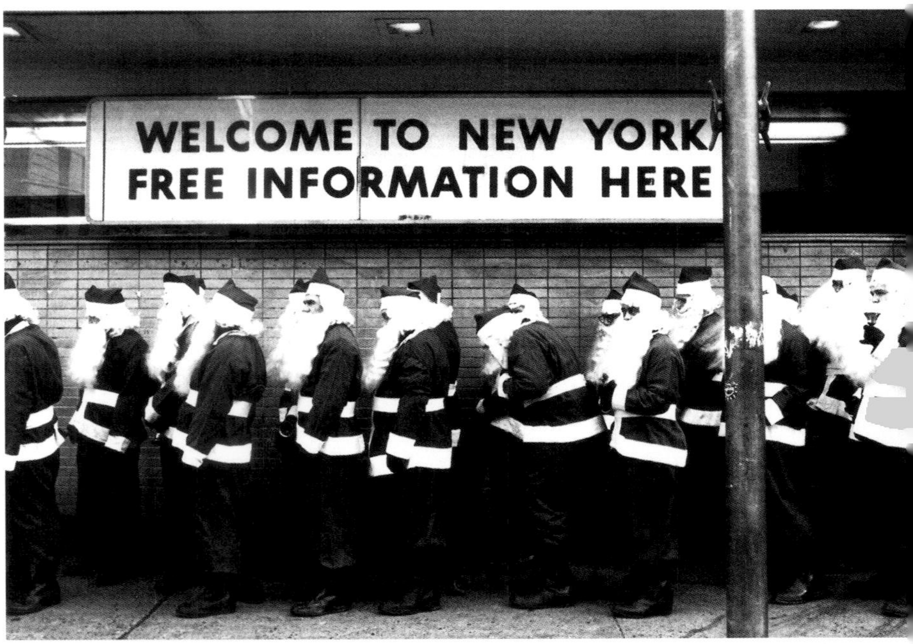

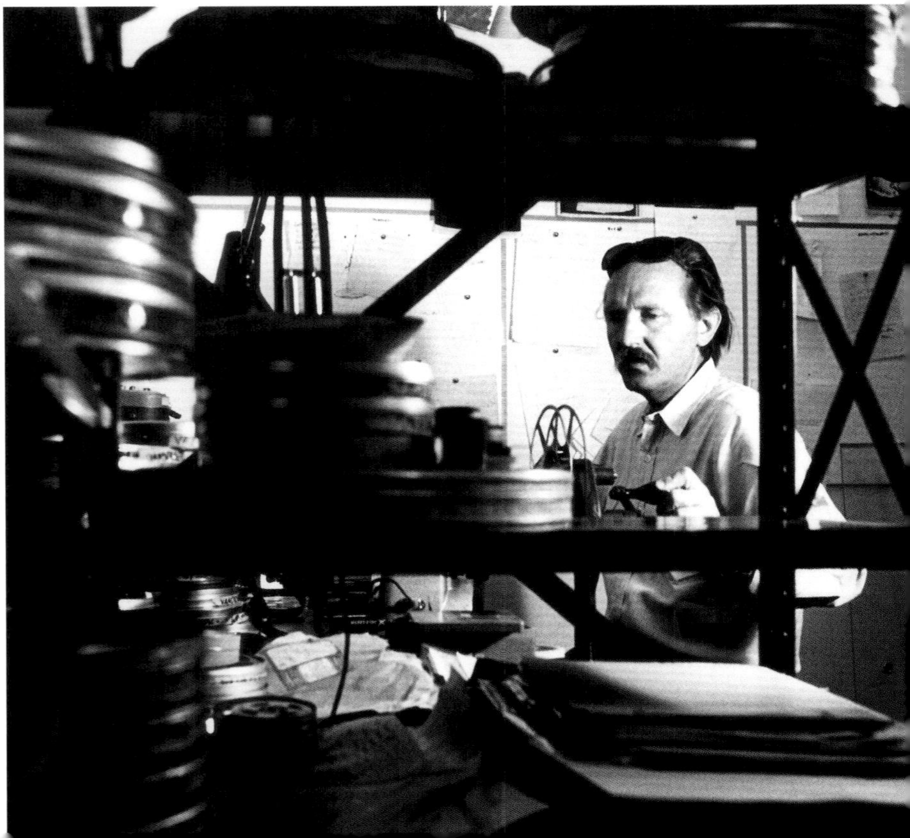

Top: Nam June Paik at CAVS, *Arttransition* conference, CAVS, MIT, 1975
Photo: Roger N. Goldstein

Middle: Miralda, "Santa Army Navy," videotape, 1979
Photo: CAVS Special Collection

Bottom: Stan VanDerBeek in his CAVS studio surrounded by film canisters, 1976
Photo: Nishan Bichajian

MIT marked CAVS's fiftieth birthday in 2018 by exhibiting artworks of CAVS fellows, graduate students and associates. Complementary events at ACT were designed to "stretch and extend the ideas and dynamics of CAVS via current and recent work of contemporary artists" (ACT website).

During the *CAVS50* installations, the Museum also mounted the two major photography exhibitions, *György Kepes Photographs: From Berlin to Chicago, 1930-1946* (September 21, 2017-March 5, 2018) which presented Kepes' time in Europe and Chicago; and *György Kepes Photographs: The MIT Years, 1946-1984* (March 21-July 15, 2018) focusing on his work created while at MIT. Together, the exhibitions traced Kepes' legacy in art, science, and technology, and the founding of the Center for Advanced Visual Studies.

Above: Wen-Ying Tsai, "Large Tuning Fork," kinetic sculpture, 1972
Facing page top: Otto Piene, "Cereus" inflatable sculpture, 2008
Facing page bottom: Mark Mendel, "Poetronics," neon poem, 1978
Exhibition views from *Celebrating the 50th Anniversary of the MIT Center for Advanced Visual Studies (CAVS50)*, MIT Museum, Cambridge, April 2018-January 2019
Photos: Ellen Sebring

Afterword

In 2009, Ute Meta Bauer, the director of MIT's Visual Arts Program, and Krzysztof Wodiczko, director of MIT's Center for Advanced Visual Studies (CAVS), joined forces to merge the academic program and the research center into the new MIT Program in Art, Culture and Technology (ACT). Their goal was to expand the notion of visual studies and push the capacity of art to enlist science and technology in cultural production and critique.

Four decades earlier, the founders of CAVS had urged artists and designers to take matters of vision, value, and imagining the future into their own hands. That future is now under siege, on all possible fronts—political, social, cultural, and ecological. Thus, CAVS's programmatic imperatives—art on the civic scale, art on the environmental scale, and art as communication—continue to live on in the DNA of the ACT program's catalyzing role during this era of escalating problems. How do art and aesthetics critically redirect technologies from serving the military-industrial complex toward decolonization and the welcoming of indigenous knowledges? How can artistic and design intelligence address extraction politics and social and ecological justice? Last but not least, how can art and artistic research contribute to the development of technologies that engage not only multi*cultural* but also multi*natural* perspectives?

Carrying on the pioneering role that CAVS had at the environmental frontier, the ACT program builds upon a pedagogical lineage aimed at changing habits of thought and developing new modes of citizenship—citizenship of the imperiled environment. By investigating the proposition that the metaphors and applications of the cybernetics model have become less productive than they once were, we should probe the usefulness of the concept of "sympoeisis" (collective production) as a new model of thought for current crises, and as an aid in imagining and working together in radical interdisciplinarity toward desirable futures.

— Gediminas Urbonas, artist and ACT director (2015-2017)

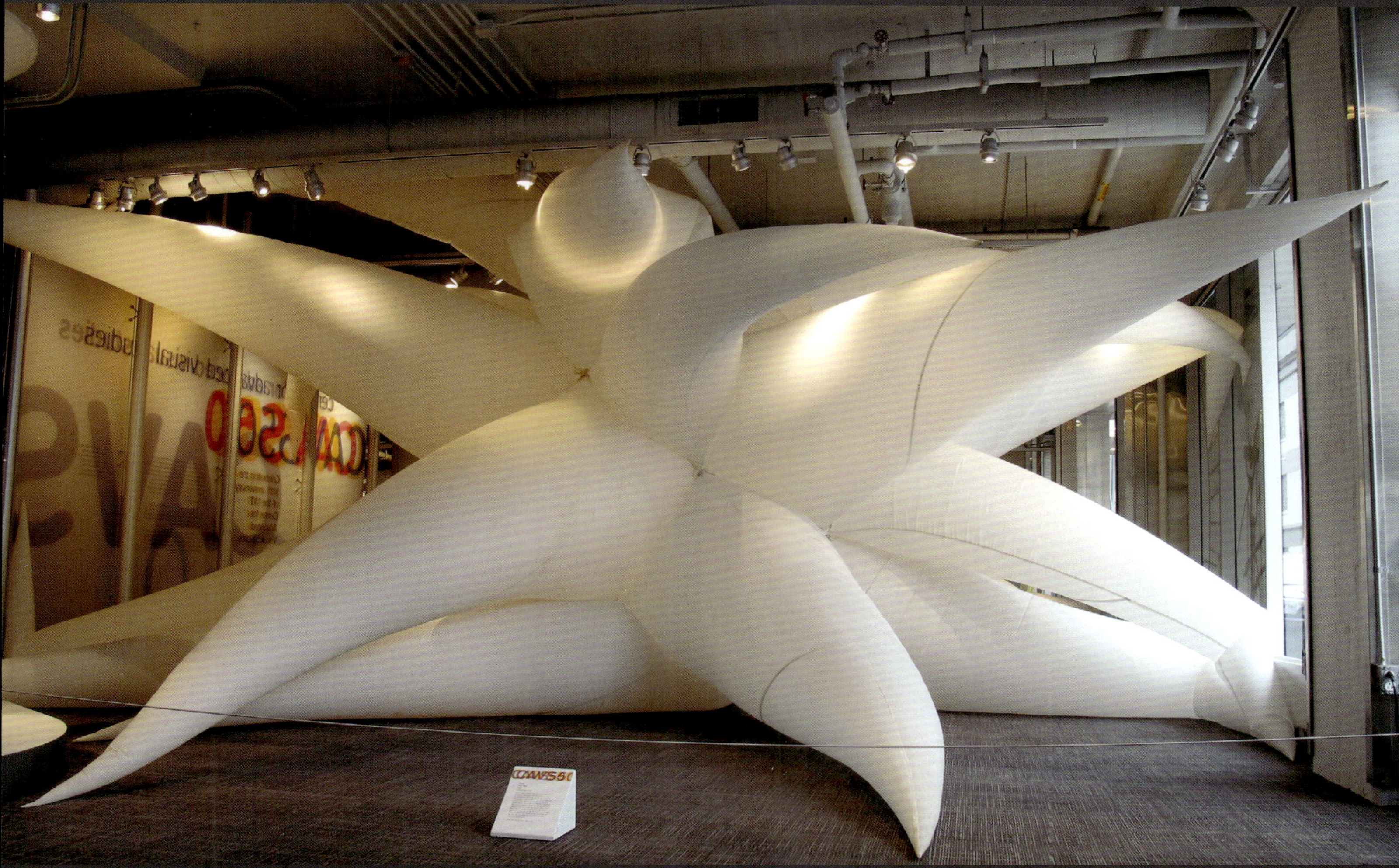

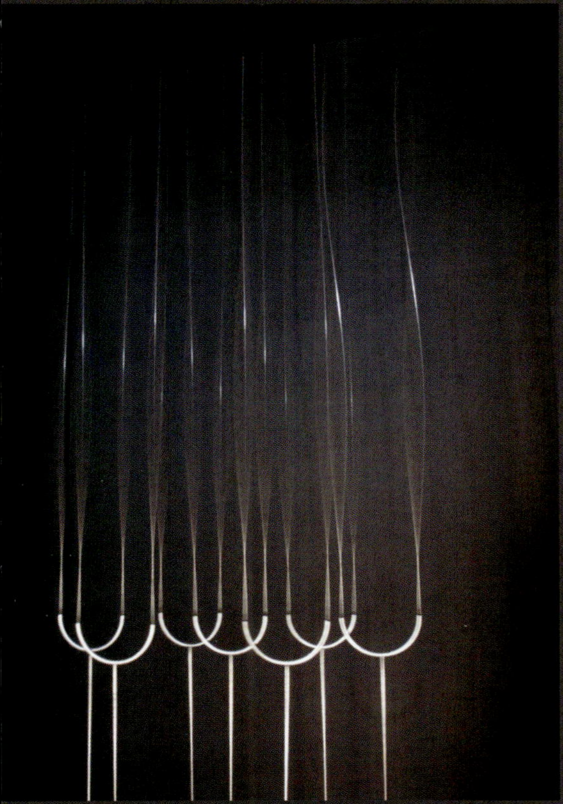

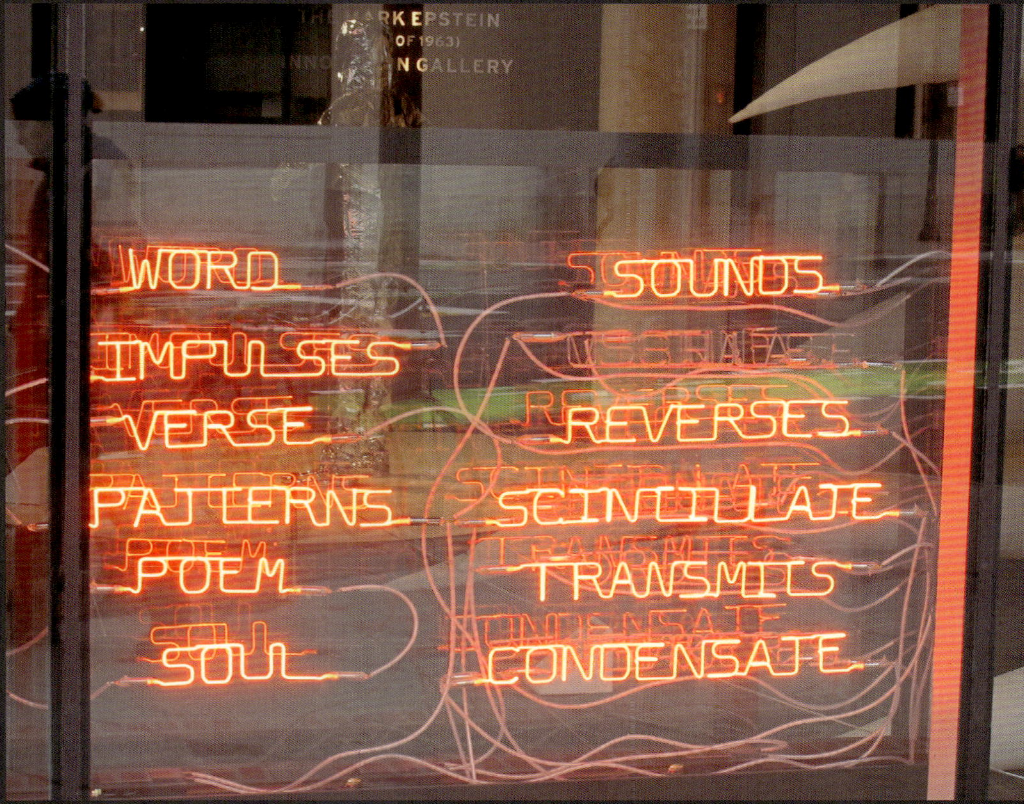
WORD
SOUNDS
IMPULSES
VERSE
REVERSES
PATTERNS
SCINTILLATE
POEM
TRANSMITS
SOUL
CONDENSATE

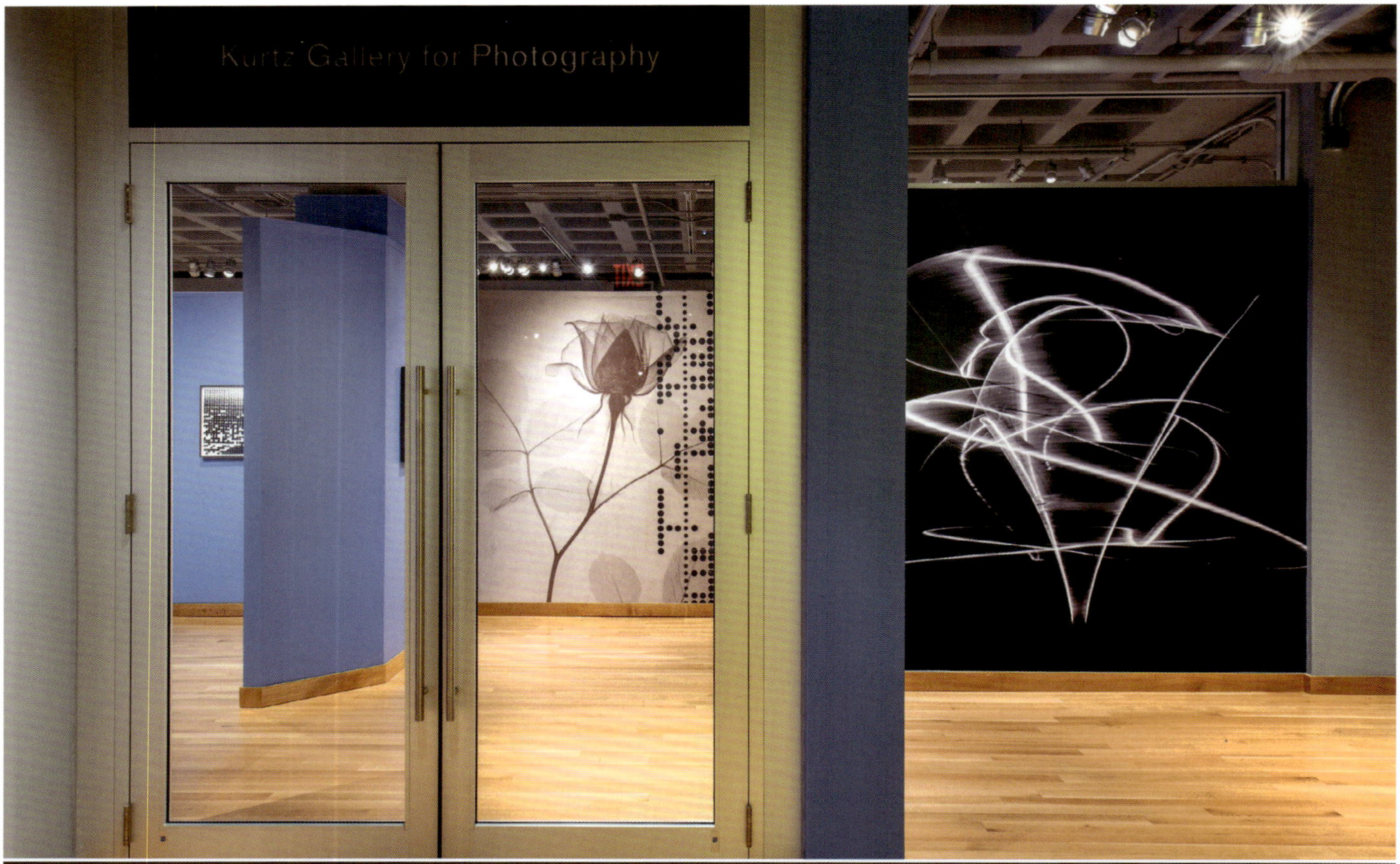

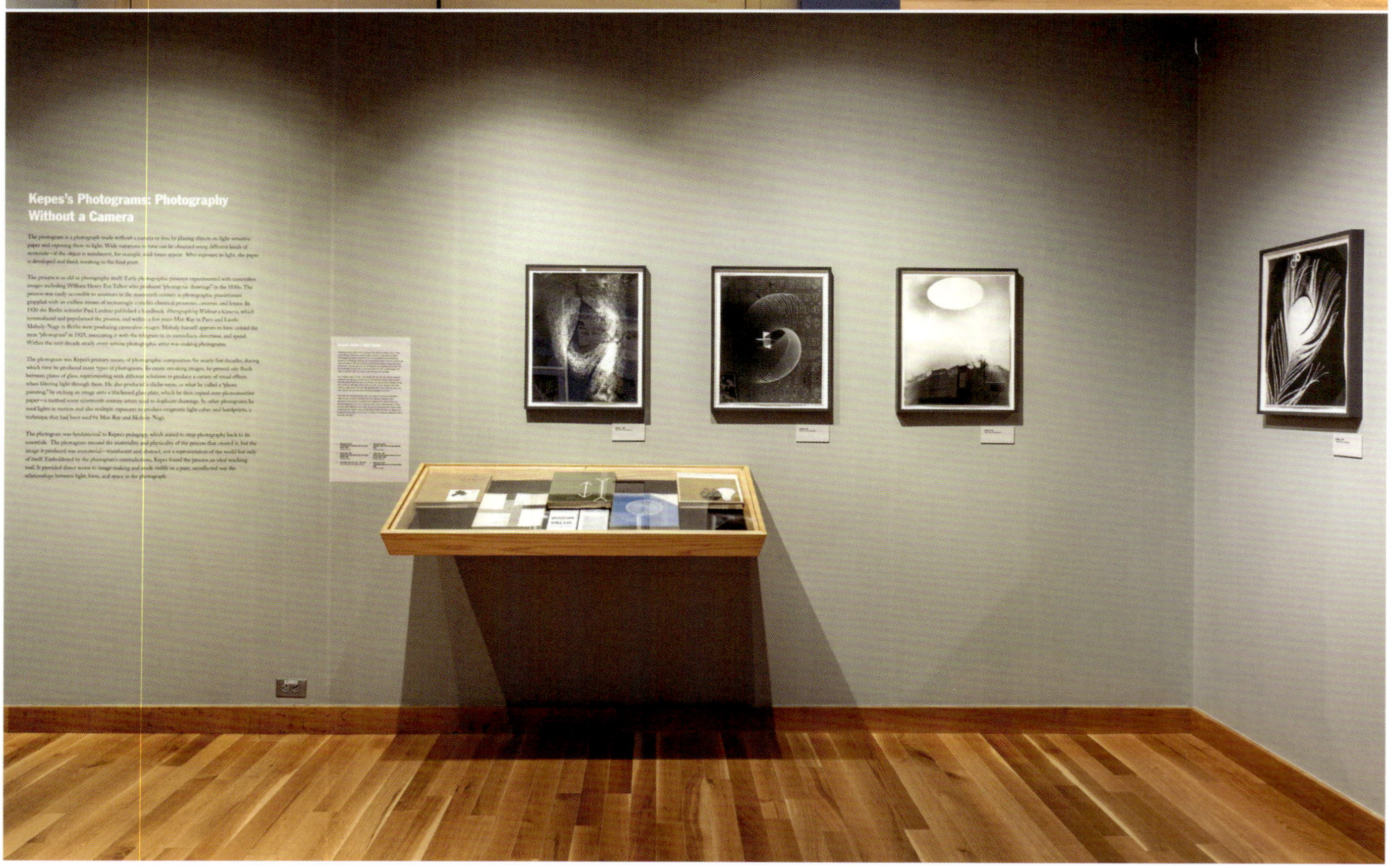

Both pages: Exhibition views, *György Kepes Photographs: The MIT Years, 1946-1984*, including (bottom left) "Kepes's Photograms: Photography Without a Camera," and his *Vision + Value* series; and (facing page bottom right) a section

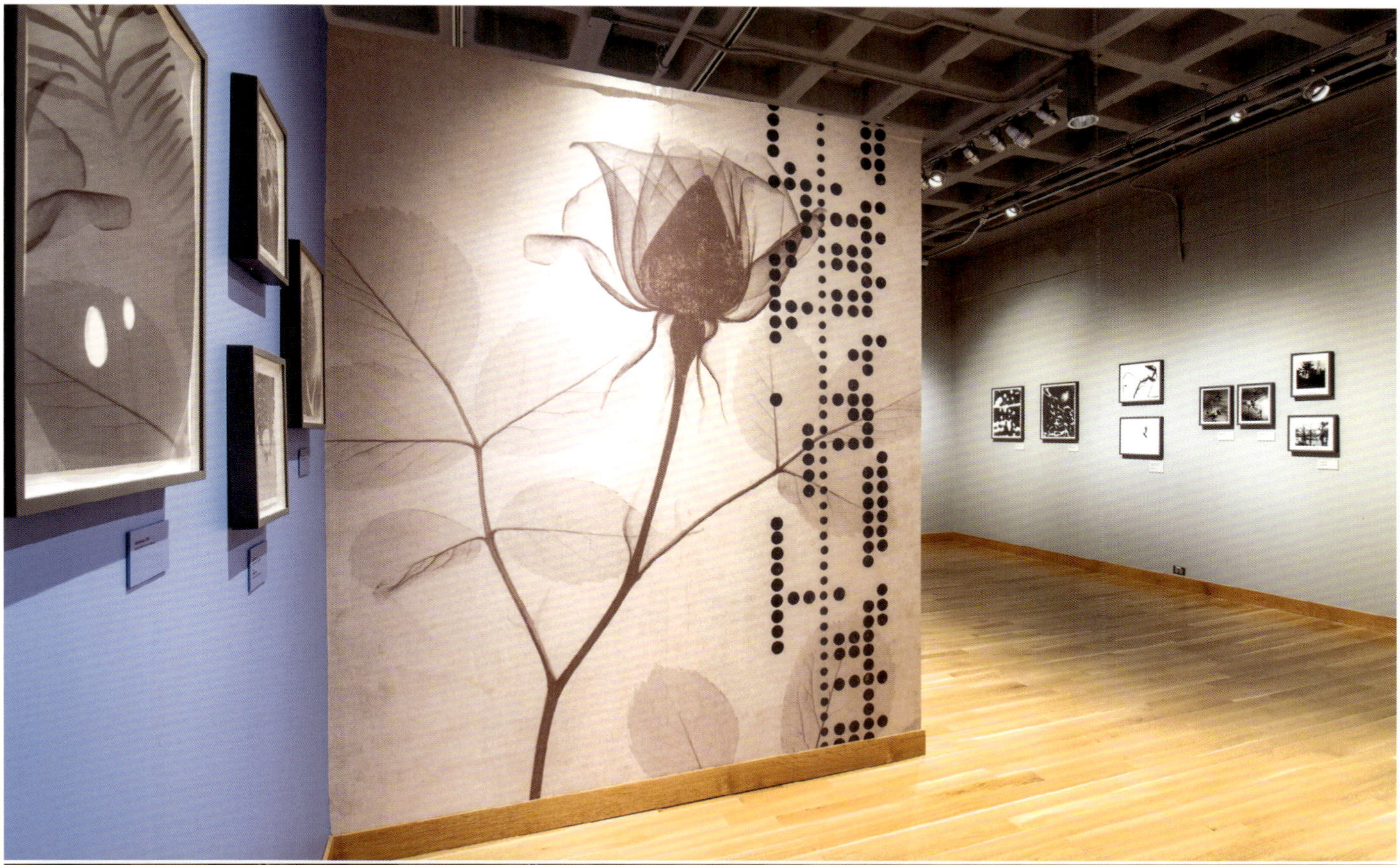

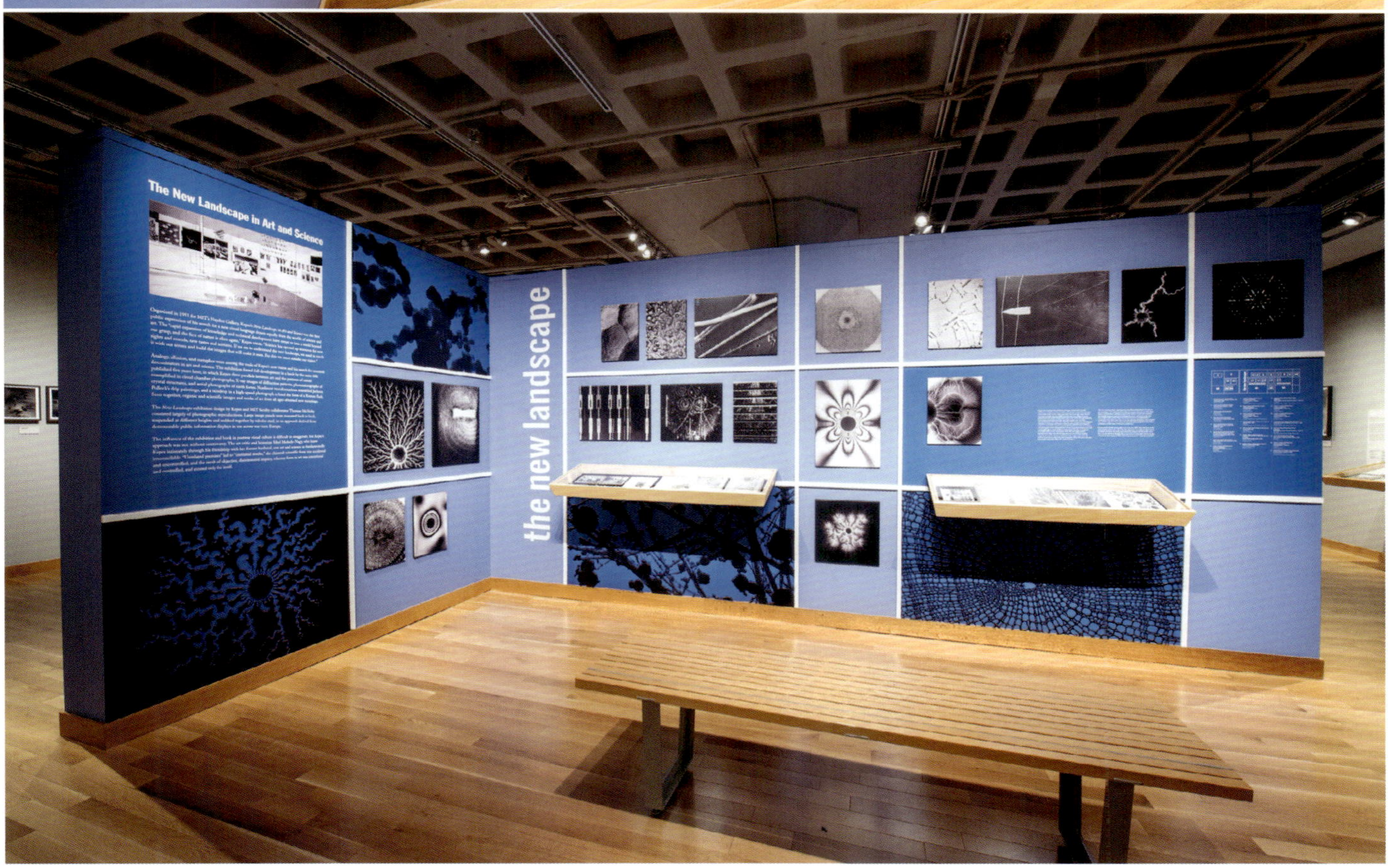

on *The New Landscape*, Kepes' 1951 exhibition exploring scientific visualizations and visual arts at MIT's Hayden Gallery; Kurtz Gallery for Photography, MIT Museum, Cambridge, March 16-July 15, 2018. Photos: Samara Vise

Bibliography

ars elecronica im Rahmen des Internationalen Brucknerfest, 1982 (exhibition catalog).

Artforum, February 25, 2019, "Jack Burnham (1931-2019)."

Arts Magazine, "Alejandro Siña," November, 1978.

Arttransition. MIT Center for Advanced Visual Studies and University Film Study Center, October 15 – 19, 1975 (exhibition catalog). MIT, Cambridge, Massachusetts, 1975.

Alexenberg, Mel and Otto Piene. *LightsOROT*. Center for Advanced Visual Studies, Yeshiva University Museum, January 31-December 31, 1988 (exhibition catalog). MIT, Cambridge, Massachusetts, 1988.

Alloway, Lawrence. *5 Artists / 5 Technologies: Environmental Light Works by Peter Campus Harriet Casdin-Silver, Paul Earls, Otto Piene, Alejandro Sina,* Grand Rapids Art Museum (exhibition catalog). Grand Rapids, Michigan, 1981.

Andersen, Wayne, *Takis: Evidence of the Unseen*, on the occasion of the Takis exhibition, Hayden Gallery, MIT, MIT Press, Cambridge, MA, 1968

Baker, Kenneth. "Review: Earth, Air, Fire, Water: Elements of Art," *Artforum*, v. 9, March, 1971, 72-74.

Blakinger, John R. "The Aesthetics of Collaboration: Complicity and Conversion at MIT's Center for Advanced Visual Studies," *Tate Papers*, no.25, Spring 2016, http://www.tate.org.uk/research/ publications/tate-papers/25/aesthetics-of-collaboration, accessed 2 July 2017.

Brigham, Joan. Interviews of CAVS Fellows and graduates, 2002-2011.

Burgess, Lowry, *Burgess: The Quiet Axis*. Montreal: Precare, 1987.

Carlock, Marty. "The Center for Advanced Visual Studies: Where Art Comes Off the Wall." *Technology Review* (July 1984) A IO-A I 5.

CAVS Portfolio, 1975, 1977.

Celebration of Light. Pyrri Art Centre, June 10 – Aug 6, 1989, Savonlinna, Finland (exhibition catalog), 1989.

Claus, Jürgen. "Sky Art-Der Himmel wird bunter." *Das Kunstwerk* (February 1985).

Coe, Ralph T. *Dale Eldred: Sculpture into Environment.* Lawrence, Kansas: Regent Press, 1978.

Conrad, Randall. "Animation and Video Art: An Interview with Betsy Connors," *Visions*, Boston Film/Video Foundation, March 1980.

Davis, Ben, Rus Gant and Vin Grabill. CAVS/MIT *Centerdisk / Skydisk*, 1983

Douglas, Deborah, Ed. *Countless Connecting Threads.* Cambridge, Massachusetts: MIT Museum, 2012.

18th Bienal Internacional de São Paulo. October 4-December 15, 1985, Fundação Bienal de São Paulo, São Paulo, Brazil, (general catalog), 1985.

Electra: Electricity and electronics in the art of the XXth century. MAM. Musée d'art moderne de la ville de Paris, December 10, 1983-February 5, 1984 (exhibition catalog).

Erickson, Ruth. *Leap Before You Look: Black Mountain College 1933-1957*. New Haven, Connecticut: Yale University Press, 2015.

Explorations, Smithsonian Institution, 1970 (brochure, catalog)

Fromm, Dorit. *Architext, Newsletter of the Department of Architecture MIT,* February 28, 1983.

Funke, Phyllis. "Science Is Blended with Art in Display," *The New York Times*, April 7, 1974.

Glibota, Ante. *Otto Piene*. Delight Edition. 2011.

Goldring, Elizabeth. "Sky Art: Art that Flies," *Flying Colors: Braniff International Travel Magazine*, Vol. 8, No. 12, 1979.

Goldring, Elizabeth and Otto Piene, Co-editors. *Centerbeam*. (Cambridge, Massachusetts, MIT Press, 1980).

Goldring, Elizabeth. "The Sky Art Conference, 1981-1986," *Lightworks*, No. 17.

Goldring, Elizabeth and Otto Piene. "Sky Art," *Parnass* (September/October 1982).

Goldring, Elizabeth and Ellen Sebring, Interview with Joan Brigham, 2016.

Goldring, Elizabeth. *Laser Treatment: Poems and Two Short Stories*. Boston: Blue Giant Press, 1983.

Goldring, Elizabeth, Otto Piene, Eds. and Rus Gant, Vin Grabill, and Ben Davis, designer/authors. *The Sky Art Interactive Video Disk Travelling Exhibition* (catalog), Center for Advanced Visual Studies, MIT. Cambridge, Massachusetts, 1984.

Goldring, Elizabeth. *Without Warning*. Helicon Nince Editions, BkMk Press, University of Missouri-Kansas City, Kansas City, Missouri, 1995.

Gunter, Virginia, Curator. *Earth, Air, Fire, Water: Elements of Art*. Boston: Museum of Fine Arts (exhibition catalog). Boston, Massachusetts, 1971.

Harvard Crimson, Cambridge, MA, 1969.

Hiemer, Astrid. "New MIT Program ACT Inauguration," *Berkshire Fine Arts*, April 10, 2010.

Holynksi, Marek. "A Spirit of Technology: An Interview with Otto Piene," *The Structurist*, Eli Bornstein, ed., No. 21/22 (1981-1982).

International Biennial Exhibition of Graphic and Visual Art, Vienna, Austria (exhibition catalog), 1979.

Janney, Chris to Elizabeth Goldring and Ellen Sebring (letter), January 2017.

Kepes, György. *The New Landscape in Art and Science*. Chicago: Paul Theobald and Co., 1956.

Kepes, György to MIT President Julius Stratton, June 17, 1965, letter.

Kepes, György. *The Center for Advanced Visual Studies inaugural catalog*. Cambridge, Massachusetts, 1967.

Kepes, György. "The Lost Pageantry of Nature," *Ars Canada*, December, 1968.

Kepes, György. *Arts of the Environment*. New York: G. Braziller, 1972.

Knott, Laura and Bernd Kracke, Eds., *Sky Art Conference 2002*, Recorded Oct 15-20, 2002, Delphi and Ikaria, Greece (conference publication), Center for Advanced Visual Studies. Cambridge, Massachusetts: MIT, 2004.

Kracke, Bernd. "Sculpture as Process," thesis, CAVS, MIT, 1981.

Kramer, Hilton. "Boston Museum Mounts an Unusual New Art Show," *The New York Times*, February 4, 1971.

Leacock, Richard and Jon Rubin, Centerbeam (film), 1978

MacCready, Paul B. "Art and Nature in the Technology of Flight." *Leonardo 19*, 1986.

Martins, Wendy. "Joe Davis and the Art of Science," *Cambridge Chronicle*, April 1, 1993.

Mehr Licht, Monocle: Tribute to György Kepes. Center for Advanced Visual Studies (exhibition catalog), Staedische Kunsthalle, Hamburg, Germany, E. Kabel Verlag, 1985.

MIT Report of the President, annual reports from 1968, 1969, 1972, 1995, 1998, Institute Archives and Special Collections, MIT.

MIT Tech Talk, May 26, 1982; October 25, 1989.

Multiple Interaction Team, Center for Advanced Visual Studies, MIT, an exhibition event to travel supported by a grant from the NEA, Museum of Science and Industry,

Chicago, San Francisco, Cincinnati, New Orleans, Philadelphia, St. Petersburg, Fl, New York, Los Angeles, 1972-74.

Muntadas, Antoni. *On Translation*, ACTAR, Museu d'Art Contemporani de Barcelona, Barcelona, July 2003.

NEA Portfolio, 1974, CAVS Archives.

News-Tribune, Waltham, MA, February 3, 1982.

Otto Piene und das CAVS. Badischer Kunstverein, Karlsruhe, Germany (exhibition catalog), September 17-October 23, 1988.

Paul, Christiane. "The Prophet's Prothesis: An Interview with Krzysztof Wodiczko," *Sculpture* Magazine, Vol. 18, No. 4 (May, 1999).

Pelham, David. *Kites.* London: Penguin, 1978.

Phillips, Patricia C. "Todd Siler at Ronald Feldman Arts," *Artforum*, May 1987.

Piene, Otto. *Elements*, Howard Wise Gallery, New York City, 1969 (Exhibition catalog).

Piene, Otto and Heinz Mack, Eds. *ZERO.* Cambridge, Massachusetts: MIT Press, 1973; original publication: ZERO 3, Diisseldorf, 1961.

Piene, Otto. "Art-and-Technology: Recent Efforts in Materials and Media," *MRS Bulletin*, 1992.

Piene, Otto. "The Development of the Group Zero," *Times Literary Supplement*, London, September 3, 1964.

Piene, Otto to György Kepes, January 28, 1966, letter.

Piene, Otto. "Sky Art: A Notebook for a Book." *Artscanada,* June 1969.

Piene, Otto. *Sky Art, Portfolio of 25 Lithographs.* Los Angeles: Tamarind Lithography Workshop, 1969.

Piene, Otto. *More Sky.* Cambridge, Massachusetts: MIT Press, 1970.

Piene, Otto and others. *Boston Celebrations, Part 1: Jubilee Projects by the Fellows of the Center for Advanced Visual Studies*, Institute of Contemporary Art (catalog). Boston, ICA, 1975; *Boston Celebrations, Part 2: You Are Here – Environmental Art*, MIT Center for Advanced Visual Studies, Cambridge, 1976.

Piene, Otto. "The Development of the Group Zero," *Lightworks*, Vol. 13, 1980

Piene, Otto. "Zero to Rainbow to Sky Art." *Lightworks*, Fall 1980.

Piene, Otto with Elizabeth Goldring and Vin Grabill, Eds. CENTERVIDEO, exhibition catalog, American Center for Students and Artists, Paris, Feb 1-6, 1981; Koelnischer Kunstverein, Cologne, March 6-10, 1981.

Piene, Otto and others, Eds. *Sky Art: Sky Art Conference '81*: Massachusetts Institute of Technology: Center for Advanced Visual Studies (conference publication). Cambridge, MIT, 1981.

Piene, Otto, Philipp Morrison, Lowry Burgess. Sky Art Conference '83, Center for Advanced Visual Studies, Sept 24 -27, 1983 (conference publication). Cambridge, Massachusetts, MIT.

Piene, Otto and Stephan Westfall. *Sky Art Alaska* (exhibition catalog), Visual Art Centre of Alaska, Anchorage, 1986.

Piene, Otto and Elizabeth Goldring, *The Light Silo*, Ante Glibota and Delight Edition, 2014.

Piene, Otto, Elizabeth Goldring and Lowry Burgess. *Sky Art Manifesto*, presented at the UNESCO Conference on the Peaceful Use of Space, Paris, Nov., 1986.

Piene, Otto and Robert Russett. "Sky, Scale and Technology in Art." *Leonardo 19*, No. 3 (1986).

Plan 75, MIT School of Architecture + Planning, 2010, "A New Program in Art, Culture and Technology."

Poleskie, Steve. "Art and Flight: Historical Origins to Contemporary Works," *Leonardo 18*, 1985.

Rawsthorn, Alice. "A Master of Image and Information," *The New York Times*, October 10, 2010.

Richard, Paul. "Of Lasers and City Lights," *The Washington Post*, May 30, 1974.

Rieveschl, Gary, *Lifeforms 1973-1976*, Jack Otto & Sons Printers, 1977

Rockne Krebs: A Retrospective of Drawings, 1965-1982, Baxter Art Gallery, California Institute of Technology, Pasadena, California, (Exhibition Catalog) February 2-March 6, 1983

Rothfuss, Joan, *Topless Cellist: The Improbable Life of Charlotte Moorman*, The MIT Press, Cambridge and London, 2014

Seaman, Bill. Discussion with Otto Piene and Elizabeth Goldring, March 6-7, 2012, Berlin.

Sebring, Ellen and Joan Brigham. Interview with Elizabeth Goldring, November 11, 2015.

Siler, Todd. *Breaking the Mind Barrier*, Simon & Schuster, 1990.

Sky Light: Kinetic Art from CAVS/ MIT. Boston, Museum of Science, 1987 (exhibition catalog).

Smee, Sebastian. "Otto Piene's 'Light Ballet' dazzles at MIT," *The Boston Globe*, November 20, 2011.

Sonfist, Alan. *Institute of Contemporary Art*, London, 1971 (catalog).

Steele, Robert. "Report on the 1975 Public-Celebration Workshop at the Center for Advanced Visual Studies, MIT," *Leonardo*, Vol. 9, No. 1, Winter 1976.

The Tech, MIT, May 3, 1974.

VanDerBeek, Stan, "Re: Vision," *The American Scholar*, Vol. 35, No. 2, Spring, 1966.

Wisnioski, Matthew. "*Centerbeam: Art of the Environment" and "Arts/Science/ Technology: Otto Piene in interview with Matthew Wisnioski"* in *A Second Modernism: MIT, Architecture, and the 'Techno-Social' Moment*, edited by Arindam Dutta. Cambridge, MA: SA+P Press/MIT Press, 2013.

Wu, Faye, PhD candidate, Department of Mechanical Engineering, MIT, email to Elizabeth Goldring, October 4, 2016.

About the Lists

A comprehensive listing of CAVS Fellows, affiliates and visiting artists with dates is difficult to compile due to diverse types of affiliations and unclear records. The following partial list is derived from information provided by CAVS special collection archivist, Jeremy Grubman, from CAVS office records, and from the Institute Archives at MIT.

CAVS Fellows and Affiliates (partial list)

The following list includes 250 CAVS fellows and affiliates in alphabetical order from 1967 through 2009. The term "affiliate" covers a variety of appointments including Research Affiliates, Visiting Scholars, Visiting Artists, Guests of the Institute, and Special Students.

Alan Abel
Marc Adrian
Werner Ahrens
Mitsunobu Akahori
Azra Akšamija
Mel Alexenberg
Rebecca Allen
Maryanne Amacher
Karin Bacon
Ken Baker
Roland Baladi
Juan Navarro Baldeweg
Judith Barry
Eric Begleiter
John Bell
Bill Bell
Mitch Benoff
James Benton
Michael Bernard
Nishan Bichajian
Hisham Bizri
Nancy Bochenek
John Boesch
Laurel Braitman
Nell Breyer
Joan Brigham
Gloria Brown Simmons
Mauricio Bueno
Lowry Burgess
Jack Burnham
James Lee Byers
Bill Cadogan
Peter Campus
Triple Candie
Mira Cantor
Ernst Caramelle
Boyd Frank Carlton
Harriet Casdin-Silver
Mel Chin
Mark Chow
Chryssa
Gillian Wise Ciobotaru
Jürgen Claus
Betsy Connors
Muriel Cooper
Luc Courchesne
Ann Craven
Attila Csáji
Dan Dailey
Dean Daderko
Peter D'Agostino
Robert D'Ancona
Ben Howell Davis
Douglas Davis
Joe Davis
Paula Dawson
Gretchen Dean
Robert Dell
Ben Denckla
Agnes Denes
Walter Dent
Juan Downey
Peter Droege
Paul Earls
Dale Eldred
Dov Eylath
Mark Faverman
Diane Fellows
David Finn
Robert Fisher
Scott Fisher
Don Foresta
Luis Frangella
Charles Frazier
Beth Galston
Susan Gamble
Rus Gant
José Wagner Garcia

William Garnett
Greg Garvey
Hope Ginsburg
Wendelin Glatzel
Derith Glover
Elon Goitein
Jon Goldman
Elizabeth Goldring
Cynthia Goodman
John Goodyear
Vin Grabill
George Greenamyer
Rosemary Grimshaw
Virginia Gunter
Fritz Haeg
Lawrence Hahn
Wolfgang Hahn
K8 Hardy
Sharon Hayes
Ron Hays
Kelly Heaton
Nathan Hemenway
Marek Holynski
Pierre-Alain Hubert
Michio Ihara
Catherine Ikam
Setsuko Ishii
Toshiro Itakura
Bertrand Ivanoff
Wendy Jacob
Christopher Janney
Jon Jeibmann
Bart Johnson
Tom Johnson
Ronald Jones
Miranda July
CM Judge
Dieter Jung
Ken Kantor
Reeta Gidwani Karmakar
Thomas B. Kass
Hidetoshi Katsuma
Harel Kedem
Sung Ho Kim
Laura Knott
Milton Komisar
Panos Kouros
Piotr Kowalski
Ellen Kozak
Bernd Kracke
Robert Kramer
Ted Kraynik
Rockne Krebs
Shelley Lake
Bruno de Lard
Leslie Larkin
Kay Larson
Pam Larson
Nicholas Lascaris
Thorbjørn Lausten
Titus Leber
Billy Lee
Richard Lerman
Xavier Le Roy
Uri Levi
Paul Levy
Joan Logue
Christina Mackie
John Malpede
Frances Masland
Paul Matisse
Mandi McIntyre
Thomas F. McNulty
Joyce Mekeel
Mark Mendel
Clara Meneres
Ioannis Michaloudis
Antoni Miralda
Mit Mitropoulos
Charlotte Moorman
Avatar Da Silva Moraes
David Curt Morris
Mike Moser
Joe Moss
Grover Mouton
Antonio Muntadas
Kevin Murphy
Michael Naimark
Carl Nesjar
John Avery Newman
José Nuño
Atsushi Ogata
Maggie Orth
Alejandro Otero
Nam June Paik
Marc Palumbo
Bill Parker
Donald Pasquella
Charles Patrick
Otto Piene
Bartolomeo Pietromarchi
Cesare Pietrouisti
Marjetica Potrč
John Powell
Jose de Prada
Seth Price
Keiko Prince
Yvonne Rainer
David Reinfurt
Stanley Resnicoff
Gary Thomas Rieveschl
Noah Riskin
Seth Riskin
David Robbins
Dario Robleto
Jenny Romaine
Bryan Rogers
Bob Rosinsky
Jon Rubin
Lees Ruoff
Jessica Rylan
Paul Rutkovsky
Narmine Sadeg
Friedrich St. Florian
Thomas Sansone
Kate Sasanoff
Ludmilla Sayenkova
Günther Schneider-Siemssen
Barbara Shawcroft
Allan Schwarz
Bill Seaman
Ellen Sebring
Leon Shiman
Todd Siler
Julean Simon
Vera Simons
Alejandro Siña
Georges Singer
Michael Smith
Frank Smullin
Alan Sonfist
Stacey Spiegel
Stelarc
Nicole Stenger
Ian Strasfogel
Tal Streeter
Peter Struycken
Steven s'Soreff
Athena Tacha
Vassilakis Takis
Aldo Tambellini
Tamiko Thiel
Don Thornton
Dennis Thurmond
Marvin Torffield
Naoko Tosa
Harold Tovish
Wen-Ying Tsai
Mary Tsouti
Stan VanDerBeek
Pam Vandiver
Gretchen Van Dyke
Tom Van Sant
Pierre Van Berkel
Tim Walters
Sally Weber
Judith Wechsler
Pantelis Xagoraris
Howard Alan York
Takeshi Yoshizumi
José Marié Yturralde
Walter Zengerle

CAVS Graduate Students

The following list includes sixty-one CAVS Master of Science in Visual Studies (SMVisS) graduates. Students who later became fellows appear on both lists.

Name	Year
Sara Baker	1986
Eric Begleiter	1984
Mitch Benoff	1993
Shawn Brixey	1988
Gloria Brown Simmons	1984
Elizabeth Cavicchi	1980
Mei-Ling Chan-Bernard	1990
Peter Codella	1982
Luc Courchesne	1984
Walter Dent	1984
Ebon Fisher	1986
Nina Frankenheim	1979
Beth Galston	1981
Lorilee Garbowski-Mallet	1982
Greg Garvey	1982
Sarah Geitz	1985
Wendelin Glatzel	1980
Derith Glover	1977
Jon Goldman	1985
Vin Grabill	1981
Jennifer Hall	1985
Gregory Haun	1992
Nathan Hemenway	1991
Christopher Janney	1978
Jon Jeibmann	1983
Caryn Johnson	1992
CM Judge	1988
Ken Kantor	1982
Harel Kedem	1978
Sue Mie Kim	1993
Laura Knott	1987
Panos Kouros	1990
Ellen Kozak	1979
Bernd Kracke	1981
Shelley Lake	1979
David Larkin	1989
Uri Levi	1985
Martha Lozano Guerra	1979
Jerry McCarty	1989
Mike Moser	1977
Mike Naimark	1979
George Numrich	1987
Atsushi Ogata	1988
Iris Oliveras	1993
Maggie Orth	1993
John Powell	1989
Seth Riskin	1989
Don Ritter	1988
Bob Rosinsky	1984
Lees Ruoff	1987
Thomas Sansone	1979
Bill Seaman	1985
Ellen Sebring	1986
Todd Siler	1981, 1986 (PhD)
Dan Spikol	1992
Don Thornton	1979
Marta Vahtar	1993
Jarlath Waldron	1991
Sally Weber	1983
Diane Willow	1992
Zafirios Xagoraris	1991

Directors

György Kepes	1967-1974
Otto Piene	1974-1994
Paul Earls and Elizabeth Goldring (Acting Co-Directors)	1994-1995
Krzysztof Wodiczko	1995-1996
Stephen Benton	1996-2003
Krzysztof Wodiczko	2004-2009
Robert Preusser (Director of Eduction)	1974-1985

Index